W9-CQO-883

McGraw-Hill series in fundamentals of physics: an undergraduate textbook program

E. U. CONDON, EDITOR, UNIVERSITY OF COLORADO

MEMBERS OF THE ADVISORY BOARD

Arthur F. Kip, University of California, Berkeley
Hugh D. Young, Carnegie Institute of Technology

INTRODUCTORY TEXTS

Young · Fundamentals of Mechanics and Heat
Kip · Fundamentals of Electricity and Magnetism
Young · Fundamentals of Optics and Modern Physics
Beiser · Concepts of Modern Physics

UPPER-DIVISION TEXTS

Kraut · Fundamentals of Mathematical Physics
Reif · Fundamentals of Statistical and Thermal Physics

Fundamentals of statistical and thermal physics

Fundamentals of

and

statistical

thermal physics

F. REIF

Professor of Physics

University of California, Berkeley

McGraw-Hill, Inc.

New York St. Louis San Francisco Auckland Bogotá
Caracas Lisbon London Madrid Mexico City Milan
Montreal New Delhi San Juan Singapore
Sydney Tokyo Toronto

Fundamentals of statistical and thermal physics

Copyright © 1965 by McGraw-Hill, Inc. All rights reserved. Typeset in the United States of America. Except as permitted under the United States Copyright Act of 1976, no part of this publication may be reproduced or distributed in any form or by any means, or stored in a data base or retrieval system, without the prior written permission of the publisher.

ISBN 07-051800-9

35 36 37 BKMBKM 9 9 8 7 6 5 4 3

Library of Congress Catalog Card Number 63-22730.

Preface

THIS BOOK is devoted to a discussion of some of the basic physical concepts and methods appropriate for the description of systems involving very many particles. It is intended, in particular, to present the disciplines of thermodynamics, statistical mechanics, and kinetic theory from a unified and modern point of view. Accordingly, the presentation departs from the historical development in which thermodynamics was the first of these disciplines to arise as an independent subject. The history of the ideas concerned with heat, work, and the kinetic theory of matter is interesting and instructive, but it does not represent the clearest or most illuminating way of developing these subjects. I have therefore abandoned the historical approach in favor of one that emphasizes the essential unity of the subject matter and seeks to develop physical insight by stressing the microscopic content of the theory.

Atoms and molecules are constructs so successfully established in modern science that a nineteenth-century distrust of them seems both obsolete and inappropriate. For this reason I have deliberately chosen to base the entire discussion on the premise that all macroscopic systems consist ultimately of atoms obeying the laws of quantum mechanics. A combination of these microscopic concepts with some statistical postulates then leads readily to some very general conclusions on a purely *macro*scopic level of description. These conclusions are valid *irrespective* of any particular models that might be assumed about the nature or interactions of the particles in the systems under consideration; they possess, therefore, the full generality of the classical laws of thermodynamics. Indeed, they are more general, since they make clear that the macroscopic parameters of a system are statistical in nature and exhibit fluctuations which are calculable and observable under appropriate conditions. Despite the microscopic point of departure, the book thus contains much general reasoning on a purely macroscopic level—probably about as much as a text on classical thermodynamics—but the microscopic content of the macroscopic arguments remains clear at all stages. Furthermore, *if* one is willing to adopt specific microscopic models concerning the particles constituting a system, then it is also apparent how one can calculate macroscopic quantities on the basis of this microscopic information. Finally, the statistical concepts used to discuss equilibrium situations constitute an appropriate preparation for their extension to the discussion of systems which are not in equilibrium.

This approach has, in my own teaching experience, proved to be no more difficult than the customary one which begins with classical thermodynamics. The latter subject, developed along purely macroscopic lines, is conceptually far from easy. Its reasoning is often delicate and of a type which seems unnatural to many physics students, and the significance of the fundamental concept of entropy is very hard to grasp. I have chosen to forego the subtleties of traditional arguments based on cleverly chosen cycles and to substitute instead the task of assimilating some elementary statistical ideas. The following gains are thereby achieved: (*a*) Instead of spending much time discussing various arguments based on heat engines, one can introduce the student at an early stage to statistical methods which are of great and recurring importance throughout all of physics. (*b*) The microscopic approach yields much better physical insight into many phenomena and leads to a ready appreciation of the meaning of entropy. (*c*) Much of modern physics is concerned with the explanation of macroscopic phenomena in terms of microscopic concepts. It seems useful, therefore, to follow a presentation which stresses at all times the interrelation between microscopic and macroscopic levels of description. The traditional teaching of thermodynamics and statistical mechanics as distinct subjects has often left students with their knowledge compartmentalized and has also left them ill-prepared to accept newer ideas, such as spin temperature or negative temperature, as legitimate and natural. (*d*) Since a unified presentation is more economical, conceptually as well as in terms of time, it permits one to discuss more material and some more modern topics.

The basic plan of the book is the following: The first chapter is designed to introduce some basic probability concepts. Statistical ideas are then applied to systems of particles in equilibrium so as to develop the basic notions of statistical mechanics and to derive therefrom the purely macroscopic general statements of thermodynamics. The *macro*scopic aspects of the theory are then discussed and illustrated at some length; the same is then done for the *micro*scopic aspects of the theory. Some more complicated equilibrium situations, such as phase transformations and quantum gases, are taken up next. At this point the text turns to a discussion of nonequilibrium situations and treats transport theory in dilute gases at varying levels of sophistication. Finally, the last chapter deals with some general questions involving irreversible processes and fluctuations. Several appendices contain mostly various useful mathematical results.

The book is intended chiefly as a text for an introductory course in statistical and thermal physics for college juniors or seniors. The mimeographed notes on which it is based have been used in this way for more than two years by myself and several of my colleagues in teaching such a course at the University of California in Berkeley. No prior knowledge of heat or thermodynamics is presupposed; the necessary prerequisites are only the equivalents of a course in introductory physics and of an elementary course in atomic physics. The latter course is merely supposed to have given the student sufficient background in modern physics (*a*) to know that quantum mechanics describes systems in terms of quantum states and wave functions, (*b*) to have encountered the energy levels of a simple harmonic oscillator and to have seen the quantum description of a free particle in a box, and (*c*) to have heard of the Heisenberg uncertainty and Pauli exclusion principles. These are essentially all the quantum ideas that are needed.

The material included here is more than can be covered in a one-semester undergraduate course. This was done purposely (*a*) to include a discussion of those basic ideas likely to be most helpful in facilitating the student's later access to more advanced works, (*b*) to allow students with some curiosity to read beyond the minimum on a given topic, (*c*) to give the instructor some possibility of selecting between alternate topics, and (*d*) to anticipate current revisions of the introductory physics course curriculum which should make upper-division students in the near future much more sophisticated and better prepared to handle advanced material than they are now. In actual practice I have successfully covered the first 12 chapters (omitting Chapter 10 and most starred sections) in a one-semester course. Chapter 1 contains a discussion of probability concepts more extensive than is needed for the understanding of subsequent chapters. In addition, the chapters are arranged in such a way that it is readily possible, after the first eight chapters, to omit some chapters in favor of others without encountering difficulties with prerequisites.

The book should also be suitable for use in an introductory graduate course if one includes the starred sections and the last three chapters, which contain somewhat more advanced material. Indeed, with students who have studied classical thermodynamics but have had no significant exposure to the ideas of statistical mechanics in their undergraduate career, one cannot hope to cover in a one-semester graduate course appreciably more subject matter than is treated here. One of my colleagues has thus used the material in our Berkeley graduate course on statistical mechanics (a course which is, as yet, mostly populated by students with this kind of preparation).

Throughout the book I have tried to keep the approach well-motivated and to strive for simplicity of presentation. It has not been my aim to pursue rigor in the formal mathematical sense. I have, however, attempted to keep the basic physical ideas in the forefront and to discuss them with care. In the process the book has become longer than it might have otherwise, for I have not hesitated to increase the ratio of words to formulas, to give illustrative examples, or to present several ways of looking at a question whenever I felt that it would enhance understanding. My aim has been to stress physical insight and important methods of reasoning, and I advise most earnestly that the student stress these aspects of the subject instead of trying to memorize various formulas meaningless in themselves. To avoid losing the reader in irrelevant details, I have often refrained from presenting the most general case of a problem and have sought instead to treat relatively simple cases by powerful and easily generalizable methods. The book is not meant to be encyclopaedic; it is merely intended to provide a basic skeleton of some fundamental ideas most likely to be useful to the student in his future work. Needless to say, some choices had to be made. For example, I thought it important to introduce the Boltzmann equation, but resisted the temptation of discussing applications of the Onsager relations to various irreversible phenomena such as thermoelectric effects.

It is helpful if a reader can distinguish material of secondary importance from that which is essential to the main thread of the argument. Two devices have been used to indicate subject matter of subsidiary importance: (*a*) Sections marked by a star (asterisk) contain material which is more advanced or more detailed; they can be omitted (and probably should be omitted in a first reading) without incurring a handicap in proceeding to subsequent sections. (*b*) Many remarks, examples, and elaborations are interspersed throughout the text and are set off on a gray background. Conversely, black marginal pointers have been used to emphasize important results and to facilitate reference to them.

The book contains about 230 problems, which should be regarded as an essential part of the text. It is indispensable that the student solve an appreciable fraction of these problems if he is to gain a meaningful understanding of the subject matter and not merely a casual hearsay acquaintance with it.

I am indebted to several of my colleagues for many valuable criticisms and suggestions. In particular, I should like to thank Prof. Eyvind H. Wichmann, who read an older version of the entire manuscript with meticulous care, Prof. Owen Chamberlain, Prof. John J. Hopfield, Dr. Allan N. Kaufman, and Dr. John M. Worlock. Needless to say, none of these people should be blamed for the flaws of the final product.

Acknowledgements are also due to Mr. Roger F. Knacke for providing the answers to the problems. Finally, I am particularly grateful to my secretary, Miss Beverly West, without whose devotion and uncanny ability to transform pages of utterly illegible handwriting into a perfectly typed technical manuscript this book could never have been written.

It has been said that "an author never finishes a book, he merely abandons it." I have come to appreciate vividly the truth of this statement and dread to see the day when, looking at the manuscript in print, I am sure to realize that so many things could have been done better and explained more clearly. If I abandon the book nevertheless, it is in the modest hope that it may be useful to others despite its shortcomings.

F. REIF

Contents

Fundamentals of statistical and thermal physics

Introduction to statistical methods 1

THIS BOOK will be devoted to a discussion of systems consisting of very many particles. Examples are gases, liquids, solids, electromagnetic radiation (photons), etc. Indeed, most physical, chemical, or biological systems do consist of many molecules; our subject encompasses, therefore, a large part of nature.

The study of systems consisting of many particles is probably the most active area of modern physics research outside the realm of high-energy physics. In the latter domain, the challenge is to understand the fundamental interactions between nucleons, neutrinos, mesons, or other strange particles. But, in trying to discuss solids, liquids, plasmas, chemical or biological systems, and other such systems involving many particles, one faces a rather different task which is no less challenging. Here there are excellent reasons for supposing that the familiar laws of quantum mechanics describe adequately the motions of the atoms and molecules of these systems; furthermore, since the nuclei of atoms are not disrupted in ordinary chemical or biological processes and since gravitational forces between atoms are of negligible magnitude, the forces between the atoms in these systems involve only well-understood electromagnetic interactions. Somebody sufficiently sanguine might therefore be tempted to claim that these systems are "understood in principle." This would, however, be a rather empty and misleading statement. For, although it might be possible to write down the equations of motion for any one of these systems, the complexity of a system containing many particles is so great that it may make the task of deducing any useful consequences or predictions almost hopeless. The difficulties involved are not just questions of quantitative detail which can be solved by the brute force application of bigger and better computers. Indeed, even if the interactions between individual particles are rather simple, the sheer complexity arising from the interaction of a large number of them can often give rise to quite unexpected *qualitative* features in the behavior of a system. It may require very deep analysis to predict the occurrence of these features from a knowledge of the individual particles. For example, it is a striking fact, and one which is difficult to understand in microscopic detail, that simple atoms forming a gas can condense abruptly to form a liquid with very different properties. It is a fantastically more difficult task to attain an understanding of how an assembly of certain kinds of molecules can lead to a system capable of biological growth and reproduction.

The task of understanding systems consisting of many particles is thus far from trivial, even when the interactions between individual atoms are well known. Nor is the problem just one of carrying out complicated computations. The main aim is, instead, to use one's knowledge of basic physical laws to develop new concepts which can illuminate the essential characteristics of such complex systems and thus provide sufficient insight to facilitate one's thinking, to recognize important relationships, and to make useful predictions. When the systems under consideration are not too complex and when the desired level of description is not too detailed, considerable progress can indeed be achieved by relatively simple methods of analysis.

It is useful to introduce a distinction between the sizes of systems whose description may be of interest. We shall call a system "*micro*scopic" (i.e. "*small* scale") if it is roughly of atomic dimensions or smaller (say of the order of 10 Å or less). For example, the system might be a molecule. On the other hand, we shall call a system "*macro*scopic" (i.e., "*large* scale") when it is large enough to be visible in the ordinary sense (say greater than 1 micron, so that it can at least be observed with a microscope using ordinary light). The system consists then of very many atoms or molecules. For example, it might be a solid or liquid of the type we encounter in our daily experience. When one is dealing with such a macroscopic system, one is, in general, not concerned with the detailed behavior of each of the individual particles constituting the system. Instead, one is usually interested in certain *macro*scopic parameters which characterize the system as a whole, e.g., quantities like volume, pressure, magnetic moment, thermal conductivity, etc. If the macroscopic parameters of an isolated system do not vary in time, then one says that the system is in equilibrium. If an isolated system is not in equilibrium, the parameters of the system will, in general, change until they attain constant values corresponding to some final equilibrium condition. Equilibrium situations can clearly be expected to be amenable to simpler theoretical discussion than more general time-dependent nonequilibrium situations.

Macroscopic systems (like gases, liquids, or solids) began first to be systematically investigated from a macroscopic phenomenological point of view in the last century. The laws thus discovered formed the subject of "thermodynamics." In the second half of the last century the theory of the atomic constitution of all matter gained general acceptance, and macroscopic systems began to be analyzed from a fundamental *micro*scopic point of view as systems consisting of very many atoms or molecules. The development of quantum mechanics after 1926 provided an adequate theory for the description of atoms and thus opened the way for an analysis of such systems on the basis of realistic microscopic concepts. In addition to the most modern methods of the "many-body problem," there have thus developed several disciplines of physics which deal with systems consisting of very many particles. Although the boundaries between these disciplines are not very sharp, it may be useful to point out briefly the similarities and differences between their respective methods of approach.

a. For a system in *equilibrium*, one can try to make some very general statements concerning relationships existing between the *macro*scopic parameters of the system. This is the approach of classical "thermodynamics," historically the oldest discipline. The strength of this method is its great generality, which allows it to make valid statements based on a minimum number of postulates *without* requiring any detailed assumptions about the *micro*scopic (i.e., molecular) properties of the system. The strength of the method also implies its weakness: only relatively few statements can be made on such general grounds, and many interesting properties of the system remain outside the scope of the method.

b. For a system in equilibrium, one can again try to make very general statements consistently based, however, on the *micro*scopic properties of the particles in the system and on the laws of mechanics governing their behavior. This is the approach of "statistical mechanics." It yields *all* the results of thermodynamics *plus* a large number of general relations for calculating the macroscopic parameters of the system from a knowledge of its microscopic constituents. This method is one of great beauty and power.

c. If the system is *not* in equilibrium, one faces a much more difficult task. One can still attempt to make very general statements about such systems, and this leads to the methods of "irreversible thermodynamics," or, more generally, to the study of "statistical mechanics of irreversible processes." But the generality and power of these methods is much more limited than in the case of systems in equilibrium.

d. One can attempt to study *in detail* the interactions of all the particles in the system and thus to calculate parameters of macroscopic significance. This is the method of "kinetic theory." It is in principle always applicable, even when the system is *not* in equilibrium so that the powerful methods of equilibrium statistical mechanics are not available. Although kinetic theory yields the most detailed description, it is by the same token also the most difficult method to apply. Furthermore, its detailed point of view may tend to obscure general relationships which are of wider applicability.

Historically, the subject of thermodynamics arose first before the atomic nature of matter was understood. The idea that heat is a form of energy was first suggested by the work of Count Rumford (1798) and Davy (1799). It was stated explicitly by the German physician R. J. Mayer (1842) but gained acceptance only after the careful experimental work of Joule (1843–1849). The first analysis of heat engines was given by the French engineer S. Carnot in 1824. Thermodynamic theory was formulated in consistent form by Clausius and Lord Kelvin around 1850, and was greatly developed by J. W. Gibbs in some fundamental papers (1876–1878).

The atomic approach to macroscopic problems began with the study of the kinetic theory of dilute gases. This subject was developed through the pioneering work of Clausius, Maxwell, and Boltzmann. Maxwell discovered the distribution law of molecular velocities in 1859, while Boltzmann formulated his fundamental integrodifferential equation (the Boltzmann equation) in 1872. The kinetic theory of gases achieved its modern form when Chapman and Enskog (1916–1917) succeeded in approaching the subject by developing systematic methods for solving this equation.

The more general discipline of statistical mechanics also grew out of the work of Boltzmann who, in 1872, succeeded further in giving a fundamental microscopic analysis of irreversibility and of the approach to equilibrium. The theory of statistical mechanics was then developed greatly in generality and power by the basic contributions of J. W. Gibbs (1902). Although the advent of quantum mechanics has brought many changes, the basic framework of the modern theory is still the one which he formulated.

In discussing systems consisting of very many particles, we shall not aim to recapitulate the historical development of the various disciplines dealing with the physical description of such systems. Instead we shall, from the outset, adopt a modern point of view based on our present-day knowledge of atomic physics and quantum mechanics. We already mentioned the fact that one can make very considerable progress in understanding systems of many particles by rather simple methods of analysis. This may seem rather surprising at first sight; for are not systems such as gases or liquids, which consist of a number of particles of the order of Avogadro's number (10^{23}), hopelessly complicated? The answer is that the very complexity of these systems contains within it the key to a successful method of attack. Since one is not concerned with the detailed behavior of each and every particle in such systems, it becomes possible to apply statistical arguments to them. But as every gambler, insurance agent, or other person concerned with the calculation of probabilities knows, statistical arguments become most satisfactory when they can be applied to large numbers. What a pleasure, then, to be able to apply them to cases where the numbers are as large as 10^{23}, i.e., Avogadro's number! In systems such as gases, liquids, or solids where one deals with very many identical particles, statistical arguments thus become particularly effective. This does not mean that all problems disappear; the physics of many-body problems does give rise to some difficult and fascinating questions. But many important problems do indeed become quite simple when approached by statistical means.

RANDOM WALK AND BINOMIAL DISTRIBUTION

1 · 1 *Elementary statistical concepts and examples*

The preceding comments make it apparent that statistical ideas will play a central role throughout this book. We shall therefore devote this first chapter to a discussion of some elementary aspects of probability theory which are of great and recurring usefulness.

The reader will be assumed to be familiar with the most rudimentary probability concepts. It is important to keep in mind that whenever it is desired to describe a situation from a statistical point of view (i.e., in terms of probabilities), it is always necessary to consider an assembly (or "ensemble") consisting of a very large number $\mathfrak{N}$ of similarly prepared systems. The probability of occurrence of a particular event is then defined with respect to this particular ensemble and is given by the fraction of systems in the ensemble which are characterized by the occurrence of this specified event. For example, in throwing a pair of dice, one can give a statistical description by considering that a very large number $\mathfrak{N}$ (in principle, $\mathfrak{N} \to \infty$) of similar pairs of dice are thrown under similar circumstances. (Alternatively one could imagine the same pair of dice thrown $\mathfrak{N}$ times in succession under similar circumstances.) The probability of obtaining a double ace is then given by the fraction of these experiments in which a double ace is the outcome of a throw.

Note also that the probability depends very much on the nature of the ensemble which is contemplated in defining this probability. For example, it makes no sense to speak simply of the probability that an individual seed will yield red flowers. But one can meaningfully ask for the probability that such a seed, regarded as a member of an ensemble of similar seeds derived from a specified set of plants, will yield red flowers. The probability depends crucially on the ensemble of which the seed is regarded as a member. Thus the probability that a given seed will yield red flowers is in general different if this seed is regarded as (*a*) a member of a collection of similar seeds which are known to be derived from plants that had produced red flowers or as (*b*) a member of a collection of seeds which are known to be derived from plants that had produced pink flowers.

In the following discussion of basic probability concepts it will be useful to keep in mind a specific simple, but important, illustrative example—the so-called "random-walk problem." In its simplest idealized form, the problem can be formulated in the following traditional way: A drunk starts out from a lamppost located on a street. Each step he takes is of equal length l. The man is, however, so drunk that the direction of each step—whether it is to the right or to the left—is completely independent of the preceding step. All one can say is that each time the man takes a step, the probability of its being to the right is p, while the probability of its being to the left is $q = 1 - p$. (In the simplest case $p = q$, but in general $p \neq q$. For example, the street might be inclined with respect to the horizontal, so that a step downhill to the right is more likely than one uphill to the left.)

Choose the x axis to lie along the street so that $x = 0$ is the position of the

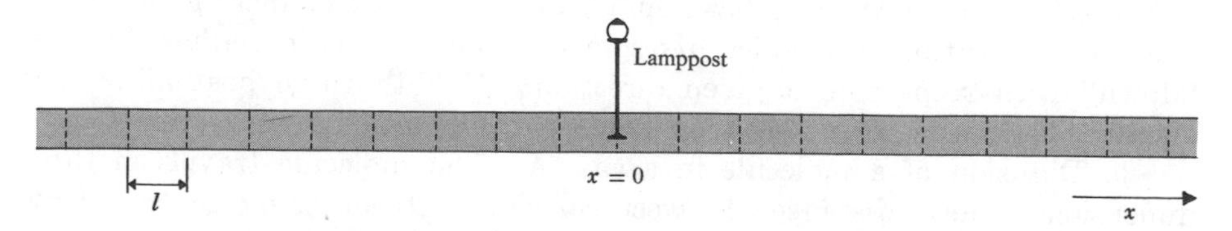

Fig. 1·1·1 The drunkard's random walk in one dimension.

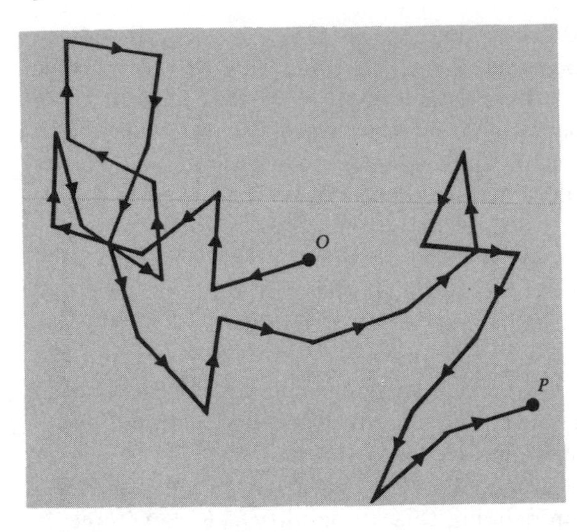

Fig. 1·1·2 Example of a random walk in two dimensions.

point of origin, the lamppost. Clearly, since each step is of length l, the location of the man along the x axis must be of the form $x = ml$, where m is an integer (positive, negative, or zero). The question of interest is then the following: After the man has taken N steps, what is the probability of his being located at the position $x = ml$?

This statistical formulation of the problem again implies that one considers a very large number $\mathfrak{N}$ of similar men starting from similar lampposts. (Alternatively, if the situation does not change in time, e.g., if the man does not gradually sober up, one could also repeat the same experiment $\mathfrak{N}$ times with the same man.) At each step, one finds that a fraction p of the men moves to the right. One then asks what fraction of the men will be located at the position $x = ml$ after N steps.

One can readily generalize this one-dimensional problem to more dimensions whether it be two (a drunk starting out from a lamppost in the middle of a parking lot), three, or more. One again asks for the probability that after N steps the man is located at a given distance from the origin (although the distance is no longer of the form ml, with m integral).

Now the main concern of physics is not with drunks who stagger home from lampposts. But the problem which this illustrates is the one of adding N vectors of equal length but of random directions (or directions specified by some probability distribution) and then asking for the probability that their resultant vector sum has a certain magnitude and direction (see Fig. 1·1·2). We mention a few physical examples where this question is relevant.

a. Magnetism: An atom has a spin $\frac{1}{2}$ and a magnetic moment μ; in accordance with quantum mechanics, its spin can therefore point either "up" or "down" with respect to a given direction. If both these possibilities are equally likely, what is the net total magnetic moment of N such atoms?

b. Diffusion of a molecule in a gas: A given molecule travels in three dimensions a mean distance l between collisions with other molecules. How far is it likely to have gone after N collisions?

c. Light intensity due to N incoherent light sources: The light amplitude due to each source can be represented by a two-dimensional vector whose direction specifies the phase of the disturbance. Here the phases are random, and the resultant amplitude, which determines the total intensity of the light from all the sources, must be computed by statistical means.

The random walk problem illustrates some very fundamental results of probability theory. The techniques used in the study of this problem are powerful and basic, and recur again and again throughout statistical physics. It is therefore very instructive to gain a good understanding of this problem.

1·2 *The simple random walk problem in one dimension*

For the sake of simplicity we shall discuss the random walk problem in one dimension. Instead of talking in terms of a drunk taking steps, let us revert to the less alcoholic vocabulary of physics and think of a particle performing successive steps, or displacements, in one dimension. After a total of N such steps, each of length l, the particle is located at

$$x = ml$$

where m is an integer lying between

$$-N \leq m \leq N$$

We want to calculate the probability $P_N(m)$ of finding the particle at the position $x = ml$ after N such steps.

Let n_1 denote the number of steps to the right and n_2 the corresponding number of steps to the left. Of course, the total number of steps N is simply

$$N = n_1 + n_2 \tag{1·2·1}$$

The net displacement (measured to the right in units of a step length) is given by

$$m = n_1 - n_2 \tag{1·2·2}$$

If it is known that in some sequence of N steps the particle has taken n_1 steps to the right, then its net displacement from the origin is determined. Indeed, the preceding relations immediately yield

$$m = n_1 - n_2 = n_1 - (N - n_1) = 2n_1 - N \tag{1·2·3}$$

This shows that if N is odd, the possible values of m must also be odd. Conversely, if N is even, m must also be even.

Our fundamental assumption was that successive steps are statistically

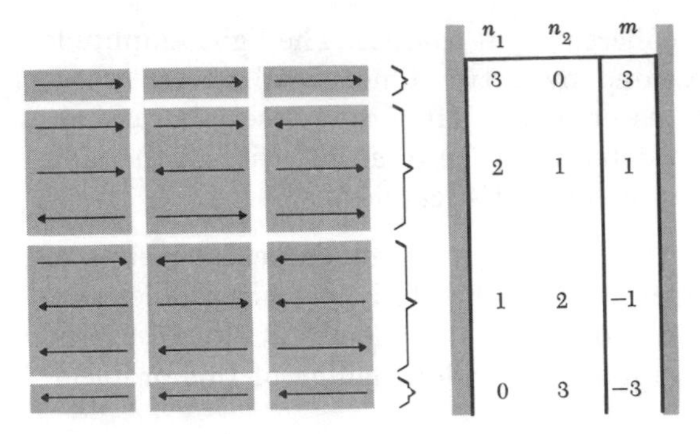

Fig. 1·2·1 Illustration showing the eight sequences of steps which are possible if the total number of steps is $N = 3$.

independent of each other. Thus one can assert simply that, irrespective of past history, each step is characterized by the respective probabilities

$$p = \text{probability that the step is to the right}$$

and

$$q = 1 - p = \text{probability that the step is to the left}$$

Now, the probability of any *one* given sequence of n_1 steps to the right and n_2 steps to the left is given simply by multiplying the respective probabilities, i.e., by

$$\underbrace{p\,p\,\cdots\,p}_{n_1\text{ factors}}\;\underbrace{q\,q\,\cdots\,q}_{n_2\text{ factors}} = p^{n_1}q^{n_2} \tag{1·2·4}$$

But there are many different possible ways of taking N steps so that n_1 of them are to the right and n_2 are to the left (see illustration in Fig. 1·2·1). Indeed, the number of distinct possibilities (as shown below) is given by

$$\frac{N!}{n_1!n_2!} \tag{1·2·5}$$

Hence the probability $W_N(n_1)$ of taking (in a total of N steps) n_1 steps to the right and $n_2 = N - n_1$ steps to the left, in any order, is obtained by multiplying the probability (1·2·4) of this sequence by the number (1·2·5) of possible sequences of such steps. This gives

▶
$$W_N(n_1) = \frac{N!}{n_1!n_2!}\,p^{n_1}q^{n_2} \tag{1·2·6}$$

Simple example Consider the simple illustration of Fig. 1·2·1, which shows the case of a total of $N = 3$ steps. There is only one way in which all three successive steps can be to the right; the corresponding probability $W(3)$ that all three steps are to the right is then simply $p \cdot p \cdot p = p^3$. On the other hand, the probability of a sequence of steps where two steps are to the right while the third step is to the left is p^2q. But there are three such possible sequences. Thus the total probability of occurrence of a situation where two steps are to the right and one is to the left is given by $3p^2q$.

Reasoning leading to Eq. (1·2·5) The problem is to count the number of distinct ways in which N objects (or steps), of which n_1 are indistinguishably of one type and n_2 of a second type, can be accommodated in a total of $N = n_1 + n_2$ possible places. In this case

the 1st place can be occupied by any one of the N objects
the 2nd place can be occupied by any one of the remaining $(N - 1)$ objects

. . .

the Nth place can be occupied only by the last 1 object

Hence all the available places can be occupied in

$$N(N-1)(N-2)\cdots 1 \equiv N!$$

possible ways. The above enumeration considers each object as distinguishable. But since the n_1 objects of the first type are indistinguishable (e.g., all are right steps), all the $n_1!$ permutations of these objects among themselves lead to the same situation. Similarly, all the $n_2!$ permutations of objects of the second type among themselves lead to the same situation. Hence, by dividing the total number $N!$ of arrangements of the objects by the number $n_1!n_2!$ of irrelevant permutations of objects of each type, one obtains the total number $N!/n_1!n_2!$ of distinct ways in which N objects can be arranged if n_1 are of one type and n_2 of another type.

Example For instance, in the previous example of three steps, there are $N = 3$ possible events (or places) designated in Fig. 1·2·2 by B_1, B_2, B_3 and capable of being filled by the three particular steps labeled A_1, A_2, A_3. The event B_1 can then occur in any of three ways, B_2 in any of two ways, and B_3 in only one way. There are thus $3 \times 2 \times 1 = 3! = 6$ possible sequences

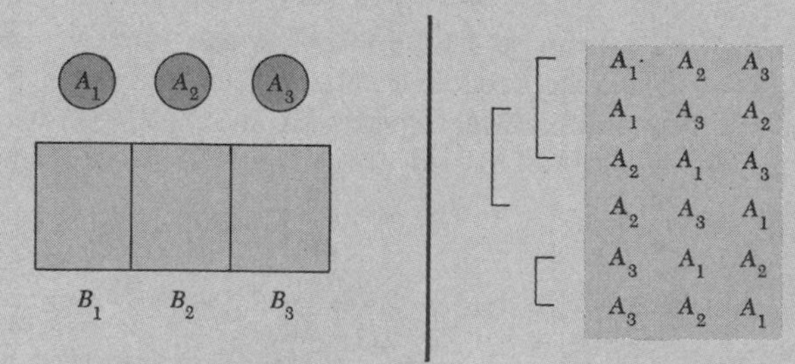

Fig. 1·2·2 Diagram illustrating the problem of distributing three objects A_1, A_2, A_3 among three places B_1, B_2, B_3. The right part of the diagram lists the possible arrangements and indicates by brackets those arrangements which are identical when A_1 and A_2 are considered indistinguishable.

of these three steps. But suppose that A_1 and A_2 denote both right steps ($n_1 = 2$), while A_3 denotes a left step ($n_2 = 1$). Then sequences of steps differing only by the two permutations of A_1 and A_2 are really identical. Thus one is left with only $\frac{6}{2} = 3$ distinct sequences for which two steps are to the right and one step is to the left.

The probability function (1·2·6) is called the binomial distribution. The reason is that (1·2·5) represents a typical term encountered in expanding $(p+q)^N$ by the binomial theorem. Indeed, we recall that the binomial expansion is given by the formula

$$(p+q)^N = \sum_{n=0}^{N} \frac{N!}{n!(N-n)!} p^n q^{N-n} \tag{1·2·7}$$

We already pointed out in (1·2·3) that if it is known that the particle has performed n_1 steps to the right in a total of N steps, then its net displacement m from the origin is determined. Thus the probability $P_N(m)$ that the particle is found at position m after N steps is the same as $W_N(n_1)$ given by (1 · 2 · 6), i.e.,

$$P_N(m) = W_N(n_1) \tag{1·2·8}$$

By (1·2·1) and (1·2·2) one finds explicitly*

$$n_1 = \tfrac{1}{2}(N+m), \qquad n_2 = \tfrac{1}{2}(N-m) \tag{1·2·9}$$

Substitution of these relations in (1·2·6) thus yields

$$P_N(m) = \frac{N!}{[(N+m)/2]![(N-m)/2]!} p^{(N+m)/2}(1-p)^{(N-m)/2} \tag{1·2·10}$$

In the special case where $p = q = \frac{1}{2}$ this assumes the symmetrical form

$$P_N(m) = \frac{N!}{[(N+m)/2]![(N-m)/2]!} \left(\frac{1}{2}\right)^N$$

Examples Suppose that $p = q = \frac{1}{2}$ and that $N = 3$ as illustrated in Fig. 1·2·1. Then the possible numbers of steps to the right are $n_1 = 0, 1, 2,$ or 3; the corresponding displacements are $m = -3, -1, 1,$ or 3; the corresponding probabilities are (as is clear from Fig. 1·2·1)

$$W_3(n_1) = P_3(m) = \tfrac{1}{8}, \tfrac{3}{8}, \tfrac{3}{8}, \tfrac{1}{8} \tag{1·2·11}$$

Figure (1·2·3) illustrates the binomial distribution for the same case where $p = q = \frac{1}{2}$, but with the total number of steps $N = 20$. The envelope of these discrete values of $P_N(m)$ is a bell-shaped curve. The physical sig-

* Note that (1·2·3) shows that $(N+m)$ and $(N-m)$ are even integers, since they equal $2n_1$ and $2n_2$, respectively.

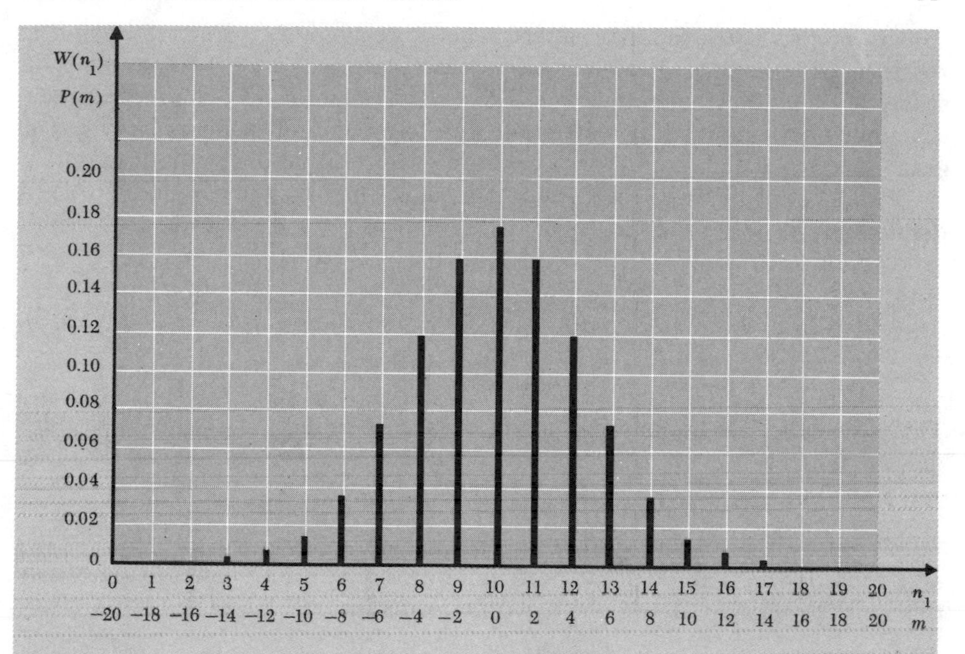

Fig. 1·2·3 Binomial probability distribution for $p = q = \frac{1}{2}$ when $N = 20$ steps. The graph shows the probability $W_N(n_1)$ of n_1 right steps, or equivalently the probability $P_N(m)$ of a net displacement of m units to the right.

nificance of this is obvious. After N random steps, the probability of the particle being a distance of N steps away from the origin is very small, while the probability of its being located in the vicinity of the origin is largest.

1·3 *General discussion of mean values*

Let u be a variable which can assume any of the M discrete values

$$u_1, u_2, \ldots, u_M$$

with respective probabilities

$$P(u_1), P(u_2), \ldots, P(u_M)$$

The mean (or average) value of u is denoted by $\bar{u}$ and is defined by

$$\bar{u} \equiv \frac{P(u_1)u_1 + P(u_2)u_2 + \cdots + P(u_M)u_M}{P(u_1) + P(u_2) + \cdots + P(u_M)}$$

or, in shorter notation, by

$$\bar{u} \equiv \frac{\sum_{i=1}^{M} P(u_i)u_i}{\sum_{i=1}^{M} P(u_i)} \tag{1·3·1}$$

This is, of course, the familiar way of computing averages. For example, if u represents the grade of a student on an examination and $P(u)$ the number of students obtaining this grade, then Eq. (1·3·1) asserts that the mean grade is computed by multiplying each grade by the number of students having this grade, adding this up, and then dividing by the total number of students.

More generally, if $f(u)$ is any function of u, then the mean value of $f(u)$ is defined by

$$\overline{f(u)} \equiv \frac{\sum_{i=1}^{M} P(u_i)f(u_i)}{\sum_{i=1}^{M} P(u_i)} \tag{1·3·2}$$

This expression can be simplified. Since $P(u_i)$ is defined as a probability, the quantity

$$P(u_1) + P(u_2) + \cdots + P(u_M) \equiv \sum_{i=1}^{M} P(u_i)$$

represents the probability that u assumes any one of its possible values and this must be unity. Hence one has quite generally

▶ $$\sum_{i=1}^{M} P(u_i) = 1 \tag{1·3·3}$$

This is the so-called "normalization condition" satisfied by every probability. As a result, the general definition (1·3·2) becomes

▶ $$\overline{f(u)} \equiv \sum_{i=1}^{M} P(u_i)f(u_i) \tag{1·3·4}$$

Note the following simple results. If $f(u)$ and $g(u)$ are any two functions of u, then

$$\overline{f(u) + g(u)} = \sum_{i=1}^{M} P(u_i)[f(u_i) + g(u_i)] = \sum_{i=1}^{M} P(u_i)f(u_i) + \sum_{i=1}^{M} P(u_i)g(u_i)$$

or

▶ $$\overline{f(u) + g(u)} = \overline{f(u)} + \overline{g(u)} \tag{1·3·5}$$

Furthermore, if c is any constant, it is clear that

▶ $$\overline{cf(u)} = c\overline{f(u)} \tag{1·3·6}$$

Some simple mean values are particularly useful for describing characteristic features of the probability distribution P. One of these is the mean value $\bar{u}$ (e.g., the mean grade of a class of students). This is a measure of the central value of u about which the various values u_i are distributed. If one measures the values of u from their mean value $\bar{u}$, i.e., if one puts

$$\Delta u \equiv u - \bar{u} \tag{1·3·7}$$

then
$$\overline{\Delta u} = \overline{(u - \bar{u})} = \bar{u} - \bar{u} = 0 \tag{1·3·8}$$

This says merely that the mean value of the deviation from the mean vanishes.

Another useful mean value is

$$\overline{(\Delta u)^2} \equiv \sum_{i=1}^{M} P(u_i)(u_i - \bar{u})^2 \geq 0 \tag{1·3·9}$$

which is called the "second moment of u about its mean," or more simply the "dispersion of u." This can never be negative, since $(\Delta u)^2 \geq 0$ so that each term in the sum contributes a nonnegative number. Only if $u_i = \bar{u}$ for *all* values u_i will the dispersion vanish. The larger the spread of values of u_i about $\bar{u}$, the larger the dispersion. The dispersion thus measures the amount of scatter of values of the variable about its mean value (e.g., scatter in grades about the mean grade of the students). Note the following general relation, which is often useful in computing the dispersion:

$$\overline{(u - \bar{u})^2} = \overline{(u^2 - 2u\,\bar{u} + \bar{u}^2)} = \overline{u^2} - 2\bar{u}\bar{u} + \bar{u}^2$$

or

$$\overline{(u - \bar{u})^2} = \overline{u^2} - \bar{u}^2 \tag{1·3·10}$$

Since the left side must be positive it also follows that

$$\overline{u^2} \geq \bar{u}^2 \tag{1·3·11}$$

One can define further mean values such as $\overline{(\Delta u)^n}$, the "nth moment of u about its mean," for integers $n > 2$. These are, however, less commonly useful.

Note that a knowledge of $P(u)$ implies complete information about the actual distribution of values of the variable u. A knowledge of a few moments, like $\bar{u}$ and $\overline{(\Delta u)^2}$, implies only partial, though useful, knowledge of the characteristics of this distribution. A knowledge of *some* mean values is *not* sufficient to determine $P(u)$ completely (unless one knows the moments $\overline{(\Delta u)^n}$ for *all* values of n). But by the same token it is often true that a calculation of the probability distribution function $P(u)$ may be quite difficult, whereas some simple mean values can be readily calculated directly without an explicit knowledge of $P(u)$. We shall illustrate some of these comments in the following pages.

1·4 *Calculation of mean values for the random walk problem*

In (1·2·6) we found that the probability, in a total of N steps, of making n_1 steps to the right (and $N - n_1 \equiv n_2$ steps to the left) is

$$W(n_1) = \frac{N!}{n_1!(N - n_1)!}\, p^{n_1} q^{N-n_1} \tag{1·4·1}$$

(For the sake of simplicity, we omit attaching the subscript N to W when no confusion is likely to arise.)

Let us first verify the normalization, i.e., the condition

$$\sum_{n_1=0}^{N} W(n_1) = 1 \tag{1·4·2}$$

which says that the probability of making any number of right steps between 0 and N must be unity. Substituting (1·4·1) into (1·4·2), we obtain

$$\begin{aligned} \sum_{n_1=0}^{N} \frac{N!}{n_1!(N-n_1)!} p^{n_1} q^{N-n_1} &= (p+q)^N && \text{by the binomial theorem} \\ &= 1^N = 1 && \text{since } q \equiv 1-p \end{aligned}$$

which verifies the result.

What is the mean number $\bar{n}_1$ of steps to the right? By definition

$$\bar{n}_1 \equiv \sum_{n_1=0}^{N} W(n_1) n_1 = \sum_{n_1=0}^{N} \frac{N!}{n_1!(N-n_1)!} p^{n_1} q^{N-n_1} n_1 \tag{1·4·3}$$

If it were not for that extra factor of n_1 in each term of the last sum, this would again be the binomial expansion and hence trivial to sum. The factor n_1 spoils this lovely situation. But there is a very useful general procedure for handling such an extra factor so as to reduce the sum to simpler form. Let us consider the purely mathematical problem of evaluating the sum occurring in (1·4·3), where p and q are considered to be any two *arbitrary* parameters. Then one observes that the extra factor n_1 can be produced by differentiation so that

$$n_1 p^{n_1} = p \frac{\partial}{\partial p} (p^{n_1})$$

Hence the sum of interest can be written in the form

$$\begin{aligned} \sum_{n_1=0}^{N} \frac{N!}{n_1!(N-n_1)!} p^{n_1} q^{N-n_1} n_1 &= \sum_{n_1=0}^{N} \frac{N!}{n_1!(N-n_1)!} \left[p \frac{\partial}{\partial p} (p^{n_1}) \right] q^{N-n_1} \\ &= p \frac{\partial}{\partial p} \left[\sum_{n_1=0}^{N} \frac{N!}{n_1!(N-n_1)!} p^{n_1} q^{N-n_1} \right] && \text{by interchanging order of summation and differentiation} \\ &= p \frac{\partial}{\partial p} (p+q)^N && \text{by the binomial theorem} \\ &= pN(p+q)^{N-1} \end{aligned}$$

Since this result is true for arbitrary values of p and q, it must also be valid in our particular case of interest where p is some specified constant and $q \equiv 1 - p$. Then $p + q = 1$ so that (1·4·3) becomes simply

▶ $$\bar{n}_1 = Np \tag{1·4·4}$$

We could have guessed this result. Since p is the probability of making a right step, the mean number of right steps in a *total* of N steps is simply given by $N \cdot p$. Clearly, the mean number of left steps is similarly equal to

$$\bar{n}_2 = Nq \tag{1·4·5}$$

Of course

$$\bar{n}_1 + \bar{n}_2 = N(p+q) = N$$

add up properly to the total number of steps.

The displacement (measured to the right in units of the step length l) is $m = n_1 - n_2$. Hence we get for the mean displacement

▶
$$\bar{m} = \overline{n_1 - n_2} = \bar{n}_1 - \bar{n}_2 = N(p-q) \tag{1·4·6}$$

If $p = q$, then $\bar{m} = 0$. This must be so since there is then complete symmetry between right and left directions.

Calculation of the dispersion Let us now calculate $\overline{(\Delta n_1)^2}$. By (1·3·10) one has

$$\overline{(\Delta n_1)^2} \equiv \overline{(n_1 - \bar{n}_1)^2} = \overline{n_1^2} - \bar{n}_1^2 \tag{1·4·7}$$

We already know $\bar{n}_1$. Thus we need to compute $\overline{n_1^2}$.

$$\begin{aligned} \overline{n_1^2} &= \sum_{n_1=0}^{N} W(n_1) n_1^2 \\ &= \sum_{n_1=0}^{N} \frac{N!}{n_1!(N-n_1)!} p^{n_1} q^{N-n_1} n_1^2 \end{aligned} \tag{1·4·8}$$

Considering p and q as arbitrary parameters and using the same trick of differentiation as before, one can write

$$n_1^2 p^{n_1} = n_1 \left(p \frac{\partial}{\partial p}\right)(p^{n_1}) = \left(p \frac{\partial}{\partial p}\right)^2 (p^{n_1})$$

Hence the sum in (1·4·8) can be written in the form

$$\sum_{n_1=0}^{N} \frac{N!}{n_1!(N-n_1)!} \left(p \frac{\partial}{\partial p}\right)^2 p^{n_1} q^{N-n_1}$$

$$= \left(p \frac{\partial}{\partial p}\right)^2 \sum_{n_1=0}^{N} \frac{N!}{n_1(N-n_1)!} p^{n_1} q^{N-n_1} \qquad \text{by interchanging order of summation and differentiation}$$

$$= \left(p \frac{\partial}{\partial p}\right)^2 (p+q)^N \qquad \text{by the binomial theorem}$$

$$= \left(p \frac{\partial}{\partial p}\right) [pN(p+q)^{N-1}]$$

$$= p[N(p+q)^{N-1} + pN(N-1)(p+q)^{N-2}]$$

The case of interest in (1·4·8) is that where $p + q = 1$. Thus (1·4·8) becomes simply

$$\begin{aligned}\overline{n_1{}^2} &= p[N + pN(N - 1)] \\ &= Np[1 + pN - p] \\ &= (Np)^2 + Npq \qquad \text{since } 1 - p = q \\ &= \bar{n}_1{}^2 + Npq \qquad \text{by } (1\cdot4\cdot4)\end{aligned}$$

Hence (1·4·7) gives for the dispersion of n_1 the result

$$\blacktriangleright \qquad \overline{(\Delta n_1)^2} = Npq \tag{1·4·9}$$

The quantity $\overline{(\Delta n_1)^2}$ is quadratic in the displacement. Its square root, i.e., the rms (root-mean-square) deviation $\Delta^* n_1 \equiv [\overline{(\Delta n_1)^2}]^{\frac{1}{2}}$, is a linear measure of the width of the range over which n_1 is distributed. A good measure of the *relative* width of this distribution is then

$$\frac{\Delta^* n_1}{\bar{n}_1} = \frac{\sqrt{Npq}}{Np} = \sqrt{\frac{q}{p}}\frac{1}{\sqrt{N}}$$

In particular,

for $p = q = \frac{1}{2}$, $$\frac{\Delta^* n_1}{\bar{n}_1} = \frac{1}{\sqrt{N}}$$

Note that as N increases, the mean value $\bar{n}_1$ increases like N, but the width $\Delta^* n_1$ increases only like $N^{\frac{1}{2}}$. Hence the *relative* width $\Delta^* n_1/\bar{n}_1$ decreases with increasing N like $N^{-\frac{1}{2}}$.

One can also compute the dispersion of m, i.e., the dispersion of the net displacement to the right. By (1·2·3)

$$m = n_1 - n_2 = 2n_1 - N \tag{1·4·10}$$

Hence one obtains

$$\Delta m \equiv m - \bar{m} = (2n_1 - N) - (2\bar{n}_1 - N) = 2(n_1 - \bar{n}_1) = 2\Delta n_1 \tag{1·4·11}$$

and $$(\Delta m)^2 = 4(\Delta n_1)^2$$

Taking averages, one gets by (1·4·9)

$$\blacktriangleright \qquad \overline{(\Delta m)^2} = 4\overline{(\Delta n_1)^2} = 4Npq \tag{1·4·12}$$

In particular,

for $p = q = \frac{1}{2}$, $$\overline{(\Delta m)^2} = N$$

Example Consider the case of $N = 100$ steps, where $p = q = \frac{1}{2}$. Then the mean number $\bar{n}_1$ of steps to the right (or to the left) is 50; the mean displacement $\bar{m} = 0$. The root-mean-square deviation of the displacement is $[\overline{(\Delta m^2)}]^{\frac{1}{2}} = 10$ steps.

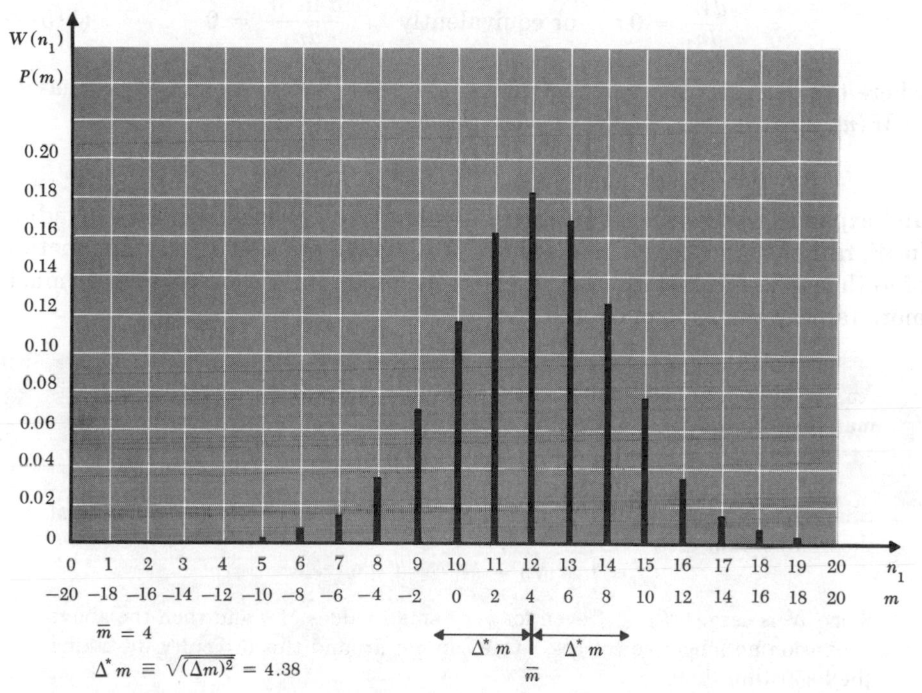

***Fig. 1·4·1** Binomial probability distribution for $p = 0.6$ and $q = 0.4$, when $N = 20$ steps. The graph shows again the probability $W(n_1)$ of n_1 right steps, or equivalently, the probability $P(m)$ of a net displacement of m units to the right. The mean values $\bar{m}$ and $\overline{(\Delta m)^2}$ are also indicated.*

1·5 *Probability distribution for large N*

When N is large, the binomial probability distribution $W(n_1)$ of (1·4·1) tends to exhibit a pronounced maximum at some value $n_1 = \tilde{n}_1$, and to decrease rapidly as one goes away from $\tilde{n}_1$ (see, for example, Fig. 1·4·1). Let us exploit this fact to find an approximate expression for $W(n_1)$ valid when N is sufficiently large.

If N is large and we consider regions near the maximum of W where n_1 is also large, the fractional change in W when n_1 changes by unity is relatively quite small, i.e.,

$$|W(n_1 + 1) - W(n_1)| \ll W(n_1) \tag{1·5·1}$$

Thus W can, to good approximation, be considered as a continuous function of the continuous variable n_1, although only integral values of n_1 are of physical relevance. The location $n_1 = \tilde{n}$ of the maximum of W is then approximately determined by the condition

$$\frac{dW}{dn_1} = 0 \qquad \text{or equivalently} \qquad \frac{d \ln W}{dn_1} = 0 \tag{1·5·2}$$

where the derivatives are evaluated for $n_1 = \tilde{n}_1$. To investigate the behavior of $W(n_1)$ near its maximum, we shall put

$$n_1 = \tilde{n}_1 + \eta \tag{1·5·3}$$

and expand $\ln W(n_1)$ in a Taylor's series about $\tilde{n}_1$. The reason for expanding $\ln W$, rather than W itself, is that $\ln W$ is a much more slowly varying function of n_1 than W. Thus the power series expansion for $\ln W$ should converge much more rapidly than the one for W.

An example may make this clearer. Suppose one wishes to find an approximate expression, valid for $y \ll 1$, of the function

$$f \equiv (1 + y)^{-N}$$

where N is large. Direct expansion in Taylor's series (or by the binomial theorem) would give

$$f = 1 - Ny + \tfrac{1}{2}N(N + 1)y^2 \cdots$$

Since N is large, $Ny \gtrsim 1$ even for very small values of y and then the above expansion no longer converges. One can get around this difficulty by taking the logarithm first:

$$\ln f = -N \ln (1 + y)$$

Expanding this in Taylor's series, one obtains

$$\ln f = -N(y - \tfrac{1}{2}y^2 \cdots)$$

or

$$f = e^{-N(y - \frac{1}{2}y^2 \cdots)}$$

which is valid as long as $y \lesssim 1$.

Expanding $\ln W$ in Taylor's series, one obtains

$$\ln W(n_1) = \ln W(\tilde{n}_1) + B_1\eta + \tfrac{1}{2}B_2\eta^2 + \tfrac{1}{6}B_3\eta^3 + \cdots \tag{1·5·4}$$

where

$$B_k \equiv \frac{d^k \ln W}{dn_1{}^k} \tag{1·5·5}$$

is the kth derivative of $\ln W$ evaluated at $n_1 = \tilde{n}_1$. Since one is expanding about a maximum, $B_1 = 0$ by (1·5·2). Also, since W is a maximum, it follows that the term $\frac{1}{2}B_2\eta^2$ must be negative, i.e., B_2 must be negative. To make this explicit, let us write $B_2 = -|B_2|$. Hence (1·5·4) yields, putting $\tilde{W} = W(\tilde{n}_1)$,

$$W(n_1) = \tilde{W}\, e^{\frac{1}{2}B_2\eta^2 + \frac{1}{6}B_3\eta^3 \cdots} = \tilde{W}\, e^{-\frac{1}{2}|B_2|\eta^2}\, e^{\frac{1}{6}B_3\eta^3 \cdots} \tag{1·5·6}$$

In the region where η is sufficiently small, higher-order terms in the expansion can be neglected so that one obtains in first approximation an expression of the simple form

$$W(n_1) = \tilde{W}\, e^{-\frac{1}{2}|B_2|\eta^2} \tag{1·5·7}$$

Let us now investigate the expansion (1·5·4) in greater detail. By (1·4·1) one has

$$\ln W(n_1) = \ln N! - \ln n_1! - \ln(N - n_1)! + n_1 \ln p + (N - n_1) \ln q \quad (1\cdot5\cdot8)$$

But, if n is any large integer so that $n \gg 1$, then $\ln n!$ can be considered an almost continuous function of n, since $\ln n!$ changes only by a small fraction of itself if n is changed by a small integer. Hence

$$\frac{d \ln n!}{dn} \approx \frac{\ln (n+1)! - \ln n!}{1} = \ln \frac{(n+1)!}{n!} = \ln (n+1)$$

Thus

for $n \gg 1$,
$$\frac{d \ln n!}{dn} \approx \ln n \quad (1\cdot5\cdot9)$$

Hence (1·5·8) yields

$$\frac{d \ln W}{dn_1} = -\ln n_1 + \ln (N - n_1) + \ln p - \ln q \quad (1\cdot5\cdot10)$$

By equating this first derivative to zero one finds the value $n_1 = \tilde{n}_1$, where W is maximum. Thus one obtains the condition

$$\ln \left[\frac{(N - \tilde{n}_1)}{\tilde{n}_1} \frac{p}{q} \right] = 0$$

or

$$(N - \tilde{n}_1)p = \tilde{n}_1 q$$

so that

▶ $$\tilde{n}_1 = Np \quad (1\cdot5\cdot11)$$

since $p + q = 1$.

Further differentiation of (1·5·10) yields

$$\frac{d^2 \ln W}{dn_1^2} = -\frac{1}{n_1} - \frac{1}{N - n_1} \quad (1\cdot5\cdot12)$$

Evaluating this for the value $n_1 = \tilde{n}_1$ given in (1·5·11), one gets

$$B_2 = -\frac{1}{Np} - \frac{1}{N - Np} = -\frac{1}{N}\left(\frac{1}{p} + \frac{1}{q}\right)$$

or

▶ $$B_2 = -\frac{1}{Npq} \quad (1\cdot5\cdot13)$$

since $p + q = 1$. Thus B_2 is indeed negative, as required for W to exhibit a maximum.

By taking further derivatives, one can examine the higher-order terms in the expansion (1·5·4). Thus, differentiating (1·5·12), one obtains

$$B_3 = \frac{1}{\tilde{n}_1^2} - \frac{1}{(N - \tilde{n}_1)^2} = \frac{1}{N^2p^2} - \frac{1}{N^2q^2}$$

or

$$|B_3| = \frac{|q^2 - p^2|}{N^2p^2q^2} < \frac{1}{N^2p^2q^2}$$

Similarly, $$B_4 = -\frac{2}{\tilde{n}_1{}^3} - \frac{2}{(N-\tilde{n}_1)^3} = -2\left(\frac{1}{N^3p^3} + \frac{1}{N^3q^3}\right)$$

or $$|B_4| = \frac{2(p^3+q^3)}{N^3p^3q^3} < \frac{4}{N^3q^3p^3}$$

Thus it is seen that the kth term in (1·5·4) is smaller in magnitude than $\eta^k/(Npq)^{k-1}$. The neglect of terms beyond $B_2\eta^2$, which leads to (1·5·7), is therefore justified if η is sufficiently small so that*

$$\eta \ll Npq \tag{1·5·14}$$

On the other hand, the factor $\exp(-\frac{1}{2}|B_2|\eta^2)$ in (1·5·6) causes W to fall off very rapidly with increasing values of $|\eta|$ if $|B_2|$ is large. Indeed, if

$$|B_2|\eta^2 = \frac{\eta^2}{Npq} \gg 1 \tag{1·5·15}$$

the probability $W(n_1)$ becomes negligibly small compared to $W(\tilde{n}_1)$. Hence it follows that if η is still small enough to satisfy (1·5·14) up to values of η so large that (1·5·15) is also satisfied, then (1·5·7) provides an excellent approximation for W throughout the *entire* region where W has an appreciable magnitude. This condition of simultaneous validity of (1·5·14) and (1·5·15) requires that

$$\sqrt{Npq} \ll \eta \ll Npq$$

i.e., that $$Npq \gg 1 \tag{1·5·16}$$

This shows that, throughout the entire domain where the probability W is not negligibly small, the expression (1·5·7) provides a very good approximation to the extent that N is large and that neither p nor q is too small.†

The value of the constant $\tilde{W}$ in (1·5·7) can be determined from the normalization condition (1·4·2). Since W and n_1 can be treated as quasicontinuous variables, the sum over all integral values of n_1 can be approximately replaced by an integral. Thus the normalization condition can be written

$$\sum_{n_1=0}^{N} W(n_1) \approx \int W(n_1)\,dn_1 = \int_{-\infty}^{\infty} W(\tilde{n}_1+\eta)\,d\eta = 1 \tag{1·5·17}$$

Here the integral over η can to excellent approximation be extended from $-\infty$ to $+\infty$, since the integrand makes a negligible contribution to the integral wherever $|\eta|$ is large enough so that W is far from its pronounced maximum value. Substituting (1·5·7) into (1·5·17) and using (A·4·2), one obtains

$$\tilde{W}\int_{-\infty}^{\infty} e^{-\frac{1}{2}|B_2|\eta^2}\,d\eta = \tilde{W}\sqrt{\frac{2\pi}{|B_2|}} = 1$$

* Note that the condition (1·5·1) is equivalent to $|\partial W/\partial n_1| \ll W$, i.e., to $|B_2\eta| = (Npq)^{-1}|\eta| \ll 1$ by virtue of (1·5·7) and (1·5·13). Thus it is also satisfied in the domain (1·5·14) where W is not too small.

† When $p \ll 1$ or $q \ll 1$, it is possible to obtain a different approximation for the binomial distribution, namely the so-called "Poisson distribution" (see Problem 1.9).

Thus (1·5·7) becomes

$$W(n_1) = \sqrt{\frac{|B_2|}{2\pi}}\, e^{-\frac{1}{2}|B_2|(n_1-\tilde{n}_1)^2} \qquad (1\cdot5\cdot18)$$

The reasoning leading to the functional form (1·5·7) or (1·5·18), the so-called "Gaussian distribution," has been very general in nature. Hence it is not surprising that Gaussian distributions occur very frequently in statistics whenever one is dealing with large numbers. In our case of the binomial distribution, the expression (1·5·18) becomes, by virtue of (1·5·11) and (1·5·13),

$$W(n_1) = (2\pi Npq)^{-\frac{1}{2}} \exp\left[-\frac{(n_1 - Np)^2}{2Npq}\right] \qquad (1\cdot5\cdot19)$$

Note that (1·5·19) is, for large values of N and n_1, much simpler than (1·4·1), since it does not require the evaluation of large factorials. Note also that, by (1·4·4) and (1·4·9), the expression (1·5·19) can be written in terms of the mean values $\bar{n}_1$ and $\overline{(\Delta n_1)^2}$ as

$$W(n_1) = [2\pi\overline{(\Delta n_1)^2}]^{-\frac{1}{2}} \exp\left[-\frac{(n_1 - \bar{n}_1)^2}{2\overline{(\Delta n_1)^2}}\right]$$

1·6 *Gaussian probability distributions*

The Gaussian approximation (1·5·19) also yields immediately the probability $P(m)$ that in a large number of N steps the net displacement is m. The corresponding number of right steps is, by (1·2·9), $n_1 = \frac{1}{2}(N + m)$. Hence (1·5·19) gives

$$P(m) = W\left(\frac{N+m}{2}\right) = [2\pi Npq]^{-\frac{1}{2}} \exp\left\{-\frac{[m - N(p-q)]^2}{8Npq}\right\} \qquad (1\cdot6\cdot1)$$

since $n_1 - Np = \frac{1}{2}[N + m - 2Np] = \frac{1}{2}[m - N(p - q)]$. By (1·2·3) one has $m = 2n_1 - N$, so that m assumes here integral values separated by an amount $\Delta m = 2$.

We can also express this result in terms of the actual displacement variable x,

$$x = ml \qquad (1\cdot6\cdot2)$$

where l is the length of each step. If l is small compared to the smallest length of interest in the physical problem under consideration,* the fact that x can only assume values in discrete increments of $2l$, rather than all values continuously, becomes unimportant. Furthermore, when N is large, the probability $P(m)$ of occurrence of a displacement m does not change significantly from

* For example, if one considers the random motion (diffusion) of an atom in a solid, the step length l is of the order of one lattice spacing, i.e., about 10^{-8} cm. But on the macroscopic scale of experimental measurement the smallest length L of relevance might be 1 micron = 10^{-4} cm.

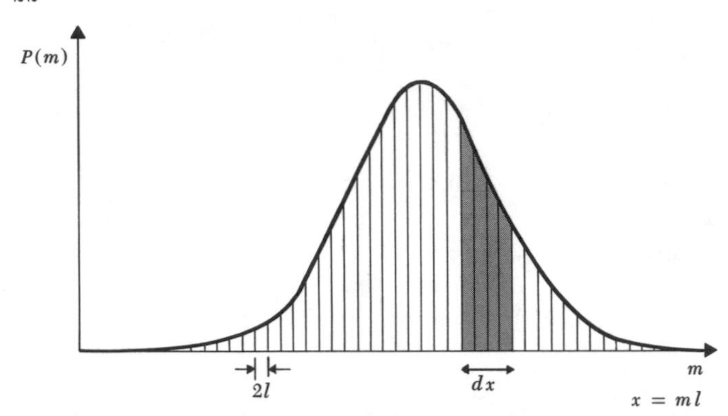

***Fig. 1·6·1** The probability $P(m)$ of a net displacement of m units when the total number N of steps is very large and the step length l is very small.*

one possible value of m to an adjacent one; i.e., $|P(m+2) - P(m)| \ll P(m)$. Therefore $P(m)$ can be regarded as a smooth function of x. A bar graph of the type shown in Fig. 1·4·1 then assumes the character illustrated in Fig. 1·6·1, where the bars are very densely spaced and their envelope forms a smooth curve.

Under these circumstances it is possible to regard x as a continuous variable on a macroscopic scale and to ask for the probability that the particle is found after N steps in the range between x and $x + dx$.* Since m assumes only integral values separated by $\Delta m = 2$, the range dx contains $dx/2l$ possible values of m, all of which occur with nearly the same probability $P(m)$. Hence the probability of finding the particle anywhere in the range between x and $x + dx$ is simply obtained by summing $P(m)$ over all values of m lying in dx, i.e., by multiplying $P(m)$ by $dx/2l$. This probability is thus proportional to dx (as one would expect) and can be written as

$$\mathcal{P}(x)\,dx = P(m)\,\frac{dx}{2l} \tag{1·6·3}$$

where the quantity $\mathcal{P}(x)$, which is independent of the magnitude of dx, is called a "probability *density*." Note that it must be multiplied by a differential element of length dx to yield a probability.

By using (1·6·1) one then obtains

$$\blacktriangleright \qquad \mathcal{P}(x)\,dx = \frac{1}{\sqrt{2\pi}\,\sigma}\,e^{-(x-\mu)^2/2\sigma^2}\,dx \tag{1·6·4}$$

where we have used the abbreviations

$$\mu \equiv (p - q)Nl \tag{1·6·5}$$

and

$$\sigma \equiv 2\sqrt{Npq}\,l \tag{1·6·6}$$

* Here dx is understood to be a differential in the *macro*scopic sense, i.e., $dx \ll L$, where L is the smallest dimension of relevance in the *macro*scopic discussion, but $dx \gg l$. (In other words, dx is *macro*scopically small, but *micro*scopically large.)

The expression (1·6·4) is the standard form of the Gaussian probability distribution. The great generality of the argument leading to (1·5·19) suggests that such Gaussian distributions occur very frequently in probability theory whenever one deals with large numbers.

Using (1·6·4), one can quite generally compute the mean values $\bar{x}$ and $\overline{(x - \bar{x})^2}$. In calculating these mean values, sums over all possible intervals dx become, of course, integrations. (The limits of x can be taken as $-\infty < x < \infty$, since $\mathcal{P}(x)$ itself becomes negligibly small whenever $|x|$ is so large as to lead to a displacement inaccessible in N steps.)

First we verify that $\mathcal{P}(x)$ is properly normalized, i.e., that the probability of the particle being somewhere is unity. Thus

$$\begin{aligned}\int_{-\infty}^{\infty} \mathcal{P}(x)\, dx &= \frac{1}{\sqrt{2\pi}\,\sigma} \int_{-\infty}^{\infty} e^{-(x-\mu)^2/2\sigma^2}\, dx \\ &= \frac{1}{\sqrt{2\pi}\,\sigma} \int_{-\infty}^{\infty} e^{-y^2/2\sigma^2}\, dy \\ &= \frac{1}{\sqrt{2\pi}\,\sigma} \sqrt{\pi 2\sigma^2} \\ &= 1 \end{aligned} \qquad (1\cdot 6\cdot 7)$$

Here we have put $y \equiv x - \mu$ and evaluated the integral by (A·4·2).

Next we calculate the mean value

$$\begin{aligned}\bar{x} &\equiv \int_{-\infty}^{\infty} x\mathcal{P}(x)\, dx \\ &= \frac{1}{\sqrt{2\pi}\,\sigma} \int_{-\infty}^{\infty} x\, e^{-(x-\mu)^2/2\sigma^2}\, dx \\ &= \frac{1}{\sqrt{2\pi}\,\sigma} \left[\int_{-\infty}^{\infty} y\, e^{-y^2/2\sigma^2}\, dy + \mu \int_{-\infty}^{\infty} e^{-y^2/2\sigma^2}\, dy \right]\end{aligned}$$

Since the integrand in the first integral is an odd function of y, the first integral

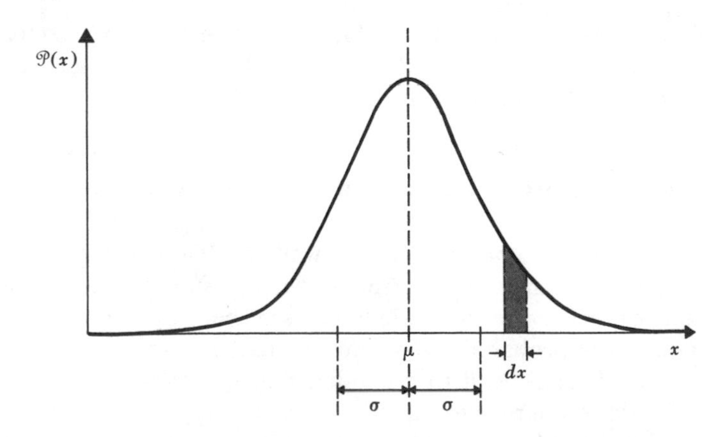

***Fig. 1·6·2** The Gaussian distribution. Here $\mathcal{P}(x)\, dx$ is the area under the curve in the interval between x and $x + dx$ and is thus the probability that the variable x lies in this range.*

vanishes by symmetry. The second integral is the same as that in (1·6·7), so that one gets

$$\bar{x} = \mu \tag{1·6·8}$$

This is a simple consequence of the fact that $\mathcal{P}(x)$ is only a function of $|x - \mu|$ and is thus symmetric about the position $x = \mu$ of its maximum. Hence this point corresponds also to the mean value of x.

The dispersion becomes

$$\begin{aligned}\overline{(x-\mu)^2} &\equiv \int_{-\infty}^{\infty} (x-\mu)^2 \mathcal{P}(x)\, dx \\ &= \frac{1}{\sqrt{2\pi}\,\sigma} \int_{-\infty}^{\infty} y^2\, e^{-y^2/2\sigma^2}\, dy \\ &= \frac{1}{\sqrt{2\pi}\,\sigma} \left[\frac{\sqrt{\pi}}{2} (2\sigma^2)^{\frac{3}{2}}\right] \\ &= \sigma^2\end{aligned}$$

where we have used the integral formulas (A·4·6). Thus

$$\overline{(\Delta x)^2} = \overline{(x-\mu)^2} = \sigma^2 \tag{1·6·9}$$

Hence σ is simply the root-mean-square deviation of x from the mean of the Gaussian distribution.

By (1·6·5) and (1·6·6), one then obtains for the random walk problem the relations

$$\bar{x} = (p - q)Nl \tag{1·6·10}$$

$$\overline{(\Delta x)^2} = 4Npql^2 \tag{1·6·11}$$

These results (derived here for the case of large N) agree, as they must, with the mean values $\bar{x} = \bar{m}l$ and $\overline{(\Delta x)^2} = \overline{(\Delta m)^2}l^2$ already calculated in (1·4·6) and (1·4·12) for the general case of arbitrary N.

GENERAL DISCUSSION OF THE RANDOM WALK

Our discussion of the random walk problem has yielded a great number of important results and introduced many fundamental concepts in probability theory. The approach we have used, based on combinatorial analysis to calculate the probability distribution, has, however, severe limitations. In particular, it is difficult to generalize this approach to other cases, for example, to situations where the length of each step is not the same or where the random walk takes place in more than one dimension. We now turn, therefore, to a discussion of more powerful methods which can readily be generalized and yet possess a basic simplicity and directness of their own.

1·7 *Probability distributions involving several variables*

The statistical description of a situation involving more than one variable requires only straightforward generalizations of the probability arguments applicable to a single variable. Let us then, for simplicity, consider the case of only two variables u and v which can assume the possible values

$$u_i \quad \text{where } i = 1, 2, \ldots, M$$

and

$$v_j \quad \text{where } j = 1, 2, \ldots, N$$

Let $P(u_i,v_j)$ be the probability that u assumes the value u_i *and* that v assumes the value v_j.

The probability that the variables u and v assume any of their possible sets of values must be unity; i.e., one has the normalization requirement

$$\sum_{i=1}^{M} \sum_{j=1}^{N} P(u_i,v_j) = 1 \tag{1·7·1}$$

where the summation extends over all possible values of u and all possible values of v.

The probability $P_u(u_i)$ that u assumes the value u_i, *irrespective* of the value assumed by the variable v, is the sum of the probabilities of all possible situations consistent with the given value of u_i; i.e.,

$$P_u(u_i) = \sum_{j=1}^{N} P(u_i,v_j) \tag{1·7·2}$$

where the summation is over all possible values of v_j. Similarly, the probability $P_v(v_j)$ that v assumes the value v_j, irrespective of the value assumed by u, is

$$P_v(v_j) = \sum_{i=1}^{M} P(u_i,v_j) \tag{1·7·3}$$

Each of the probabilities P_u and P_v is, of course, properly normalized. For example, by (1·7·2) and (1·7·1) one has

$$\sum_{i=1}^{M} P_u(u_i) = \sum_{i=1}^{M} \left[\sum_{j=1}^{N} P(u_i,v_j) \right] = 1 \tag{1·7·4}$$

An important special case occurs when the probability that one variable assumes a certain value does not depend on the value assumed by the other variable. The variables are then said to be "statistically independent" or "uncorrelated." The probability $P(u_i,v_j)$ can then be expressed very simply in terms of the probability $P_u(u_i)$ that u assumes the value u_i and the probability $P_v(v_j)$ that v assumes the value v_j. Indeed, in this case [the number of instances in the ensemble where $u = u_i$ *and* where simultaneously $v = v_j$] is simply obtained by multiplying [the number of instances where $u = u_i$] by [the number of instances where $v = v_j$]; hence

$$P(u_i,v_j) = P_u(u_i)P_v(v_j) \tag{1·7·5}$$

if u and v are statistically independent.

Let us now mention some properties of mean values. If $F(u,v)$ is any function of u and v, then its mean value is defined by

$$\overline{F(u,v)} \equiv \sum_{i=1}^{M} \sum_{j=1}^{N} P(u_i,v_j)F(u_i,v_j) \tag{1·7·6}$$

Note that if $f(u)$ is a function of u only, it also follows by (1 · 7 · 2) that

$$\overline{f(u)} = \sum_i \sum_j P(u_i,v_j)f(u_i) = \sum_i P_u(u_i)f(u_i) \tag{1·7·7}$$

If F and G are any functions of u and v, then one has the general result

$$\begin{aligned}\overline{F + G} &\equiv \sum_i \sum_j P(u_i,v_j)[F(u_i,v_j) + G(u_i,v_j)] \\ &= \sum_i \sum_j P(u_i,v_j)F(u_i,v_j) + \sum_i \sum_j P(u_i,v_j)G(u_i,v_j)\end{aligned}$$

or

▶ $$\overline{F + G} = \overline{F} + \overline{G} \tag{1·7·8}$$

i.e., the average of a sum equals simply the sum of the averages.

Given any two functions $f(u)$ and $g(v)$, one can also make a general statement about the mean value of their product *if* u and v are statistically independent variables. Indeed, one then finds

$$\begin{aligned}\overline{f(u)g(v)} &\equiv \sum_i \sum_j P(u_i,v_j)f(u_i)g(v_j) \\ &= \sum_i \sum_j P_u(u_i)P_v(v_j)f(u_i)g(v_j) \qquad \text{by (1·7·5)} \\ &= \Big[\sum_i P_u(u_i)f(u_i)\Big] \Big[\sum_j P_v(v_j)g(v_j)\Big]\end{aligned}$$

Thus

▶ $$\overline{f(u)g(v)} = \overline{f(u)}\;\overline{g(v)} \tag{1·7·9}$$

i.e., the average of a product equals the product of the averages *if* u and v are *statistically independent.* If u and v are statistically *not* independent, the statement (1 · 7 · 9) is in general *not* true.

The generalization of the definitions and results of this section to the case of more than two variables is immediate.

1·8 *Comments on continuous probability distributions*

Consider first the case of a single variable u which can assume any value in the continuous range $a_1 < u < a_2$. To give a probability description of such a situation, one can focus attention on any infinitesimal range of the variable between u and $u + du$ and ask for the probability that the variable assumes a value in this range. One expects that this probability is proportional to the magnitude of du if this interval is sufficiently small; i.e., one expects that this probability can be written in the form $\mathcal{P}(u)\,du$, where $\mathcal{P}(u)$ is independent of the size of du.* The quantity $\mathcal{P}(u)$ is called a "probability density." Note that it must be multiplied by du to yield an actual probability.

It is readily possible to reduce the problem dealing with a continuous variable to an equivalent discrete problem where the number of possible values of the variable becomes countable. It is only necessary to subdivide the accessible range $a_1 < u < a_2$ of the variable into (arbitrarily small) equal intervals of fixed size δu. Each such interval can then be labeled by some index i. The value of u in this interval can be denoted simply by u_i and the probability of finding the variable in this range by $P(u_i)$. One can then deal with a denumerable set of values of the variable u (each of them corresponding to one of the chosen fixed infinitesimal intervals). It also becomes clear that relations involving probabilities of discrete variables are equally valid for probabilities of continuous variables. For example, the simple properties (1·3·5) and (1·3·6) of mean values are also applicable if u is a continuous variable.

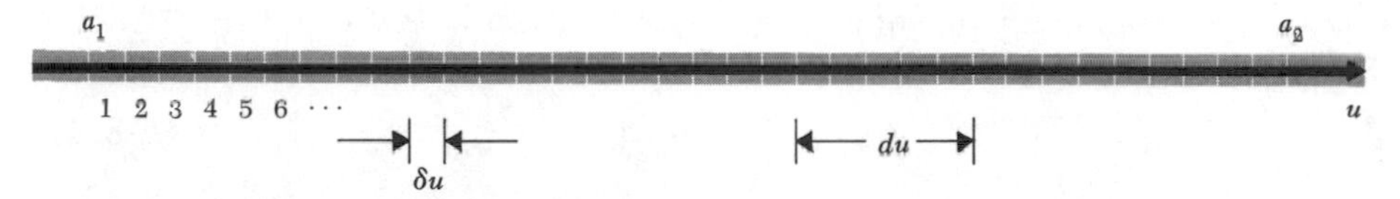

***Fig. 1·8·1** Subdivision of the range $a_1 < u < a_2$ of a continuous variable u into a countable number of infinitesimal intervals δu of fixed size.*

To make the connection between the continuous and discrete points of view quite explicit, note that in terms of the original infinitesimal subdivision interval δu,

$$P(u) = \mathcal{P}(u)\,\delta u$$

Similarly, if one considers any interval between u and $u + du$ which is such that du is macroscopically small although $du \gg \delta u$, then this interval contains $du/\delta u$ possible values of u_i for which the probability $P(u_i)$ has essentially the same value—call it simply $P(u)$. Then the probability $\mathcal{P}(u)\,du$ of

* Indeed, the probability must be expressible as a Taylor's series in powers of du and must vanish as $du \to 0$. Hence the leading term must be of the form $\mathcal{P}\,du$, while terms involving higher powers of du are negligible if du is sufficiently small.

the variable assuming a value between u and $u + du$ should be given by multiplying the probability $P(u_i)$ for assuming any discrete value in this range by the number $du/\delta u$ of discrete values in this range; i.e., one has properly

$$\mathcal{P}(u)\, du = P(u_i) \frac{du}{\delta u} = \frac{P(u)}{\delta u}\, du \tag{1·8·1}$$

Note that the sums involved in calculating normalization conditions or mean values can be written as integrals if the variable is continuous. For example, the normalization condition asserts that the sum of the probabilities over all possible values of the variable must equal unity; in symbols

$$\sum_i P(u_i) = 1 \tag{1·8·2}$$

But if the variable is continuous, one can first sum over all values of the variable in a range between u and $u + du$, thus obtaining the probability $\mathcal{P}(u)\, du$ that the variable lies in this range, and then complete the sum (1·8·2) by summing (i.e., integrating) over all such possible ranges du. Thus (1·8·2) is equivalent to

$$\int_{a_1}^{a_2} \mathcal{P}(u)\, du = 1 \tag{1·8·3}$$

which expresses the normalization condition in terms of the probability density $\mathcal{P}(u)$. Similarly, one can calculate mean values in terms of $\mathcal{P}(u)$. The general definition of the mean value of a function f was given in terms of discrete variables by (1·3·4) as

$$\overline{f(u)} = \sum_i P(u_i) f(u_i) \tag{1·8·4}$$

In a continuous description one can again sum first over all values between u and $u + du$ (this contributes to the sum an amount $\mathcal{P}(u)\, du\, f(u)$) and then integrate over all possible ranges du. Thus (1·8·4) is equivalent to the relation

$$\overline{f(u)} = \int_{a_1}^{a_2} \mathcal{P}(u) f(u)\, du \tag{1·8·5}$$

Remark Note that in some cases it is possible that the probability *density* $\mathcal{P}(u)$ itself becomes infinite for certain values of u. This does not lead to any difficulties as long as the *integral* $\int_{c_1}^{c_2} \mathcal{P}(u)\, du$, which measures the probability that u assumes any value in some arbitrary range between c_1 and c_2, always remains finite.

The extension of these comments to probabilities involving several variables is immediate. Consider, for example, the case of two variables u and v which can assume all values in the continuous respective ranges $a_1 < u < a_2$

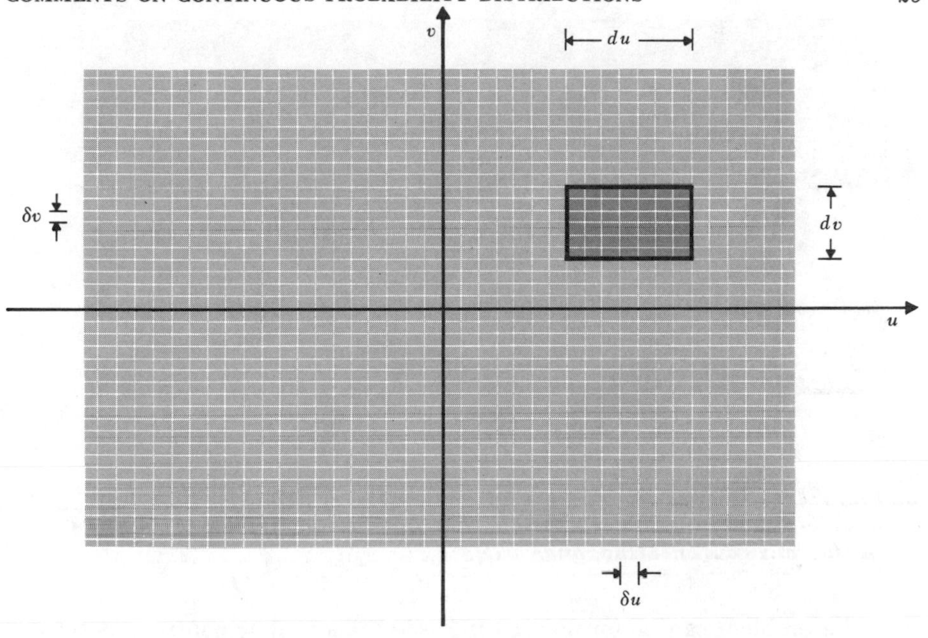

Fig. 1·8·2 Subdivision of the continuous variables u and v into small intervals of magnitude δu and δv.

and $b_1 < v < b_2$. One can then talk of the probability $\mathcal{P}(u,v)\,du\,dv$ that the variables lie in the ranges between u and $u + du$ and between v and $v + dv$, respectively, where $\mathcal{P}(u,v)$ is a probability density independent of the size of du and dv. It is again possible to reduce the problem to an equivalent one involving discrete countable values of the variables. It is only necessary to subdivide the possible values of u into fixed infinitesimal intervals of size δu and labeled by i, and those of v into fixed infinitesimal intervals of size δv and labeled by j. Then one can speak of the probability $P(u_i,v_j)$ that $u_i = u$ and that simultaneously $v = v_j$. Analogously to (1·8·1) one then has the relation

$$\mathcal{P}(u,v)\,du\,dv = P(u,v)\,\frac{du\,dv}{\delta u\;\delta v}$$

where the factor multiplying $P(u,v)$ is simply the number of infinitesimal cells of magnitude $\delta u\;\delta v$ contained in the range lying between u and $u + du$ and between v and $v + dv$.

The normalization condition (1·7·2) can then be written in terms of the probability density $\mathcal{P}(u,v)$ as

$$\int_{a_1}^{a_2}\int_{b_1}^{b_2} du\,dv\;\mathcal{P}(u,v) = 1 \tag{1·8·6}$$

Analogously to (1·7·7) one can also write

$$\overline{F(u,v)} = \int_{a_1}^{a_2}\int_{b_1}^{b_2} du\,dv\;\mathcal{P}(u,v)F(u,v) \tag{1·8·7}$$

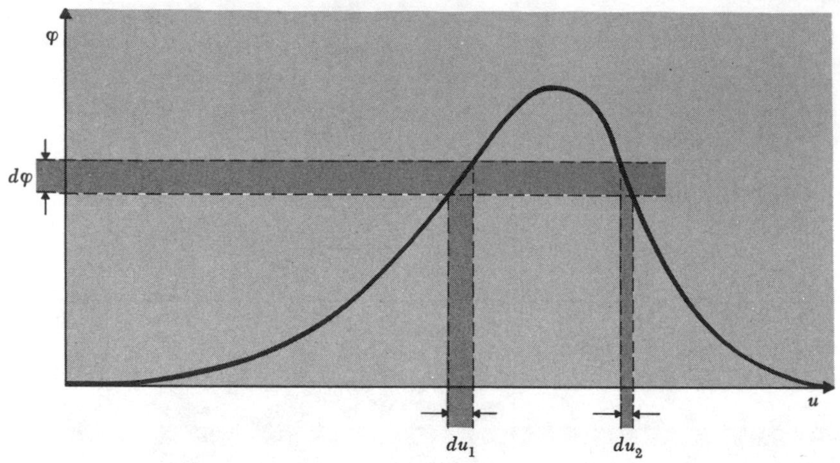

Fig. 1·8·3 Illustration showing a function $\varphi(u)$ which is such that $u(\varphi)$ is a double-valued function of φ. Here the range $d\varphi$ corresponds to u being either in the range du_1 or in the range du_2.

Since the problem can be formulated in discrete as well as continuous terms, the general properties (1·7·8) and (1·7·9) of mean values remain, of course, valid in the continuous case.

Functions of random variables Consider the case of a single variable u and suppose that $\varphi(u)$ is some continuous function of u. The following question arises quite frequently. If $\mathcal{P}(u)\,du$ is the probability that u lies in the range between u and $u + du$, what is the corresponding probability $W(\varphi)\,d\varphi$ that φ lies in the range between φ and $\varphi + d\varphi$? Clearly, the latter probability is obtained by adding up the probabilities for all those values u which are such that φ lies in the range between φ and $\varphi + d\varphi$; in symbols

$$W(\varphi)\,d\varphi = \int_{d\varphi} \mathcal{P}(u)\,du \tag{1·8·8}$$

Here u can be considered a function of φ and the integral extends over all those values of u lying in the range between $u(\varphi)$ and $u(\varphi + d\varphi)$. Thus (1·8·8) becomes simply

$$W(\varphi)\,d\varphi = \int_{\varphi}^{\varphi+d\varphi} \mathcal{P}(u) \left|\frac{du}{d\varphi}\right| d\varphi = \mathcal{P}(u) \left|\frac{du}{d\varphi}\right| d\varphi \tag{1·8·9}$$

The last step assumes that u is a single-valued function of φ and follows, since the integral is extended only over an infinitesimal range $d\varphi$. Since $u = u(\varphi)$, the right side of (1·8·9) can, of course, be expressed completely in terms of φ. If $u(\varphi)$ is not a single-valued function of φ, then the integral (1·8·8) may consist of several contributions similar to those of (1·8·9) (see Fig. 1·8·3).

Similar arguments can be used to find the probabilities for functions of *several* variables when the probabilities of the variables themselves are known.

Example Suppose that a two-dimensional vector $\boldsymbol{B}$ of constant length $B = |\boldsymbol{B}|$ is equally likely to point in any direction specified by the angle θ (see Fig. 1·8·4). The probability $\mathcal{P}(\theta)\,d\theta$ that this angle lies in the range between θ and $\theta + d\theta$ is then given by the ratio of the angular range $d\theta$ to the total angular range 2π subtended by a full circle; i.e.,

$$\mathcal{P}(\theta)\,d\theta = \frac{d\theta}{2\pi} \tag{1·8·10}$$

If the vector makes an angle θ with the x axis, its x component is given by

$$B_x = B\cos\theta \tag{1·8·11}$$

What is the probability $W(B_x)\,dB_x$ that the x component of this vector lies between B_x and $B_x + dB_x$? Clearly, B_x is always such that $-B \le B_x \le B$. In this interval an infinitesimal range between B_x and $B_x + dB_x$ corresponds to *two* possible infinitesimal ranges of $d\theta$ (see Fig. 1·8·4), each of magnitude $d\theta$ connected to dB_x through the relation (1·8·11) so that $dB_x = |B\sin\theta|\,d\theta$. By virtue of (1·8·10) the probability $W(B_x)\,dB_x$ is then given by

$$W(B_x)\,dB_x = 2\left[\frac{1}{2\pi}\frac{dB_x}{|B\sin\theta|}\right] = \frac{1}{\pi B}\frac{dB_x}{|\sin\theta|}$$

But, by (1·8·11),

$$|\sin\theta| = (1-\cos^2\theta)^{\frac{1}{2}} = \left[1-\left(\frac{B_x}{B}\right)^2\right]^{\frac{1}{2}}$$

Hence

$$W(B_x)\,dB_x = \begin{cases} \dfrac{dB_x}{\pi\sqrt{B^2 - B_x^2}} & \text{for } -B \le B_x \le B \\ 0 & \text{otherwise} \end{cases} \tag{1·8·12}$$

The probability density is maximum (indeed infinite) as $|B_x| \to B$ and is

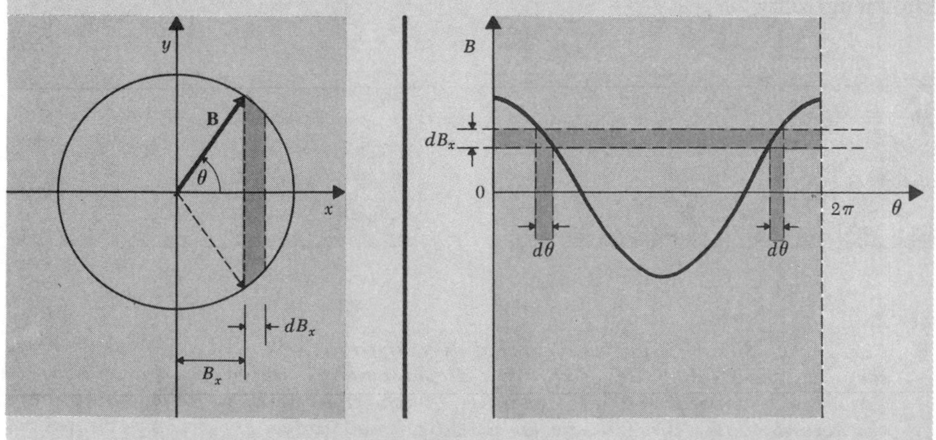

Fig. 1·8·4 Dependence of the x component $B_x = B\cos\theta$ of a two-dimensional vector B on its polar angle θ.

minimum when $B_x = 0$. This result is apparent from the geometry of Fig. (1·8·4), since a given narrow range of dB_x corresponds to a relatively large range of the angle θ when $B_x \approx B$ and to a very much narrower range of the angle θ when $B_x \approx 0$.

1 · 9 *General calculation of mean values for the random walk*

The comments of Sec. 1·7 allow one to calculate mean values for very general situations by very simple and transparent methods. Consider a quite general form of the one-dimensional random walk problem. Let s_i denote the displacement (positive or negative) in the ith step

> Let $w(s_i)\,ds_i$ be the probability that the ith displacement lies in the range between s_i and $s_i + ds_i$.

We assume again that this probability is independent of what displacements occur in any other steps. For simplicity we further assume that the probability distribution w is the same for each step i. Nevertheless, it is clear that the situation envisaged here is considerably more general than before, for we no longer necessarily assume a fixed magnitude l of displacement at each step, but a distribution of possible step lengths with relative probability specified by w.

We are interested in the total displacement x after N steps. We can ask for $\mathcal{P}(x)\,dx$, the probability that x lies in the range between x and $x + dx$. We can also ask for mean values (or moments) of x. In this section we show that the calculation of these moments can be achieved very simply without prior knowledge of $\mathcal{P}(x)$.

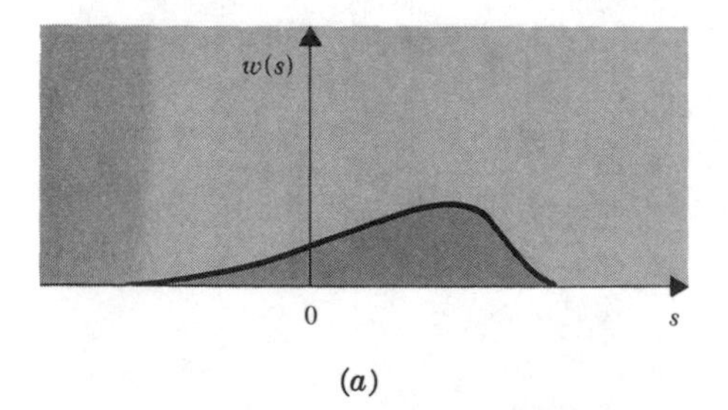

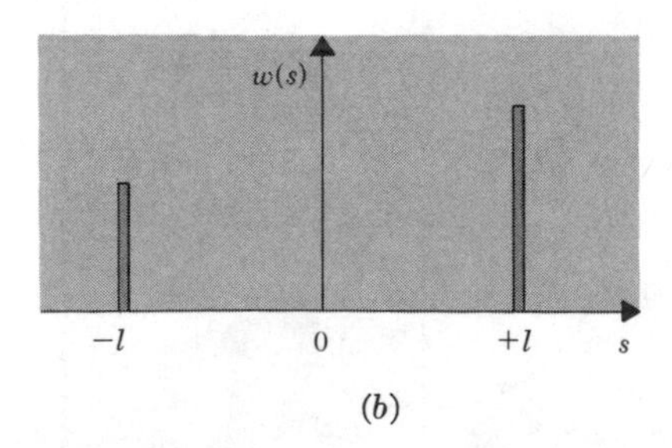

***Fig. 1·9·1** Some examples of probability distributions giving, for any one step, the probability $w(s)\,ds$ that the displacement is between s and $s + ds$. (a) A rather general case, displacements to the right being more probable than those to the left. (b) The special case discussed in Sec. 1·2. Here the peaks, centered about $+l$ and $-l$, respectively, are very narrow; the area under the right peak is p, that under the left one is q. (The curves (a) and (b) are not drawn to the same scale; the total area under each should be unity.)*

The total displacement x is equal to

$$x = s_1 + s_2 + \cdots + s_N = \sum_{i=1}^{N} s_i \tag{1·9·1}$$

Taking mean values of both sides,

$$\bar{x} = \overline{\sum_{i=1}^{N} s_i} = \sum_{i=1}^{N} \bar{s}_i \tag{1·9·2}$$

where we have used the property (1·7·8). But since $w(s_i)$ is the same for each step, independent of i, each mean value $\bar{s}_i$ is the same. Thus (1·9·2) is simply the sum of N equal terms and becomes

$$\bar{x} = N\bar{s} \tag{1·9·3}$$

where

$$\bar{s} \equiv \bar{s}_i = \int ds\, w(s)s \tag{1·9·4}$$

is merely the mean displacement per step.

Next we calculate the dispersion

$$\overline{(\Delta x)^2} \equiv \overline{(x - \bar{x})^2} \tag{1·9·5}$$

By (1·9·1) and (1·9·2) one has

$$x - \bar{x} = \sum_i (s_i - \bar{s})$$

or

$$\Delta x = \sum_{i=1}^{N} \Delta s_i \tag{1·9·6}$$

where

$$\Delta s = s_i - \bar{s} \tag{1·9·7}$$

By squaring (1·9·6) one obtains

$$(\Delta x)^2 = \left(\sum_{i=1}^{N} \Delta s_i\right)\left(\sum_{j=1}^{N} \Delta s_j\right) = \sum_i (\Delta s_i)^2 + \sum_i \sum_{\substack{j \\ i \neq j}} (\Delta s_i)(\Delta s_j) \tag{1·9·8}$$

Here the first term on the right represents all the square terms, and the second term all the cross terms, originating from the multiplication of the sum by itself. Taking the mean value of (1·9·8) yields, by virtue of (1·7·8),

$$\overline{(\Delta x)^2} = \sum_i \overline{(\Delta s_i)^2} + \sum_i \sum_{\substack{j \\ i \neq j}} \overline{\Delta s_i \, \Delta s_j} \tag{1·9·9}$$

In the cross terms we make use of the fact that different steps are statistically independent and apply the relation (1·7·9) to write for $i \neq j$

$$\overline{(\Delta s_i)(\Delta s_j)} = \overline{(\Delta s_i)}\;\overline{(\Delta s_j)} = 0 \tag{1·9·10}$$

since

$$\overline{\Delta s_i} = \bar{s}_i - \bar{s} = 0$$

In short, each cross term vanishes on the average, being as often positive as negative. Thus (1·9·9) reduces simply to a sum of square terms

$$\overline{(\Delta x)^2} = \sum_{i=1}^{N} \overline{(\Delta s_i)^2} \tag{1·9·11}$$

Of course, none of these square terms can be negative. Since the probability distribution $w(s_i)$ is the same for each step, independent of i, it again follows that $\overline{(\Delta s_i)^2}$ must be the same for each step. Thus the sum in (1·9·11) consists merely of N equal terms and becomes simply

▶ $$\overline{(\Delta x)^2} = N\overline{(\Delta s)^2} \tag{1·9·12}$$

where $$\overline{(\Delta s)^2} \equiv \overline{(\Delta s_i)^2} = \int ds\, w(s)(\Delta s)^2 \tag{1·9·13}$$

is just the dispersion of the displacement per step.

Despite their great simplicity, the relations (1·9·3) and (1·9·12) are very general and important results. The dispersion $\overline{(\Delta x)^2} = \overline{(x - \bar{x})^2}$ is a measure of the *square* of the width of the distribution of the net displacement about its mean $\bar{x}$. The square root $\Delta^* x \equiv [\overline{(\Delta x)^2}]^{\frac{1}{2}}$, i.e., the "root-mean-square (rms) deviation from the mean," thus provides a direct measure of the width of the distribution of the displacement about the mean $\bar{x}$. The results (1·9·3) and (1·9·12) thus make possible the following interesting statements about the sum (1·9·1) of statistically independent variables. If $\bar{s} \neq 0$ and the number N of these variables (e.g., of displacements) increases, the mean value $\bar{x}$ of their sum tends to increase proportionaly to N. The width $\Delta^* x$ of the distribution about the mean also increases, but only proportionately to $N^{\frac{1}{2}}$. Hence the *relative* magnitude of the width Δx^* compared to the mean $\bar{x}$ itself *decreases* like $N^{-\frac{1}{2}}$; explicitly, one has by (1·9·3) and (1·9·12), if $\bar{s} \neq 0$,

$$\frac{\Delta^* x}{\bar{x}} = \frac{\Delta^* s}{\bar{s}} \frac{1}{\sqrt{N}}$$

where $\Delta^* s \equiv [\overline{(\Delta s)^2}]^{\frac{1}{2}}$. This means that the percentage deviation of the distribution of values of x about their mean $\bar{x}$ becomes increasingly negligible as the number N becomes large. This is a characteristic feature of statistical distributions involving large numbers.

Example Let us apply the general results (1·9·3) and (1·9·12) of this section to the special case of the random walk with fixed step length l previously discussed in Sec. 1·2. There the probability of a step to the right is p, that of a step to the left is $q = 1 - p$. The mean displacement per step is then given by

$$\bar{s} = pl + q(-l) = (p - q)l = (2p - 1)l \tag{1·9·14}$$

As a check, note that $\bar{s} = 0$ if $p = q$, as required by symmetry.

Also $$\overline{s^2} = pl^2 + q(-l)^2 = (p + q)l^2 = l^2$$

Hence $$\overline{(\Delta s)^2} = \overline{s^2} - \bar{s}^2 = l^2[1 - (2p - 1)^2]$$
$$= l^2[1 - 4p^2 + 4p - 1] = 4l^2p(1 - p)$$

or $$\overline{(\Delta s)^2} = 4pql^2 \tag{1·9·15}$$

Hence the relations (1·9·3) and (1·9·12) yield

$$\left.\begin{aligned}\bar{x} &= (p-q)Nl \\ \overline{(\Delta x)^2} &= 4pqNl^2\end{aligned}\right\} \tag{1·9·16}$$

Since $x = ml$, these agree precisely with the previously computed results (1·4·6) and (1·4·12).

*1·10 *Calculation of the probability distribution*

For the problem discussed in the last section, the total displacement x in N steps is given by

$$x = \sum_{i=1}^{N} s_i \tag{1·10·1}$$

We now want to find the probability $\mathcal{P}(x)\,dx$ of finding x in the range between x and $x + dx$. Since the steps are statistically independent, the probability of *a particular* sequence of steps where

the 1st displacement lies in the range between s_1 and $s_1 + ds_1$
the 2nd displacement lies in the range between s_2 and $s_2 + ds_2$
. . .
the Nth displacement lies in the range between s_N and $s_N + ds_N$

is simply given by the product of the respective probabilities, i.e., by

$$w(s_1)\,ds_1 \cdot w(s_2)\,ds_2 \cdots w(s_N)\,ds_N$$

If we sum this probability over all the possible individual displacements which are consistent with the condition that the total displacement x in (1·10·1) always lies in the range between x and $x + dx$, then we obtain the total probability $\mathcal{P}(x)\,dx$, irrespective of the sequence of steps producing this total displacement. In symbols we can write

$$\mathcal{P}(x)\,dx = \underset{(dx)}{\int\int_{-\infty}^{\infty}\cdots\int} w(s_1)w(s_2)\cdots w(s_N)\,ds_1\,ds_2\cdots ds_N \tag{1·10·2}$$

where the integration is over all possible values of the variables s_i, subject to the restriction that

$$x < \sum_{i=1}^{N} s_i < x + dx \tag{1·10·3}$$

In principle, evaluation of the integral (1·10·2) solves completely the problem of finding $\mathcal{P}(x)$.

In practice, the integral (1·10·2) is difficult to evaluate because the condition (1·10·3) makes the limits of integration very awkward; i.e., one faces the complicated geometrical problem of determining over what subspace, consistent with (1·10·3), one has to integrate. A powerful way of handling this kind of problem is to eliminate the geometrical problem by integrating over *all* values of the variables s_i *without* restriction, while shifting the complication introduced by the condition of constraint (1·10·3) to the *integrand.* This can readily be done by multiplying the integrand in (1·10·2) by a factor which is equal to unity when the s_i are such that (1·10·3) is satisfied, but which equals zero otherwise. The Dirac δ function $\delta(x - x_0)$, discussed in Appendix A · 7, has precisely the selective property that it vanishes whenever $|x - x_0| > \frac{1}{2}|dx|$, while it becomes infinite like $(dx)^{-1}$ in the infinitesimal range where $|x - x_0| < \frac{1}{2}|dx|$; i.e., $\delta(x - x_0)\,dx = 1$ in this latter range. Hence (1·10·2) can equally well be written

$$\mathcal{P}(x)\,dx = \iint_{-\infty}^{\infty}\cdots\int w(s_1)w(s_2)\cdots w(s_N)\left[\delta\left(x - \sum_{i=1}^{N} s_i\right)dx\right]ds_1\,ds_2\cdots ds_N \tag{1·10·4}$$

where there is now *no* further restriction on the domain of integration. At this point we can use the convenient analytical representation of the δ function in terms of the integral (A·7·14); i.e., we can write

$$\delta(x - \Sigma s_i) = \frac{1}{2\pi}\int_{-\infty}^{\infty} dk\, e^{ik[\Sigma s_i - x]} \tag{1·10·5}$$

Substituting this result in (1·10·4) yields:

$$\mathcal{P}(x) = \iint\cdots\int w(s_1)w(s_2)\cdots w(s_N)\,\frac{1}{2\pi}\int_{-\infty}^{\infty} dk\, e^{ik(s_1+\cdots+s_N-x)}\,ds_1\,ds_2\cdots ds_N$$

$$\text{or}\quad \mathcal{P}(x) = \frac{1}{2\pi}\int_{-\infty}^{\infty} dk\, e^{-ikx}\int_{-\infty}^{\infty} ds_1\, w(s_1)\, e^{iks_1}\cdots\int_{-\infty}^{\infty} ds_N\, w(s_N)\, e^{iks_N} \tag{1·10·6}$$

where we have interchanged the order of integration and used the multiplicative property of the exponential function. Except for the irrelevant symbol used as variable of integration, each of the last N integrals is identical and equal to

$$\blacktriangleright \qquad Q(k) \equiv \int_{-\infty}^{\infty} ds\, e^{iks} w(s) \tag{1·10·7}$$

Hence (1·10·6) becomes

$$\blacktriangleright \qquad \mathcal{P}(x) = \frac{1}{2\pi}\int_{-\infty}^{\infty} dk\, e^{-ikx} Q^N(k) \tag{1·10·8}$$

Thus the evaluation of two simple (Fourier) integrals solves the problem completely.

Example Let us again apply the present results to the case of equal step lengths l discussed in Sec. 1·2. There the probability of a displacement $+l$ is equal to p, that of a displacement $-l$ is equal to $q = 1 - p$; i.e., the corresponding probability *density* w is given by

$$w(s) = p\delta(s - l) + q\delta(s + l)$$

The quantity (1.10.7) becomes

$$Q(k) \equiv \overline{e^{iks}} = p\, e^{ikl} + q\, e^{-ikl}$$

Using the binomial expansion, one then obtains

$$\begin{aligned} Q^N(k) &= (p\, e^{ikl} + q\, e^{-ikl})^N \\ &= \sum_{n=0}^{N} \frac{N!}{n!(N-n)!} (p\, e^{ikl})^n (q\, e^{-ikl})^{N-n} \\ &= \sum_{n=0}^{N} \frac{N!}{n!(N-n)!} p^n q^{N-n}\, e^{ikl(2n-N)} \end{aligned}$$

Thus (1·10·8) yields

$$\begin{aligned} \mathcal{P}(x) &= \frac{1}{2\pi} \int_{-\infty}^{\infty} dk\, e^{-ikx} Q^N(k) \\ &= \sum_{n=0}^{N} \frac{N!}{n!(N-n)!} p^n q^{N-n} \left\{ \frac{1}{2\pi} \int_{-\infty}^{\infty} dk\, e^{ik[(2n-N)l-x]} \right\} \end{aligned}$$

$$\text{or} \quad \mathcal{P}(x) = \sum_{n=0}^{N} \frac{N!}{n!(N-n)!} p^n q^{N-n} \delta[x - (2n-N)l] \qquad (1\cdot10\cdot9)$$

This says that the probability density $\mathcal{P}(x)$ vanishes unless

$$x = (2n - N)l, \qquad \text{where } n = 0, 1, 2, \ldots, N$$

The probability $P(2n - N)$ of finding the particle at such a position is then given by

$$P(2n - N) = \int_{(2n-N)l-\epsilon}^{(2n-N)l+\epsilon} \mathcal{P}(x)\, dx = \frac{N!}{n!(N-n)!} p^n q^{N-n}$$

where ϵ is some sufficiently small quantity; i.e., P is given by the coefficient of the corresponding δ function in (1·10·9). Thus we regain the result of Sec. 1·2 as a special case of the present more general formulation and *without* the need to use any combinatorial reasoning.

*1·11 *Probability distribution for large N*

We consider the integrals (1·10·7) and (1·10·8) which solve the problem of finding $\mathcal{P}(x)$ and ask what approximations become appropriate when N is large. The argument used here will be similar to the method detailed in Appendix A·6 to derive Stirling's formula.

The integrand in (1·10·7) contains the factor e^{iks}, which is an oscillatory function of s and oscillates the more rapidly with increasing magnitude of k. Hence the quantity $Q(k)$ given by the integral in (1·10·7) tends in general to be increasingly small as k becomes large. (See remark below.) If Q is raised to a large power N, it thus follows that $Q^N(k)$ tends to decrease very rapidly with increasing k. To compute $\mathcal{P}(x)$ by Eq. (1·10·8), a knowledge of $Q^N(k)$ for *small* values of k is then sufficient for calculating the integral, since for large values of k the contribution of $Q^N(k)$ to this integral is negligibly small. But for small values of k, it should be possible to approximate $Q^N(k)$ by a suitable expansion in powers of k. Since $Q^N(k)$ is a rapidly varying function of k, it is preferable (as in Sec. 1·5) to seek the more readily convergent power series expansion of its slowly varying logarithm $\ln Q^N(k)$.

Remark To the extent that $w(s)$ varies slowly over a period of oscillation, the integral $Q(k) = \int ds\, e^{iks}w(s) \approx 0$. The reason is that in any range $a < s < b$ in which w varies slowly so that $|dw/ds|(b - a) \ll w$, but which contains many oscillations so that $(b - a)k \gg 1$, the integral

$$\int_a^b ds\, e^{iks}w(s) \approx w(a) \int_a^b ds\, e^{iks} \approx 0$$

Combining these two inequalities one can say that

$$\int_{-\infty}^{\infty} ds\, e^{iks}w(s) \approx 0$$

to the extent that k is large enough so that everywhere

$$\left|\frac{dw}{ds}\right| \frac{1}{k} \ll w$$

The actual calculation is straightforward. We want first to compute $Q(k)$ for small values of k. Expanding e^{iks} in Taylor's series, Eq. (1·10·7) becomes

$$Q(k) \equiv \int_{-\infty}^{\infty} ds\, w(s)\, e^{iks} = \int_{-\infty}^{\infty} ds\, w(s)(1 + iks - \tfrac{1}{2}k^2s^2 + \cdots)$$

or
$$Q(k) = 1 + i\bar{s}k - \tfrac{1}{2}\overline{s^2}k^2 \cdots \tag{1·11·1}$$

where
$$\overline{s^n} \equiv \int_{-\infty}^{\infty} ds\, w(s)s^n \tag{1·11·2}$$

is a constant which represents the usual definition of the nth moment of s. Here we assume that $|w(s)| \to 0$ rapidly enough as $|s| \to \infty$ so that these moments are finite. Hence (1·11·1) yields

$$\ln Q^N(k) = N \ln Q(k) = N \ln [1 + i\bar{s}k - \tfrac{1}{2}\overline{s^2}k^2 \cdots] \tag{1·11·3}$$

Using the Taylor's series expansion valid for $y \ll 1$,

$$\ln (1 + y) = y - \tfrac{1}{2}y^2 \cdots$$

Eq. (1 · 11 · 3) becomes, up to terms quadratic in k,

$$\begin{aligned}\ln Q^N &= N[i\bar{s}k - \tfrac{1}{2}\overline{s^2}k^2 - \tfrac{1}{2}(i\bar{s}k)^2 \cdots] \\ &= N[i\bar{s}k - \tfrac{1}{2}(\overline{s^2} - \bar{s}^2)k^2 \cdots] \\ &= N[i\bar{s}k - \tfrac{1}{2}\overline{(\Delta s)^2}k^2 \cdots]\end{aligned}$$

where

$$\overline{(\Delta s)^2} \equiv \overline{s^2} - \bar{s}^2 \tag{1·11·4}$$

Hence we obtain

$$Q^N(k) = e^{iN\bar{s}k - \frac{1}{2}N\overline{(\Delta s)^2}k^2} \tag{1·11·5}$$

Thus (1 · 10 · 8) becomes

$$\mathcal{P}(x) = \frac{1}{2\pi}\int_{-\infty}^{\infty} dk\, e^{i(N\bar{s}-x)k - \frac{1}{2}N\overline{(\Delta s)^2}k^2} \tag{1·11·6}$$

The integral here is of the following form, where a is real and positive:

$$\begin{aligned}\int_{-\infty}^{\infty} du\, e^{-au^2+bu} &= \int_{-\infty}^{\infty} du\, e^{-a[u^2-(b/a)u]} \\ &= \int_{-\infty}^{\infty} du\, e^{-a(u-b/2a)^2+b^2/4a} && \text{by completing the square} \\ &= e^{b^2/4a}\int_{-\infty}^{\infty} dy\, e^{-ay^2} && \text{by putting } y = u - \frac{b}{2a} \\ &= e^{b^2/4a}\sqrt{\frac{\pi}{a}} && \text{by (A·4·2)}\end{aligned}$$

Thus

$$\int_{-\infty}^{\infty} du\, e^{-au^2+bu} = \sqrt{\frac{\pi}{a}}\, e^{b^2/4a} \tag{1·11·7}$$

Applying this integral formula to (1·11·6) we get, with $b = i(N\bar{s} - x)$ and $a = \frac{1}{2}N\overline{(\Delta s)^2}$, the result

$$\mathcal{P}(x) = \frac{1}{\sqrt{2\pi\sigma^2}}\, e^{-(x-\mu)^2/2\sigma^2} \tag{1·11·8}$$

where

$$\left.\begin{aligned}\mu &\equiv N\bar{s} \\ \sigma^2 &\equiv N\overline{(\Delta s)^2}\end{aligned}\right\} \tag{1·11·9}$$

Thus the distribution has the Gaussian form previously encountered in Sec. 1·6. Note, however, the extreme generality of this result. *No matter what* the probability distribution $w(s)$ for each step may be, as long as the steps are *statistically independent* and $w(s)$ falls off rapidly enough as $|s| \to \infty$, the total displacement x will be distributed according to the Gaussian law *if N is sufficiently large.* This very important result is the content of the so-called "central limit theorem," probably the most famous theorem in mathematical probability theory.* The generality of the result also accounts for the fact that so many phenomena in nature (e.g., errors in measurement) obey approximately a Gaussian distribution.

* A proof of the theorem with attention to fine points of mathematical rigor can be found in A. I. Khinchin, "Mathematical Foundations of Statistical Mechanics," p. 166, Dover Publications, New York, 1949.

We already showed that for the Gaussian distribution (1·6·4)

$$\left.\begin{aligned}\bar{x} &= \mu \\ \text{and} \qquad \overline{(\Delta x)^2} &= \sigma^2\end{aligned}\right\}$$

Hence (1·11·9) implies

$$\left.\begin{aligned}\bar{x} &= N\bar{s} \\ \text{and} \qquad \overline{(\Delta x)^2} &= N\overline{(\Delta s)^2}\end{aligned}\right\} \qquad (1\cdot 11\cdot 10)$$

which agree with the results obtained from our general moment calculations (1·9·3) and (1·9·12).

SUGGESTIONS FOR SUPPLEMENTARY READING

Probability theory

F. Mosteller, R. E. K. Rourke, and G. B. Thomas: "Probability and Statistics," Addison-Wesley Publishing Company, Reading, Mass., 1961. (An elementary introduction.)

W. Feller: "An Introduction to Probability Theory and its Applications," 2d ed., John Wiley & Sons, Inc., New York, 1959.

H. Cramer: "The Elements of Probability Theory," John Wiley & Sons, Inc., New York, 1955.

Random walk problem

S. Chandrasekhar: "Stochastic Problems in Physics and Astronomy, *Rev. Mod. Phys.*, vol. 15, pp. 1–89 (1943). This article is also reprinted in M. Wax, "Selected Papers on Noise and Stochastic Processes," Dover Publications, New York, 1954.

R. B. Lindsay: "Introduction to Physical Statistics," chap. 2, John Wiley & Sons, Inc., New York, 1941. (An elementary discussion of the random walk and related physical problems.)

PROBLEMS

1.1 What is the probability of throwing a total of 6 points or less with three dice?

1.2 Consider a game in which six true dice are rolled. Find the probability of obtaining

(*a*) exactly one ace
(*b*) at least one ace
(*c*) exactly two aces

1.3 A number is chosen at random between 0 and 1. What is the probability that exactly 5 of its first 10 decimal places consist of digits less than 5?

1.4 A drunk starts out from a lamppost in the middle of a street, taking steps of equal length either to the right or to the left with equal probability. What is the probability that the man will again be at the lamppost after taking N steps

(a) if N is even?
(b) if N is odd?

1.5 In the game of Russian roulette (*not* recommended by the author), one inserts a single cartridge into the drum of a revolver, leaving the other five chambers of the drum empty. One then spins the drum, aims at one's head, and pulls the trigger.

(a) What is the probability of being still alive after playing the game N times?

(b) What is the probability of surviving $(N - 1)$ turns in this game and then being shot the Nth time one pulls the trigger?

(c) What is the mean number of times a player gets the opportunity of pulling the trigger in this macabre game?

1.6 Consider the random walk problem with $p = q$ and let $m = n_1 - n_2$ denote the net displacement to the right. After a total of N steps, calculate the following mean values: $\overline{m}$, $\overline{m^2}$, $\overline{m^3}$, and $\overline{m^4}$.

1.7 Derive the binomial distribution in the following algebraic way, which does not involve any explicit combinatorial analysis. One is again interested in finding the probability $W(n)$ of n successes out of a total of N independent trials. Let $w_1 \equiv p$ denote the probability of a success, $w_2 = 1 - p = q$ the corresponding probability of a failure. Then $W(n)$ can be obtained by writing

$$W(n) = \sum_{i=1}^{2} \sum_{j=1}^{2} \sum_{k=1}^{2} \cdots \sum_{m=1}^{2} w_i w_j w_k \cdots w_m \tag{1}$$

Here each term contains N factors and is the probability of a particular combination of successes and failures. The sum over all combinations is then to be taken only over those terms involving w_1 exactly n times, i.e., only over those terms involving $w_1{}^n$.

By rearranging the sum (1), show that the *un*restricted sum can be written in the form

$$W(n) = (w_1 + w_2)^N$$

Expanding this by the binomial theorem, show that the sum of all terms in (1) involving $w_1{}^n$, i.e., the desired probability $W(n)$, is then simply given by the one binomial expansion term which involves $w_1{}^n$.

1.8 Two drunks start out together at the origin, each having equal probability of making a step to the left or right along the x axis. Find the probability that they meet again after N steps. It is to be understood that the men make their steps simultaneously. (It may be helpful to consider their relative motion.)

1.9 The probability $W(n)$ that an event characterized by a probability p occurs n times in N trials was shown to be given by the binomial distribution

$$W(n) = \frac{N!}{n!(N-n)!} p^n (1-p)^{N-n} \tag{1}$$

Consider a situation where the probability p is small ($p \ll 1$) and where one is interested in the case $n \ll N$. (Note that if N is large, $W(n)$ becomes very small if $n \to N$ because of the smallness of the factor p^n when $p \ll 1$. Hence $W(n)$ is indeed only appreciable when $n \ll N$.) Several approximations can then be made to reduce (1) to simpler form.

(a) Using the result $\ln(1 - p) \approx -p$, show that $(1 - p)^{N-n} \approx e^{-Np}$.
(b) Show that $N!/(N - n)! \approx N^n$.
(c) Hence show that (1) reduces to

$$W(n) = \frac{\lambda^n}{n!} e^{-\lambda} \tag{2}$$

where $\lambda \equiv Np$ is the mean number of events. The distribution (2) is called the "Poisson distribution."

1.10 Consider the Poisson distribution of the preceding problem.

(a) Show that it is properly normalized in the sense that $\sum_{n=0}^{N} W_n = 1$. (The sum can be extended to infinity to an excellent approximation, since W_n is negligibly small when $n \gtrsim N$.)

(b) Use the Poisson distribution to calculate $\bar{n}$.

(c) Use the Poisson distribution to calculate $\overline{(\Delta n)^2} \equiv \overline{(n - \bar{n})^2}$.

1.11 Assume that typographical errors committed by a typesetter occur completely at random. Suppose that a book of 600 pages contains 600 such errors. Use the Poisson distribution to calculate the probability

(a) that a page contains no errors

(b) that a page contains at least three errors

1.12 Consider the α particles emitted by a radioactive source during some time interval t. One can imagine this time interval to be subdivided into many small intervals of length Δt. Since the α particles are emitted at random times, the probability of a radioactive disintegration occurring during any such time Δt is completely independent of whatever disintegrations occur at other times. Furthermore, Δt can be imagined to be chosen small enough so that the probability of more than one disintegration occurring in a time Δt is negligibly small. This means that there is some probability p of one disintegration occurring during a time Δt (with $p \ll 1$, since Δt was chosen small enough) and probability $1 - p$ of no disintegration occurring during this time. Each such time interval Δt can then be regarded as an independent trial, there being a total of $N = t/\Delta t$ such trials during a time t.

(a) Show that the probability $W(n)$ of n disintegrations occurring in a time t is given by a Poisson distribution.

(b) Suppose that the strength of the radioactive source is such that the mean number of disintegrations per minute is 24. What is the probability of obtaining n counts in a time interval of 10 seconds? Obtain numerical values for all integral values of n from 0 to 8.

1.13 A metal is evaporated in vacuum from a hot filament. The resultant metal atoms are incident upon a quartz plate some distance away and form there a thin metallic film. This quartz plate is maintained at a low temperature so that any metal atom incident upon it sticks at its place of impact without further migration. The metal atoms can be assumed equally likely to impinge upon any element of area of the plate.

If one considers an element of substrate area of size b^2 (where b is the metal atom diameter), show that the number of metal atoms piled up on this area should be distributed approximately according to a Poisson distribution. Suppose that one evaporates enough metal to form a film of mean thickness corresponding to 6 atomic layers. What fraction of the substrate area is then not covered by metal at all? What fraction is covered, respectively, by metal layers 3 atoms thick and 6 atoms thick?

1.14 A penny is tossed 400 times. Find the probability of getting 215 heads. (Suggestion: use the Gaussian approximation.)

1.15 A set of telephone lines is to be installed so as to connect town A to town B. The town A has 2000 telephones. If each of the telephone users of A were to be guaranteed instant access to make calls to B, 2000 telephone lines would be needed. This would be rather extravagant. Suppose that during the busiest hour of the day each subscriber in A requires, on the average, a telephone connection to B for two minutes, and that these telephone calls are made at random. Find the minimum number M of telephone lines to B which must be installed so that at most only 1 percent of the callers of town A will fail to have immediate access to a telephone line to B. (Suggestion: approximate the distribution by a Gaussian distribution to facilitate the arithmetic.)

1.16 Consider a gas of N_0 noninteracting molecules enclosed in a container of volume V_0. Focus attention on any subvolume V of this container and denote by N the number of molecules located within this subvolume. Each molecule is equally likely to be located anywhere within the container; hence the probability that a given molecule is located within the subvolume V is simply equal to V/V_0.

(*a*) What is the mean number $\bar{N}$ of molecules located within V? Express your answer in terms N_0, V_0, and V.

(*b*) Find the relative dispersion $\overline{(N - \bar{N})^2}/\bar{N}^2$ in the number of molecules located within V. Express your answer in terms of $\bar{N}$, V, and V_0.

(*c*) What does the answer to part (*b*) become when $V \ll V_0$?

(*d*) What value should the dispersion $\overline{(N - \bar{N})^2}$ assume when $V \to V_0$? Does the answer to part (*b*) agree with this expectation?

1.17 Suppose that in the preceding problem the volume V under consideration is such that $0 \ll V/V_0 \ll 1$. What is the probability that the number of molecules in this volume is between N and $N + dN$?

1.18 A molecule in a gas moves equal distances l between collisions with equal probability in any direction. After a total of N such displacements, what is the mean square displacement $\overline{R^2}$ of the molecule from its starting point?

1.19 A battery of total emf V is connected to a resistor R; as a result an amount of power $P = V^2/R$ is dissipated in this resistor. The battery itself consists of N individual cells connected in series so that V is just equal to the sum of the emf's of all these cells. The battery is old, however, so that not all cells are in perfect condition. Thus there is only a probability p that the emf of any individual cell has its normal value v; and a probability $1 - p$ that the emf of any individual cell is zero because the cell has become internally shorted. The individual cells are statistically independent of each other. Under these conditions, calculate the *mean* power $\bar{P}$ dissipated in the resistor, expressing the result in terms of N, v, and p.

1.20 Consider N similar antennas emitting linearly polarized electromagnetic radiation of wavelength λ and velocity c. The antennas are located along the x axis at a separation λ from each other. An observer is located on the x axis at a great distance from the antennas. When a *single* antenna radiates, the observer measures an *intensity* (i.e., mean-square electric-field amplitude) equal to I.

(*a*) If all the antennas are driven in phase by the same generator of frequency $\nu = c/\lambda$, what is the total intensity measured by the observer?

(*b*) If the antennas all radiate at the same frequency $\nu = c/\lambda$ but with completely random phases, what is the mean intensity measured by the observer? (*Hint:* Represent the amplitudes by vectors, and deduce the observed intensity from the resultant amplitude.)

1.21 Radar signals have recently been reflected from the planet Venus. Suppose that in such an experiment a pulse of electromagnetic radiation of duration τ is sent from the earth toward Venus. A time t later (which corresponds to the time necessary for light to go from the earth to Venus and back again) the receiving antenna on the earth is turned on for a time τ. The returning echo ought then to register on the recording meter, placed at the output of the electronic equipment following the receiving antenna, as a very faint signal of definite amplitude a_s. But a fluctuating random signal (due to the inevitable fluctuations in the radiation field in outer space and due to current fluctuations always existing in the sensitive receiving equipment itself) also registers as a signal of amplitude a_n on the recording meter. This meter thus registers a total amplitude $a = a_s + a_n$.

Although $\bar{a}_n = 0$ on the average, since a_n is as likely to be positive as negative, there is considerable probability that a_n attains values considerably in excess of a_s; i.e., the root-mean-square amplitude $(\overline{a_n^2})^{\frac{1}{2}}$ can be considerably greater than the signal a_s of interest. Suppose that $(\overline{a_n^2})^{\frac{1}{2}} = 1000\ a_s$. Then the fluctuating signal a_n constitutes a background of "noise" which makes observation of the desired echo signal essentially impossible.

On the other hand, suppose that N such radar pulses are sent out in succession and that the total amplitudes a picked up at the recording equipment after each pulse are all added together before being displayed on the recording meter. The resulting amplitude must then have the form $A = A_s + A_n$, where A_n represents the resultant noise amplitude (with $\bar{A}_n = 0$) and $\bar{A} = A_s$ represents the resultant echo-signal amplitude. How many pulses must be sent out before $(\overline{A_n^2})^{\frac{1}{2}} = A_s$ so that the echo signal becomes detectable?

1.22 Consider the random walk problem in one dimension, the probability of a displacement between s and $s + ds$ being

$$w(s)\,ds = (2\pi\sigma^2)^{-\frac{1}{2}}\,e^{-(s-l)^2/2\sigma^2}ds$$

After N steps,

(*a*) What is the mean displacement $\bar{x}$ from the origin?

(*b*) What is the dispersion $\overline{(x - \bar{x})^2}$?

1.23 Consider the random walk problem for a particle in one dimension. Assume that in each step its displacement is always positive and equally likely to be anywhere in the range between $l - b$ and $l + b$ where $b < l$. After N steps, what is

(*a*) the mean displacement $\bar{x}$?

(*b*) the dispersion $\overline{(x - \bar{x})^2}$?

1.24 (*a*) A particle is equally likely to lie anywhere on the circumference of a circle. Consider as the z axis any straight line in the plane of the circle and passing through its center. Denote by θ the angle between this z axis and the straight line connecting the center of the circle to the particle. What is the probability that this angle lies between θ and $\theta + d\theta$?

(*b*) A particle is equally likely to lie anywhere on the surface of a sphere. Consider any line through the center of this sphere as the z axis. Denote by θ the angle between this z axis and the straight line connecting the center of the sphere to the particle. What is the probability that this angle lies between θ and $\theta + d\theta$?

1.25 Consider a polycrystalline sample of $CaSO_4 \cdot 2H_2O$ in an external magnetic field $\mathbf{B}$ in the z direction. The internal magnetic field (in the z direction) produced at the position of a given proton in the H_2O molecule by the neighboring proton is given by $(\mu/a^3)(3\cos^2\theta - 1)$ if the spin of this neighboring proton points along the applied field; it is given by $-(\mu/a^3)(3\cos^2\theta - 1)$ if this neighboring spin points in a direction opposite to the applied field. Here μ is the magnetic moment of the proton and a is the distance between the two protons, while θ denotes the angle between the line joining the protons and the z axis. In this sample of randomly oriented crystals the neighboring proton is equally likely to be located anywhere on the sphere of radius a surrounding the given proton.

(*a*) What is the probability $W(b)\,db$ that the internal field b lies between b and $b + db$ if the neighboring proton spin is parallel to $\mathbf{B}$?

(*b*) What is this probability $W(b)\,db$ if the neighboring proton spin is equally likely to be parallel or antiparallel to $\mathbf{B}$? Draw a sketch of $W(b)$ as a function of b.

(In a nuclear magnetic resonance experiment the frequency at which energy is absorbed from a radio-frequency magnetic field is proportional to the local magnetic field existing at the position of a proton. The answer to part (*b*) gives, therefore, the shape of the absorption line observed in the experiment.)

**1.26* Consider the random walk problem in one dimension and suppose that the probability of a single displacement between s and $s + ds$ is given by

$$w(s)\,ds = \frac{1}{\pi}\frac{b}{s^2 + b^2}\,ds$$

Calculate the probability $\mathcal{P}(x)\,dx$ that the total displacement after N steps lies between x and $x + dx$. Does $\mathcal{P}(x)$ become Gaussian when N becomes large? If not, does this violate the central limit theorem of Sec. 1·11?

**1.27* Consider a very general one-dimensional random walk, where the probability that the ith displacement lies between s_i and $s_i + ds_i$ is given by $w_i(s_i)\,ds_i$. Here the probability density w_i characterizing each step may be different and thus dependent on i. It is still true, however, that different displacements are statistically independent, i.e., w_i for any one step does not depend on the displacements performed by the particle in any other step. Use arguments similar to those of Sec. 1·11 to show that when the number N of displacements becomes large, the probability $\mathcal{P}(x)\,dx$ that the total displacement lies between x and $x + dx$ will still tend to approach the Gaussian form with a mean value $\bar{x} = \Sigma\bar{s}_i$ and a dispersion $\overline{(\Delta x)^2} = \Sigma\overline{(\Delta s_i)^2}$. This result constitutes a very general form of the central limit theorem.

**1.28* Consider the random walk of a particle in three dimensions and let $w(\mathbf{s})\,d^3\mathbf{s}$ denote the probability that its displacement $\mathbf{s}$ lies in the range between $\mathbf{s}$ and $\mathbf{s} + d\mathbf{s}$ (i.e., that s_x lies between s_x and $s_x + ds_x$, s_y between s_y and $s_y + ds_y$, and s_z between s_z and $s_z + ds_z$). Let $\mathcal{P}(\mathbf{r})\,d^3\mathbf{r}$ denote the probability that the total displacement $\mathbf{r}$ of the particle after N steps lies in the range between $\mathbf{r}$ and $\mathbf{r} + d\mathbf{r}$. By generalizing the argument of Sec. 1·10 to three dimensions, show that

$$\mathcal{P}(\mathbf{r}) = \frac{1}{(2\pi)^3}\int_{-\infty}^{\infty} d^3\mathbf{k}\, e^{-i\mathbf{k}\cdot\mathbf{r}}Q^N(\mathbf{k})$$

where

$$Q(\mathbf{k}) = \int_{-\infty}^{\infty} d^3\mathbf{s}\, e^{i\mathbf{k}\cdot\mathbf{s}}w(\mathbf{s})$$

*1.29 (a) Using an appropriate Dirac-delta function, find the probability density $w(\mathbf{s})$ for displacements of uniform length l, but in any random direction of three-dimensional space. (Hint: Remember that the function $w(\mathbf{s})$ must be such that $\int\int\int w(\mathbf{s})\, d\mathbf{s} = 1$ when integrated over all space.)

(b) Use the result of part (a) to calculate $Q(\mathbf{k})$. (Perform the integration in spherical coordinates.)

(c) Using this value of $Q(\mathbf{k})$, compute $\mathcal{P}(\mathbf{r})$ for $N = 3$, thus solving the random walk problem in three dimensions for the case of three steps.

Statistical description of systems of particles

2

NOW THAT we are familiar with some elementary statistical ideas, we are ready to turn our attention to the main subject of this book, the discussion of systems consisting of very many particles. In analyzing such systems, we shall attempt to combine some statistical ideas with our knowledge of the laws of mechanics applicable to the particles. This approach forms the basis of the subject of "statistical mechanics" and is quite similar to that which would be used in discussing a game of chance. To make the analogy explicit consider, for example, a system consisting of 10 dice which are thrown from a cup onto a table in a gambling experiment. The essential ingredients necessary for an analysis of this situation are the following:

1. *Specification of the state of the system:* One needs a detailed method for describing the outcome of each experiment. For example, in this case a specification of the state of the system after the throw requires a statement as to which face is uppermost for each of the 10 dice.

2. *Statistical ensemble:* In principle it may be true that the problem is deterministic in the following sense: if we really knew the initial positions and orientations as well as corresponding velocities of all the dice in the cup at the beginning of the experiment, and if we knew just how the cup is manipulated in the act of throwing the dice, then we could indeed predict the outcome of the experiment by applying the laws of classical mechanics and solving the resulting differential equations. But we do not have any such detailed information available to us. Hence we proceed to describe the experiment in terms of probabilities. That is, instead of focusing our attention on a single experiment, we consider an ensemble consisting of many such experiments, all carried out under similar conditions where 10 dice are thrown from the cup. The outcome of each such experiment will, in general, be different. But we can ask for the *probability* of occurrence of a particular outcome, i.e., we can determine the fraction of cases (in this set of similar experiments) which are characterized by a particular final state of the dice. This procedure shows how the prob-

ability is determined experimentally. Our theoretical aim is to predict this probability on the basis of some fundamental postulates.

3. *Basic postulate about a priori probabilities:* To make theoretical progress, we must introduce some basic postulates. Our knowledge of the physical situation leads us to expect that there is nothing in the laws of mechanics which, for regular dice of uniform density, would result in the preferred appearance uppermost of any one face of a die compared to any other face. Hence we may introduce the postulate that *a priori* (i.e., based on our prior notions as yet unverified by actual observations) the probabilities are equal that any of the six faces of a die land uppermost. This postulate is eminently reasonable and certainly does not contradict any of the laws of mechanics. That the postulate is actually valid can only be decided by making theoretical predictions based on this postulate and checking that these predictions are confirmed by experimental observations. To the extent that such predictions are repeatedly verified, the validity of this postulate can be accepted with increasing confidence.

4. *Probability calculations:* Once the basic postulate has been adopted, the theory of probability allows the theoretical calculation of the probability of the outcome for any experiment with these dice.

In studying systems consisting of a large number of particles, our considerations will be similar to those used in formulating the preceding problem of several dice.

STATISTICAL FORMULATION OF THE MECHANICAL PROBLEM

2 · 1 *Specification of the state of a system*

Consider any system of particles, no matter how complicated (e.g., an assembly of weakly interacting harmonic oscillators, a gas, a liquid, an automobile). We know that the particles in any such system (i.e., the electrons, atoms, or molecules composing the system) can be described in terms of the laws of quantum mechanics. Specifically, the system can then be described by a wave function $\psi(q_1, \ldots, q_f)$ which is a function of some set of f coordinates (including possible spin variables) required to characterize the system. The number f is the "number of degrees of freedom" of the system. A particular quantum state of the system is then specified by giving the values of some set of f quantum numbers. This description is complete since, if ψ is thus specified at any time t, the equations of motion of quantum mechanics allow prediction of ψ at any other time.

Example 1 Consider a system consisting of a single particle, considered fixed in position, but having a spin $\frac{1}{2}$ (i.e., intrinsic spin angular momentum $\frac{1}{2}\hbar$). In a quantum-mechanical description the state of this particle is specified by the projection m of its spin along some fixed axis (which we shall choose to call the z axis). The quantum number m can then assume the two values $m = \frac{1}{2}$ or $m = -\frac{1}{2}$; i.e., roughly speaking, one can say that the spin can point either "up" or "down" with respect to the z axis.

Example 2 Consider a system consisting of N particles considered fixed in position, but each having spin $\frac{1}{2}$. Here N may be large, say of the order of Avogadro's number $N_a = 6 \times 10^{23}$. The quantum number m of each particle can then assume the two values $\frac{1}{2}$ or $-\frac{1}{2}$. The state of the entire system is then specified by stating the values of the N quantum numbers $m_1, \ldots, m_N$ which specify the orientation of the spin of each particle.

Example 3 Consider a system consisting of a one-dimensional simple harmonic oscillator whose position coordinate is x. The possible quantum states of this oscillator can be specified by a quantum number n such that the energy of the oscillator can be expressed as

$$E_n = (n + \tfrac{1}{2})\hbar\omega$$

where ω is the classical angular frequency of oscillation. Here the quantum number n can assume any integral value $n = 0, 1, 2, \ldots$.

Example 4 Consider a system consisting of N weakly interacting one-dimensional simple harmonic oscillators. The quantum state of this system can be specified by the set of numbers $n_1, \ldots, n_N$, where the quantum number n_i refers to the ith oscillator and can assume any value $0, 1, 2, \ldots$.

Example 5 Consider a system consisting of a single particle (without spin) confined within a rectangular box (so that the particle's coordinates lie within the ranges $0 \leq x \leq L_x$, $0 \leq y \leq L_y$, $0 \leq z \leq L_z$), but otherwise subject to no forces. The wave function ψ of the particle (of mass m) must then satisfy the Schrödinger equation

$$-\frac{\hbar^2}{2m}\left(\frac{\partial^2}{\partial x^2} + \frac{\partial^2}{\partial y^2} + \frac{\partial^2}{\partial z^2}\right)\psi = E\psi \tag{2·1·1}$$

and, to guarantee confinement of the particle within the box, ψ must vanish at the walls. The wave function having these properties has the form

$$\psi = \sin\left(\pi\frac{n_x x}{L_x}\right)\sin\left(\pi\frac{n_y y}{L_y}\right)\sin\left(\pi\frac{n_z z}{L_z}\right) \tag{2·1·2}$$

This satisfies (2·1·1), provided that the energy E of the particle is related to n_x, n_y, n_z by

$$E = \frac{\hbar^2}{2m}\pi^2\left(\frac{n_x^2}{L_x^2} + \frac{n_y^2}{L_y^2} + \frac{n_z^2}{L_z^2}\right) \tag{2·1·3}$$

Also $\psi = 0$ properly when $x = 0$ or $x = L_x$, when $y = 0$ or $y = L_y$, and when $z = 0$ or $z = L_z$, provided that the three numbers n_x, n_y, n_z assume any discrete integral values. The state of the particle is then specified by stating the values assumed by these three quantum numbers n_x, n_y, n_z, while Eq. (2·1·3) gives the corresponding value of the quantized energy.

Comments on the classical description Atoms and molecules are properly described in terms of quantum mechanics. Throughout this book our theoretical discussion of systems of many such particles will, therefore, be based consistently on quantum ideas. A description in terms of classical mechanics, although inadequate in general, may nevertheless sometimes be a useful approximation. It is, therefore, worth making a few comments about the specification of the state of a system in classical mechanics.

Let us start with a very simple case—a single particle in one dimension. This system can be completely described in terms of its position coordinate q and its corresponding momentum p. (This specification is complete, since the laws of classical mechanics are such that a knowledge of q and p at any one time permits prediction of the values of q and p at any other time.) It is possible to represent the situation geometrically by drawing cartesian axes labeled by q and p, as shown in Fig. 2·1·1. Specification of q and p is then equivalent to specifying a point in this two-dimensional space (commonly called "phase space"). As the coordinate and momentum of the particle change in time, this representative point moves through this phase space.

In order to describe the situation in terms where the possible states of the particle are countable, it is convenient to subdivide the ranges of the variables q and p into arbitrarily small discrete intervals. For example, one can choose fixed intervals of size δq for the subdivision of q, and fixed intervals of size δp for the subdivision of p. Phase space is then subdivided into small cells of equal size and of two-dimensional volume (i.e., area)

$$\delta q \, \delta p = h_0$$

where h_0 is some small constant having the dimensions of angular momentum. The state of the system can then be specified by stating that its coordinate lies

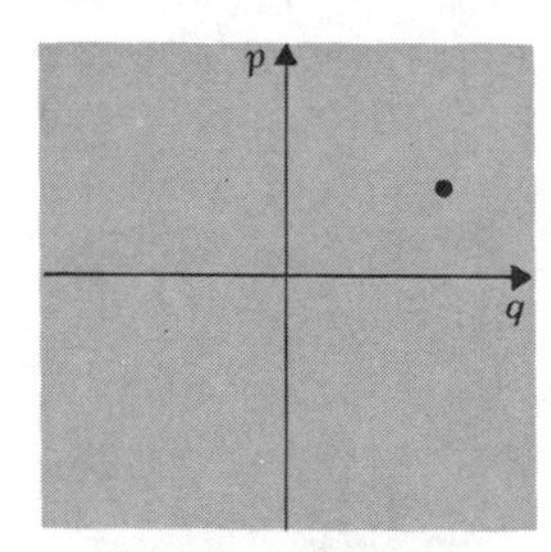

Fig. 2·1·1 Classical phase space for a single particle in one dimension.

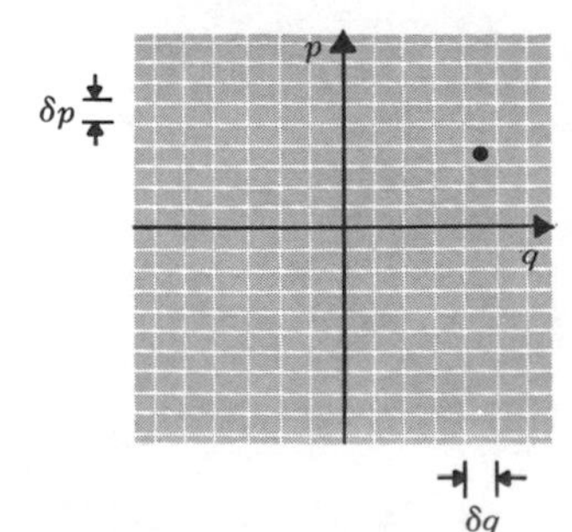

Fig. 2·1·2 The phase of Fig. 2·1·1 is here shown subdivided into equal cells of volume $\delta q \, \delta p = h_0$.

in some interval between q and $q + \delta q$ and that its momentum lies in some interval between p and $p + \delta p$, i.e., by stating that the representative point (q,p) lies in a particular cell of phase space. The specification of the state of the system clearly becomes more precise as one decreases the size chosen for the cells into which phase space has been divided, i.e., as one decreases the magnitude chosen for h_0. Of course, h_0 can be chosen arbitrarily small in this classical description.

It may be remarked that quantum theory imposes a limitation on the accuracy with which a simultaneous specification of a coordinate q and its cooresponding momentum p is possible. Indeed, this limitation is expressed by the Heisenberg uncertainty principle, which states that the uncertainties δq and δp in these two quantities are such that $\delta q \, \delta p \gtrsim \hbar$, where $\hbar$ is Planck's constant (divided by 2π.) Thus subdivision of the phase space into cells of volume less than $\hbar$ is really physically meaningless; i.e., a choice of $h_0 < \hbar$ would lead to a specification of the system more precise than is allowed by quantum theory.

The generalization of the above remarks to an arbitrarily complex system is immediate. Such a system can be described by some set of f coordinates $q_1, q_2, \ldots, q_f$ and f corresponding momenta $p_1, p_2, \ldots, p_f$, i.e., by a total of $2f$ parameters. The number f of independent coordinates needed for the description of the system is called the "number of degrees of freedom" of the system. (For example, if we are dealing with a system of N point particles, then each particle is characterized by three position coordinates so that $f = 3N$.) The set of numbers $\{q_1, \ldots, q_f, p_1, \ldots, p_f\}$ can again be regarded as a "point" in a "phase space" of $2f$ dimensions in which each cartesian-coordinate axis is labeled by one of the coordinates or momenta. (Except for the fact that this space is not so readily visualized by our provincial three-dimensional minds, this space is completely analogous to the two-dimensional diagram of Fig. 2·1·1). Once again this space can be subdivided into little cells (of volume $\delta q_1 \cdots \delta q_f \, \delta p_1 \cdots \delta p_f = h_0{}^f$ if one always chooses the interval of subdivision δq_k of the kth coordinate and δp_k of the kth momentum such that $\delta q_k \, \delta p_k = h_0$). The state of the system can then again be specified by stating in which particular range, or cell in phase space, the coordinates $q_1, \ldots, q_f, p_1, \ldots, p_f$ of the system can be found.

Summary The microscopic state, or "microstate," of a system of particles can be simply specified in the following way:

Enumerate in some convenient order, and label with some index r ($r = 1, 2, 3 \ldots$), all the possible quantum states of the system. The state of the system is then described by specifying the particular state r in which the system is found.

If it is desired to use the approximation of classical mechanics, the situation is quite analogous. After the phase space for the system has been subdivided into suitable cells of equal size, one can enumerate these cells in some convenient order and label them with some index r ($r = 1, 2, 3, \ldots$). The state of the system is then described by specifying the particular cell r in which the representative point of the system is located.

The quantum mechanical and classical descriptions are thus very similar, a cell in phase space in the classical discussion being analogous to a quantum state in the quantum-mechanical discussion.

2 · 2 *Statistical ensemble*

In principle the problem of a system consisting of many particles is completely deterministic in the sense that a complete specification of the quantum state ψ of the system at any one time allows calculation of all physical quantities, as well as prediction of the state ψ of the system at all other times. (Similarly, in classical mechanics, complete specification of the state of the system by all of its coordinates q and momenta p at any one time allows calculation of all physical quantities, as well as prediction of the coordinates and momenta at all other times.) But in general we neither have available to us, nor are we interested in, such a complete specification of the system. Hence we proceed to a discussion of the system in terms of probability concepts. For this purpose we consider *not* an isolated instance of a single system, but instead imagine attention focused on an ensemble consisting of a very large number of identical systems, all prepared subject to whatever conditions are specified as known. The systems in this ensemble will, in general, be in different states and will, therefore, also be characterized by different macroscopic parameters (e.g., by different values of pressure or magnetic moment). But we can ask for the probability of occurrence of a particular value of such a parameter, i.e., we can determine the fraction of cases in the ensemble when the parameter assumes this particular value. The aim of theory will be to predict the probability of occurrence in the ensemble of various values of such a parameter on the basis of some basic postulates.

Example Consider a system of three fixed particles, each having spin $\frac{1}{2}$ so that each spin can point either up or down (i.e., along or opposite some direction chosen as the z axis). Each particle has a magnetic moment along the z axis of μ when it points up, and $-\mu$ when it points down. The system is placed in an external magnetic field H pointing along this z axis.

The state of the particle i can be specified by its magnetic quantum number m_i which can assume the two values $m_i = \pm\frac{1}{2}$. The state of the whole system is specified by giving the values of the three quantum numbers m_1, m_2, m_3. A particle has energy $-\mu H$ when its spin points up, and energy μH when its spin points down.

We list in the table below all the possible states of the system. We also list some parameters, such as total magnetic moment and total energy, which characterize the system as a whole. (For the sake of brevity $m = \frac{1}{2}$ is denoted simply by $+$, and $m = -\frac{1}{2}$ by $-$.)

State index r	*Quantum numbers* m_1, m_2, m_3	*Total magnetic moment*	*Total energy*
1	$+ + +$	3μ	$-3\mu H$
2	$+ + -$	μ	$-\mu H$
3	$+ - +$	μ	$-\mu H$
4	$- + +$	μ	$-\mu H$
5	$+ - -$	$-\mu$	μH
6	$- + -$	$-\mu$	μH
7	$- - +$	$-\mu$	μH
8	$- - -$	-3μ	$3\mu H$

One usually has available some partial knowledge about the system under consideration. (For example, one might know the total energy and the volume of a gas.) The system can then only be in any of its states which are compatible with the available information about the system. These states will be called the "states accessible to the system." In a statistical description the representative ensemble thus contains only systems all of which are consistent with the specified available knowledge about the system; i.e., the systems in the ensemble must all be distributed over the various accessible states.

Example Suppose that in the previous example of a system consisting of three spins the total energy of the system is known to be equal to $-\mu H$. If this is the only information available, then the system can be in only one of the following three states:

$$(+ + -) \qquad (+ - +) \qquad (- + +)$$

Of course, we do not know in which of these states the system may actually be, nor do we necessarily know the relative probability of finding the system in any one of these states.

2·3 *Basic postulates*

In order to make theoretical progress, it is necessary to introduce some postulate about the relative probability of finding a system in any of its accessible states. Suppose that the system under consideration is *isolated* and thus cannot exchange energy with its surroundings. The laws of mechanics then imply that the total energy of the system is conserved. Thus it is known that the system must always be characterized by this value of the energy and that the states accessible to the system must all have this energy. But there are usually a great many states of this kind, and the system can be in any one of them. What can one say about the relative probability of finding the system in any such state?

One can hope to make some general statements in the simple case where the isolated system is in *equilibrium*. Such an equilibrium situation is characterized by the fact that the probability of finding the system in any one state is independent of time (i.e., the representative ensemble is the same irrespective of time). All macroscopic parameters describing the isolated system are then also time-independent. When one considers such an isolated system in equilibrium, the only information one has available about the system is that it must be in one of its accessible states consistent with the constant value of its energy. But there is nothing in the laws of mechanics which would lead one to expect that the system should be found more frequently in one of its accessible states rather than in another. Hence it seems eminently reasonable to *assume* that the system is equally likely to be found in any one of its accessible states. Indeed, one can show explicitly from the laws of mechanics that if one considers a representative ensemble of such isolated systems where these systems are distributed uniformly (i.e., with equal probability) over all their accessible states at *any one time*, then they will remain uniformly distributed over these states *forever*.* This fact shows that such a uniform distribution of systems in the ensemble over their accessible states corresponds indeed to a possible equilibrium situation which does not change in time. It also suggests that there is nothing intrinsic in the laws of mechanics which favors some states at the expense of others, because there exists no tendency to destroy the uniform distribution by populating some states preferentially while depleting other states.

The foregoing considerations suggest that all accessible states of an isolated system have intrinsically the same probability of being occupied by this system. One is thus led to introduce the following fundamental *postulate* of equal a priori probabilities:

▶ An isolated system in equilibrium is equally likely to be in any of its accessible states.

The same postulate is made in classical mechanics where state refers to a cell in phase space. That is, if phase space is subdivided into small cells of equal size, then an isolated system in equilibrium is equally likely to be in any of its accessible cells.†

This fundamental postulate is eminently reasonable and certainly does not contradict any of the laws of mechanics. Whether the postulate is actually

* This is a consequence of what is called "Liouville's theorem." A proof of this theorem in classical mechanics is given in Appendix A·13 and discussed more fully in R. C. Tolman, "The Principles of Statistical Mechanics," chap. 3, Oxford University Press, Oxford, 1938. A discussion of this theorem in quantum mechanics can be found in chap. 9 of the same book.

† Further comments about the postulate can be found at the end of this section.

valid can, of course, only be decided by making theoretical predictions based on it and by checking whether these predictions are confirmed by experimental observations. A large body of calculations based on this postulate have indeed yielded results in very good agreement with observations. The validity of this postulate can therefore be accepted with great confidence as the basis of our theory.

We illustrate this postulate with a few simple examples.

Example 1 In the previous example of a system of three spins, assume that the system is isolated. Its total energy is then known to have some constant value; suppose that it is known to be equal to $-\mu H$. As already mentioned, the system can then be in any of the following three states

$$(+ + -) \qquad (+ - +) \qquad (- + +)$$

The postulate asserts that when the system is in equilibrium it is equally likely to be found in any of these three states.

Note, incidentally, that it is *not* true that a given spin is equally likely to point up or down, i.e., to be in any of its two possible states. (There is, of course, no paradox here, since a given spin is not an isolated system, but interacts with the other two spins.) Indeed, it is seen that in the present example it is twice as probable that a given spin points up (is in a state of lower energy) than that it points down (is in a state of higher energy).

Example 2 An example more representative of situations encountered in practice is that of a macroscopic system consisting of N magnetic atoms, where N is of the order of Avogadro's number. If these atoms have spin $\frac{1}{2}$ and are placed in an external magnetic field, the situation is, of course, completely analogous to that of the preceding case of only three spins. But now there exists, in general, an extremely large number of possible states of the system for each specified value of its total energy.

Example 3 Consider a one-dimensional harmonic oscillator of mass m and spring constant κ, and let us discuss it in terms of classical mechanics. Denote the displacement coordinate of the oscillator by x and its linear momentum by p. Phase space is then two dimensional. The energy E of the oscillator is given by

$$E = \frac{p^2}{2m} + \frac{1}{2}\kappa x^2 \qquad (2 \cdot 3 \cdot 1)$$

where the first term on the right is its kinetic, the second term its potential energy. For a constant energy E, Eq. $(2 \cdot 3 \cdot 1)$ describes an ellipse in phase space, i.e., in the px plane. Suppose one knows that the energy of the oscillator lies in the small range* between E and $E + \delta E$. Then there are still

* The energy of a system can never be physically known to infinite precision (in quantum physics not even in principle, unless one spends an infinite amount of time in the measurement), and working with an infinitely sharply defined energy can also lead to unnecessary conceptual difficulties of a purely mathematical kind.

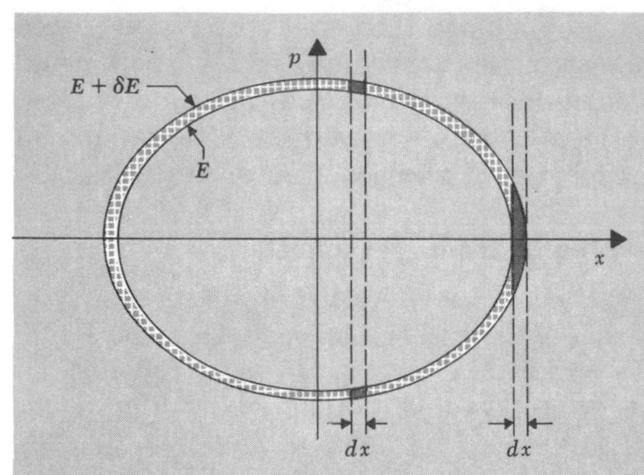

Fig. 2·3·1 Classical phase space for a one-dimensional harmonic oscillator with energy between E and $E + \delta E$. The accessible region of phase space consists of the area lying between the two ellipses.

many cells in phase space contained between the two ellipses corresponding to the respective energies E and $E + \delta E$, i.e., many different corresponding sets of values of x and p are possible for the oscillators in a representative ensemble. If the only information available about the oscillator is that it is in equilibrium with an energy in the specified range, then our statistical postulate asserts that it is equally probable that the oscillator has values of x and p lying within any one of these cells.

Another way of looking at the situation is the following. The time dependence of x and p for the oscillator is, by elementary mechanics, of the form

$$x = A \cos (\omega t + \varphi)$$
$$p = m\dot{x} = -mA\omega \sin (\omega t + \varphi)$$

where $\omega = \sqrt{\kappa/m}$, while A and φ are constants. By (2·3·1) the total energy is then

$$E = \frac{m\omega^2}{2} A^2 \sin^2 (\omega t + \varphi) + \frac{\kappa}{2} A^2 \cos^2 (\omega t + \varphi) = \tfrac{1}{2} m\omega^2 A^2$$

This is indeed equal to a constant, and the above relation determines the amplitude A in terms of E. But the phase angle φ is still quite arbitrary, depending on unknown initial conditions, and can assume any value in the range $0 < \varphi < 2\pi$. This gives rise to the many possible sets of values of x and p which correspond to the same energy.

Note that a given interval dx corresponds to a larger number of cells (i.e., a larger area) lying between the two ellipses when $x \approx A$ than when $x \approx 0$. Hence it is more probable that an oscillator in the ensemble is found with its position x close to A than close to 0. This result is, of course, also obvious from the fact that near the extremes of its position, where $x \approx A$, the oscillator has small velocity; hence it spends a longer time there than near $x \approx 0$, where it moves rapidly.

The approach to equilibrium Consider a situation where it is known that an isolated system is *not* equally likely to be found in any of the states accessible to it. Our fundamental postulate asserts that this situation cannot be one where equilibrium prevails. Thus one expects the situation to change with time. This means that in the representative statistical ensemble the distribution of systems over the accessible states will change in time; correspondingly, the mean values of various macroscopic parameters describing the system will also change.

Before discussing this nonequilibrium situation in greater detail, it is worth making a few comments about the nature of the states used in our theory to describe an isolated system of many particles. These states are *not* rigorously exact quantum states of the perfectly isolated system with all interactions between particles taken into account.* It would be prohibitively complicated to attempt any such utterly precise description; nor does one have available sufficiently detailed information about a macroscopic system to make such a precise description of any experimental interest. Instead one describes the system in terms of some complete set of approximate quantum states which take into account substantially all of its predominant dynamical features without being rigorously exact. When the system is known to be in such a state at any one time, it will not remain in this state indefinitely. Instead, there exists a finite probability that the system will at some later time be found in some of the other approximate states accessible to it, the transitions to these other states being caused by the presence of small residual interactions between the particles (interactions not taken into account in defining the approximate quantum states of the system).

Suppose then that at some initial time t the system is known to be in some *sub*set of the states actually accessible to it. There are no restrictions which would prevent the system from being found in *any* of its accessible states at some later time since all these states satisfy the conservation of energy and are consistent with the other constraints to which the system is known to be subject; nor is there anything in the laws of mechanics which would make any of these states intrinsically preferable to any other one. It is therefore exceedingly unlikely that the system remains indefinitely in the restricted subset of states in which it finds itself at the initial time t. Instead, the system will in the course of time always make transitions between all its various accessible states as a result of small interactions between its constituent particles. What then is the probability of finding the system in any of these states at some much later time?†

To see what happens it is only necessary to consider a statistical ensemble

* If the system were known to be in such an exact eigenstate at any one time, it would remain in this state forever.

† In principle, one could ask more detailed questions about subtle correlations existing between states of the system, i.e., questions about the quantum mechanical phases as well as amplitudes of the relevant wave functions. But it is generally meaningless to seek such a precise description in a theory where completely detailed information about any system is neither available nor of interest.

of such systems. Suppose that these systems are initially distributed over their accessible states in some arbitrary way; e.g., that they are only found in some particular subset of these accessible states. The systems in the ensemble will then constantly continue making transitions between the various accessible states, each system passing ultimately through practically all the states in which it can possibly be found. One expects that the net effect of these constant transitions will be analogous to the effect of repeated shufflings of a deck of cards. In the latter case, if one keeps on shuffling long enough, the cards get so mixed up that each one is equally likely to occupy any position in the deck irrespective of how the deck was arranged initially. Similarly, in the case of the ensemble of systems, one expects that ultimately the systems will become randomly (i.e., uniformly) distributed over all their accessible states. The distribution of systems over these states then remains uniform, i.e., it corresponds to a final time-independent equilibrium situation. In other words, one expects that, no matter what the initial conditions may be, an isolated system left to itself will ultimately attain a final equilibrium situation in which it is equally likely to be found in any one of its accessible states. One can look upon the expectation discussed in the preceding sentences as a highly plausible basic hypothesis very well confirmed by experience. From a more fundamental point of view this hypothesis can be regarded as a consequence of the so-called "H theorem," which can be established on the basis of the laws of mechanics and certain approximations inherent in a statistical description.*

Example 1 Consider again the very simple example of an isolated system of three spins $\frac{1}{2}$ in a large external magnetic field H. The approximate quantum states of the system can be labeled by the orientation of each spin with respect to this field ("up" or "down"). Suppose that this system has been prepared in such a way that it is known to be in the state $(+ + -)$ at some initial time; the system is then left to itself. Small interactions exist between the spins because the magnetic moment of one spin produces a small field H_m ($H_m \ll H$) with which the moment of some other spin can interact. These interactions between the magnetic moments of the spins bring about transitions in which one spin flips from the "up" direction to the "down" direction while some other spin does the reverse; of course, such a mutual spin-flip leaves the total energy of the system unchanged. The net result is that, after a sufficiently long time, the system will be found with equal probability in any of its three accessible states $(+ + -)$, $(+ - +)$, and $(- + +)$.

Example 2 Another vivid example is shown in Fig. 2·3·2, in which a gas of molecules is originally confined to the left half of a box, the right half being empty. Suppose now that the partition is removed at some initial time t. Immediately after this act, the molecules are certainly not distributed with equal probability over all their accessible states, since the molecules are all localized in the left half of the box whereas the right half, although now per-

* The interested reader can find a discussion of the H theorem in Appendix A·12.

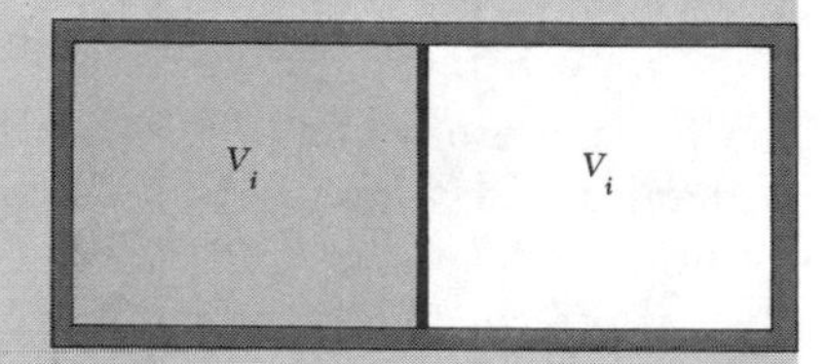

***Fig. 2·3·2** A system consisting of a box divided by a partition into two equal parts, each of volume V_i. The left side is filled with gas; the right side is empty.*

fectly accessible, is empty. But it is clearly fantastically improbable that this situation will prevail for any length of time. Indeed, as a result of collisions with the walls and with each other, the molecules will very quickly redistribute themselves over the entire volume of the box. The final equilibrium situation, where the density of molecules is uniform throughout the entire box, is thus attained quite rapidly.

Note that the preceding comments say nothing about how long one has to wait before the ultimate equilibrium situation is reached. If a system is initially not in equilibrium, the time necessary to attain equilibrium, (the so-called "relaxation time") might be shorter than a microsecond or longer than a century. It all depends on the detailed nature of the interactions between the particles of the particular system and on the resultant rate at which transitions actually occur between the accessible states of this system. The problem of calculating the *rate* of approaching equilibrium is a difficult one. On the other hand, one knows that isolated systems *do* tend to approach equilibrium if one waits long enough. The task of calculating the properties of systems in such *time-independent* situations is then quite straightforward (in principle), since it requires only arguments based on the fundamental statistical postulate of equal a priori probabilities.

Remark on classical phase space Phase space is defined in terms of generalized coordinates and *momenta* because it is in terms of these variables that Liouville's theorem holds. In cartesian coordinates it is usually true that $p_i = mv_i$, and phase space could therefore have been defined equally well in terms of coordinates and *velocities*. But more generally, for example in the presence of a magnetic field, the relation between p_i and v_i is more complicated.

Remark on the fundamental postulate in quantum mechanics The probability P_r that a quantum-mechanical system is in a state r (which is an eigenstate of the Hamiltonian) is given by $P_r = |a_r|^2$, where a_r is the complex "probability amplitude" which characterizes the state r of the system. Strictly speaking, the fundamental postulate asserts that in equilibrium the probabilities P_r are equal for all accessible states *and* that the corresponding amplitudes a_r have *random* phase factors.*

* R. C. Tolman, "The Principles of Statistical Mechanics," Oxford University Press, Oxford, 1938, pp. 349–356.

2 · 4 *Probability calculations*

The postulate of equal a priori probabilities is fundamental to all statistical mechanics and allows a complete discussion of the properties of systems in equilibrium. In principle, the calculations are very simple. For purposes of illustration, consider a system in equilibrium which is isolated so that its total energy is known to have a constant value in some range between E and $E + \delta E$. To make statistical predictions, we focus attention on an ensemble of such systems, all of which satisfy the condition that their total energy lies in this energy range. Let $\Omega(E)$ denote the *total* number of states of the system in this range. Suppose that there are among these states a certain number $\Omega(E; y_k)$ of states for which some parameter y of the system assumes the value y_k. The parameter might be the magnetic moment of the system, or the pressure exerted by the system, etc. (We label the possible values which y may assume by the index k; if the possible values which y can assume are continuous instead of discrete, we think of successive values of k as corresponding to values of y which differ by infinitesimal amounts.) Our fundamental postulate tells us that among the states accessible to the system, i.e., among the $\Omega(E)$ states which satisfy the condition that the energy of the system lies in the specified range, all states are equally likely to occur in the ensemble. Hence we can simply write for the probability $P(y_k)$ that the parameter y of the system assumes the value y_k

$$P(y_k) = \frac{\Omega(E; y_k)}{\Omega(E)} \tag{2·4·1}$$

Also, to calculate the *mean* value of the parameter y for this system, we simply take the average over the systems in the ensemble; i.e.,

$$\bar{y} = \frac{\sum_k \Omega(E; y_k) y_k}{\Omega(E)} \tag{2·4·2}$$

Here the summation over k denotes a sum over all possible values which the parameter y can assume.

Calculations of this kind are in principle quite straightforward. It is true that purely mathematical difficulties of computation may be encountered unless one is dealing with very simple systems. The reason is that, although it is quite easy to count states when there are no restrictions, it may be a formidable problem to pick out only those particular $\Omega(E)$ states which satisfy the condition that they have an energy near some specified value E. Mathematical complications of this sort are, however, not prohibitive and there are methods for overcoming them quite readily.

Example Let us illustrate these general comments in the case of the extremely simple example of the system consisting of three spins in equilibrium in a magnetic field H. If the total energy of this system is known to be $-\mu H$, then the system is equally likely to be in any of the three states

$$(++-) \qquad (+-+) \qquad (-++)$$

Focus attention on one of these spins, say the first. What is the probability P_+ that this spin points up? Since there are two cases where it points up, one has

$$P_+ = \tfrac{2}{3}$$

What is the mean magnetic moment $\bar{\mu}_z$ (in the $+z$ direction) of such a spin? Since the probability of occurrence of each state of the entire system is $\frac{1}{3}$, one has simply

$$\bar{\mu}_z = \tfrac{1}{3}\mu + \tfrac{1}{3}\mu + \tfrac{1}{3}(-\mu) = \tfrac{1}{3}\mu$$

2 · 5 *Behavior of the density of states*

A *macro*scopic system is one which has very many degrees of freedom (e.g., a copper block, a bottle of wine, etc.). Denote the energy of the system by E. Subdivide the energy scale into equal small ranges of magnitude δE, the magnitude of δE determining the precision within which one chooses to measure the energy of the system. For a macroscopic system, even a physically very small interval δE contains many possible states of the system. We shall denote by $\Omega(E)$ the number of states whose energy lies between E and $E + \delta E$.

The number of states $\Omega(E)$ depends on the magnitude δE chosen as the subdivision interval in a given discussion. Suppose that δE, while being large compared to the spacing between the possible energy levels of the system, is *macro*scopically sufficiently small. Then $\Omega(E)$ must be proportional* to δE, i.e., one can write

$$\Omega(E) = \omega(E)\,\delta E \qquad (2\cdot5\cdot1)$$

where $\omega(E)$ is independent of the size of δE. Thus $\omega(E)$ is a characteristic property of the system which measures the number of states per unit energy range, i.e., the "density of states." Since all statistical calculations involve the counting of states, it is worth examining how sensitively $\Omega(E)$ (or equivalently $\omega(E)$) depends on the energy E of a macroscopic system.

We are not interested in any exact results, but rather in a rough estimate adequate to reveal the essential behavior of Ω as a function of E. A simple argument can then be given along the following lines. Consider a system of f degrees of freedom so that f quantum numbers are required to specify each of its possible states. Let E be the energy of the system measured from its lowest possible energy (i.e., measured from the energy of its quantum-mechanical ground state) and let $\Phi(E)$ denote the total number of possible quantum

* The number of states $\Omega(E)$ must vanish when $\delta E \to 0$ and must be expressible as a Taylor's series in powers of δE. When δE is sufficiently small, all terms involving higher powers of δE are negligibly small and one is left with an expression of the form (2.5.1).

states of the system which are characterized by energies less than E Clearly $\Phi(E)$ increases as the energy E increases. Let us examine how *rapidly* $\Phi(E)$ increases.

Consider first *one* typical degree of freedom of the system. Denote by $\Phi_1(\epsilon)$ the *total* number of possible values which can be assumed by the quantum number associated with this particular degree of freedom when it contributes to the system an amount of energy ϵ or less. Again $\Phi_1(\epsilon)$ must clearly increase as ϵ increases (the smallest value of Φ_1 being unity when ϵ has its lowest possible value). If ϵ is not too small, Φ_1 is of the order of $\epsilon/\Delta\epsilon$ where $\Delta\epsilon$ denotes the mean spacing between the possible quantized energies associated with a typical degree of freedom and may itself depend on the magnitude of ϵ. But without getting involved in insignificant details, one can say that Φ_1 ought to increase roughly proportionally to ϵ; or in symbols that

$$\Phi_1 \propto \epsilon^\alpha \qquad (\alpha \approx 1) \tag{2·5·2}$$

where α is some number of the order of unity.

Let us now return to the whole system having an energy E and described by f quantum numbers $\{s_1, s_2, \ldots, s_f\}$. Then the energy ϵ per degree of freedom is of the order of

$$\epsilon \approx \frac{E}{f} \tag{2·5·3}$$

and, corresponding to this amount of energy or less, there are roughly $\Phi_1(\epsilon)$ possible values which can be assumed by the quantum number describing this degree of freedom. Corresponding to a total energy of E or less of the entire system, there are then approximately $\Phi_1(\epsilon)$ possible values which can be assumed by the quantum number s_1 associated with the first degree of freedom, approximately $\Phi_1(\epsilon)$ possible values which can be assumed by s_2, approximately $\Phi_1(\epsilon)$ possible values which can be assumed by s_3, etc. Hence the total number $\Phi(E)$ of possible sets of values of the f quantum numbers is approximately given by

$$\Phi(E) \approx [\Phi_1(\epsilon)]^f, \qquad \text{where } \epsilon = \frac{E}{f} \tag{2·5·4}$$

This gives the total number of states of the system when it has energy E or less. The number of states $\Omega(E)$ in the range between E and $E + \delta E$ is then

$$\Omega(E) = \Phi(E + \delta E) - \Phi(E) = \frac{\partial\Phi}{\partial E}\,\delta E \tag{2·5·5}$$

Thus

$$\Omega(E) \approx f\Phi_1{}^{f-1}\,\frac{\partial\Phi_1}{\partial\epsilon}\,\frac{1}{f}\,\delta E = \Phi_1{}^{f-1}\,\frac{\partial\Phi_1}{\partial\epsilon}\,\delta E \tag{2·5·6}$$

When the energy E of the system increases, the number of states Φ_1 per degree of freedom increases slowly and, by (2·5·3), roughly proportionately to $\epsilon = E/f$. But when one is dealing with a macroscopic system, f is very large—of the order of Avogadro's number—so that $f \approx 10^{24}$. Since the exponent in (2·5·6) is so very large, it follows that the number of possible states $\Omega(E)$ accessible to the entire system is an *extremely* rapidly increasing function of the energy E of the system. This is a general characteristic of the number of states $\Omega(E)$, or equivalently of the density of states $\omega = \Omega/\delta E$, of all ordinary macroscopic systems.

The relation (2·5·6) allows one to make some statements about orders of magnitude. Thus it implies that

$$\ln \Omega = (f - 1) \ln \Phi_1 + \ln \left(\frac{\partial \Phi_1}{\partial \epsilon} \delta E \right) \tag{2·5·7}$$

We recall that δE is supposed to be large compared to the spacing between the energy levels of the system. The quantity $(\partial \Phi_1/\partial \epsilon)\, \delta E$ is thus of the order of unity in the widest sense; i.e., it is certainly not greater than f nor much less than f^{-1}. Hence its logarithm is certainly of the order of unity in a strict sense (i.e., it lies between $\ln f$ and $-\ln f$ or between 55 and -55, if $f \approx 10^{24}$). On the other hand, the first term of (2·5·7) is of the order of f itself (if the energy ϵ of the system is not so very close to its ground-state energy that $\Phi_1 = 1$ for all degrees of freedom) and is thus fantastically larger than the second term which is only of order $\ln f$. (That is, $\ln f <<< f$, if f is very large.) Thus to an excellent approximation, (2·5·7) becomes

$$\ln \Omega \approx f \ln \Phi_1 \tag{2·5·8}$$

and

$$\ln \Omega \approx \mathcal{O}(f) \qquad \text{if } E > 0 \tag{2·5·9}$$

That is, $\ln \Omega$ is of the order of f if the energy of the system is not too close to the energy of its ground state. Furthermore, it follows by (2·5·8) and (2·5·2) that

$$\Omega \approx \Phi_1{}^f \propto E^f \tag{2·5·10}$$

This last relation is intended to be only an order-of-magnitude relation showing roughly how rapidly $\Omega(E)$ varies with the energy E of the system. Thus we have put $\alpha = 1$ for simplicity, since we are not particularly interested in whether the exponent in (2·5·10) should be f, $\frac{1}{2}f$, or any other number of the order of f.

Special case: ideal gas in the classical limit Consider the case of a gas of N identical molecules enclosed in a container of volume V. The energy of this system can be written

$$E = K + U + E_{\text{int}} \tag{2·5·11}$$

Here K denotes the total kinetic energy of translation of the molecules. If the momentum of the center of mass of the ith molecule is denoted by $\mathbf{p}_i$, then K depends only on these momenta and is given by

$$K = K(\mathbf{p}_1, \mathbf{p}_2, \ldots, \mathbf{p}_N) = \frac{1}{2m} \sum_{i=1}^{N} \mathbf{p}_i{}^2 \tag{2·5·12}$$

The quantity $U = U(\mathbf{r}_1, \mathbf{r}_2, \ldots, \mathbf{r}_N)$ represents the potential energy of mutual interaction between the molecules. It depends on the relative separations between the molecules, i.e., on their center-of-mass positions $\mathbf{r}_i$.

Finally, if the molecules are not monatomic, the atoms of each molecule can also rotate and vibrate relative to its center of mass. Let $Q_1, Q_2, \ldots, Q_M$ and $P_1, P_2, \ldots, P_M$ denote the coordinates and momenta describing this intramolecular motion. Then E_{int} represents the *total* energy of the system due to such intramolecular motion; it depends only on the internal coordinates Q_i and internal momenta P_i of all the molecules. Of course, if the molecules are monatomic, $E_{\text{int}} = 0$.

A particularly simple case is that where the mutual energy of interaction between the molecules is negligibly small. Then $U \approx 0$ and the molecules are said to form an "ideal gas." This situation can be achieved physically in the limit where the concentration N/V of the molecules is made sufficiently small, for then the mean separation between molecules becomes so large that their mutual interaction becomes negligibly small.

What is $\Omega(E)$ for such an ideal gas? Let us consider the situation in the classical limit, i.e., under circumstances where the energy E of the gas is much greater than its ground-state energy so that all quantum numbers are large. A description in terms of classical mechanics is then expected to be a good approximation. The number of states $\Omega(E)$ lying between the energies E and $E + \delta E$ is then equal to the number of cells in phase space contained between these energies; i.e., it is proportional to the volume of phase space contained therein. In symbols,

$$\Omega(E) \propto \int_E^{E+\delta E} \cdots \int d^3\mathbf{r}_1 \cdots d^3\mathbf{r}_N \, d^3\mathbf{p}_1 \cdots d^3\mathbf{p}_N \, dQ_1 \cdots dQ_M \, dP_1 \cdots dP_M \tag{2·5·13}$$

Here the integrand is simply the element of volume of phase space where we have used the abbreviations

$$d^3\mathbf{r}_i \equiv dx_i \, dy_i \, dz_i$$
$$d^3\mathbf{p}_i = dp_{ix} \, dp_{iy} \, dp_{iz}$$

to express three-dimensional volume elements in terms of the three cartesian components of the respective position and momentum vectors. The integration extends over all coordinates and momenta which are such that the total energy given by $(2 \cdot 5 \cdot 11)$ lies in the range between E and $E + \delta E$.

Since $U = 0$ for an ideal gas, the expression E in $(2 \cdot 5 \cdot 11)$ is independent of the center-of-mass positions $\mathbf{r}_i$ of the molecules.* Hence the integration over the position vectors $\mathbf{r}_i$ can be performed immediately. Since each integral over $\mathbf{r}_i$ extends over the volume V of the container, $\int d^3\mathbf{r}_i = V$. But there are N such integrals. Hence $(2 \cdot 5 \cdot 13)$ becomes simply

▶ $$\Omega(E) \propto V^N \chi(E) \tag{2·5·14}$$

* This is true as long as each molecule remains within the container. Of course, the walls of the container serve to confine the molecules within its volume by making $E \to \infty$ whenever a molecule tends to penetrate into a wall.

$$\text{where} \quad \chi(E) \propto \int_E^{E+\delta E} \cdots \int d^3\mathbf{p}_1 \cdots d^3\mathbf{p}_N \, dQ_1 \cdots dQ_M \, dP_1 \cdots dP_M \tag{2·5·15}$$

is independent of V, since neither K nor E_{int} in (2·5·11) depends on the coordinates $\mathbf{r}_i$, so that the integral (2·5·15) does not depend on the volume of the container. The relation (2·5·14) expresses a physically reasonable result for noninteracting molecules. It makes the obvious assertion that if the kinetic energy of each molecule is kept fixed while the volume of the box is doubled, twice as many states become available to each molecule; the number of states accessible to the N molecules is then simply increased by a factor $2 \times 2 \times 2 \times 2 \cdots = 2^N$.

Consider now the particularly simple case where the molecules are monatomic so that $E_{\text{int}} = 0$ and no intramolecular coordinates Q_i and P_i appear in the problem. Then (2·5·11) reduces simply to the kinetic energy and becomes

$$2mE = \sum_{i=1}^{N} \sum_{\alpha=1}^{3} p_{i\alpha}^2 \tag{2·5·16}$$

where the sum contains the square of each momentum component $p_{i\alpha}$ of each particle (since $\mathbf{p}_i^2 = p_{i1}^2 + p_{i2}^2 + p_{i3}^2$, denoting x, y, z components by 1, 2, 3, respectively). The sum in (2·5·16) thus contains $3N = f$ square terms. For E = constant, Eq. (2·5·16) then describes, in the f-dimensional space of the momentum components, a sphere of radius $R(E) = (2mE)^{\frac{1}{2}}$. Hence $\Omega(E)$, or $\chi(E)$ in (2·5·15), is proportional to the volume of phase space contained in the spherical shell lying between the sphere of radius $R(E)$ and that of slightly larger radius $R(E + \delta E)$ (see Fig. 2·5·1). But the volume of a sphere in f dimensions is proportional to R^f, since it is essentially obtained (just as the volume of a cube in f dimensions) by multiplying f linear dimensions by each other. Thus the *total* number of state $\Phi(E)$ of energy *less* than E is proportional

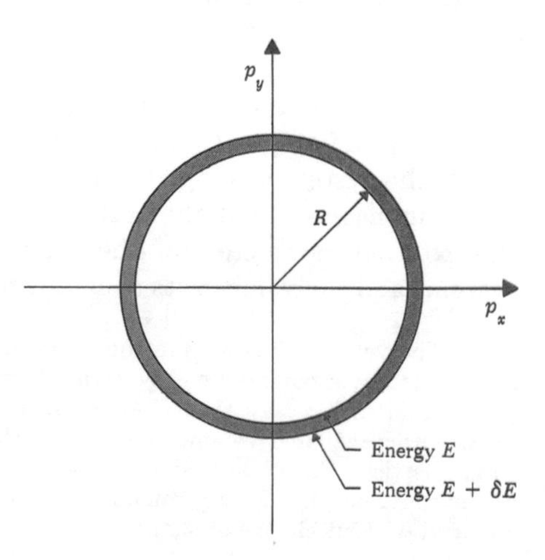

Fig. 2·5·1 Illustration in two dimensions of the "sphere" in momentum space for a single particle (of mass m) moving in two dimensions. Here $(2m)^{-1}(p_x^2 + p_y^2) = E$, the energy of the particle. The radius of the shell is $R = (2mE)^{\frac{1}{2}}$.

to this volume, i.e.,

$$\Phi(E) \propto R^f = (2mE)^{f/2} \tag{2·5·17}$$

The number of states $\Omega(E)$ lying in the spherical *shell* between energies E and $E + \delta E$ is then given by (2·5·5), so that

$$\Omega(E) \propto E^{(f/2)-1} \propto E^{(3N/2)-1} \tag{2·5·18}$$

which is properly proportional to R^{f-1}, i.e., to the *area* of the sphere in phase space. Combining this result with (2·5·14) one obtains for the classical *monatomic* ideal gas

► $$\Omega(E) = BV^N E^{3N/2} \tag{2·5·19}$$

where B is a constant independent of V and E, and where we have neglected 1 compared to N. Note again that since N is of the order of Avogadro's number and thus very large, $\Omega(E)$ is an extremely rapidly increasing function of the energy E of the system.

INTERACTION BETWEEN MACROSCOPIC SYSTEMS

2 · 6 *Thermal interaction*

In describing a macroscopic system it is, in general, possible to specify some macroscopically measurable independent parameters $x_1, x_2, \ldots, x_n$ which are known to affect the equations of motion (i.e., to appear in the Hamiltonian) of this system. These parameters are known as the "external parameters" of the system. Examples of such parameters are the applied magnetic or electric fields in which the system is located, or the volume V of the system (e.g., the volume V of the container confining a gas).* The energy levels of the system depend then, of course, on the values of the external parameters. If a particular quantum state r of the system is characterized by an energy E_r, one can thus write the functional relation

$$E_r = E_r(x_1, x_2, \ldots, x_n) \tag{2·6·1}$$

The "macroscopic state," or "macrostate," of the system is defined by specifying the external parameters of the system and any other conditions to which the system is subject. For example, if one deals with an isolated system, the *macro*state of the system might be specified by stating the values of the external parameters of the system (e.g., the value of the volume of the system) and the value of its constant total energy. The representative ensem-

* The volume V enters the equations of motion because the walls of the container are represented by a potential energy term U which depends on the position coordinates of the particles in such a way that $U \to \infty$ whenever the position coordinate of a molecule lies outside the available volume, i.e., inside the wall itself. For example, in the case of a single particle, Eq. (2·1·3) shows explicitly that its energy levels depend on the dimensions of the container; i.e., for a given quantum state, $E \propto V^{-\frac{2}{3}}$ if the volume V of the container is changed without change of shape.

ble for the system is prepared in accordance with the specification of this macrostate; e.g., all systems in the ensemble are characterized by the given values of the external parameters and of the total energy. Of course, corresponding to this given *macro*state, the system can be in any one of a very large number of possible *micro*states (i.e., quantum states).

Let us now consider two macroscopic systems A and A' which can interact with each other so that they can exchange energy. (Their total energy remains constant, of course, since the combined system $A^{(0)}$ consisting of A and A' is isolated.) In a macroscopic description it is useful to distinguish between two types of possible interactions between such systems. In one case all the external parameters remain fixed so that the possible energy levels of the systems do not change; in the other case the external parameters are changed and some of the energy levels are thereby shifted. We shall discuss these types of interaction in greater detail.

The first kind of interaction is that where the external parameters of the system remain unchanged. This represents the case of purely "thermal interaction."

Example As a trivial illustration, suppose that a bottle of beer is removed from a refrigerator and placed in the trunk of a car, where it remains for a while. No external parameters are changed; i.e., neither the volume of the bottle nor that of the air in the trunk is changed. But energy is transferred from the air in the trunk to the beer and results in a change of the latter's properties (e.g., the beer tastes less good).

As a result of the purely thermal interaction, energy is transferred from one system to the other. In a statistical description where one focuses attention on an ensemble of similar systems $(A + A')$ in interaction (see Fig. 2·6·1), the energy of every A system (or every A' system) does not change by precisely the same amount. One can, however, describe the situation conveniently in terms of the change in *mean* energy of each of the systems. The mean energy transferred from one system to the other as a result of purely thermal interaction is called "heat." More precisely, the change $\Delta\bar{E}$ of the mean energy of system A is called the "heat Q absorbed" by this system; i.e., $Q \equiv \Delta\bar{E}$. This heat can, of course, be negative as well as positive; the quantity $(-Q)$ is called the "heat given off" by the system. Since the combined energy of $(A + A')$ is unchanged, it follows that

$$\Delta\bar{E} + \Delta\bar{E}' = 0 \tag{2·6·2}$$

where $\Delta\bar{E}$ denotes the change of mean energy of A and $\Delta\bar{E}'$ that of A'. In terms of the definition of heat, one can write correspondingly

$$Q + Q' = 0 \qquad \text{or} \qquad Q = -Q' \tag{2·6·3}$$

This merely expresses the conservation of energy by the statement that the

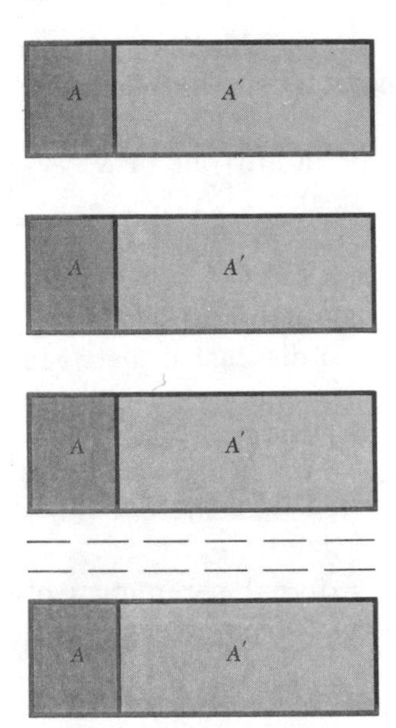

Fig. 2·6·1 Diagram illustrating schematically a representative statistical ensemble of similar isolated systems $A^{(0)}$, each consisting of two systems A and A' in interaction with each other.

heat absorbed by one system must be equal to the heat given off by the other system.

Since the external parameters do not change in a purely thermal interaction, the energy levels of neither system are in any way affected. The change of mean energy of a system comes about because the interaction results in a change in the relative number of systems in the ensemble which are distributed over the fixed energy levels (see Fig. 2·7·3*a* and *b*).

2 · 7 *Mechanical interaction*

A system which cannot interact thermally with any other system is said to be "thermally isolated," (or "thermally insulated"). It is easy to prevent thermal interaction between any two systems by keeping them spatially sufficiently separated, or by surrounding them with "thermally insulating" (sometimes also called "adiabatic") envelopes. These names are applied to an envelope provided that it has the following defining property: if it separates *any* two systems A and A' whose external parameters are fixed and each of which is initially in internal equilibrium, then these systems will remain in their respective equilibrium macrostates indefinitely (see Fig. 2·7·1). This definition implies physically that the envelope is such that no energy transfer is possible through it. (In practice, envelopes made of asbestos or fiberglass might approximate adiabatic envelopes reasonably well.)

When two systems are thermally insulated, they are still capable of interacting with each other through changes in their respective external parameters. This represents the second kind of simple macroscopic interaction, the

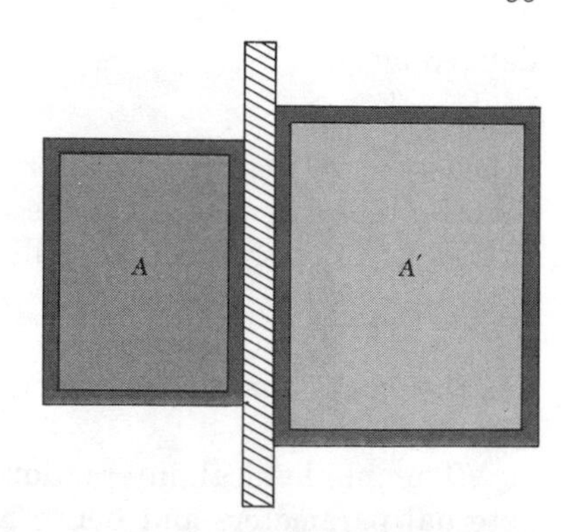

***Fig. 2·7·1** Two systems A and A′, each consisting of a gas in a container of fixed volume, are separated by a partition. If the partition is adiabatic, each system can independently remain in equilibrium for any value of its mean pressure. If the partition is not adiabatic, the gas pressures will, in general, change in time until they attain mutually compatible values in the final equilibrium situation.*

case of purely "mechanical interaction." The systems are then said to exchange energy by doing "macroscopic work" on each other.

Example Consider the situation shown in Fig. 2·7·2 in which a gas is enclosed in a vertical cylinder by a piston of weight w, the piston being thermally insulated from the gas. Initially the piston is clamped in position at a height s_i. When the piston is released, it oscillates for a while and finally comes to rest at a greater height s_f. Let A denote the system consisting of the gas and cylinder, and A' the system consisting of the piston (including the weight) and the earth. Here the interaction involves changes in the external parameters of the system, i.e., a change in the volume of the gas and in the height of the piston. In this process the gas does a net amount of work in lifting the weight.

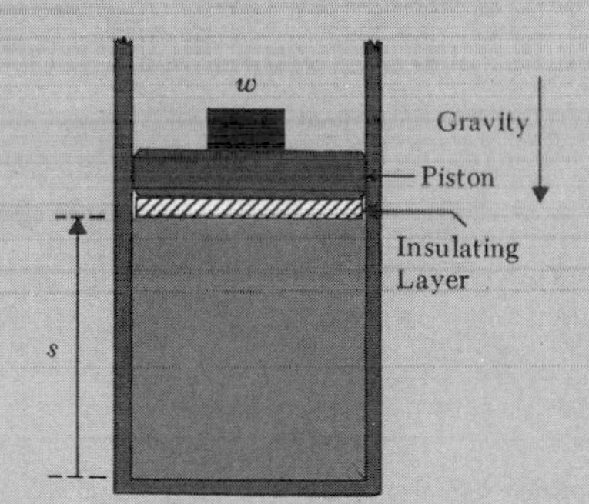

***Fig. 2·7·2** A gas contained in a cylinder closed by a piston of weight w. A layer of thermally insulating material (of negligible weight) is attached to the bottom of the piston to separate it from the gas.*

In a statistical description one again focuses attention on an ensemble of similar systems $(A + A')$ in interaction. Not every system in the ensemble has its energy changed by exactly the same amount as a result of the change of external parameters, but one can again describe the situation in terms of the change in *mean* energy of the systems. Consider, for example, system A. If the change in its mean energy due to the change of external parameters is denoted by $\Delta_x\bar{E}$, then the "macroscopic work" $\mathcal{W}$ done *on* the system is defined as

$$\mathcal{W} = \Delta_x\bar{E} \qquad (2\cdot7\cdot1)$$

The macroscopic work W done *by* the system is the negative of this and is thus defined as

$$W \equiv -\mathcal{W} \equiv -\Delta_x \bar{E} \tag{2·7·2}$$

Whenever we shall use the term "work" without further qualifications, we shall be referring to the macroscopic work just defined. The conservation of energy (2·6·2) is, of course, still valid and can be written in the form

$$W + W' = 0 \quad \text{or} \quad W = -W' \tag{2·7·3}$$

i.e., the work done by one system must be equal to the work done on the other system.

The mechanical interaction between systems involves changes in the external parameters and hence results in changes of the energy levels of the systems. Note that, even if the energies E_r of different quantum states are originally equal, a change of external parameters usually shifts these energy levels by different amounts for different states r. In general, the change in mean energy of a system depends on how the external parameters are changed and on how rapidly they are changed. When these parameters are changed in some arbitrary way, the energy levels of the possible states of the system change; in addition, transitions are produced between various states of the system. (Thus, if the system is initially in a particular state, it will in general be distributed over many of its states after the parameter change.) Thus the situation may be quite complicated while the parameters are changed and shortly thereafter, even when equilibrium conditions prevail initially and finally.

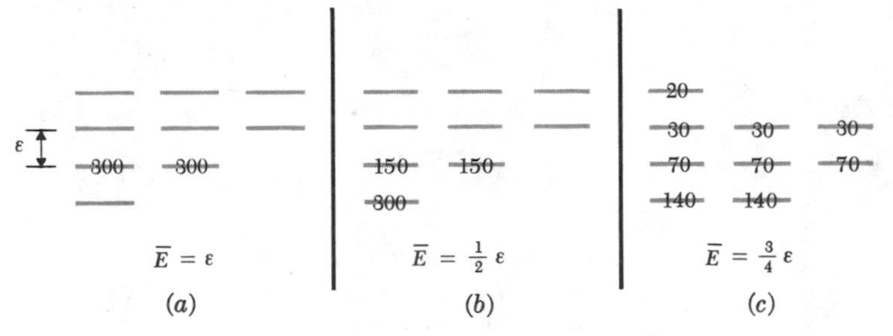

***Fig. 2·7·3** Schematic illustration of heat and work. The diagram shows the energy levels (separated by an amount ε) of a hypothetical system which can be in any of nine possible states. There are 600 systems in the statistical ensemble and the numbers indicate the number of these systems in each state. (a) Initial equilibrium situation. (b) Final equilibrium situation after the system has given off heat $\frac{1}{2}\epsilon$ to some other system. (c) Final equilibrium situation after the system in (a) has (in some arbitrary way) done work $\frac{1}{4}\epsilon$ on some other system. (The very small numbers used for simplicity in this illustration are, of course, not representative of real macroscopic systems.)*

Example The complications can be illustrated by the previous example of Fig. 2·7·2, or perhaps even more directly by the example of Fig. 2·7·4. Sup-

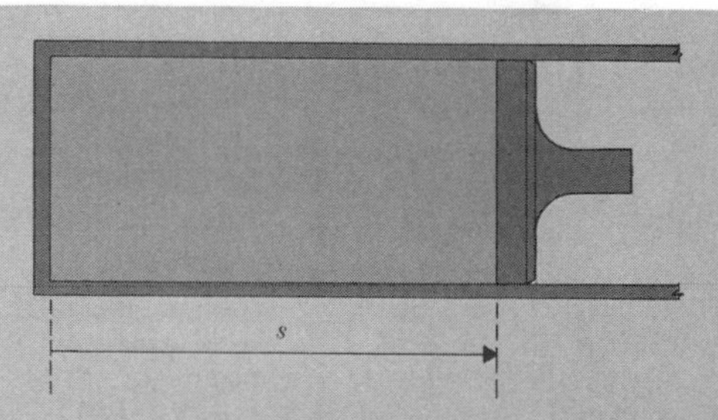

Fig. 2·7·4 ***A system consisting of a fluid (gas or liquid) contained in a cylinder closed off by a movable piston. The one external parameter is the distance s, which is the distance of the piston from the end wall of the cylinder.***

pose that this system is initially in equilibrium, the piston being clamped at a distance s_i from the end wall of the cylinder. The system is then equally likely to be in any of its possible states compatible with the initial value $s = s_i$ and with the initial energy E_i of the system. Suppose that some external device now rapidly moves the piston to a new position $s = s_f$, thus compressing the fluid. In this process the external device does work and the mean energy of the system is increased by some amount $\Delta_x\bar{E}$. But all kinds of pressure nonuniformities and turbulence are also set up in the fluid; during this time the system is *not* equally likely to be in any of its accessible states. Of course, if one keeps s at the value s_f and waits long enough, a new equilibrium situation will again be reached, where there is equal probability that the system is in any of its states compatible with the new value $s = s_f$ and the new mean energy $E_i + \Delta_x\bar{E}$.

Macroscopic work is, nevertheless, a quantity which can be readily measured experimentally. Suppose that in the mechanical interaction between two systems A and A' at least one of them, say A', is a relatively simple system whose change in mean energy can readily be computed from a change in its external parameters by using considerations based on mechanics. For example, one may know that A' exerts a measurable mean force on A and that the change of external parameters corresponds simply to a definite displacement of the center of mass of A'. Then the mean work W' done by A' on A is immediately obtained as the product of a mean force multiplied by the corresponding displacement; by (2·7·3), the work done by A is then given by $W = -W'$.

Example 1 Consider the previously mentioned illustration of Fig. 2·7·2, where the piston is initially at a height s_i and finally comes to rest at a height s_f. Here the center of mass of the piston is simply displaced by a net amount $(s_f - s_i)$, and one can neglect any change in the internal energy of motion of the molecules in the piston relative to its center of mass. Then the entire change in the energy of the system A', consisting of the piston and the earth, is due to the change of potential energy $w(s_f - s_i)$ of the center of mass of the piston in the gravitational field of the earth. Hence it follows that in the process here contemplated the system A, consisting of gas and cylinder, does an amount of work $W = w(s_f - s_i)$ on the system A'.

Example 2 In Fig. 2·7·5 the falling weight w is connected through a string to a paddle wheel, which is thus made to rotate and to churn the liquid in which it is immersed. Suppose that the weight descends with uniform speed a distance s. Then the energy of the system A', consisting of the weight and the earth, is decreased by an amount ws; this is then also the work done on the system A consisting of the paddle wheel and the liquid.

Fig. 2·7·5 A system consisting of a vessel containing a liquid and a paddle wheel. The falling weight can perform work on the system by rotating the paddle wheel.

Example 3 Figure 2·7·6 illustrates a similar situation where a battery of emf $\mathcal{V}$ is connected electrically to a resistor immersed in a liquid. When a charge q flows through the circuit, the energy stored in the battery decreases by an amount $q\mathcal{V}$. Hence the battery does an amount of work $q\mathcal{V}$ on the system A consisting of the resistor and the liquid.

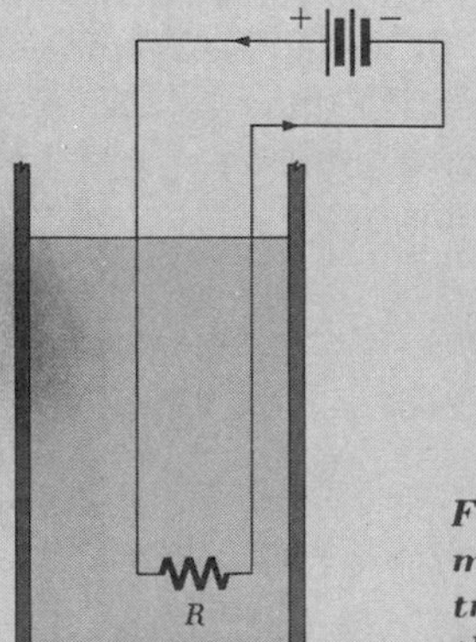

Fig. 2·7·6 A system consisting of a resistor immersed in a liquid. The battery can perform electrical work on the system by sending current through the resistor.

2·8 *General interaction*

In the most general case of interaction between two systems their external parameters do *not* remain fixed and the systems are *not* thermally insulated. As a result of such a general interaction the mean energy of a system is changed by some amount $\Delta\bar{E}$, but not all of this change is due to the change of its external parameters. Let $\Delta_x\bar{E} = \mathcal{W}$ denote the increase of its mean energy calculable from the change of external parameters (i.e., due to the macroscopic work $\mathcal{W}$ done *on* the system). Then the total change in mean energy of the system can be written in the form

$$\Delta\bar{E} \equiv \Delta_x\bar{E} + Q = \mathcal{W} + Q \tag{2·8·1}$$

where the quantity Q thus introduced is simply a measure of the mean energy change *not* due to the change of external parameters. In short, (2·8·1) *defines* the quantity Q by the relation

$$Q \equiv \Delta\bar{E} - \mathcal{W} = \Delta\bar{E} + W \tag{2·8·2}$$

where $W \equiv -\mathcal{W}$ is the work done *by* the system. The relation (2·8·2) constitutes the general *definition* of the heat absorbed by a system. When the external parameters are kept fixed, (2·8·2) reduces, of course, to the definition already introduced in Sec. 2·6 for the case of purely thermal interaction.

The relation (2·8·1) simply splits the total mean energy change into a part $\mathcal{W}$ due to mechanical interaction and a part Q due to thermal interaction. One of the fundamental aims of our study will be to gain a better understanding of the relationship between thermal and mechanical interactions. This is the reason for the name "thermodynamics" applied to the classical discipline dealing with such questions.

Note that, by virtue of (2·8·1), both heat and work have the dimensions of energy and are thus measured in units of ergs or joules.

Example Consider Fig. 2·8·1 where two gases A and A' are contained in a cylinder and separated by a movable piston.

a. Suppose first that the piston is clamped in a fixed position and that it is thermally insulating. Then the gases A and A' do not interact.

b. If the piston is *not* insulating but is clamped in position, energy will in general flow from one gas to the other (although no macroscopic work gets

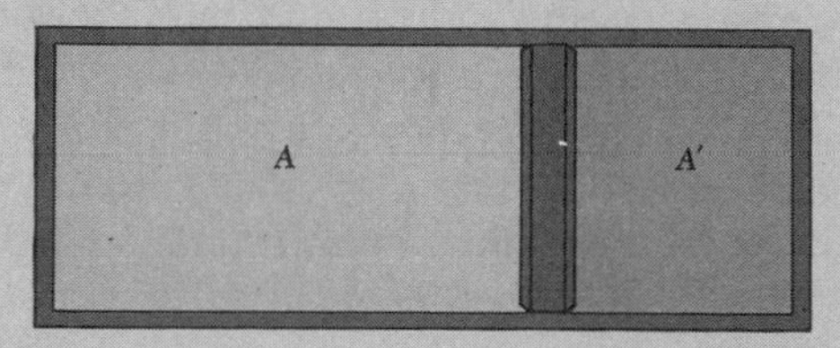

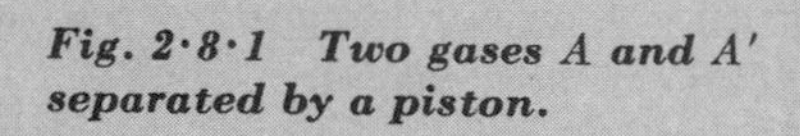
Fig. 2·8·1 ***Two gases A and A' separated by a piston.***

done) and the pressures of the gases will change as a result. This is an example of purely thermal interaction.

c. If the piston is insulating but free to move, then it will in general move so that the volumes and pressures of the gases change, one gas doing mechanical work on the other. This is an example of purely mechanical interaction.

d. Finally, if the piston is noninsulating and free to move, both thermal and mechanical interaction can take place between the two gases A and A'.

If one contemplates infinitesimal changes, the small increment of mean energy resulting from the interaction can be written as the differential $d\bar{E}$. The infinitesimal amount of work done by the system in the process will be denoted by $đW$; similarly, the infinitesimal amount of heat absorbed by the system in the process will be denoted by $đQ$. In terms of the above notation, the definition (2·8·2) becomes for an infinitesimal process

$$đQ \equiv d\bar{E} + đW \tag{2·8·3}$$

Remark The special symbol $đW$ is introduced, instead of W, merely as a convenient notation to emphasize that the work itself is infinitesimal. It does *not* designate any difference between works. Indeed, the work done is a quantity referring to the interaction *process* itself. Thus it makes no sense to talk of the work in the system before and after the process, or of the difference between these. Similar comments apply to $đQ$, which again denotes just the infinitesimal amount of heat absorbed in the process, *not* any meaningless difference between heats.

2 · 9 *Quasi-static processes*

In the last few sections we have considered quite general processes whereby systems can interact with each other. An important, and much simpler, special case is that where a system A interacts with some other system in a process (involving the performance of work, exchange of heat, or both) which is carried out so slowly that A remains arbitrarily close to equilibrium at all stages of the process. Such a process is said to be "quasi-static" for the system A. Just *how* slowly one must proceed to keep a situation quasi-static depends on the time τ (the "relaxation time") that the system requires to attain equilibrium if it is suddenly disturbed. To be slow enough to be quasi-static implies that one proceeds slowly compared to the time τ. For example, if the gas in Fig. 2·7·4 returns to equilibrium within a time $\tau \approx 10^{-3}$ seconds after the distance s is suddenly halved, then a process wherein the piston is moved so as to halve the volume of the gas in 0.1 second can be considered quasi-static to a good approximation.

If the external parameters of a system have values $x_1, \ldots, x_n$, then the energy of the system in a definite quantum state r has some value

$$E_r = E_r(x_1, \ldots, x_n) \tag{2·9·1}$$

When the values of the external parameters are changed, the energy of this state r changes in accordance with the functional relation (2·9·1). In particular, when the parameters are changed by infinitesimal amounts so that $x_\alpha \rightarrow x_\alpha + dx_\alpha$ for each α, then (2·9·1) gives for the corresponding change in energy

$$dE_r = \sum_{\alpha=1}^{n} \frac{\partial E_r}{\partial x_\alpha} dx_\alpha \tag{2·9·2}$$

The work $đW$ done *by* the system when it remains in this particular state r is then defined as

$$đW_r \equiv -dE_r = \Sigma X_{\alpha,r}\, dx_\alpha \tag{2·9·3}$$

where we have introduced the definition

$$X_{\alpha,r} \equiv -\frac{\partial E_r}{\partial x_\alpha} \tag{2·9·4}$$

This is called the "generalized force" (conjugate to the external parameter x_α) in the state r. Note that if x_α denotes a distance, then X_α is simply an ordinary force.

Consider now the statistical description where one focuses attention on an ensemble of similar systems. When the external parameters of the system are changed quasi-statically, then the generalized forces $X_{\alpha,r}$ have at any time well-defined mean values; these are calculable from the distribution of systems in the ensemble corresponding to the equilibrium situation consistent with the values of these external parameters at that time. (For example, if the system is thermally isolated, then the systems in the ensemble are at any time equally likely to be in any of their accessible states which are compatible with the values of the external parameters at that time.) The macroscopic work $đW$ resulting from an infinitesimal quasi-static change of external parameters is then obtained by calculating the decrease in mean energy resulting from this parameter change. Calculating the mean value of (2·9·3) averaged over all accessible states r then gives

$$đW = \sum_{\alpha=1}^{n} \bar{X}_\alpha\, dx_\alpha \tag{2·9·5}$$

where

$$\bar{X}_\alpha \equiv -\overline{\frac{\partial E_r}{\partial x_\alpha}} \tag{2·9·6}$$

is the *mean* generalized force conjugate to x_α. Here the mean values are to be calculated with the equilibrium distribution of systems in the ensemble corresponding to the external parameter values x_α. The macroscopic work W resulting from a *finite* quasi-static change of external parameters can then be obtained by integration.

Remark If one were dealing with an isolated system in a state r which is an *exact* stationary quantum state of the *entire* Hamiltonian (including *all* interactions between particles), then this system would remain in this state of energy E_r indefinitely when the external parameters are kept fixed; it would also remain in this state (its energy E_r varying in accordance with (2·9·1)) when the external parameters are changed infinitely slowly. Thus no transitions to other states would occur in the course of time. But in a statistical description one does not deal with such precisely defined situations. Instead, one contemplates a system which can be in any one of a large number of accessible quantum states which are not exact stationary quantum states of the entire Hamiltonian (including all the interactions), so that transitions between these states do occur. Indeed, if one waits long enough, these transitions bring about the final equilibrium situation where the system, if isolated, is equally likely to be found in any of its accessible states. When the external parameters of the system are changed quasi-statically, a given system in the ensemble does then not always remain in the same state. Instead, there occurs a continual redistribution of systems over their accessible states so as to maintain always a distribution consistent with an equilibrium situation, i.e., a uniform distribution over all accessible states in an ensemble of isolated systems (see Fig. 2·9·1).

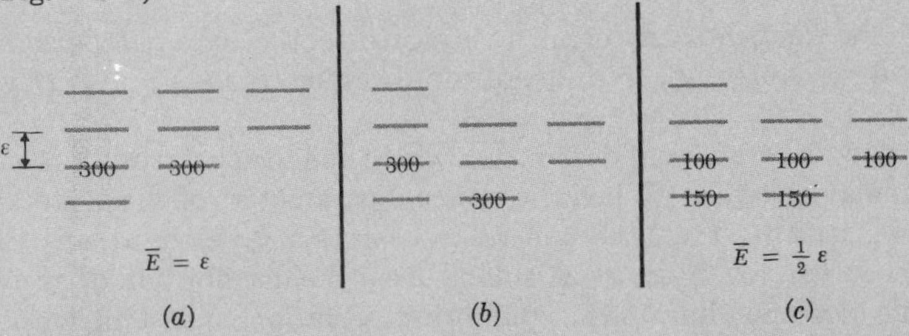

Fig. 2·9·1 ***Schematic illustration of quasi-static work done by the thermally isolated system of Fig. 2·7·3 as a result of the same change of external parameter as that shown in part (c) of that figure. The diagram shows again the energy levels of this system, and the numbers indicate the number of systems present in each state in the ensemble. (a) Initial equilibrium situation. (b) Hypothetical situation which would prevail after the quasi-static external parameter change if each system remained in its former state. (c) Actual final equilibrium situation resulting from the quasi-static external parameter change, work $\frac{1}{2}\epsilon$ having been done by the system.***

2 · 10 *Quasi-static work done by pressure*

As an important example of quasi-static work, consider the case in which there is only one external parameter of significance, the volume V of the system. Then the work done in changing the volume from V to $V + dV$ can be calculated from elementary mechanics as the product of a force multiplied by a displacement. Suppose that the system under consideration (see Fig. 2·7·4) is contained in a cylinder. If the system is in state r, let its pressure on the piston of area A be denoted by p_r. The force exerted by the system on the piston is then p_rA. The volume of the system is specified by the distance s of the piston from the end wall of the cylinder; thus $V = As$. If the distance s is now changed very slowly by an amount ds, the system remains in the state r and performs an amount of work

$$đW_r = (p_rA)\, ds = p_r(A\, ds) = p_r\, dV \qquad (2 \cdot 10 \cdot 1)$$

Since $đW_r = -dE_r$, it follows from this that

$$p_r = -\frac{\partial E_r}{\partial V} \qquad (2 \cdot 10 \cdot 2)$$

Thus p_r is the generalized force conjugate to the volume V.

If the volume of the system is changed quasi-statically, the system remains always in internal equilibrium so that its pressure has a well-defined mean value $\bar{p}$. The macroscopic work done by the system in a quasi-static change of volume is then, by (2·10·1); related to the mean pressure by the relation*

$$đW = \bar{p}\, dV \qquad (2 \cdot 10 \cdot 3)$$

Remark The expression (2·10·3) for the work done is much more general than the derivation based on the simple cylinder would indicate. To show this, consider an arbitrary slow expansion of the system from the volume enclosed by the solid boundary to that enclosed by the dotted boundary in Fig. 2·10·1. If the mean pressure is $\bar{p}$, the mean force on an element of area dA is $\bar{p}\, dA$ in the direction of the normal $\mathbf{n}$. If the displacement of this element of area is by an amount ds in the direction making an angle θ with the normal,

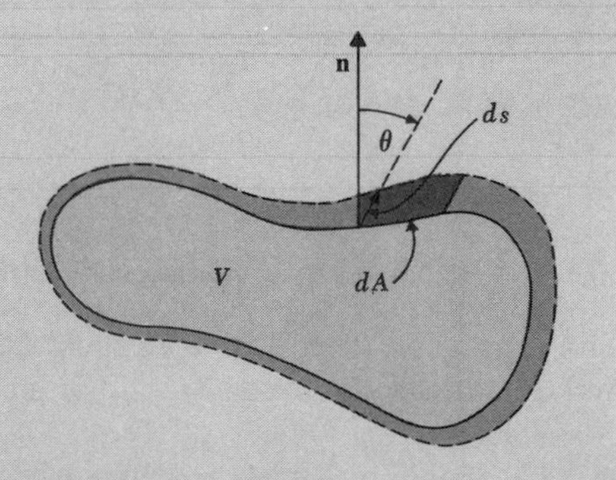

Fig. 2·10·1 Arbitrary expansion of a system of volume V.

* When a system is in one definite state r, the work $đW_r$ and the corresponding pressure p_r may, in general, depend on just how the volume is changed. (For example, if the system is in the shape of a rectangular parallelepiped, the work $đW_r$ may depend on which wall is moved and the force per unit area on different walls may be different.) But after averaging over all the states r, the macroscopic work and mean pressure $\bar{p}$ become insensitive to the precise mode of deformation contemplated for the volume.

then the work done by the pressure on this area is $(\bar{p}\, dA)\, ds \cos\theta = \bar{p}\, dv$, where $dv \equiv (dA\, ds \cos\theta)$ is the volume of the parallelepiped swept out by the area element dA in its motion through ds. Summing over all the elements of area of the boundary surface gives then for the total work

$$đW = \Sigma\bar{p}\, dv = \bar{p}\Sigma\, dv = \bar{p}\, dV$$

where $dV = \Sigma\, dv$ is the sum of all the little volumes swept out, i.e., the increase in volume of the total system. Thus one regains (2·10·3).

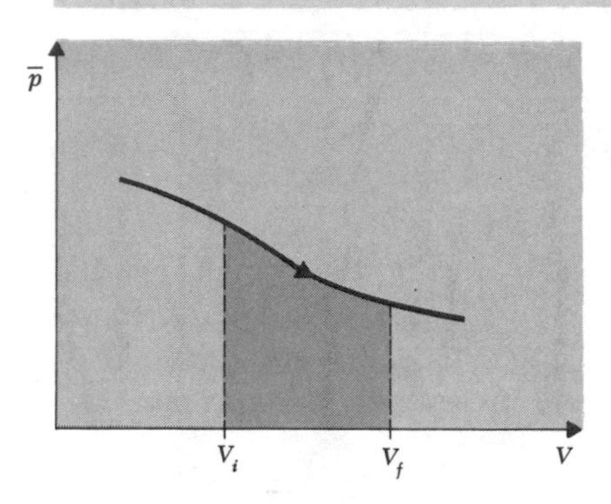

***Fig. 2·10·2** Dependence of the mean pressure $\bar{p}$ on the volume V of a system. The shaded area under the curve represents the work done by the system when its volume changes quasi-statically from V_i to V_f.*

Suppose that a quasi-static process is carried out in which the volume is changed from V_i to V_f. For example, this process might be carried out in such a way that for all volumes $V_i \le V \le V_f$ the mean pressure $\bar{p} = \bar{p}(V)$ assumes the values indicated by the curve of Fig. 2·10·2. In this process the macroscopic work done by the system is given by

$$W_{if} = \int_{V_i}^{V_f} đW = \int_{V_i}^{V_f} \bar{p}\, dV \qquad (2\cdot10\cdot4)$$

Note that this integral represents geometrically just the shaded area contained below the curve of Fig. 2·10·2.

2 · 11 *Exact and "inexact" differentials*

The expression (2·8·3) relates the differential $d\bar{E}$ of the energy to the infinitesimal quantities $đW$ and $đQ$. It is instructive to examine these infinitesimals more closely.

Consider the purely mathematical problem where $F(x,y)$ is some function of the two independent variables x and y. This means that the value of F is determined when the values of x and y are specified. If one goes to a neighboring point corresponding to $x + dx$ and $y + dy$, the function F changes by an amount

$$dF = F(x + dx,\, y + dy) - F(x,y) \qquad (2\cdot11\cdot1)$$

This can also be written in the form

$$dF = A(x,y)\,dx + B(x,y)\,dy \tag{2·11·2}$$

where $A = \partial F/\partial x$ and $B = \partial F/\partial y$. Clearly dF in (2·11·1) is simply the infinitesimal difference between two adjacent values of the function F. The infinitesimal quantity dF is here just an ordinary differential; it is also called an "exact differential" to distinguish it from other kinds of infinitesimal quantities to be discussed presently. Note that if one goes from an initial point i corresponding to (x_i,y_i) to a final point f corresponding to (x_f,y_f), the corresponding change in F is simply given by

$$\Delta F = F_f - F_i = \int_i^f dF = \int_i^f (A\,dx + B\,dy) \tag{2·11·3}$$

Since the difference on the left side depends only on the initial and final points, the integral on the right can only depend on these end points; it thus *cannot* depend on the path along which it is evaluated in going from the initial point i to the final point f.

On the other hand, not every infinitesimal quantity is an exact differential. Consider, for example, the infinitesimal quantity

$$A'(x,y)\,dx + B'(x,y)\,dy \equiv đG \tag{2·11·4}$$

where A' and B' are some functions of x and y, and where $đG$ has been introduced merely as an abbreviation for the expression on the left side. Although $đG$ is certainly an infinitesimal quantity, it does *not* follow that it is necessarily an exact differential; i.e., it is in general *not* true that there exists some function $G = G(x,y)$ whose value is determined when x and y are given, and which is such that $dG = G(x + dx,\ y + dy) - G(x,y)$ is equal to the expression (2·11·4). Equivalently, it is in general *not* true that, if one sums (i.e., integrates) the infinitesimal quantities $đG$ in going from the point i to the point f along a certain path, that the integral

$$\int_i^f đG = \int_i^f (A'\,dx + B'\,dy) \tag{2·11·5}$$

is independent of the particular path used. When an infinitesimal quantity is not an exact differential it is called an "inexact differential."

Example Consider the infinitesimal quantity

$$đG = \alpha\,dx + \beta\frac{x}{y}\,dy = \alpha\,dx + \beta x\,d(\ln y)$$

where α and β are constants. Let i denote the initial point (1,1) and f the final point (2,2). In Fig. 2·11·1 one can then, for example, calculate the integral of $đG$ along the path $i \to a \to f$ passing through the point a with coordinates (2,1); this gives

$$\int_{iaf} đG = \alpha + 2\beta\ln 2$$

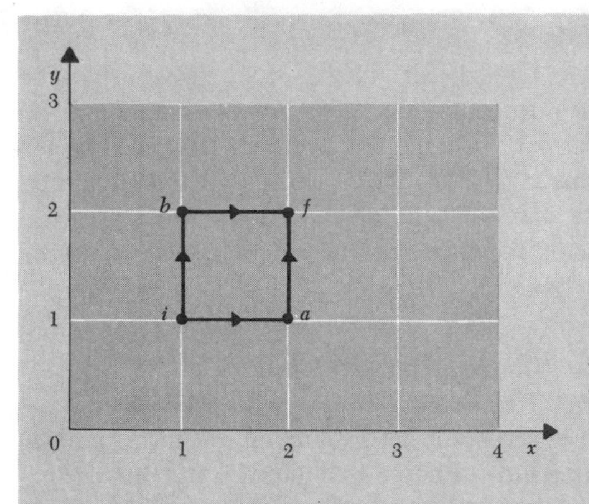

Fig. 2·11·1 Alternative paths connecting the points i and f in the xy plane.

Alternatively, one can calculate it along the path $i \to b \to f$ passing through the point b with coordinates (1,2); this gives

$$\int_{ibf} đG = \beta \ln 2 + \alpha$$

Thus the integrals are different and the quantity $đG$ is not an exact differential.

On the other hand, the infinitesimal quantity

$$dF \equiv \frac{đG}{x} = \frac{\alpha}{x}\,dx + \frac{\beta}{y}\,dy$$

is an exact differential of the well-defined function $F = \alpha \ln x + \beta \ln y$. The integral of dF from i to f is thus always equal to

$$\int_i^f dF = \int_i^f \frac{đG}{x} = (\alpha + \beta) \ln 2$$

irrespective of the path chosen to go from i to f.

After this purely mathematical illustration, let us return to the physical situation of interest. The macrostate of a macroscopic system can be specified by the values of its external parameters (e.g., of its volume V) and of its mean energy $\bar{E}$; other quantities, such as its mean pressure $\bar{p}$, are then determined. Alternatively, one can choose the external parameters and the pressure $\bar{p}$ as the independent variables describing the macrostate; the mean energy $\bar{E}$ is then determined. Quantities such as $d\bar{p}$ or $d\bar{E}$ are thus infinitesimal differences between well-defined quantities, i.e., they are just ordinary (i.e., exact) differentials. For example, $d\bar{E} = \bar{E}_f - \bar{E}_i$ is simply the difference between the well-defined mean energy $\bar{E}_f$ of the system in a final macrostate f and its well-defined mean energy $\bar{E}_i$ in an initial macrostate i when these two states are only infinitesimally different. It also follows that, if the system is taken from any initial macrostate i to any final macrostate f, its mean energy change is simply given by

$$\Delta\bar{E} = \bar{E}_f - \bar{E}_i = \int_i^f d\bar{E} \tag{2·11·6}$$

But since a quantity like $\bar{E}$ is just a function of the macrostate under consideration, $\bar{E}_i$ and $\bar{E}_f$ depend only on the particular initial and final macrostates; thus the integral $\int_i^f d\bar{E}$ over all the energy increments gained in the process depends *only* on the initial and final macrostates. In particular, therefore, the integral does *not* depend on what particular process is chosen to go from i to f in evaluating the integral.

On the other hand, consider the infinitesimal work $đW$ done by the system in going from some initial macrostate i to some neighboring final macrostate f. In general $đW = \Sigma\bar{X}_\alpha\, dx_\alpha$ is *not* the difference between two numbers referring to the two neighboring macrostates, but is merely an infinitesimal quantity characteristic of the *process* of going *from* state i *to* state f. (It is meaningless to talk of the work *in* a given state; one can only talk of the work done in going *from* one state *to* another state.) The work $đW$ is then in general an *inexact* differential. The total work done by the system in going from any macrostate i to some other macrostate f can be written as

$$W_{if} = \int_i^f đW \tag{2·11·7}$$

where the integral represents simply the sum of the infinitesimal amounts of work $đW$ performed at each stage of the process. But, in general, the value of the integral *does* depend on the particular process which is used in going from macrostate i to macrostate f.

Example Consider a system, e.g., a gas, whose volume V is the only relevant external parameter (see Fig. 2·7·4). Assume that the system is brought quasi-statically from its initial macrostate of volume V_i to its final macrostate of volume V_f. (During this process the system may be allowed to exchange heat with some other system.) We can describe the particular process used by specifying the mean pressure $\bar{p}(V)$ of the system for all values assumed by its volume in the course of the process. This functional relation can be repre-

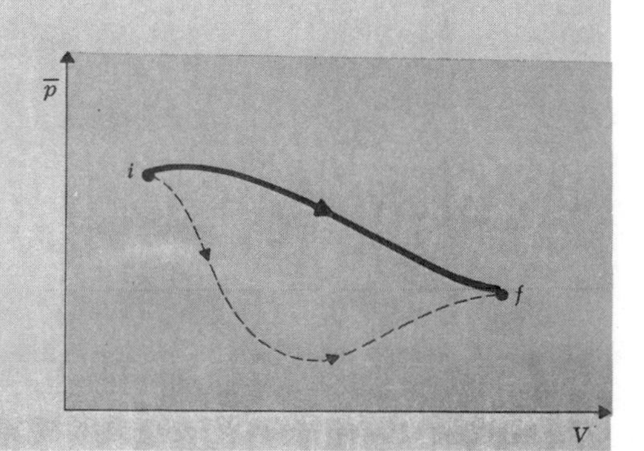

Fig. 2·11·2 Relation between mean pressure $\bar{p}$ and volume V for two different quasi-static processes.

sented by the curve of Fig. 2·11·2 and the corresponding work is given by (2·10·4), i.e., by the area under the curve. If two different processes are used in going from i to f, described respectively by the solid and dotted $\bar{p}$ versus V curves in Fig. 2·11·2, then the areas under these two curves will be different. Thus the work W_{if} done by the system certainly depends on the particular process used in going from i to f.

In going from macrostate i to macrostate f the change $\Delta\bar{E}$ does *not* depend on the process, while the work W in general *does*. Hence it follows by (2·8·2) that the heat Q in general also *does* depend on the process used. Thus $đQ$ denotes just an infinitesimal amount of heat absorbed during a process; like $đW$, it is, in general, *not* an exact differential.

Of course, if the system is thermally insulated so that $Q = 0$, Eq. (2·8·2) implies that

$$W_{if} = -\Delta\bar{E} \tag{2·11·8}$$

Then the work done depends *only* on the energy difference between initial and final macrostates and *is* independent of the process. Thus we have a result which is sometimes referred to as the "first law of thermodynamics":

> If a *thermally isolated* system is brought from some initial to some final macrostate, the work done by the system is independent of the process used. (2·11·9)

Remark This statement is an expression of conservation of energy and is subject to direct experimental verification. For example, one may conceive of the following type of experiment. A thermally insulated cylinder is closed by a piston. The cylinder contains a system which consists of a liquid in which there is immersed a small paddle wheel which can be rotated from outside by a falling weight. Work can be done on this system by either (*a*) moving the piston, or (*b*) rotating the paddle wheel. The respective amounts of work can be measured in terms of mechanical quantities by knowing (*a*) the mean pressure on, and displacement of, the piston, and (*b*) the distance

Fig. 2·11·3 A thermally insulated system on which work can be done in various ways.

by which the known weight descends. By doing such work the system can be brought from its initial macrostate of volume V_i and pressure $\bar{p}_i$ to a final state of volume V_f and pressure $\bar{p}_f$. But this can be done in many ways: e.g., by rotating the paddle wheel first and then moving the piston the required amount; or by moving the piston first and then rotating the paddle wheel through the requisite number of revolutions; or by performing these two types of work alternately in smaller amounts. The statement (2·11·9) asserts that if the *total* work performed in each such procedure is measured, the result is always the same.*

Similarly, it follows that if the external parameters of a system are kept fixed so that it does no work, then $đW = 0$ and (2·8·3) reduces to

$$đQ = d\bar{E}$$

so that $đQ$ becomes an exact differential. The amount of heat Q absorbed in going from one macrostate to another is then independent of the process used and depends only on the mean energy difference between them.

SUGGESTIONS FOR SUPPLEMENTARY READING

Statistical formulation

R. C. Tolman: "The Principles of Statistical Mechanics," chaps. 3 and 9, Oxford University Press, Oxford, 1938. (This book is a classic in the field of statistical mechanics and is entirely devoted to a careful exposition of fundamental ideas. The chapters cited discuss ensembles of systems and the fundamental statistical postulate in classical and quantum mechanics, respectively.)

Work and heat—macroscopic discussion

M. W. Zemansky: "Heat and Thermodynamics," 4th ed, chaps. 3 and 4, McGraw-Hill Book Company, New York, 1957.

H. B. Callen: "Thermodynamics," secs. 1.1–1.7, John Wiley & Sons, Inc., New York, 1960. (The analogy mentioned on pp. 19 and 20 is particularly instructive.)

PROBLEMS

2.1 A particle of mass m is free to move in one dimension. Denote its position coordinate by x and its momentum by p. Suppose that this particle is confined within a box so as to be located between $x = 0$ and $x = L$, and suppose that its energy is known to lie between E and $E + \delta E$. Draw the classical phase space

* Paddle wheels such as this were historically used by Joule in the last century to establish the equivalence of heat and mechanical energy. In the experiment just mentioned we might equally well replace the paddle wheel by an electric resistor on which electrical work can be done by sending through it a known electric current.

of this particle, indicating the regions of this space which are accessible to the particle.

2.2 Consider a system consisting of two weakly interacting particles, each of mass m and free to move in one dimension. Denote the respective position coordinates of the two particles by x_1 and x_2, their respective momenta by p_1 and p_2. The particles are confined within a box with end walls located at $x = 0$ and $x = L$. The total energy of the system is known to lie between E and $E + \delta E$. Since it is difficult to draw a four-dimensional phase space, draw separately the part of the phase space involving x_1 and x_2 and that involving p_1 and p_2. Indicate on these diagrams the regions of phase space accessible to the system.

2.3 Consider an ensemble of classical one-dimensional harmonic oscillators.

(*a*) Let the displacement x of an oscillator as a function of time t be given by $x = A \cos(\omega t + \varphi)$. Assume that the phase angle φ is equally likely to assume any value in its range $0 < \varphi < 2\pi$. The probability $w(\varphi)\,d\varphi$ that φ lies in the range between φ and $\varphi + d\varphi$ is then simply $w(\varphi)\,d\varphi = (2\pi)^{-1}\,d\varphi$. For any fixed time t, find the probability $P(x)\,dx$ that x lies between x and $x + dx$ by summing $w(\varphi)\,d\varphi$ over all angles φ for which x lies in this range. Express $P(x)$ in terms of A and x.

(*b*) Consider the classical phase space for such an ensemble of oscillators, their energy being known to lie in the small range between E and $E + \delta E$. Calculate $P(x)\,dx$ by taking the ratio of that volume of phase space lying in this energy range *and* in the range between x and $x + dx$ to the total volume of phase space lying in the energy range between E and $E + \delta E$ (see Fig. 2·3·1). Express $P(x)$ in terms of E and x. By relating E to the amplitude A, show that the result is the same as that obtained in part (*a*).

2.4 Consider an isolated system consisting of a large number N of very weakly interacting localized particles of spin $\frac{1}{2}$. Each particle has a magnetic moment μ which can point either parallel or antiparallel to an applied field H. The energy E of the system is then $E = -(n_1 - n_2)\mu H$, where n_1 is the number of spins aligned parallel to H and n_2 the number of spins aligned antiparallel to H.

(*a*) Consider the energy range between E and $E + \delta E$ where δE is very small compared to E but is microscopically large so that $\delta E \gg \mu H$. What is the total number of states $\Omega(E)$ lying in this energy range?

(*b*) Write down an expression for $\ln \Omega(E)$ as a function of E. Simplify this expression by applying Stirling's formula in its simplest form (A·6·2).

(*c*) Assume that the energy E is in a region where $\Omega(E)$ is appreciable, i.e., that it is not close to the extreme possible values $\pm N\mu H$ which it can assume. In this case apply a Gaussian approximation to part (*a*) to obtain a simple expression for $\Omega(E)$ as a function of E.

2.5 Consider the infinitesimal quantity

$$A\,dx + B\,dy \equiv đF$$

where A and B are both functions of x and y.

(*a*) Suppose that $đF$ is an exact differential so that $F = F(x,y)$. Show that A and B must then satisfy the condition

$$\frac{\partial A}{\partial y} = \frac{\partial B}{\partial x}$$

(*b*) If $đF$ is an exact differential, show that the integral $\int đF$ evaluated along *any* closed path in the xy plane must vanish.

2.6 Consider the infinitesimal quantity

$$(x^2 - y)\,dx + x\,dy \equiv đF \tag{1}$$

(*a*) Is this an exact differential?

(*b*) Evaluate the integral $\int đF$ between the points (1,1) and (2,2) of Fig. 2·11·1 along the straight-line paths connecting the following points:

$$(1,1) \to (1,2) \to (2,2)$$
$$(1,1) \to (2,1) \to (2,2)$$
$$(1,1) \to (2,2)$$

(*c*) Suppose that both sides of (1) are divided by x^2. This yields the quantity $đG = đF/x^2$. Is $đG$ an exact differential?

(*d*) Evaluate the integral $\int đG$ along the three paths of part (*b*).

2.7 Consider a particle confined within a box in the shape of a cube of edges $L_x = L_y = L_z$. The possible energy levels of this particle are then given by (2·1·3).

(*a*) Suppose that the particle is in a given state specified by particular values of the three integers n_x, n_y, and n_z. By considering how the energy of this state must change when the length L_x of the box is changed quasistatically by a small amount dL_x, show that the force exerted by the particle in this state on a wall perpendicular to the x axis is given by $F_x = -\partial E/\partial L_x$.

(*b*) Calculate explicitly the force per unit area (or pressure) on this wall. By averaging over all possible states, find an expression for the mean pressure on this wall. (Exploit the property that the average values $\overline{n_x^2} = \overline{n_y^2} = \overline{n_z^2}$ must all be equal by symmetry.) Show that this mean pressure can be very simply expressed in terms of the mean energy $\bar{E}$ of the particle and the volume $V = L_xL_yL_z$ of the box.

2.8 A system undergoes a quasi-static process which appears in a diagram of mean pressure $\bar{p}$ versus volume V as a closed curve. (See diagram. Such a process is called "cyclic" since the system ends up in a final macrostate which is identical to its initial macrostate.) Show that the work done by the system is given by the area contained within the closed curve.

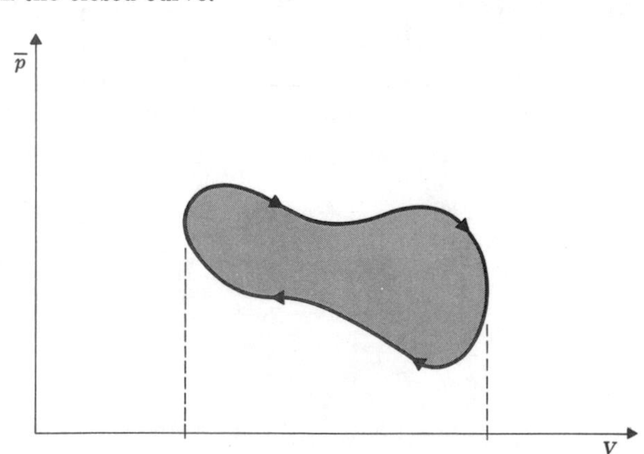

2.9 The tension in a wire is increased quasi-statically from F_1 to F_2. If the wire has length L, cross-sectional area A, and Young's modulus Y, calculate the work done.

2.10 The mean pressure $\bar{p}$ of a thermally insulated amount of gas varies with its volume V according to the relation

$$\bar{p}V^{\gamma} = K$$

where γ and K are constants. Find the work done by this gas in a quasistatic process from a macrostate with pressure $\bar{p}_i$ and volume V_i to one with pressure $\bar{p}_f$ and volume V_f. Express your answer in terms of $\bar{p}_i$, V_i, $\bar{p}_f$, V_f, and γ.

2.11 In a quasi-static process $A \rightarrow B$ (see diagram) in which no heat is exchanged with the environment, the mean pressure $\bar{p}$ of a certain amount of gas is found to change with its volume V according to the relation

$$\bar{p} = \alpha V^{-\frac{5}{3}}$$

where α is a constant. Find the quasi-static work done and the net heat absorbed by this system in each of the following three processes, all of which take the system from macrostate A to macrostate B.

(*a*) The system is expanded from its original to its final volume, heat being added to maintain the pressure constant. The volume is then kept constant, and heat is extracted to reduce the pressure to 10^6 dynes cm^{-2}.

(*b*) The volume is increased and heat is supplied to cause the pressure to decrease linearly with the volume.

(*c*) The two steps of process (*a*) are performed in the opposite order.

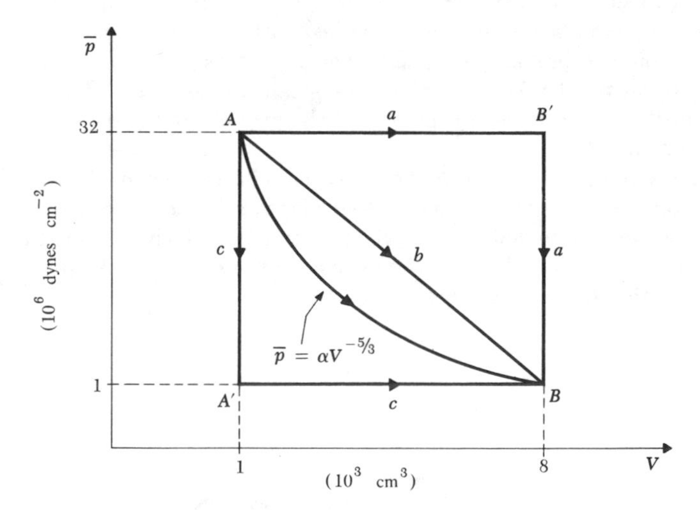

Statistical thermodynamics

3

THE FUNDAMENTAL statistical postulate of equal a priori probabilities can be used as the basis of the entire theory of systems in equilibrium. In addition, the hypothesis mentioned at the end of Sec. 2·3 (and based on the assumed validity of the H theorem) also makes a statement about isolated systems not in equilibrium, asserting that these tend to approach ultimate equilibrium situations (characterized by the uniform statistical distribution over accessible states which is demanded by the fundamental postulate).

In this chapter we shall show how these basic statements lead to some very general conclusions concerning all macroscopic systems. The important results and relationships thus established constitute the basic framework of the discipline of "equilibrium statistical mechanics" or, as it is sometimes called, "statistical thermodynamics." Indeed, the major portion of this book will deal with systems in equilibrium and will therefore be an elaboration of the fundamental ideas developed in this chapter.

IRREVERSIBILITY AND THE ATTAINMENT OF EQUILIBRIUM

3·1 *Equilibrium conditions and constraints*

Consider an isolated system whose energy is specified to lie in a narrow range. As usual, we denote by Ω the number of states accessible to this system. By our fundamental postulate we know that, in equilibrium, such a system is equally likely to be found in any one of these states.

We recall briefly what we mean by "accessible states." There are in general some specified conditions which the system is known to satisfy. These act as constraints which limit the number of states in which the system can possibly be found without violating these conditions. The accessible states are then all the states consistent with these constraints.

The constraints can be described more quantitatively by specifying the values of some parameters* $y_1, y_2, \ldots, y_n$ which characterize the system on a macroscopic scale. The number of states accessible to the system depends then on the values of these parameters; i.e., one can write the functional relation

$$\Omega = \Omega(y_1, \ldots, y_n)$$

for the number of states accessible to the system when each parameter labeled by α lies in the range between y_α and $y_\alpha + \delta y_\alpha$. For example, a parameter y_α might denote the volume or the energy of some subsystem. We give some concrete illustrations.

Example 1 Consider the system shown in Fig. 2.3.2 where a box is divided by a partition into two equal parts, each of volume V_i. The left half of the box is filled with gas, while the right one is empty. Here the partition acts as a constraint which specifies that only those states of the system are accessible for which the coordinates of all the molecules lie in the left half of the box. In other words, the volume V accessible to the gas is a parameter which has the prescribed value $V = V_i$.

Example 2 Consider a system $A^{(0)}$ consisting of two subsystems A and A' separated by a fixed thermally insulating partition (see Fig. 2·7·1). This partition acts as a constraint which specifies that no energy can be exchanged between A and A'. Hence only those states of $A^{(0)}$ are accessible which have the property that the energy of A remains constant at some specified value $E = E_i$, while that of A' remains constant at some other specified value $E' = E_i'$.

Example 3 Consider the system of Fig. 2·8·1 where a thermally insulated piston separates two gases A and A'. If the piston is clamped in position, then this piston acts as a constraint which specifies that only those states of the total system are accessible which are such that the A molecules lie within a given fixed volume V_i, while the A' molecules lie within a given fixed volume V_i'.

Suppose that the initial situation with the given constraints is one of equilibrium where the isolated system is equally likely to be found in any of its Ω_i accessible states. Consider that some of the constraints are now removed. Then all the states formerly accessible to the system still remain accessible to it; but many more additional states will, in general, also become accessible. A removal of constraints can then only result in increasing, or possibly leaving unchanged, the number of states accessible to the system. Denoting the final number of accessible states by Ω_f, one can write

$$\Omega_f \geq \Omega_i \qquad (3\cdot 1\cdot 1)$$

* These are not necessarily *external* parameters.

Focus attention on a representative ensemble of systems similar to the one under consideration and suppose that, when the constraints are removed, $\Omega_f > \Omega_i$. *Immediately* after the constraints are removed, the systems in the ensemble will not be in any of the states from which they were previously excluded. But the systems occupy then only a fraction

$$P_i = \frac{\Omega_i}{\Omega_f} \tag{3·1·2}$$

of the Ω_f states now accessible to them. This is *not* an equilibrium situation. Indeed, our fundamental postulate asserts that in the final equilibrium situation consistent with the absence of constraints, it is equally likely that each of the Ω_f states be occupied by the systems. If $\Omega_f \gg \Omega_i$, the particular situation where the systems are distributed only over the Ω_i original states becomes thus a very unlikely one; to be precise, its probability of occurrence is given by (3·1·2). In accordance with the hypothesis discussed at the end of Sec. 2·3, there is then a pronounced tendency for the situation to change in time until the much more probable final equilibrium situation is reached where the systems in the ensemble are distributed equally over all the possible Ω_f states.

Let us illustrate these statements with the examples mentioned previously.

Example 1 Suppose that the partition in Fig. 2·3·2 is removed. There is now no longer any constraint preventing a molecule from occupying the right half of the box. It is therefore exceedingly improbable that all the molecules will remain concentrated in the left half. Instead, they move about until they become randomly distributed throughout the entire box. In this final equilibrium situation each molecule is then equally likely to be found anywhere inside the box.

Suppose that this final equilibrium situation in the absence of the partition has been attained. What then is the probability P_i of encountering a situation where all the molecules are again concentrated in the left half of the box? The probability of finding one given molecule in the left half of the box is $\frac{1}{2}$. Hence the probability P_i of simultaneously finding all N molecules in the left half of the box is obtained by simply multiplying the respective probabilities of each molecule being in the left half, i.e.,

$$P_i = (\tfrac{1}{2})^N$$

When N is of the order of Avogadro's number, so that $N \approx 6 \times 10^{23}$, this probability is *fantastically* small; i.e.,

$$P_i \approx 10^{-2\times 10^{23}}$$

Example 2 Imagine that the partition in Fig. 2·7·1 is made thermally conducting. This removes the former constraint because the systems A and A' are now free to exchange energy with each other. The number of states accessible to the combined system $A^{(0)} = A + A'$ will, in general, be much greater if A adjusts its energy to some new value (and if A' correspondingly adjusts its energy so as to keep the energy of the isolated total system $A^{(0)}$ unchanged). Hence the most probable final equilibrium situation results when such an adjustment has taken place by virtue of heat transfer between the two systems.

Example 3 Imagine that the piston in Fig. 2·8·1 is unclamped so that it is free to move. Then the number of states accessible to the combined system $A + A'$ is, in general, much increased if the volumes of A and A' assume new values significantly different from their original ones. Thus a much more probable final equilibrium situation for the combined system $A + A'$ is attained if the piston moves so as to bring the volumes of A and A' to these new values.* As one would expect (and as we shall prove later), this final equilibrium situation corresponds to one where the mean pressures of both gases are equal so that the piston is in mechanical equilibrium.

This discussion can be phrased in terms of the relevant parameters $y_1, \ldots, y_n$ of the system. Suppose that a constraint is removed; for example, one of the parameters (call it simply y), which originally had the value $y = y_i$, is now allowed to vary. Since all the states accessible to the system are a priori equally likely, the equilibrium probability distribution $P(y)$ of finding the system in the range between y and $y + \delta y$ is proportional to the number of states accessible to the system when the parameter lies in this range; i.e.,

$$P(y) \propto \Omega(y) \tag{3·1·3}$$

This probability implies an occurrence of possible values of y, which is, in general, extremely different from the original situation where all systems in the ensemble were characterized by the value $y = y_i$ (see Fig. 3·1·1). In the absence of constraints the value $y = y_i$ of the parameter represents, therefore, a very improbable configuration. Hence the situation tends to change in time until the uniform equilibrium distribution of systems over

* The piston may oscillate back and forth several times before settling down in its final equilibrium position.

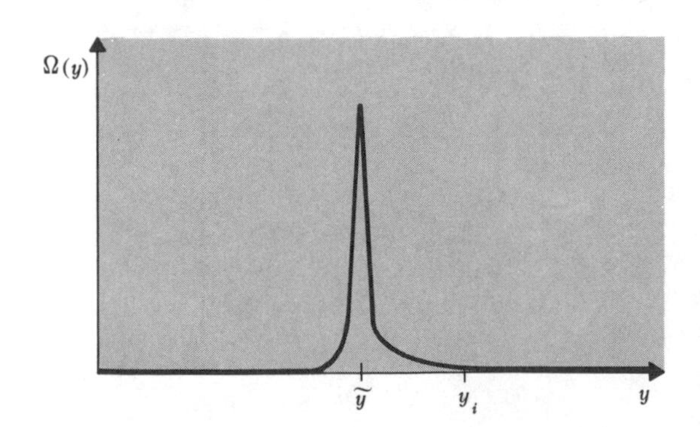

Fig. 3·1·1 ***Schematic diagram showing the number of states $\Omega(y)$ accessible to a system as a function of a parameter y. The initial value of this parameter is denoted by y_i.***

accessible states is attained, i.e., until the various values of y occur with respective probabilities given by (3·1·3). Usually $\Omega(y)$ has a very pronounced maximum at some value $\tilde{y}$. In that case practically all systems in the final equilibrium situation will have values corresponding to the most probable situation where y is very close to $\tilde{y}$. Hence, if initially $y_i \neq \tilde{y}$, the parameter y will change after the constraint is removed until it attains values close to $\tilde{y}$ where Ω is maximum. This discussion can be summarized by the following statement:

> If some constraints of an isolated system are removed, the parameters of the system tend to readjust themselves in such a way that $\Omega(y_1, \ldots, y_n)$ approaches a maximum. In symbols
>
> $$\Omega(y_1, \ldots, y_n) \rightarrow \text{maximum} \tag{3·1·4}$$

3·2 *Reversible and irreversible processes*

Suppose that the final equilibrium situation has been reached so that the systems in the ensemble are uniformly distributed over the Ω_f accessible final states. If the constraints are now simply restored, the systems in the ensemble will still occupy these Ω_f states with equal probability. Thus, if $\Omega_f > \Omega_i$, simply restoring the constraints does *not* restore the initial situation. Once the systems are randomly distributed over the Ω_f states, simply imposing or reimposing a constraint cannot cause the systems to move spontaneously out of some of their possible states so as to occupy a more restricted class of states. Nor could the removal of any other constraints make the situation any better; it could only lead to a situation where even more states become accessible to the system so that the system could be found in them as well.

Consider an isolated system (i.e., one which cannot exchange energy in the form of heat or work with any other system) and suppose that some process occurs in which the system goes from some initial situation to some final situation. If the final situation is such that the imposition or removal of constraints of this *isolated* system cannot restore the initial situation, then the process is said to be "irreversible." On the other hand, if it is such that the imposition or removal of constraints *can* restore the initial situation, then the process is said to be "reversible."

In terms of these definitions, the original removal of the constraints in the case where $\Omega_f > \Omega_i$ can be said to be an irreversible process. Of course, it is possible to encounter the special case where the original removal of the constraints does not change the number of accessible states so that $\Omega_f = \Omega_i$. Then the system, originally in equilibrium and equally likely to be in any of its Ω_i states, will simply remain distributed with equal probability over these states. The equilibrium of the system is then completely undisturbed so that this special process is reversible.

We again illustrate these comments with the previous examples:

Example 1 Once the molecules are in equilibrium and uniformly distributed throughout the box, the simple act of replacing the partition does not change the essential situation. The molecules still remain uniformly distributed throughout the box. The original removal of the partition thus constitutes an irreversible process.

This does *not* mean that the original situation of this system can *never* be restored. It can, provided that the system is *not* kept isolated but is allowed to interact with other systems. For example, in the present situation one can take a thin piston, which is originally coincident with the right wall of the box. One can now use some outside device A' (e.g., a falling weight) to move the piston to the center of the box, thus doing work on the gas in order to recompress it into the left half against the pressure exerted by the gas. The volume of the gas has now been restored to its original value V_i and the right half of the box is empty as before. But the energy of the gas is greater than originally because of the work done on it during recompression. To restore the energy of the gas to its previous value, one can now let the gas give off just the right amount of heat by bringing it into thermal contact with some suitable system A''. The gas has thus been restored to its original situation, its volume and energy being the same as initially.

Of course, the *isolated* system $A^{(0)}$ consisting of the gas *and* the systems A' and A'' has *not* been restored to its original situation, since the systems A' and A'' have been changed in the process. The process is still irreversible for the entire system $A^{(0)}$. Indeed, in releasing the weight to move the piston and in eliminating the thermal insulation to allow heat exchange with A'', we have removed constraints of $A^{(0)}$ and increased the number of states accessible to this isolated system.

Example 2 Suppose that thermal interaction between A and A' has taken place and that the systems are in equilibrium. Simply making the partition again thermally insulating does not change the new energies of A and A'. One cannot restore the system $A + A'$ to its original situation by making heat flow in a direction opposite to the original direction of spontaneous heat transfer (unless one introduces interaction with suitable outside systems). The original heat transfer is thus an irreversible process.

Of course, a special case may arise where the initial energies of the systems A and A' are such that making the partition originally thermally conducting does *not* increase the number of states accessible to the combined system $A + A'$. Then no net energy exchange takes place between A and A', and the process is reversible.

Example 3 This again is, in general, an irreversible process. Simply clamping the piston in its new position, so that it is again not free to move, does not restore the initial volumes of the gases.

The discussion of this section can be summarized by the statement that if some constraints of an isolated system in equilibrium are removed, the number of states accessible to the system can only increase or remain the same, i.e., $\Omega_f \geq \Omega_i$.

If $\Omega_f = \Omega_i$, then the systems in the representative ensemble are already distributed with equal probability over all their accessible states. The system remains, therefore, always in equilibrium and the process is *reversible*.

If $\Omega_f > \Omega_i$, then the distribution of systems over the possible states in the representative ensemble is a very improbable one. The system will therefore tend to change in time until the most probable final equilibrium situation of uniform distribution of systems over accessible states is reached. Equilibrium does not prevail at all stages of the process and the process is *irreversible*.

Remarks on significant time scales Note that we have nowhere made any statements about the *rate* of a process, i.e., about the relaxation time τ required by a system to reach the final equilibrium situation. An answer to this kind of question could only be obtained by a *detailed* analysis of the interactions between particles, since these interactions are responsible for bringing about the change of the system in time which results in the attainment of the final equilibrium state. The beauty of our general probability arguments is precisely the fact that they yield information about equilibrium situations, *without* the necessity of getting involved in the difficult detailed analysis of interactions between the very many particles of a system.

The general probability arguments of statistical mechanics are thus basically restricted to a consideration of equilibrium situations which do not change in time. But this limitation is not quite as severe as it might appear at first sight. The important parameter is really the time t_{exp} of experimental interest compared to the significant relaxation times τ of the system under consideration. There are really three cases which may arise.

1. $\tau \ll t_{\text{exp}}$: In this case the system comes to equilibrium very quickly compared to times of experimental interest. Hence probability arguments concerning the resulting equilibrium situation are certainly applicable.

2. $\tau \gg t_{\text{exp}}$: This is the opposite limit where equilibrium is achieved very slowly compared to experimental times. Here the situation would not be changed significantly if one imagined constraints to be introduced which would prevent the system from ever reaching equilibrium at all. But in the presence of these constraints the system would be in equilibrium; hence it can again be treated by general probability arguments.

Example 1 Imagine that in Fig. 2·7·1 the partition has a very small thermal conductivity so that the amount of energy transferred between A and A' during a time t_{exp} of experimental interest is very small. Then the situation would be substantially the same if the partition were made thermally insulating; in that case both A and A' can separately be considered to be in thermal equilibrium and can be discussed accordingly.

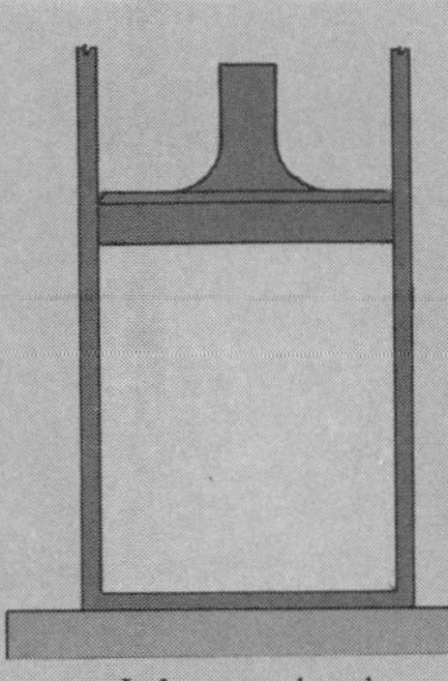

***Fig. 3·2·1** Experimental setup in which a gas is contained in a cylinder closed by a piston which is free to oscillate. The gas is in thermal contact with the laboratory bench.*

Example 2 As a second example, consider the situation of Fig. 3·2·1. Here a gas is contained in a cylinder closed by a movable piston, and the whole apparatus is sitting on a bench in the laboratory. When the piston is pushed down and then released, it will oscillate with a period t_{osc} about its equilibrium position. There are two significant relaxation times in the problem. If the piston is suddenly displaced, it takes a time τ_{th} before the gas will again come to thermal equilibrium with the laboratory bench by exchanging heat with it; it also takes a time τ_{int} before the gas of molecules will regain internal thermal equilibrium so that it is again uniformly distributed over all its accessible states. Ordinarily $\tau_{int} \ll \tau_{th}$. If the time of experimental interest (which is here the period of oscillation t_{osc}) is such that

$$\tau_{int} \ll t_{osc} \ll \tau_{th}$$

one can treat the problem to good approximaton by considering the gas to be always in internal equilibrium in a macrostate corresponding to the instantaneous position of the piston, and by considering the walls of the cylinder to be thermally insulating.

3. $\tau \approx t_{exp}$: In this case the time required to reach equilibrium is comparable to times of experimental significance. The statistical distribution of the system over its accessible states is then not uniform and keeps on changing during the time under consideration. One is then faced with a difficult problem which cannot be reduced to a discussion of equilibrium situations.

THERMAL INTERACTION BETWEEN MACROSCOPIC SYSTEMS

3 · 3 *Distribution of energy between systems in equilibrium*

Let us now discuss in greater detail the thermal interaction between two macroscopic systems A and A'. We shall denote the respective energies of these systems by E and E'. For convenience, we imagine these energy scales to be subdivided into equal small intervals of respective magnitudes δE and $\delta E'$; then we shall denote by $\Omega(E)$ the number of states of A in the range

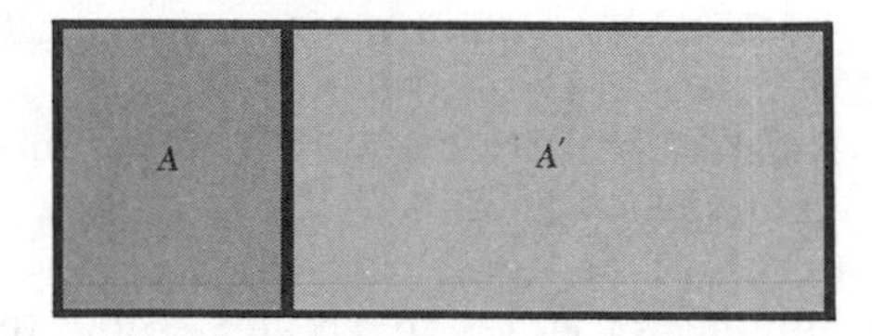

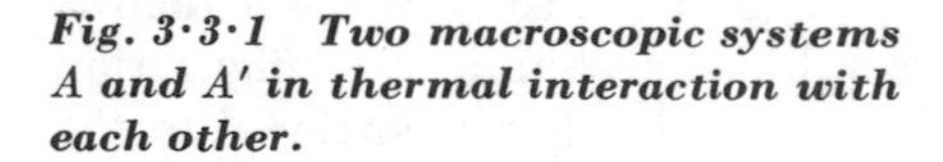
***Fig. 3·3·1** Two macroscopic systems A and A′ in thermal interaction with each other.*

between E and $E + \delta E$, and by $\Omega'(E')$ the number of states of A' in the range between E' and $E' + \delta E'$.

We assume that the systems are not thermally insulated from each other so that they are free to exchange energy. (The external parameters of the systems are supposed to remain fixed; thus the energy transfer is in the form of heat.) The combined system $A^{(0)} \equiv A + A'$ is isolated and its total energy $E^{(0)}$ is therefore constant. The energy of each system separately is, however, not fixed, since it can exchange energy with the other system. We assume always, when speaking of thermal contact between two systems, that the interaction between the systems is weak so that their energies are simply additive. Thus we can write

$$E + E' = E^{(0)} = \text{constant} \tag{3·3·1}$$

Remark The Hamiltonian (or energy) $\mathcal{H}$ of the combined system can always be written in the form

$$\mathcal{H} = \mathcal{H} + \mathcal{H}' + \mathcal{H}^{(\text{int})}$$

where $\mathcal{H}$ depends only on the variables describing A, $\mathcal{H}'$ only on the variables describing A', and the interaction term $\mathcal{H}^{(\text{int})}$ on the variables of both systems.* This last term $\mathcal{H}^{(\text{int})}$ cannot be zero, because then the two systems would not interact at all and would have no way of exchanging energy and thus coming to equilibrium with each other. But the assumption of *weak* interaction is that $\mathcal{H}^{(\text{int})}$, although finite, is negligibly small compared to $\mathcal{H}$ and $\mathcal{H}'$.

Suppose that the systems A and A' are in equilibrium with each other, and focus attention on a representative ensemble such as that shown in Fig. 2·6·1. Then the energy of A can assume a large range of possible values, but these values occur by no means with equal probability. Indeed, suppose that A has an energy E (i.e., more precisely an energy between E and $E + \delta E$.) Then the corresponding energy of A' is by (3·3·1) known to be

$$E' = E^{(0)} - E \tag{3·3·2}$$

The number of states accessible to the entire system $A^{(0)}$ can thus be regarded as a function of a single parameter, the energy E of system A. Let us denote

* For example, for two particles moving in one dimension

$$\mathcal{H} = \frac{p^2}{2m} + \frac{p'^2}{2m'} + U(x,x')$$

where the first terms describe their kinetic energies and the last one describes their potential energy of mutual interaction which depends on their positions x and x'.

by $\Omega^{(0)}(E)$ the number of states accessible to $A^{(0)}$ when A has an energy between E and $E + \delta E$. Our fundamental postulate asserts that in equilibrium $A^{(0)}$ must be equally likely to be found in any one of its states. Hence it follows that the probability $P(E)$ of finding this combined system in a configuration where A has an energy between E and $E + \delta E$ is simply proportional to the number of states $\Omega^{(0)}(E)$ accessible to the total system $A^{(0)}$ under these circumstances. In symbols this can be written

$$P(E) = C\Omega^{(0)}(E) \tag{3·3·3}$$

where C is a constant of proportionality independent of E.

More explicitly, this probability could also be written as

$$P(E) = \frac{\Omega^{(0)}(E)}{\Omega^{(0)}{}_{\text{tot}}}$$

where $\Omega^{(0)}{}_{\text{tot}}$ denotes the *total* number of states accessible to $A^{(0)}$. Of course, $\Omega^{(0)}{}_{\text{tot}}$ can be obtained by summing $\Omega^{(0)}(E)$ over all possible energies E of the system A. Similarly, the constant C in (3·3·3) can be determined by the normalization requirement that the probability $P(E)$ summed over all possible energies of A must yield unity. Thus

$$C^{-1} = \Omega^{(0)}{}_{\text{tot}} = \sum_E \Omega^{(0)}(E)$$

But when A has an energy E it can be in any one of its $\Omega(E)$ possible states. At the same time A' must then have an energy $E' = E^{(0)} - E$ so that it can be in any one of its $\Omega'(E') = \Omega'(E^{(0)} - E)$ possible states. Since every possible state of A can be combined with every possible state of A' to give a different state of the total system $A^{(0)}$, it follows that the number of distinct states accessible to $A^{(0)}$ when A has energy E is simply given by the product

$$\Omega^{(0)}(E) = \Omega(E)\Omega'(E^{(0)} - E) \tag{3·3·4}$$

Correspondingly, the probability (3·3·3) of system A having an energy near E is simply given by

▶ $$P(E) = C\Omega(E)\Omega'(E^{(0)} - E) \tag{3·3·5}$$

Illustrative example with very small numbers Consider the two systems A and A' having the characteristics illustrated in Fig. 3·3·2. Suppose that the total energy $E^{(0)}$ of both systems is, in the arbitrary units used, equal to 15. One possible situation, for example, would be that $E = 4$ and $E' = 11$. In this case A could be any one of its two possible states and A' in any one of its 40 states. There are then a total of $\Omega^{(0)} = 2 \times 40 = 80$ different possible states for the combined system $A + A'$. Let us enumerate systematically some of the conceivable situations in a table when the total energy of the system is $E^{(0)} = 15$.

Suppose $E =$	; then $E' =$.	Here $\Omega(E) =$	, and $\Omega'(E') =$.	Hence $\Omega^{(0)}(E) =$.
4	11	2	40	80
5	10	5	26	130
6	9	10	16	160
7	8	17	8	136
8	7	25	3	75

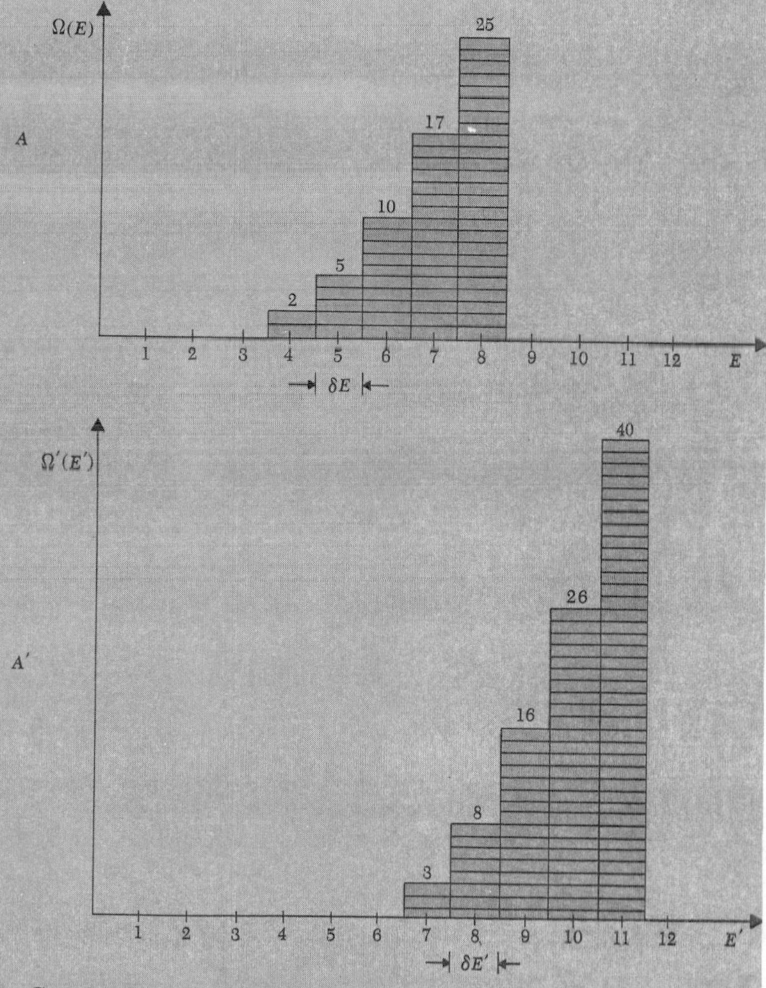

Fig. 3·3·2 Graph showing, in the case of two special very small systems A and A', the number of states $\Omega(E)$ accessible to A when its energy is E and the number of states $\Omega'(E')$ accessible to A' when its energy is E'. (The energies are measured in terms of an arbitrary unit.)

Note that it would be most probable in the ensemble to find the combined system in a state where A has energy $E = 6$ and A' has energy $E' = 9$. This situation would be likely to occur twice as frequently as the situation where $E = 4$ and $E' = 11$.

Let us now investigate the dependence of $P(E)$ on the energy E. Since A and A' are both systems of very many degrees of freedom, we know by (2·5·10) that both $\Omega(E)$ and $\Omega'(E')$ are extremely rapidly increasing functions of their respective arguments. Hence it follows that if one considers the expression (3·3·5) as a function of increasing energy E, the factor $\Omega(E)$ *increases* extremely rapidly while the factor $\Omega'(E^{(0)} - E)$ *decreases* extremely rapidly. The result is that the product of these two factors, i.e., the probability $P(E)$, exhibits an extremely sharp maximum for some particular value $\tilde{E}$ of the energy E. Thus the dependence of $P(E)$ on E must show the general behavior illustrated in Fig. 3·3·3 where the width Δ^*E of the region where $P(E)$ has appreciable magnitude is such that $\Delta^*E <<< \tilde{E}$.

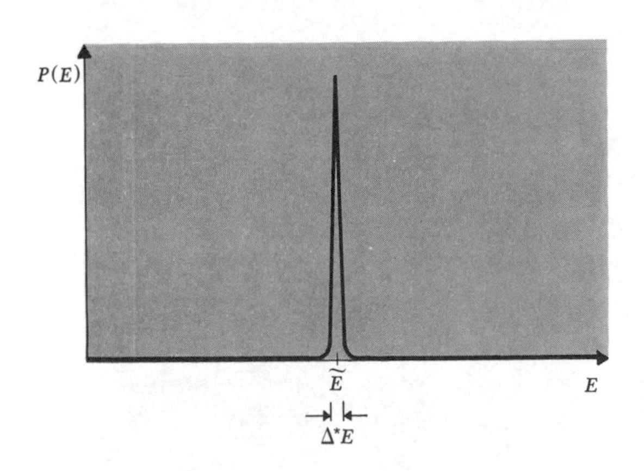

***Fig. 3·3·3** Schematic illustration of the functional dependence of the probability $P(E)$ on the energy E.*

Remark More explicitly, if the number of states exhibits the behavior discussed in (2·5·10) so that $\Omega \propto E^f$ and $\Omega' \propto E'^{f'}$, then (3·3·5) gives

$$\ln P \approx f \ln E + f' \ln (E^{(0)} - E) + \text{constant}$$

Thus $\ln P$ exhibits a unique maximum as a function of E, and this maximum of the logarithm corresponds to a *very* pronounced maximum of P itself. Except for the fact that this maximum is *enormously* sharper for these macroscopic systems where Ω and Ω' are such rapidly varying functions of energy, the situation is analogous to the simple example discussed above. We shall postpone until Sec. 3·6 a more quantitative estimate of the width Δ^*E of the maximum in the case of macroscopic systems.

To locate the position of the maximum of $P(E)$, or equivalently, of the maximum of its logarithm, we need to find the value $E = \tilde{E}$, where*

$$\frac{\partial \ln P}{\partial E} = \frac{1}{P}\frac{\partial P}{\partial E} = 0 \tag{3·3·6}$$

But by (3·3·5)

$$\ln P(E) = \ln C + \ln \Omega(E) + \ln \Omega'(E') \tag{3·3·7}$$

where $E' = E^{(0)} - E$. Hence (3·3·6) becomes

$$\frac{\partial \ln \Omega(E)}{\partial E} + \frac{\partial \ln \Omega'(E')}{\partial E'}(-1) = 0$$

or

▶ $$\beta(\tilde{E}) = \beta'(\tilde{E}') \tag{3·3·8}$$

where $\tilde{E}$ and $\tilde{E}'$ denote the corresponding energies of A and A' at the maximum, and where we have introduced the definition

▶ $$\beta(E) \equiv \frac{\partial \ln \Omega}{\partial E} \tag{3·3·9}$$

with a corresponding definition for β'. The relation (3·3·8) is the equation which determines the value $\tilde{E}$ where $P(E)$ is maximum.

By its definition, the parameter β has the dimensions of a reciprocal energy. It is convenient to introduce a *dimensionless* parameter T defined by writing

$$kT \equiv \frac{1}{\beta} \tag{3·3·10}$$

where k is some positive constant having the dimensions of energy and whose magnitude can be chosen in some convenient arbitrary way. The parameter T is then by (3·3·9) defined as

$$\frac{1}{T} = \frac{\partial S}{\partial E} \tag{3·3·11}$$

where we have introduced the definition

▶ $$S \equiv k \ln \Omega \tag{3·3·12}$$

This quantity S is given the name of "entropy." The condition of maximum probability $P(E)$ is then, by (3·3·7), expressible as the condition that the total entropy

$$S + S' = \text{maximum} \tag{3·3·13}$$

The condition that this occurs can, by (3·3·8), be written as

$$T = T' \tag{3·3·14}$$

* We write this as a *partial* derivative to emphasize that all external parameters of the system are considered to remain unchanged in this discussion. The reason that it is somewhat more convenient to work with $\ln P$ instead of P itself is that the logarithm is a much more slowly varying function of the energy E, and that it involves the numbers Ω and Ω' as a simple sum rather than as a product.

Remark Note that the number Ω of accessible states in the energy range δE (and hence also the entropy $S = k \ln \Omega$ of (3·3·12)) depends on the size δE chosen as the fixed small energy-subdivision interval in a given discussion. This dependence of S is, however, utterly negligible for a macroscopic system and does not at all affect the parameter β.

Indeed, by (2·5·1), $\Omega(E)$ is simply proportional to δE; i.e., $\Omega(E) = \omega(E)\,\delta E$, where ω is the *density* of states which is independent of δE. Since δE is a fixed interval independent of E, it follows by (3·3·9) that

$$\beta = \frac{\partial}{\partial E}(\ln \omega + \ln \delta E) = \frac{\partial \ln \omega}{\partial E} \tag{3·3·15}$$

which is independent of δE. Furthermore, suppose that one had chosen instead of δE a different energy subdivision interval $\delta^* E$. The corresponding number of states $\Omega^*(E)$ in the range between E and $E + \delta^* E$ would then be given by

$$\Omega^*(E) = \frac{\Omega(E)}{\delta E}\,\delta^* E$$

The corresponding entropy defined by (3·3·12) would then be

$$S^* = k \ln \Omega^* = S + k \ln \frac{\delta^* E}{\delta E} \tag{3·3·16}$$

Now by (2·5·9), $S = k \ln \Omega$ is of the order of kf, where f is the number of degrees of freedom of the system. Imagine then an extreme situation where the interval δE^* would be chosen to be so fantastically different from δE as to differ from it by a factor as large as f (e.g., by as much as 10^{24}). Then the second term on the right side of (3·3·16) would at most be of the order of $k \ln f$. But when f is a large number, $\ln f <<< f$. (For example, if $f = 10^{24}$, $\ln f = 55$, which is certainly utterly negligible compared to f itself.) The last term in (3·3·16) is therefore completely negligible compared to S, so that one has to excellent approximation

$$S^* = S$$

The value of the entropy $S = k \ln \Omega$ calculated by (3·3·12) is thus essentially independent of the interval δE chosen for the subdivision of the energy scale.

3 · 4 *The approach to thermal equilibrium*

We pointed out already that the maximum exhibited by $P(E)$ at the energy $E = \tilde{E}$ is extremely sharp. There is, therefore, an overwhelmingly large probability that, in an equilibrium situation where A and A' are in thermal contact, the system A has an energy E very close to $\tilde{E}$, while the system A' has correspondingly an energy very close to $E^{(0)} - \tilde{E} \equiv \tilde{E}'$. The respective *mean* energies of the systems in thermal contact must therefore also be equal to these energies; i.e., when the systems are in thermal contact,

$$\bar{E} = \tilde{E} \qquad \text{and} \qquad \bar{E}' = \tilde{E}' \tag{3·4·1}$$

Consider then the situation where A and A' are initially separately in equilibrium and isolated from each other, their respective energies being very close to E_i and E_i'. (Their respective mean energies are accordingly $\bar{E}_i = E_i$ and $\bar{E}_i' = E_i'$.) The systems A and A' are now placed in thermal contact so that they are free to exchange energy with each other. The resulting situation is then an extremely improbable one, unless it happens that the systems initially have energies very close to $\tilde{E}$ and $\tilde{E}'$, respectively. The situation will therefore tend to change in time until the systems attain final mean energies $\bar{E}_f$ and $\bar{E}_f'$ which are such that

$$\bar{E}_f = \tilde{E} \quad \text{and} \quad \bar{E}_f' = \tilde{E}' \tag{3·4·2}$$

so that the probability $P(E)$ becomes maximum. By (3·3·8) the β parameters of the systems are then equal, i.e.,

$$\beta_f = \beta_f' \tag{3·4·3}$$

where $\qquad \beta_f \equiv \beta(\bar{E}_f) \quad \text{and} \quad \beta_f' \equiv \beta(\bar{E}_f')$

The final probability is maximum and thus never less than the original one. By virtue of (3·3·7) this statement can be expressed in terms of the definition (3·3·12) of the entropy as

$$S(\bar{E}_f) + S'(\bar{E}_f') \geq S(\bar{E}_i) + S'(\bar{E}_i') \tag{3·4·4}$$

When A and A' exchange energy in attaining the final equilibrium, their total energy is, of course, always conserved. Thus

$$\bar{E}_f + \bar{E}_f' = \bar{E}_i + \bar{E}_i' \tag{3·4·5}$$

Let us denote the entropy changes of the systems by

$$\begin{aligned} \Delta S &\equiv S_f - S_i \equiv S(\bar{E}_f) - S(\bar{E}_i) \\ \Delta S' &\equiv S_f' - S_i' \equiv S(\bar{E}_f') - S(\bar{E}_i') \end{aligned} \tag{3·4·6}$$

Then the condition (3·4·4) can be written more compactly as

▶ $$\Delta S + \Delta S' \geq 0 \tag{3·4·7}$$

Similarly, the mean energy changes are, by definition, simply the respective heats absorbed by the two systems. Thus

$$\begin{aligned} Q &\equiv \bar{E}_f - \bar{E}_i \\ Q' &\equiv \bar{E}_f' - \bar{E}_i' \end{aligned} \tag{3·4·8}$$

The conservation of energy (3·4·5) can then be written more compactly as

▶ $$Q + Q' = 0 \tag{3·4·9}$$

Hence $Q' = -Q$, so that if Q is positive then Q' is negative, and vice versa. A negative heat absorbed is simply a heat given off, and (3·4·9) expresses the obvious fact that the heat absorbed by one system must be equal to the heat given off by the other system.

By *definition*, we shall call the system which absorbs heat the "colder" system and the system which gives off heat the "warmer," or "hotter," system.

There are thus basically two cases which can arise:

1. The initial energies of the system may be such that $\beta_i = \beta_i'$, where $\beta_i = \beta(\bar{E}_i)$ and $\beta_i' \equiv \beta(\bar{E}_i')$. Then $\bar{E}_i = \tilde{E}$, and the condition of maximum probability (or entropy) is already fulfilled. (Equation (3·4·7) becomes an equality.) The systems remain, therefore, in equilibrium. There is then also no net exchange of energy (i.e., of heat) between the systems.

2. More generally, the initial energies of the systems are such that $\beta_i \neq \beta_i'$. Then $\bar{E}_i \neq \tilde{E}$, and the systems are in a very improbable nonequilibrium situation. (Equation (3·4·7) is an inequality.) This situation will therefore change in time. Transfer of heat between the systems takes place until the condition of maximum probability (or entropy) is achieved where $\bar{E}_f = \tilde{E}$ and where $\beta_f' = \beta_f$.

3 · 5 *Temperature*

In the preceding section we saw that the parameter β (or equivalently, $T = (k\beta)^{-1}$) has the following two properties:

1. If two systems separately in equilibrium are characterized by the *same* value of the parameter, then the systems will remain in equilibrium when brought into thermal contact with each other.

2. If the systems are characterized by *different* values of the parameter, then they will *not* remain in equilibrium when brought into thermal contact with each other.

In addition, suppose we have three systems A, B, and C. We know that if A and C remain in equilibrium when brought into thermal contact, then $\beta_A = \beta_C$. Similarly, we know that if B and C also remain in equilibrium when brought into thermal contact, then $\beta_B = \beta_C$. But then we can conclude that $\beta_A = \beta_B$, so that systems A and B will also remain in equilibrium when brought into thermal contact. We thus arrive at the following statement, sometimes known as the "zeroth law of thermodynamics":

> If two systems are in thermal equilibrium with a third system, then they must be in thermal equilibrium with each other. (3·5·1)

This property makes possible the use of test systems, called "thermometers," which allow measurements to decide whether any two systems will or will not remain in equilibrium when brought into thermal contact with each other. Such a thermometer is any macroscopic system M chosen in accordance with the following two specifications:

1. Among the many macroscopic parameters characterizing the system M, select one (call it ϑ) which varies by appreciable amounts when M is brought

into thermal contact with the various systems to be tested. All the other macroscopic parameters of M are held fixed. The parameter ϑ, which is allowed to vary, is called the "thermometric parameter" of M.

2. The system M is chosen to be much smaller (i.e., to have many fewer degrees of freedom) than the systems which it is designed to test. This is desirable in order to minimize the possible energy transfer to the systems under test so as to reduce the disturbance of the systems under test to a minimum.

Examples of thermometers

a. Mercury in a glass tube. The height of the mercury in the tube is taken as the thermometric parameter ϑ. This is the familiar "mercury-in-glass thermometer."

b. Gas in a bulb, its volume being maintained constant. The mean pressure of the gas is taken as the thermometric parameter ϑ. This is called a "constant-volume gas thermometer."

c. Gas in a bulb, its pressure being maintained constant. The volume of the gas is taken as the thermometric parameter ϑ. This is called a "constant-pressure gas thermometer."

d. An electrical conductor maintained at constant pressure and carrying a current. The electrical resistance of the conductor is the thermometric parameter ϑ. This is called a "resistance thermometer."

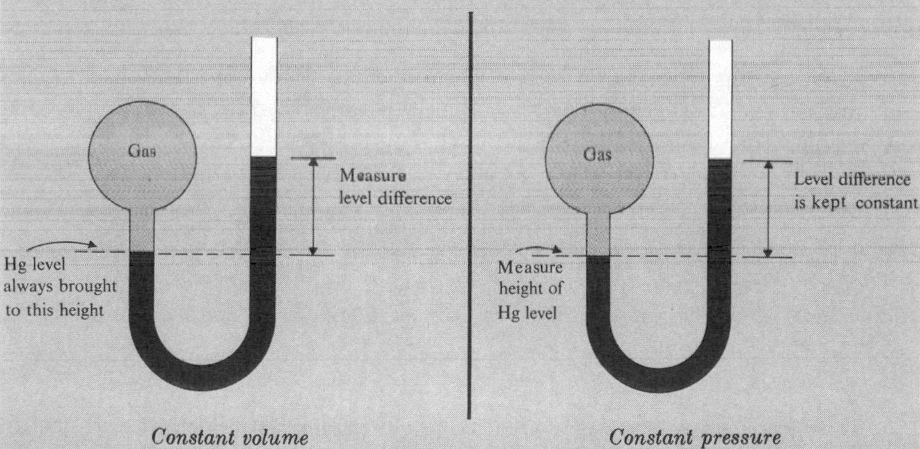

Fig. 3·5·1 Constant-volume and constant-pressure gas thermometers.

A thermometer M is used in the following way. It is successively placed in thermal contact with the systems under test, call them A and B, and is allowed to come to equilibrium with each.

1. If the thermometric parameter ϑ of the thermometer M (e.g., the height of the Hg column of the mercury-in-glass thermometer) has the same value in both cases, one knows that after M has come to equilibrium with A, it remains in equilibrium after being placed in thermal contact with B. Hence the zeroth law allows one to conclude that A and B will remain in equilibrium if brought into contact with each other.

2. If the thermometric parameter of M does *not* have the same value in both cases, then one knows that A and B will *not* remain in equilibrium if brought into thermal contact with each other. For suppose they did remain in equilibrium; then, after M attains thermal equilibrium with A, it would by the zeroth law have to remain in equilibrium when brought into thermal contact with B. Thus the parameter ϑ could not change when M is brought into thermal contact with B.

Consider *any* thermometer M with *any one* parameter ϑ chosen as its thermometric parameter. The value assumed by ϑ when the thermometer M has come to thermal equilibrium with some system A will, by definition, be called the "temperature" of the system A with respect to the particular thermometric parameter ϑ of the particular thermometer M.

According to this definition the temperature can be a length, a pressure, or any other quantity. Note that, even if two different thermometers have parameters of the same dimensions, it is in general not true that they will yield the same value of temperature for the same body. Furthermore, if a body C has a temperature halfway between the temperatures of bodies A and B when measured by one thermometer, this statement is not necessarily true with respect to the temperatures measured by some other thermometer.

Nevertheless, the somewhat arbitrary temperature concept which we have defined has, according to our discussion, the following fundamental and useful property:

> Two systems will remain in equilibrium when placed in thermal contact with each other if and only if they have the same temperature (referred to the same thermometer). (3·5·2)

Note that if $\psi(\vartheta)$ is *any* single-valued function of ϑ, then it can be used as a thermometric parameter just as well as ϑ itself. This function $\psi(\vartheta)$ also satisfies the property (3·5·2) and can be equally well designated as the temperature of the system with respect to the particular thermometer chosen for the measurement.

The temperature concept which we have defined is important and useful, but is rather arbitrary in the sense that the temperature assigned to a system depends in an essential way on the peculiar properties of the particular system M used as the thermometer.

On the other hand, we can exploit the properties of the parameter β and use the particular parameter β_M of the thermometer M as its thermometric parameter. Then we know that when the thermometer is in thermal equilibrium with a system A, $\beta_M = \beta_A$. The thermometer measures then, by virtue of (3·3·9), a fundamental property of the system A, namely, the variation of its density of states with energy. Furthermore, if one uses *any other* thermometer M', it too will read a value $\beta_{M'} = \beta_A$ when brought into thermal contact with system A. Thus we see that

If the parameter β is used as a thermometric parameter, then *any* thermometer yields the *same* temperature reading when used to measure the temperature of a particular system. Furthermore, this temperature measures a fundamental property of the density of states of the system under test.

The parameter β is, therefore, a particularly useful and fundamental temperature parameter. The corresponding dimensionless quantity $T \equiv (k\beta)^{-1}$ is accordingly called the "absolute temperature." We shall postpone until later a discussion of practical procedures for finding numerical values of β or T by appropriate measurements.

Some properties of the absolute temperature By (3·3·9) the absolute temperature is given by

$$\frac{1}{kT} \equiv \beta \equiv \frac{\partial \ln \Omega}{\partial E} \tag{3·5·3}$$

We saw in Sec. 2·5 that $\Omega(E)$ is ordinarily a very rapidly increasing function of the energy E. Hence (3·5·3) shows that ordinarily

$$\beta > 0 \qquad \text{or} \qquad T > 0 \tag{3·5·4}$$

****Remark*** This is true for all ordinary systems where one takes into account the kinetic energy of the particles. Such systems have no upper bound on their possible energy (a lower bound, of course, always exists—namely, the quantum mechanical ground-state energy of the system); and as we have seen in Sec. 2·5, $\Omega(E)$ increases then roughly like E^f, where f is the number of degrees of freedom of the system. Exceptional situations may arise, however, where one does not want to take into account the translational degrees of freedom of a system (i.e., coordinates and momenta), but focuses attention *only* on its spin degrees of freedom. In that case the system has an upper bound to its possible energy (e.g., all spins lined up antiparallel to the field) as well as a lower bound (e.g., all spins lined up parallel to the field). Correspondingly, the *total* number of states (irrespective of energy) available to the system is finite. In this case the number of possible spin states $\Omega_{spin}(E)$ at first increases, as usual, with increasing energy; but then it reaches a maximum and decreases again. Thus it is possible to get absolute *spin* temperatures which are negative as well as positive.

If one disregards such exceptional cases where systems have an upper bound to their possible total energy, T is always positive and some further general statements can readily be made. In (2·5·10) it was shown that the functional dependence of $\Omega(E)$ is roughly given by

$$\Omega(E) \propto E^f$$

where f is the number of degrees of freedom of the system and the energy E is measured with respect to its ground state. Thus

$$\ln \Omega \approx f \ln E + \text{constant}$$

Thus, when $E = \tilde{E} \approx \bar{E}$, one gets

$$\beta = \frac{\partial \ln \Omega(E)}{\partial E} \approx \frac{f}{\bar{E}} \tag{3·5·5}$$

and

$$kT \approx \frac{\bar{E}}{f} \tag{3·5·6}$$

Thus the quantity kT is a rough measure of the mean energy, above the ground state, per degree of freedom of a system.

The condition of equilibrium (3·3·8) between two systems in thermal contact asserts that their respective absolute temperatures must be equal. By virtue of (3·5·6) we see that this condition is roughly equivalent to the fairly obvious statement that the total energy of the interacting systems is shared between them in such a way that the mean energy per degree of freedom is the same for both systems.

In addition, the absolute temperature T has the general property of indicating the direction of heat flow between two systems in thermal contact. This is most readily seen in the case where an infinitesimal amount of heat Q is transferred between two systems A and A', originally at slightly different initial temperatures β_i and β_i'. Using the notation of Sec. 3·3, the condition that the probability (3·3·7) must increase in the process (or equivalently, the condition (3·4·7) for the entropy) can be written as

$$\frac{\partial \ln \Omega(\bar{E}_i)}{\partial E} (\bar{E}_f - \bar{E}_i) + \frac{\partial \ln \Omega'(\bar{E}_i')}{\partial E'} (\bar{E}_f' - \bar{E}_i') \geq 0$$

Using the definition of β as well as the relations (3·4·8), this becomes

$$(\beta_i - \beta_i')Q \geq 0$$

Thus, if $Q > 0$, then

$$\beta_i \geq \beta_i'$$

and

$$T_i \leq T_i'$$

if T_i and T_i' are positive. Hence positive heat is always absorbed by the system with higher β and given off by the system with lower β. Or, in the ordinary case where the absolute temperatures are positive, heat is absorbed by the system at the lower absolute temperature T and given off by the system at the higher absolute temperature T. Since the words "colder" and "warmer" were defined in Sec. 3·4 in terms of the direction of heat flow, one can say (in the case of ordinary positive absolute temperatures) that the warmer system has a higher absolute temperature than the colder one. (We shall see later that the same conclusions are true in situations where a *finite* amount of heat is transferred.)

3·6 *Heat reservoirs*

The thermal interaction between two systems is particularly simple if one of them is very much larger than the other one (i.e., if it has many more degrees of freedom). To be precise, suppose that A' denotes the large system and A any relatively small system with which it may interact. The system A' is then said to act as a "heat reservoir," or "heat bath," with respect to the smaller system if it is so large that its temperature parameter remains essentially unchanged irrespective of any amount of heat Q' which it may absorb from the smaller system. In symbols, this condition says that A' is such that

$$\left| \frac{\partial \beta'}{\partial E'} Q' \right| \ll \beta' \tag{3·6·1}$$

Here $\partial\beta'/\partial E'$ is of the order of $\beta'/\bar{E}'$, where $\bar{E}'$ is the mean energy of A' measured from its ground state,* while the heat Q' absorbed by A' is at most of the order of the mean energy $\bar{E}$ of the small system A above its ground state. Hence one expects (3·6·1) to be valid if

$$\frac{\bar{E}}{\bar{E}'} \ll 1$$

i.e., if A' is sufficiently large compared to A.

Note that the concept of a heat reservoir is a relative one. A glass of tea acts approximately as a heat reservoir with respect to a slice of lemon immersed in it. On the other hand, it is certainly not a heat bath with respect to the whole room; indeed, the relationship there is the opposite.

If the macroscopic system A' has $\Omega'(E')$ accessible states and absorbs heat $Q' = \Delta\bar{E}'$, one can express the resulting change in $\ln \Omega'$ by a Taylor's expansion. Thus

$$\begin{aligned} \ln \Omega'(E' + Q') - \ln \Omega'(E') &= \left(\frac{\partial \ln \Omega'}{\partial E'}\right) Q' + \frac{1}{2}\left(\frac{\partial^2 \ln \Omega'}{\partial E'^2}\right) Q'^2 + \cdots \\ &= \beta' Q' + \frac{1}{2}\frac{\partial \beta'}{\partial E'} Q'^2 + \cdots \end{aligned} \tag{3·6·2}$$

where we have used the definition (3·3·9). But if A' acts as a heat reservoir so that (3·6·1) is satisfied, then β' does not change appreciably and higher-order terms on the right of (3·6·2) are negligible. Thus (3·6·2) reduces simply to

$$\ln \Omega'(E' + Q') - \ln \Omega'(E') = \beta' Q' = \frac{Q'}{kT'} \tag{3·6·3}$$

The left side expresses, by the definition (3·3·12), the entropy change of the heat reservoir. Thus one arrives at the simple result that, if a heat reservoir at temperature T' absorbs heat Q', its resulting entropy change is given by

► $$\Delta S' = \frac{Q'}{T'} \qquad \text{(for a heat reservoir)} \tag{3·6·4}$$

A similar relation holds for any system which is at absolute temperature $T = (k\beta)^{-1}$ and which absorbs an *infinitesimal* amount of heat $đQ$ from some

* Indeed, assuming the approximate dependence $\Omega' \propto E'^{f'}$ of (2·5·10), it follows for $E' = \bar{E}'$ that $\beta' \approx (\partial \ln \Omega'/\partial \bar{E}) \approx f'/\bar{E}'$; hence $|\partial\beta'/\partial E'| \approx f'/\bar{E}'^2 \approx \beta'/\bar{E}'$.

other system at a slightly different temperature. Since $đQ \ll E$, where E is the energy of the system under consideration, it follows that

$$\ln \Omega\,(E + đQ) - \ln \Omega(E) = \frac{\partial \ln \Omega}{\partial E}\,đQ = \beta đQ$$

or, since $S = k \ln \Omega$, that

$$dS = \frac{đQ}{T} \tag{3·6·5}$$

where dS is the increase in entropy of the system.

3 · 7 *Sharpness of the probability distribution*

In Sec. 3·3 we argued that the probability $P(E)$ that A has an energy E exhibits a very sharp maximum. Let us now investigate more quantitatively just how sharp this maximum really is.

Our method of approach is identical to that used in Sec. 1·5. To investigate the behavior of $P(E)$ near its maximum $E = \tilde{E}$, we consider the more slowly varying function $\ln P(E)$ of (3·3·7) and expand it in a power series of the energy difference

$$\eta \equiv E - \tilde{E} \tag{3·7·1}$$

Expanding $\ln \Omega(E)$ in Taylor's series about $\tilde{E}$, one gets

$$\ln \Omega(E) = \ln \Omega(\tilde{E}) + \left(\frac{\partial \ln \Omega}{\partial E}\right)\eta + \frac{1}{2}\left(\frac{\partial^2 \ln \Omega}{\partial E^2}\right)\eta^2 + \cdots \tag{3·7·2}$$

Here the derivatives are evaluated at $E = \tilde{E}$. Let us use the abbreviations

$$\beta \equiv \left(\frac{\partial \ln \Omega}{\partial E}\right) \tag{3·7·3}$$

and

$$\lambda \equiv -\left(\frac{\partial^2 \ln \Omega}{\partial E^2}\right) = -\left(\frac{\partial \beta}{\partial E}\right) \tag{3·7·4}$$

The minus sign has been introduced for convenience since we shall see that the second derivative is intrinsically negative.

Hence (3·7·2) can be written

$$\ln \Omega(E) = \ln \Omega(\tilde{E}) + \beta\eta - \tfrac{1}{2}\lambda\eta^2 + \cdots \tag{3·7·5}$$

One can write down a corresponding expression for $\ln \Omega'(E')$ near $E' = \tilde{E}'$. By conservation of energy $E' = E^{(0)} - E$, so that

$$E' - \tilde{E}' = -(E - \tilde{E}) = -\eta \tag{3·7·6}$$

Analogously to (3·7·5) one thus obtains

$$\ln \Omega'(E') = \ln \Omega'(\tilde{E}') + \beta'(-\eta) - \tfrac{1}{2}\lambda'(-\eta)^2 + \cdots \qquad (3\cdot7\cdot7)$$

where β' and λ' are the parameters (3·7·3) and (3·7·4) correspondingly defined for system A' and evaluated at the energy $E' = \tilde{E}'$. Adding (3·7·5) and (3·7·7) gives

$$\ln [\Omega(E)\Omega'(E')] = \ln [\Omega(\tilde{E})\Omega'(\tilde{E}')] + (\beta - \beta')\eta - \tfrac{1}{2}(\lambda + \lambda')\eta^2 \qquad (3\cdot7\cdot8)$$

At the maximum of $\Omega(E)\Omega'(E')$ it follows by (3·3·8) that $\beta = \beta'$, so that the term linear in η vanishes as it should. Hence (3 · 7 · 8) yields for (3 · 3 · 8) the result

$$\ln P(E) = \ln P(\tilde{E}) - \tfrac{1}{2}\lambda_0\eta^2$$

or

$$P(E) = P(\tilde{E})\, e^{-\frac{1}{2}\lambda_0(E-\tilde{E})^2} \qquad (3\cdot7\cdot9)$$

where $$\lambda_0 \equiv \lambda + \lambda' \qquad (3\cdot7\cdot10)$$

Note that λ_0 cannot be negative, since then the probability $P(E)$ would not exhibit a maximum value, i.e., the combined system $A^{(0)}$ would not attain a well-defined final equilibrium situation as, physically, we know it must. Furthermore, neither λ nor λ' can be negative. Indeed, one could choose for A' a system for which $\lambda' \ll \lambda$; in that case $\lambda \approx \lambda_0$ and, since we already argued that this last quantity cannot be negative, it follows that $\lambda \geq 0$. Similar reasoning shows that $\lambda' \geq 0$.*

The same conclusion follows also from the argument that ordinarily $\Omega \propto E^f$. Indeed, using the definition (3·7·4), one obtains from (3·5·5)

$$\lambda = -\left(-\frac{f}{\tilde{E}^2}\right) = \frac{f}{\tilde{E}^2} > 0 \qquad (3\cdot7\cdot11)$$

The preceding discussion leads to several interesting remarks. In Sec. 3·3 we concluded from the general behavior of the densities of states of the interacting systems that the probability $P(E)$ has a unique maximum at some energy $\tilde{E}$. We have now shown more specifically that, for E not too far from $\tilde{E}$, the probability $P(E)$ is described by the Gaussian distribution (3·7·9). It then follows by (1·6·8) that the mean energy $\bar{E}$ is given by

$$\bar{E} = \tilde{E} \qquad (3\cdot7\cdot12)$$

Thus the mean energy of A is indeed equal to the energy $\tilde{E}$ corresponding to the situation of maximum probability. Furthermore, (3·7·9) shows that $P(E)$ becomes negligibly small compared to its maximum value when

* The equals sign corresponds to exceptional circumstances. An example might be a system consisting of a mixture of ice and water in equilibrium. The addition of energy to this system results in melting some of the ice, but does not change its temperature parameter. Thus $\lambda = -\partial\beta/\partial E = 0$.

$\frac{1}{2}\lambda_0(E - \tilde{E})^2 \gg 1$, i.e., when $|E - \tilde{E}| \gg \lambda_0^{-\frac{1}{2}}$. In other words, it is very improbable that the energy of A lies far outside the range $\tilde{E} \pm \Delta^*E$ where *

$$\Delta^*E = \lambda_0^{-\frac{1}{2}} \tag{3·7·13}$$

Suppose now that A is the system with the larger value of the parameter λ. Then

$$\lambda_0 \approx \lambda \approx \frac{f}{\tilde{E}^2} = \frac{f}{\bar{E}^2}$$

and

$$\Delta^*E \approx \frac{\bar{E}}{\sqrt{f}}$$

where $\bar{E}$ is the mean energy of A above its ground state. The fractional width of the maximum in $P(E)$ is then given by

► $$\frac{\Delta^*E}{\bar{E}} \approx \frac{1}{\sqrt{f}} \tag{3·7·14}$$

If A contains a mole of particles, $f \approx N_a \approx 10^{24}$ and $(\Delta^*E/\bar{E}) \approx 10^{-12}$.

Hence the probability distribution has indeed an exceedingly sharp maximum when one is dealing with macroscopic systems containing very many particles. In our example, the probability $P(E)$ becomes already negligibly small if the energy E differs from its mean value by more than 1 part in 10^{12}! This is an example of a general characteristic of macroscopic systems. Since the number of particles is so very large, fluctuations in any macroscopic parameter y (e.g., energy or pressure) are ordinarily utterly negligible. This means that one nearly always observes the mean value $\bar{y}$ of the parameter and tends, therefore, to remain unaware of the statistical aspects of the macroscopic world. It is only when one is making very precise measurements or when one is dealing with very small systems that the existence of fluctuations becomes apparent.

The condition $\lambda \geq 0$ implies by (3·7·4) that

$$\lambda = -\frac{\partial^2 \ln \Omega}{\partial E^2} = -\frac{\partial \beta}{\partial E} \geq 0$$

or

$$\frac{\partial \beta}{\partial E} \leq 0 \tag{3·7·15}$$

Using $\beta = (kT)^{-1}$, the equivalent condition for T becomes

$$\frac{\partial \beta}{\partial T}\frac{\partial T}{\partial E} = -\frac{1}{kT^2}\frac{\partial T}{\partial E} \leq 0$$

Thus

$$\frac{\partial T}{\partial E} \geq 0 \tag{3·7·16}$$

i.e., the absolute temperature of any system increases with its energy.

The relation (3·7·15) allows one to establish the general connection between absolute temperature and the direction of heat flow. In the situation of Sec. 3·4, suppose that initially $\beta_i \neq \beta_i'$. If A absorbs positive heat Q,

* The result (1·6·9) applied to the Gaussian distribution (3·7·9) shows that λ_0^{-1} is indeed the dispersion of the energy.

(3·7·15) implies that its value of β must *de*crease. At the same time, A' must give off heat so that its value of β' must *in*crease. Since β is a continuous function of E for each system, the β values of the systems change in this way continuously until they reach the common final value β_f; this must therefore be such that $\beta_f < \beta_i$ and $\beta_f > \beta_i'$. Thus $\beta_i > \beta_i'$ and the positive heat Q gets absorbed by the system with the higher value of β. Correspondingly, for ordinary positive absolute temperatures, positive heat gets absorbed by the system at the lower absolute temperature T.*

Remark on the total number of accessible states It is of some interest to calculate the total number of states $\Omega^{(0)}{}_{tot}$ accessible to the entire system $A^{(0)}$. Since the probability distribution is so sharply peaked, practically all states lie in a range within a width $\Delta^*E = \lambda_0^{-\frac{1}{2}}$ of $\tilde{E}$ (see Fig. 3·3·3). Since the density of states near $E = \tilde{E}$ is equal to $\Omega^{(0)}(\tilde{E})/\delta E$, the total number of states is approximately given by

$$\Omega^{(0)}{}_{tot} \approx \frac{\Omega^{(0)}(\tilde{E})}{\delta E}\Delta^*E = K\Omega^{(0)}(\tilde{E}) \tag{3·7·17}$$

where

$$K \equiv \frac{\Delta^*E}{\delta E} \tag{3·7·18}$$

By (3·7·17) it follows that

$$\ln \Omega^{(0)}{}_{tot} = \ln \Omega^{(0)}(\tilde{E}) + \ln K \approx \ln \Omega(\tilde{E}) \tag{3·7·19}$$

The last result is true to excellent approximation because $\ln K$ is utterly negligible. This is another striking consequence of the fact that we are dealing with such large numbers. The reason is that, no matter what reasonable energy subdivision interval we may choose, the number K in (3·7·18) should certainly not be much bigger than f, where f may be of the order of Avogadro's number N_a. Thus $\ln K$ is of order $\ln f$ or less. On the other hand, (2·5·9) shows that $\ln \Omega$ is some number of the order of f. But when f is large, say $f \approx N_a \approx 10^{24}$, $f >>> \ln f$ (10^{24} compared to only 55). Thus $\ln K$ is completely negligible compared to $\ln \Omega(\tilde{E})$. The relation (3·7·19) asserts then that the probability distribution is so sharply peaked around its maximum that, for purposes of calculating logarithms, the total number of states is equal to the maximum number of states. From this it also follows that, if in analogy to (3·3·13), one defines the total entropy of $A^{(0)}$ in terms of the total number of its accessible states so that

$$S^{(0)} = k \ln \Omega^{(0)}{}_{tot}$$

Then

$$S^{(0)} = k \ln \Omega^{(0)}(\tilde{E}) = k \ln [\Omega(\tilde{E})\Omega'(\tilde{E}')] = k \ln \Omega(\tilde{E}) + k \ln \Omega'(\tilde{E}')$$

or

$$S^{(0)} = S(\tilde{E}) + S'(\tilde{E}') \tag{3·7·20}$$

Thus the entropy so defined has the above simple additive property.

* The statement relating to T is restricted to the ordinary case of positive temperatures, since otherwise T is not a continuous function of E. Indeed, for spin systems with an upper bound of possible energies, Ω has a maximum, and hence β passes continuously through the value $\beta = 0$. But correspondingly, $T \equiv (k\beta)^{-1}$ jumps from ∞ to $-\infty$.

GENERAL INTERACTION BETWEEN MACROSCOPIC SYSTEMS

3 · 8 *Dependence of the density of states on the external parameters*

Now that we have examined in detail the thermal interaction between systems, let us turn to the general case where mechanical interaction can also take place, i.e., where the external parameters of the systems are also free to change. We begin, therefore, by investigating how the density of states depends on the external parameters.

For the sake of simplicity, consider the situation where only a single external parameter x of the system is free to vary; the generalization to the case where there are several such parameters will be immediate. The number of states accessible to this system in the energy range between E and $E + \delta E$ will also depend on the particular value assumed by this external parameter; we shall denote it by $\Omega(E,x)$. We are interested in examining how Ω depends on x.

When x is changed by an amount dx, the energy $E_r(x)$ of each microstate r changes by an amount $(\partial E_r/\partial x)\, dx$. The energies of different states are, in general, changed by different amounts. Let us denote by $\Omega_Y(E,x)$ the number of those states which have an energy in the range between E and $E + \delta E$ when the external parameter has the value x, *and* which are such that their derivative $\partial E_r/\partial x$ has a value in the range between Y and $Y + \delta Y$. The total number of states is then given by

$$\Omega(E,x) = \sum_Y \Omega_Y(E,x) \tag{3·8·1}$$

where the summation is over all possible values of Y.

Consider a particular energy E. When the external parameter is changed, some states which originally had an energy less than E will acquire an energy greater than E, and vice versa. What then is the total number of states $\sigma(E)$ whose energy is changed from a value less than E to a value greater than E when the parameter changes from x to $x + dx$? Those states for which $\partial E_r/\partial x$ has the particular value Y change their energy by the infinitesimal amount $Y\, dx$. Hence all these states located within an energy $Y\, dx$ below E will change their energy from a value smaller to one greater than E (see Fig. 3 · 8 · 1). The number $\sigma_Y(E)$ of such states is thus given by the number per unit

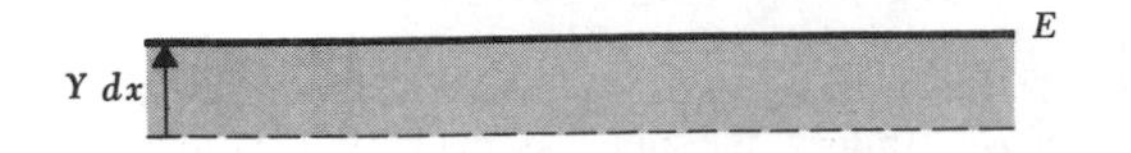

***Fig. 3·8·1** The shaded area indicates the energy range occupied by states with a value of $\partial E_r/\partial x = Y$, whose energy changes from a value smaller than E to one greater than E when the external parameter is changed from x to $x + dx$.*

energy multiplied by the energy range $Y\,dx$, i.e., by

$$\sigma_Y(E) = \frac{\Omega_Y(E,x)}{\delta E} Y dx \tag{3·8·2}$$

Different states have their energy changed by different amounts $Y\,dx$ (positive or negative). Hence the *total* number of states $\sigma(E)$ whose energy is changed from a value less to a value greater than E is given by summing (3·8·2) over all possible values of $\partial E_r/\partial x = Y$; thus

$$\sigma(E) = \sum_Y \frac{\Omega_Y(E,x)}{\delta E} Y\,dx = \frac{\Omega(E,x)}{\delta E} \bar{Y}\,dx \tag{3·8·3}$$

where we have used the definition

$$\bar{Y} = \frac{1}{\Omega(E,x)} \sum_Y \Omega_Y(E,x) Y \tag{3·8·4}$$

as the mean value of Y over all accessible states, each state being considered equally likely. The mean value thus defined is, of course, a function of E and x, i.e., $\bar{Y} = \bar{Y}(E,x)$. Since $Y = \partial E_r/\partial x$, one has

$$\bar{Y} = \overline{\frac{\partial E_r}{\partial x}} \equiv -\bar{X} \tag{3·8·5}$$

where $\bar{X}$ is, by the definition (2·9·6), the mean generalized force conjugate to the external parameter x.

Let us now consider the total number of states $\Omega(E,x)$ between E and $E + \delta E$ (see Fig. 3·8·2). When the parameter changes from x to $x + dx$, the number of states in this energy range changes by an amount $[\partial\Omega(E,x)/\partial x]\,dx$ which must be due to [the net number of states which enter this range by having their energy changed from a value less than E to one greater than E] minus [the net number of states which leave this range by having their energy changed from a value less than $E + \delta E$ to one greater than $E + \delta E$]. In symbols this can be written

$$\frac{\partial\Omega(E,x)}{\partial x} dx = \sigma(E) - \sigma(E + \delta E) = -\frac{\partial\sigma}{\partial E}\delta E$$

Using (3·8·3) this becomes

$$\frac{\partial\Omega}{\partial x} = -\frac{\partial}{\partial E}(\Omega\bar{Y}) \tag{3·8·6}$$

or

$$\frac{\partial\Omega}{\partial x} = -\frac{\partial\Omega}{\partial E}\bar{Y} - \Omega\frac{\partial\bar{Y}}{\partial E}$$

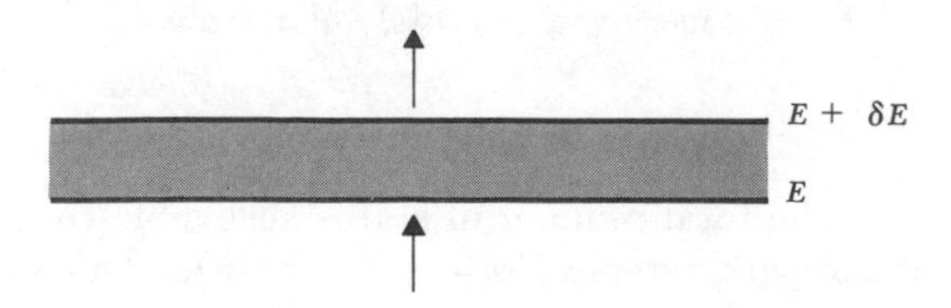

***Fig. 3·8·2** The number of states in the shaded energy range changes when the external parameter is varied because states enter and leave this energy range.*

By dividing both sides by Ω, this can be written

$$\frac{\partial \ln \Omega}{\partial x} = -\frac{\partial \ln \Omega}{\partial E}\bar{Y} - \frac{\partial \bar{Y}}{\partial E} \tag{3·8·7}$$

But for a large system the first term on the right is of the order of $(f/E)\bar{Y}$, since one has approximately $\Omega \propto E^f$. On the other hand, the second term on the right is only of the order of $\partial \bar{Y}/\partial E \approx \bar{Y}/E$ and is therefore smaller than the first term by a factor f. For macroscopic systems where f is of the order of Avogadro's number, the second term on the right is thus utterly negligible. Hence (3·8·7) becomes to excellent approximation

$$\frac{\partial \ln \Omega}{\partial x} = -\frac{\partial \ln \Omega}{\partial E}\bar{Y} = \beta \bar{X} \tag{3·8·8}$$

where we have used (3·8·5) and the definition (3·3·9) of the temperature parameter β.

When there are several external parameters $x_1, \ldots, x_n$ (so that $\Omega = \Omega(E, x_1, \ldots, x_n)$) the above relation is clearly valid for each one of them. One obtains then for each external parameter x_α and its corresponding mean generalized force $\bar{X}_\alpha$ the general relation

▶ $$\frac{\partial \ln \Omega}{\partial x_\alpha} = \beta \bar{X}_\alpha \tag{3·8·9}$$

3·9 *Equilibrium between interacting systems*

Consider two systems A and A' which can interact both by exchanging heat and by doing work on each other. A specific example might be the situation illustrated in Fig. 2·8·1 where the piston is free to move and is capable of conducting heat. System A has energy E and is characterized by some external parameters $x_1, \ldots, x_n$ which are free to vary. Similarly, A' has energy E' and is characterized by some external parameters $x_1', \ldots, x_n'$. The combined system $A^{(0)} \equiv A + A'$ is isolated. Hence

$$E + E' = E^{(0)} = \text{constant} \tag{3·9·1}$$

The energy E' of A' is then determined if the energy E of A is known. Furthermore, if A and A' can interact mechanically, this implies that the parameters x' are some functions of the parameters x.

Example In Fig. 2·8·1, the gas A is described by one external parameter x, its volume V; similarly, the gas A' is described by its volume V'. But as the piston moves, the total volume remains unchanged, i.e.,

$$V + V' = V^{(0)} = \text{constant} \tag{3·9·2}$$

The total number of states accessible to $A^{(0)}$ is thus a function of E and of the parameters $x_\alpha (\alpha = 1, \ldots, n)$. Once again $\Omega^{(0)}(E; x_1, \ldots, x_n)$ will have a very sharp maximum for some values $E = \tilde{E}$ and $x_\alpha = \tilde{x}_\alpha$. The equilibrium situation corresponds then to the one of maximum probability where practically all systems $A^{(0)}$ have values of E and x_α very close to $\tilde{E}$ and $\tilde{x}_\alpha$. The mean values of these quantities in equilibrium will thus be equal to $\bar{E} = \tilde{E}$ and $\bar{x}_\alpha = \tilde{x}_\alpha$.

Infinitesimal quasi-static process Consider a quasi-static process in which the system A, by virtue of its interaction with system A', is brought from an equilibrium state described by $\bar{E}$ and $\bar{x}_\alpha (\alpha = 1, \ldots, n)$ to an infinitesimally different equilibrium state described by $\bar{E} + d\bar{E}$ and $\bar{x}_\alpha + d\bar{x}_\alpha$. What is the resultant change in the number of states Ω accessible to A?

Since $\Omega = \Omega(E; x_1, \ldots, x_n)$, one can write for the resultant change in $\ln \Omega$ the purely mathematical result

$$d \ln \Omega = \frac{\partial \ln \Omega}{\partial E} d\bar{E} + \sum_{\alpha=1}^{n} \frac{\partial \ln \Omega}{\partial x_\alpha} d\bar{x}_\alpha \tag{3·9·3}$$

By using the relation (3·8·9), this can be written

$$d \ln \Omega = \beta(d\bar{E} + \sum_\alpha \bar{X}_\alpha \, d\bar{x}_\alpha) \tag{3·9·4}$$

By (2·9·5) the last term in the parentheses is just the macroscopic work $đW$ done by A in this infinitesimal process. Hence (3·9·4) can be written

$$d \ln \Omega = \beta(d\bar{E} + đW) \equiv \beta \, đQ \tag{3·9·5}$$

where we have used the definition (2·8·3) for the infinitesimal heat absorbed by A. Equation (3·9·5) is a fundamental relation valid for any quasi-static infinitesimal process. By (3·3·10) and (3·3·12) it can also be written in the form

► $$đQ = T \, dS = d\bar{E} + đW \tag{3·9·6}$$

or equivalently

► $$dS = \frac{đQ}{T} \tag{3·9·7}$$

Thus the relation (3·6·5) remains valid even if the external parameters of the system are changed quasi-statically. Note that in the special case when the system is thermally insulated (i.e., when the process is "adiabatic") the absorbed heat $đQ = 0$ and (3·9·7) asserts that

$$dS = 0$$

This shows that S, or $\ln \Omega$, does not change even if the external parameters are varied *quasi-statically* by a *finite* amount. Hence one has the important result that

> If the external parameters of a *thermally isolated* system are changed *quasi-statically* by any amount, $\Delta S = 0$. (3·9·8)

Thus the performance of quasi-static work changes the energy of a thermally isolated system, but does not affect the number of states accessible to it. In accordance with the discussion of Sec. 3·2, such a process is thus reversible.

It is worth emphasizing that even if a system is thermally isolated so that it absorbs *no* heat, its entropy will *increase* if processes take place which are *not* quasi-static. For example, each total system $A^{(0)}$ is thermally isolated in the three examples discussed at the beginning of Sec. 3·1, yet the number of states accessible to it, and hence its entropy, increases.

Equilibrium conditions Consider the equilibrium between the systems A and A' in the simple case where the external parameters are the volumes V and V' of the two systems. The number of states available to the combined system $A^{(0)}$ is, as in (3·3·5), given by the simple product

$$\Omega^{(0)}(E,V) = \Omega(E,V)\Omega'(E',V') \tag{3·9·9}$$

where E' and V' are related to E and V by (3·9·1) and (3·9·2).

Taking logarithms of (3·9·9), one gets

$$\ln \Omega^{(0)} = \ln \Omega + \ln \Omega' \tag{3·9·10}$$

or

$$S^{(0)} = S + S' \tag{3·9·11}$$

The maximum value of $\Omega^{(0)}$ or $S^{(0)}$ is then determined by the condition that

$$d \ln \Omega^{(0)} = d(\ln \Omega + \ln \Omega') = 0 \tag{3·9·12}$$

for arbitrary changes of dE and dV.

But

$$d \ln \Omega = \frac{\partial \ln \Omega}{\partial E} dE + \frac{\partial \ln \Omega}{\partial V} dV = \beta\, dE + \beta \bar{p}\, dV \tag{3·9·13}$$

where we have used (3·8·8) and the generalized force $\bar{X} = -(\overline{\partial E_r/\partial V})$ is, by (2·10·2), just the mean pressure $\bar{p}$ exerted by A. Similarly, one has for A'

$$d \ln \Omega' = \beta'\, dE' + \beta'\bar{p}'\, dV' = -\beta'\, dE - \beta'\bar{p}'\, dV \tag{3·9·14}$$

since the conservation conditions (3·9·1) and (3·9·2) imply that $dE' = -dE$ and $dV' = -dV$, respectively. Hence the condition of maximum entropy (3·9·12) becomes

$$(\beta - \beta')\, dE + (\beta\bar{p} - \beta'\bar{p}')\, dV = 0 \tag{3·9·15}$$

Since this must be satisfied for arbitrary values of dE and dV, it follows that the coefficients of both of these differentials must separately vanish.

Hence and

$$\left.\begin{aligned} \beta - \beta' &= 0 \\ \beta\bar{p} - \beta'\bar{p}' &= 0 \end{aligned}\right\}$$

or and

$$\left.\begin{aligned} \beta &= \beta' \\ \bar{p} &= \bar{p}' \end{aligned}\right\} \tag{3·9·16}$$

As might be expected, these conditions simply assert that the temperatures of the systems must be equal to guarantee their thermal equilibrium and that their mean pressures must be equal to guarantee their mechanical equilibrium.

As a particularly simple example consider the mechanical interaction of a system A with a purely mechanical device A' whose energy E' is a function only of some external parameter x. The situation might be the one illustrated in Fig. 3·9·1 where A' is a spring whose elongation is measured by the distance x. The total number $\Omega^{(0)}$ of states accessible to the system $A + A'$ is then simply proportional to the number of states $\Omega(E,x)$ accessible to A.* But if $E^{(0)}$ is the constant *total* energy, the energy of A is

$$E = E^{(0)} - E'(x) \tag{3·9·17}$$

and is thus a function of x. If x is free to adjust itself, it will tend to change so that the system approaches an equilibrium situation where Ω is a maximum, i.e., where

$$\frac{\partial}{\partial x} \ln \Omega(E,x) = 0$$

This means that

$$\frac{\partial \ln \Omega}{\partial E}\frac{\partial E}{\partial x} + \frac{\partial \ln \Omega}{\partial x} = 0$$

By (3·9·17) and (3·8·9) this becomes

$$\beta\left(-\frac{\partial E'}{\partial x}\right) + \beta \bar{X} = 0$$

or

$$\bar{X} = \frac{\partial E'}{\partial x}$$

This condition asserts simply that in equilibrium the mean force $\bar{X}$ exerted by the gas A must be equal to the force $\partial E'/\partial x$ exerted by the spring.

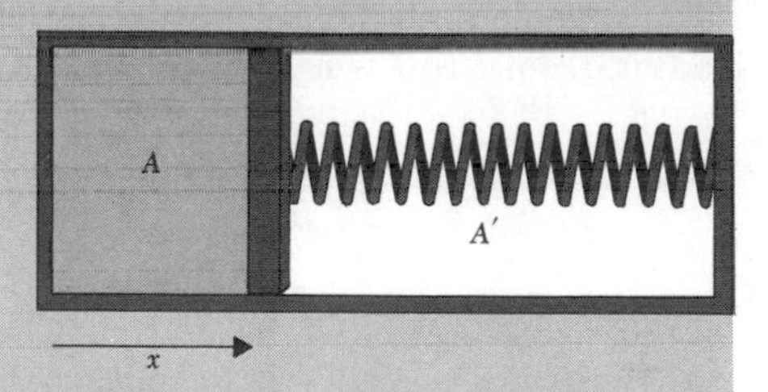

***Fig. 3·9·1** A gas A interacting through a movable piston with a spring A′.*

3·10 *Properties of the entropy*

Entropy and exact differentials In Sec. 2·9 we discussed the fact that the infinitesimal heat $đQ$ is *not* an exact differential. The relation (3·9·7) establishes, however, the following remarkable result: Although the heat $đQ$ absorbed in a quasi-static infinitesimal process is *not* an exact differential, the quantity

* The number of states accessible to A' remains unchanged—this is what one means by a purely mechanical device which is completely described by its external parameters; i.e., A' itself is thermally insulated, and its external parameters are changed sufficiently slowly (compared to its significant relaxation time) that its entropy remains constant.

$đQ/T = dS$ obtained by dividing $đQ$ by the absolute temperature T of the system *is* an exact differential. This is so because the entropy S is a function characteristic of each macrostate of the system, and dS is simply the difference between two such functions for adjacent macrostates.

Remark If multiplication by a factor converts an inexact differential into an exact differential, then this factor is in mathematical terminology said to be an "integrating factor" for the inexact differential. Thus one can say that the absolute temperature T is characterized by the property that T^{-1} in an integrating factor for $đQ$.

It follows that, given any two macrostates i and f of a system, the entropy difference between them can be written as

$$S_f - S_i = \int_i^f dS = \int_{i\ (\text{eq})}^f \frac{đQ}{T} \qquad (3 \cdot 10 \cdot 1)$$

where we have used the result (3 · 9 · 7) in the last integral. We have labeled this integral by the subscript "eq" (standing for "equilibrium") to emphasize explicitly the fact that it is to be evaluated for any process by which the system is brought *quasi-statically* through a sequence of near-equilibrium situations from its initial to its final macrostates. The temperature in the integrand is thus well defined at all stages of the process. Since the left side of (3 · 10 · 1) refers only to the initial and final macrostates, it follows that the integral on the right side must be independent of whatever particular quasi-static process leading from i to f is chosen to evaluate the integral. In symbols,

$$\int_{i\ (\text{eq})}^f \frac{đQ}{T} \quad \text{is independent of the process} \qquad (3 \cdot 10 \cdot 2)$$

Example Consider the two quasi-static processes indicated by the solid and dashed lines of the $\bar{p}$ versus V diagram in Fig. 2 · 11 · 2. The integral $\int_i^f đQ$ giving the total heat absorbed in going from i to f will be different when evaluated for the two processes. But the integral $\int_i^f (đQ/T)$ will yield the *same* result when evaluated for these processes.

Let us be clear as to how such an integral is to be evaluated. At any stage of the process the system is characterized by a certain value of V and corresponding value $\bar{p}$ given by the graph. This information is adequate to determine a definite temperature T for this macrostate of the system. In going to an adjacent value of V, an amount of heat $đQ$ is absorbed. Hence one knows both T and $đQ$, and can sum all the successive quantities $đQ/T$ as one proceeds to increase the volume from V_i to V_f.

Implications of the statistical definition of entropy In (3·3·12) we defined the entropy in terms of the number Ω of accessible states in the range between E and $E + \delta E$ by the relation

$$S \equiv k \ln \Omega \tag{3·10·3}$$

It is important to note that if the macrostate of a system is specified, i.e., if one knows the external parameters and the energy E of the system, then the number Ω of states accessible to the system is completely determined if the system is discussed in terms of quantum mechanics. Hence the entropy S has, by (3·10·3), a *unique* value calculable from a knowledge of the microscopic constitution of the system.*

Remark This last statement would *not* be true if the system were described in terms of classical mechanics. If the system has f degrees of freedom, phase space is subdivided (as in Sec. 2·1) into cells of arbitrarily chosen volume $h_0{}^f$. The total number of cells, or states, available to the system is then obtained by dividing the accessible volume of phase space (contained between the energies E and $E + \delta E$) by the volume per cell. Thus

$$\Omega = \frac{1}{h_0{}^f} \int \cdots \int dq_1 \cdots dq_f \, dp_1 \cdots dp_f$$

or

$$S = k \ln \left(\int \cdots \int dq_1 \cdots dp_f \right) - kf \ln h_0 \tag{3·10·4}$$

These relations show that Ω depends in an essential way on the size of the cells into which phase space is subdivided. Correspondingly, S contains an additive constant which depends on this cell size. In a classical description the value of the entropy is thus not unique, but only defined to within an arbitrary additive constant. What happens effectively in quantum mechanics is that there exists a natural unit of cell size according to which h_0 is to be put equal to Planck's constant; the additive constant becomes then uniquely determined.

Limiting behavior of the entropy As one goes to lower energy, every system described by quantum mechanics approaches the lowest possible energy E_0 of its ground state. Corresponding to this energy there exists usually only one possible state of the system; or there may be a relatively small number of such states, all of the same energy E_0 (the ground state is then said to be "degenerate"). When one considers energies somewhat greater than E_0, the number of states $\Omega(E)$ increases, of course, very rapidly; by (2·5·10), it then behaves roughly like $\Omega \propto (E - E_0)^f$ if f is the number of degrees of freedom of the system. The general dependence of $\ln \Omega$ on the energy E of the system is thus of the form sketched in Fig. 3·10·1.

* The value of S is certainly unique for a given choice of the energy subdivision interval δE. In addition, we showed at the end of Sec. 3·3 that the value of the entropy is also utterly insensitive to the exact choice of the magnitude of δE.

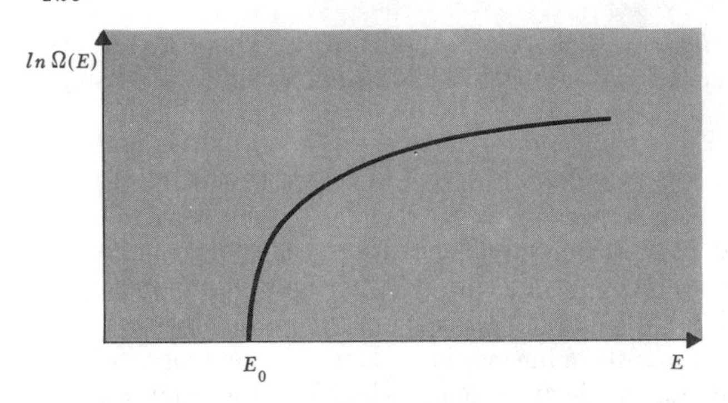

Fig. 3·10·1 Behavior of ln $\Omega(E)$ *for energies* $E > E_0$. *Note that* β, *the slope of the curve, becomes very large for* $E \to E_0$ *and that* $\partial\beta/\partial E < 0$.

Whenever the energy of the system is significantly greater than its ground state energy E_0, its entropy S is the order of $k \ln \Omega$, i.e., by (2·5·9), of the order of kf. As its energy approaches E_0, the number of states $\Omega(E)$ within the given interval δE falls off rapidly; it ultimately becomes a number of the order of f itself or smaller since the ground state itself consists of only one state, or at most a relatively small number of states. But then $S = k \ln \Omega$ approaches a number of the order of $k \ln f$ or less, and this is utterly negligible compared to the magnitude kf of the entropy at higher energies. Thus one can assert to excellent approximation that the entropy becomes vanishingly small as the system approaches its ground-state energy. In symbols,

$$\text{as } E \to E_0, \qquad S \to 0 \tag{3·10·5}$$

Remark We note again the quantum-mechanical basis for the validity of the above discussion. In the framework of classical mechanics there would not exist a situation of lowest energy with an associated definite small number of states.

This limiting behavior of S can also be expressed in terms of the temperature of the system. We know by (3·7·15) that $\partial\beta/\partial E < 0$, or equivalently that $\partial T/\partial E > 0$. Hence it follows that as E decreases toward E_0, β increases and becomes very large, while $T = (k\beta)^{-1}$ decreases and becomes very small. In the limiting case as $T \to 0$, E must increasingly approach its ground-state value E_0. By virtue of (3·10·5), the entropy must then become negligibly small. Thus,

$$\text{as } T \to 0, \qquad S \to 0 \tag{3·10·6}$$

In applying (3·10·6) to situations of experimental interest one must, as usual, be sure to ascertain that equilibrium arguments are applicable to the system under consideration. At very low temperatures one must be particularly careful in this respect since the *rate* of attaining equilibrium may then become quite slow. Another question which may arise in practice concerns the extent to which the limiting situation $T \to 0$ has indeed been reached. In other words, how small a temperature is sufficiently small to apply (3·10·6)?

It is only possible to answer this question by knowing something about the particular system under consideration. A case of very frequent occurrence is that where the nuclei of the atoms in the system have nuclear spins. If one brings such a system to some sufficiently low temperature T_0, the entropy (or number of states) associated with the degrees of freedom *not* involving nuclear spins may have become quite negligible. Yet the number of states Ω_s corresponding to the possible nuclear spin orientations may be very large, indeed just as large as at very much higher temperatures. The reason is that nuclear magnetic moments are very small; interactions affecting the spin orientation of a nucleus are thus also very small. Hence each nuclear spin is ordinarily oriented completely at random, even at temperatures as low as T_0.

Suppose, for example, that the system consists of atoms having nuclear spin $\frac{1}{2}$ (e.g., the system might be a silver spoon). Each spin can have two possible orientations. If spin-independent interactions are very small, these two orientations do not differ significantly in energy and each spin is equally likely to point "up" or "down." If there are N nuclei in the system, there are then $\Omega_S = 2^N$ possible spin states accessible to the system, even at a temperature as low as T_0. Only at temperatures very much less than T_0 will the interactions involving nuclear spins become of significance. For instance, the situation where all nuclear spins are aligned parallel to each other ("nuclear ferromagnetism") may have slightly lower energy than other spin orientations. The system will then (if it attains equilibrium) settle down at the very lowest temperatures ($T <<< T_0$) into this *one* very lowest state where all nuclei are aligned.

In the case under discussion, where one knows that the nuclear spins are still randomly oriented at a temperature as low as T_0 and do not depart from this random configuration until one goes to temperatures very much less than T_0, one can still make a useful statement. Thus one can assert that as T decreases towards T_0, the entropy approaches a value S_0 which is simply given by the number of possible nuclear spin orientations, i.e., $S_0 = k \ln \Omega_S$. In symbols this yields the result that

$$\text{as } T \rightarrow 0_+, \qquad S \rightarrow S_0 \tag{3·10·7}$$

Here $T \rightarrow 0_+$ denotes a limiting temperature which is very small, yet large enough so that spins remain randomly oriented. Furthermore, S_0 is a definite constant which depends only on the kinds of atomic nuclei of which the system is composed, but which is completely independent of any details concerned with the energy levels of the system. One can say, in short, that S_0 is independent of all parameters of the system in the widest sense, i.e., independent of the spatial arrangement of its atoms or of the interactions between them. The statement (3·10·7) is a useful one because it can be applied at temperatures which are not prohibitively low.

SUMMARY OF FUNDAMENTAL RESULTS

3 · 11 *Thermodynamic laws and basic statistical relations*

The entire discussion of this chapter has been based on the fundamental statistical postulates of Sec. 2 · 3. We have elaborated some details and given various illustrative examples to gain familiarity with the main properties of macroscopic systems. The main ideas have, however, been quite simple; most of them are contained in Secs. 3 · 1, 3 · 3, and 3 · 9. The discussion of this chapter encompasses, nevertheless, all the fundamental results of classical thermodynamics and all the essential results of statistical mechanics. Let us now summarize these results in the form of very general statements and group them into two categories. The first of these will consist of purely *macro*scopic statements which make *no* reference to the *micro*scopic properties of the systems, i.e., to the molecules of which they consist. These statements we shall call "thermodynamic laws." The other class of statements will refer to the microscopic properties of the systems and will be designated as "statistical relations."

Thermodynamic laws The first statement is the fairly trivial one discussed in Sec. 3 · 5.

▶ *Zeroth law:* If two systems are in thermal equilibrium with a third system, they must be in thermal equilibrium with each other.

Next there is a statement expressing the conservation of energy and discussed in Sec. 2 · 8.

▶ *First law:* An equilibrium macrostate of a system can be characterized by a quantity $\bar{E}$ (called "internal energy") which has the property that

for an isolated system, $$\bar{E} = \text{constant} \tag{3·11·1}$$

If the system *is* allowed to interact and thus goes from one macrostate to another, the resulting change in $\bar{E}$ can be written in the form

$$\Delta\bar{E} = -W + Q \tag{3·11·2}$$

where W is the macroscopic work done *by* the system as a result of the system's change in external parameters. The quantity Q, *defined* by (3 · 11 · 2), is called the "heat absorbed by the system."

Next we were led to introduce the entropy S which had the simple properties discussed in Secs. 3 · 1 and 3 · 9. Thus we obtain the following results:

▶ *Second law:* An equilibrium macrostate of a system can be characterized by a quantity S (called "entropy"), which has the properties that

a. In any process in which a thermally *isolated* system goes from one macrostate to another, the entropy tends to increase, i.e.,

$$\Delta S \geq 0 \tag{3·11·3}$$

b. If the system is not isolated and undergoes a quasi-static infinitesimal process in which it absorbs heat $đQ$, then

$$dS = \frac{đQ}{T} \tag{3·11·4}$$

where T is a quantity characteristic of the macrostate of the system. (T is called the "absolute temperature" of the system.)

Finally, there is a last statement based on (3·10·7).

▶ *Third law:* The entropy S of a system has the limiting property that

$$\text{as } T \to 0_+, \qquad S \to S_0 \tag{3·11·5}$$

where S_0 is a constant independent of all parameters of the particular system.

Note again that the above four laws are completely *macro*scopic in content. We have introduced three quantities ($\bar{E}$, S, and T) which are asserted to be defined for each macrostate of the system, and we have made some statements about properties relating these quantities. But *nowhere* in these four laws have we made any explicit reference to the microscopic nature of the system (e.g., to the properties of the molecules constituting the system or to the forces of interaction between them).

Statistical relations There is first the general connection (3 · 3 · 12)

▶
$$S = k \ln \Omega \tag{3·11·6}$$

This allows one to relate the quantities entering in the thermodynamic laws to one's microscopic knowledge of the system. Indeed, if one knows the nature of the particles constituting the system and the interactions between them, then one can in principle use the laws of mechanics to compute the possible quantum states of the system and thus to find Ω.

Furthermore, one can use the fundamental statistical postulate of Sec. 2·3 to make statements about the probability P of finding an isolated system in a situation characterized by certain parameters $y_1, \ldots, y_n$. If the corresponding number of accessible states is Ω, then in equilibrium

▶
$$P \propto \Omega \propto e^{S/k} \tag{3·11·7}$$

A large number of conclusions follow from the purely *macro*scopic statements which we have called the laws of thermodynamics. Indeed, the whole discipline of classical thermodynamics assumes these laws as basic postulates and then proceeds to deduce their consequences in a macroscopic discussion which never refers to the microscopic description of matter in terms of atoms or molecules. The approach is sufficiently fruitful to have given rise to a large body of important results. We shall discuss it further, particularly in Chapter 5. This approach was also historically the oldest, since it could arise in a context where the atomic constitution of matter was not yet known or sufficiently understood.* If one does make use of *microscopic* information and uses statistical mechanics to calculate Ω, one's powers of prediction are, of course, tremendously increased. Not only can one then calculate thermodynamic quantities from first principles by using (3·11·6), but one can also calculate probabilities and thus the fluctuations of physical quantities about their mean values. Statistical mechanics is thus the more inclusive discipline which encompasses all of classical thermodynamics; to emphasize this fact it is sometimes called "statistical thermodynamics."

3 · 12 *Statistical calculation of thermodynamic quantities*

It is worth pointing out explicitly how a knowledge of the number of states $\Omega = \Omega(E; x_1, \ldots, x_n)$ of a system allows one to calculate important macroscopic quantities characterizing the system in equilibrium. The quantity Ω is a function of the energy of the system under consideration and of its external parameters. The relations of particular interest are (3·3·9) and (3·8·9), namely,

$$\beta = \frac{\partial \ln \Omega}{\partial E} \qquad \text{and} \qquad \bar{X}_\alpha = \frac{1}{\beta}\frac{\partial \ln \Omega}{\partial x_\alpha} \tag{3·12·1}$$

These allow one to compute the absolute temperature and the mean generalized forces of the system from a knowledge of Ω. For example, in the particular case when $x_\alpha = V$ is the volume V of the system, the corresponding mean generalized force $\bar{X}_\alpha$ is, by (2·10·2), the mean pressure $\bar{p}$ given by

$$\bar{p} = \frac{1}{\beta}\frac{\partial \ln \Omega}{\partial V} \tag{3·12·2}$$

The equations (3·12·1) permit one to find relations connecting the generalized forces, the external parameters, and the absolute temperature T. Such relations are called "equations of state" and are important since they relate parameters that are readily measured by experiment. For example, one can find how the mean pressure $\bar{p}$ depends on the temperature T and volume V of the system; the relation $\bar{p} = \bar{p}(T,V)$ would be the corresponding "equation of state."

Remark Note that the relations (3·12·1) are implied by the statement (3·11·4). The latter yields for an infinitesimal quasi-static process the entropy change

$$dS = \frac{đQ}{T} = \frac{1}{T}\left(d\bar{E} + \sum_{\alpha=1}^{n} \bar{X}_\alpha \, d\bar{x}_\alpha\right) \tag{3·12·3}$$

* See the references at the end of this chapter for books treating the subject entirely from a macroscopic point of view.

But since the entropy S is a function of the energy and the external parameters, one can also write the purely mathematical result

$$dS = \left(\frac{\partial S}{\partial E}\right) d\bar{E} + \sum_{\alpha=1}^{n} \left(\frac{\partial S}{\partial x_\alpha}\right) d\bar{x}_\alpha \qquad (3 \cdot 12 \cdot 4)$$

Since $(3 \cdot 12 \cdot 3)$ and $(3 \cdot 12 \cdot 4)$ must be identically equal for arbitrary values of $d\bar{E}$ and $d\bar{x}_\alpha$, the corresponding coefficients multiplying the differentials must be equal. Hence one obtains

$$\frac{1}{T} = \frac{\partial S}{\partial E} \quad \text{and} \quad \frac{\bar{X}_\alpha}{T} = \left(\frac{\partial S}{\partial x_\alpha}\right) \qquad (3 \cdot 12 \cdot 5)$$

where the derivatives are to be evaluated corresponding to the energy $\tilde{E} = \bar{E}$ and parameters $\tilde{x}_\alpha = \bar{x}_\alpha$ of the equilibrium state under consideration. The relations $(3 \cdot 12 \cdot 5)$ are identical to $(3 \cdot 12 \cdot 1)$, since $S = k \ln \Omega$ and $T = (k\beta)^{-1}$.

Let us illustrate the preceding comments by applying them to a very simple but important system, an ideal gas. In $(2 \cdot 5 \cdot 14)$ we showed that for an ideal gas of N molecules in a volume V the quantity Ω is of the form

$$\Omega \propto V^N \chi(E) \qquad (3 \cdot 12 \cdot 6)$$

where $\chi(E)$ is independent of V and depends only on the energy E of the gas. Hence

$$\ln \Omega = N \ln V + \ln \chi(E) + \text{constant} \qquad (3 \cdot 12 \cdot 7)$$

Thus $(3 \cdot 12 \cdot 2)$ yields immediately for the mean pressure of the gas the simple relation

$$\bar{p} = \frac{N}{\beta} \frac{1}{V} = \frac{N}{V} kT \qquad (3 \cdot 12 \cdot 8)$$

or

▶ $$\bar{p} = nkT \qquad (3 \cdot 12 \cdot 9)$$

where $n \equiv N/V$ is the number of molecules per unit volume. This is the equation of state for an ideal gas. Alternatively one can write $N = \nu N_a$, where ν is the number of moles of gas present and N_a is Avogadro's number. Then $(3 \cdot 12 \cdot 8)$ becomes

▶ $$\bar{p}V = \nu RT \qquad (3 \cdot 12 \cdot 10)$$

where $R \equiv N_a k$ is called the "gas constant." Note that neither the equation of state nor the constant R depends on the *kind* of molecules constituting the ideal gas.

By $(3 \cdot 12 \cdot 1)$ and $(3 \cdot 12 \cdot 7)$ one obtains further

$$\beta = \frac{\partial \ln \chi(E)}{\partial E}$$

evaluated for the mean energy $E = \bar{E}$ of the gas. Here the right side is only a function of E, but *not* of V. Thus it follows that, for an ideal gas $\beta = \beta(\bar{E})$

or
$$\bar{E} = \bar{E}(T) \tag{3·12·11}$$

Hence one reaches the important conclusion that the mean energy of an ideal gas depends only on its temperature and is independent of its volume. This result is physically plausible. An increase in volume of the container increases the mean distance between the molecules and thus changes, in general, their mean potential energy of mutual interaction. But in the case of an *ideal* gas this interaction energy is negligibly small, while the kinetic and internal energies of the molecules do not depend on the distances between them. Hence the total energy of the gas remains unchanged.

SUGGESTIONS FOR SUPPLEMENTARY READING

The following books give a statistical discussion somewhat similar to the one of this text:

C. Kittel: "Elementary Statistical Physics," secs. 1–10, John Wiley & Sons, Inc., New York, 1958.

R. Becker: "Theorie der Wärme," secs. 32–35, 45, Springer-Verlag, Berlin, 1955. (A good book, but in German.)

L. Landau and E. M. Lifshitz: "Statistical Physics," secs. 1–13, Addison-Wesley, Reading, Mass., 1963.

The following books are good introductions to classical thermodynamics from a completely macroscopic point of view:

M. W. Zemansky: "Heat and Thermodynamics," 4th ed, McGraw-Hill Book Company, New York, 1957.

E. Fermi: "Thermodynamics," Dover Publications, New York, 1957.

H. B. Callen: "Thermodynamics," John Wiley & Sons, Inc., New York, 1960. (More sophisticated in approach than the preceding books.)

PROBLEMS

3.1 A box is separated by a partition which divides its volume in the ratio 3:1. The larger portion of the box contains 1000 molecules of Ne gas; the smaller, 100 molecules of He gas. A small hole is punctured in the partition, and one waits until equilibrium is attained.

(*a*) Find the mean number of molecules of each type on either side of the partition.

(*b*) What is the probability of finding 1000 molecules of Ne gas in the larger portion and 100 molecules of He gas in the smaller (i.e., the same distribution as in the initial system)?

3.2 Consider a system of N localized weakly interacting particles, each of spin $\frac{1}{2}$ and magnetic moment μ, located in an external magnetic field H. This system was already discussed in Problem 2.4.

(a) Using the expression for $\ln \Omega(E)$ calculated in Problem 2.4b and the definition $\beta = \partial \ln \Omega / \partial E$, find the relation between the absolute temperature T and the total energy E of this system.

(b) Under what circumstances is T negative?

(c) The total magnetic moment M of this system is related to its energy E. Use the result of part (a) to find M as a function of H and the absolute temperature T.

3.3 Consider two spin systems A and A' placed in an external field H. System A consists of N weakly interacting localized particles of spin $\frac{1}{2}$ and magnetic moment μ. Similarly, system A' consists of N' weakly interacting localized particles of spin $\frac{1}{2}$ and magnetic moment μ'. The two systems are initially isolated with respective total energies $bN\mu H$ and $b'N'\mu' H$. They are then placed in thermal contact with each other. Suppose that $|b| \ll 1$ and $|b'| \ll 1$ so that the simple expressions of Problem 2.4c can be used for the densities of states of the two systems.

(a) In the most probable situation corresponding to the final thermal equilibrium, how is the energy $\tilde{E}$ of system A related to the energy $\tilde{E}'$ of system A'?

(b) What is the value of the energy $\tilde{E}$ of system A?

(c) What is the heat Q absorbed by system A in going from the initial situation to the final situation when it is in equilibrium with A'?

(d) What is the probability $P(E)\,dE$ that A has its final energy in the range between E and $E + dE$?

(e) What is the dispersion $(\Delta^*E)^2 \equiv \overline{(E - \bar{E})^2}$ of the energy E of system A in the final equilibrium situation?

(f) What is the value of the relative energy spread $|\Delta^*E/\tilde{E}|$ in the case when $N' \gg N$?

3.4 Suppose that a system A is placed into thermal contact with a heat reservoir A' which is at an absolute temperature T' and that A absorbs an amount of heat Q in this process. Show that the entropy increase ΔS of A in this process satisfies the inequality $\Delta S \geq Q/T'$, where the equals sign is only valid if the initial temperature of A differs infinitesimally from the temperature T' of A'.

3.5 A system consists of N_1 molecules of type 1 and N_2 molecules of type 2 confined within a box of volume V. The molecules are supposed to interact very weakly so that they constitute an ideal gas mixture.

(a) How does the total number of states $\Omega(E)$ in the range between E and $E + \delta E$ depend on the volume V of this system? You may treat the problem classically.

(b) Use this result to find the equation of state of this system, i.e., to find its mean pressure $\bar{p}$ as a function of V and T.

3.6 A glass bulb contains air at room temperature and at a pressure of 1 atmosphere. It is placed in a chamber filled with helium gas at 1 atmosphere and at room temperature. A few months later, the experimenter happens to read in a journal article that the particular glass of which the bulb is made is quite permeable to helium, although not to any other gases. Assuming that equilibrium has been attained by this time, what gas pressure will the experimenter measure inside the bulb when he goes back to check?

Macroscopic parameters and their measurement 4

SECTION 3·11 CONTAINS all the results necessary for an extensive discussion of systems in equilibrium. We shall begin this discussion by exploring a few of the purely *macro*scopic consequences of the theory. The present chapter will consider briefly some of the parameters which are commonly used in the description of macroscopic systems. Many of these parameters, such as heat, absolute temperature, and entropy, have already been introduced. They have been defined in terms of the *micro*scopic mechanical concepts applicable to the particles of a system, and their properties and interrelations have already been established on the basis of the microscopic theory. But we have yet to examine how these quantities are to be determined operationally by suitable macroscopic measurements on a system. An examination of this kind of question is, of course, essential to any physical theory, since one must show how theoretical constructs and predictions can be compared with well-defined experimental measurements. In this chapter we shall discuss how the theory suggests *what* quantities are to be studied experimentally and *how* they are to be measured. In Chapter 5 we then shall show how the theory is capable of predicting various important relationships between such measurable macroscopic quantities.

4·1 *Work and internal energy*

The macroscopic work done by a system is very easily determined, since one can readily measure the external parameters of the system and the associated mean generalized forces. For example, if the volume of a system is changed quasi-statically from V_i to V_f and throughout this process the mean pressure of the system has the measurable value $\bar{p}(V)$, the macroscopic work done by the system is given by calculating the integral (2·10·4)

$$W = \int_{V_i}^{V_f} \bar{p}(V)\, dV \tag{4·1·1}$$

The determination of the internal energy $\bar{E}$ of a system is, by (3·11·2), reducible to a measurement of macroscopic work. If one considers a system which is *thermally insulated* so that it cannot absorb any heat, then $Q = 0$ and one has simply

$$\Delta \bar{E} = -W$$

or

$$\bar{E}_b - \bar{E}_a = -W_{ab} = -\int_a^b đW \tag{4·1·2}$$

This relation defines internal energy *differences* in terms of the macroscopic work W_{ab} done by the system in going from macrostate a to macrostate b. Only such energy differences are of physical significance; i.e., the mean energy is defined only to within an arbitrary additive constant (just as potential energy in mechanics is defined only to within an arbitrary constant). Thus one can choose one particular macrostate a of a system as a standard state from which the mean energy is measured. For example, one can adopt the convention of putting $\bar{E}_a = 0$. To determine the internal energy $\bar{E}_b$ of any other macrostate b of the system, it is only necessary to insulate the system thermally and to go from state a to state b (or conversely, from state b to state a) by the performance of a suitable amount of macroscopic work. Equation (4·1·2) shows that the work thus done is independent of the particular process used in going from a to b. (This was the essential content of the first law of thermodynamics discussed in Sec. 2·11.) Hence the work thus done by the system is guaranteed to yield a unique measurable number. The internal energy of macrostate b is thus uniquely determined by (4·1·2) as

$$\bar{E}_b = -W_{ab} = W_{ba} \tag{4·1·3}$$

The above procedure can be applied to all macrostates b of the system and allows one, therefore, to characterize each such state by a definite operationally measurable value of the internal energy parameter $\bar{E}_b$. Note that the units of energy are the same as those of work, i.e., ergs or joules.

Example 1 Consider a system consisting of a vessel which contains a liquid and a paddle wheel which is free to rotate (see Fig. 2·7·5). If the system is kept at fixed pressure, its macrostate is completely specified by its internal energy $\bar{E}$. Equivalently, it can be specified by its temperature ϑ measured with respect to any *arbitrary* thermometer, since $\bar{E}$ and ϑ are functionally related. A falling weight can do macroscopic work on the system by rotating the paddle wheel.

Consider some standard macrostate a, where $\vartheta = \vartheta_a$ and $\bar{E} = \bar{E}_a$. By doing some measurable amount of work $\mathcal{W}$ *on* the system one can attain a different macrostate characterized by a different temperature ϑ and a larger internal energy $\bar{E} = E_a + \mathcal{W}$. Similarly, one can determine the internal energy of a macrostate of internal energy lower than $\bar{E}_a$ by starting from this macrostate, characterized by some temperature ϑ, and measuring the amount of work $\mathcal{W}$ which must be done *on* the system to bring its temperature to ϑ_a.

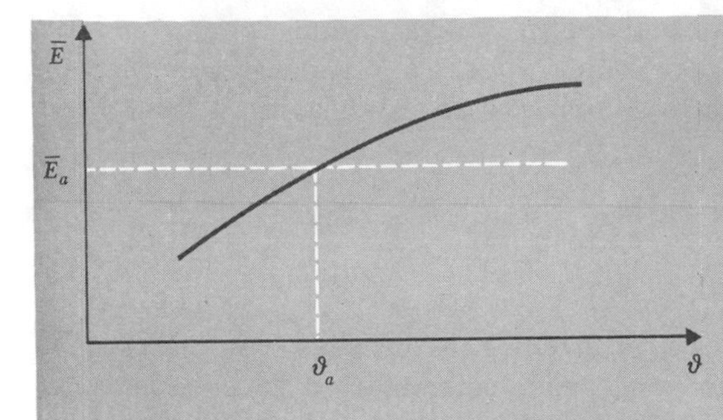

Fig. 4·1·1 Schematic curve showing the dependence of the measured mean energy $\bar{E}$ on the arbitrary temperature parameter ϑ characterizing the system of Fig. 2·7·5.

The energy of the initial macrostate is then $\bar{E}_a - \mathcal{W}$. In this way one can construct a curve of $\bar{E}$ versus ϑ (see Fig. 4·1·1). The energy $\bar{E}_a$ of the standard state can, of course, be set equal to zero.

Example 2 Consider a system consisting of an electric resistor (e.g., a coil of platinum wire). If this system is kept at fixed pressure, its macrostate can again be specified completely by an arbitrary temperature parameter ϑ. Here one can determine values of the internal energy of this system corresponding to various values of ϑ by connecting a battery to the resistor and doing electrical work on it. Except for the fact that electrical measurements are usually more convenient and accurate than mechanical ones, the analysis of this example is identical to that of the preceding one.

Example 3 Consider the system of Fig. 4·1·2 consisting of a cylinder containing a gas. The macrostate of this system can be specified by two parameters, e.g., its volume V and internal energy $\bar{E}$. The mean gas pressure $\bar{p}$ is then determined. (Alternatively, one can specify the macrostate by specifying V and $\bar{p}$ as independent variables; the mean energy $\bar{E}$ is then determined.) Consider a standard macrostate a of volume V_a and mean pressure $\bar{p}_a$, where $\bar{E} = \bar{E}_a$. How would one determine the mean energy $\bar{E}_b$ of any other macrostate b of volume V_b and mean pressure $\bar{p}_b$?

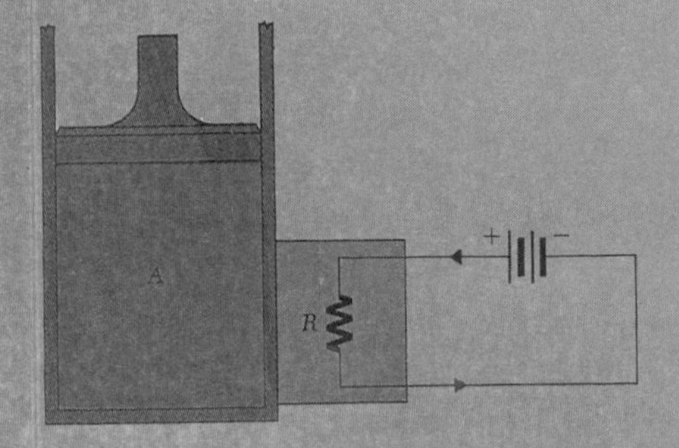

Fig. 4·1·2 A system consisting of a cylinder containing a gas. The volume V of the gas is determined by the position of the movable piston. The resistor R can be brought into thermal contact with this system.

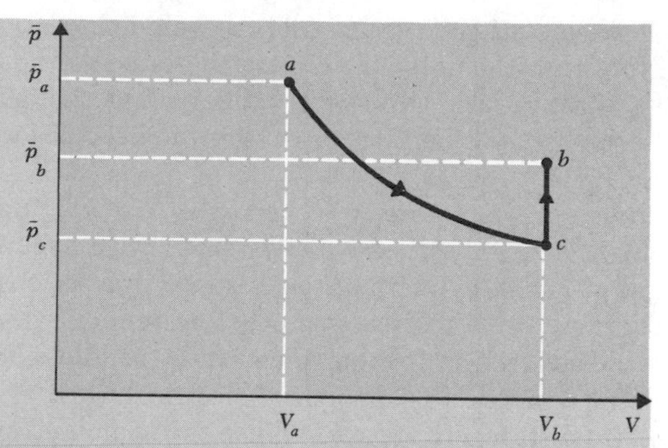

Fig. 4·1·3 Diagram illustrating the volumes and mean pressures describing different macrostates of the gas of Fig. 4·1·2.

Each macrostate can be represented by a point on a $\bar{p}\,V$ diagram (see Fig. 4·1·3). One could proceed as follows.

1. Let the gas expand against the piston until its volume changes from its initial value V_a to its desired final value V_b. As a result, the mean pressure of the gas decreases to some value $\bar{p}_c$. Denote by W_{ac} the work done *by* the gas on the piston in this process.

2. To change the mean pressure to the desired final value $\bar{p}_b$ while keeping the volume constant, bring the system into thermal contact with another system whose internal energy is already known; e.g., bring the system into thermal contact with the electrical resistor of the preceding example. Do electric work $\mathcal{W}_R$ *on* this resistor of an amount just sufficient to bring the gas pressure to $\bar{p}_b$. In this process the internal energy of the resistor changes by a measurable amount $\Delta\bar{\epsilon}$, and an amount of energy $\mathcal{W} - \Delta\bar{\epsilon}$ (in the form of heat) is transferred to the system of interest.

The total internal energy of the system in state b is then given by*

$$\bar{E}_b = \bar{E}_a - W_{ac} + (\mathcal{W} - \Delta\bar{\epsilon})$$

4·2 *Heat*

The heat Q_{ab} absorbed by a system in going from a macrostate a to another macrostate b is, by (2·8·2), defined as

$$Q_{ab} = (\bar{E}_b - \bar{E}_a) + W_{ab} \tag{4·2·1}$$

where W_{ab} is the macroscopic work done *by* the system in this process. Since we have already discussed how to measure the work W_{ab} and the internal energy

* Since the performance of work on the resistor results in an increase in the pressure of the gas, the above procedure could not be used if $\bar{p}_b < \bar{p}_c$. But then one could proceed in the reverse direction by measuring the work required to go from state b to state a.

$\bar{E}$, the relation (4·2·1) yields a well-defined number for the heat absorbed. Note that the units of heat are the same as those of work, i.e., ergs or joules.

In practice, two slightly different methods are commonly used for doing "calorimetry," i.e., for measuring the heat absorbed by a system.

Direct measurement in terms of work Suppose it is desired to measure the heat Q_{ab} absorbed by a system A with fixed external parameters (e.g., the gas of Fig. 4·1·2 with piston clamped in position). Then one can bring A into thermal contact with some other system on which work can be done, e.g., an electrical resistor (or equivalently, the paddle-wheel system of Fig. 2·7·5). By doing a measurable amount of electrical work $\mathcal{W}$ *on* the resistor, one can bring A from state a to state b. Since the combined system consisting of A and the resistor does not exchange heat with any outside system, Eq. (4·2·1) applied to this combined system yields

$$\mathcal{W} = \Delta\bar{E} + \Delta\bar{\epsilon}$$

where $\Delta\bar{E}$ is the change in mean energy of A and $\Delta\bar{\epsilon}$ is the change in mean energy of the resistor. Since A itself does no work, the same Eq. (4·2·1) applied to A implies that

$$Q_{ab} = \Delta\bar{E}$$

Hence

$$Q_{ab} = \mathcal{W} - \Delta\bar{\epsilon} \qquad (4\cdot 2\cdot 2)$$

If the resistor system is sufficiently small compared to A, then $\Delta\bar{\epsilon} \ll \Delta\bar{E}$ and the term $\Delta\bar{\epsilon}$ in (4·2·2) is negligibly small. Otherwise, the internal energy $\bar{\epsilon}$ of the resistor can be considered known as a function of its macrostate (e.g. of its temperature ϑ) from prior measurements involving the performance of electrical work on the isolated resistor system. Equation (4·2·2) thus determines the heat absorbed by A in going from macrostate a to b.

Comparison method (sometimes called the "method of mixtures") While keeping all external parameters fixed, bring the system A into thermal contact with a reference system B whose internal energy is already known as a function of its parameters. No work is done in this process. The conservation of energy for the isolated combined system implies then that in the process of going [from the initial situation, where the systems are in equilibrium and isolated from each other] to [the final situation, where the systems are in equilibrium and in thermal contact], the changes in internal energies satisfy the condition

$$\Delta\bar{E}_A + \Delta\bar{E}_B = 0$$

In terms of the heats absorbed by the two systems, this can be written

$$Q_A + Q_B = 0 \qquad (4\cdot 2\cdot 3)$$

Since $Q_B = \Delta\bar{E}_B$ is known for the reference system B in terms of the changes of its parameters in the process, one has thus measured

$$Q_A = -Q_B \qquad (4\cdot 2\cdot 4)$$

A familiar example of this method is that where water is used as the reference system B. The system A is immersed in the water. The resulting temperature change of the water can be measured and determines the internal energy change of the water. The heat absorbed by A is then known.

4 · 3 *Absolute temperature*

We discussed in Sec. 3 · 5 the measurement of temperature with respect to an arbitrary thermometric parameter of some arbitrary thermometer. We now want to consider the operational determination of the *absolute temperature* T of a system. Compared to an arbitrary temperature parameter, the absolute temperature has the following two important properties:

1. As discussed in Sec. 3 · 5, the absolute temperature provides one with a temperature parameter which is completely independent of the nature of the particular thermometer used to perform the temperature measurement.
2. The absolute temperature T is a parameter of fundamental significance which enters all the theoretical equations. Hence all theoretical predictions will involve this particular temperature parameter.

Any theoretical relation involving the absolute temperature T can be used as the basis of an experimental determination of T. We can distinguish between two classes of relations on which such a determination can be based.

a. Theoretical relations involving *micro*scopic aspects of the theory. For example, one can apply statistical mechanics to a particular system to calculate from microscopic considerations the equation of state of this system. The equation of state (discussed in Sec. 3 · 12) is a relation between macroscopic parameters of the system and the absolute temperature T. Hence it can be used as a basis for measuring T.

b. Theoretical relations based on the purely *macro*scopic statements of the theory. For example, the second law states that $dS = đQ/T$ for an infinitesimal quasi-static process. This relation involves the absolute temperature T and can thus be used for measuring T. (An example of this procedure will be discussed in Sec. 11 · 3.)

The simplest and most important illustration of the first method is that based on the equation of state for an ideal gas. In (3 · 12 · 8) we found that this equation of state can be written in the form

$$\bar{p}V = NkT \tag{4·3·1}$$

or equivalently,

$$\bar{p}V = \nu RT \tag{4·3·2}$$

where

$$R \equiv N_a k \tag{4·3·3}$$

Here ν is the number of moles of gas, and N_a is Avogadro's number. In practice, the ideal gas conditions of negligible interaction between molecules can be achieved by working with gases in the limit of very high dilution so that the mean intermolecular separation is large.

In the limit of sufficiently high dilution the mean intermolecular separation also becomes large compared to the mean De Broglie wavelength corresponding to the mean momentum of a gas molecule. Quantum mechanical effects thus become unimportant in this limit and the equation of state (4·3·1), derived on the basis of classical statistical mechanics, must also be valid. (The strictly quantum-mechanical derivation of (4·3·1) will be given in Chapter 9.)

The equation of state (4·3·1) makes some definite predictions. For example, it asserts that, if the temperature is kept fixed, one has the relation

$$\bar{p}V = \text{constant}$$

This result is the familiar and historically important "Boyle's law." Another consequence of (4·3·1) is that this equation of state is the same *irrespective* of the particular gas considered; e.g., it applies equally well to helium, hydrogen, or methane, as long as these gases are sufficiently dilute that they can be considered ideal.

To use the equation of state (4·3·1) as a means for determining the absolute temperature T, one can proceed as follows. Keep the volume V of the given amount of gas fixed. Then one has a constant-volume gas thermometer of the type described in Sec. 3·5. Its thermometric parameter is the pressure $\bar{p}$. By virtue of (4·3·1) one knows that $\bar{p}$ is directly proportional to the absolute temperature T of the gas. Indeed, once one chooses the value of the arbitrary constant k still at one's disposal (or equivalently, the value of the constant R), Eq. (4·3·1) determines a definite value of T. We now describe how this constant k is conventionally chosen, or equivalently, how the absolute temperature *scale* is chosen.

When the constant-volume gas thermometer is brought into thermal contact with some system A with which it is allowed to come to equilibrium, its pressure will attain some definite value $\bar{p}_A$. When it is brought into thermal contact with some other system B (this may be a system of the same kind but in some other macrostate) with which it is allowed to come to equilibrium, its pressure will attain some other definite value $\bar{p}_B$. By (4·3·1), the pressure ratio is then given by

$$\frac{\bar{p}_A}{\bar{p}_B} = \frac{T_A}{T_B} \tag{4·3·4}$$

where T_A and T_B are the absolute temperatures of the systems A and B. Thus any ideal gas thermometer can be used to measure absolute temperature ratios. In particular, if system B is chosen as some standard system in some standard macrostate, then the gas thermometer can be used to measure the ratio of the absolute temperature T of any system to the temperature T_B of this standard system.

The relation (4·3·4) is another consequence of the equation of state (4·3·1) which can be checked experimentally, for it implies that this pressure ratio ought to be the same no matter what gas is used in the gas thermometer, provided that the gas is sufficiently dilute. In symbols,

$$\lim_{\nu \to 0} \frac{\bar{p}_A}{\bar{p}_B} \rightarrow \text{constant independent of nature of gas} \qquad (4 \cdot 3 \cdot 5)$$

where ν denotes the number of moles of gas used in the bulb of the thermometer.

This comment also provides an experimental criterion for deciding when a gas is sufficiently dilute to be considered ideal. The pressure ratio $\bar{p}_A/\bar{p}_B$ can be measured with a given amount of gas in the bulb. This measurement can then be repeated with successively smaller amounts of gas in the bulb; the ratio $\bar{p}_A/\bar{p}_B$ must then, by (4·3·5), reach a limiting constant value. When this is the case, the gas is known to be sufficiently dilute to exhibit ideal behavior.

By international convention one chooses as the standard system pure water, and as the standard macrostate of this system the situation where the solid, liquid, and gas phases of this system (i.e., ice, water, and water vapor) are in equilibrium with each other. (This macrostate is called the "triple

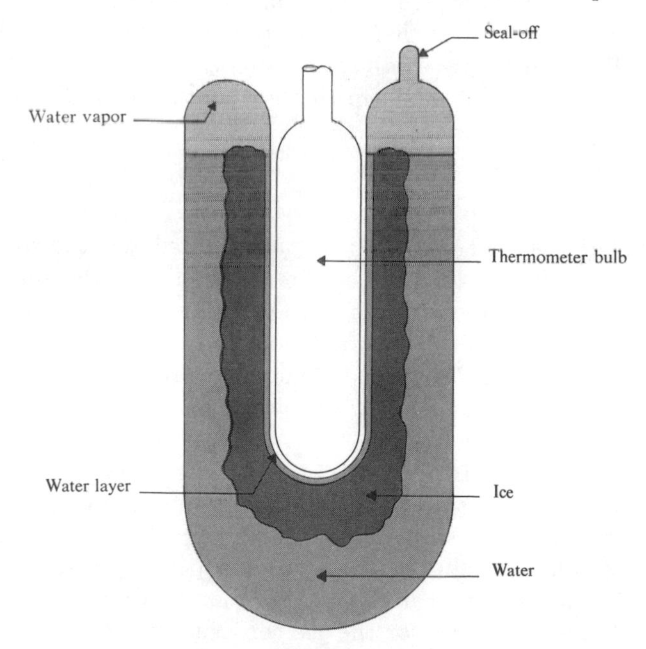

***Fig. 4·3·1** Diagram illustrating a triple-point cell designed to calibrate a thermometer at the triple point of water. A freezing mixture is first introduced into the central well to produce some of the ice. After the freezing mixture is removed, the thermometer bulb is placed in the well and the system is allowed to come to thermal equilibrium.*

point" of water.) The reason for this choice is that there is only one definite value of pressure and temperature at which all these three phases can coexist in equilibrium and, as can readily be verified experimentally, the temperature of this system is unaffected by any changes in the relative amounts of solid, liquid, and gas present under these circumstances. The triple point provides, therefore, a very reproducible standard of temperature. By international convention one chooses to assign to the absolute temperature T_t of water at its triple point the value

$$T_t \equiv 273.16 \qquad \textit{exactly} \tag{4·3·6}$$

This particular choice was motivated by the desire to keep the modern temperature scale, adopted by international convention in 1954, as nearly identical as possible with a historically older temperature scale.

The choice (4·3·6) fixes a scale factor for T which we can indicate by assigning to T the unit* of "degree K." This stands for "degree Kelvin" (so named after the famous British physicist of the last century), and is commonly abbreviated as "°K." Whenever we use the unit "degree" without further qualifications, we shall always mean "degree K." We mentioned in Sec. 3·3 that kT has the dimensions of energy. It thus follows that the constant k has the units of ergs/degree.

Once the above conventions have been adopted, the absolute temperature T_A of any system A is completely determinable by the gas thermometer. If the thermometer exhibits pressure $\bar{p}_A$ when in thermal contact with this system, and pressure $\bar{p}_t$ when in thermal contact with water at the triple point, then

$$T_A = 273.16 \frac{\bar{p}_A}{\bar{p}_t} \tag{4·3·7}$$

Here the pressure ratio is to be evaluated in the ideal gas limit, i.e., in the limit when the gas used in the thermometer is made sufficiently dilute. Thus the absolute temperature of any system can be directly determined by measuring the pressure of a constant volume gas thermometer. This is in practice a relatively simple method of measuring the absolute temperature, provided that the temperature is not so low or so high that the use of gas thermometers becomes impracticable.

Once the temperature scale has been fixed by (4·3·6), one can return to the equation of state (4·3·2) and determine the constant R. Taking ν moles of any gas at the triple-point temperature $T_t = 273.16$°K, one need only measure its volume V (in cm^3) and its corresponding mean pressure $\bar{p}$ (in dynes/cm^2); this information permits computation of R by (4·3·2). Careful measurements of this type yield for the gas constant the value

$$R = (8.3143 \pm 0.0012) \text{ joules mole}^{-1} \text{ deg}^{-1} \tag{4·3·8}$$

* This unit is a unit in the same sense as the degree of angular measure; it does not involve length, mass, or time.

(1 joule = 10^7 ergs). Knowing Avogadro's number ("unified scale," atomic weight of C^{12} = 12 exactly)

$$N_a = (6.02252 \pm 0.00028) \times 10^{23} \text{ molecules mole}^{-1} \qquad (4 \cdot 3 \cdot 9)$$

one can use (4·3·3) to find the value of k. This important constant is called "Boltzmann's constant" in honor of the Austrian physicist who contributed so significantly to the development of kinetic theory and statistical mechanics. Its value is found to be*

$$k = (1.38054 \pm 0.00018) \times 10^{-16} \text{ ergs degree}^{-1} \qquad (4 \cdot 3 \cdot 10)$$

On the absolute-temperature scale defined by the choice of (4·3·6), an energy of 1 ev (electron volt) corresponds to an energy kT, where $T \approx 11\,600°$K. Also, room temperature is approximately 300°K on this scale and corresponds to an energy $kT \approx \frac{1}{40}$ ev.

Another temperature scale sometimes used is the Celsius (or centigrade) temperature θ *defined* in terms of the absolute temperature T by the relation

$$\theta \equiv T - 273.15 \qquad \text{degrees Celsius} \qquad (4 \cdot 3 \cdot 11)$$

(abbreviated as "°C"). On this scale water at atmospheric pressure freezes at *approximately* 0°C and boils at *approximately* 100°C.

Historical remark We wish to mention briefly the historical context which motivates the particular choice of numerical value in (4·3·6). The basic reason was the adoption, before the full significance of the absolute temperature concept had become clear, of the Celsius (or centigrade) temperature scale based on *two* fixed standard temperatures. In this scheme the Celsius temperature θ was chosen as a linear function of the thermometric parameter. For example, in the case of the constant-volume gas thermometer where the thermometric parameter is the pressure $\bar{p}$, the Celsius temperature was taken to be

$$\theta = a\bar{p} + b \qquad (4 \cdot 3 \cdot 12)$$

where a and b are constants to be determined in terms of the two fixed standard temperatures. The latter were again based on water as the standard system, but the two standard macrostates of this substance were chosen as follows:

1. The state where ice is in equilibrium with air-saturated water at atmospheric pressure. This is the so-called "ice point" of water. By definition one assigned to the temperature of this state the value $\theta = 0$.
2. The state where water is in equilibrium with water vapor at a pressure of 1 atmosphere. This is the so-called "steam point" of water. By definition one assigned to the temperature of this state the value $\theta = 100$.

* The values of these physical constants are those of the least-squares adjustment of E. R. Cohen and J. W. M. DuMond and approved for adoption by the National Research Council in April, 1963. See the Table of Numerical Constants at the end of the book.

(We note parenthetically that these points are more difficult to reproduce experimentally than the triple point of water, which involves only pure water (without the presence of air) and needs no specification of the applied pressure.)

If the pressure readings of the gas thermometer at the ice and steam points are denoted by $\bar{p}_i$ and $\bar{p}_s$, respectively, then Eq. (4 · 3 · 12) yields the following two relations when applied to these two standard states:

$$0 = a\bar{p}_i + b$$
$$100 = a\bar{p}_s + b$$

These two equations can be solved for a and b in terms of $\bar{p}_i$ and $\bar{p}_s$. The relation (4 · 3 · 12) then becomes

$$\theta = 100\,\frac{\bar{p} - \bar{p}_i}{\bar{p}_s - \bar{p}_i} \tag{4·3·13}$$

Alternatively, this can be used to express $\bar{p}$ in terms of θ. One finds

$$\bar{p} = \bar{p}_i\left(1 + \frac{\theta}{\theta_0}\right) \tag{4·3·14}$$

where

$$\theta_0 \equiv 100\left(\frac{\bar{p}_s}{\bar{p}_i} - 1\right)^{-1} \tag{4·3·15}$$

depends only on a pressure *ratio* and is thus independent of the nature of the gas used. Thus θ_0 is a universal constant for all gases and can be *measured* by using a gas thermometer at the ice and steam points. One finds thus

$$\theta_0 = 273.15 \tag{4·3·16}$$

By (4 · 3 · 14), measurements of two systems at the respective temperatures θ_A and θ_B yield for the corresponding pressure ratio

$$\frac{\bar{p}_A}{\bar{p}_B} = \frac{\theta_0 + \theta_A}{\theta_0 + \theta_B} \tag{4·3·17}$$

This is of the same form as (4 · 3 · 4) if one *defines* the absolute temperature T by the relation

$$T \equiv \theta_0 + \theta \tag{4·3·18}$$

If the triple point of water is measured on this temperature scale, one finds $\theta \approx 0.01°C$, or by (4 · 3 · 18), $T \approx 273.16°K$ *approximately*.

It is clear that this old-fashioned procedure for establishing a temperature scale is cumbersome, logically not very satisfying, and not of the highest possible accuracy. The modern convention using a single fixed point is far more satisfactory in all these respects. But by *choosing* T_t to be *exactly* 273.16°K rather than some other number, one gains the convenience that all the older temperature measurements based on the former temperature scale will (within the limits of accuracy with which the triple point of water was measured on that scale) agree numerically with the values based on the modern convention.

4·4 *Heat capacity and specific heat*

Consider a macroscopic system whose macrostate can be specified by its absolute temperature T and some other macroscopic parameter (or set of macroscopic parameters) y. For example, y might be the volume or the mean pressure of the system. Suppose that, starting with the system at temperature T, an infinitesimal amount of heat $đQ$ is added to the system while all its parameters y are kept fixed. The resulting change dT in the temperature of the system depends on the nature of the system as well as on the parameters T and y specifying the macrostate of the system. We define the ratio

$$\left(\frac{đQ}{dT}\right)_y \equiv C_y \qquad (4\cdot4\cdot1)$$

in the limit as $đQ \to 0$ (or $dT \to 0$) as the "heat capacity" of the system. Here we have used the subscript y to denote explicitly the parameters kept constant in the process of adding heat. The quantity C_y depends, of course, on the nature of the system and on the particular macrostate under consideration, i.e., in general

$$C_y = C_y(T,y) \qquad (4\cdot2\cdot2)$$

The amount of heat $đQ$ which needs to be added to produce a given temperature change dT of a homogeneous system will be proportional to the quantity of matter contained therein. Hence it is convenient to define a quantity, the "specific heat," which depends only on the nature of the substance under consideration, not on the amount present. This can be achieved by dividing the heat capacity C_y of ν moles (or m grams) of the substance by the corresponding number of moles (or of grams). The "specific heat per mole" or "heat capacity per mole" is thus defined as

$$c_y \equiv \frac{1}{\nu} C_y = \frac{1}{\nu}\left(\frac{đQ}{dT}\right)_y \qquad (4\cdot4\cdot3)$$

Equivalently, the "specific heat per gram" is defined as

$$c_y' \equiv \frac{1}{m} C_y = \frac{1}{m}\left(\frac{đQ}{dT}\right)_y \qquad (4\cdot4\cdot4)$$

The cgs units of the molar specific heat are, by (4·4·3), ergs degree $^{-1}$ mole^{-1}.

It should be noted from the operational definition (4·4·1) of the heat capacity C_y that this quantity does depend on which particular parameters y of the system are kept constant in the process of adding the heat. Suppose we consider a substance, e.g., a gas or liquid, whose macrostate can be specified by two parameters, say the temperature T and the volume V (see Fig. 4·4·1). When the system is in a given macrostate, we can ask for the following two quantities: (1) c_V, the molar specific heat at constant volume of the system in this state, and (2) c_p, the molar specific heat at constant pressure of the system in this state.

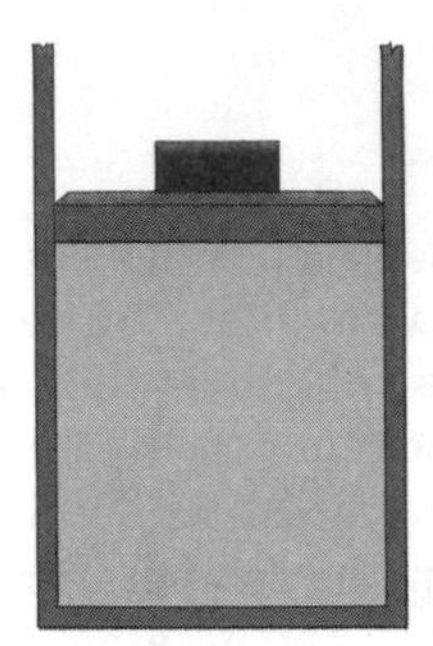

Fig. 4·4·1 Diagram illustrating specific heat measurements of a gas kept at constant volume or at constant pressure.

1. To determine c_V, we clamp the piston in position so that the volume of the system is kept fixed. In this case the system cannot do any work, and the heat $đQ$ added to the system goes entirely to increase the internal energy of the system

$$đQ = d\bar{E} \qquad (4\cdot4\cdot5)$$

2. To determine c_p, the piston is left completely free to move, the weight of the piston being equal to the constant force per unit area (or mean pressure $\bar{p}$) on the system. In this case the piston will move when heat $đQ$ is added to the system; as a result, the system does also mechanical work. Thus the heat $đQ$ is used *both* to increase the internal energy of the system *and* to do mechanical work on the piston; i.e.

$$đQ = d\bar{E} + \bar{p}\, dV \qquad (4\cdot4\cdot6)$$

For a *given* amount of heat absorbed, the internal energy $\bar{E}$ will therefore increase by a smaller amount (and hence the temperature T will also increase by a smaller amount) in the second case compared to the first. By (4·4·1) one expects, therefore, that the heat capacity is greater in the second case; i.e., one expects that

$$c_p > c_V \qquad (4\cdot4\cdot7)$$

Remark Note that the specific heat at constant volume may itself still be a function of the volume V; i.e., $c_V = c_V(T,V)$ in general. For example, the heat required to raise the temperature of a gas from 300 to 301 degrees is in general not the same if its volume is kept constant at 50 cm³ in the process of adding heat, as it is if its volume is kept constant at 1000 cm³ in the process of adding heat.

Since the second law allows us to write $đQ = T\, dS$, the heat capacity (4·4·1) can be written in terms of the entropy as

$$C_y = T\left(\frac{\partial S}{\partial T}\right)_y \qquad (4\cdot4\cdot8)$$

If S in this expression is the entropy per mole of substance, then C_y is the molar specific heat.

If one contemplates a situation where all the *external* parameters of a system are kept fixed, then the system does no macroscopic work, $đW = 0$, and the first law reduces simply to the statement $đQ = d\bar{E}$. For example, if the volume V is the only external parameter, one can write

$$C_V = T\left(\frac{\partial S}{\partial T}\right)_V = \left(\frac{\partial \bar{E}}{\partial T}\right)_V \tag{4·4·9}$$

By virtue of (3·7·16) it follows that this quantity is always positive.

Measurements of the specific heat involve measurements of heat of the type discussed in Sec. 4·2. In measuring heat by the comparison method (or method of mixtures), it used to be popular to select water as the reference substance. Hence a knowledge of its specific heat was of particular importance. Such a measurement can be made directly in terms of work and was first performed by Joule during the years 1843–1849. The specific heat of water at a pressure of 1 atmosphere and a temperature of 15°C (288.2°K) is found to be 4.18 joules deg^{-1} gram^{-1}.

Before the nature of heat as a form of energy was understood, it was customary to define a unit of heat, the "calorie," as the heat required to raise the temperature of water at 1 atmosphere from 14.5 to 15.5°C. Joule's measurement of the heat capacity of water in terms of work allowed expression of the calorie in terms of absolute energy units. The calorie unit is gradually becoming obsolete and is now *defined* in terms of energy units by the relation

$$1 \text{ calorie} \equiv 4.1840 \text{ joules} \tag{4·4·10}$$

Example Let us consider heat measurements by the method of mixtures in terms of the specific heats of the substances involved. Consider that two substances A and B, of respective masses m_A and m_B, are brought into thermal contact under conditions where the pressure is kept constant. (For example, a copper block is immersed in water, the pressure being atmospheric.) Suppose that at this pressure the specific heats per gram of the respective substances are $c_A'(T)$ and $c_B'(T)$. Assume that before the substances are brought into thermal contact their respective equilibrium temperatures are T_A and T_B, respectively. Denote their final common temperature, after equilibrium is reached, by T_f. No work gets done in this process, so that the conservation of energy is expressed by (4·2·3) as

$$Q_A + Q_B = 0 \tag{4·4·11}$$

But by (4·4·4) the heat absorbed by a substance when its temperature is increased by an amount dT is given by $đQ = mc'\,dT$. Hence the heat absorbed by A in going from temperature T_A to temperature T_f is given by

$$Q_A = \int_{T_A}^{T_f} m_A c_A'(T')\,dT'$$

or

$$Q_A = m_A c_A'(T_f - T_A)$$

if the temperature dependence of c_A' is negligible. Similar expressions hold for B. Hence the fundamental condition (4·4·11) can be written

$$m_A \int_{T_A}^{T_f} c_A' \, dT' + m_B \int_{T_B}^{T_f} c_B' \, dT' = 0 \tag{4·4·12}$$

This relation allows, for example, computation of the final temperature T_f if the other quantities are known. The situation is particularly simple if the specific heats c_A' and c_B' are temperature independent. In this case (4·4·12) becomes simply

$$m_A c_A'(T_f - T_A) + m_B c_B'(T_f - T_B) = 0 \tag{4·4·13}$$

If desired, this can be solved explicitly for the final temperature T_f to give

$$T_f = \frac{m_A c_A' T_A + m_B c_B' T_B}{m_A c_A' + m_B c_B'}$$

4 · 5 *Entropy*

The entropy can readily be determined by using the second-law statement (3·11·4) that $dS = đQ/T$ for an infinitesimal quasi-static process. Given any macrostate b of the system, one can find the entropy difference between this state and some standard state a by considering *any quasi-static* process which carries the system from state a to state b and calculating for this process the integral

$$S_b - S_a = \int_{a\,(\text{eq})}^{b} \frac{đQ}{T} \tag{4·5·1}$$

The evaluation of this integral yields, as discussed in Sec. 3·10, the same unique value of $S_b - S_a$ *irrespective* of what quasi-static process may be chosen to bring the system from state a to state b.

Let us emphasize again that the process chosen for calculating the integral in (4·5·1) *must* be quasi-static. This means that we must somehow bring the system from state a to state b by continuously changing its parameters so slowly (compared to significant relaxation times) that it is at all times very close to an equilibrium situation. In general, this will require the use of other auxiliary systems on which the system can do work and from which it can absorb heat. For example, if we need to change the volume of the system, we could do it by successively moving a piston by small amounts, proceeding sufficiently slowly to allow the system to reach equilibrium at all stages. Or, if we need to change the temperature of the system, we could do it by bringing the system successively into contact with a large number of heat reservoirs of slightly different temperatures, again proceeding sufficiently slowly to allow the system to reach equilibrium at all stages. Clearly, the temperature T is a well-defined quantity as we go through this succession of equilibrium situations, and the heat $đQ$ absorbed in going from one macrostate to an adjacent one is also a measurable quantity. The evaluation of the entropy difference by (4·5·1) thus presents no conceptual difficulties. Note that the units of entropy are, by (4·5·1), ergs/degree or joules/degree.

Suppose that the macrostate of a body is only specified by its temperature, since all its other parameters y (e.g., its volume V or its mean pressure $\bar{p}$) are kept constant. If one knows the heat capacity $C_y(t)$ of the body under these conditions, then its entropy difference (for the given values of the parameters y) is given by

$$S(T_b) - S(T_a) = \int_a^b \frac{đQ}{T} = \int_{T_a}^{T_b} \frac{C_y(T')\, dT'}{T'} \qquad (4\cdot5\cdot2)$$

In the particular case that $C_y(T)$ is independent of T, this becomes simply

$$S(T_b) - S(T_a) = C_y \ln \frac{T_b}{T_a} \qquad (4\cdot5\cdot3)$$

Example Consider the example discussed at the end of Sec. 4·4 where two systems A and B, with constant specific heats c_A' and c_B' and originally at respective temperatures T_A and T_B, are brought into thermal contact with each other. After the systems come to equilibrium, they reach a common final temperature T_f. What is the entropy change of the entire system in this process? The process which occurred here was certainly *not* a quasi-static one (unless $T_B = T_A$); nonequilibrium conditions prevailed between the initial and final situations. To calculate the entropy change of system A, we can imagine that it is brought from its initial temperature T_A to its final temperature T_f by a succession of infinitesimal heat additions whereby the system at any intermediate equilibrium temperature T absorbs from a heat reservoir at an infinitesimally higher temperature $(T + dT)$ a small amount of heat $đQ = m_A c_A'\, dT$. Thus the entropy change of A is given by

$$\Delta S_A = S_A(T_f) - S_A(T_A) = \int_{T_A}^{T_f} \frac{m_A c_A'\, dT}{T} = m_A c_A' \ln \frac{T_f}{T_A}$$

A similar expression holds for system B. Hence the entropy change of the total system is given by

$$\Delta S_A + \Delta S_B = m_A c_A' \ln \frac{T_f}{T_A} + m_B c_B' \ln \frac{T_f}{T_B} \qquad (4\cdot5\cdot4)$$

Since this represents the total entropy change of the isolated system $(A + B)$, we know by the second law (3·11·3) that this can never be negative. To verify this explicitly, we make use of the simple inequality (proved in Appendix A·8)

$$\ln x \le x - 1 \qquad (= \text{sign for } x = 1) \qquad (4\cdot5\cdot5)$$

Hence

$$-\ln x \ge -x + 1$$

or, putting $y = 1/x$,

$$\ln y \geq 1 - \frac{1}{y} \qquad (= \text{sign for } y = 1) \tag{4·5·6}$$

Equation (4·5·4) implies, therefore, the inequality

$$\begin{aligned} \Delta S_A + \Delta S_B &\geq m_A c_A' \left(1 - \frac{T_A}{T_f}\right) + m_B c_B' \left(1 - \frac{T_B}{T_f}\right) \\ &= T_f^{-1}[m_A c_A'(T_f - T_A) + m_B c_B'(T_f - T_B)] \\ &= 0 \qquad \text{by (4·4·13)} \end{aligned}$$

Thus $$\Delta S_A + \Delta S_B \geq 0 \tag{4·5·7}$$

The equals sign holds only if $T_A/T_f = 1$ and $T_B/T_f = 1$, i.e., if $T_B = T_A$; then equilibrium is indeed preserved and no irreversible process occurs after the systems are brought into thermal contact with each other.

The relation (4·5·2) is actually quite interesting because it gives an explicit connection between two different types of information about the system under consideration. On the one hand, (4·5·2) involves the heat capacity $C(T)$ obtainable from macroscopic measurements of absorbed heat. On the other hand, it involves the entropy which is related to a microscopic knowledge of the quantum states of the system and which can be calculated either from first principles or from experimental information obtainable from spectroscopic data.

Example As a simple illustration, suppose that one is dealing with a simple system of N magnetic atoms, each with spin $\frac{1}{2}$. If this system is known to be ferromagnetic at sufficiently low temperatures, all spins must be completely aligned as $T \to 0$ so that the number of accessible states $\Omega \to 1$, or $S = k \ln \Omega \to 0$ (in accordance with the third law). But at sufficiently high temperatures all spins must be completely randomly oriented so that $\Omega = 2^N$ and $S = kN \ln 2$. Hence it follows that this system must have a heat capacity $C(T)$ which satisfies, by (4·5·2), the equation

$$\int_0^\infty \frac{C(T')\,dT'}{T'} = kN \ln 2$$

This relation must be valid irrespective of the details of the interactions which bring about ferromagnetic behavior and irrespective of the temperature dependence of $C(T)$.

These comments should make it apparent that the measuring of heat capacities is not just a dull activity undertaken to fill up handbooks with data useful to engineers concerned with the properties of materials. Accurate measurements of heat capacities may be of considerable interest because they can provide some important information about the nature of the energy levels of physical systems.

4 · 6 *Consequences of the absolute definition of entropy*

In many applications it is true that only entropy *differences*, i.e., values of the entropy measured with respect to some chosen standard state, are of importance. In this respect the entropy is similar to the internal energy $\bar{E}$ of a system, and the entropy of any state with respect to the standard state can be determined by the integral (4·5·1). But we do know, as discussed in Sec. 3·10, that the entropy S is a completely calculable number and is *not* merely defined to within an arbitrary additive constant. This reflects itself in the third law statement that the entropy approaches, as $T \to 0$, a definite value S_0 (usually $S_0 = 0$) independent of all parameters of the system. To obtain an absolute value of the entropy one can either use statistical mechanics to *calculate* the absolute value of the entropy in the standard state, or one can measure entropy differences from a standard state chosen at $T \to 0$ where $S = S_0$ is known to have a definite value independent of all parameters of the system.

The fact that the entropy has a definite value (without any arbitrary additive constant) makes possible physically significant statements which could not be made otherwise. The following two examples will serve to illustrate this point.

Example 1 Consider the case of a solid which can exist in two different crystal structures. A classical example of this kind is tin, which exists in two rather different forms: one of these is "white" tin, which is a metal; the other is "gray" tin, which is a semiconductor. Gray tin is the stable form at temperatures below $T_0 \equiv 292°K$, while white tin is the stable form above this temperature. At the temperature T_0 the two forms are in equilibrium with each other. They can then coexist indefinitely in arbitrary proportions, and a positive amount of heat Q_0 must be absorbed to transform one mole of gray tin into the white tin modification.

Although white tin is the unstable form at temperatures below T_0, the speed with which the transformation proceeds to the gray form is very slow compared to times of experimental interest. It is thus very easy to work with white tin, the ordinary metal, down to very low temperatures. (One is, in practice, scarcely aware that there is a tendency for the metal to transform to the gray form.) A sample of white tin readily achieves internal equilibrium, although it exhibits a negligible tendency to transform to the gray form. One can thus easily measure the molar specific heat $C^{(w)}(T)$ of white tin in the temperature range $T < T_0$. There is, of course, no difficulty in working with a sample of gray tin in this temperature range; one can thus also make measurements of the molar specific heat $C^{(g)}(T)$ of gray tin at temperatures $T < T_0$.*

* All quantities in this discussion refer to measurements at the same constant pressure.

Since the transformation from the white to the gray form proceeds at a negligible rate, the situation would not be changed significantly if one imagined imposing a constraint which would prevent the transformation altogether. In that case white tin can be considered in a genuine equilibrium situation statistically distributed over all states consistent with the crystal structure of white tin; similarly, gray tin can be considered in an equilibrium situation statistically distributed over all states consistent with the crystal structure of gray tin. Arguments of equilibrium statistical mechanics can then be applied to each of these systems. In particular, let us consider the limit as the temperature $T \to 0$. By this limit we shall mean a reasonably low temperature (say 0.1°K), but not one so extremely low (say, less than 10^{-6}°K) that the random spin orientation of the tin nuclei would be affected.* In accordance with the discussion of Sec. 3 · 10, a sample consisting of a mole of white tin approaches a ground-state configuration consistent with the white-tin crystal structure. Correspondingly its entropy $S^{(w)}$ tends to zero except for the contribution $S_0 = k \ln \Omega_S$ associated with the Ω_S states of possible nuclear spin orientations. Similarly, a sample consisting of a mole of gray tin approaches a ground state configuration consistent with the gray tin crystal structure. Correspondingly, its entropy $S^{(g)}$ also tends to zero except for the contribution due to the possible nuclear spin orientations. Since one is dealing with the *same* number of the *same* kind of nuclei, there are again Ω_S possible spin states, and this nuclear spin contribution to the entropy is again $S_0 = k \ln \Omega_S$. Thus, as $T \to 0$,

$$S^{(w)}(T) \to S_0 \qquad \text{and} \qquad S^{(g)}(T) \to S_0$$

i.e.,

$$S^{(w)}(0) = S^{(g)}(0) \tag{4·6·1}$$

The relation (4 · 6 · 1) expresses just the content of the third law that the entropy approaches, as $T \to 0$, a value *independent* of all parameters of the system (in this case independent of crystal structure). We shall now show how this statement can be combined with a knowledge of the specific heats to calculate the heat of transformation Q_0 from gray to white tin at the transition temperature T_0. For suppose that it is desired to calculate the entropy $S^{(w)}(T_0)$ of a mole of white tin at $T = T_0$. One can use two different quasi-static processes in going from $T = 0$ to the same final macrostate.

1. Bring a mole of white tin quasi-statically from $T = 0$ to $T = T_0$. This yields a final entropy

$$S^{(w)}(T_0) = S^{(w)}(0) + \int_0^{T_0} \frac{C^{(w)}(T')}{T'}\, dT' \tag{4·6·2}$$

2. Take a mole of gray tin at $T = 0$ and bring it first quasi-statically to temperature T_0. Then transform it quasi-statically (at this equilibrium transition temperature) to white tin; its entropy change in this transformation is

* The magnetic field H produced by a nuclear moment μ at a neighboring nucleus at an interatomic distance r is of the order of μ/r^3, i.e., of the order of 5 gauss if μ is a nuclear magneton (5×10^{-24} ergs/gauss) and $r = 10^{-8}$ cm. Departures from random nuclear spin orientation can therefore only be expected at a temperature T low enough so that $kT \lesssim \mu H$, the interaction energy between nuclei.

simply Q_0/T_0. Hence one can write*

$$S^{(w)}(T_0) = S^{(g)}(0) + \int_0^{T_0} \frac{C^{(g)}(T')}{T'} dT' + \frac{Q_0}{T_0} \tag{4·6·3}$$

Making use of the result (4·6·1), one thus obtains the relation

$$\frac{Q_0}{T_0} = \int_0^{T_0} \frac{C^{(w)}(T')}{T'} dT' - \int_0^{T_0} \frac{C^{(g)}(T')}{T'} dT' \tag{4·6·4}$$

By using the experimental specific heat measurements and performing the integrations numerically, one finds that the first integral has the value 51.4 joules/deg, while the second integral has the value 44.1 joules/deg. Using $T_0 = 292°K$, one then calculates $Q_0 = (292)(7.3) = 2130$ joules, which compares favorably with the value of 2240 joules obtained by direct measurement of the heat of transformation. Note that this calculation of Q_0 would have been impossible if the third law had not allowed the comparison (4·6·1) between the entropies at $T = 0$.

Example 2 As a second example illustrating the significance of the third law for calculations of the entropy, consider a system A consisting of a mole of solid lead (Pb) and a mole of solid sulfur (S) separated by a partition. Consider also another system B consisting of a mole of the solid compound lead sulfide (PbS). Although the systems A and B are very different, they consist of the same atoms. Hence the third law asserts that the entropies of both systems approach the same value as $T \to 0$, a value corresponding merely to the number of possible orientations of the nuclear spins. In symbols,

$$S^{(\mathrm{Pb+S})}(0) = S^{(\mathrm{PbS})}(0) \tag{4·6·5}$$

if $S^{(\mathrm{Pb+S})}(T)$ denotes the entropy of system A and $S^{(\mathrm{PbS})}$ denotes the entropy of system B.

Suppose that the systems A and B are both at atmospheric pressure. Suppose further that one knows as a function of temperature the heat capacity per mole (at constant atmospheric pressure) $C^{(\mathrm{Pb})}$ of solid lead, $C^{(\mathrm{S})}$ of solid sulfur, and $C^{(\mathrm{PbS})}$ of lead sulfide (PbS). Then one can write for the entropy of the system A consisting of Pb and S separately

$$S^{(\mathrm{Pb+S})}(T) = S^{(\mathrm{Pb+S})}(0) + \int_0^{T} \frac{C^{(\mathrm{Pb})}(T')}{T'} dT' + \int_0^{T} \frac{C^{(\mathrm{S})}(T')}{T'} dT' \tag{4·6·6}$$

For the entropy of the system B consisting of PbS one can write at the same temperature

$$S^{(\mathrm{PbS})}(T) = S^{(\mathrm{PbS})}(0) + \int_0^{T} \frac{C^{(\mathrm{PbS})}(T')}{T'} dT' \tag{4·6·7}$$

* The integrals in (4·6·2) and (4·6·3) must converge properly, since all other quantities are finite. Hence the specific heats must approach zero as $T \to 0$. This is a general property of the specific heats of all substances and one that is well verified experimentally.

By virtue of (4·6·5) the last two relations provide then a quite unique value for the entropy difference $[S^{(Pb+S)}(T) - S^{(PbS)}(T)]$, even though the calculations involve only a knowledge of the specific heats and no information whatever about how Pb and S might react to form PbS. This implies also a very definite prediction as to the value of the integral $\int đQ/T$ which one would obtain by carrying out a quasi-static process in which PbS at a temperature T would be transformed through successive equilibrium states into separated Pb and S at the same temperature. Such a quasi-static process could be performed as follows: Heat the PbS slowly until it evaporates; then heat it further until all the PbS molecules are completely dissociated into Pb and S atoms; then separate the gases slowly with the aid of a semipermeable membrane; then lower the temperature of the whole system back to the temperature T while keeping the membrane in position.

Remark on the separation of gases by semipermeable membranes It is possible to conceive of a membrane which is completely permeable to molecules of one type and completely impermeable to all other molecules. (One can also realize such membranes in practice; e.g., hot palladium metal is permeable to hydrogen (H_2) gas but to no other gases.) With the aid of such membranes one can unmix gases as illustrated in Fig. 4·6·1. For example, to unmix the A and B molecules in a quasi-static way, one needs only to move the two membranes slowly until they meet somewhere in the container. Then all the A molecules will be in the left part and all the B molecules in the right part of the container.

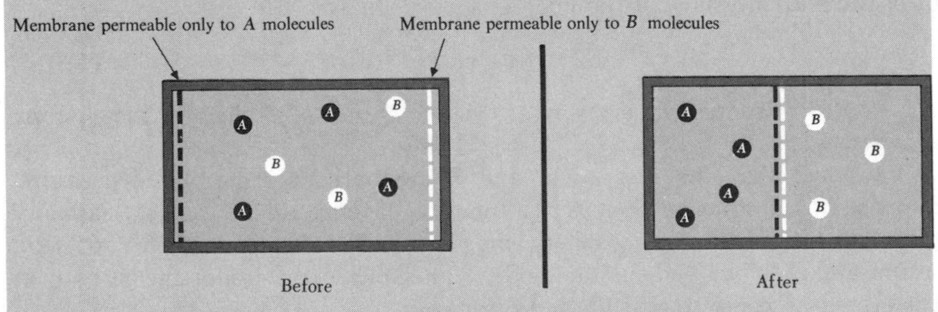

Fig. 4·6·1 Separation of two gases A and B by means of semipermeable membranes.

4 · 7 *Extensive and intensive parameters*

The macroscopic parameters specifying the macrostate of a homogeneous system can be classified into two types. Let y denote such a parameter. Consider that the system is divided into two parts, say by introducing a partition, and denote by y_1 and y_2 the values of this parameter for the two subsystems.

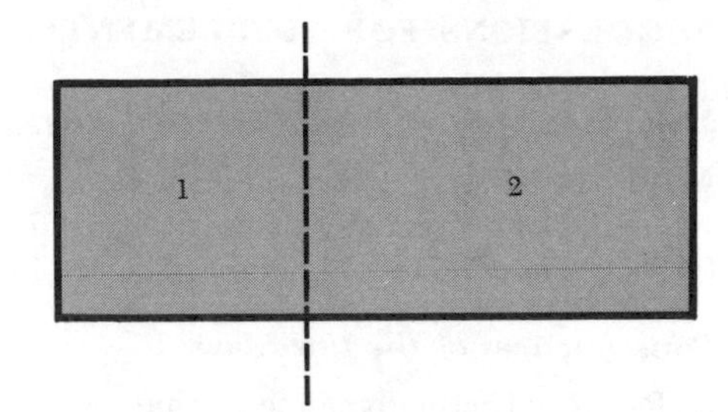

Fig. 4·7·1 Division of a homogeneous system into two parts by a partition.

Then two cases can arise:

1. One has $y_1 + y_2 = y$, in which case the parameter y is said to be *extensive.*
2. One has $y_1 = y_2 = y$, in which case the parameter y is said to be *intensive.*

In simple terms one can say that an extensive parameter gets doubled if the size of the system is doubled, while an intensive parameter remains unchanged.

Thus the volume V of a system is an extensive parameter, as is the total mass M of a system. On the other hand, the density ρ of a system, $\rho = M/V$, is an intensive parameter. Indeed, it is clear that the ratio of any two extensive parameters is an intensive parameter.

The mean pressure of a system is an intensive parameter, since both parts of a system, after subdivision, will have the same pressure as before. Similarly, the temperature T of a system is an intensive parameter.

The internal energy $\bar{E}$ of a system is an extensive quantity. Indeed, no work is required to subdivide the system into two parts (*if* one neglects the work involved in creating the two new surfaces; this is negligible for large systems for which the ratio of the number of molecules near the boundary to the number of molecules in the bulk of the system is very small). Thus the total energy of the system is the same after subdivision as it was before, i.e., $\bar{E}_1 + \bar{E}_2 = \bar{E}$.

The heat capacity C, being the ratio of an energy increase divided by a fixed small temperature increment, is similarly an extensive quantity. The specific heat per mole (or per gram) is, by its definition C/ν (where ν is the number of moles in the system), obviously an intensive quantity.

The entropy S is also an extensive quantity. This follows from the relation $\Delta S = \int đQ/T$, since the heat absorbed $đQ = C\,dT$ is an extensive quantity. It also follows from the statistical definition, e.g., from (3·7·20).

When dealing with extensive quantities such as the entropy S, it is often convenient to talk in terms of the quantity per mole S/ν which is an intensive parameter independent of the size of the system. It is sometimes convenient to denote the quantity per mole by a small letter, e.g., to denote the entropy per mole by s. Thus $S = \nu s$.

SUGGESTIONS FOR SUPPLEMENTARY READING

Macroscopic discussion of internal energy, heat, and temperature

M. W. Zemansky: "Heat and Thermodynamics," 4th ed., chaps. 1 and 4, McGraw-Hill Book Company, New York, 1957.

H. B. Callen: "Thermodynamics," chap. 1, John Wiley & Sons, Inc., New York, 1960.

Consequences of the third law

E. Fermi: "Thermodynamics," chap. 8, Dover Publications, New York, 1956.

J. Wilks: "The Third Law of Thermodynamics," Oxford University Press, Oxford, 1961. (A more advanced book.)

PROBLEMS

4.1 (*a*) One kilogram of water at 0°C is brought into contact with a large heat reservoir at 100°C. When the water has reached 100°C, what has been the change in entropy of the water? of the heat reservoir? of the entire system consisting of both water and heat reservoir?

(*b*) If the water had been heated from 0°C to 100°C by first bringing it in contact with a reservoir at 50°C and then with a reservoir at 100°C, what would have been the change in entropy of the entire system?

(*c*) Show how the water might be heated from 0°C to 100°C with no change in the entropy of the entire system.

4.2 A 750-g copper calorimeter can containing 200 g of water is in equilibrium at a temperature of 20°C. An experimenter now places 30 g of ice at 0°C in the calorimeter and encloses the latter with a heat-insulating shield.

(*a*) When all the ice has melted and equilibrium has been reached, what will be the temperature of the water? (The specific heat of copper is 0.418 joules g^{-1} deg^{-1}. Ice has a specific gravity of 0.917 and its heat of fusion is 333 joules g^{-1}; i.e., it requires 333 joules of heat to convert 1 g of ice to water at 0°C.)

(*b*) Compute the total entropy change resulting from the process of part (*a*).

(*c*) After all the ice has melted and equilibrium has been reached, how much work, in joules, must be supplied to the system (e.g., by means of a stirring rod) to restore all the water to 20°C?

4.3 The heat absorbed by a mole of ideal gas in a quasi-static process in which its temperature T changes by dT and its volume V by dV is given by

$$đQ = c\,dT + \bar{p}\,dV$$

where c is its constant molar specific heat at constant volume and $\bar{p}$ is its mean pressure, $\bar{p} = RT/V$. Find an expression for the change of entropy of this gas in a quasi-static process which takes it from initial values of temperature T_i and volume V_i to the final values T_f and V_f. Does your answer depend on the process involved in going from the initial to the final state?

4.4 A solid contains N magnetic atoms having spin $\frac{1}{2}$. At sufficiently high temperatures, each spin is completely randomly oriented, i.e., equally likely to be in either of its two possible states. But at sufficiently low temperatures the interactions between the magnetic atoms causes them to exhibit ferromagnetism, with the result that all their spins become oriented along the same direction as $T \to 0$. A very crude approximation suggests that the spin-dependent contribution $C(T)$ to the heat capacity of this solid has an approximate temperature dependence given by

$$\begin{aligned} C(T) &= C_1\left(2\frac{T}{T_1} - 1\right) \qquad &\text{if } \tfrac{1}{2}T_1 < T < T_1 \\ &= 0 &\text{otherwise} \end{aligned}$$

The abrupt increase in specific heat as T is reduced below T_1 is due to the onset of ferromagnetic behavior.

Use entropy considerations to find an explicit expression for the maximum value C_1 of the heat capacity.

4.5 A solid contains N magnetic iron atoms having spin S. At sufficiently high temperatures, each spin is completely randomly oriented, i.e., equally likely to be in any of its $2S + 1$ possible states. But at sufficiently low temperatures the interactions between the magnetic atoms causes them to exhibit ferromagnetism, with the result that all their spins become oriented along the same direction as $T \to 0$. The magnetic atoms contribute to the solid a heat capacity $C(T)$ which has, very crudely, an approximate temperature dependence given by

$$\begin{aligned} C(T) &= C_1\left(2\frac{T}{T_1} - 1\right) \qquad &\text{if } \tfrac{1}{2}T_1 < T < T_1 \\ &= 0 &\text{otherwise} \end{aligned}$$

The abrupt increase in specific heat as T is reduced below T_1 is due to the onset of ferromagnetic behavior.

If one dilutes the magnetic atoms by replacing 30 percent of the iron atoms by nonmagnetic zinc atoms, then the remaining 70 percent of the iron atoms still become ferromagnetic at sufficiently low temperatures. The magnetic atoms now contribute to the solid a heat capacity $C(T)$ with the different temperature dependence given, very crudely, by

$$\begin{aligned} C(T) &= C_2\frac{T}{T_2} \qquad &\text{if } 0 < T < T_2 \\ &= 0 &\text{otherwise} \end{aligned}$$

Since the interactions between magnetic ions have been reduced, the ferromagnetic behavior now sets in at a temperature T_2 lower than the previous temperature T_1, and the heat capacity falls off more slowly below the temperature T_2.

Use entropy considerations to compare the magnitude of the specific heat maximum C_2 in the dilute case with the magnitude of the specific heat maximum C_1 in the undiluted case. Find an explicit expression for C_2/C_1.

Simple applications of macroscopic thermodynamics 5

IN THIS chapter we shall explore the purely macroscopic consequences of our theory in order to derive various important relationships between macroscopic quantities. The whole chapter will be based solely on the general statements, called "thermodynamic laws," which were derived in Chapter 3 and summarized in Sec. 3·11. Despite their apparent innocuousness, these statements allow one to draw an impressive number of remarkable conclusions which are completely *independent* of any specific models assumed to describe the microscopic constituents of a system.

Since the discussion of this chapter will be completely macroscopic, quantities such as energy E and pressure p will always refer to their respective mean values. For simplicity, we shall therefore omit the averaging bar symbols above the letters designating these quantities.

Most of the systems considered in this chapter will be characterized by a single external parameter, the volume V. The macrostate of such a system can then be specified completely by two macroscopic variables: its external parameter V and its internal energy E.* The other macroscopic parameters, like temperature T or pressure p, are then determined. But the quantities V and E do not always represent the most convenient choice of independent variables. Any two other macroscopic parameters, e.g., E and p, or T and V, might equally well be chosen as independent variables. In either case, E and V would then be determined.

Most of the mathematical manipulations encountered in making thermodynamic calculations involve changing variables and taking partial derivatives. To avoid ambiguity, it is customary to indicate explicitly by subscripts which of the independent variables are kept constant in evaluating a given partial derivative. For example, if T and V are chosen as independent variables, $(\partial E/\partial T)_V$ denotes a partial derivative where the other independent variable V is kept constant. On the other hand, if T and p are chosen as independent

* These are, of course, the same variables that specify the number of states $\Omega(E,V)$ accessible to the system.

variables, $(\partial E/\partial T)_p$ denotes a partial derivative where the other independent variable p is kept constant. These two partial derivatives are, in general, not equal. If one simply wrote the partial derivative $(\partial E/\partial T)$ without a subscript, it would not be clear which is the other independent variable kept constant in the differentiation.*

The first law (3·11·2) applied to any infinitesimal process yields the relation

$$đQ = dE + đW$$

where dE is the change of internal energy of the system under consideration. If the process is quasi-static, the second law (3·11·2) allows one to express the heat $đQ$ absorbed by the system in this process in terms of the change of entropy of the system, i.e., $đQ = T\,dS$; furthermore, the work done by the system when its volume is changed by an amount dV in the process is simply given by $đW = p\,dV$. Hence one obtains the fundamental thermodynamic relation

► $$T\,dS = dE + p\,dV$$

Most of this chapter will be based on this one equation. Indeed, it is usually simplest to make this fundamental relation the starting point for discussing any problem.

PROPERTIES OF IDEAL GASES

5·1 *Equation of state and internal energy*

Macroscopically, an ideal gas is described by the equation of state relating its pressure p, volume V, and absolute temperature T. For ν moles of gas, this equation of state is given by

$$pV = \nu RT \tag{5·1·1}$$

We derived this relation in (3·12·10) by *micro*scopic arguments of statistical mechanics applied to an ideal gas in the classical limit. But from the point of view of the present macroscopic discussion, Eq. (5·1·1) merely characterizes the kind of system we are talking about; thus (5·1·1) might equally well be considered as a purely phenomenological relation summarizing experimental measurements on the system.

An ideal gas has a second important property already proved in (3·12·11) on the basis of microscopic statistical mechanics: its internal energy does not depend on its volume, but only on its temperature. Thus

$$E = E(T) \qquad \text{independent of } V \tag{5·1·2}$$

This property is actually a direct consequence of the equation of state (5·1·1). Thus, even if (5·1·1) were to be considered a purely empirical equation of state describing a particular gas, the thermodynamic laws would allow

* Further discussion of partial derivatives can be found in Appendix A · 9.

us to conclude immediately that this gas must satisfy the property (5·1·2). Let us show this explicitly.

Quite generally, the internal energy E of ν moles of any gas can be considered a function of T and V,

$$E = E(T,V) \tag{5·1·3}$$

Thus we can write the purely mathematical statement

$$dE = \left(\frac{\partial E}{\partial T}\right)_V dT + \left(\frac{\partial E}{\partial V}\right)_T dV \tag{5·1·4}$$

But the fundamental thermodynamic relation for a quasi-static continuous change of parameters can be written as

▶ $$T\,dS = đQ = dE + p\,dV \tag{5·1·5}$$

Using (5·1·1) to express p in terms of V and T, (5·1·5) becomes

$$dS = \frac{1}{T}\,dE + \frac{\nu R}{V}\,dV$$

or by (5·1·4)

$$dS = \frac{1}{T}\left(\frac{\partial E}{\partial T}\right)_V dT + \left[\frac{1}{T}\left(\frac{\partial E}{\partial V}\right)_T + \frac{\nu R}{V}\right] dV \tag{5·1·6}$$

The mere fact that dS on the left side of (5·1·6) is the exact differential of a well-defined function allows us to draw an important conclusion. We can consider S dependent on T and V. Thus $S = S(T,V)$, and we can write the mathematical statement

$$dS = \left(\frac{\partial S}{\partial T}\right)_V dT + \left(\frac{\partial S}{\partial V}\right)_T dV \tag{5·1·7}$$

Since this expression must be true for all values of dT and dV, comparison with (5·1·4) shows immediately that

$$\begin{aligned}\left(\frac{\partial S}{\partial T}\right)_V &= \frac{1}{T}\left(\frac{\partial E}{\partial T}\right)_V \\ \left(\frac{\partial S}{\partial V}\right)_T &= \frac{1}{T}\left(\frac{\partial E}{\partial V}\right)_T + \frac{\nu R}{V}\end{aligned} \tag{5·1·8}$$

But the equality of the second derivatives, irrespective of order of differentiation,

$$\frac{\partial^2 S}{\partial V\,\partial T} = \frac{\partial^2 S}{\partial T\,\partial V} \tag{5·1·9}$$

implies a definite connection between the expressions on the right side of (5·1·8). Thus

$$\left(\frac{\partial}{\partial V}\right)_T\left(\frac{\partial S}{\partial T}\right)_V = \left(\frac{\partial}{\partial T}\right)_V\left(\frac{\partial S}{\partial V}\right)_T$$

by (5·1·9), and

$$\frac{1}{T}\left(\frac{\partial^2 E}{\partial V\,\partial T}\right) = \left[-\frac{1}{T^2}\left(\frac{\partial E}{\partial V}\right)_T + \frac{1}{T}\left(\frac{\partial^2 E}{\partial T\,\partial V}\right)\right] + 0$$

using (5·1·8). Since the second derivatives of E are again (analogously to (5·1·9)) equal irrespective of the order of differentiation, this last relation shows immediately that

$$\left(\frac{\partial E}{\partial V}\right)_T = 0 \tag{5·1·10}$$

This establishes that E is independent of V and completes the proof that (5·1·2) follows from (5·1·1).

Historical remark on the "free expansion" experiment The fact that the internal energy E of a gas does not depend on its volume (if the gas is sufficiently dilute that it can be considered ideal) was verified in a classical experiment by Joule. He made use of the "free expansion" of an ideal gas as illustrated in Fig. 5·1·1.

A container consisting of two compartments separated by a valve is immersed in water. Initially, the valve is closed and one compartment is filled with the gas under investigation, while the other compartment is evacuated. Suppose that the valve is now opened so that the gas is free to expand and fill both compartments. In this process no work gets done by the system consisting of the gas and container. (The container walls are rigid and nothing moves.) Hence one can say, by the first law, that the heat Q absorbed by this system equals its increase in internal energy,

$$Q = \Delta E \tag{5·1·11}$$

Assume that the internal energy change of the (thin-walled) container is negligibly small. Then ΔE measures simply the energy change of the gas.

Joule found that the temperature of the water did not change in this experiment. (Because of the large heat capacity of the water, any anticipated temperature change is, however, quite small; Joule's actual sensitivity of temperature measurement was, in retrospect, rather inadequate.) Thus the water absorbed no heat from the gas; consequently, the heat Q absorbed by the gas

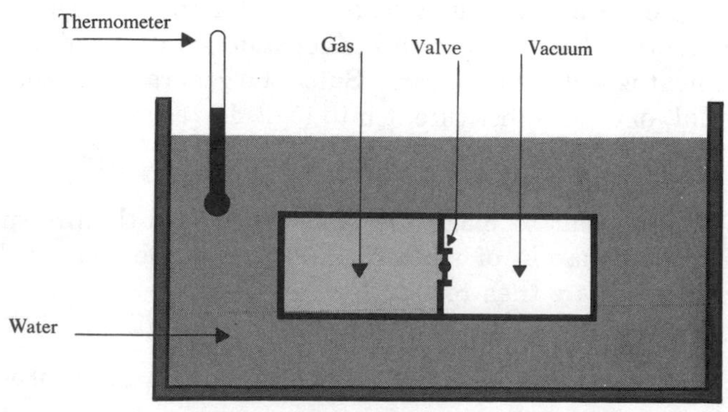

Fig. 5·1·1 Experimental setup for studying the free expansion of a gas.

also vanished. All that happens in the experiment is that the temperature of the gas remains unchanged while its volume changes from its initial value V_i to its final value V_f. Since $Q = 0$, Joule's experiment leads by virtue of (5·1·11) to the conclusion

$$E(T,V_f) - E(T,V_i) = 0$$

which verifies that $E(T,V)$ is independent of the volume V.

5 · 2 *Specific heats*

The heat absorbed in an infinitesimal process is given by the first law as

$$đQ = dE + p\,dV \tag{5·2·1}$$

Let us first obtain an expression for the molar specific heat c_V at constant volume. Then $dV = 0$ and (5·2·1) reduces simply to

$$đQ = dE$$

Hence one obtains

$$c_V \equiv \frac{1}{\nu}\left(\frac{đQ}{dT}\right)_V = \frac{1}{\nu}\left(\frac{\partial E}{\partial T}\right)_V \tag{5·2·2}$$

The specific heat c_V may itself, of course, be a function of T. But by virtue of (5·1·2), it is independent of V for an ideal gas.

Since E is independent of V, so that E is only a function of T, the general relation (5·1·4) reduces to

$$dE = \left(\frac{\partial E}{\partial T}\right)_V dT \tag{5·2·3}$$

i.e., the change of energy depends *only* on the temperature change of the gas, even if the volume of the gas also changes. Using (5·2·2), one can then write quite generally

$$dE = \nu c_V\,dT \tag{5·2·4}$$

for an *ideal* gas.

Let us now obtain an expression for the molar specific heat c_p at constant pressure. Here the pressure is constant, but the volume V changes, in general, as heat is added to the gas. Since the general expression (5·2·4) for dE is still valid, one can substitute it into (5·2·1) to get

$$đQ = \nu c_V\,dT + p\,dV \tag{5·2·5}$$

We now want to make use of the fact that the pressure p is kept constant· By the equation of state (5·1·1), a volume change dV and a temperature change dT are then related by

$$p\,dV = \nu R\,dT \tag{5·2·6}$$

Substituting this into (5·2·5) yields for the heat absorbed at constant pressure

$$đQ = \nu c_V\,dT + \nu R\,dT \tag{5·2·7}$$

But, by definition,

$$c_p = \frac{1}{\nu}\left(\frac{đQ}{dT}\right)_p$$

Using (5·2·7), this becomes

$$c_p = c_V + R \qquad (5\cdot2\cdot8)$$

Thus $c_p > c_V$, in general agreement with (4·4·7), and these molar specific heats of an ideal gas differ precisely by the gas constant R.

The ratio γ of the specific heats is then given by

$$\gamma \equiv \frac{c_p}{c_V} = 1 + \frac{R}{c_V} \qquad (5\cdot2\cdot9)$$

The quantity γ can be determined from the velocity of sound in the gas and can also be measured directly by other methods. Table 5·2·1 lists some representative experimental values of c_V for a few gases. It also illustrates the extent of agreement between the values of γ computed by (5·2·9) and the experimentally measured values of this quantity.

Table 5·2·1 Specific heats of some gases (at 15°C and 1 atm)*

Gas	*Symbol*	c_V (*experimental*) (*joules mole*$^{-1}$ deg^{-1})	γ (*experimental*)	γ (*computed by* (5·2·9))
Helium	He	12.5	1.666	1.666
Argon	Ar	12.5	1.666	1.666
Nitrogen	N_2	20.6	1.405	1.407
Oxygen	O_2	21.1	1.396	1.397
Carbon dioxide	CO_2	28.2	1.302	1.298
Ethane	C_2H_6	39.3	1.220	1.214

* Experimental values taken from J. R. Partington and W. G. Shilling, "The Specific Heats of Gases," p. 201, Benn, London, 1924.

Microscopic calculation of specific heats If one is willing to make use of *micro*scopic information, one can, of course, make many more interesting statements. The situation of a *monatomic* ideal gas is particularly simple. In (2·5·19) we found for the number of states of such a gas in some small energy range δE the expression

$$\Omega(E,V) = BV^N E^{3N/2}$$

where N is the number of molecules in the gas and B is some constant independent of E and V. Hence

$$\ln \Omega = \ln B + N \ln V + \frac{3N}{2} \ln E$$

The temperature parameter $\beta = (kT)^{-1}$ is then given by

$$\beta = \frac{\partial \ln \Omega}{\partial E} = \frac{3N}{2}\frac{1}{E}$$

Thus

$$E = \frac{3N}{2\beta} = \frac{3N}{2} kT \qquad (5\cdot 2\cdot 10)$$

This expresses directly the relation between the internal energy and the absolute temperature of the gas. If N_a denotes Avogadro's number, $N = \nu N_a$ and (5 · 2 · 10) can also be written

$$E = \tfrac{3}{2}\nu(N_a k)T = \tfrac{3}{2}\nu RT \qquad (5\cdot 2\cdot 11)$$

where $R = N_a k$ is the gas constant.

The molar specific heat at constant volume of a monatomic ideal gas is then, by (5 · 2 · 2) and (5 · 2 · 11),

$$c_V = \frac{1}{\nu}\left(\frac{\partial E}{\partial T}\right)_V = \frac{3}{2} R \qquad (5\cdot 2\cdot 12)$$

By (4 · 3 · 8) this has the numerical value

$$c_V = 12.47 \text{ joules deg}^{-1} \text{ mole}^{-1} \qquad (5\cdot 2\cdot 13)$$

Furthermore, (5 · 2 · 8) then gives

$$c_p = \tfrac{3}{2}R + R = \tfrac{5}{2}R \qquad (5\cdot 2\cdot 14)$$

and

$$\gamma \equiv \frac{c_p}{c_v} = \tfrac{5}{3} = 1.667 \qquad (5\cdot 2\cdot 15)$$

These simple microscopic arguments lead thus to very definite quantitative predictions. The experimental values given in Table 5 · 2 · 1 for the monatomic gases helium and argon show very satisfactory agreement with the theoretical values (5 · 2 · 13) and (5 · 2 · 15).

5 · 3 *Adiabatic expansion or compression*

Suppose that the temperature of a gas is maintained constant by being kept in thermal contact with a heat reservoir. If the gas is allowed to expand quasi-statically under such "isothermal" (i.e., "same temperature") conditions, the pressure p and volume V satisfy, by virtue of the equation of state (5 · 1 · 1), the relation

$$pV = \text{constant} \qquad (5\cdot 3\cdot 1)$$

Suppose, however, that the gas is thermally insulated from its surroundings (i.e., that it is maintained under adiabatic conditions). If the gas is allowed to expand under these conditions, it will do work at the expense of its internal energy; as a result its temperature will also change. In a quasi-static adiabatic process of this kind, how is the pressure p of the gas related to its volume V?

Our starting point is again the first law (5·2·1). Since no heat is absorbed in the adiabatic process here contemplated, $đQ = 0$. Using Eq. (5·2·4) for an ideal gas, (5·2·1) becomes

$$0 = \nu c_V \, dT + p \, dV \tag{5·3·2}$$

This relation involves the three variables p, V, and T. By the equation of state (5·1·1), one can express one of these in terms of the other two. Thus (5·1·1) yields

$$p \, dV + V \, dp = \nu R \, dT \tag{5·3·3}$$

Let us solve this for dT and substitute the result into (5·3·2). This gives a relation between dp and dV.

$$0 = \frac{c_V}{R} (p \, dV + V \, dp) + p \, dV = \left(\frac{c_V}{R} + 1 \right) p \, dV + \frac{c_V}{R} V \, dp$$

or

$$(c_V + R) p \, dV + c_V V \, dp = 0$$

Dividing both sides of this equation by the quantity $c_V pV$ yields the equation

$$\gamma \frac{dV}{V} + \frac{dp}{p} = 0 \tag{5·3·4}$$

where, by (5·2·9),

$$\gamma \equiv \frac{c_V + R}{c_V} = \frac{c_p}{c_V} \tag{5·3·5}$$

Now c_V is temperature independent for most gases. In other cases it may be a *slowly* varying function of T. Thus it is always an excellent approximation to assume that the specific heat ratio γ is independent of T in a limited temperature range. Then (5·3·4) can be immediately integrated to give

$$\gamma \ln V + \ln p = \text{constant}$$

or

▶ $$pV^\gamma = \text{constant} \tag{5·3·6}$$

Since $\gamma > 1$ by virtue of (5·3·5), p will vary more rapidly with V than in the isothermal case (5·3·1) where $pV = \text{constant}$.

From (5·3·6) one can, of course, also obtain corresponding relations between V and T, or between p and T. For example, since $p = \nu RT/V$, (5·3·6) implies that

$$V^{\gamma-1} T = \text{constant} \tag{5·3·7}$$

5 · 4 *Entropy*

The entropy of an ideal gas can readily be computed from the fundamental thermodynamic relation (5 · 1 · 5) by the procedure of Sec. 4 · 5. By virtue of (5 · 2 · 4) and the equation of state (5 · 1 · 1), the relation (5 · 1 · 5) becomes

$$T\,dS = \nu c_V(T)\,dT + \frac{\nu RT}{V}\,dV$$

or

$$dS = \nu c_V(T)\,\frac{dT}{T} + \nu R\,\frac{dV}{V} \tag{5·4·1}$$

This allows one to find, by integration, the entropy of ν moles of this gas at any arbitrary temperature T and volume V compared to the entropy of the gas in some standard macrostate.

Let us choose as the standard macrostate of this kind of gas one where ν_0 moles of the gas occupy a volume V_0 at the temperature T_0. We denote the *molar* entropy of the gas in this standard state by s_0. To calculate the entropy $S(T,V;\,\nu)$ of ν moles of this gas at temperature T and volume V, we need merely to consider *any* quasi-static process whereby we bring these ν moles of gas from the standard state to the final state of interest. Let us then first divide off by a partition ν moles of gas in the standard state; these will have entropy νs_0 and occupy a volume $V_0(\nu/\nu_0)$. Take these ν moles of gas and slowly increase the temperature to the value T while keeping the volume constant at $V_0(\nu/\nu_0)$. Then change the volume slowly to the value V while keeping the temperature constant at T. For the process just described, integration of (5 · 4 · 1) gives the result

$$S(T,V;\,\nu) - \nu s_0 = \nu \int_{T_0}^{T} \frac{c_V(T')\,dT'}{T'} + \nu R \int_{V_0(\nu/\nu_0)}^{V} \frac{dV'}{V'} \tag{5·4·2}$$

The last integration is immediate:

$$\int_{V_0(\nu/\nu_0)}^{V} \frac{dV'}{V'} = [\ln V']_{V_0(\nu/\nu_0)}^{V} = \ln V - \ln\left(V_0\,\frac{\nu}{\nu_0}\right) = \ln\frac{V}{\nu} - \ln\frac{V_0}{\nu_0}$$

Hence (5 · 4 · 2) becomes

$$S(T,V;\,\nu) = \nu\left[\int_{T_0}^{T} \frac{c_V(T')}{T'}\,dT' + R\ln\frac{V}{\nu} - R\ln\frac{V_0}{\nu_0} + s_0\right] \tag{5·4·3}$$

or

$$S(T,V;\,\nu) = \nu\left[\int \frac{c_V(T')}{T'}\,dT' + R\ln V - R\ln\nu + \text{constant}\right] \tag{5·4·4}$$

In this last expression we have lumped all the quantities referring to the standard state into a single constant. The expressions (5 · 4 · 3) or (5 · 4 · 4) give the dependence of the entropy S on T, V, and ν. In the special case when c_V is temperature independent, the integral over temperature becomes, of course, trivial; i.e.,

$$\int \frac{c_V}{T'}\,dT' = c_V \ln T$$

if c_V is constant.

GENERAL RELATIONS FOR A HOMOGENEOUS SUBSTANCE

5 · 5 *Derivation of general relations*

We consider a homogeneous system whose volume V is the only external parameter of relevance. The starting point of our whole discussion is again the fundamental thermodynamic relation for a quasi-static infinitesimal process

$$đQ = T\,dS = dE + p\,dV \tag{5·5·1}$$

This equation gives rise to a wealth of other relations which we want to exhibit presently.

Independent variables S and V Equation (5 · 5 · 1) can be written

$$dE = T\,dS - p\,dV \tag{5·5·2}$$

This shows how E depends on independent variations of the parameters S and V. If these are considered the two independent parameters specifying the system, then

$$E = E(S,V)$$

and one can write the corresponding purely mathematical statement

$$dE = \left(\frac{\partial E}{\partial S}\right)_V dS + \left(\frac{\partial E}{\partial V}\right)_S dV \tag{5·5·3}$$

Since (5·5·2) and (5·5·3) must be equal for all possible values of dS and dV, it follows that the corresponding coefficients of dS and dV must be the same. Hence

$$\left(\frac{\partial E}{\partial S}\right)_V = T \qquad \left(\frac{\partial E}{\partial V}\right)_S = -p \tag{5·5·4}$$

The important content of the relation (5·5·2) is that the combination of parameters on the right side is always equal to the exact differential of a quantity, which in this case is the energy E. Hence the parameters T, S, p, and V which occur on the right side of (5·5·2) cannot be varied completely arbitrarily; there must exist some connection between them to guarantee that their combination yields the differential dE. To obtain this connection, it is only necessary to note that the second derivatives of E must be independent of the order of differentiation, i.e.,

$$\frac{\partial^2 E}{\partial V\,\partial S} = \frac{\partial^2 E}{\partial S\,\partial V}$$

or

$$\left(\frac{\partial}{\partial V}\right)_S \left(\frac{\partial E}{\partial S}\right)_V = \left(\frac{\partial}{\partial S}\right)_V \left(\frac{\partial E}{\partial V}\right)_S$$

Hence one obtains by (5·5·4) the result

$$\left(\frac{\partial T}{\partial V}\right)_S = -\left(\frac{\partial p}{\partial S}\right)_V \tag{5·5·5}$$

This useful relation reflects merely the fact that dE is the exact differential of a well-defined quantity E characteristic of the macrostate of the system.

Independent variables S and p Equation (5·5·2) exhibits the effect of independent variations of S and V. One might equally well exhibit the effect of independent variations of S and p. One can easily pass from the expression $p\,dV$, where the variation dV appears, to an equivalent expression where dp appears by the simple transformation

$$p\,dV = d(pV) - V\,dp$$

Let us then return to (5·5·2) and transform this into an expression involving dp rather than dV. We get

$$dE = T\,dS - p\,dV = T\,dS - d(pV) + V\,dp$$

or

$$d(E + pV) = T\,dS + V\,dp$$

We can write this as

$$dH = T\,dS + V\,dp \tag{5·5·6}$$

where we have introduced the definition

$$H \equiv E + pV \tag{5·5·7}$$

The function H is called the "enthalpy."

Considering S and p as independent variables, one can write

$$H = H(S,p)$$

and

$$dH = \left(\frac{\partial H}{\partial S}\right)_p dS + \left(\frac{\partial H}{\partial p}\right)_S dp \tag{5·5·8}$$

Comparison between (5·5·6) and (5·5·8) yields the relations

$$\left(\frac{\partial H}{\partial S}\right)_p = T \qquad \left(\frac{\partial H}{\partial p}\right)_S = V \tag{5·5·9}$$

The important aspect of (5·5·6) is again the fact that the combination of parameters on the right side is equal to the exact differential of a quantity which we happen to have designated by the letter H. The equality of the cross derivatives of this quantity, i.e., the equality

$$\frac{\partial^2 H}{\partial p\,\partial S} = \frac{\partial^2 H}{\partial S\,\partial p}$$

then implies immediately the relation

$$\left(\frac{\partial T}{\partial p}\right)_S = \left(\frac{\partial V}{\partial S}\right)_p \tag{5·5·10}$$

This equation is analogous to (5·5·5) and represents again a necessary connection between the parameters T, S, p, V. By this time it should be clear what kind of game one plays to get thermodynamic relations of this sort. All we need do is play it to the bitter end by performing all other possible changes of variable in the fundamental equation (5·5·2).

Independent variables T and V We transform (5·5·2) into an expression involving dT rather than dS. Thus we can write

$$dE = T\,dS - p\,dV = d(TS) - S\,dT - p\,dV$$

or

$$dF = -S\,dT - p\,dV \tag{5·5·11}$$

where we have introduced the definition

$$F \equiv E - TS \tag{5·5·12}$$

The function F is called the "Helmholtz free energy."

Considering T and V as independent variables,

$$F = F(T,V)$$

and

$$dF = \left(\frac{\partial F}{\partial T}\right)_V dT + \left(\frac{\partial F}{\partial V}\right)_T dV \tag{5·5·13}$$

Comparison of (5·5·11) with (5·5·13) yields

$$\left(\frac{\partial F}{\partial T}\right)_V = -S$$
$$\left(\frac{\partial F}{\partial V}\right)_T = -p \tag{5·5·14}$$

Equality of the cross derivatives

$$\frac{\partial^2 F}{\partial V\,\partial T} = \frac{\partial^2 F}{\partial T\,\partial V}$$

then implies

$$\left(\frac{\partial S}{\partial V}\right)_T = \left(\frac{\partial p}{\partial T}\right)_V \tag{5·5·15}$$

Independent variables T and p We finally transform (5·5·2) into an expression involving dT and dp rather than dS and dV. Thus we can write

$$dE = T\,dS - p\,dV = d(TS) - S\,dT - d(pV) + V\,dp$$

or

$$dG = -S\,dT + V\,dp \tag{5·5·16}$$

where we have introduced the definition

$$G \equiv E - TS + pV \tag{5·5·17}$$

The function G is called the "Gibbs free energy." In terms of the previous definitions (5·5·7) or (5·5·12), we could also write $G = H - TS$, or $G = F + pV$.

Considering T and p as independent variables,

$$G = G(T,p)$$

and

$$dG = \left(\frac{\partial G}{\partial T}\right)_p dT + \left(\frac{\partial G}{\partial p}\right)_T dp \tag{5·5·18}$$

Comparison of (5·5·16) with (5·5·18) yields

$$\left(\frac{\partial G}{\partial T}\right)_p = -S \qquad \left(\frac{\partial G}{\partial p}\right)_T = V \tag{5·5·19}$$

Equality of the cross derivatives

$$\frac{\partial^2 G}{\partial p\,\partial T} = \frac{\partial^2 G}{\partial T\,\partial p}$$

then implies

$$-\left(\frac{\partial S}{\partial p}\right)_T = \left(\frac{\partial V}{\partial T}\right)_p \tag{5·5·20}$$

5 · 6 *Summary of Maxwell relations and thermodynamic functions*

Maxwell relations The entire discussion of the preceding section was based upon the fundamental thermodynamic relation

▶
$$dE = T\,dS - p\,dV \tag{5·6·1}$$

From this statement we derived the important relations (5·5·5), (5·5·10), (5·5·15), and (5·5·20), which are repeated below:

$$\left(\frac{\partial T}{\partial V}\right)_S = -\left(\frac{\partial p}{\partial S}\right)_V \tag{5·6·2}$$

$$\left(\frac{\partial T}{\partial p}\right)_S = \left(\frac{\partial V}{\partial S}\right)_p \tag{5·6·3}$$

$$\left(\frac{\partial S}{\partial V}\right)_T = \left(\frac{\partial p}{\partial T}\right)_V \tag{5·6·4}$$

$$\left(\frac{\partial S}{\partial p}\right)_T = -\left(\frac{\partial V}{\partial T}\right)_p \tag{5·6·5}$$

These are known as "Maxwell's relations." They are a direct consequence of the fact that the variables T, S, p, and V are not completely independent, but are related through the fundamental thermodynamic relation (5·6·1). All the Maxwell relations are basically equivalent;* any one of them can be derived from any other one by a simple change of independent variables.

It is worth recalling explicitly why there exists this connection between variables which is expressed by the Maxwell relations. The basic reason is as follows: It is possible to give a complete macroscopic description of a system in equilibrium if one knows the number of states Ω accessible to the system (or equivalently, its entropy $S = k \ln \Omega$) as a function of its energy E and its one external parameter V. But both the temperature T and mean pressure p of the system can be expressed in terms of $\ln \Omega$ or S; in Chapter 3 we found the explicit expressions (3·12·1) or, equivalently, (3·12·5)

$$\frac{1}{T} = \left(\frac{\partial S}{\partial E}\right)_V \quad \text{and} \quad p = T\left(\frac{\partial S}{\partial V}\right)_E \tag{5·6·6}$$

It is the fact that both T and p are expressible in terms of the same function S which leads to the connection (5·6·1) and hence to the Maxwell relations.

Indeed, one has

$$\begin{aligned} dS &= \left(\frac{\partial S}{\partial E}\right)_V dE + \left(\frac{\partial S}{\partial V}\right)_E dV \\ &= \frac{1}{T}dE + \frac{p}{T}dV \qquad \text{by (5·6·6)} \end{aligned}$$

and the latter expression is simply the fundamental relation (5·6·1)

Note that the fundamental relation (5·6·1) involves the variables on the right side in pairs, one pair consisting of T and S, the other of p and V. In these two pairs

$$(T,S) \qquad \text{and} \qquad (p,V)$$

the first involves quantities like entropy and temperature which describe the density of accessible states of the system, whereas the second involves an external parameter and its corresponding generalized force. The essential content of the Maxwell relations is the existence of a connection between the *cross* derivatives of these two kinds of quantities. Specifically, each of the Maxwell relations is a statement asserting that [the derivative of a variable of the first pair with respect to a variable of the second pair] is (except for sign) equal to [the corresponding derivative of the *other* variable of the second pair with respect to the *other* variable of the first pair].

The above property characterizing the Maxwell relations makes it very easy to read them off directly from the fundamental relation (5·6·1). The proper sign can be obtained in the following way: If the two variables with respect to which one differentiates are the same variables S and V which occur as differentials in (5·6·1), then the minus sign that occurs in (5·6·1) also

* They can, for example, all be summarized by the single statement that the Jacobian determinant $\partial(T,S)/\partial(p,V) = 1$.

occurs in the Maxwell relation. Any one permutation away from these particular variables introduces a change of sign.

Example A minus sign occurs in (5·6·2) because the variables S and V with respect to which one differentiates are the same as those appearing as differentials in (5·6·1). On the other hand, in (5·6·3) the derivatives are with respect to S and p, whereas S and V appear as differentials in (5·6·1). The switch from p to V implies *one* sign change with respect to the minus sign in (5·6·1); hence there is a plus sign in (5·6·3).

Thermodynamic functions The Maxwell relations constitute the most important result of the last section. It is, however, also worth summarizing for future reference the various thermodynamic functions that were introduced in that section. We list them below, together with the most convenient independent variables used in conjunction with each of these functions (i.e., the variables in terms of which the fundamental relation (5·6·1) is expressed most simply):

$$\begin{array}{ll} E & E = E(S,V) \\ H \equiv E + pV & H = H(S,p) \\ F \equiv E - TS & F = F(T,V) \\ G \equiv E - TS + pV & G = G(T,p) \end{array} \tag{5·6·7}$$

Next we summarize the thermodynamic relations satisfied by each of these functions

$$dE = T\,dS - p\,dV \tag{5·6·8}$$
$$dH = T\,dS + V\,dp \tag{5·6·9}$$
$$dF = -S\,dT - p\,dV \tag{5·6·10}$$
$$dG = -S\,dT + V\,dp \tag{5·6·11}$$

The relations (5·5·4), (5·5·9), (5·5·14), and (5·5·19), involving derivatives of the functions E, H, F, and G, respectively, can immediately be read off from these equations.

The equations (5·6·9) through (5·6·11) are very simply related to the fundamental equation (5·6·8) or (5·6·1). It is only necessary to note that all of them involve the same variable pairs (T,S) and (p,V), the variables entering as differentials being the independent variables listed in (5·6·7); and that any change of variable away from those used in (5·6·8) introduces a change of sign.

5 · 7 *Specific heats*

We consider any homogeneous substance whose volume V is the only relevant external parameter. Let us first investigate the general relation existing between the molar specific heat c_V at constant volume and the molar specific heat c_p at constant pressure. This relation has practical importance, since calculations by statistical mechanics are usually more easily performed for an assumed fixed volume, while experimental measurements are more readily carried out under conditions of constant (say atmospheric) pressure. Thus, to compare the theoretically calculated quantity c_V with the experimentally measured parameter c_p, a knowledge of the relation between these quantities is necessary.

The heat capacity at constant volume is given by

$$C_V = \left(\frac{đQ}{dT}\right)_V = T\left(\frac{\partial S}{\partial T}\right)_V \qquad (5\cdot7\cdot1)$$

The heat capacity at constant pressure is similarly given by

$$C_p = \left(\frac{đQ}{dT}\right)_p = T\left(\frac{\partial S}{\partial T}\right)_p \qquad (5\cdot7\cdot2)$$

We seek a general relation between these two quantities.

Experimentally, the parameters which can be controlled most conveniently are the temperature T and pressure p. Let us consider these as independent variables. Then $S = S(T,p)$ and one obtains the following general expression for the heat $đQ$ absorbed in an infinitesimal quasi-static process

$$đQ = T\,dS = T\left[\left(\frac{\partial S}{\partial T}\right)_p dT + \left(\frac{\partial S}{\partial p}\right)_T dp\right] \qquad (5\cdot7\cdot3)$$

One can use (5·7·2) to write this in the form

$$đQ = T\,dS = C_p\,dT + T\left(\frac{\partial S}{\partial p}\right)_T dp \qquad (5\cdot7\cdot4)$$

If the pressure is maintained constant, $dp = 0$, and (5·7·4) reduces to (5·7·2). But in calculating C_V by (5·7·1), T and V are used as the independent variables. To express $đQ$ in (5·7·4) in terms of dT and dV, it is only necessary to express dp in terms of these differential quantities. This gives

$$đQ = T\,dS = C_p\,dT + T\left(\frac{\partial S}{\partial p}\right)_T \left[\left(\frac{\partial p}{\partial T}\right)_V dT + \left(\frac{\partial p}{\partial V}\right)_T dV\right] \qquad (5\cdot7\cdot5)$$

The heat $đQ$ absorbed under conditions when V is constant is then immediately obtained by putting $dV = 0$. Dividing this heat by dT gives C_V. Thus

$$C_V = T\left(\frac{\partial S}{\partial T}\right)_V = C_p + T\left(\frac{\partial S}{\partial p}\right)_T \left(\frac{\partial p}{\partial T}\right)_V \qquad (5\cdot7\cdot6)$$

This is a relation between C_V and C_p, but it involves on the right side quantities which are not readily measured. For example, what is $(\partial S/\partial p)_T$? It is not readily measured, but since it is the derivative of a variable from the (T,S) pair with respect to a variable from the (p,V) pair, we can use one of the Maxwell relations. By (5·6·5)

$$\left(\frac{\partial S}{\partial p}\right)_T = -\left(\frac{\partial V}{\partial T}\right)_p \qquad (5\cdot7\cdot7)$$

Here the quantity on the right *is* a readily measured and familiar quantity, since it is simply the change of volume with temperature under conditions of constant pressure. Indeed, one defines the intensive quantity

$$\alpha \equiv \frac{1}{V}\left(\frac{\partial V}{\partial T}\right)_p \tag{5·7·8}$$

as the "volume coefficient of expansion" of the substance. Thus

$$\left(\frac{\partial S}{\partial p}\right)_T = -V\alpha \tag{5·7·9}$$

The derivative $(\partial p/\partial T)_V$ is also not very readily determined, since it implies a measurement where the volume V is kept constant.* It is usually more convenient to control the temperature T and pressure p. But we can express V in terms of T and p. Thus

$$dV = \left(\frac{\partial V}{\partial T}\right)_p dT + \left(\frac{\partial V}{\partial p}\right)_T dp$$

and, under conditions of constant volume where $dV = 0$, this gives for the desired ratio dp/dT at constant volume the result

$$\left(\frac{\partial p}{\partial T}\right)_V = -\frac{\left(\frac{\partial V}{\partial T}\right)_p}{\left(\frac{\partial V}{\partial p}\right)_T} \tag{5·7·10}$$

(This is simply the result (A · 9 · 5), which we could have written down without rederiving it.) Here the numerator is again related to α by (5 · 7 · 8). The denominator is another familiar quantity, since it measures the change in volume of the substance with increasing pressure at a constant temperature. (The change of volume will be negative, since the volume decreases with increasing pressure.) One defines the positive intensive quantity

$$\kappa \equiv -\frac{1}{V}\left(\frac{\partial V}{\partial p}\right)_T \tag{5·7·11}$$

as the "isothermal compressibility" of the substance. Hence (5 · 7 · 10) becomes

$$\left(\frac{\partial p}{\partial T}\right)_V = \frac{\alpha}{\kappa} \tag{5·7·12}$$

Substitution of (5 · 7 · 9) and (5 · 7 · 12) into (5 · 7 · 6) yields then

$$C_V = C_p + T(-V\alpha)\left(\frac{\alpha}{\kappa}\right)$$

or

▶ $$C_p - C_V = VT\frac{\alpha^2}{\kappa} \tag{5·7·13}$$

* In the case of solids and liquids a small temperature increase at constant volume tends to produce a very large pressure increase. This imposes rather severe demands on the strength of the vessel containing the substance.

If C_p and C_V in this expression are the heat capacities per mole, then the corresponding volume V is the volume per mole.

Equation (5·7·13) provides the desired relation between C_p and C_V in terms of quantities which can be either readily measured directly or computed from the equation of state. For solids and liquids the right side of (5·7·13) is fairly small, so that C_p and C_V do not differ very much.

Numerical example Consider the case of copper at room temperature (298°K) and atmospheric pressure. The density of the metal is 8.9 g cm^{-3} and its atomic weight is 63.5; hence its molar volume is $V = 63.5/8.9 = 7.1$ cm^3 $mole^{-1}$. The other observed values* are $\alpha = 5 \times 10^{-5}$ deg^{-1}, $\kappa = 4.5 \times 10^{-13}$ cm^2 $dyne^{-1}$, and $c_p = 24.5$ joules deg^{-1} $mole^{-1}$. Then one computes by (5·7·13) that $c_p - c_V = 1.2 \cdot 10^7$ ergs deg^{-1} $mole^{-1}$. Thus $c_V = 23.3$ joules deg^{-1} $mole^{-1}$ and $\gamma \equiv c_p/c_V = 1.05$.

Simple application: ideal gas Let us apply (5·7·13) to the special case of the ideal gas discussed in Sec. 5·2. Then the equation of state (5·1·1) is

$$pV = \nu RT \tag{5·7·14}$$

We calculate first the expansion coefficient α defined by (5·7·8). For constant p

$$p\,dV = \nu R\,dT$$

Hence

$$\left(\frac{\partial V}{\partial T}\right)_p = \frac{\nu R}{p}$$

and

$$\alpha = \frac{1}{V}\left(\frac{\nu R}{p}\right) = \frac{\nu R}{\nu RT} = \frac{1}{T} \tag{5·7·15}$$

We calcaulate next the compressibility κ defined in (5·7·11). For constant T, (5·7·14) yields

$$p\,dV + V\,dp = 0$$

Hence

$$\left(\frac{\partial V}{\partial p}\right)_T = -\frac{V}{p}$$

and

$$\kappa = -\frac{1}{V}\left(-\frac{V}{p}\right) = \frac{1}{p} \tag{5·7·16}$$

Thus (5·7·13) becomes

$$C_p - C_V = VT\,\frac{(1/T)^2}{1/p} = \frac{Vp}{T} = \nu R$$

or, per mole,

$$c_p - c_V = R \tag{5·7·17}$$

which agrees with our previous result (5·2·8).

Limiting properties near absolute zero The third law of thermodynamics (3·11·5) asserts that, as the temperature $T \to 0$, the entropy S of a system

* Data taken from "American Institute of Physics Handbook," 2d ed., McGraw-Hill Book Company, New York, 1963.

approaches smoothly some limiting constant value S_0 independent of all parameters of the system. In symbols*

$$\text{as } T \to 0, \qquad S \to S_0 \tag{5·7·18}$$

In the limit as $T \to 0$, the derivatives $\partial S/\partial T$ appearing in (5·7·1) and (5·7·2) remain thus finite, and one can conclude from these relations that

$$\text{as } T \to 0, \qquad C_V \to 0 \quad \text{and} \quad C_p \to 0 \tag{5·7·19}$$

The argument leading to (5·7·19) becomes particularly clear if one writes (5·7·1) or (5·7·2) in integrated form. For example, if the volume is kept constant,

$$S(T) - S(0) = \int_0^T \frac{C_V(T')}{T'} \, dT'$$

But since the entropy difference on the left must be finite, it must be true that $C_V(T) \to 0$ as $T \to 0$ in order to guarantee proper convergence of the integral on the right. The limiting behavior (5·7·19) of the heat capacities is not surprising. It merely reflects the fact that as $T \to 0$, the system tends to settle down in its ground state. The mean energy of the system then becomes essentially equal to its ground-state energy, and no further reduction of temperature can result in a further change of mean energy to a smaller value.

Since the limiting value approached by the entropy as $T \to 0$ is independent of all parameters of the system, it is also independent of volume or pressure variations in this limit. Hence $(\partial S/\partial p)_T \to 0$, and the Maxwell relation (5·7·7) applied to the definition (5·7·8) allows one to make a statement about the limiting behavior of the coefficient of expansion α; i.e.,

$$\text{as } T \to 0, \qquad \alpha \to 0 \tag{5·7·20}$$

On the other hand, the compressibility κ is a purely mechanical property and a system (e.g., a solid) in its ground state has a well-defined compressibility. Thus κ remains finite as $T \to 0$.

Since the product $T\alpha^2$ on the right side of (5·7·13) approaches zero very rapidly as $T \to 0$, it follows that the difference $C_p - C_V$ becomes increasingly negligible compared to C_V itself as one goes to very low temperatures; i.e.,

$$\text{as } T \to 0, \qquad \frac{C_p - C_V}{C_V} \to 0 \tag{5·7·21}$$

This statement in no way contradicts the relation (5·7·17), according to which $C_p - C_V$ is a constant for an ideal gas. The reason is that when $T \to 0$ and the system approaches its ground state, quantum mechanical effects become very important. Hence the classical equation of state $pV = \nu RT$ is no longer valid, even if the interactions between the particles in a gas are so small that the gas can be treated as ideal.

* As usual, this low-temperature limit may be understood to be sufficiently high that the nuclear spin orientations are still completely random.

5·8 *Entropy and internal energy*

Consider the temperature T and volume V of a substance as the independent variables. Then one can write its entropy as

$$S = S(T,V)$$

so that

$$dS = \left(\frac{\partial S}{\partial T}\right)_V dT + \left(\frac{\partial S}{\partial V}\right)_T dV \tag{5·8·1}$$

But the first of the derivatives is simply related to the heat capacity at constant volume, while the second can be reexpressed in terms of a Maxwell relation. Specifically, by (5·7·1) and (5·6·4) one has

$$\left(\frac{\partial S}{\partial T}\right)_V = \frac{1}{T} C_V \tag{5·8·2}$$

$$\left(\frac{\partial S}{\partial V}\right)_T = \left(\frac{\partial p}{\partial T}\right)_V \tag{5·8·3}$$

Hence (5·8·1) becomes

$$dS = \frac{C_V}{T} dT + \left(\frac{\partial p}{\partial T}\right)_V dV \tag{5·8·4}$$

Note that the right side of (5·8·3) can be evaluated if one knows the equation of state. The quantity C_V is in general a function of both T and V. Its dependence on V can, however, also be calculated from the equation of state. Indeed, from its definition,

$$C_V = T\left(\frac{\partial S}{\partial T}\right)_V$$

Differentiation at a fixed temperature T then yields

$$\begin{aligned}\left(\frac{\partial C_V}{\partial V}\right)_T = \left(\frac{\partial}{\partial V}\right)_T \left[T\left(\frac{\partial S}{\partial T}\right)_V\right] &= T\frac{\partial^2 S}{\partial V\,\partial T}\\ &= T\frac{\partial^2 S}{\partial T\,\partial V} = T\left(\frac{\partial}{\partial T}\right)_V\left(\frac{\partial S}{\partial V}\right)_T\\ &= T\left(\frac{\partial}{\partial T}\right)_V\left(\frac{\partial p}{\partial T}\right)_V \qquad \text{by (5·8·3)}\end{aligned}$$

Thus

▶ $$\left(\frac{\partial C_V}{\partial V}\right)_T = T\left(\frac{\partial^2 p}{\partial T^2}\right)_V \tag{5·8·5}$$

and the right side can be evaluated from a knowledge of the equation of state.

We have already pointed out, most recently in Sec. 5·6, that all thermodynamic properties of a system can be calculated from a knowledge of its entropy. Let us then ask what experimental knowledge is necessary in order to calculate the entropy S and thus also all other thermodynamic functions. It is readily seen that what is required is only a knowledge of

1. The heat capacity as a function of T for some *one* fixed value $V = V_1$ of the volume
2. The equation of state

For example, the equation of state can be used in (5 · 8 · 5) to calculate $(\partial C_V/\partial V)_T$ as a function of T and V. This information is then sufficient to relate the heat capacity $C_V(T,V)$ at any volume V to the known heat capacity $C_V(T,V_1)$ at the volume V_1 and the same temperature T; i.e., one has simply

$$C_V(T,V) = C_V(T,V_1) + \int_{V_1}^{V} \left(\frac{\partial C_V(T,V')}{\partial V}\right)_T dV' \qquad (5 \cdot 8 \cdot 6)$$

Knowing $C_V(T,V)$ and using the knowledge of $(\partial p/\partial T)_V$ as a function of T and V which is provided by the equation of state, one can immediately use (5 · 8 · 4) to find $S(T,V)$ at any temperature T and volume V compared to its value in some standard state of temperature T_0 and volume V_0. One needs only to integrate (5 · 8 · 4) by writing

$$S(T,V) - S(T_0,V_0) = [S(T,V) - S(T_0,V)] + [S(T_0,V) - S(T_0,V_0)] \qquad (5 \cdot 8 \cdot 7)$$

where the first term on the right represents the entropy change at the constant volume V and the second term the entropy change at the constant temperature T_0. Thus one gets

$$S(T,V) - S(T_0,V_0) = \int_{T_0}^{T} \frac{C_V(T',V)}{T'}\, dT' + \int_{V_0}^{V} \left(\frac{\partial p(T_0,V')}{\partial T}\right)_V dV' \qquad (5 \cdot 8 \cdot 8)$$

Remark One could, of course, have equally well integrated (5 · 8 · 4) in opposite order by writing, instead of (5 · 8 · 7),

$$\begin{aligned} S(T,V) - S(T_0,V_0) &= [S(T,V) - S(T,V_0)] + [S(T,V_0) - S(T_0,V_0)] \\ &= \int_{V_0}^{V} \left(\frac{\partial p(T,V')}{\partial T}\right)_V dV' + \int_{T_0}^{T} \frac{C_V(T',V_0)}{T'}\, dT' \end{aligned} \qquad (5 \cdot 8 \cdot 9)$$

This expression involves C_V at the volume V_0 instead of the volume V, and $(\partial p/\partial T)_V$ at the temperature T instead of the temperature T_0. Nevertheless (5 · 8 · 9) must yield the result (5 · 8 · 8). The reason is, of course, the fundamental one that the entropy is a quantity characteristic of a particular macrostate, so that the entropy difference calculated is independent of the process used to go from the macrostate T_0, V_0 to that corresponding to T, V.

Let us now turn to the internal energy E of the substance and consider it as a function of T and V. The fundamental thermodynamic relation asserts that

$$dE = T\, dS - p\, dV$$

By expressing dS in terms of T and V as we already did in (5·8·4), this can be written as

$$dE = C_V\,dT + \left[T\left(\frac{\partial p}{\partial T}\right)_V - p\right]dV \tag{5·8·10}$$

Comparing this with the purely mathematical result

$$dE = \left(\frac{\partial E}{\partial T}\right)_V dT + \left(\frac{\partial E}{\partial V}\right)_T dV$$

we obtain the relations

► $$\left(\frac{\partial E}{\partial T}\right)_V = C_V \tag{5·8·11}$$

► $$\left(\frac{\partial E}{\partial V}\right)_T = T\left(\frac{\partial p}{\partial T}\right)_V - p \tag{5·8·12}$$

Equation (5·8·12) shows that the dependence of the internal energy on the volume can again be calculated from the equation of state. A knowledge of this equation of state and of the heat capacity permits one thus to integrate (5·8·10) to find $E(T,V)$ at any temperature T and volume V compared to the energy $E(T_0,V_0)$ of some standard macrostate.

Example: The van der Waals gas Consider a gas whose equation of state is

$$\left(p + \frac{a}{v^2}\right)(v - b) = RT \tag{5·8·13}$$

where $v \equiv V/\nu$ is the molar volume. This is an empirical equation known as the van der Waals equation. (With suitable approximations it can also be derived from statistical mechanics; see Chapter 10.) It represents the behavior of real gases more accurately than the ideal gas law by introducing two additional positive constants a and b characteristic of the particular gas under consideration. (Indeed, it is approximately valid even at temperatures and molar volumes so low that the gas has become a liquid.)

From a qualitative microscopic point of view, long-range attractive forces between molecules tend to keep them closer together than would be the case for noninteracting molecules. These forces thus have the same effect as a slight compression of the gas; the term a/v^2 represents this additional positive pressure. On the other hand, there are also short-range repulsive forces between the molecules which keep them apart sufficiently to prevent them from occupying the same place at the same time. The term b represents the volume occupied by the molecules themselves and which must thus be subtracted from the volume available to any one molecule in the container.

For $a = b = 0$, or in the limit where the gas becomes very dilute (so that $v \to \infty$), Eq. (5·8·13) reduces to the ideal gas equation

$$pv = RT$$

as it must.

We first calculate by Eq. (5·8·12) the volume dependence of the molar energy ϵ. We need to find $(\partial p/\partial T)_V$. Solving (5·8·13) for p, one gets

$$p = \frac{RT}{v - b} - \frac{a}{v^2} \tag{5·8·14}$$

Hence

$$\left(\frac{\partial p}{\partial T}\right)_v = \frac{R}{v - b} \tag{5·8·15}$$

Thus (5·8·12) yields

$$\left(\frac{\partial \epsilon}{\partial v}\right)_T = T\left(\frac{\partial p}{\partial T}\right)_v - p = \frac{RT}{v - b} - p$$

or, by (5·8·14),

$$\left(\frac{\partial \epsilon}{\partial v}\right)_T = \frac{a}{v^2} \tag{5·8·16}$$

For an ideal gas, $a = 0$ so that $(\partial \epsilon/\partial v)_T = 0$ in agreement with our earlier result (5·1·10).

Also we have by (5·8·5) and (5·8·15)

$$\left(\frac{\partial c_V}{\partial v}\right)_T = T\left(\frac{\partial^2 p}{\partial T^2}\right) = T\left(\frac{\partial}{\partial T}\right)_v\left(\frac{R}{v - b}\right) = 0$$

Hence c_V is *independent* of the molar volume and thus only a function of T, i.e.,

$$c_V = c_V(T) \tag{5·8·17}$$

(The same result is, of course, true *a fortiori* for an ideal gas.) Equation (5·8·10) can then be written

$$d\epsilon = c_V(T)\, dT + \frac{a}{v^2}\, dv \tag{5·8·18}$$

If some standard macrostate of the gas is chosen to have temperature T_0 and molar volume v_0, then integration of (5·8·18) gives

$$\epsilon(T,v) - \epsilon(T_0,v_0) = \int_{T_0}^{T} c_V(T')\, dT' - a\left(\frac{1}{v} - \frac{1}{v_0}\right)$$

or

$$\epsilon(T,v) = \int_{T_0}^{T} c_V(T')\, dT' - \frac{a}{v} + \text{constant} \tag{5·8·19}$$

If c_V is independent of temperature, this becomes simply

$$\epsilon(T,v) = c_V T - \frac{a}{v} + \text{constant} \tag{5·8·20}$$

Note that here ϵ *does* depend on the molar volume v. As v increases, ϵ also increases. This makes physical sense because the intermolecular separation increases as v increases, and thus the attractive (i.e., negative) potential energy of interaction between the molecules is decreased in magnitude.

Finally, let us compute the entropy per mole of gas. By using (5·8·15), the relation (5·8·4) becomes

$$ds = \frac{c_V(T)}{T}\, dT + \frac{R}{v - b}\, dv \tag{5·8·21}$$

Integrating, one gets

$$s(T,v) - s(T_0,v_0) = \int_{T_0}^{T} \frac{c_V(T')\,dT'}{T'} + R \ln \left(\frac{v - b}{v_0 - b} \right) \qquad (5 \cdot 8 \cdot 22)$$

If c_V is independent of temperature, this can be written

$$s(T,v) = c_V \ln T + R \ln (v - b) + \text{constant} \qquad (5 \cdot 8 \cdot 23)$$

FREE EXPANSION AND THROTTLING PROCESSES

5·9 *Free expansion of a gas*

This experiment is one we have mentioned before. Consider a rigid container which is thermally insulated. It is divided into two compartments separated by a valve which is initially closed (see Fig. 5·9·1). One compartment of volume V_1 contains the gas under investigation, the other compartment is empty. The initial temperature of the system is T_1. The valve is now opened and the gas is free to expand so as to fill the entire container of volume V_2. What is the temperature T_2 of the gas after the final equilibrium state has been reached?

Since the system consisting of gas and container is adiabatically insulated, no heat flows into the system; i.e.,

$$Q = 0$$

Furthermore, the system does no work in the process; i.e.,

$$W = 0$$

Thus it follows by the first law that the total energy of the system is conserved; i.e.,

$$\Delta E = 0 \qquad (5 \cdot 9 \cdot 1)$$

Assume that the container itself has negligible heat capacity so that the internal energy of the container itself does not change. (This is a condition difficult to realize in practice; we shall come back to this point later.) Then the energy

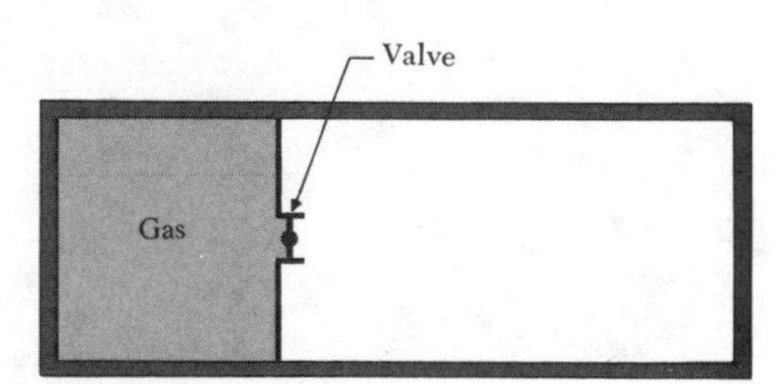

Fig. 5·9·1 ***Free expansion of a gas.***

change is simply that of the gas, and the conservation of energy (5 · 9 · 1) can be written

$$E(T_2, V_2) = E(T_1, V_1) \tag{5·9·2}$$

To predict the outcome of the experiment it is only necessary to know the internal energy of the gas $E(T,V)$ as a function of T and V; for if the initial parameters T_1 and V_1 and the final volume V_2 are known, Eq. (5 · 9 · 2) provides one with an equation for the unknown final temperature T_2.

Remark The actual free expansion is, of course, a complicated irreversible process involving turbulence and gross nonuniformities of pressure and temperature (to the extent that these quantities can be defined at all for such a marked nonequilibrium situation). Equilibrium conditions prevail only in the initial and final situations. Nevertheless, to predict the outcome of the process, the only knowledge required is that of the energy function E characteristic of equilibrium macrostates of the system.

For an *ideal* gas, E is independent of the volume V; i.e., $E = E(T)$. Then (5 · 9 · 2) becomes simply $E(T_2) = E(T_1)$, so that one must have $T_2 = T_1$. There is then *no* temperature change in the free expansion of an ideal gas.

More generally, the energy $E(T,V)$ is a function of both T and V. It can be represented in a two-dimensional graph by plotting E versus T for various values of the parameter V, as shown schematically in Fig. 5 · 9 · 2. From such a diagram the result of the experiment can be immediately predicted. Given T_1 and V_1, one can read off the value $E = E_1$. By (5 · 9 · 2) the intersection of the horizontal line $E = E_1$ with the curve V_2 yields then the final temperature T_2. If the curves are as drawn, $T_2 < T_1$.

Alternatively, and somewhat more directly, one can use the knowledge of $E(T,V)$ shown in Fig. 5 · 9 · 2 to draw curves of T versus V for various values of the energy E. On such a plot (illustrated in Fig. 5 · 9 · 3) one knows by (5 · 9 · 2) that the initial values of T and V determine a given energy curve, say $E = E_1$, and that as a result of the free expansion one must always end up

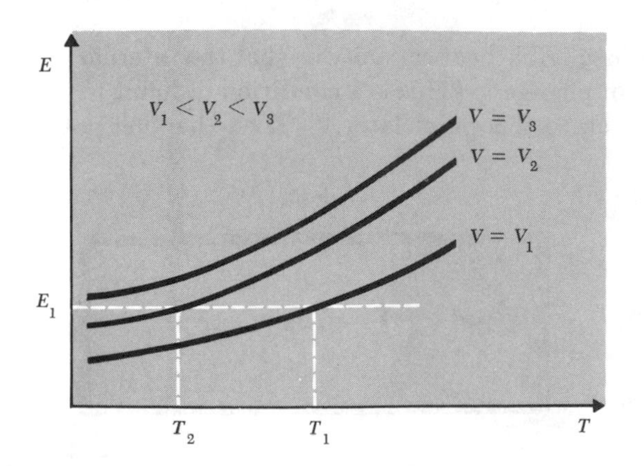

Fig. 5 · 9 · 2 Schematic diagram showing the dependence of the internal energy E of a gas on its temperature T for various values of its volume V.

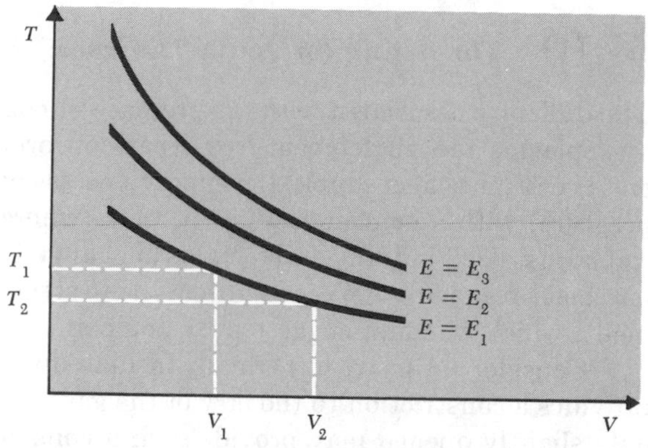

***Fig. 5·9·3** Schematic diagram showing curves of constant internal energy E. Each curve describes corresponding values of T and V which yield the given energy E.*

somewhere on this *same* curve. The final temperature can thus be read off immediately from this curve for any value of the final volume V_2.

Example: van der Waals gas Let us calculate the temperature change in the case of the free expansion of one mole of a van der Waals gas. If we denote the molar internal energy by $\epsilon(T,v)$, the energy-conservation condition (5·9·2) is

$$\epsilon(T_2,v_2) = \epsilon(T_1,v_1)$$

This becomes by (5·8·19)

$$\int_{T_0}^{T_2} c_V(T')\,dT' - \frac{a}{v_2} = \int_{T_0}^{T_1} c_V(T')\,dT' - \frac{a}{v_1}$$

Hence $$\int_{T_0}^{T_2} c_V(T')\,dT' - \int_{T_0}^{T_1} c_V(T')\,dT' = a\left(\frac{1}{v_2} - \frac{1}{v_1}\right)$$

or $$\int_{T_1}^{T_2} c_V(T')\,dT' = a\left(\frac{1}{v_2} - \frac{1}{v_1}\right) \tag{5·9·3}$$

Over the small temperature range $T_1 < T' < T_2$, any possible temperature dependence of c_V is negligibly small. Thus c_V can be regarded as substantially constant, and (5·9·3) becomes simply

$$c_V(T_2 - T_1) = a\left(\frac{1}{v_2} - \frac{1}{v_1}\right)$$

or $$T_2 - T_1 = -\frac{a}{c_V}\left(\frac{1}{v_1} - \frac{1}{v_2}\right) \tag{5·9·4}$$

For an expansion where $v_2 > v_1$, or $1/v_1 > 1/v_2$, one gets thus (since $c_V > 0$)

$$T_2 < T_1 \tag{5·9·5}$$

Hence the temperature is reduced as a result of the free expansion.

In principle, it appears that the free expansion of a gas could provide a method of cooling the gas to low temperatures. In practice, a difficulty is encountered because of the appreciable heat capacity C_c of the container. Since its internal energy also changes by an amount $C_c(T_2 - T_1)$, a given volume change of the gas results in a much smaller net temperature change when C_c is finite than when it is zero. (If the container is taken into account, the heat capacity c_V in (5·9·4) must be replaced by the total heat capacity $c_V + C_c$.)

5 · 10 *Throttling (or Joule-Thomson) Process*

The difficulty associated with the presence of containing walls can be overcome by replacing the single-event free-expansion process just discussed (where this one event must also supply the energy necessary to change the container temperature) with a continuous-flow process (where the temperature of the walls can adjust itself initially and remains unchanged after the steady-state situation has been reached). We now discuss this steady-state experimental arrangement, which was first suggested by Joule and Thomson.

Consider a pipe with thermally insulated walls. A porous plug in the pipe provides a constriction to the flow of the gas. (Alternatively, a valve which is only slightly opened may provide such a constriction.) A continuous stream of gas flows from left to right. The presence of the constriction results in a constant pressure difference being maintained across this constriction. Thus the gas pressure p_1 to the left of the constriction is greater than the gas pressure p_2 to the right of the constriction. Let T_1 denote the temperature of the gas on the left side of the constriction. What then is the gas temperature T_2 on the right side?

Let us analyze the situation. Focus attention on the system consisting of the mass M of gas lying between the dashed planes A and B shown in Fig. 5·10·2. (We suppose that the planes A and B are chosen so far apart that the volume occupied by the constriction itself is negligible compared to the volume contained between A and B.) At some initial time the plane B coincides with the constriction, and virtually the entire mass M of gas lies to the left of the constriction (see Fig. 5·10·2a). Then it occupies some volume V_1 corresponding to the pressure p_1. As this mass M of gas flows down the pipe, the planes A and B which define its geometrical boundaries also move down the pipe. After some time has elapsed, the plane A will have moved so as to coincide with the constriction, and virtually the entire mass M of the gas will lie to the right of the constriction. There it occupies some different volume V_2 corresponding to the lower pressure p_2. This is the final situation illustrated in Fig. 5·10·2b.

In the process just described, the difference in internal energy of the mass M of gas between the final situation when it is to the right and the initial

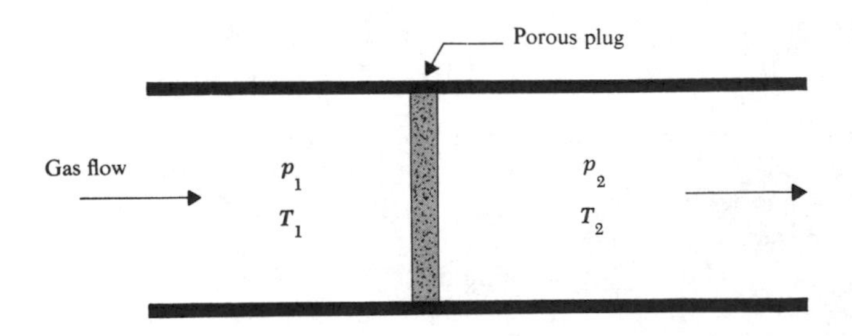

***Fig. 5·10·1** A steady-state throttling process in which a gas is flowing through a porous plug.*

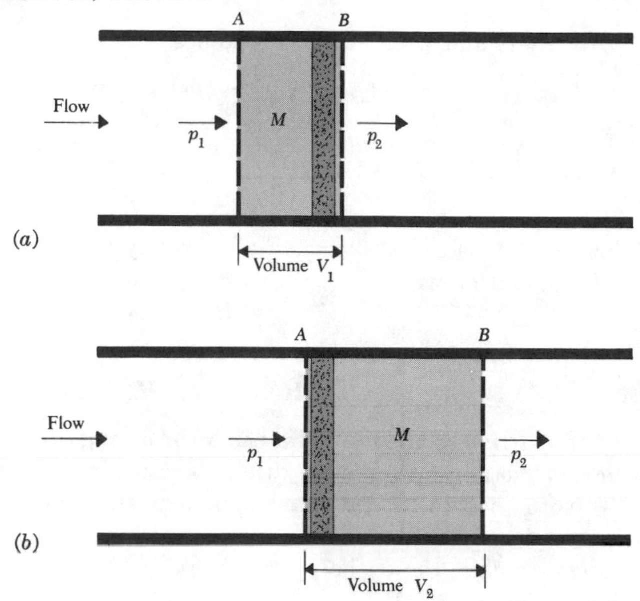

***Fig. 5·10·2** Diagram showing a mass M of gas passing through a constriction (a porous plug in this case) (a) before passing through the constriction (b) after passing through the constriction.*

situation when it is to the left of the constriction is simply

$$\Delta E = E_2 - E_1 = E(T_2,p_2) - E(T_1,p_1) \tag{5·10·1}$$

In this process the mass M of gas also does work. Indeed, it does work p_2V_2 in displacing the gas to the right of the constriction by the volume V_2 against the *constant* pressure p_2. Furthermore, the gas to the left of the constriction does work p_1V_1 *on* the mass M of gas by displacing it by the volume V_1 with a *constant* pressure p_1. Hence the *net* work done *by* the mass M of gas in the process is simply

$$W = p_2V_2 - p_1V_1 \tag{5·10·2}$$

But no heat is absorbed by the mass M of gas in the process we have described. This is not just because the walls are adiabatically insulated so that no heat enters them from the outside; more importantly, after the steady state situation has been established, there is no temperature difference between the walls and the adjacent gas, so that no heat flows from the walls into the gas. Thus

$$Q = 0 \tag{5·10·3}$$

Application of the first law to the mass M of gas yields then for the process under consideration the relation

$$\Delta E + W = Q = 0 \tag{5·10·4}$$

By (5 · 10 · 1) and (5 · 10 · 2) this becomes

$$(E_2 - E_1) + (p_2V_2 - p_1V_1) = 0$$

or

$$E_2 + p_2V_2 = E_1 + p_1V_1 \tag{5·10·5}$$

Let us define the quantity

$$H \equiv E + pV \tag{5·10·6}$$

This is the so-called "enthalpy" already encountered in (5 · 5 · 7). Then (5 · 10 · 5) can be written

$$H_2 = H_1$$

or

▶ $$H(T_2,p_2) = H(T_1,p_1) \tag{5·10·7}$$

Thus we arrive at the result that in a throttling process the gas passes through the constriction in such a way that its enthalpy H remains constant.

Note that (5 · 10 · 7) is analogous to the condition (5 · 9 · 2) for the free expansion case. The difference is that the gas *does* work in the throttling process, so that the enthalpy rather than the internal energy is the quantity which is conserved.

Remark Here again the actual passage of the gas through the constriction involves complicated irreversible nonequilibrium processes. Equilibrium situations prevail only to the left and to the right of the constriction. But a knowledge of the enthalpy function $H(T,p)$ characteristic of equilibrium macrostates of the system is sufficient to predict the outcome of the process.

Suppose that $H(T,p)$ is known as a function of T and p. Then, given T_1 and p_1 and the final pressure p_2, (5 · 10 · 7) provides an equation to determine the unknown final temperature T_2. In the case of an ideal gas,

$$H = E + pV = E(T) + \nu RT$$

so that $H = H(T)$ is a function of the temperature only. Then the condition (5 · 10 · 7) implies immediately

$$H(T_2) = H(T_1)$$

so that $T_2 = T_1$. Thus the temperature of an *ideal* gas does *not* change in a throttling process.

In the more general case the method of analysis is similar to that used for the discussion of the free expansion in Sec. 5 · 9. From a knowledge of $H(T,p)$ one can construct curves of T versus p for various fixed values of the enthalpy H (see Fig. 5 · 10 · 3). On such a plot the initial values T_1 and p_1 determine a particular enthalpy curve. By virtue of (5 · 10 · 7) one must end up somewhere on this *same* curve as a result of the throttling process. The final temperature T_2 can then be read off immediately from this curve for any value of the final pressure p_2.

The curves of Fig. 5 · 10 · 3 do in general exhibit maxima. Thus it is possible to attain, as a result of a throttling process with $p_2 < p_1$, conditions where

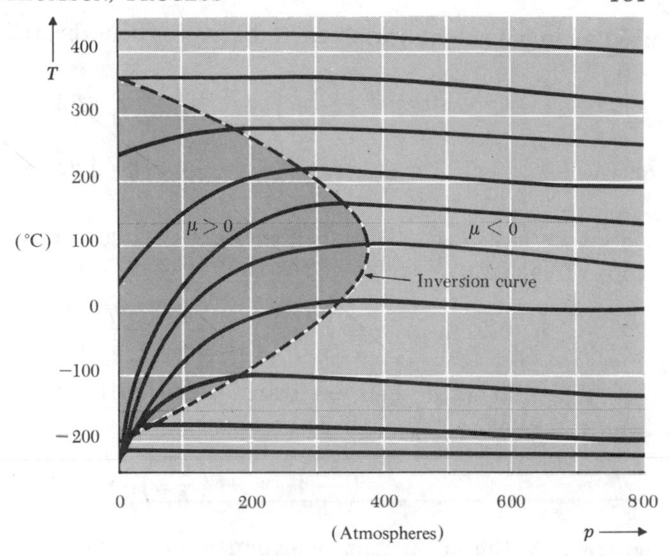

***Fig. 5·10·3** Curves of constant enthalpy H in the pT plane of a gas. The numerical values are for nitrogen (N_2). The dashed line is the inversion curve.*

the temperature T either increases, decreases, or remains the same. A significant parameter in this context is the slope μ of these curves,

$$\mu \equiv \left(\frac{\partial T}{\partial p}\right)_H \tag{5·10·8}$$

called the Joule-Thomson coefficient.* This quantity gives the change of temperature produced in a throttling process involving an infinitesimal pressure differential. For an infinitesimal pressure drop, T will decrease if $\mu > 0$. The condition $\mu = 0$ implies that no temperature change occurs and locates the maxima of the curves in Fig. 5·10·3. The locus of the maxima forms a curve (shown dashed in Fig. 5·10·3) which is called the "inversion curve." It separates on the diagram the region of positive slope μ (where the temperature tends to fall) from the region of negative slope μ (where the temperature tends to rise).

Let us find an expression for μ in terms of readily measured parameters of the gas. By (5·10·7) we are interested in a situation where H is constant. Starting from the fundamental thermodynamic relation

$$dE = T\,dS - p\,dV$$

we get for the enthalpy change the result

$$dH \equiv d(E + pV) = T\,dS + V\,dp \tag{5·10·9}$$

(This was already obtained in (5·5·6).) In our case, where H is constant, $dH = 0$. Writing (5·10·9) in terms of the quantities T and p which we have

* It is sometimes also called the Joule-Kelvin coefficient. The reason is that the names Thomson and Kelvin refer to the same person, William Thomson, who was raised to the peerage and thus became Lord Kelvin.

used as independent variables in discussing the throttling process, we get

$$0 = T\left[\left(\frac{\partial S}{\partial T}\right)_p dT + \left(\frac{\partial S}{\partial p}\right)_T dp\right] + V\,dp$$

or

$$C_p\,dT + \left[T\left(\frac{\partial S}{\partial p}\right)_T + V\right] dp = 0$$

where we have used $C_p = T(\partial S/\partial T)_p$. Using this result, valid under conditions of constant H, to solve for the ratio dT/dp, we get

$$\mu \equiv \left(\frac{\partial T}{\partial p}\right)_H = -\frac{T(\partial S/\partial p)_T + V}{C_p} \tag{5·10·10}$$

The numerator can be transformed into more convenient form by a Maxwell relation; by (5·6·5) one has

$$\left(\frac{\partial S}{\partial p}\right)_T = -\left(\frac{\partial V}{\partial T}\right)_p = -V\alpha$$

where α is the coefficient of expansion defined in (5·7·8). Thus (5·10·10) becomes

▶
$$\mu = \frac{V}{C_p}(T\alpha - 1) \tag{5·10·11}$$

Of course, this is properly an intensive quantity, since both the volume V and the heat capacity C_p are extensive quantities.

For an ideal gas we found in (5·7·15) that $\alpha = T^{-1}$. Then $\mu = 0$ and, as mentioned previously, no temperature change results from a throttling process.

More generally, $\mu > 0$ if $\alpha > T^{-1}$, and conversely $\mu < 0$ if $\alpha < T^{-1}$. The locus of points in the pT plane, where α is such that $\alpha = T^{-1}$, gives the inversion curve.

The Joule-Thomson effect constitutes a practical method for cooling gases and is often used in processes designed to liquefy gases. In order to achieve a lower temperature as a result of throttling a gas, it is necessary to work in that region of pressure and temperature where $\mu > 0$; in particular, the initial temperature must be less than the maximum temperature on the inversion curve (see Fig. 5·10·3). For example, this maximum inversion temperature is 34°K for helium, 202°K for hydrogen, and 625°K for nitrogen. An attempt to throttle helium gas starting from room temperature would thus result in an *increase*, rather than a decrease, of the gas temperature. To use the Joule-Thomson effect for cooling helium gas to very low temperatures it is then necessary to precool it first to temperatures below 34°K. This can be done by using liquid hydrogen to precool the helium gas. Alternatively, one can keep the helium gas thermally insulated and let it do mechanical work at the expense of its internal energy. After this procedure has resulted in a sufficiently large temperature decrease of the helium gas, the Joule-Thomson effect can be used as the final stage in the cooling process.

Joule-Thomson effect and molecular forces Neither free-expansion nor throttling processes result in a temperature change in the case of an ideal gas.

Both of these processes become interesting only if the gas is not ideal, i.e., when the mutual interaction between molecules is of importance. The equation of state of any gas can be written in the general form of a series

$$p = kT[n + B_2(T)n^2 + B_3(T)n^3 \cdots] \qquad (5\cdot 10\cdot 12)$$

which is an expansion in powers of the number of molecules per unit volume $n \equiv N/V$. The expression (5·10·12) is called the "virial expansion," and the coefficients B_2, B_3, . . . are called virial coefficients. For an ideal gas $B_2 = B_3 = \cdots = 0$. If n is not too large, only the first few terms in (5·10·12) are important. The first correction to the ideal gas consists of retaining the term B_2n^2 and neglecting all higher-order terms. In this case (5·10·12) becomes

$$p = \frac{N}{V} kT \left(1 + \frac{N}{V} B_2\right) \qquad (5\cdot 10\cdot 13)$$

One can readily make some qualitative statements about the behavior of B_2 as a function of T on the basis of some simple microscopic considerations. The interaction between two gas molecules is weakly attractive when their mutual separation is relatively large, but becomes strongly repulsive when their separation becomes of the order of a molecular diameter.* At low temperatures the mean kinetic energy of a molecule is small. The weak long-range attraction between molecules is then quite significant and tends to make the mean intermolecular separation less than would be the case in the absence of interaction. This attraction thus tends to reduce the gas pressure below that for an ideal gas; i.e., B_2 in (5·10·13) is then negative. But at higher temperatures the mean kinetic energy of a molecule becomes so large that the weak intermolecular attractive potential energy becomes comparatively negligible. In that case it is the strong short-range repulsive interaction between molecules which is most significant. This repulsion tends to increase the gas pressure above that expected for an ideal gas; i.e., B_2 is then positive. These qualitative considerations lead one to expect that B_2 is an increasing function of T, being negative for sufficiently low temperatures and becoming positive at higher temperatures. (These arguments will be made more quantitative in Sec. 10·4; they lead to a curve of $B_2(T)$ versus T of the type shown in Fig. 10·4·1.)

Let us now apply these considerations to a discussion of the Joule-Thomson effect by evaluating (5·10·11). Let us use the equation of state (5·10·13) to express V as a function of T and p. This is most readily done by noting that the term $(N/V)B_2$ is a correction term which is small compared to unity; hence one commits negligible error by replacing the ratio N/V in that term by the value $p/(kT)$ which this ratio assumes in first approximation. Thus (5·10·13) becomes

$$p = \frac{NkT}{V}\left(1 + \frac{p}{kT} B_2\right) = \frac{N}{V}(kT + pB_2)$$

or

$$V = N\left(\frac{kT}{p} + B_2\right) \qquad (5\cdot 10\cdot 14)$$

* Figure 10·3·1 illustrates a curve of potential energy of mutual interaction as a function of intermolecular distance.

Hence (5 · 10 · 11) yields the result

$$\mu = \frac{1}{C_p}\left[T\left(\frac{\partial V}{\partial T}\right)_p - V\right] = \frac{N}{C_p}\left(T\frac{\partial B_2}{\partial T} - B_2\right) \qquad (5 \cdot 10 \cdot 15)$$

The previously discussed temperature dependence of B_2 allows us now to draw some interesting conclusions. Since B_2 is an increasing function of T, the term $T(\partial B_2/\partial T)$ is positive. At low temperatures where molecular attraction is predominant, B_2 itself is negative; hence (5 · 10 · 15) shows that $\mu > 0$ in this temperature range. But if one goes to sufficiently high temperatures where molecular repulsion becomes predominent, B_2 becomes positive and sufficiently large to make $\mu < 0$ in (5 · 10 · 15). The existence of the inversion curve where $\mu = 0$ reflects, therefore, the competing effects between molecular attraction and repulsion.

HEAT ENGINES AND REFRIGERATORS

5 · 11 *Heat engines*

Historically, the subject of thermodynamics began with a study of the basic properties of heat engines. Since the subject has not only great technological importance (sufficient to have been responsible for the industrial revolution) but also intrinsic physical interest, we shall devote some time to a discussion of it.

It is very easy to do mechanical work w upon a device M, and then to extract from it an equivalent amount of heat q, which goes to increase the internal energy of some heat reservoir B.* For example, the device M might be a paddle wheel rotated in a liquid by a falling weight, or an electric resistor upon which electrical work is done.

The fundamentally significant question is: To what extent is it possible to proceed in the reverse way, i.e., to build a device (called a "heat engine") which can extract some of the internal energy from a heat reservoir in the form of heat and convert it into macroscopic work? The situation would then be as diagrammed in Fig. 5 · 11 · 3.

It is necessary to keep in mind the following points. The work should not be provided at the expense of the heat engine itself; otherwise one could not continue the process of heat-to-work conversion indefinitely. Thus one wants the engine to be in the same macrostate at the end of the process as it was at the beginning (i.e., to have gone through a cycle) so that it is ready to start again converting more heat into work in the next cycle. Clearly, steam engines and gasoline engines all go through such cycles. Furthermore, the work put out by the heat engine should be capable of simply changing an external parameter of some outside device (e.g., of lifting a weight) without doing it at the expense of affecting the other degrees of freedom (or entropy) of that device. One can

* We use the small letters w and q to denote intrinsically *positive* amounts of work and heat.

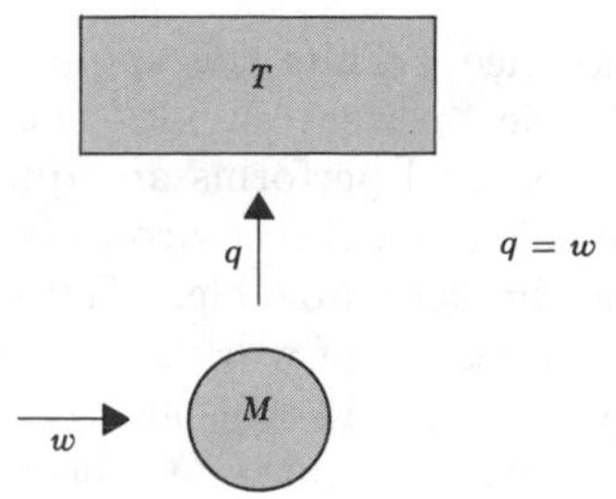

Fig. 5·11·1 Conversion of work w into heat q given off to a heat reservoir at temperature T.

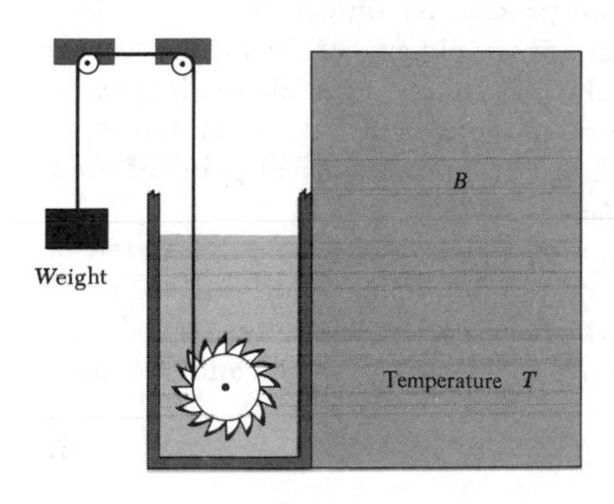

Fig. 5·11·2 A physical illustration showing the conversion of mechanical work into heat.

thus phrase the essential problem of constructing a heat engine in the following way: To what extent is it possible to extract a net amount of energy from one (or more) heat reservoirs, where that energy is randomly distributed over very many degrees of freedom, so as to transform it into energy associated with the single degree of freedom connected with the external parameter of an outside device?

Figure 5·11·3 would then be the prototype of the most desirable type of engine. After a cycle, M is back in the same macrostate as at the beginning, so that its internal energy is the same. Hence the first law of thermodynamics implies that

$$w = q \tag{5·11·1}$$

i.e., to conserve energy, the work put out by the engine must be equal to the heat extracted from the reservoir. One can certainly *not* build an engine which violates this condition.

But one may not be able to construct an engine even when this condition

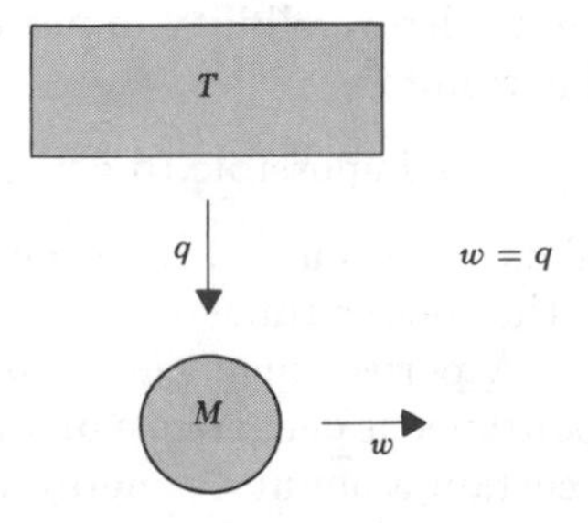

Fig. 5·11·3 A perfect engine.

is satisfied. Thus the engine illustrated in Fig. 5·11·3 is indeed a highly desirable "perfect engine"; i.e., working in a cycle, it extracts heat from a reservoir and performs an equivalent amount of work *without* producing any other effect on the environment. But a perfect engine of this kind is, unfortunately, not realizable. Indeed we know from our discussion of Sec. 3·2 that the conversion of work into heat illustrated in Fig. 5·11·2, or more schematically in Fig. 5·11·1, is an irreversible process in which the distribution of systems over accessible states becomes more random, so that the entropy increases. One cannot, therefore, simply reverse the process as shown in Fig. 5·11·3. In the concrete case of Fig. 5·11·2, one cannot simply expect the heat reservoir B to convert its internal energy, randomly distributed over all its degrees of freedom, into a systematic upward motion of the weight. It is, of course, in principle *possible* that this might happen, but from a statistical point of view such an occurrence is *fantastically improbable.*

Equivalently, we can show that an ideal engine of the type illustrated in Fig. 5·11·3 violates the second law of thermodynamics. Indeed, we must require that the total entropy change ΔS of the complete system (consisting of the heat engine, the outside device on which it does work, and the heat reservoir) be such that in a cycle

$$\Delta S \geq 0 \tag{5·11·2}$$

Now the engine itself returns to its previous state after a cycle; its entropy is thus unchanged after the cycle. Furthermore, we have already pointed out that no entropy change is associated with the outside device on which work is done. On the other hand, the entropy change of the heat reservoir at absolute temperature T_1 is, by (3·6·4), given by $-q/T_1$, since the reservoir absorbs heat $(-q)$. Hence (5·11·2) becomes

$$\frac{-q}{T_1} \geq 0$$

or, by (5·11·1),

$$\frac{q}{T_1} = \frac{w}{T_1} \leq 0 \tag{5·11·3}$$

Since we want the work w done *by* the engine to be positive, (5·11·3) cannot be satisfied. The inverse process of Fig. 5·11·1 where $w < 0$ is, of course, feasible. There is no objection to converting any amount of work into heat, but the converse is not possible. The second law thus again implies a fundamental irreversibility of natural processes. By (5·11·3) it specifically implies this result:

It is impossible to construct a perfect heat engine. (5·11·4)

(This statement is sometimes known as Kelvin's formulation of the second law of thermodynamics.)

A perfect heat engine is thus not realizable because it would require the spontaneous occurrence of a process which goes from an initial situation, where a certain amount of energy is distributed randomly over the many degrees of freedom of a heat reservoir, to a much more special and enormously less probable final situation, where this energy is all associated with the motion of a single degree of freedom capable of performing macroscopic work; in short, because it would require a process where the entropy S decreases. But this kind of process, in which the system consisting of heat reservoir and engine goes to a less random situation, *can* take place if this system is coupled to some other auxiliary system whose degree of randomness (or entropy) is increased in this process by a compensating amount (i.e., by an amount large enough that the *entire* system does increase in randomness). The simplest such auxiliary system is a second heat reservoir at some temperature T_2 lower than T_1. One then obtains a nonperfect, but realizable, heat engine which not only absorbs heat q_1 from a reservoir at temperature T_1, but also *rejects* heat to some second reservoir at some lower temperature T_2. Thus a real engine can be diagrammed as shown in Fig. 5·11·4.

In this case the first law requires that in a cycle

$$q_1 = w + q_2 \tag{5·11·5}$$

On the other hand, the second law is satisfied if in this cycle the total entropy change of these reservoirs satisfies the inequality

$$\Delta S = \frac{(-q_1)}{T_1} + \frac{q_2}{T_2} \geq 0 \tag{5·11·6}$$

Equations (5·11·5) and (5·11·6) *can* be satisfied with positive work w performed by the engine on the outside world. By combining these equations one gets

$$\frac{-q_1}{T_1} + \frac{q_1 - w}{T_2} \geq 0$$

$$\frac{w}{T_2} \leq q_1\left(\frac{1}{T_2} - \frac{1}{T_1}\right)$$

or

$$\eta \equiv \frac{w}{q_1} \leq 1 - \frac{T_2}{T_1} = \frac{T_1 - T_2}{T_1} \tag{5·11·7}$$

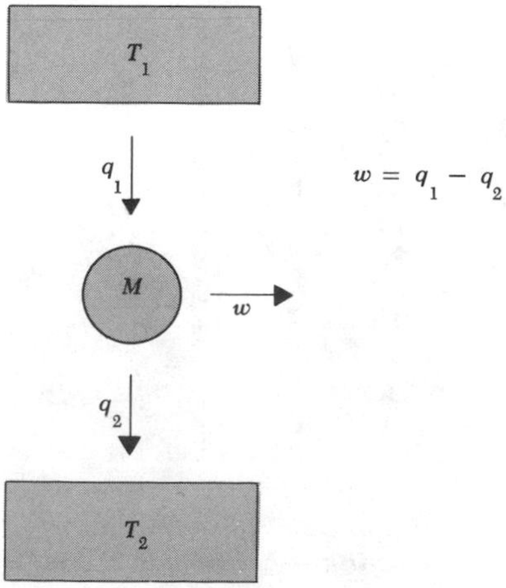

Fig. 5·11·4 A real engine.

For a perfect engine one would have $w/q_1 = 1$. For a real engine this ratio is less than 1, i.e.,

$$\eta \equiv \frac{w}{q_1} = \frac{q_1 - q_2}{q_1} < 1 \tag{5·11·8}$$

since some heat does not get transformed into work but is instead rejected to some other heat reservoir. The quantity $\eta = w/q_1$ is called the "efficiency" of the engine. Equation (5·11·7) provides us then with a relation for the maximum possible efficiency of an engine operating between two reservoirs of given absolute temperatures. Since the equals sign in the second statement (5·11·6) holds only for a quasi-static process, (5·11·7) implies also that no engine operating between the two given heat reservoirs can have an efficiency greater than that of an engine which operates between the same two reservoirs in a quasi-static manner. Furthermore, (5·11·7) implies that *any* engine which operates between these two reservoirs in a quasi-static manner has the *same* efficiency:

if quasi-static, $$\eta = \frac{T_1 - T_2}{T_1} \tag{5·11·9}$$

Carnot engines It is of interest to exhibit explicitly how such an engine operating quasi-statically between two heat reservoirs can be constructed. Such an engine is the simplest conceivable engine and is called a "Carnot engine" (named after Carnot, the French engineer who was the first to examine theoretically the operation of heat engines). Let x denote the external parameter of the engine M; changes in this parameter give rise to the work performed by the engine. Let the engine initially be in a state where $x = x_a$ and its temperature $T = T_2$, the temperature of the colder heat reservoir. The Carnot engine then goes through a cycle consisting of four steps, all per-

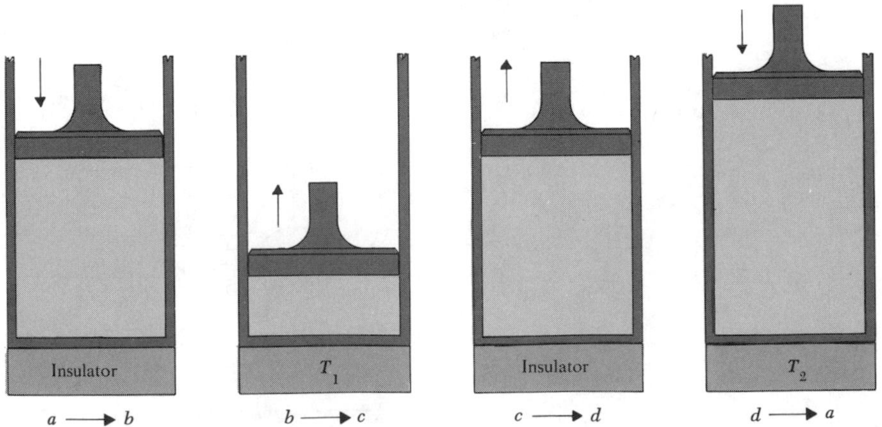

Fig. 5·11·5 The four steps of a Carnot cycle in which a gas is used as the working substance. The external parameter x is the volume V of the gas.

formed in a quasi-static fashion. Label macrostates of the engine by small letters a, b, c, d.

1. $a \rightarrow b$: The engine is *thermally insulated.* Its external parameter is changed slowly until the engine temperature reaches T_1. Thus $x_a \rightarrow x_b$ such that $T_2 \rightarrow T_1$.

2. $b \rightarrow c$: The engine is now placed in *thermal contact* with the heat reservoir at temperature T_1. Its external parameter is changed further, the engine remaining at temperature T_1 and absorbing some heat q_1 from the reservoir. Thus $x_b \rightarrow x_c$ such that heat q_1 is absorbed by the engine.

3. $c \rightarrow d$: The engine is again *thermally insulated.* Its external parameter is changed in such a direction that its temperature goes back to T_2. Thus $x_c \rightarrow x_d$ such that $T_1 \rightarrow T_2$.

4. $d \rightarrow a$: The engine is now placed in *thermal contact* with the heat reservoir at temperature T_2. Its external parameter is then changed until it returns to its initial value x_a, the engine remaining at temperature T_2 and rejecting some heat q_2 to this reservoir. Thus $x_d \rightarrow x_a$ and heat q_2 is given off by the engine.

The engine is now back in its initial state and the cycle is completed.

Example Let us illustrate a Carnot cycle with a particular kind of system. Take, for example, a gas (not necessarily ideal) contained in a cylinder closed off by a piston. The external parameter is the volume V of the gas. The four steps of the Carnot cycle are illustrated in Fig. 5·11·5. The area enclosed by the quadrilateral figure in Fig. 5·11·6 represents the total work

$$w = \int_a^b p\,dV + \int_b^c p\,dV + \int_c^d p\,dV + \int_d^a p\,dV$$

performed by the engine in a cycle.

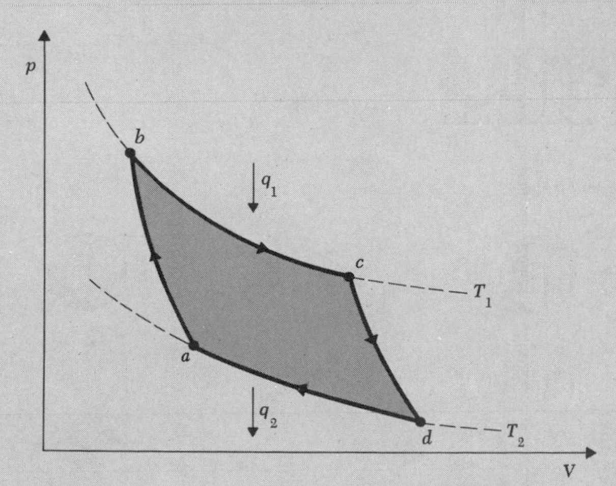

Fig. 5·11·6 The Carnot cycle of Fig. 5·11·5 illustrated on a pV diagram.

Practical engines, such as steam or gasoline engines, are somewhat more complicated than Carnot engines. But, like all heat engines, they cannot be perfect and are well known to have mechanisms (such as condensers or exhausts) by which they reject heat to some low-temperature reservoir, usually the surrounding atmosphere.

5 · 12 *Refrigerators*

A refrigerator is a device which, operating in a cycle, removes heat from a reservoir at lower absolute temperature and rejects it to a reservoir at higher absolute temperature. It can be represented by the diagram of Fig. 5 · 12 · 1, which is similar to that of Fig. 5 · 11 · 4 except that the directions of all the arrows have been reversed. The first law, applied to the refrigerator of Fig. 5 · 12 · 1, requires that

$$w + q_2 = q_1 \tag{5·12·1}$$

Since a Carnot engine operates quasi-statically by passing continuously through a series of near-equilibrium states, one could run it equally well quasi-statically in the reverse direction. In this case it would operate like a particularly simple special kind of refrigerator.

Needless to say, Fig. (5 · 12 · 1) does not represent the ideal refrigerator one might like to have. After all, if the two heat reservoirs were placed in thermal contact with each other, some amount of heat q would flow spontaneously from the reservoir at higher temperature T_1 to the reservoir at lower temperature T_2. The "perfect refrigerator" would just reverse the process, i.e., it would simply remove heat q from the reservoir at lower temperature and transfer it to the reservoir at higher temperature *without* affecting the environment in any other way; i.e., the perfect refrigerator would not require any work

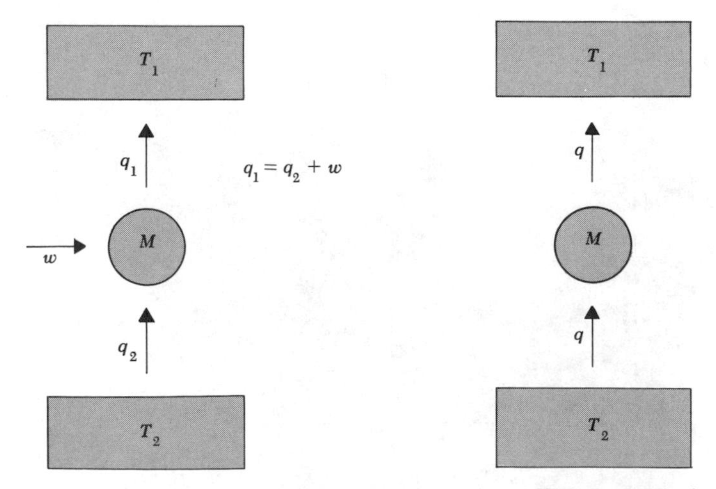

Fig. 5·12·1 A real refrigerator. *Fig. 5·12·2 A perfect refrigerator.*

to be done on it and would be represented by the diagram of Fig. 5·12·2. But a perfect refrigerator would again violate the second law. Indeed, the total entropy change in Fig. (5·12·2) has to satisfy the inequality

$$\Delta S = \frac{q}{T_1} + \frac{(-q)}{T_2} \geq 0$$

or

$$q\left(\frac{1}{T_1} - \frac{1}{T_2}\right) \geq 0 \qquad (5\cdot12\cdot2)$$

which is impossible for $q > 0$ and $T_1 > T_2$. Thus we arrive at the following statement:

It is impossible to construct a perfect refrigerator. (5·12·3)

(This statement is sometimes known as the Clausius formulation of the second law of thermodynamics.)

This result is, of course, only too familiar. Kitchen refrigerators have a nasty habit of requiring an external source of power.

A real refrigerator is then properly represented by Fig. 5·12·1, where some amount of work w must be done on the refrigerator to make it function. In that case one has, by (5·12·1),

$$q_2 = q_1 - w \qquad (5\cdot12\cdot4)$$

i.e., the heat removed from the colder reservoir is *less* than that given off to the warmer reservoir. The second law imposes then the requirement that

$$\Delta S = \frac{q_1}{T_1} + \frac{(-q_2)}{T_2} \geq 0$$

or

$$\frac{q_2}{q_1} \leq \frac{T_2}{T_1} \qquad (5\cdot12\cdot5)$$

where the equals sign holds only for a refrigerator operating between the two reservoirs in a quasi-static manner.

Remark It can be shown that the Kelvin and Clausius statements of the second law are equivalent and that either one implies that there must exist a function with the properties of entropy. This was the basis of the historical macroscopic approach to classical thermodynamics. The interested reader is referred to the bibliography at the end of this chapter for books developing this point of view.

SUGGESTIONS FOR SUPPLEMENTARY READING

Thermodynamic relations and properties of pure substances

M. W. Zemansky: "Heat and Thermodynamics," 4th ed., chaps. 6, 11, and 13, McGraw-Hill Book Company, New York, 1957.

H. B. Callen: "Thermodynamics," chaps. 5 and 6, John Wiley & Sons, Inc., New York, 1960.

Practical engines and refrigerators

M. W. Zemansky: "Heat and Thermodynamics," 4th ed., secs. 7.2–7.4, 12.1–12.10, McGraw-Hill Book Company, New York, 1957.

J. K. Roberts and A. R. Miller: "Heat and Thermodynamics," 5th ed., chap. 14, Interscience Publishers, New York, 1960.

Liquefaction of gases

M. W. Zemansky: "Heat and Thermodynamics," 4th ed., secs. 14.1–14.2, 16.1, McGraw-Hill Book Company, New York, 1957.

J. K. Roberts and A. R. Miller: "Heat and Thermodynamics," 5th ed., pp. 130–139, Interscience Publishers, New York, 1960.

Completely macroscopic development of thermodynamics based on the Kelvin or Clausius statements of the second law; discussion of Carnot engines

M. W. Zemansky: "Heat and Thermodynamics," 4th ed., chaps. 7, 9, and 10, McGraw-Hill Book Company, New York, 1957.

E. Fermi: "Thermodynamics," secs. 7–13, Dover Publications, New York, 1957.

Jacobian method of manipulating partial derivatives

L. D. Landau and E. M. Lifshitz: "Statistical Physics," sec. 16, Addison-Wesley Puslishing Company, Reading, Mass., 1958.

F. W. Crawford: "Heat, Thermodynamics, and Statistical Physics," secs. 11.13–11.17, Harcourt, Brace and World, New York, 1963.

PROBLEMS

5.1 An ideal gas has a temperature-independent molar specific heat c_V at constant volume. Let $\gamma \equiv c_p/c_V$ denote the ratio of its specific heats. The gas is thermally insulated and is allowed to expand quasi-statically from an initial volume V_i at temperature T_i to a final volume V_f.

(*a*) Use the relation $pV^\gamma =$ constant to find the final temperature T_f of this gas.

(*b*) Use the fact that the entropy remains constant in this process to find the final temperature T_f.

5.2 The molar specific heat at constant volume of a monatomic ideal gas is known to be $\frac{3}{2}R$. Suppose that one mole of such a gas is subjected to a cyclic quasi-static process which appears as a circle on the diagram of pressure p versus volume V shown in the figure. Find the following quantities:

(*a*) The net work (in joules) done by the gas in one cycle.

(*b*) The internal energy difference (in joules) of the gas between state C and state A.

(*c*) The heat absorbed (in joules) by the gas in going from A to C via the the path ABC of the cycle.

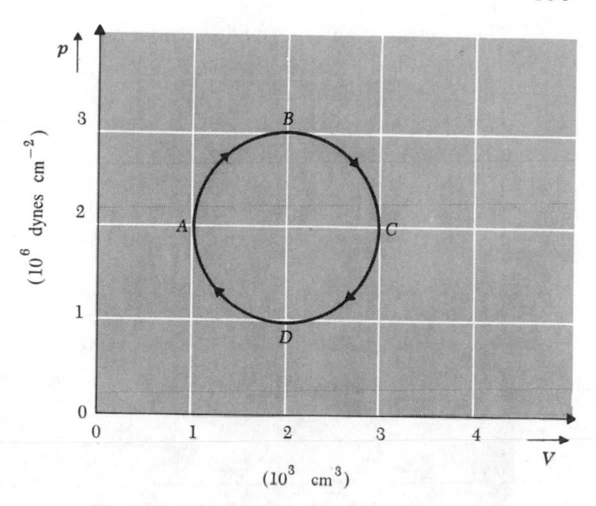

5.3 An ideal diatomic gas has a molar internal energy equal to $E = \frac{5}{2}RT$ which depends only on its absolute temperature T. A mole of this gas is taken quasi-statically first from state A to state B, and then from state B to state C along the straight line paths shown in the diagram of pressure p versus volume V.

(*a*) What is the molar heat capacity at constant volume of this gas?
(*b*) What is the work done by the gas in the process $A \to B \to C$?
(*c*) What is the heat absorbed by the gas in this process?
(*d*) What is its change of entropy in this process?

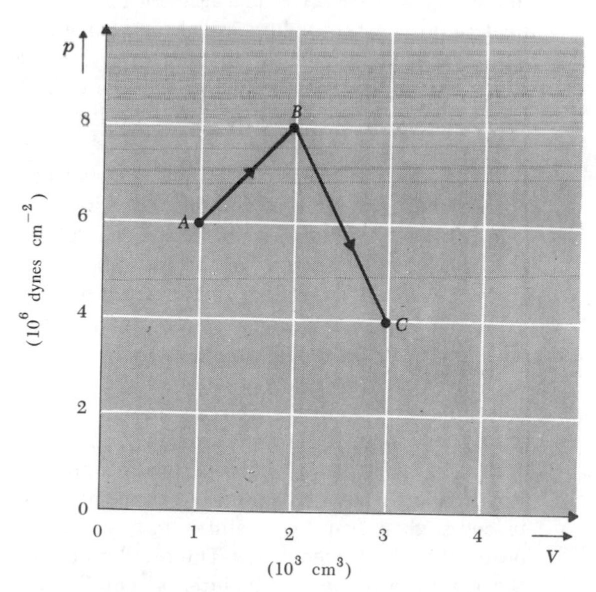

5.4 A cylindrical container 80 cm long is separated into two compartments by a thin piston, originally clamped in position 30 cm from the left end. The left compartment is filled with one mole of helium gas at a pressure of 5 atmospheres; the right compartment is filled with argon gas at 1 atmosphere of pressure.

These gases may be considered ideal. The cylinder is submerged in 1 liter of water, and the entire system is initially at the uniform temperature of 25°C. The heat capacities of the cylinder and piston may be neglected. When the piston is unclamped, a new equilibrium situation is ultimately reached with the piston in a new position.

(*a*) What is the increase in temperature of the water?

(*b*) How far from the left end of the cylinder will the piston come to rest?

(*c*) What is the increase of total entropy of the system?

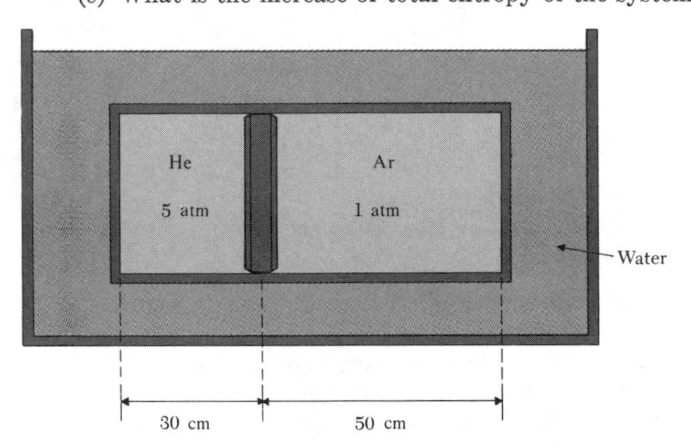

5.5 A vertical cylinder contains ν moles of an ideal gas and is closed off by a piston of mass M and area A. The acceleration due to gravity is g. The molar specific heat c_V (at constant volume) of the gas is a constant independent of temperature. The heat capacities of the piston and cylinder are negligibly small and any frictional forces between the piston and the cylinder walls can be neglected. The whole system is *thermally insulated*. Initially, the piston is clamped in position so that the gas has a volume V_0 and a temperature T_0. The piston is now released and, after some oscillations, comes to rest in a final equilibrium situation corresponding to a larger volume of the gas.

(*a*) Does the temperature of the gas increase, decrease, or remain the same?

(*b*) Does the entropy of the gas increase, decrease, or remain the same?

(*c*) Calculate the final temperature of the gas in terms of T_0, V_0, and the other parameters mentioned in the statement of the problem.

5.6 The following describes a method used to measure the specific heat ratio $\gamma \equiv c_p/c_V$ of a gas. The gas, assumed ideal, is confined within a vertical cylindrical container and supports a freely moving piston of mass m. The piston and cylinder both have the same cross-sectional area A. Atmospheric pressure is p_0, and when the piston is in equilibrium under the influence of gravity (acceleration g) and the gas pressure, the volume of the gas is V_0. The piston is now displaced slightly from its equilibrium position and is found to oscillate about this position with frequency ν. The oscillations of the piston are slow enough that the gas always remains in internal equilibrium, but fast enough that the gas cannot exchange heat with the outside. The variations in gas pressure and volume are thus adiabatic. Express γ in terms of m, g, A, p_0, V_0, and ν.

5.7 Consider the earth's atmosphere as an ideal gas of molecular weight μ in a uniform gravitational field. Let g denote the acceleration due to gravity.

(*a*) If z denotes the height above sea level, show that the change of atmospheric pressure p with height is given by

$$\frac{dp}{p} = -\frac{\mu g}{RT}\,dz$$

where T is the absolute temperature at the height z.

(*b*) If the decrease of pressure in (*a*) is due to an adiabatic expansion, show that

$$\frac{dp}{p} = \frac{\gamma}{\gamma - 1}\frac{dT}{T}$$

(*c*) From (*a*) and (*b*) calculate dT/dz in degrees per kilometer. Assume the atmosphere to consist mostly of nitrogen (N_2) gas for which $\gamma = 1.4$.

(*d*) In an isothermal atmosphere at temperature T, express the pressure p at height z in terms of the pressure p_0 at sea level.

(*e*) If the sea-level pressure and temperature are p_0 and T_0, respectively, and the atmosphere is regarded as adiabatic as in part (*b*), find again the pressure p at height z.

5.8 When a sound wave passes through a fluid (liquid or gas), the period of vibration is short compared to the relaxation time necessary for a macroscopically small element of volume of the fluid to exchange energy with the rest of the fluid through heat flow. Hence compressions of such an element of volume can be considered adiabatic.

By analyzing one-dimensional compressions and rarefactions of the system of fluid contained in a slab of thickness dx, show that the pressure $p(x,t)$ in the fluid depends on the position x and the time t so as to satisfy the wave equation

$$\frac{\partial^2 p}{\partial t^2} = u^2\frac{\partial^2 p}{\partial x^2}$$

where the velocity of sound propagation u is a constant given by $u = (\rho\kappa_S)^{-\frac{1}{2}}$. Here ρ is the equilibrium density of the fluid and κ_S is its *adiabatic* compressibility $\kappa_S = -V^{-1}(\partial V/\partial p)_S$, i.e., its compressibility measured under conditions where the fluid is thermally insulated.

5.9 Refer to the results of the preceding problem.

(*a*) Calculate the adiabatic compressibility κ_S of an ideal gas in terms of its pressure p and specific heat ratio γ.

(*b*) Find an expression for the velocity of sound in an ideal gas in terms of γ, its molecular weight μ, and its absolute temperature T.

(*c*) How does the sound velocity depend on the gas temperature T at a fixed pressure? How does it depend on the gas pressure p at a fixed temperature?

(*d*) Calculate the velocity of sound in nitrogen (N_2) gas at room temperature and pressure. Take $\gamma = 1.4$.

5.10 Liquid mercury at atmospheric pressure and 0°C (i.e., 273°K) has a molar volume of 14.72 cm³/mole and a specific heat at constant pressure of $c_p = 28.0$ joules mole⁻¹ deg⁻¹. Its coefficient of expansion is $\alpha = 1.81 \times 10^{-4}$ deg⁻¹, and its compressibility is $\kappa = 3.88 \times 10^{-12}$ cm² dyne⁻¹. Find its specific heat c_V at constant volume and the ratio $\gamma \equiv c_p/c_V$.

5.11 Consider an isotropic solid of length L. Its coefficient of linear expansion α_L is defined as $\alpha_L \equiv L^{-1}(\partial L/\partial T)_p$ and is a measure of the change in length of this solid produced by a small change of temperature. By considering an infinitesimal rectangular parallelepiped of this solid, show that the coefficient of volume expansion $\alpha \equiv V^{-1}(\partial V/\partial T)_p$ for this solid is simply equal to $\alpha = 3\alpha_L$.

5.12 The following problem arises when experiments are done on solids at high pressures. If the pressure is increased by an amount Δp, this being done under circumstances where the sample is thermally insulated and at a sufficiently slow rate that the process can be regarded as quasi-static, what is the resulting change of temperature ΔT of the sample? If Δp is fairly small, derive an expression for ΔT in terms of Δp, the absolute temperature T of the sample, its specific heat at constant pressure c_p (in ergs g^{-1} deg^{-1}), its density ρ (in g/cm^3), and its volume coefficient of thermal expansion α (in deg^{-1}).

5.13 A homogeneous substance at temperature T and pressure p has a molar volume v and a molar specific heat (measured at constant pressure) given by c_p. Its coefficient of volume expansion α is known as a function of temperature. Calculate how c_p depends on the pressure at a given temperature; i.e., calculate $(\partial c_p/\partial p)_T$, expressing the result in terms of T, v, and the properties of α.

5.14 In a temperature range near absolute temperature T, the tension force F of a stretched plastic rod is related to its length L by the expression

$$F = aT^2(L - L_0)$$

where a and L_0 are positive constants, L_0 being the unstretched length of the rod. When $L = L_0$, the heat capacity C_L of the rod (measured at constant length) is given by the relation $C_L = bT$, where b is a constant.

(*a*) Write down the fundamental thermodynamic relation for this system, expressing dS in terms of dE and dL.

(*b*) The entropy $S(T,L)$ of the rod is a function of T and L. Compute $(\partial S/\partial L)_T$.

(*c*) Knowing $S(T_0,L_0)$, find $S(T,L)$ at *any* other temperature T and length L. (It is most convenient to calculate first the change of entropy with temperature at the length L_0 where the heat capacity is known.)

(*d*) If one starts at $T = T_i$ and $L = L_i$ and stretches the thermally insulated rod quasi-statically until it attains the length L_f, what is the final temperature T_f? Is T_f larger or smaller than T_i?

(*e*) Calculate the heat capacity $C_L(L,T)$ of the rod when its length is L instead of L_0.

(*f*) Calculate $S(T,L)$ by writing $S(T,L) - S(T_0,L_0) = [S(T,L) - S(T_0,L)] + [S(T_0,L) - S(T_0,L_0)]$ and using the result of part (*e*) to compute the first term in the square brackets. Show that the final answer agrees with that found in (*c*).

5.15 The figure illustrates a soap film (shown in gray) supported by a wire frame. Because of surface tension the film exerts a force $2\sigma l$ on the cross wire. This force is in such a direction that it tends to move this wire so as to decrease the area of the film. The quantity σ is called the "surface tension" of the film and the factor 2 occurs because the film has two surfaces. The temperature dependence of σ is given by

$$\sigma = \sigma_0 - \alpha T$$

where σ_0 and α are constants independent of T or x.

(*a*) Suppose that the distance x (or equivalently, the total film area $2lx$) is the only external parameter of significance in the problem. Write a relation expressing the change dE in mean energy of the film in terms of the heat $đQ$

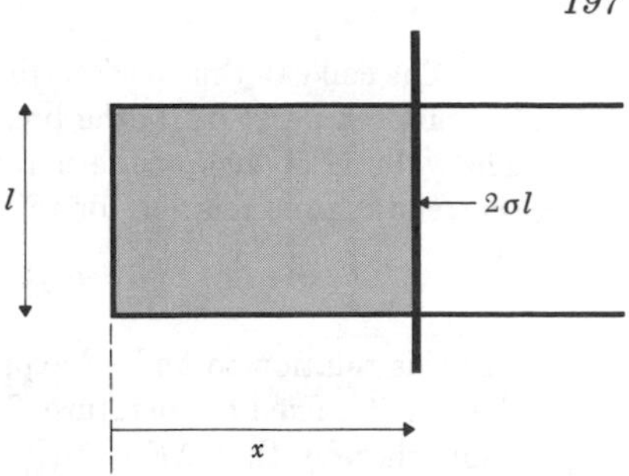

absorbed by it and the work done by it in an infinitesimal quasi-static process in which the distance x is changed by an amount dx.

(b) Calculate the change in mean energy $\Delta E = E(x) - E(0)$ of the film when it is stretched at a constant temperature T_0 from a length $x = 0$ to a length x.

(c) Calculate the work $\mathcal{W}(0 \rightarrow x)$ done on the film in order to stretch it at this constant temperature from a length $x = 0$ to a length x.

5.16 Consider an electrochemical cell of the type illustrated in the figure. The cell can be maintained in equilibrium by connecting a potentiometer across its terminals in such a way that the emf $\mathcal{V}$ produced by the cell is precisely compensated and no net current flows through the outside circuit. The following chemical reaction can take place in the cell:

$$Zn + CuSO_4 \rightleftarrows Cu + ZnSO_4 \qquad (1)$$

Suppose that the equilibrium is shifted quasi-statically by an infinitesimal amount so that the reaction proceeds from left to right, dN atoms of copper (Cu) being produced in the process. Then a charge $ze\,dN$ flows from the Cu to the zinc (Zn) electrode through the outside circuit (where $z = 2$ is the valence of copper) and the cell does an amount of work $\mathcal{V}ze\,dN$. Expressed in terms of moles, when $d\nu = dN/N_a$ moles of Cu are produced, the charge transferred is $zeN_a\,d\nu = zf\,d\nu$ (where $f \equiv N_ae$ is called the Faraday constant), and the work done by the cell is $\mathcal{V}zf\,d\nu$.

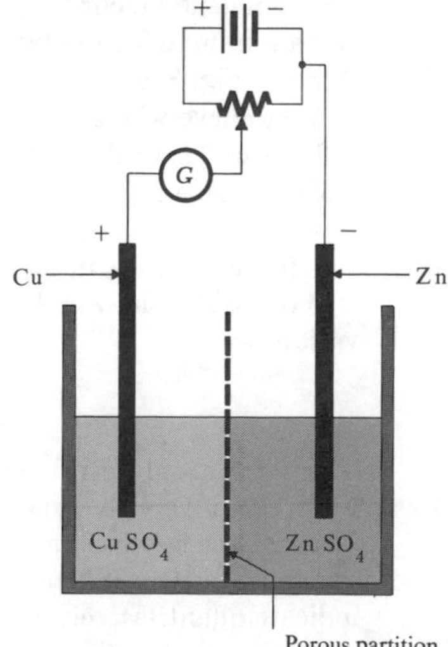

The cell can thus be described in terms of the following independent parameters: its temperature T, the pressure p, and the number of moles ν of Cu metal. The volume change of the material in the cell is negligible. The fundamental thermodynamic relation for this cell becomes then

$$T\,dS = dE + zf\mathcal{V}\,d\nu \tag{2}$$

Use this relation to find an expression for the change ΔE in the mean energy of the cell at a fixed temperature T and pressure p when one mole of Cu is produced. Show thereby that ΔE (which is the heat of reaction involved in the chemical transformation (1)) can be determined solely by measurements of the emf $\mathcal{V}$ of the cell without the necessity of doing any calorimetry.

5.17 The equation of state of a gas can be written in the form

$$p = nkT(1 + B_2 n)$$

where p is the mean pressure of the gas, T its absolute temperature, $n \equiv N/V$ the number of molecules per unit volume, and $B_2 = B_2(T)$ is the second virial coefficient. The discussion of Sec. 5·10 showed that B_2 is an increasing function of the temperature.

Find how the mean internal energy E of this gas depends on its volume V, i.e., find an expression for $(\partial E/\partial V)_T$. Is it positive or negative?

5.18 The free expansion of a gas is a process where the total mean energy E remains constant. In connection with this process, the following quantities are of interest.

(*a*) What is $(\partial T/\partial V)_E$? Express the result in terms of p, T, $(\partial p/\partial T)_V$, and C_V.

(*b*) What is $(\partial S/\partial V)_E$? Express the result in terms of p and T.

(*c*) Using the results (*a*) and (*b*), calculate the temperature change $\Delta T = T_2 - T_1$ in a free expansion of a gas from volume V_1 to volume V_2. Give explicit results for ν moles of a van der Waals gas, assuming C_V to be temperature independent.

5.19 The van der Waals equation for 1 mole of gas is given by $(p + av^{-2})(v - b) = RT$. In general, curves of p versus v for various values of T exhibit a maximum and a minimum at the two points where $(\partial p/\partial v)_T = 0$ (the curves are similar to those of Fig. 8·6·1). The maximum and minimum coalesce into a single point on that curve where $(\partial^2 p/\partial v^2)_T = 0$ in addition to $(\partial p/\partial v)_T = 0$. This point is called the "critical point" of the substance and its temperature, pressure, and molar volume are denoted by T_c, p_c, and v_c, respectively.

(*a*) Express a and b in terms of T_c and v_c.

(*b*) Express p_c in terms of T_c and v_c.

(*c*) Write the van der Waals equation in terms of the reduced dimensionless variables

$$T' \equiv \frac{T}{T_c}, \qquad v' \equiv \frac{v}{v_c}, \qquad p' \equiv \frac{p}{p_c}$$

This form should involve neither a nor b.

5.20 Find the inversion curve for a van der Waals gas. Express your result in terms of the reduced variables p' and T', and find p' as a function of T' along the inversion curve. Sketch this curve on a graph of p' versus T', being sure to indicate quantitatively the intercepts on the T' axis and the location of the maximum pressure on this curve.

5.21 The Joule-Kelvin coefficient is given by

$$\mu \equiv \left(\frac{\partial T}{\partial p}\right)_H = \frac{V}{C_p}\left[\frac{T}{V}\left(\frac{\partial V}{\partial T}\right)_p - 1\right] \tag{1}$$

Since it involves the absolute temperature T, this relation can be used to determine the absolute temperature T.

Consider *any* readily measurable *arbitrary* temperature parameter ϑ (e.g., the height of a mercury column). All that is known is that ϑ is some (unknown) function of T; i.e., $\vartheta = \vartheta(T)$.

(*a*) Express (1) in terms of the various directly measurable quantities involving the temperature parameter ϑ instead of the absolute temperature T, i.e., in terms of $\mu' \equiv (\partial\vartheta/\partial p)_H$, $C_p' \equiv (đQ/d\vartheta)_p$, $\alpha' \equiv V^{-1}(\partial V/\partial\vartheta)_p$, and the derivative $d\vartheta/dT$.

(*b*) Show that, by integrating the resulting expression, one can find T for any given value of ϑ if one knows that $\vartheta = \vartheta_0$ when $T = T_0$ (e.g., if one knows the value of $\vartheta = \vartheta_0$ at the triple point where $T_0 = 273.16$).

5.22 Refrigeration cycles have been developed for heating buildings. The procedure is to design a device which absorbs heat from the surrounding earth or air outside the house and then delivers heat at a higher temperature to the interior of the building. (Such a device is called a "heat pump.")

(*a*) If a device is used in this way, operating between the outside absolute temperature T_0 and an interior absolute temperature T_i, what would be the maximum number of kilowatt-hours of heat that could be supplied to the building for every kilowatt-hour of electrical energy needed to operate the device?

(*b*) Obtain a numerical answer for the case that the outside temperature is 0°C and the interior temperature is 25°C.

5.23 Two identical bodies, each characterized by a heat capacity at constant pressure C which is independent of temperature, are used as heat reservoirs for a heat engine. The bodies remain at constant pressure and undergo no change of phase. Initially, their temperatures are T_1 and T_2, respectively; finally, as a result of the operation of the heat engine, the bodies will attain a common final temperature T_f.

(*a*) What is the total amount of work W done by the engine? Express the answer in terms of C, T_1, T_2, and T_f.

(*b*) Use arguments based upon entropy considerations to derive an inequality relating T_f to the initial temperatures T_1 and T_2.

(*c*) For given initial temperatures T_1 and T_2, what is the maximum amount of work obtainable from the engine?

5.24 The latent heat of melting of ice is L per unit mass. A bucket contains a mixture of water and ice at the ice point (absolute temperature T_0). It is desired to use a refrigerator in order to freeze an additional mass m of water in the bucket. The heat rejected by the refrigerator goes to warm up a body of constant heat capacity C and, initially, also at temperature T_0. What is the minimum amount of heat which the refrigerator must reject to this body in the process?

5.25 Consider the physical situation illustrated in Fig. 5·11·2. Suppose that under the influence of gravity ($g = 980$ cm sec^{-2}) the weight, having a mass $m = 50$ grams, is allowed to descend a distance $L = 1$ cm before coming to rest on a platform. In this process the weight turns the paddle wheel and raises the temperature of the liquid by a slight amount above its original temperature of 25°C.

Calculate the probability that, as a result of a spontaneous fluctuation, the water gives off its energy to the weight and raises it again so as to restore it to a height of 1 cm or more.

5.26 A gasoline engine can be *approximately* represented by the idealized cyclic process *abcd* shown in the accompanying diagram of pressure p versus volume V of the gas in the cylinder. Here $a \to b$ represents the adiabatic compression of the air-gasoline mixture, $b \to c$ the rise in pressure at constant volume due to the explosion of the mixture, $c \to d$ the adiabatic expansion of the mixture during which the engine performs useful work, and $d \to a$ the final cooling down of the gas at constant volume.

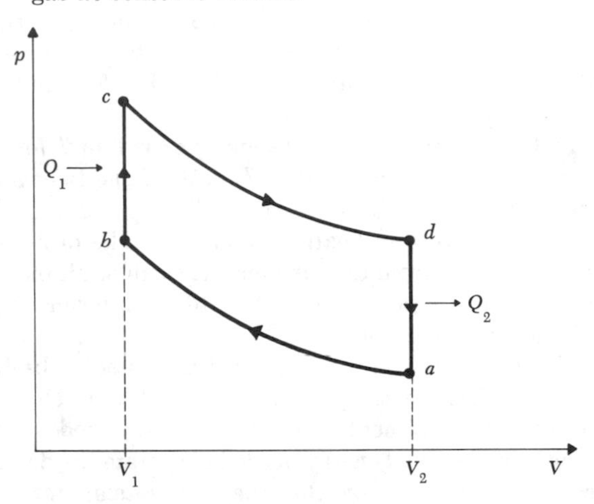

Assume this cycle to be carried out quasi-statically for a fixed amount of ideal gas having a constant specific heat. Denote the specific heat ratio by $\gamma \equiv c_p/c_v$. Calculate the efficiency η (ratio of work performed to heat intake Q_1) for this process, expressing your answer in terms of V_1, V_2, and γ.

Basic methods and results of statistical mechanics

6

ONE CAN readily discuss many more important applications based on the macroscopic aspects of the general theory of Chapter 3. But appreciably greater insight and power are gained by considering also the microscopic aspects of the theory. In this chapter we shall therefore turn our attention to the statistical relations summarized in the latter part of Sec. 3·11. Our aim will be (1) to derive general probability statements for a variety of situations of physical interest and (2) to describe practical methods for calculating macroscopic quantities (such as entropies or specific heats) from a knowledge of the purely microscopic properties of a system. In Chapter 7 we shall then apply these methods to a discussion of some important physical situations.

ENSEMBLES REPRESENTATIVE OF SITUATIONS OF PHYSICAL INTEREST

6 · 1 *Isolated system*

In giving a statistical description of a system, one always has some information available about the physical situation under consideration. The representative statistical ensemble is then constructed in such a way that all the systems in the ensemble satisfy conditions consistent with one's information about the system. Since it is possible to visualize a variety of physical situations, one is led to consider a corresponding number of representative ensembles. We shall describe some of the most important cases in the following sections.

An isolated system represents a situation of fundamental importance, one which we have discussed at length in Chapters 2 and 3. Indeed, whenever one is dealing with a situation where a system A is not isolated but is allowed to interact with some other system A', it is always possible to reduce the situation to the case of an isolated system by focusing attention on the combined system $A + A'$.

For the sake of simplicity, suppose that the volume V of the system is its only relevant external parameter. An isolated system of this kind consists then of a given number N of particles in a specified volume V, the constant energy of the system being known to lie in some range between E and $E + \delta E$. Probability statements are then made with reference to an ensemble which consists of many such systems, all consisting of this number N of particles in this volume V, and all with their energy lying in the range between E and $E + \delta E$. The fundamental statistical postulate asserts that in an equilibrium situation the system is equally likely to be found in any one of its accessible states. Thus, if the energy of a system in state r is denoted by E_r, the probability P_r of finding the system in state r is given by

$$P_r = \begin{cases} C & \text{if } E < E_r < E + \delta E \\ 0 & \text{otherwise} \end{cases} \tag{6·1·1}$$

where C is a constant. It can be determined by the normalization condition that $\Sigma P_r = 1$ when summed over all accessible states in the range between E and $E + \delta E$.

An ensemble representing an isolated system in equilibrium consists then of systems distributed in accordance with (6·1·1). It is sometimes called a "microcanonical" ensemble.

6 · 2 *System in contact with a heat reservoir*

We consider the case of a small system A in thermal interaction with a heat reservoir A'. This is the situation already discussed in Sec. 3·6 where $A \ll A'$, i.e., where A has many fewer degrees of freedom than A'. The system A may be any relatively small *macro*scopic system. (For example, it may be a bottle of wine immersed in a swimming pool, the pool acting as a heat reservoir.) Sometimes it may also be a distinguishable *micro*scopic system which can be clearly identified.* (For example, it may be an atom at some lattice site in a solid, the solid acting as a heat reservoir.) We ask the following question: Under conditions of equilibrium, what is the probability P_r of finding the system A in any *one* particular microstate r of energy E_r?

This question is immediately answered by the same reasoning as was used in Sec. 3·3. We again assume weak interaction between A and A' so that their energies are additive. The energy of A is, of course, not fixed. It is only the total energy of the combined system $A^{(0)} = A + A'$ which has a constant value in some range between $E^{(0)}$ and $E^{(0)} + \delta E$. The conservation of energy can then be written as

$$E_r + E' = E^{(0)} \tag{6·2·1}$$

where E' denotes the energy of the reservoir A'. When A has an energy E_r,

* The qualifying remark is introduced because it may not always be possible to label the identity of an individual atomic particle in a quantum mechanical description.

the reservoir A' must then have an energy near $E' = E^{(0)} - E_r$. Hence, if A is in the *one* definite state r, the number of states accessible to the combined system $A^{(0)}$ is just the number of states $\Omega'(E^{(0)} - E_r)$ accessible to A' when its energy lies in a range δE near the value $E' = E^{(0)} - E_r$. But, according to the fundamental statistical postulate, the probability of occurrence in the ensemble of a situation where A is in state r is simply proportional to the number of states accessible to $A^{(0)}$ under these conditions. Hence

$$P_r = C'\Omega'(E^{(0)} - E_r) \tag{6·2·2}$$

where C' is a constant of proportionality independent of r. As usual, it can be determined from the normalization condition for probabilities, i.e.,

$$\sum_r P_r = 1 \tag{6·2·3}$$

where the sum extends over all possible states of A irrespective of energy.

Up to now, our discussion has been completely general. Let us now make use of the fact that A is a very much smaller system than A'. Then $E_r \ll E^{(0)}$ and (6·2·2) can be approximated by expanding the slowly varying logarithm of $\Omega'(E')$ about the value $E' = E^{(0)}$. Thus

$$\ln \Omega'(E^{(0)} - E_r) = \ln \Omega'(E^{(0)}) - \left[\frac{\partial \ln \Omega'}{\partial E'}\right]_0 E_r \cdots \tag{6·2·4}$$

Since A' acts as a heat reservoir, $E_r \lll E^{(0)}$ and higher-order terms in the expansion can be neglected. The derivative

$$\left[\frac{\partial \ln \Omega'}{\partial E'}\right]_0 \equiv \beta \tag{6·2·5}$$

is evaluated at the fixed energy $E' = E^{(0)}$ and is thus a constant independent of the energy E_r of A. By (3·3·10) it is just the constant temperature parameter $\beta = (kT)^{-1}$ characterizing the *heat reservoir* A'. (Physically, this means that the reservoir A' is so large compared to A that its temperature remains unaffected by whatever small amount of energy it gives to A.) Hence (6·2·4) becomes

$$\ln \Omega'(E^{(0)} - E_r) = \ln \Omega'(E^{(0)}) - \beta E_r$$

or

$$\Omega'(E^{(0)} - E_r) = \Omega'(E^{(0)})\, e^{-\beta E_r} \tag{6·2·6}$$

Since $\Omega'(E^{(0)})$ is just a constant independent of r, (6·2·2) becomes then simply

$$P_r = C\, e^{-\beta E_r} \tag{6·2·7}$$

where C is some constant of proportionality independent of r. Using the normalization condition (6·2·3), C is determined by the relation

$$C^{-1} = \sum_r e^{-\beta E_r}$$

so that (6·2·7) can also be written explicitly in the form

$$P_r = \frac{e^{-\beta E_r}}{\sum_r e^{-\beta E_r}} \tag{6·2·8}$$

Let us discuss the results (6·2·2) or (6·2·7) more fully. If A is known to be in a definite one of its states, the reservoir can be in any one of the large number $\Omega'(E^{(0)} - E_r)$ of states accessible to it. Remember that ordinarily the number of states $\Omega'(E')$ accessible to the reservoir is a very rapidly increasing function of its energy (i.e., β in (6·2·5) is positive). Thus, if A is in a state r where its energy E_r is higher, the conservation of energy for the total system implies that the energy of the reservoir is correspondingly lower so that the number of states accessible to the reservoir is markedly reduced. The probability of encountering this situation in the ensemble is accordingly very much less. The exponential dependence of P_r on E_r in (6·2·7) just expresses this state of affairs in mathematical terms.

Example A simple numerical illustration is provided by Fig. 6·2·1 where the bar graphs show the number of states accessible to A and A' for various values of their respective energies. Assume that the total energy of the combined system is known to be 1007. Suppose that A is in one of its states, call it r, of energy 6. Then the energy of the reservoir A' must be 1001 so that it can be in any one of 400,000 possible states. In an ensemble

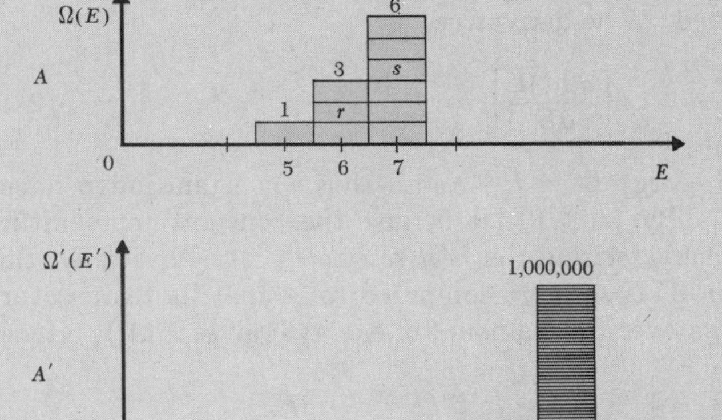

Fig. 6·2·1 Schematic illustration (not drawn to scale) showing the number of states accessible to a system A and to a heat reservoir A′ as a function of their respective energies. (The energy scale is in terms of an arbitrary unit.)

consisting of many systems $A^{(0)}$, there will then be 400,000 different possible examples of systems for which A is in state r. Suppose, however, that we consider the situation where A is in a state, such as s, where its energy is 7. Here the reservoir must have energy 1000 so that only 100,000 states are accessible to it. The ensemble contains then merely 100,000 different possible examples of systems for which A is in state s.

The probability (6·2·7) is a very general result and is of fundamental importance in statistical mechanics. The exponential factor $e^{-\beta E_r}$ is called the "Boltzmann factor"; the corresponding probability distribution (6·2·7) is known as the "canonical distribution." An ensemble of systems all of which are in contact with a heat reservoir of known temperature T, i.e., all of which are distributed over states in accordance with (6·2·7), is called a "canonical ensemble."

The fundamental result (6·2·7) gives the probability of finding A in *one* particular state r of energy E_r. The probability $P(E)$ that A has an energy in a small *range* between E and $E + \delta E$ is then simply obtained by adding the probabilities for all states whose energy lies in this range; i.e.,

$$P(E) = \sum_r P_r$$

where r is such that $E < E_r < E + \delta E$. But all these states are, by (6·2·7), equally probable and are characterized by essentially the same exponential factor $e^{-\beta E}$; hence one needs simply to multiply the probability of finding A in any one of these states by the number $\Omega(E)$ of its states in this energy range, i.e.,

$$P(E) = C\Omega(E)\, e^{-\beta E} \tag{6·2·9}$$

To the extent that A itself is a large system (although very much smaller than A'), $\Omega(E)$ is a rapidly increasing function of E. The presence of the rapidly *decreasing* factor $e^{-\beta E}$ in (6·2·9) results then in a maximum of the product $\Omega(E)\, e^{-\beta E}$. The larger A is, the sharper is this maximum in $P(E)$; i.e., the more

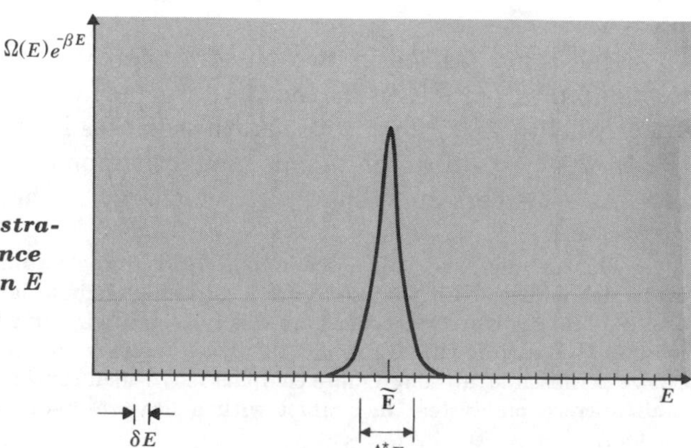

***Fig. 6·2·2** Schematic illustration showing the dependence of the function $\Omega(E)e^{-\beta E}$ on E for a macroscopic system.*

rapidly $\Omega(E)$ increases with E, the shaper this maximum becomes. Thus we arrive again at the conclusions of Sec. 3·7. We emphasize, however, that (6·2·9) is valid no matter how small A is. It may even be a system of atomic size, provided that it can be treated as a distinguishable system satisfying the additivity of energy (6·2·1).

Once the probability distribution (6·2·7) is known, various mean values can readily be computed. For example, let y be any quantity assuming the value y_r in state r of the system A. Then

$$\bar{y} = \frac{\sum_r e^{-\beta E_r} y_r}{\sum_r e^{-\beta E_r}} \qquad (6\cdot2\cdot10)$$

where the summation is over all states r of the system A.

6 · 3 *Simple applications of the canonical distribution*

The canonical distribution (6·2·7) yields a host of conclusions. Here we mention only a few illustrative applications where the canonical distribution leads immediately to physically very important results. Most of these will be discussed more fully in Chapter 7.

Paramagnetism Consider a substance which contains N_0 magnetic atoms per unit volume and which is placed in an external magnetic field $\mathbf{H}$. Assume that each atom has spin $\frac{1}{2}$ (corresponding to one unpaired electron) and an intrinsic magnetic moment μ. In a quantum-mechanical description the magnetic moment of each atom can then point either parallel or antiparallel to the external field $\mathbf{H}$. If the substance is at absolute temperature T, what is the mean magnetic moment $\bar{\mu}_H$ (in the direction of $\mathbf{H}$) of such an atom? We assume that each atom interacts only weakly with the other atoms and with the other degrees of freedom of the substance. It is then permissible to focus attention on a single atom as the small system under consideration and to regard all the other atoms and other degrees of freedom as constituting a heat reservoir.*

Each atom can be in two possible states: the state (+) where its spin points up (i.e., parallel to $\mathbf{H}$) and the state (−) where its spin points down (i.e., antiparallel to $\mathbf{H}$). Let us discuss these states in turn.

In the (+) state, the atomic magnetic moment $\boldsymbol{\mu}$ is parallel to $\mathbf{H}$ so that $\mu_H = \mu$. The corresponding magnetic energy of the atom is then $\epsilon_+ = -\mu H$.

* This assumes that it is possible to identify a single atom unambiguously, an assumption which is justified if the atoms are localized at definite lattice sites of a solid or if they form a dilute gas where the atoms are widely separated. In a concentrated gas the assumption might break down. It would then be necessary to adopt a point of view (which is always permissible, although more complicated) which considers the *entire* gas of atoms as a small microscopic system in contact with a heat reservoir provided by other degrees of freedom.

The probability of finding the atom in this state is thus

$$P_+ = C\,e^{-\beta\epsilon_+} = C\,e^{\beta\mu H} \tag{6·3·1}$$

where C is a constant of proportionality and $\beta = (kT)^{-1}$. This is the state of lower energy (if μ is positive) and is thus the state in which the atom is more likely to be found.

In the $(-)$ state, $\boldsymbol{\mu}$ is antiparallel to $\boldsymbol{H}$ so that $\mu_H = -\mu$. The corresponding energy of the atom is then $\epsilon_- = +\mu H$. The probability of finding the atom in this state is thus

$$P_- = C\,e^{-\beta\epsilon_-} = C\,e^{-\beta\mu H} \tag{6·3·2}$$

This is the state of higher energy (if μ is positive) and is thus the state in which the atom is less likely to be found.

Since the first state where $\boldsymbol{\mu}$ is parallel to $\boldsymbol{H}$ is more probable, it is clear that the mean magnetic moment $\bar{\mu}_H$ must point in the direction of the external field $\boldsymbol{H}$. By virtue of (6·3·1) and (6·3·2), the significant parameter in this problem is the quantity

$$y \equiv \beta\mu H = \frac{\mu H}{kT}$$

which measures the ratio of a typical magnetic energy to a typical thermal energy. It is apparent that if T is very large, i.e., if $y \ll 1$, the probability that $\boldsymbol{\mu}$ is parallel to $\boldsymbol{H}$ is almost the same as that of its being antiparallel. In this case $\boldsymbol{\mu}$ is almost completely randomly oriented so that $\bar{\mu}_H \approx 0$. On the other hand, if T is very small, i.e., if $y \gg 1$, then it is much more probable that $\boldsymbol{\mu}$ is parallel to $\boldsymbol{H}$ rather than antiparallel to it. In this case $\bar{\mu}_H \approx \mu$.

All these qualitative conclusions can readily be made quantitative by actually calculating the mean value $\bar{\mu}_H$. Thus we have

$$\bar{\mu}_H = \frac{P_+\mu + P_-(-\mu)}{P_+ + P_-} = \mu\,\frac{e^{\beta\mu H} - e^{-\beta\mu H}}{e^{\beta\mu H} + e^{-\beta\mu H}}$$

or

$$\bar{\mu}_H = \mu \tanh \frac{\mu H}{kT} \tag{6·3·3}$$

Here we have used the definition of the hyperbolic tangent

$$\tanh y \equiv \frac{e^y - e^{-y}}{e^y + e^{-y}}$$

The "magnetization" $\bar{M}_0$, or mean magnetic moment per unit volume, is then in the direction of H and is given by

$$\bar{M}_0 = N_0\bar{\mu}_H \tag{6·3·4}$$

One can easily check that $\bar{\mu}_H$ exhibits the qualitative behavior already discussed. If $y \ll 1$, then $e^y = 1 + y + \cdots$ and $e^{-y} = 1 - y + \cdots$. Hence

$$\text{for } y \ll 1, \qquad \tanh y = \frac{(1 + y + \cdots) - (1 - y - \cdots)}{2} = y$$

On the other hand, if $y \gg 1$, then $e^y \gg e^{-y}$. Hence

$$\text{for } y \gg 1, \qquad \tanh y = 1$$

Thus (6·3·3) leads to the limiting behavior that

$$\text{for } \mu H/kT \ll 1, \qquad \bar{\mu}_H = \frac{\mu^2 H}{kT} \tag{6·3·5a}$$

$$\text{for } \mu H/kT \gg 1, \qquad \bar{\mu}_H = \mu \tag{6·3·5b}$$

By (6·3·4) and (6·3·5a) it then follows that

$$\text{if } \mu H/kT \ll 1, \qquad \bar{M}_0 = \chi H \tag{6·3·6}$$

where χ is a constant of proportionality independent of H. This parameter χ is called the "magnetic susceptibility" of the substance. Equation (6·3·5a) provides an explicit expression for χ in terms of microscopic quantities, i.e.,

$$\chi = \frac{N_0 \mu^2}{kT} \tag{6·3·7}$$

The fact that $\chi \propto T^{-1}$ is known as Curie's law. On the other hand,

$$\text{if } \mu H/kT \gg 1, \qquad \bar{M}_0 \to N_0 \mu \tag{6·3·8}$$

becomes independent of H and equal to the maximum (or "saturation") magnetization which the substance can exhibit. The complete dependence of the magnetization $\bar{M}_0$ on temperature T and magnetic field H is shown in Fig. 6·3·1.

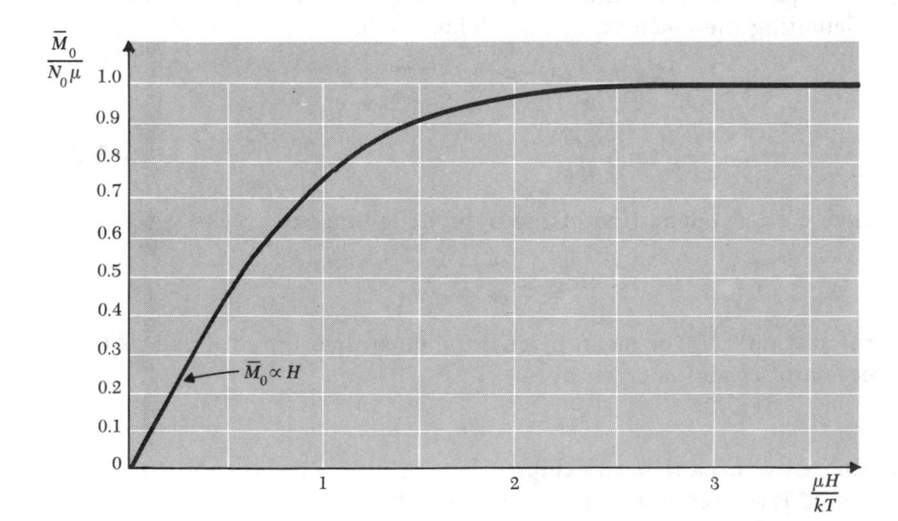

***Fig. 6·3·1** Dependence of the magnetization $\bar{M}_0$ on magnetic field H and temperature T for noninteracting magnetic atoms of spin $\frac{1}{2}$ and magnetic moment μ.*

Molecule in an ideal gas Consider a monatomic gas at absolute temperature T confined in a container of volume V. Assume that the number of molecules per unit volume is small enough that the interaction between molecules is very weak; then the total energy of the gas equals the sum of the energies of each molecule. We treat the problem classically so that it is permissible to focus attention on a given distinct molecule (without having to be concerned about the essential indistinguishability of the molecules in the gas). All the remaining molecules can then be regarded as a heat reservoir at temperature T.

The molecule can only be located somewhere inside the container. There its energy is purely kinetic; i.e.,

$$E = \frac{1}{2} m\mathbf{v}^2 = \frac{1}{2}\frac{\mathbf{p}^2}{m} \tag{6·3·9}$$

where m is the mass of the molecule and $\mathbf{v} = \mathbf{p}/m$ is its velocity. If the molecule's position lies in the range between $\mathbf{r}$ and $\mathbf{r} + d\mathbf{r}$ (i.e., if its x coordinate lies between x and $x + dx$, its y coordinate between y and $y + dy$, and its z coordinate between z and $z + dz$) and if its momentum lies in the range between $\mathbf{p}$ and $\mathbf{p} + d\mathbf{p}$ (i.e., if its x component of momentum lies between p_x and $p_x + dp_x$, . . .), then the volume of phase space corresponding to this range of $\mathbf{r}$ and $\mathbf{p}$ is

$$d^3\mathbf{r}\, d^3\mathbf{p} \equiv (dx\, dy\, dz)(dp_x\, dp_y\, dp_z) \tag{6·3·10}$$

To find the probability $P(\mathbf{r},\mathbf{p})\, d^3\mathbf{r}\, d^3\mathbf{p}$ that the molecule has position lying in the range between $\mathbf{r}$ and $\mathbf{r} + d\mathbf{r}$ and momentum in the range between $\mathbf{p}$ and $\mathbf{p} + d\mathbf{p}$, one need only multiply the number $(d^3\mathbf{r}\, d^3\mathbf{p})/h_0^3$ of cells in phase space corresponding to this range by the probability that the molecule is found in a particular cell. Thus

$$P(\mathbf{r},\mathbf{p})\, d^3\mathbf{r}\, d^3\mathbf{p} \propto \left(\frac{d^3\mathbf{r}\, d^3\mathbf{p}}{h_0^3}\right) e^{-\beta(p^2/2m)} \tag{6·3·11}$$

where $\beta \equiv (kT)^{-1}$.

Note that the probability density P does not depend on the position $\mathbf{r}$ of the molecule in the box. This reflects merely the fact that, in the absence of external forces, the symmetry of the physical situation is such that there can be no preferred location of a molecule within the box.

To find the probability $P(\mathbf{p})\, d^3\mathbf{p}$ that a molecule has momentum lying in the range between $\mathbf{p}$ and $\mathbf{p} + d\mathbf{p}$, irrespective of its location $\mathbf{r}$, one need only sum the probability (6·3·11) over all possible positions $\mathbf{r}$, i.e., integrate it over the volume of the container

$$P(\mathbf{p})\, d^3\mathbf{p} = \int_{(r)} P(\mathbf{r},\mathbf{p})\, d^3\mathbf{r}\, d^3\mathbf{p} \propto e^{-\beta(p^2/2m)}\, d^3\mathbf{p} \tag{6·3·12}$$

One can equally well express this in terms of the velocity $\mathbf{v} = \mathbf{p}/m$. The probability $P'(\mathbf{v})\, d^3\mathbf{v}$ that a molecule has a velocity between $\mathbf{v}$ and $\mathbf{v} + d\mathbf{v}$ is then

$$P'(\mathbf{v})\, d^3\mathbf{v} = P(\mathbf{p})\, d^3\mathbf{p} = C\, e^{-\beta m v^2/2}\, d^3\mathbf{v} \tag{6·3·13}$$

where C is a constant of proportionality which can be determined by the normalization condition that the integral of the probability (6·3·13) over all possible velocities of the molecule must be equal to unity. The result (6·3·13) is the famous "Maxwell distribution" of molecular velocities.

Molecule in an ideal gas in the presence of gravity Consider the situation of the preceding example, but suppose now that a uniform gravitational field acts in the $-z$ direction. Then, instead of (6·3·9), the energy of a molecule in the gas becomes

$$E = \frac{\mathbf{p}^2}{2m} + mgz \tag{6·3·14}$$

where g is the constant acceleration due to gravity. Analogously to (6·3·11), one then has

$$\begin{aligned} P(\mathbf{r},\mathbf{p})\, d^3\mathbf{r}\, d^3\mathbf{p} &\propto \frac{d^3\mathbf{r}\, d^3\mathbf{p}}{h_0^3}\, e^{-\beta[(p^2/2m)+mgz]} \\ &\propto d^3\mathbf{r}\, d^3\mathbf{p}\, e^{-\beta(p^2/2m)}\, e^{-\beta mgz} \end{aligned} \tag{6·3·15}$$

The probability now does depend on the z coordinate of the molecule. The probability $P(\mathbf{p})\, d^3\mathbf{p}$ that a molecule has momentum in the range between $\mathbf{p}$ and $\mathbf{p} + d\mathbf{p}$, irrespective of its location, is given as before by

$$P(\mathbf{p})\, d^3\mathbf{p} = \int_{(\mathbf{r})} P(\mathbf{r},\mathbf{p})\, d^3\mathbf{r}\, d^3\mathbf{p} \tag{6·3·16}$$

where the integration over $\mathbf{r}$ extends over the volume V of the container. Since (6·3·15) factors into the product of two exponentials, (6·3·16) becomes simply

$$P(\mathbf{p})\, d^3\mathbf{p} = C\, e^{-\beta(p^2/2m)}\, d^3\mathbf{p} \tag{6·3·17}$$

where C is a constant of proportionality. This means that the momentum distribution function, and thus also the velocity distribution function, is exactly the same as that obtained in (6·3·12) in the absence of a gravitational field.

Finally we can find the probability $P(z)\, dz$ that a molecule is located at a height between z and $z + dz$, irrespective of its momentum or x and y position components. This is found from (6·3·15) by integration:

$$P(z)\, dz = \int_{(x,y)} \int_{(p)} P(\mathbf{r},\mathbf{p})\, d^3\mathbf{r}\, d^3\mathbf{p} \tag{6·3·18}$$

where one integrates over all momenta (from $-\infty$ to $+\infty$ for each momentum component) and over all possible x and y values lying within the container, (i.e., over the cross-sectional area of the container). Again, since (6·3·15) factors into a product of exponentials, (6·3·18) becomes simply for a container of constant cross section

$$P(z)\, dz = C'\, e^{-\beta mgz}\, dz \tag{6·3·19}$$

where C' is a constant of proportionality. This implies that

$$P(z) = P(0)\, e^{-mgz/kT} \tag{6·3·20}$$

i.e., the probability of finding a molecule at height z decreases exponentially with the height. The result (6·3·20) is sometimes called the "law of atmospheres," since it would describe the density variation of the air near the surface of the earth if the atmosphere were at a constant temperature (which it is *not*).

6·4 *System with specified mean energy*

Another situation of physical interest is that where a system A consists of a fixed number N of particles in a given volume V, but where the only information available about the energy of the system is its *mean* energy $\bar{E}$. This is a very common situation. Suppose, for example, that a system A is brought to some final macrostate as a result of interaction with other macroscopic systems. Then the measurement of the macroscopic work done or the heat absorbed in the process does not tell one the energy of each system in the ensemble, but provides information only about the *mean* energy of the final macrostate of A.

A system A with specified *mean* energy $\bar{E}$ is also described by a canonical distribution. For, if such a system were placed in thermal contact with a heat reservoir at some temperature β, the mean energy of the system would be determined. Thus a proper choice of β would guarantee that the mean energy of the system assumes the specified value $\bar{E}$.

A more direct argument is readily given. Denote the energy of the system A in state r by E_r. Suppose that the statistical ensemble consists of a very large number a of such systems, a_r of which are in state r. Then the information available to us is that

$$\frac{1}{a}\sum_s a_s E_s = \bar{E} \tag{6·4·1}$$

equals the specified mean energy. Thus it follows that

$$\Sigma a_s E_s = a\bar{E} = \text{constant}$$

This implies that the situation is equivalent to one where a fixed total amount of energy $a\bar{E}$ is to be distributed over all the systems in the ensemble, each such system being equally likely to be in any one state. If a system in the ensemble is in state r, the remaining $(a - 1)$ systems must then have a combined energy $(a\bar{E} - E_r)$. These $(a - 1)$ systems can be distributed over some very large number $\Phi(E')$ of accessible states if their combined energy is E'. If the one system under consideration is in state r, the remaining $(a - 1)$ systems can then be with equal probability in any of the $\Phi(a\bar{E} - E_r)$ states accessible to them. Since $E_r \ll a\bar{E}$, the mathematical problem is here exactly the same as that of Sec. 6·2, dealing with a system in thermal contact with a heat reservoir, except that the role of energy reservoir is now played not by any physical heat reservoir of specified temperature parameter β, but by the totality of all the other systems in the ensemble. Accordingly, one gets again the canonical distribution

$$P_r \propto e^{-\beta E_r} \qquad (6\cdot4\cdot2)$$

The parameter $\beta = (\partial \ln \Phi/\partial E')$ does not here have any immediate physical significance in terms of the temperature of a real heat bath. Rather, it is to be determined by the condition that the mean energy calculated with the distribution (6·4·2) is indeed equal to the specified mean value $\bar{E}$, i.e., by the condition

$$\frac{\sum_r e^{-\beta E_r} E_r}{\sum_r e^{-\beta E_r}} = \bar{E} \qquad (6\cdot4\cdot3)$$

In short, when one is dealing with a system in contact with a heat reservoir of temperature $\beta = (kT)^{-1}$, the canonical distribution (6·4·2) is valid and the mean energy $\bar{E}$ can be calculated by (6·4·3) from the known value of β. If one is dealing with a system of specified mean energy $\bar{E}$, the canonical distribution (6·4·2) is again valid, but the parameter β is to be calculated by (6·4·3) from the known value of $\bar{E}$.

6 · 5 *Calculation of mean values in a canonical ensemble*

When a system A is in thermal contact with a heat reservoir as in Sec. 6·2, or when only its mean energy is known as in Sec. 6·4, the systems in the representative statistical ensemble are distributed over their accessible states in accordance with the canonical distribution

$$P_r = C\, e^{-\beta E_r} = \frac{e^{-\beta E_r}}{\sum_r e^{-\beta E_r}} \qquad (6\cdot5\cdot1)$$

In these physical situations the energy of the system is not precisely specified and the calculation of important mean values becomes particularly simple.

By (6·5·1) the mean energy is given by

$$\bar{E} = \frac{\sum_r e^{-\beta E_r} E_r}{\sum_r e^{-\beta E_r}} \qquad (6\cdot5\cdot2)$$

where the sums are over *all* accessible states r of the system, irrespective of their energy. The relation (6·5·2) can be reduced to much simpler form by noting that the sum in the numerator can be readily expressed in terms of the sum appearing in the denominator. Thus

$$\sum_r e^{-\beta E_r} E_r = -\sum_r \frac{\partial}{\partial\beta}(e^{-\beta E_r}) = -\frac{\partial}{\partial\beta} Z$$

where

$$Z \equiv \sum_r e^{-\beta E_r} \tag{6·5·3}$$

is just the sum in the denominator of (6·5·2). Hence one obtains

$$\bar{E} = -\frac{1}{Z}\frac{\partial Z}{\partial \beta} = -\frac{\partial \ln Z}{\partial \beta} \tag{6·5·4}$$

The quantity Z defined in (6·5·3) is called the "sum over states" or the "partition function." (The letter Z is used because the German name is "*Zustandsumme.*")*

The canonical distribution implies a distribution of systems over possible energies; the resulting dispersion of the energy is also readily computed. We can use the general statistical relation (1·3·10), i.e.,

$$\overline{(\Delta E)^2} \equiv \overline{(E - \bar{E})^2} = \overline{E^2 - 2\bar{E}E + \bar{E}^2} = \overline{E^2} - \bar{E}^2 \tag{6·5·5}$$

Here

$$\overline{E^2} = \frac{\sum_r e^{-\beta E_r} E_r^2}{\sum_r e^{-\beta E_r}} \tag{6·5·6}$$

But

$$\sum_r e^{-\beta E_r} E_r^2 = -\frac{\partial}{\partial \beta}\left(\sum_r e^{-\beta E_r} E_r\right) = \left(-\frac{\partial}{\partial \beta}\right)^2 \left(\sum_r e^{-\beta E_r}\right)$$

Hence (6·5·6) becomes

$$\overline{E^2} = \frac{1}{Z}\frac{\partial^2 Z}{\partial \beta^2} \tag{6·5·7}$$

This can be written in a form involving the mean energy $\bar{E}$ of (6·5·4). Thus

$$\overline{E^2} = \frac{\partial}{\partial \beta}\left(\frac{1}{Z}\frac{\partial Z}{\partial \beta}\right) + \frac{1}{Z^2}\left(\frac{\partial Z}{\partial \beta}\right)^2 = -\frac{\partial \bar{E}}{\partial \beta} + \bar{E}^2$$

Hence (6·5·5) yields

$$\overline{(\Delta E)^2} = -\frac{\partial \bar{E}}{\partial \beta} = \frac{\partial^2 \ln Z}{\partial \beta^2} \tag{6·5·8}$$

Since $\overline{(\Delta E)^2}$ can never be negative, it follows that $\partial \bar{E}/\partial \beta \leq 0$ (or equivalently, that $\partial \bar{E}/\partial T \geq 0$). These results agree with those of (3·7·15) and (3·7·16).

Suppose that the system is characterized by a single external parameter x. (The generalization of all results to the case when there are several such parameters will be immediate.) Consider a quasi-static change of the external parameter from x to $x + dx$. In this process the energy of the system in state r changes by the amount

$$\Delta_x E_r = \frac{\partial E_r}{\partial x}\, dx$$

The macroscopic work $đW$ done *by* the system as a result of this parameter

* Note that since there are in general very many states of the same energy, the sum Z contains very many terms which are equal.

change is then, corresponding to (2·9·5), given by

$$đW = \frac{\sum_r e^{-\beta E_r}\left(-\frac{\partial E_r}{\partial x}\,dx\right)}{\sum_r e^{-\beta E_r}} \tag{6·5·9}$$

where the mean value has been calculated with the canonical distribution (6·5·1). Once again the numerator can be written in terms of Z. Thus

$$\sum e^{-\beta E_r}\frac{\partial E_r}{\partial x} = -\frac{1}{\beta}\frac{\partial}{\partial x}\left(\sum_r e^{-\beta E_r}\right) = -\frac{1}{\beta}\frac{\partial Z}{\partial x}$$

and (6·5·9) becomes

$$đW = \frac{1}{\beta Z}\frac{\partial Z}{\partial x}\,dx = \frac{1}{\beta}\frac{\partial \ln Z}{\partial x}\,dx \tag{6·5·10}$$

Since one can express $đW$ in terms of the mean generalized force X

$$đW = \bar{X}\,dx, \qquad \bar{X} \equiv -\overline{\frac{\partial E_r}{\partial x}}$$

it follows also, by (6·5·10), that

▶
$$\bar{X} = \frac{1}{\beta}\frac{\partial \ln Z}{\partial x} \tag{6·5·11}$$

For example, if $x = V$, the volume of the system, Eq. (6·5·11) provides an expression for its mean pressure. That is,

$$đW = \bar{p}\,dV = \frac{1}{\beta}\frac{\partial \ln Z}{\partial V}\,dV$$

or

▶
$$\bar{p} = \frac{1}{\beta}\frac{\partial \ln Z}{\partial V} \tag{6·5·12}$$

Now Z is a function of β and V (since the energies E_r depend on V). Hence (6·5·12) is an equation relating $\bar{p}$ to $T = (k\beta)^{-1}$ and V, i.e., it gives the equation of state of the system.

6 · 6 *Connection with thermodynamics*

Note that all the important physical quantities can be expressed completely in terms of $\ln Z$.* In particular, the fact that both the mean energy $\bar{E}$ and the work $đW$ are expressible in terms of $\ln Z$ implies immediately the intimate connection between $d\bar{E}$ and $đW$ which is the content of the second law of thermodynamics. To show this explicitly, we recall that Z in (6·5·3) is a function of both

* The situation is completely analogous to that encountered in (3 · 12 · 1), where all physical quantities could be expressed in terms of $\ln \Omega$. The physical consequence (the validity of the second law in the form (6·6·4)) is the same in both cases.

β and x, since $E_r = E_r(x)$. Hence one has $Z = Z(\beta,x)$ and can write for a small change of this quantity

$$d \ln Z = \frac{\partial \ln Z}{\partial x} dx + \frac{\partial \ln Z}{\partial \beta} d\beta \qquad (6\cdot6\cdot1)$$

Consider a quasi-static process where x and β change so slowly that the system is always very close to equilibrium and thus always distributed according to the canonical distribution; then (6·6·1) implies, by virtue of (6·5·4) and (6·5·10), the relation

$$d \ln Z = \beta\, đW - \bar{E}\, d\beta \qquad (6\cdot6\cdot2)$$

The last term can be rewritten in terms of the change in $\bar{E}$ rather than the change in β. Thus

$$d \ln Z = \beta\, đW - d(\bar{E}\beta) + \beta\, d\bar{E}$$

or

$$d(\ln Z + \beta\bar{E}) = \beta(đW + d\bar{E}) \equiv \beta\, đQ \qquad (6\cdot6\cdot3)$$

where we have used the definition (2·8·3) for the heat $đQ$ absorbed by the system. Equation (6·6·3) shows again that although $đQ$ is not an exact differential, an exact differential results when $đQ$ is multiplied by the temperature parameter β. This is, of course, the content of the second law of thermodynamics, previously derived in (3·9·5) and expressed there in the form

$$dS = \frac{đQ}{T} \qquad (6\cdot6\cdot4)$$

The identification of (6·6·3) and (6·6·4) becomes complete if one puts

$$S \equiv k(\ln Z + \beta\bar{E}) \qquad (6\cdot6\cdot5)$$

It can readily be verified that this result agrees with the general definition $S \equiv k \ln \Omega(\bar{E})$ introduced in (3·3·12) for the entropy of a macroscopic system of mean energy $\bar{E}$. The partition function (6·5·3) is a sum over all states r, very many of which have the same energy. One can perform the sum by first summing over all the $\Omega(E)$ states in the energy range between E and $E + \delta E$, and then summing over all such possible energy ranges. Thus

$$Z = \sum_r e^{-\beta E_r} = \sum_E \Omega(E)\, e^{-\beta E} \qquad (6\cdot6\cdot6)$$

The summand here is just proportional to the probability (6·2·9) that the system A has an energy between E and $E + \delta E$. Since $\Omega(E)$ increases very rapidly while $e^{-\beta E}$ decreases very rapidly with increasing E, the summand $\Omega(E)e^{-\beta E}$ exhibits a *very* sharp maximum at some value $\tilde{E}$ of the energy (see Fig. 6·2·2). The mean value of the energy must then be equal to $\tilde{E}$ (i.e., $\bar{E} = \tilde{E}$), and the summand is only appreciable in some narrow range Δ^*E surrounding $\bar{E}$. The subsequent argument is similar to that used in (3·7·17). The sum in (6·6·6) must be equal to the value $\Omega(\bar{E})e^{-\beta\bar{E}}$ of the summand at its maximum multiplied by a number of the order of $(\Delta^*E/\delta E)$, this being the number of energy intervals δE contained in the range Δ^*E. Thus

$$Z = \Omega(\bar{E})\, e^{-\beta \bar{E}} \frac{\Delta^*E}{\delta E}$$

and

$$\ln Z = \ln \Omega(\bar{E}) - \beta \bar{E} + \ln \frac{\Delta^*E}{\delta E}$$

But, if the system has f degrees of freedom, the last term on the right is at most of the order of $\ln f$ and is thus utterly negligible compared to the other terms which are of the order of f. Hence

$$\ln Z = \ln \Omega(\bar{E}) - \beta \bar{E} \tag{6·6·7}$$

so that (6·6·5) reduces indeed to

$$S = k \ln \Omega(\bar{E}) \tag{6·6·8}$$

Since $k\beta = T^{-1}$, (6·6·5) can be written in the form

$$TS = kT \ln Z + \bar{E}$$

or

▶ $$F \equiv \bar{E} - TS = -kT \ln Z \tag{6·6·9}$$

Thus $\ln Z$ is very simply related to the Helmholtz free energy F already encountered in (5·5·12). Indeed, the relations (6·5·12) and (6·5·4) expressing $\bar{p}$ and $\bar{E}$ in terms of derivatives of $\ln Z$ are equivalent to the relations (5·5·14) expressing $\bar{p}$ and S in terms of derivatives of F. They express a connection between these macroscopic quantities and the partition function Z, which is calculable from microscopic information about the system. They are thus analogous to the relations (3·12·1) or (3·12·5) which connect T and $\bar{E}$ with the quantity $\ln \Omega$ or S.

Let us examine the partition function (6·5·3) in the limit as $T \to 0$ or $\beta \to \infty$. Then the only terms of appreciable magnitude in the sum are those with the lowest possible value of the energy E_r, i.e., the Ω_0 states corresponding to the ground state energy E_0. Hence

as $T \to 0$, $$Z \to \Omega_0\, e^{-\beta E_0}$$

In this limit the mean energy $\bar{E} \to E_0$, and the entropy S defined in (6·6·5) becomes,

as $T \to 0$, $$S \to k[(\ln \Omega_0 - \beta E_0) + \beta E_0] = k \ln \Omega_0 \tag{6·6·10}$$

Thus we regain the statement (known as the "third law of thermodynamics") that the entropy has the limiting property already discussed in Sec. 3·10; i.e., the entropy approaches a value (equal to zero in the absence of randomness of nuclear spin orientations) independent of all parameters of the system.

Suppose that one is dealing with a system $A^{(0)}$ consisting of two systems A and A' which are weakly interacting with each other. Let each state of A be denoted by an index r and its corresponding energy by E_r. Similarly, let each state of A' be denoted by an index s and its corresponding energy by E_s'. A state of the combined system $A^{(0)} = A + A'$ can then be denoted by the pair of indices r,s; since A and A' interact only weakly, the corresponding energy of this state is simply given by

$$E_{rs}^{(0)} = E_r + E_s' \tag{6·6·11}$$

The partition function of $A^{(0)}$ is then, by definition,

$$\begin{aligned} Z^{(0)} &= \sum_{r,s} e^{-\beta E_{rs}^{(0)}} \\ &= \sum_{r,s} e^{-\beta(E_r + E_s')} \\ &= \sum_{r,s} e^{-\beta E_r} e^{-\beta E_s'} \\ &= \left(\sum_r e^{-\beta E_r}\right)\left(\sum_s e^{-\beta E_s'}\right) \end{aligned}$$

that is,

$$Z^{(0)} = ZZ' \tag{6·6·12}$$

or

$$\ln Z^{(0)} = \ln Z + \ln Z' \tag{6·6·13}$$

where Z and Z' are the partition functions of A and A', respectively. By virtue of (6·5·4), the respective mean energies of $A^{(0)}$, A, and A' are then related by

$$\bar{E}^{(0)} = \bar{E} + \bar{E}' \tag{6·6·14}$$

It then also follows that the respective entropies of these systems are, by virtue of the definition (6·6·5), related by

$$S^{(0)} = S + S' \tag{6·6·15}$$

Hence (6·6·12) or (6·6·13) reflect the obvious fact that the extensive thermodynamic functions of two weakly interacting systems are simply additive.

Suppose, finally, that two systems A and A' are each separately in internal equilibrium with specified mean energies, or equivalently, with specified temperature parameters β and β', respectively. Then the probability P_r of finding system A in state r and the probability P_s' of finding A' in state s are given by the canonical distributions

$$P_r = \frac{e^{-\beta E_r}}{\sum_r e^{-\beta E_r}} \qquad \text{and} \qquad P_s = \frac{e^{-\beta' E_s'}}{\sum_s e^{-\beta' E_s'}} \tag{6·6·16}$$

If these systems are placed in thermal contact so that they interact only weakly with each other, then their respective probabilities are statistically independent and the probability P_{rs} of finding system A in state r *and* system s in state s is given by $P_{rs} = P_r P_s$. Immediately after the systems are brought into thermal contact, it then follows by (6·6·16) that

$$P_{rs} = \frac{e^{-\beta E_r}}{\sum_r e^{-\beta E_r}} \frac{e^{-\beta' E_s'}}{\sum_s e^{-\beta' E_s'}} \tag{6·6·17}$$

If $\beta = \beta'$, this becomes simply

$$P_{rs} = \frac{e^{-\beta(E_r+E_s')}}{\sum_r \sum_s e^{-\beta(E_r+E_s')}} \tag{6·6·18}$$

which is the canonical distribution (corresponding to temperature β) characterizing the equilibrium of the combined system $A + A'$ whose energy levels are given by (6·6·11). Hence the systems A and A' do remain in equilibrium after being joined. On the other hand, if $\beta \neq \beta'$, then (6·6·17) does not correspond to a canonical distribution of the combined system and thus does *not* describe an equilibrium situation. Hence a redistribution of systems over states tends to occur until an ultimate equilibrium situation is reached where P_{rs} is given by a canonical distribution of the form (6·6·18) with some common temperature parameter β. These comments show directly that the parameter β occurring in the canonical distribution has the familiar properties of a temperature.

The discussion of this section makes it apparent that the canonical distribution implies all the thermodynamic relations already familiar from Chapter 3. The particular definition (6·6·5) of the entropy is actually quite convenient since it involves a knowledge of $\ln Z$ rather than of $\ln \Omega$. But a computation of Z by (6·5·3) is relatively simple since it involves an unrestricted sum over *all* states, whereas a computation of $\Omega(E)$ involves the more difficult problem of counting only those states lying between the energies E and $E + \delta E$. The definition (6·6·5) for the entropy of a system at specified temperature β has the further advantages that it does not depend, even in principle, on the size δE of any arbitrary energy interval; and that it can be used to define the entropy of an arbitrarily small system. These are distinct mathematical advantages, although the physical significance of the original (and, for large systems, equivalent) definition (6·6·8) of the entropy is more transparent.

Remark It is instructive to express the physical quantities of interest directly in terms of the canonical probability P_r of (6·5·1). By (6·5·3), one can write P_r in the form

$$P_r = \frac{e^{-\beta E_r}}{Z} \tag{6·6·19}$$

The mean energy of the system is then given by

$$\bar{E} = \Sigma P_r E_r \tag{6·6·20}$$

In a general quasi-static process this energy changes because both E_r and P_r change. Thus

$$d\bar{E} = \sum_r (E_r\, dP_r + P_r\, dE_r) \tag{6·6·21}$$

The work done by the system in this process is

$$đW = \sum_r P_r(-dE_r) = -\sum P_r\, dE_r \tag{6·6·22}$$

In doing work, the energy of each state, occupied with the given probability P_r, is thus simply changed by dE_r by virtue of the change of external parameters.

The heat absorbed in this process is, by definition,

$$đQ \equiv d\bar{E} + đW$$

so that

$$đQ = \sum_r E_r \, dP_r \qquad (6 \cdot 6 \cdot 23)$$

In absorbing heat, the energy of each state is thus unaffected, but its probability of occurrence is changed.

The entropy (6·6·5) can be written

$$\begin{aligned} S &= k\left[\ln Z + \beta \sum_r P_r E_r\right] \\ &= k\left[\ln Z - \sum_r P_r \ln (ZP_r)\right] \\ &= k\left[\ln Z - \ln Z \left(\sum_r P_r\right) - \sum_r P_r \ln P_r\right] \end{aligned}$$

or

▶ $$S = -k \sum_r P_r \ln P_r \qquad (6 \cdot 6 \cdot 24)$$

since

$$\sum_r P_r = 1$$

APPROXIMATION METHODS

6 · 7 *Ensembles used as approximations*

Suppose that one is interested in discussing an isolated system with a given number N of particles in a given volume V, the energy of the system being known to lie in the range between E and $E + \delta E$. The physical equilibrium situation is then such that the system is described statistically in terms of a microcanonical ensemble where all states in the given energy range are equally probable. If a parameter y assumes the value y_r in state r, then the mean value of $\bar{y}$ is given by

$$\bar{y} = \frac{\sum_r y_r}{\Omega(E)} \qquad (6 \cdot 7 \cdot 1)$$

Here all summations are subject to the condition that one sums *only* over those states for which the energy E_r lies in the small range

$$E < E_r < E + \delta E \qquad (6 \cdot 7 \cdot 2)$$

and $\Omega(E)$ is the number of states in this particular range. The calculation of such sums and of $\Omega(E)$ may be quite difficult because of the equation of constraint (6·7·2). The trouble is that one cannot simply sum indiscriminately over all states without restriction as we did in Sec. 6·5 in calculating mean values with the canonical distribution. Instead, one must pick out only those particular states which satisfy the restriction (6·7·2). This difficulty can, however, be readily overcome by the use of quite accurate approximation methods.

One way of circumventing the difficulties presented by the condition (6·7·2) is to replace it with the weaker condition that only the *mean* energy $\bar{E}$ of the system is specified, with $\bar{E}$ chosen to be equal to the given energy E. Then the canonical distribution (6·4·2) is applicable and the probability of the system being in any one of its $\Omega(E_1)$ states of energy between E_1 and $E_1 + \delta E_1$ is given by

$$P(E_1) \propto \Omega(E_1)\, e^{-\beta E_1} \tag{6·7·3}$$

Since the number of states $\Omega(E_1)$ for a large system is a very rapidly increasing function of E_1 while $e^{-\beta E_1}$ is rapidly decreasing, the expression (6·7·3) has the usual very sharp maximum at the energy $\bar{E} = E$ (see Fig. 6·2·2). Indeed, the sharpness of this maximum can be explicitly calculated by using the canonical distribution to compute the dispersion $\overline{(E_1 - \bar{E})^2}$ by (6·5·8). The width $\Delta^* E_1$ of the maximum, given by the square root of this dispersion, is very small relative to $\bar{E}$ for a macroscopic system. (By the arguments of (3·7·14), $\Delta^* E/\bar{E}$ is ordinarily of the order of $f^{-\frac{1}{2}}$ where f is the number of degrees of freedom.) Thus, even if the energy of the system should be so precisely known that δE in (6·7·2) is very small (say $\delta E/E \approx 10^{-11}$), it is yet true that $\Delta^* E_1 < \delta E$ for a system consisting of a mole of particles. Thus values of the energy E_1 lying outside the range (6·7·2) occur with negligible probability in the canonical distribution. A specification of the mean energy $\bar{E}$ is then almost equivalent to a specification of the total energy E by (6·7·2). Hence one expects that mean values can be computed with negligible error by using the canonical distribution; i.e., instead of (6·7·1) one can write

$$\bar{y} = \frac{\sum_r e^{-\beta E_r} y_r}{\sum_r e^{-\beta E_r}} \tag{6·7·4}$$

where there appears now no further complicating restriction on the domain of summation, since one sums over *all* states.

The foregoing comments can be phrased in more physical terms. If a macroscopic system A is in contact with a heat reservoir, the relative fluctuations in the energy of A are exceedingly small. Suppose now that A is removed from contact with the heat reservoir and is thermally insulated; then its total energy cannot change at all. But the distinction between this situation and the previous one is so small that it is really utterly irrelevant for most purposes; in particular, the mean values of all physical quantities (e.g., of the mean pressure or the mean magnetic moment of A) remain quite unaffected. Hence it makes no difference whether these mean values are calculated by considering the system to be isolated so that it has equal probability of being in any one of its states of accurately specified fixed energy, or by considering it to be in contact with a heat reservoir so that it is distributed over all its states in accordance with a canonical distribution. But the latter procedure is mathematically simpler.

Calculating the *dispersion* $\overline{(y - \bar{y})^2}$ of some quantity y is a much more delicate matter. There is no guarantee that the dispersion is the same when calculated under conditions where E is precisely specified (i.e., $\delta E \to 0$ in (6·6·2)) or under conditions where only the mean energy $\bar{E}$ is specified. As a matter of fact, one would expect the dispersion to be greater in the second case. In particular, if y were the energy E of the system, its dispersion would vanish in the first case where E is precisely specified, but would not vanish in the second case where only the mean value $\bar{E}$ is specified.

When one is dealing with a macroscopic system of very precisely specified energy, the mathematical difficulties encountered in the evaluation of (6·7·1) can therefore be circumvented to excellent approximation. For purposes of calculating mean values, the situation is quite equivalent to one where the system is described by a canonical distribution with a mean energy corresponding to its actual energy.

*6·8 *Mathematical approximation methods*

The use of a canonical ensemble as an approximation method for handling the difficulties caused by the restrictive condition (6·7·2) can also be considered as a purely mathematical approximation method. This point of view is instructive both because it makes apparent how to find approximations for related situations, and because it permits one to make estimates of the errors involved.

To calculate physical quantities for an isolated system by the relations (3·12·1), one needs to know the function $\ln \Omega(E)$. Simply counting states is not very difficult if one can simply proceed in any order and add them up one at a time to get $1 + 1 + 1 + 1 + \cdots$ But the difficulty is that among all these states one wants to count only those which have an energy E_r lying in the range

$$E < E_r < E + \delta E \tag{6·8·1}$$

Thus the sum to be performed is of the form

$$\Omega(E) = \sum_r{}' u_r, \qquad u_r = 1 \text{ for all } r \tag{6·8·2}$$

where the prime on Σ denotes that the sum is to be performed subject to the restriction (6·8·1).

The basic problem is again that of handling the constraint (6·8·1). There are several ways of doing this conveniently.

Method 1 This is the mathematical analogue of the physical approximation used in the preceding section. By virtue of (6·8·1), the sum (6·8·2) depends on the particular energy E. If the energy of interest were not E, but E_1, the sum would be quite different. Indeed the sum, i.e., $\Omega(E_1)$, is a very rapidly increasing function of E_1. We wish to calculate it for the particular value $E_1 = E$. We can exploit the rapidly increasing property of the sum $\Omega(E_1)$ by noting that multiplication by the rapidly decreasing function $e^{-\beta E_1}$ produces a function $\Omega(E_1)\, e^{-\beta E_1}$ with a very sharp maximum near some value $E_1 = \tilde{E}_1$. Here β is some arbitrary positive parameter which (for the time being) has no connection whatever with temperature. By proper choice of β one can make the maximum occur at the desired value $\tilde{E}_1 = E$; one need only choose β so that

$$\frac{\partial}{\partial E_1} \ln [\Omega(E_1)\, e^{-\beta E_1}] = \frac{\partial \ln \Omega}{\partial E_1} - \beta = 0 \tag{6·8·3}$$

when $E_1 = E$.

The sharp maximum property of $\Omega(E_1)\, e^{-\beta E_1}$ implies that when this quantity is summed indiscriminately over all possible energies E_1, only those terms in some narrow range Δ^*E near E will contribute appreciably. Thus one selects only those terms of interest; i.e.,

$$\sum_{E_1} \Omega(E_1)\, e^{-\beta E_1} = \Omega(E)\, e^{-\beta E} K, \qquad K \equiv \frac{\Delta^* E}{\delta E_1}$$

where the sum is expressed in terms of the value of the summand at the maximum, multiplied by the number K of terms in the sum contained in the range Δ^*E (see Fig. 6·2·2). Taking logarithms, one gets

$$\ln \Big[\sum_{E_1} \Omega(E_1)\, e^{-\beta E_1}\Big] = \ln \Omega(E) - \beta E$$

since $\ln K$ is utterly negligible compared to the other terms. Hence

▶ $$\ln \Omega(E) = \ln Z + \beta E \tag{6·8·4}$$

where
$$Z \equiv \sum_{E_1} \Omega(E_1)\, e^{-\beta E_1} = \sum_r e^{-\beta E_r} \tag{6·8·5}$$

The last form on the right is obtained by summing over all individual states, whereas in the first sum one first groups together all terms with a given energy E and then sums over all energies. The relation (6·8·4) represents the desired approximate evaluation of $\ln \Omega$ in terms of the *unrestricted* sum Z over *all* states.

The parameter β is to be determined by the maximum condition (6·8·3) which is an equation expressing β in terms of E. Thus Z is a function of E through its dependence on β. By (6·8·4) the condition (6·8·3) becomes for $E_1 = E$

$$\left[\frac{\partial \ln Z}{\partial \beta}\frac{\partial \beta}{\partial E} + \left(E\frac{\partial \beta}{\partial E} + \beta\right)\right] - \beta = 0$$

or
$$\frac{\partial \ln Z}{\partial \beta} + E = 0 \tag{6·8·6}$$

Using (6·8·5), this equation for determining β is simply

$$\frac{\sum_r e^{-\beta E_r} E_r}{\sum_r e^{-\beta E_r}} = E \tag{6·8·7}$$

It is clear from (6·8·3) that the parameter β introduced in this approximation method is just the temperature of the system. Similarly, the entropy can be calculated by (6·8·4) as

$$S = k \ln \Omega = k(\ln Z + \beta E)$$

where the sum Z defined in (6·8·5) is simply the partition function already encountered in (6·5·3).

Method 2 It is possible to handle the restrictive condition in the sum (6·8·2) in a very straightforward fashion by a method similar to that used in Sec. 1·10. Let us shift the complication introduced by the restriction from the summation to the summand by multiplying each term in the sum by the function $\delta(E_r - E)\,\delta E$, which is equal to unity whenever E_r lies in a range δE about E, but which vanishes otherwise. By Appendix A·7 the function $\delta(E_r - E)$ is just the Dirac δ function. Then one can write

$$\Omega(E) = \sum_r \delta(E_r - E)\,\delta E \tag{6·8·8}$$

where the sum is now over all states *without* any restriction, but where the δ function in the summand guarantees that only those terms in (6·8·8) which satisfy the condition (6·8·1) contribute to the sum.

But at this point one can make use of the simple analytic representation for the δ function given in (A·7·16). Thus

$$\delta(E - E_r) = \frac{1}{2\pi}\int_{-\infty}^{\infty} d\beta'\, e^{i(E-E_r)\beta'}\, e^{(E-E_r)\beta}$$

or in more compact form

$$\delta(E - E_r) = \frac{1}{2\pi}\int_{-\infty}^{\infty} d\beta'\, e^{(E-E_r)\underline{\beta}} \tag{6·8·9}$$

where

$$\underline{\beta} \equiv \beta + i\beta' \tag{6·8·10}$$

is complex, the integration is only over its imaginary part, and β is an arbitrary parameter which can be chosen at will.

The sum (6·8·8) is now readily evaluated. One has simply

$$\Omega(E) = \frac{\delta E}{2\pi}\sum_r \int_{-\infty}^{\infty} d\beta'\, e^{(E-E_r)\underline{\beta}}$$

or

$$\Omega(E) = \frac{\delta E}{2\pi}\int_{-\infty}^{\infty} d\beta'\, e^{\underline{\beta}E} Z(\underline{\beta}) \tag{6·8·11}$$

where

$$Z(\underline{\beta}) \equiv \sum_r e^{-\underline{\beta}E_r} = \sum_r e^{-(\beta+i\beta')E_r} \tag{6·8·12}$$

This last sum is over all states without restriction and is thus relatively simple to evaluate.

These results are *exact*. We note that if $\beta' = 0$, all the terms in the sum (6 · 8 · 12) are positive. On the other hand, if $\beta' \neq 0$, the oscillatory factors $e^{-i\beta' E_r}$ cause the terms in the sum not to add in phase, but to add with more or less random signs in the real and imaginary parts. Since there are so very many terms in the sum, the result is that the absolute value $|e^{\underline{\beta} E} Z(\underline{\beta})|$ is *very* much larger for $\beta' = 0$ than for $\beta' \neq 0$. Because of this very sharp maximum, only the region of integration near $\beta' = 0$ contributes appreciably to the integral (6 · 8 · 11). Hence we expect that the integral is very well approximated by

$$\Omega(E) = K' \, e^{\beta E} Z(\beta) \tag{6·8·13}$$

where K' is some constant which is certainly small compared to the number of degrees of freedom. Thus

▶ $$\ln \Omega(E) = \beta E + \ln Z(\beta) \tag{6·8·14}$$

since $\ln K'$ is negligibly small compared to the other terms which are of order f. Thus we regain the result (6 · 8 · 4).

It is worth doing the argument leading to (6 · 8 · 13) more carefully. Because the integrand in (6 · 8 · 11) is only appreciable for $\beta' \approx 0$, one can in the significant domain of integration expand its logarithm in a power series about $\beta' = 0$. Thus

$$\begin{aligned} \ln [e^{\underline{\beta} E} Z(\underline{\beta})] &= \underline{\beta} E + \ln Z(\underline{\beta}) \\ &= (\beta + i\beta')E + \ln Z(\beta) + B_1(i\beta') + \tfrac{1}{2} B_2 (i\beta')^2 + \cdots \end{aligned}$$

or

$$\ln [e^{\underline{\beta} E} Z(\underline{\beta})] = \beta E + \ln Z(\beta) + i(E + B_1)\beta' - \tfrac{1}{2} B_2 \beta'^2 + \cdots \tag{6·8·15}$$

where
$$B_k \equiv \left[\frac{\partial^k \ln Z}{\partial \underline{\beta}^k} \right]_{\beta' = 0} = \frac{\partial^k \ln Z}{\partial \beta^k} \tag{6·8·16}$$

Hence

$$e^{\underline{\beta} E} Z(\underline{\beta}) = e^{\beta E} Z(\beta) e^{-\frac{1}{2} B_2 \beta'^2} e^{i(E + B_1)\beta'} \tag{6·8·17}$$

The parameter β is still at our disposal, and we can choose it so as to optimize our approximation. Irrespective of the choice of β, we already know (and (6 · 8 · 17) shows this explicitly) that $|e^{\underline{\beta} E} Z(\underline{\beta})|$ is always maximum for $\beta' = 0$. We should like the integrand to contribute most significantly to the integral in the immediate vicinity of $\beta' = 0$ where the expansion (6 · 8 · 15) is most nearly valid. Because of the oscillatory behavior of the integrand $e^{\underline{\beta} E} Z(\underline{\beta})$ caused by the imaginary part β', this integrand contributes to the integral most importantly in that region where it oscillates *least* rapidly, i.e., where

$$\frac{\partial}{\partial \beta'} [e^{\underline{\beta} E} Z(\underline{\beta})] = 0$$

so that the integrand is stationary with respect to the phase β'. Choosing the region of least rapid oscillation to lie at $\beta' = 0$ means choosing β so that

$$E + B_1 = 0$$

or

$$E + \frac{\partial \ln Z}{\partial \beta} = 0 \tag{6·8·18}$$

Then (6·8·17) reduces to

$$e^{\underline{\beta} E} Z(\underline{\beta}) = e^{\beta E} Z(\beta)\, e^{-\frac{1}{2}B_2\beta'^2} \tag{6·8·19}$$

The argument which led us to expect a sharp maximum of $e^{\underline{\beta} E} Z(\underline{\beta})$ at $\beta' = 0$ implies that B_2 must be such that $B_2 \gg 1$. Hence (6·8·11) becomes simply

$$\Omega(E) = \frac{\delta E}{2\pi} e^{\beta E} Z(\beta) \int_{-\infty}^{\infty} d\beta'\, e^{-\frac{1}{2}B_2\beta'^2}$$

or

$$\Omega(E) = e^{\beta E} Z(\beta) \frac{\delta E}{\sqrt{2\pi B_2}} \tag{6·8·20}$$

Thus

$$\ln \Omega(E) \approx \beta E + \ln Z$$

These are the results (6·8·13) and (6·8·14). Note also that the condition (6·8·18) which determines β is the same as that of (6·8·6), i.e., it is again equivalent to (6·8·7).*

GENERALIZATIONS AND ALTERNATIVE APPROACHES

*6·9 *Grand canonical and other ensembles*

System with an indefinite number of particles The discussion of the last few sections can be readily generalized to a variety of other situations. Consider, for example, the case where a system A of fixed volume V is in contact with a large reservoir A' with which it can exchange not only energy, but particles (see Fig. 6·9·1). Then neither the energy E of A nor the number N of particles in A are fixed, but the *total* energy $E^{(0)}$ and the total number of particles $N^{(0)}$ of the combined system $A^{(0)} \equiv A + A'$ *are* fixed; i.e.,

$$\begin{aligned} E + E' &= E^{(0)} = \text{constant} \\ N + N' &= N^{(0)} = \text{constant} \end{aligned} \tag{6·9·1}$$

where E' and N' denote the energy and number of particles in the reservoir A'. In this situation one can ask for the probability in the ensemble of finding the

* This method based on the integral (6·8·11) and its approximate evaluation by the method of stationary phase is equivalent to the so-called "Darwin-Fowler" method which employs contour integration in the complex plane and the method of steepest descents. See, for example, R. H. Fowler, "Statistical Mechanics," 2d ed., chap. 2, Cambridge University Press, Cambridge, 1955, or E. Schrödinger, "Statistical Thermodynamics," 2d ed., chap. 6, Cambridge University Press, Cambridge, 1952.

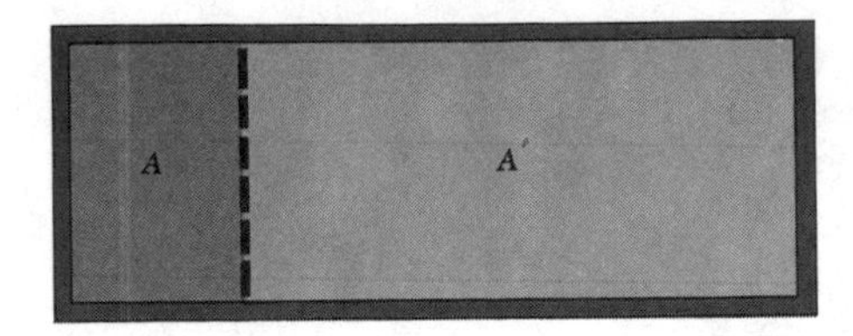

Fig. 6·9·1 A small system A separated from a much larger system A′ by a perforated partition. The systems can exchange both energy and particles.

system A in any one particular state r where it contains N_r particles and has an energy E_r.

The argument which answers this question is identical to that of Sec. 6·2. Let $\Omega'(E',N')$ denote the number of states accessible to the reservoir A' when it contains N' particles and has an energy in the range near E'. If A is in the particular state r, the number of states accessible to the combined system $A^{(0)}$ is just the number of states accessible to the reservoir. The probability P_r of finding A in this state is then proportional to this; i.e.,

$$P_r(E_r,N_r) \propto \Omega'(E^{(0)} - E_r, N^{(0)} - N_r) \tag{6·9·2}$$

where we have used the conservation equations (6·9·1). Since A is very small compared to A', $E_r \ll E^{(0)}$ and $N_r \ll N^{(0)}$. Thus

$$\ln \Omega'(E^{(0)} - E_r, N^{(0)} - N_r) = \ln \Omega'(E^{(0)}, N^{(0)}) - \left[\frac{\partial \ln \Omega}{\partial E'}\right]_0 E_r - \left[\frac{\partial \ln \Omega}{\partial N'}\right]_0 N_r$$

Here the derivatives are evaluated for $E' = E^{(0)}$ and $N' = N^{(0)}$; they are, therefore, constants characterizing the reservoir A'. Denote them by

$$\beta \equiv \left[\frac{\partial \ln \Omega}{\partial E'}\right]_0 \quad \text{and} \quad \alpha \equiv \left[\frac{\partial \ln \Omega}{\partial N'}\right]_0 \tag{6·9·3}$$

Then
$$\Omega'(E^{(0)} - E_r, N^{(0)} - N_r) = \Omega'(E^{(0)}, N^{(0)})\, e^{-\beta E_r - \alpha N_r}$$
and

▶
$$P_r \propto e^{-\beta E_r - \alpha N_r} \tag{6·9·4}$$

This is called the "grand canonical" distribution. An ensemble of systems distributed according to this probability distribution is called a "grand canonical ensemble." The parameter β is, by (6·9·3), the temperature parameter of the reservoir; thus $T \equiv (k\beta)^{-1}$ is the absolute temperature of the reservoir. The quantity $\mu \equiv -kT\alpha$ is called the "chemical potential" of the reservoir.

It is obvious from the discussion of Sec. 6·4 that if one considers a physical situation where only the *mean* energy $\bar{E}$ and the *mean* number $\bar{N}$ of particles of a system A are known, the distribution over systems in the ensemble is again described by a grand canonical distribution of the form (6·9·4). But then the parameters β and α no longer characterize any reservoir. Instead, they are to be determined by the conditions that the system A has the specified mean energy $\bar{E}$ and mean number $\bar{N}$ of particles, i.e., by the equations

$$\bar{E} = \frac{\sum_r e^{-\beta E_r - \alpha N_r} E_r}{\sum_r e^{-\beta E_r - \alpha N_r}}$$

$$\bar{N} = \frac{\sum_r e^{-\beta E_r - \alpha N_r} N_r}{\sum_r e^{-\beta E_r - \alpha N_r}} \tag{6·9·5}$$

Here the sums are over all possible states of the system A irrespective of its number of particles or of its energy.

When A is a macroscopic system in contact with a reservoir as illustrated in Fig. 6·9·1, it is again clear that the relative fluctuations of its energy about its mean energy $\bar{E}$, and of its number of particles about its mean number $\bar{N}$, are very small. Thus the physical properties of A would not be appreciably affected if it were removed from contact with the reservoir so that both its energy and number of particles would be rigorously fixed. Thus, for purposes of calculating mean values of physical quantities, it makes no noticeable difference whether a macroscopic system is isolated, or in contact with a reservoir with which it can only exchange energy, or in contact with a reservoir with which it can exchange both energy and particles. Hence these mean values can equally well be calculated by considering the system to be distributed with equal probability over all states of given energy and number of particles (microcanonical distribution), or to be distributed according to the canonical distribution (6·2·7) over all its states with a given number of particles irrespective of energy, or to be distributed according to the grand canonical distribution (6·9·4) over all its states irrespective of energy and number. In some problems where the constraint of a fixed number of particles is cumbersome, one can thus readily circumvent the complication by approximating the actual situation with one where only the mean number of particles is fixed, i.e., by using the grand canonical distribution (6·9·4). This is sometimes a useful procedure in practical calculations.

System in macroscopic motion Up to now we have always been careful to satisfy the condition of conservation of total energy for an isolated system. But what about other constants of motion like the total linear momentum or total angular momentum? The reason that we have not paid attention to these quantities is that we have always effectively considered the system A of interest to be enclosed in a container A' of very large mass. This container can take up arbitrary amounts of momentum from the system A with negligible effect on the velocity of its center of mass. The system A can thus have arbitrary amounts of momentum, and one need not be concerned about satisfying any momentum conservation conditions for it. What is effectively specified in the problem is then the velocity $\boldsymbol{v}_0$ of the container A', and we have chosen $\boldsymbol{v}_0 = 0$ with respect to the laboratory. The system A itself can then have arbitrary momentum; the condition of equilibrium is only that its *mean* velocity be the same as the specified velocity $\boldsymbol{v}_0$ of the container.

These comments show that the system A' acts like a momentum reservoir with a mass M' much larger than that of A. The analogy to the case of energy reservoirs discussed in Sec. 6·2 is apparent. It is, indeed, of some interest to discuss briefly the situation where the combined system $A^{(0)} \equiv A + A'$ is in macroscopic motion with respect to the laboratory. Consider the case where A can exchange both energy and momentum with the much larger system A'. If A is in a state r where its total energy is ϵ_r and its momentum is $\boldsymbol{p}_r$, then the conservation conditions for the combined system $A^{(0)}$ of total energy ϵ_0 and momentum $\boldsymbol{p}_0$ are

$$\begin{aligned} \epsilon_r + \epsilon' &= \epsilon_0 = \text{constant} \\ \boldsymbol{p}_r + \boldsymbol{p}' &= \boldsymbol{p}_0 = \text{constant} \end{aligned} \tag{6·9·6}$$

Here ϵ' denotes the total energy and $\boldsymbol{p}'$ the total momentum of the reservoir A'.

Up to now we have always considered systems whose center of mass is at rest with respect to the laboratory; then the total energy ϵ of a system consists only of the internal energy E of particle motion with respect to the center of mass. In the present problem the situation is different. The number of states $\Omega'(E')$ accessible to A' depends on its internal energy E' with respect to its center of mass. Since the latter moves with velocity $\boldsymbol{p}'/M'$, the internal energy of A' differs from its total energy ϵ' by the macroscopic kinetic energy of center-of-mass motion. Thus

$$E' = \epsilon' - \frac{\boldsymbol{p}'^2}{2M'} \tag{6·9·7}$$

When A is in state r, it follows by (6·9·6) that the internal energy of A' is

$$\begin{aligned} E' &= \epsilon_0 - \epsilon_r - \frac{1}{2M'}(\boldsymbol{p}_0 - \boldsymbol{p}_r)^2 \\ &= \epsilon_0 - \epsilon_r - \frac{\boldsymbol{p}_0{}^2}{2M'} + \frac{\boldsymbol{p}_0 \cdot \boldsymbol{p}_r}{M'} - \frac{\boldsymbol{p}_r{}^2}{2M'} \end{aligned}$$

or

$$E' \approx \left(\epsilon_0 - \frac{\boldsymbol{p}_0{}^2}{2M'}\right) - (\epsilon_r - \boldsymbol{v}_0 \cdot \boldsymbol{p}_r) \tag{6·9·8}$$

Since M' is very large, we have neglected the term $\boldsymbol{p}_r{}^2/M'$. Also M' is then nearly the mass of the combined system $A^{(0)}$ so that $\boldsymbol{v}_0 = \boldsymbol{p}_0/M'$ is the velocity of the center of mass of the total system $A^{(0)}$ (or equivalently, of A').

The probability P_r that A is in state r is

$$P_r \propto \Omega'(E')$$

with E' given by (6·9·8). Expanding $\ln \Omega'(E')$ in the usual way this becomes

$$▶ \qquad P_r \propto e^{-\beta(\epsilon_r - \boldsymbol{v}_0 \cdot \boldsymbol{p}_r)} \tag{6·9·9}$$

where $\beta = \partial \ln \Omega'/\partial E'$ is the temperature parameter of the reservoir evaluated when its internal energy $E' = \epsilon_0 - p_0{}^2/2M'$.

Example Consider a molecule A in an ideal gas A', the center of mass of the whole gas moving with constant velocity $\mathbf{v}_0$. Suppose that A is in a state with momentum between $\mathbf{p}$ and $\mathbf{p} + d\mathbf{p}$, or velocity between $\mathbf{v}$ and $\mathbf{v} + d\mathbf{v}$, where $\mathbf{p} = m\mathbf{v}$ and m is the mass of the molecule. Then

$$\epsilon_r - \mathbf{v}_0 \cdot \mathbf{p}_r = \tfrac{1}{2}mv^2 - \mathbf{v}_0 \cdot m\mathbf{v} = \tfrac{1}{2}m(\mathbf{v} - \mathbf{v}_0)^2 - \tfrac{1}{2}mv_0^2$$

Since $\mathbf{v}_0$ is just a constant, it follows by (6·9·9) that the probability of the molecule's velocity being in the range between $\mathbf{v}$ and $\mathbf{v} + d\mathbf{v}$ is simply

$$P(\mathbf{v})\, d^3\mathbf{v} \propto e^{-\frac{1}{2}\beta m(\mathbf{v}-\mathbf{v}_0)^2}\, d^3\mathbf{v}$$

This is, of course, what one would expect. The molecule has simply a Maxwellian velocity distribution *relative* to the frame of reference moving with the constant velocity $\mathbf{v}_0$.

*6·10 *Alternative derivation of the canonical distribution*

The canonical distribution is so important that it is worth deriving it by an alternative method. Although this derivation is more cumbersome than the one given in Sec. 6·4, it has some instructive features.

We use the notation introduced in Sec. 6·4 and consider a system A of constant specified mean energy $\bar{E}$. The representative ensemble is supposed to consist of a very large number a of such systems, a_r of which are in state r. Then we know that

$$\sum_r a_r = a \tag{6·10·1}$$

while

$$\frac{1}{a}\sum a_r E_r = \bar{E} \tag{6·10·2}$$

The number $\Gamma(a_1, a_2, \ldots)$ of distinct possible ways of selecting a total of a distinct systems in such a way that a_1 of them are in state $r = 1$, a_2 in state $r = 2$ etc., is given by the same combinatorial reasoning as that used in Sec. 1·2, i.e., by

$$\Gamma = \frac{a!}{a_1!a_2!a_3! \cdots} \tag{6·10·3}$$

Thus

$$\ln \Gamma = \ln a! - \sum_r \ln a_r! \tag{6·10·4}$$

Since the representative ensemble is supposed to consist of a very large number of systems, all the numbers a and a_r are *very* large so that Stirling's approximation can be used in its simplest form (A·6·2):

$$\ln a_r! = a_r \ln a_r - a_r$$

Hence (6·10·4) becomes

$$\ln \Gamma = a \ln a - a - \sum_r a_r \ln a_r + \sum_r a_r$$

or

$$\ln \Gamma = a \ln a - \Sigma a_r \ln a_r \qquad (6\cdot10\cdot5)$$

where we have used (6·10·1). We can now ask: For what distribution of systems over the possible states will the total number Γ of possible ways of achieving this distribution be a maximum? That is, for what set of numbers $a_1, a_2, a_3, \ldots$, subject to conditions (6·10·1) and (6·10·2), will Γ (or $\ln \Gamma$) be a maximum?

The condition that $\ln \Gamma$ has an extremum is that for small changes* δa_r of the various numbers there is no change in Γ. Thus we require that

$$\delta \ln \Gamma = -\sum_r (\delta a_r + \ln a_r\, \delta a_r) = 0 \qquad (6\cdot10\cdot6)$$

subject to the conditions (6·10·1) and (6·10·2), i.e.,

$$\sum_r \delta a_r = 0 \qquad (6\cdot10\cdot7)$$

and

$$\sum_r E_r\, \delta a_r = 0 \qquad (6\cdot10\cdot8)$$

By virtue of (6·10·7) the condition (6·10·6) becomes

$$\sum_r \ln a_r\, \delta a_r = 0 \qquad (6\cdot10\cdot9)$$

In (6·10·9) all the changes δa_r are not independent of each other since they must satisfy the equations of constraint (6·10·7) and (6·10·8). The situation is most expeditiously handled by the method of Lagrange multipliers (see Appendix A·10). Thus one can multiply (6·10·7) by a parameter α and (6·10·8) by a parameter β and then add these equations to (6·10·9) to obtain

$$\sum_r (\ln a_r + \alpha + \beta E_r)\, \delta a_r = 0 \qquad (6\cdot10\cdot10)$$

With a proper choice of α and β, all the δa_r in (6·10·10) can now be regarded as independent. Hence we can conclude that each coefficient of δa_r must separately vanish. If we denote by $\tilde{a}_r$ the value of a_r when Γ is maximum, we have then

$$\ln \tilde{a}_r + \alpha + \beta E_r = 0$$

or

$$\tilde{a}_r = e^{-\alpha}\, e^{-\beta E_r} \qquad (6\cdot10\cdot11)$$

Here the parameter α is to be determined by the normalization condition

* The numbers a_r are integers, but they are so large that even a change δa_r of many integer units is negligibly small compared to a_r itself and can be considered infinitesimal. Thus differential calculus methods are applicable.

(6·10·1); i.e.,

$$e^{-\alpha} = a(\Sigma\, e^{-\beta E_r})^{-1}$$

The parameter β is to be determined by the condition (6·10·2), i.e., by the relation

$$\frac{\Sigma\, e^{-\beta E_r} E_r}{\Sigma\, e^{-\beta E_r}} = \bar{E} \tag{6·10·12}$$

Putting

▶ $$\tilde{P}_r \equiv \frac{\tilde{a}_r}{a} = \frac{e^{-\beta E_r}}{\sum_r e^{-\beta E_r}} \tag{6·10·13}$$

we have in (6·10·13) regained the canonical distribution (6·4·2) as corresponding to that distribution of systems in the ensemble which makes the number Γ of possible configurations a maximum.

In terms of the probability $P_r = a_r/a$, the expression (6·10·5) for Γ can be written

$$\begin{aligned}\ln \Gamma &= a \ln a - \sum_r a P_r \ln (a P_r) \\ &= a \ln a - a \sum P_r (\ln a + \ln P_r) \\ &= a \ln a - a \ln a \left(\sum_r P_r\right) - a \sum P_r \ln P_r\end{aligned}$$

or

$$\ln \Gamma = -a \sum_r P_r \ln P_r \tag{6·10·14}$$

since $\Sigma P_r = 1$. (Note that the right side is properly positive, since $0 < P_r < 1$, so that $\ln P_r < 0$.)

Hence the canonical distribution P_r is characterized by the fact that it makes the quantity $-\sum_r P_r \ln P_r$ a maximum subject to a given value $\Sigma P_r E_r = \bar{E}$ of the mean energy. Comparison with the expression (6·6·24) for the entropy in terms of the canonical distribution shows that

$$\ln \tilde{\Gamma} = \frac{a}{k} S \tag{6·10·15}$$

An increase in $\ln \Gamma$ reflects a more random distribution of systems over the available states, i.e., a loss of specific information about the distribution of systems over the available states. The maximum possible value of Γ gives, by (6·10·15), the entropy of the final equilibrium state.

The quantity $-\ln \Gamma$, i.e., the function $\sum_r P_r \ln P_r$, can be used as a measure of nonrandomness, or information, available about systems in the ensemble. This function plays a key role as a measure of information in problems of communication and general "information theory."*

* See, for example, L. Brillouin, "Science and Information Theory," 2d ed., Academic Press, New York, 1962; or J. R. Pierce, "Symbols, Signals, and Noise," Harper, New York, 1961. Statistical mechanics is considered from the point of view of information theory by E. T. Jaynes in *Phys. Rev.*, vol. 106, p. 620 (1957).

Remark One can easily check that Γ is indeed a *maximum* by expanding (6·10·5) about the value $\tilde{a}_r$. Thus

$$\ln \Gamma = a \ln a - \Sigma(\tilde{a}_r + \delta a_r) \ln (\tilde{a}_r + \delta a_r) \tag{6·10·16}$$

But

$$\ln (\tilde{a}_r + \delta a_r) = \ln \tilde{a}_r + \ln \left(1 + \frac{\delta a_r}{\tilde{a}_r}\right) \approx \ln \tilde{a}_r + \frac{\delta a_r}{\tilde{a}_r} - \frac{1}{2}\left(\frac{\delta a_r}{\tilde{a}_r}\right)^2 \cdots$$

Then (6·10·16) becomes

$$\ln \Gamma = a \ln a - \sum_r \tilde{a}_r \ln \tilde{a}_r - \sum_r (1 + \ln \tilde{a}_r)\, \delta a_r - \sum_r \frac{1}{2} \frac{(\delta a_r)^2}{\tilde{a}_r}$$

The terms in δa_r vanish, as they must for an extremum, since

$$\Sigma \ln \tilde{a}_r\, \delta a_r = -\Sigma(\alpha + \beta E_r)\, \delta a_r = 0$$

by (6·10·7) and (6·10·8). Hence one is left with

$$\ln \Gamma = \left(a \ln a - \sum_r \tilde{a}_r \ln \tilde{a}_r\right) - \frac{1}{2} \sum_r \frac{(\delta a_r)^2}{\tilde{a}_r}$$

or

$$\Gamma = \tilde{\Gamma} \exp \left[-\frac{1}{2} \sum_r \frac{(\delta a_r)^2}{\tilde{a}_r} \right] \tag{6·10·17}$$

The argument of the exponential function is a very large positive number unless practically all the δa_r vanish. Hence Γ exhibits a very sharp maximum.

SUGGESTIONS FOR SUPPLEMENTARY READING

C. Kittel: "Elementary Statistical Physics," secs. 11–14, John Wiley & Sons, Inc., New York, 1958.

T. L. Hill: "An Introduction to Statistical Thermodynamics," chaps. 1 and 2, Addison-Wesley Publishing Company, Reading, Mass., 1960.

E. Schrödinger: "Statistical Thermodynamics," 2d ed., chaps. 2 and 6, Cambridge University Press, Cambridge, 1952.

R. Becker: "Theorie der Wärme," secs. 36–41, 46, Springer-Verlag, Berlin, 1955.

PROBLEMS

6.1 A simple harmonic one-dimensional oscillator has energy levels given by $E_n = (n + \frac{1}{2})\hbar\omega$, where ω is the characteristic (angular) frequency of the oscillator and where the quantum number n can assume the possible integral values $n = 0, 1, 2, \ldots$. Suppose that such an oscillator is in thermal contact with a heat reservoir at temperature T low enough so that $kT/(\hbar\omega) \ll 1$.

(*a*) Find the ratio of the probability of the oscillator being in the first excited state to the probability of its being in the ground state.

(*b*) Assuming that only the ground state and first excited state are appreciably occupied, find the mean energy of the oscillator as a funcion of the temperature T.

6.2 Consider again the system of Problem 3.2, i.e., N weakly interacting particles, each of spin $\frac{1}{2}$ and magnetic moment μ, located in an external field H. Suppose that this system is in thermal contact with a heat reservoir at the absolute temperature T. Calculate its mean energy $\bar{E}$ as a function of T and H. Compare the result with the answer to Problem 3.2*a*.

6.3 A solid at absolute temperature T is placed in an external magnetic field $H =$ 30,000 gauss. The solid contains weakly interacting paramagnetic atoms of spin $\frac{1}{2}$ so that the energy of each atom is $\pm\mu H$.

(*a*) If the magnetic moment μ is equal to one Bohr magneton, i.e., $\mu = 0.927 \times 10^{-20}$ ergs/gauss, below what temperature must one cool the solid so that more than 75 percent of the atoms are polarized with their spins parallel to the external magnetic field?

(*b*) Suppose that one considered instead a solid which is free of paramagnetic atoms but contains many protons (e.g., paraffin). Each proton has spin $\frac{1}{2}$ and a magnetic moment $\mu = 1.41 \times 10^{-23}$ ergs/gauss. Below what temperature must one cool this solid so that more than 75 percent of the protons have their spins aligned parallel to the external magnetic field?

6.4 A sample of mineral oil is placed in an external magnetic field H. Each proton has spin $\frac{1}{2}$ and a magnetic moment μ; it can, therefore, have two possible energies $\epsilon = \mp\mu H$, corresponding to the two possible orientations of its spin. An applied radio-frequency field can induce transitions between these two energy levels if its frequency ν satisfies the Bohr condition $h\nu = 2\mu H$. The power absorbed from this radiation field is then proportional to the *difference* in the number of nuclei in these two energy levels. Assume that the protons in the mineral oil are in thermal equilibrium at a temperature T which is so high that $\mu H \ll kT$. How does the absorbed power depend on the temperature T of the sample?

6.5 Consider an ideal gas at the absolute temperature T in a uniform gravitational field described by acceleration g. By writing the condition of hydrostatic equilibrium for a slice of the gas located between heights z and $z + dz$, derive an expression for $n(z)$, the number of molecules per cm^3 at height z. Compare this with Eq. (6·3·20), which was derived from statistical mechnics.

6.6 A system consists of N weakly interacting particles, each of which can be in either of two states with respective energies ϵ_1 and ϵ_2, where $\epsilon_1 < \epsilon_2$.

(*a*) Without explicit calculation, make a qualitative plot of the mean energy $\bar{E}$ of the system as a function of its temperature T. What is $\bar{E}$ in the limit of very low and very high temperatures? Roughly near what temperature does $\bar{E}$ change from its low to its high temperature limiting values?

(*b*) Using the result of (*a*), make a qualitative plot of the heat capacity C_V (at constant volume) as a function of the temperature T.

(*c*) Calculate explicitly the mean energy $\bar{E}(T)$ and heat capacity $C_V(T)$ of this system. Verify that your expressions exhibit the qualitative features discussed in (*a*) and (*b*).

6.7 The nuclei of atoms in a certain crystalline solid have spin one. According to quantum theory, each nucleus can therefore be in any one of three quantum states labeled by the quantum number m, where $m = 1$, 0, or -1. This quantum number measures the projection of the nuclear spin along a crystal axis of the solid. Since the electric charge distribution in the nucleus is not spherically symmetrical, but ellipsoidal, the energy of a nucleus depends on its spin orientation with respect to the internal electric field existing at its location. Thus a nucleus has the same energy $E = \epsilon$ in the state $m = 1$ and the state $m = -1$, compared with an energy $E = 0$ in the state $m = 0$.

(*a*) Find an expression, as a function of absolute temperature T, of the nuclear contribution to the molar internal energy of the solid.

(*b*) Find an expression, as a function of T, of the nuclear contribution to the molar entropy of the solid.

(*c*) By directly counting the total number of accessible states, calculate the nuclear contribution to the molar entropy of the solid at very low temperatures. Calculate it also at very high temperatures. Show that the expression in part (*b*) reduces properly to these values as $T \to 0$ and $T \to \infty$.

(*d*) Make a qualitative graph showing the temperature dependence of the nuclear contribution to the molar heat capacity of the solid. Calculate its temperature dependence explicitly. What is its temperature dependence for large values of T?

6.8 The following describes a simple two-dimensional model of a situation of actual physical interest. A solid at absolute temperature T contains N negatively charged impurity ions per cm^3, these ions replacing some of the ordinary atoms of the solid. The solid as a whole is, of course, electrically neutral. This is so because each negative ion with charge $-e$ has in its vicinity one positive ion with charge $+e$. The positive ion is small and thus free to move between lattice sites. In the absence of an external electric field it will, therefore, be found with equal probability in any one of the four equidistant sites surrounding the stationary negative ion (see diagram; the lattice spacing is a).

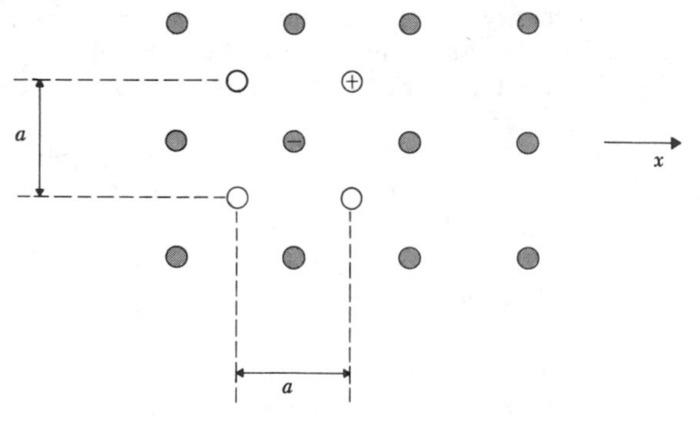

If a small electrical field $\mathcal{E}$ is applied along the x direction, calculate the electric polarization, i.e., the mean electric dipole moment per unit volume along the x direction.

6.9 A wire of radius r_0 is coincident with the axis of a metal cylinder of radius R and length L. The wire is maintained at a positive potential V (statvolts) with respect to the cylinder. The whole system is at some high absolute temperature T. As a result, electrons emitted from the hot metals form a dilute gas filling the cylindrical container and in equilibrium with it. The density of these electrons is so low that their mutual electrostatic interaction can be neglected.

(*a*) Use Gauss's theorem to obtain an expression for the electrostatic field which exists at points at a radial distance r from the wire ($r_0 < r < R$). The cylinder of length L may be assumed very long so that end effects are negligible.

(*b*) In thermal equilibrium, the electrons form a gas of variable density which fills the entire space between the wire and cylinder. Using the result of part (*a*), find the dependence of the electric charge density on the radial distance r.

6.10 A dilute solution of macromolecules (large molecules of biological interest) at temperature T is placed in an ultracentrifuge rotating with angular velocity ω. The centripetal acceleration $\omega^2 r$ acting on a particle of mass m may then be replaced by an equivalent centrifugal force $m\omega^2 r$ in the rotating frame of reference.

(*a*) Find how the relative density $\rho(r)$ of molecules varies with their radial distance r from the axis of rotation.

(*b*) Show quantitatively how the molecular weight of the macromolecules can be determined if the density ratio ρ_1/ρ_2 at the radii r_1 and r_2 is measured by optical means.

6.11 Two atoms of mass m interact with each other by a force derivable from a mutual potential energy of the form

$$U = U_0\left[\left(\frac{a}{x}\right)^{12} - 2\left(\frac{a}{x}\right)^{6}\right]$$

where x is the separation between the two particles. The particles are in contact with a heat reservoir at a temperature T low enough so that $kT \ll U_0$, but high enough so that classical statistical mechanics is applicable. Calculate the mean separation $\bar{x}(T)$ of the particles and use it to compute the quantity

$$\alpha \equiv \frac{1}{\bar{x}}\frac{\partial \bar{x}}{\partial T}$$

(This illustrates the fundamental procedure for calculating the coefficient of linear expansion of a solid.) Your calculation should make approximations based on the fact that the temperature is fairly low; thus retain only the lowest order terms which yield a value of $\alpha \neq 0$. (Hint: Expand the potential function about its minimum in a power series in x. To evaluate some of the integrals, use approximations similar to those used in evaluating the integral (A·6·12).)

6.12 Consider a rectangular box with four walls and a bottom (but no top). The total area of the walls and bottom is A. Find the dimensions of the box which give a maximum volume using

(*a*) the methods of straightforward calculus;

(*b*) Lagrange multipliers (see Appendix A·10).

****6.13*** Suppose that the expression

$$S \equiv -k \sum_r P_r \ln P_r$$

is accepted as the general definition of the entropy of a system. The following problems illustrate that the entropy so defined has indeed some very interesting properties showing that S is a measure of disorder or randomness in a system.

Imagine that a system A_1 has probability $P_r^{(1)}$ of being found in a state r and a system A_2 has probability $P_s^{(2)}$ of being found in a state s. Then one has

$$S_1 = -k \sum_r P_r^{(1)} \ln P_r^{(1)} \qquad \text{and} \qquad S_2 = -k \sum_s P_s^{(2)} \ln P_s^{(2)}$$

Each state of the composite system A consisting of A_1 and A_2 can then be labeled by the pair of numbers r, s; let the probability of A being found in this state be denoted by P_{rs}. Then its entropy is defined by

$$S = -k \sum_r \sum_s P_{rs} \ln P_{rs}$$

If A_1 and A_2 are weakly interacting so that they are statistically independent, then $P_{rs} = P_r^{(1)}P_s^{(2)}$. Show that under these circumstances the entropy is simply additive, i.e., $S = S_1 + S_2$.

***6.14** In the preceding problem, assume that A_1 and A_2 are *not* weakly interacting so that $P_{rs} \neq P_r^{(1)}P_s^{(2)}$. One has, of course, the general relations

$$P_r^{(1)} = \sum_s P_{rs} \qquad \text{and} \qquad P_s^{(2)} = \sum_r P_{rs}$$

Furthermore, all the probabilities are properly normalized so that

$$\sum_r P_r^{(1)} = 1, \qquad \sum_s P_s^{(2)} = 1, \qquad \sum_r \sum_s P_{rs} = 1$$

(*a*) Show that

$$S - (S_1 + S_2) = k \sum_{r,s} P_{rs} \ln \left(\frac{P_r^{(1)}P_s^{(2)}}{P_{rs}} \right)$$

(*b*) By using the inequality of Appendix A·8, $-\ln x \geq -x + 1$, show that

$$S \leq S_1 + S_2$$

where the equals sign holds only if $P_{rs} = P_r^{(1)}P_s^{(2)}$ holds for all r and s. This means that the existence of correlations between the systems, due to the interaction between them, leads to a situation less random than that where the systems are completely independent of each other.

***6.15** Consider a system distributed over its accessible states r in accordance with an *arbitrary* probability distribution P_r and let its entropy be defined by the relation $S = -k \sum_r P_r \ln P_r$. The distribution is properly normalized so that $\sum_r P_r = 1$. Compare this distribution with the canonical distribution

$$P_r^{(0)} = \frac{e^{-\beta E_r}}{Z}, \qquad Z \equiv \sum_r e^{-\beta E_r}$$

corresponding to the *same* mean energy $\bar{E}$, i.e.,

$$\Sigma P_r^{(0)} E_r = \Sigma P_r E_r = \bar{E}$$

The entropy for this canonical distribution is given by $S_0 = -k \sum_r P_r^{(0)} \ln P_r^{(0)}$.

(*a*) Show that

$$S - S_0 = k \sum_r [-P_r \ln P_r + P_r \ln P_r^{(0)} - P_r \ln P_r^{(0)} + P_r^{(0)} \ln P_r^{(0)}]$$

$$= k \sum_r P_r \ln \frac{P_r^{(0)}}{P_r}$$

(*b*) Using again the inequality $\ln x \leq x - 1$ of Appendix A·8, go on to show that $S_0 \geq S$, the equals sign holding only if $P_r = P_r^{(0)}$ for all states r.

This shows (in agreement with the discussion of Sec. 6·10) that, for a specified value of the mean energy, the entropy S is a maximum for the canonical distribution.

Simple applications of statistical mechanics

THE DISCUSSION of the preceding chapter dealt with some detailed microscopic aspects of the general theory of Chapter 3. As a result of this discussion, we have acquired some very powerful tools for calculating the macroscopic properties of any system in equilibrium from a knowledge of its microscopic constituents. The range of applicability of these conceptual tools is very wide indeed. In the present chapter, we shall illustrate their usefulness by discussing some rather simple, but very important, physical situations.

GENERAL METHOD OF APPROACH

7 · 1 *Partition functions and their properties*

The procedure for calculating macroscopic properties by statistical mechanics is, in principle, exceedingly simple. If the system under consideration is at a specified temperature T, i.e., if it is in thermal contact with some heat reservoir at this temperature, then one need only calculate the partition function Z of (6·5·3). Other physical quantities such as $\bar{E}$, $\bar{p}$, S, or even dispersions such as $\overline{(\Delta E)^2}$, can then be immediately obtained from the relations of Sec. 6·5 by simply taking suitable derivatives of $\ln Z$. Nor is the situation significantly different if the system is not in contact with a heat reservoir. Even if the system is isolated and has fixed energy, the mean values of the macroscopic parameters of the system are still related to its temperature T as though it were in thermal contact with a heat reservoir of this temperature. Thus the calculation is again reduced to the evaluation of the partition function Z.

Thus one arrives at the near-universal prescription for calculating macroscopic properties by statistical mechanics: evaluate the partition function*

$$Z \equiv \sum_r e^{-\beta E_r} \tag{7·1·1}$$

* An alternative prescription would, of course, be to evaluate $\Omega(E)$ and then to use relations such as (3·12·1) to find other quantities. But, for reasons already discussed, a direct calculation of $\Omega(E)$ is in general more difficult than a calculation of Z.

This is an unrestricted sum over all states of the system. If one knows the particles which constitute the system and the interactions between them, it is possible to find the quantum states of this system and to evaluate the sum (7·1·1). The statistical mechanical problem is then solved. In principle there is no difficulty in formulating the problem, no matter how complex the system may be. The difficulties are reduced to the mathematical ones of carrying out these prescriptions. Thus it is an easy task to find the quantum states and the partition function for an ideal gas of noninteracting atoms; but it is a formidable task to do the same for a liquid where all the molecules interact strongly with each other.

If the system can be treated in the classical approximation, then its energy $E(q_1, \ldots, q_f, p_1, \ldots, p_f)$ depends on some f generalized coordinates and f momenta. If phase space is subdivided into cells of volume $h_0{}^f$, the partition function in Eq. (7·1·1) can be evaluated by first summing over the number $(dq_1 \cdots dq_f\, dp_1 \cdots dp_f)/h_0{}^f$ of cells of phase space which lie in the element of volume $(dq_1 \cdots dq_f\, dp_1 \cdots dp_f)$ at the point $\{q_1, \ldots, q_f, p_1, \ldots, p_f\}$ and which have nearly the same energy $E(q_1, \ldots, q_f, p_1, \ldots, p_f)$; and then summing (or integrating) over all such elements of volume. Thus one obtains in the classical approximation

$$Z = \int \cdots \int e^{-\beta E(q_1, \ldots, p_f)} \frac{dq_1 \cdots dp_f}{h_0{}^f} \tag{7·1·2}$$

It is worth keeping in mind the following remarks concerning the partition function Z. The first remark pertains to the energy scale used in evaluating Z. The energy of a system is only defined to within an arbitrary additive constant. If one changes by a constant amount ϵ_0 the standard state with respect to which the energy is measured, the energy of each state r becomes $E_r{}^* = E_r + \epsilon_0$. Correspondingly, the partition function becomes

$$Z^* = \sum_r e^{-\beta(E_r + \epsilon_0)} = e^{-\beta\epsilon_0} \sum_r e^{-\beta E_r} = e^{-\beta\epsilon_0} Z \tag{7·1·3}$$

or

$$\ln Z^* = \ln Z - \beta\epsilon_0$$

Thus the partition function is also changed. By (6·5·4) the new mean energy is then given by

$$\bar{E}^* = -\frac{\partial \ln Z^*}{\partial\beta} = -\frac{\partial \ln Z}{\partial\beta} + \epsilon_0 = \bar{E} + \epsilon_0$$

i.e., it is properly shifted by the amount ϵ_0. On the other hand, the entropy is properly unchanged, since by (6·6·5)

$$S^* = k(\ln Z^* + \beta\bar{E}^*) = k(\ln Z + \beta\bar{E}) = S$$

Similarly, all expressions for generalized forces (i.e., all equations of state) are unchanged, since they involve only derivatives of $\ln Z$ with respect to an external parameter.

The second remark concerns the decomposition of the partition function for a system A when the latter consists of two parts A' and A'' which interact only weakly with each other. If the states of A' and A'' are labeled respectively by r and s, then a state of A can be specified by the pair of numbers r,s and its corresponding energy E_{rs} is simply additive, i.e.,

$$E_{rs} = E_r' + E_s'' \tag{7·1·4}$$

Here A' and A'' may refer to two different distinguishable groups of particles which interact weakly with each other (e.g., He and Ne molecules in an ideal-gas mixture of these two gases). Alternatively, they may refer to two different sets of degrees of freedom of the *same* group of particles (e.g., in a diatomic gas, they may refer to (1) the degrees of freedom describing the translational motion of the centers of mass of the molecules and (2) the degrees of freedom describing the rotation of these molecules about their respective centers of mass).

The important point is only the additivity of the energies in (7·1·4); for then the partition function Z for the total system A is a sum over all states labeled by rs, i.e.,

$$Z = \sum_{r,s} e^{-\beta(E_r'+E_s'')} = \sum_{r,s} e^{-\beta E_r'} e^{-\beta E_s''} = \left(\sum_r e^{-\beta E_r'}\right)\left(\sum_s e^{-\beta E_s''}\right)$$

Thus

$$Z = Z'Z'' \tag{7·1·5}$$

and

$$\ln Z = \ln Z' + \ln Z'' \tag{7·1·6}$$

where Z' and Z'' are the partition functions of A' and A'', respectively. Thus we have shown that if a system consists of distinct noninteracting parts, the partition function factors into a simple product.* This is a useful result and one which clearly is equally valid when one is dealing with more than two weakly interacting parts.

IDEAL MONATOMIC GAS

7·2 *Calculation of thermodynamic quantities*

Consider a gas consisting of N identical monatomic molecules of mass m enclosed in a container of volume V. Denote the position vector of the ith molecule by $\mathbf{r}_i$, its momentum by $\mathbf{p}_i$. Then the total energy of the gas is given by

$$E = \sum_{i=1}^{N} \frac{\mathbf{p}_i^2}{2m} + U(\mathbf{r}_1, \mathbf{r}_2, \ldots, \mathbf{r}_N) \tag{7·2·1}$$

Here the first term on the right represents the total kinetic energy of all the molecules. The term U represents the potential energy of interaction between the molecules. If the gas is sufficiently dilute that the interaction between molecules is negligible, $U \to 0$ and we obtain the simple case of an ideal gas.

* We already established this result in (6·6·13) where we showed that it implies the additivity of all extensive thermodynamic functions.

In writing (7·2·1), we assume the constraining condition that all the position vectors $\mathbf{r}_i$ lie inside the volume of the container.

Let us treat the problem classically; the validity of this approximation will be examined in Sec. 7·4. Then we can immediately use (7·1·2) to write the *classical* partition function (denote it by Z')

$$Z' = \int \exp \left\{ -\beta \left[\frac{1}{2m} (\mathbf{p}_1^2 + \cdots + \mathbf{p}_N^2) + U(\mathbf{r}_1, \ldots, \mathbf{r}_N) \right] \right\} \frac{d^3\mathbf{r}_1 \cdots d^3\mathbf{r}_N \, d^3\mathbf{p}_1 \cdots d^3\mathbf{p}_N}{h_0^{3N}}$$

or

$$Z' = \frac{1}{h_0^{3N}} \int e^{-(\beta/2m)p_1^2} \, d^3\mathbf{p}_1 \cdots \int e^{-(\beta/2m)p_N^2} \, d^3\mathbf{p}_N \int e^{-\beta U(\mathbf{r}_1, \ldots, \mathbf{r}_N)} \, d^3\mathbf{r}_1 \cdots d^3\mathbf{r}_N \quad (7\cdot2\cdot2)$$

where the second expression follows from the first by using the multiplicative property of the exponential function. Since the kinetic energy is a sum of terms, one for each molecule, the corresponding part of the partition function breaks up into a product of N integrals, each identical except for the irrelevant variable of integration, and equal to

$$\int_{-\infty}^{\infty} e^{-(\beta/2m)p^2} \, d^3\mathbf{p}$$

Since U is *not* in the form of a simple sum of terms for individual molecules, the integral over the coordinates $\mathbf{r}_1, \ldots, \mathbf{r}_N$ is very difficult to carry out. This is why the treatment of nonideal gases is complicated. But if the gas is sufficiently dilute to be ideal, then $U = 0$ and the integral becomes trivial; i.e.,

$$\int d^3\mathbf{r}_1 \cdots d^3\mathbf{r}_N = \int d^3\mathbf{r}_1 \int d^3\mathbf{r}_2 \cdots \int d^3\mathbf{r}_N = V^N$$

since each integration extends over the volume of the container. Then Z' factors into a simple product

$$Z' = \zeta^N \quad (7\cdot2\cdot3)$$

or

$$\ln Z' = N \ln \zeta \quad (7\cdot2\cdot4)$$

where

$$\zeta \equiv \frac{V}{h_0^3} \int_{-\infty}^{\infty} e^{-(\beta/2m)p^2} \, d^3\mathbf{p} \quad (7\cdot2\cdot5)$$

is the partition function for a single molecule.

Remark It would be possible to formulate this problem in a slightly different way by not imposing the condition that the position coordinate $\mathbf{r}$ of each molecule lie within the container. In that case, to write down an expression for the total energy valid everywhere, one would have to add to (7·2·1) a term

$$U' = \sum_i u(\mathbf{r}_i)$$

where $u(\mathbf{r}_i)$ represents the potential energy of a molecule due to the container, i.e.,

$$u(\mathbf{r}) = \begin{cases} 0 & \text{if } \mathbf{r} \text{ lies inside the container} \\ \infty & \text{if } \mathbf{r} \text{ lies outside the container} \end{cases}$$

In this case the partition function (7·2·2) would contain a factor $e^{-\beta U'}$ which would equal unity whenever all molecules are within the container and would equal zero whenever any one molecule is outside the container. Thus the integration over all coordinates without restriction would again immediately reduce to the form (7·2·2) of integration over the volume of the container only.

The integral in (7·2·5) is readily evaluated.

$$\int_{-\infty}^{\infty} e^{-(\beta/2m)p^2}\, d^3\mathbf{p} = \iiint_{-\infty}^{\infty} e^{-(\beta/2m)(p_x^2+p_y^2+p_z^2)}\, dp_x\, dp_y\, dp_z$$

$$= \int_{-\infty}^{\infty} e^{-(\beta/2m)p_x^2}\, dp_x \int_{-\infty}^{\infty} e^{-(\beta/2m)p_y^2}\, dp_y \int_{-\infty}^{\infty} e^{-(\beta/2m)p_z^2}\, dp_z$$

$$= \left(\sqrt{\frac{\pi 2m}{\beta}}\right)^3 \qquad \text{by (A·4·2)}$$

Hence

$$\zeta = V\left(\frac{2\pi m}{h_0^2\beta}\right)^{\frac{3}{2}} \tag{7·2·6}$$

and

$$\ln Z' = N\left[\ln V - \frac{3}{2}\ln\beta + \frac{3}{2}\ln\left(\frac{2\pi m}{h_0^2}\right)\right] \tag{7·2·7}$$

From this partition function one can immediately calculate a host of other physical quantities. By (6·5·12) the mean gas pressure $\bar{p}$ is given by

$$\bar{p} = \frac{1}{\beta}\frac{\partial \ln Z'}{\partial V} = \frac{1}{\beta}\frac{N}{V}$$

Thus

▶
$$\bar{p}V = NkT \tag{7·2·8}$$

and one regains the equation of state already derived in (3·12·8) under more general conditions (gas not necessarily monatomic).

By (6·5·4), the total mean energy of the gas is

$$\bar{E} = -\frac{\partial}{\partial\beta}\ln Z' = \frac{3}{2}\frac{N}{\beta} = N\bar{\epsilon} \tag{7·2·9}$$

where

▶
$$\bar{\epsilon} = \tfrac{3}{2}kT \tag{7·2·10}$$

is the mean energy per molecule. The heat capacity at constant volume of the gas is then given by

$$C_V = \left(\frac{\partial\bar{E}}{\partial T}\right)_V = \frac{3}{2}Nk = \frac{3}{2}\nu N_a k \tag{7·2·11}$$

where ν is the number of moles and N_a is Avogadro's number. Hence the *molar* specific heat at constant volume of a monatomic gas is

$$c_V = \tfrac{3}{2}R \tag{7·2·12}$$

where $R = N_a k$ is the gas constant. These results agree with those already obtained in (5·2·10) and (5·2·12).

Remark on fluctuations The fluctuation in the total energy of the gas in contact with a heat reservoir at temperature T can also be readily calculated. By (6·5·8) the dispersion in energy is given by

$$\overline{(\Delta E)^2} = -\frac{\partial \bar{E}}{\partial \beta}$$

Here the volume V is, of course, kept constant in taking the derivative. Putting $\beta = (kT)^{-1}$, this becomes

$$\overline{(\Delta E)^2} = -\left(\frac{\partial \bar{E}}{\partial T}\right)_V \frac{\partial T}{\partial \beta} = kT^2 \left(\frac{\partial \bar{E}}{\partial T}\right)_V$$

or

▶ $$\overline{(\Delta E)^2} = kT^2 C_V \tag{7·2·13}$$

Thus the fluctuation in energy of *any* system is quite generally related to its heat capacity at constant volume. In the particular case of a monatomic ideal gas consisting of N molecules one obtains, by (7·2·11),

$$\overline{(\Delta E)^2} = \tfrac{3}{2}Nk^2T^2 \tag{7·2·14}$$

The root-mean-square fluctuation in energy $\Delta^* E = [\overline{(\Delta E)^2}]^{\frac{1}{2}}$ can be compared to the mean energy $\bar{E}$ of the gas. Thus

$$\frac{\Delta^* E}{\bar{E}} = \frac{\sqrt{\frac{3}{2}Nk^2T^2}}{\frac{3}{2}NkT} = \sqrt{\frac{2}{3N}} \tag{7·2·15}$$

This is very small when N is of the order of Avogadro's number.

The entropy of the gas can be calculated by (6·6·5). Using (7·2·7) and (7·2·9) one obtains

$$S = k(\ln Z' + \beta \bar{E}) = Nk\left[\ln V - \frac{3}{2}\ln \beta + \frac{3}{2}\ln\left(\frac{2\pi m}{h_0^2}\right) + \frac{3}{2}\right]$$

or

$$S = Nk[\ln V + \tfrac{3}{2}\ln T + \sigma] \tag{7·2·16}$$

where

$$\sigma \equiv \frac{3}{2}\ln\left(\frac{2\pi m k}{h_0^2}\right) + \frac{3}{2}$$

is a constant independent of T, V, or N. This expression for the entropy is, however, *not* correct.

7·3 *Gibbs paradox*

The challenging statement at the end of the last section suggests that the expression (7·2·16) for the entropy merits some discussion. First, note that our calculation was carried out within the framework of classical mechanics which certainly is not valid at very low temperatures where the system is in the (relatively few) states of very low energy close to the quantum-mechanical ground state and where a quantum description is certainly needed. Hence the circumstance that Eq. (7·2·16) yields $S \to -\infty$ as $T \to 0$ (in apparent contradiction to the third law of thermodynamics) is no cause for alarm. In accordance with its classical derivation, (7·2·16) is not expected to be valid at such low temperatures.

Nevertheless, the expression (7·2·16) for S is clearly wrong since it implies that the entropy does not behave properly as an extensive quantity. Quite generally, one must require that all thermodynamic relations remain valid if the size of the whole system under consideration is simply increased by a scale factor α, i.e., if all its extensive parameters are multiplied by the same factor α. In our case, if the independent extensive parameters V and N are multiplied by α, the mean energy $\bar{E}$ in (7·2·9) is indeed properly increased by this same factor, but the entropy S in (7·2·16) is *not* increased by α because of the term $N \ln V$.

Indeed, (7·2·16) asserts that the entropy S of a fixed volume V of gas is simply proportional to the number N of molecules. But this dependence on N is *not* correct, as can readily be seen in the following way. Imagine that a partition is introduced which divides the container into two parts. This is a reversible process which does not affect the distribution of systems over accessible states. Thus the total entropy ought to be the same with, or without, the partition in place; i.e.,

$$S = S' + S'' \tag{7·3·1}$$

where S' and S'' are the entropies of the two parts. But the expression (7·2·16) does *not* yield the simple additivity required by (7·3·1). This is easily verified. Suppose, for example, that the partition divides the gas into two *equal* parts, each containing N' molecules of gas in a volume V'. Then the entropy of each part is given by (7·2·16) as

$$S' = S'' = N'k[\ln V' + \tfrac{3}{2} \ln T + \sigma]$$

while the entropy of the whole gas without partition is by (7·2·16)

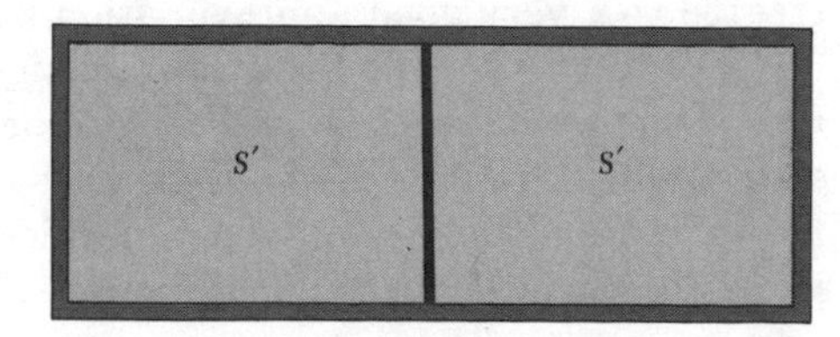

***Fig. 7·3·1** A container of gas divided into two equal parts by a partition.*

$$S = 2N'k[\ln (2V') + \tfrac{3}{2} \ln T + \sigma]$$

Hence $$S - 2S' = 2N'k \ln (2V') - 2N'k \ln V' = 2N'k \ln 2 \qquad (7\cdot 3\cdot 2)$$

and is *not* equal to zero as required by (7·3·1).

This paradox was first discussed by Gibbs and is commonly referred to as the "Gibbs paradox." Something is obviously wrong in our discussion; the question is what. Did we not prove quite generally in (6·6·15) that the entropies of two weakly interacting systems are additive? How then can we fail to satisfy the condition (7·3·1)? The answer is quite simple. Our general argument in Sec. 6·6 was based on the premise that the external parameters of each subsystem remain the *same.* If we brought the two gases in our example together and left them separated by a partition, then the volume V' of each subsystem would remain the same and their entropies would satisfy the additivity (7·3·1). But we did more than that—we also removed the partition. In that case Eq. (6·6·11) is no longer valid because the energies E_r' and E_s'' are both calculated with the volume V' as the external parameter, while for the combined system (with partition removed) the possible states of energy E_{rs} are to be calculated with total volume $2V'$ as the external parameter.

The act of removing the partition has thus very definite physical consequences. Whereas before removal of the partition a molecule of each subsystem could only be found within a volume V', after the partition is removed it can be located anywhere within the volume $V = 2V'$. *If* the two subsystems consisted of different gases, the act of removing the partition would lead to diffusion of the molecules throughout the whole volume $2V'$ and consequent random mixing of the different molecules. This is clearly an irreversible process; simply replacing the partition would not unmix the gases. In this case the increase of entropy in (7·3·2) would make sense as being simply a measure of the irreversible increase of disorder resulting from the mixing of *un*like gases.

But if the gases in the subsystems are identical, such an increase of entropy does *not* make physical sense. The root of the difficulty embodied in the Gibbs paradox is that we treated the gas molecules as individually distinguishable, as though interchanging the positions of two like molecules would lead to a physically distinct state of the gas. This is not so. Indeed, if we treated the gas by quantum mechanics (as we shall do in Chapter 9), the molecules would, as a matter of principle, have to be regarded as completely indistinguishable. A calculation of the partition function would then automatically yield the correct result, and the Gibbs paradox would never arise. Our mistake has been to take the classical point of view too seriously. Even though one may be in a temperature and density range where the motion of molecules can be treated to a very good approximation by classical mechanics, one cannot go so far as to disregard the essential indistinguishability of the molecules; one cannot observe and label individual atomic particles as though they were macroscopic billiard balls. If one does want to use the classical approximation, then the indistinguishability of the molecules must be taken into account explicitly in calculating the partition function (7·2·2). This can be done by noting that the $N!$ possible permutations of the molecules among themselves do not lead to physically distinct situations, so that the number of distinct states over which one sums in (7·2·2) is too large by a factor of $N!$. The correct partition function Z, which does take into account the essential indistinguishability of the molecules and does not lead to the Gibbs paradox difficulties, is then given by dividing (7·2·3) by this factor, i.e.,

$$Z = \frac{Z'}{N!} = \frac{\zeta^N}{N!} \tag{7·3·3}$$

Note that in a *strictly* classical description it would be permissible to consider every particle as distinguishable. If one agrees to consider identical molecules as essentially indistinguishable so as to avoid the Gibbs paradox, then the following question arises: Just how different must molecules be before they should be considered distinguishable (i.e., before their mutual mixing leads to a finite, instead of no, increase of entropy)? In a classical view of nature two molecules could, of course, differ by infinitesimal amounts (e.g., the nuclei of two atoms could have infinitesimally different masses). In a quantum description this troublesome question does not arise because of the quantized discreteness of nature (e.g., the nuclei of two isotopes differ by at least one nucleon mass). Hence the distinction between identical and non-identical molecules is completely unambiguous in a quantum-mechanical description. The Gibbs paradox thus foreshadowed already in the last century conceptual difficulties that were resolved satisfactorily only by the advent of quantum mechanics.

By (7·3·3) one then gets

$$\ln Z = N \ln \zeta - \ln N!$$

or

$$\ln Z = N \ln \zeta - N \ln N + N \tag{7·3·4}$$

where we have used Stirling's formula. Equation (7·3·4) differs from the corresponding expression (7·2·4) only by the additive term $(-N \ln N + N)$. Since the pressure $\bar{p}$ and energy $\bar{E}$ depend only on *derivatives* of $\ln Z$ with respect to V or β, the previous results (7·2·8) and (7·2·9) for these quantities are unaffected. But the expression for S, which does involve $\ln Z$ itself rather than only its derivatives, is changed by this additive term. Thus (7·3·4) yields, instead of (7·2·16), the result

$$S = kN[\ln V + \tfrac{3}{2} \ln T + \sigma] + k(-N \ln N + N)$$

or

$$▶ \quad S = kN\left[\ln \frac{V}{N} + \frac{3}{2} \ln T + \sigma_0\right] \tag{7·3·5}$$

where

$$\sigma_0 \equiv \sigma + 1 = \frac{3}{2} \ln \left(\frac{2\pi mk}{h_0^2}\right) + \frac{5}{2} \tag{7·3·6}$$

It is apparent that the extra term involving $\ln N$ avoids the difficulties of the Gibbs paradox. The entropy S in (7·3·5) behaves properly like an extensive quantity; i.e., it does get multiplied by a factor α if both V and N are multiplied by α.

Since h_0 is an arbitrary constant in the present classical calculation, σ_0 is some arbitrary additive constant in the entropy. Note that (7·3·5) agrees exactly with the entropy expression derived by macroscopic reasoning in (5·4·4). It is only necessary to put $N = \nu N_a$, where ν is the number of moles of gas, and to use the relation (7·2·12), according to which $c_V = \frac{3}{2}N_a k$ for a monatomic ideal gas.

7 · 4 *Validity of the classical approximation*

We saw that the essential indistinguishability of identical molecules cannot be disregarded even if the motion of the molecules can be treated by classical mechanics. But to what extent is the latter procedure itself valid? That is, to what extent is it permissible to evaluate the partition function (7·2·2) in terms of coordinates $\mathbf{r}_i$ and momenta $\mathbf{p}_i$ which can be simultaneously specified?

An approximate criterion for the validity for this classical description can be obtained by appealing to the Heisenberg uncertainty principle

$$\Delta q\, \Delta p \gtrsim \hbar \tag{7·4·1}$$

This relates the uncertainties Δq and Δp introduced by quantum effects in any attempt at simultaneous specification of a position q and corresponding momentum p of a particle. Suppose that one tries to describe the motion of the gas molecules by classical mechanics. Denote the magnitude of the mean momentum of a molecule by* $\bar{p}$ and the mean separation between molecules by $\bar{R}$. Then one would certainly expect a classical description to be applicable if

$$\bar{R}\bar{p} \gg \hbar \tag{7·4·2}$$

when (7·4·1) implies that quantum mechanical effects are not important. Equivalently (7·4·2) expresses the condition that

▶
$$\bar{R} \gg \bar{\lambda} \tag{7·4·3}$$

i.e., that the mean separation between particles is much greater than their mean de Broglie wavelength

$$\bar{\lambda} = 2\pi \frac{\hbar}{\bar{p}} = \frac{h}{\bar{p}} \tag{7·4·4}$$

When (7·4·3) is satisfied so that $\bar{R} \gg \bar{\lambda}$, the quantum description ought to be equivalent to the motion of wave packets describing individual particles which move independently in a quasi-classical manner. In the opposite limit, where $\bar{R} \ll \bar{\lambda}$, a state of the whole gas will be shown in Chapter 9 to be described by a single wave function which cannot be decomposed in any simple way; it thus

* The symbol $\bar{p}$ should not be confused with the mean pressure of the gas. We shall denote the latter quantity by capital $\bar{P}$ later in this section.

results in correlations between the motions of the particles even if no forces exist between them.

The mean intermolecular separation $\bar{R}$ can be estimated by imagining each molecule at the center of a little cube of side $\bar{R}$, these cubes filling the available volume V. Then

$$\bar{R}^3 N = V$$

or

$$\bar{R} = \left(\frac{V}{N}\right)^{\frac{1}{3}} \tag{7·4·5}$$

The mean momentum $\bar{p}$ can be estimated from the known mean energy $\bar{\epsilon}$ of a molecule in the gas at temperature T. By (7·2·10)

$$\frac{1}{2m}\bar{p}^2 \approx \bar{\epsilon} = \frac{3}{2}kT$$

Thus

$$\bar{p} \approx \sqrt{3mkT}$$

and

$$\bar{\lambda} \approx \frac{h}{\sqrt{3mkT}} \tag{7·4·6}$$

Hence the condition (7·4·3) becomes

$$\left(\frac{V}{N}\right)^{\frac{1}{3}} \gg \frac{h}{\sqrt{3mkT}} \tag{7·4·7}$$

This shows that the classical approximation ought to be applicable if the concentration N/V of molecules in the gas is sufficiently small, if the temperature T is sufficiently high, and if the mass of the molecules is not too small.

Numerical estimates Consider, for example, helium (He) gas at room temperature and pressure. Then one has

mean pressure $\bar{P} = 760$ mm Hg $\approx 10^6$ dynes/cm^2
temperature $T \approx 300°$K; hence $kT \approx 4 \times 10^{-14}$ ergs
molecular mass $m = \dfrac{4}{6 \times 10^{23}} \approx 7 \times 10^{-24}$ grams

The equation of state gives

$$\frac{N}{V} = \frac{\bar{P}}{kT} = 2.5 \times 10^{19} \text{ molecules/cm}^3$$

Thus $\bar{R} \approx 34 \times 10^{-8}$ cm by (7·4·5)
and $\bar{\lambda} \approx 0.6 \times 10^{-8}$ cm by (7·4·6)

Here the criterion (7·4·3) is quite well satisfied, and the classical evaluation of the partition function ought to be a very good approximation if the indistinguishability of the particles is taken into account. Most gases have larger molecular weights and thus smaller de Broglie wavelengths; the criterion (7·4·3) is then even better satisfied.

On the other hand, consider the conduction electrons in a typical metal. In a first approximation, interactions between these electrons can be neglected so that they can be treated as an ideal gas. But the numerical values of the significant parameters are then quite different. First, the mass of the electron is very small, about 10^{-27} g or 7000 times less than that of the He atom. This makes the de Broglie wavelength of the electron much longer,

$$\bar{\lambda} \approx (0.6 \times 10^{-8}) \sqrt{7000} \approx 50 \times 10^{-8} \text{ cm}$$

In addition, there is about one conduction electron per atom in the metal. Since there is roughly one atom in a cube 2×10^{-8} cm on a side,

$$\bar{R} \approx 2 \times 10^{-8} \text{ cm}$$

This is much smaller than for the He gas case; i.e., the electrons in a metal form a very dense gas. Hence the criterion (7·4·3) is certainly not satisfied. Thus there exists no justification for discussing electrons in a metal by classical statistical mechanics; indeed, a completely quantum-mechanical treatment is essential.

THE EQUIPARTITION THEOREM

7 · 5 *Proof of the theorem*

In *classical* statistical mechanics there exists a very useful general result which we shall now establish. As usual, the energy of a system is a function of some f generalized coordinates q_k and corresponding f generalized momenta p_k; i.e.,

$$E = E(q_1, \ldots, q_f, p_1, \ldots, p_f) \tag{7·5·1}$$

The following is a situation of frequent occurrence:

a. The total energy splits additively into the form

$$E = \epsilon_i(p_i) + E'(q_1, \ldots, p_f) \tag{7·5·2}$$

where ϵ_i involves only the one variable p_i and the remaining part E' does *not* depend on p_i.

b. The function ϵ_i is quadratic in p_i; i.e., it is of the form

$$\epsilon_i(p_i) = bp_i^2 \tag{7·5·3}$$

where b is some constant.

The most common situation is one where p_i is a momentum. The reason is that the kinetic energy is usually a quadratic function of each momentum component, while the potential energy does not involve the momenta.

If in assumptions (*a*) and (*b*) the variable were not a momentum p_i but a coordinate q_i satisfying the same two conditions, the theorem we want to establish would be exactly the same.

We ask the question: What is the mean value of ϵ_i in thermal equilibrium if conditions (*a*) and (*b*) are satisfied?

If the system is in equilibrium at the absolute temperature $T = (k\beta)^{-1}$, it is distributed in accordance with the canonical distribution; the mean value

$\bar{\epsilon}_i$ is then, by definition, expressible in terms of integrals over all phase space

$$\bar{\epsilon}_i = \frac{\int_{-\infty}^{\infty} e^{-\beta E(q_1, \ldots, p_f)} \epsilon_i \, dq_1 \cdots dp_f}{\int_{-\infty}^{\infty} e^{-\beta E(q_1, \ldots, p_f)} \, dq_1 \cdots dp_f} \tag{7·5·4}$$

By condition (*a*) this becomes

$$\bar{\epsilon}_i = \frac{\int e^{-\beta(\epsilon_i + E')} \epsilon_i \, dq_1 \cdots dp_f}{\int e^{-\beta(\epsilon_i + E')} \, dq_1 \cdots dp_f}$$
$$= \frac{\int e^{-\beta \epsilon_i} \epsilon_i \, dp_i \int e^{-\beta E'} \, dq_1 \cdots dp_f}{\int e^{-\beta \epsilon_i} \, dp_i \int e^{-\beta E'} \, dq_1 \cdots dp_f}$$

where we have used the multiplicative property of the exponential function and where the last integrals in both numerator and denominator extend over all terms q and p except p_i. These integrals are equal and thus cancel; hence only the one-dimensional integrals survive:

$$\bar{\epsilon}_i = \frac{\int e^{-\beta \epsilon_i} \epsilon_i \, dp_i}{\int e^{-\beta \epsilon_i} \, dp_i} \tag{7·5·5}$$

This can be simplified further by reducing the integral in the numerator to that in the denominator. Thus

$$\bar{\epsilon}_i = \frac{-\frac{\partial}{\partial \beta} \left(\int e^{-\beta \epsilon_i} \, dp_i \right)}{\int e^{-\beta \epsilon_i} \, dp_i}$$

or

$$\bar{\epsilon}_i = -\frac{\partial}{\partial \beta} \ln \left(\int_{-\infty}^{\infty} e^{-\beta \epsilon_i} \, dp_i \right) \tag{7·5·6}$$

Up to now we have made use only of the assumption (7·5·2). Let us now use the second assumption (7·5·3). Then the integral in (7·5·6) becomes

$$\int_{-\infty}^{\infty} e^{-\beta \epsilon_i} \, dp_i = \int_{-\infty}^{\infty} e^{-\beta b p_i^2} \, dp_i = \beta^{-\frac{1}{2}} \int_{-\infty}^{\infty} e^{-b y^2} \, dy$$

where we have introduced the variable $y \equiv \beta^{\frac{1}{2}} p_i$. Thus

$$\ln \int_{-\infty}^{\infty} e^{-\beta \epsilon_i} \, dp_i = -\tfrac{1}{2} \ln \beta + \ln \int_{-\infty}^{\infty} e^{-b y^2} \, dy$$

But here the integral on the right does *not* involve β at all. Hence (7·5·6) becomes simply

$$\bar{\epsilon}_i = -\frac{\partial}{\partial \beta} \left(-\frac{1}{2} \ln \beta \right) = \frac{1}{2\beta}$$

or

▶ $$\bar{\epsilon}_i = \tfrac{1}{2} kT \tag{7·5·7}$$

Note the great generality of this result and that we obtained it *without* needing to evaluate a single integral.

Equation (7·5·7) is the so-called "equipartition theorem" of classical statistical mechanics. In words it states that the mean value of each independent quadratic term in the energy is equal to $\frac{1}{2}kT$.

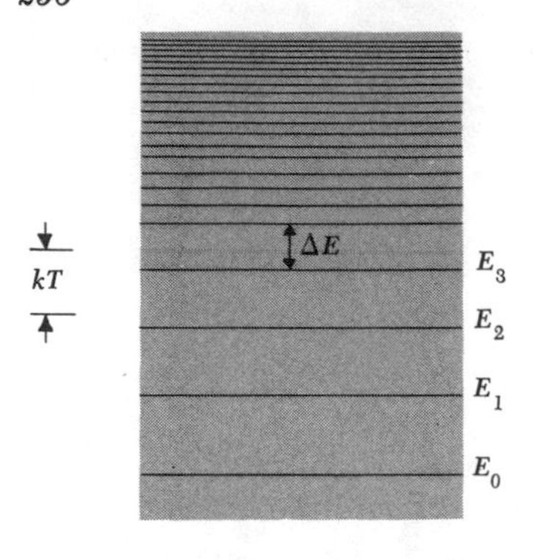

Fig. 7·5·1 Schematic diagram of the energy levels of a system.

It should be emphasized that the equipartition theorem is valid only in *classical* statistical mechanics. In the correct quantum-mechanical description a system has a set of possible energy levels, as indicated in Fig. 7·5·1, where E_0 is the ground-state energy and where at higher energies the levels are, in general, increasingly closely spaced. When the absolute temperature is sufficiently high (and thus the mean energy of the system is sufficiently high) the spacing ΔE between levels around the mean energy $\bar{E}$ is small compared to the thermal energy kT; i.e., $\Delta E \ll kT$. In this case the fact that there are discrete energy levels is not particularly important, and the classical description (and equipartition theorem where applicable) can be expected to be a good approximation. On the other hand, when the temperature is sufficiently low so that $kT \lesssim \Delta E$, the classical description must certainly break down.

7 · 6 *Simple applications*

Mean kinetic energy of a molecule in a gas Consider a molecule in a gas (not necessarily an ideal gas) at temperature T. If this molecule has mass m and a center-of-mass momentum $\mathbf{p} = m\mathbf{v}$, its kinetic energy of translation is

$$K = \frac{1}{2m}(p_x^2 + p_y^2 + p_z^2) \tag{7·6·1}$$

The kinetic energy of the other molecules does not involve the momentum $\mathbf{p}$ of this particular molecule. The potential energy of interaction between molecules depends only on their position coordinates and thus certainly does not involve $\mathbf{p}$. Finally, if the molecule is polyatomic, the internal energy of vibration or rotation of its atoms relative to its center of mass also does not involve $\mathbf{p}$. Hence the essential conditions of the equipartition theorem are satisfied. Since (7·6·1) contains three quadratic terms, the equipartition theorem allows one to conclude immediately that

$$\bar{K} = \tfrac{3}{2}kT \tag{7·6·2}$$

if the motion of the center of mass can be treated classically.

For an ideal monatomic gas the *entire* energy is kinetic, so that the mean energy per mole of gas is simply

$$\bar{E} = N_a(\tfrac{3}{2}kT) = \tfrac{3}{2}RT$$

The molar specific heat at constant volume is then

$$c_V = \left(\frac{\partial \bar{E}}{\partial T}\right)_V = \frac{3}{2}R \tag{7·6·3}$$

Brownian motion Consider a macroscopic particle of mass m immersed in a liquid at temperature T. Let the z axis point in the direction of the gravitational field (if one is present) and focus attention on v_x, the x component of the center-of-mass velocity of the particle. The mean value of v_x must vanish by symmetry; i.e.,

$$\bar{v}_x = 0$$

But it is, of course, not true that v_x itself is always found to vanish if one observes a collection of such particles; velocity fluctuations *do* occur. Indeed, the equipartition theorem can be applied to the center-of-mass energy terms just as in the preceding example; thus one can conclude that

$$\overline{\tfrac{1}{2}mv_x^2} = \tfrac{1}{2}kT \qquad \text{or} \qquad \overline{v_x^2} = \frac{kT}{m}$$

The dispersion $\overline{v_x^2}$ in this velocity component is thus negligibly small when m is large. For example, when the particle is the size of a golf ball, fluctuations in its velocity are essentially unobservable and the particle appears to be at rest. But when m is small (e.g., when the particle has a diameter of about a micron), $\overline{v_x^2}$ becomes appreciable and velocity fluctuations can readily be observed under a microscope. The fact that small particles of this kind perpetually move about in a random manner was first observed by Brown, a botanist, in the last century. The phenomenon is, therefore, called "Brownian motion." It was explained theoretically by Einstein in 1905 on the basis of the intrinsic thermal fluctuations resulting from the interaction of the small particle with the heat bath, i.e., from the random collisions of the particle with the molecules of the liquid. The phenomenon was historically important in helping to gain acceptance for the atomic theory of all matter and for the validity of the statistical description thereof.

Harmonic oscillator Consider a one-dimensional harmonic oscillator which is in equilibrium with a heat reservoir at absolute temperature T. The energy of such an oscillator is given by

$$E = \frac{p^2}{2m} + \frac{1}{2}\kappa_0 x^2 \tag{7·6·4}$$

where the first term on the right is the kinetic energy involving the momentum p and mass m, and the second term on the right is the potential energy involving the position coordinate x and spring constant κ_0. Each of these terms is quadratic in the respective variable. Hence the equipartition theorem leads immediately to the following conclusions, valid in the classical approximation:

$$\text{mean kinetic energy} = \frac{1}{2m}\overline{p^2} = \tfrac{1}{2}kT$$

$$\text{mean potential energy} = \tfrac{1}{2}\kappa_0\overline{x^2} = \tfrac{1}{2}kT$$

Hence the mean total energy is

$$\bar{E} = \tfrac{1}{2}kT + \tfrac{1}{2}kT = kT \tag{7·6·5}$$

It is instructive to treat this example by quantum mechanics as an illustration of the limits of validity of the classical description. According to quantum mechanics the possible energy levels of the harmonic oscillator are given by

$$E_n = (n + \tfrac{1}{2})\hbar\omega \tag{7·6·6}$$

where the possible states of the oscillator are labeled by the quantum number n which can assume all integral values

$$n = 0, 1, 2, 3, \ldots$$

Here $\hbar$ is Planck's constant (divided by 2π) and

$$\omega = \sqrt{\frac{\kappa_0}{m}} \tag{7·6·7}$$

is the classical angular frequency of oscillation of the oscillator. The mean energy of the oscillator is then given by

$$\bar{E} = \frac{\sum_{n=0}^{\infty} e^{-\beta E_n} E_n}{\sum_{n=0}^{\infty} e^{-\beta E_n}} = -\frac{1}{Z}\frac{\partial Z}{\partial \beta} = -\frac{\partial}{\partial \beta}\ln Z \tag{7·6·8}$$

where $$Z \equiv \sum_{n=0}^{\infty} e^{-\beta E_n} = \sum_{n=0}^{\infty} e^{-(n+\frac{1}{2})\beta\hbar\omega} \tag{7·6·9}$$

or $$Z = e^{-\frac{1}{2}\beta\hbar\omega} \sum_{n=0}^{\infty} e^{-n\beta\hbar\omega} = e^{-\frac{1}{2}\beta\hbar\omega}(1 + e^{-\beta\hbar\omega} + e^{-2\beta\hbar\omega} + \cdots)$$

This sum is just an infinite geometric series where each term is obtained from the preceding one as a result of multiplication by $e^{-\beta\hbar\omega}$. The sum can thus immediately be evaluated to give

$$Z = e^{-\frac{1}{2}\beta\hbar\omega} \frac{1}{1 - e^{-\beta\hbar\omega}} \tag{7·6·10}$$

or $$\ln Z = -\tfrac{1}{2}\beta\hbar\omega - \ln(1 - e^{-\beta\hbar\omega}) \tag{7·6·11}$$

Thus one obtains, by (7·6·8),

$$\bar{E} = -\frac{\partial}{\partial \beta}\ln Z = -\left(-\frac{1}{2}\hbar\omega - \frac{e^{-\beta\hbar\omega}\hbar\omega}{1 - e^{-\beta\hbar\omega}}\right)$$

or

$$\blacktriangleright \qquad \bar{E} = \hbar\omega\left(\frac{1}{2} + \frac{1}{e^{\beta\hbar\omega} - 1}\right) \tag{7·6·12}$$

Let us now investigate some limiting cases. When

$$\beta\hbar\omega = \frac{\hbar\omega}{kT} \ll 1 \tag{7·6·13}$$

the temperature is so high that the thermal energy kT is large compared to the separation $\hbar\omega$ between energy levels. Then one expects the classical description to be a good approximation. Indeed, if (7·6·13) is valid, the exponential function can be expanded in Taylor's series so that (7·6·12) becomes

$$\bar{E} = \hbar\omega\left[\frac{1}{2} + \frac{1}{(1 + \beta\hbar\omega + \cdots) - 1}\right] \approx \hbar\omega\left[\frac{1}{2} + \frac{1}{\beta\hbar\omega}\right]$$

$$\approx \hbar\omega\left[\frac{1}{\beta\hbar\omega}\right] \qquad \text{by virtue of (7·6·13)}$$

or

$$\bar{E} = \frac{1}{\beta} = kT \tag{7·6·14}$$

in agreement with the classical result (7·6·5).

On the other hand, at low temperatures where

$$\beta\hbar\omega = \frac{\hbar\omega}{kT} \gg 1 \tag{7·6·15}$$

one has $e^{\beta\hbar\omega} \gg 1$, so that (7·6·12) becomes

$$\bar{E} = \hbar\omega(\tfrac{1}{2} + e^{-\beta\hbar\omega}) \tag{7·6·16}$$

This is quite different from the equipartition result (7·6·5) and approaches properly the ("zero point") energy $\frac{1}{2}\hbar\omega$ of the ground state as $T \to 0$.

7·7 *Specific heats of solids*

Consider any simple solid with Avogadro's number N_a of atoms per mole. Examples might be copper, gold, aluminum, or diamond. These atoms are free to vibrate about their equilibirum positions. (Such vibrations are called "lattice vibrations.") Each atom is specified by three position coordinates and three momentum coordinates. Since the vibrations are supposed to be small, the potential energy of interaction between atoms can be expanded about their equilibrium positions and is therefore quadratic in the atomic displacements from their equilibrium positions. The net result is that the total energy of lattice vibrations can be written (when expressed in terms of appropriate "normal mode coordinates") in the simple form

$$E = \sum_{i=1}^{3N_a}\left(\frac{p_i^2}{2m} + \frac{1}{2}\kappa_i q_i^2\right) \tag{7·7·1}$$

Here the first term is the total kinetic energy involving the $3N_a$ (normal-mode) momenta of the atoms, while the second term is the total potential energy involving their $3N_a$ (normal-mode) coordinates. The coefficients κ_i are positive constants. Thus the total energy is the same as that of $3N_a$ independent one-dimensional harmonic oscillators. If the temperature T is high enough so that classical mechanics is applicable (and room temperature is usually sufficiently high for that), application of the equipartition theorem allows one to conclude immediately that the total mean energy per mole is

$$\bar{E} = 3N_a[(\tfrac{1}{2}kT) \times 2]$$

or

$$\bar{E} = 3N_akT = 3RT \qquad (7\cdot7\cdot2)$$

Thus the molar specific heat at constant volume becomes

$$c_V = \left(\frac{\partial \bar{E}}{\partial T}\right)_V = 3R \qquad (7\cdot7\cdot3)$$

This result asserts that at sufficiently high temperatures all simple solids have the same molar specific heat equal to $3R$ (25 joules mole^{-1} deg^{-1}). Historically, the validity of this result was first discovered empirically and is known as the law of Dulong and Petit. Table 7·7·1 lists directly measured values of the molar specific heat c_p at constant *pressure* for some solids at room temperature. The molar specific heat c_V at constant volume is somewhat less (by about 5 percent, as calculated in the numerical example of Sec. 5·7).

Table 7·7·1 Values* of c_p (joules mole^{-1} deg^{-1}) for some solids at T = 298°K

Solid	c_p	*Solid*	c_p
Copper	24.5	Aluminum	24.4
Silver	25.5	Tin (white)	26.4
Lead	26.4	Sulfur (rhombic)	22.4
Zinc	25.4	Carbon (diamond)	6.1

* "American Institute of Physics Handbook," 2d ed., McGraw-Hill Book Company, New York, 1963, p. 4–48.

Of course, the preceding arguments are not valid for solids at appreciably lower temperatures. Indeed, the third law leads to the general result (5·7·19), which requires that c_V must approach zero as $T \to 0$. One can obtain an approximate idea of the behavior of c_V at all temperatures by making the crude assumption (first introduced by Einstein) that all atoms in the solid vibrate with the *same* angular frequency ω. Then $\kappa_i = m\omega^2$ for all terms i in (7·7·1), and the mole of solid is equivalent to an assembly of $3N_a$ independent one-dimensional harmonic oscillators. These can be treated by quantum mechanics so that their total mean energy is just $3N_a$ times that of the single oscillator discussed in (7·6·12); i.e.,

$$\bar{E} = 3N_a\hbar\omega\left(\frac{1}{2} + \frac{1}{e^{\beta\hbar\omega} - 1}\right) \qquad (7\cdot7\cdot4)$$

Hence the molar specific heat of the solid on the basis of this simple Einstein model is given by

$$c_V = \left(\frac{\partial \bar{E}}{\partial T}\right)_V = \left(\frac{\partial \bar{E}}{\partial \beta}\right)_V \frac{\partial \beta}{\partial T} = -\frac{1}{kT^2}\left(\frac{\partial \bar{E}}{\partial \beta}\right)_V$$

$$= -\frac{3N_a\hbar\omega}{kT^2}\left[-\frac{e^{\beta\hbar\omega}\hbar\omega}{(e^{\beta\hbar\omega}-1)^2}\right]$$

or

$$c_V = 3R\left(\frac{\Theta_E}{T}\right)^2 \frac{e^{\Theta_E/T}}{(e^{\Theta_E/T}-1)^2} \tag{7·7·5}$$

where $R \equiv N_a k$ and where we have written

$$\beta\hbar\omega = \frac{\hbar\omega}{kT} \equiv \frac{\Theta_E}{T}$$

by introducing the characteristic "Einstein temperature"

$$\Theta_E \equiv \frac{\hbar\omega}{k} \tag{7·7·6}$$

If the temperature is so high that $kT \gg \hbar\omega$ or $T \gg \Theta_E$, then $\Theta_E/T \ll 1$ and expansion of the exponentials in (7·7·5) yields again the classical result

for $T \gg \Theta_E$,
$$c_V \to 3R \tag{7·7·7}$$

On the other hand, if the temperature is so low that $kT \ll \hbar\omega$ or $T \ll \Theta_E$ then $\Theta_E/T \gg 1$ and the exponential factor becomes very large compared to unity. The specific heat then becomes quite small; more precisely,

for $T \ll \Theta_E$,
$$c_V \to 3R\left(\frac{\Theta_E}{T}\right)^2 e^{-\Theta_E/T} \tag{7·7·8}$$

Thus the specific heat should approach zero exponentially as $T \to 0$. Experimentally the specific heat approaches zero more slowly than this, indeed $c_V \propto T^3$ as $T \to 0$. The reason for this discrepancy is the crude assumption that all atoms vibrate with the same characteristic frequency. In reality this is, of course, not the case (even if all the atoms are identical). The reason is that each atom does not vibrate separately as though it were experiencing a force due to stationary neighboring atoms; instead, there exist many different modes of motion in which various *groups* of atoms oscillate in phase at the same frequency. It is thus necessary to know the many different possible frequencies of these modes of oscillation (i.e., the values of all the coefficients κ_i in (7·7·1)). This problem will be considered in greater detail in Secs. 10·1 and 10·2. But it is qualitatively clear that, although T may be quite small, there are always some modes of oscillation (those corresponding to sufficiently large groups of atoms moving together) with a frequency ω so low that $\hbar\omega \ll kT$. These modes still contribute appreciably to the specific heat and thus prevent c_V from decreasing quite as rapidly as indicated by (7·7·8).

Nevertheless, the very simple Einstein approximation does give a reasonably good description of the specific heats of solids. It also makes clear the existence of a characteristic parameter Θ_E which depends on the properties of

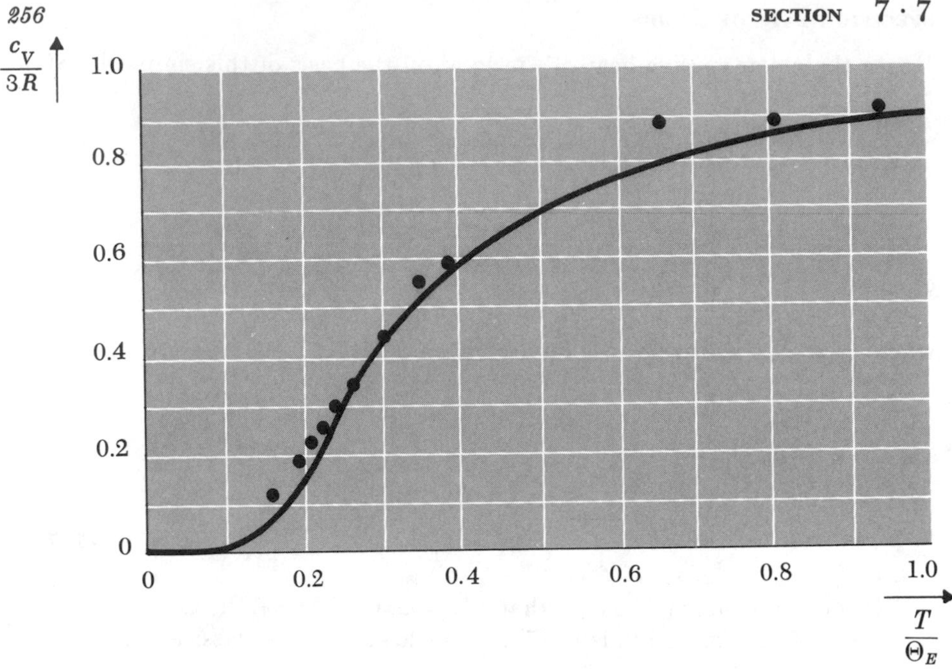

***Fig. 7·7·1** Temperature dependence of c_V according to the Einstein model. The points are experimental values of c_V for diamond, the fit to the curve being achieved by choosing Θ_E = 1320°K (after A. Einstein, Ann. Physik, vol. 22, p. 186 (1907)).*

the solid under consideration. For example, if a solid has atoms of low molecular weight and is hard (i.e., relatively incompressible), this implies that each oscillator has a small mass m and a large spring constant κ_0 (i.e., the spring is stiff). Then (7·6·7) shows that the angular frequency of vibration ω of the atoms is large, or that Θ_E defined in (7·7·6) is large. Thus one must go to higher temperatures before the classical limit $c_V = 3R$ is reached. This explains why a solid such as diamond, which consists of relatively light carbon atoms and is quite hard, has at room temperature a specific heat c_V which is still considerably smaller than the classical value $3R$ (see Table 7·7·1). Thus for diamond a reasonably good fit with experiment can be obtained by choosing $\Theta_E = 1320°\text{K}$ (see Fig. 7·7·1). For most other solids Θ_E lies closer to $\Theta_E \approx 300°\text{K}$. This corresponds to a frequency of vibration $\omega/2\pi \approx k\Theta_E/(2\pi\hbar)$ of about 6×10^{12} cycles/sec, i.e., to a frequency in the infrared region of the electromagnetic spectrum.

> Before the introduction of quantum ideas it was not possible to understand why the molar specific heats of solids should fall below the classical equipartition value $3R$ at low temperatures. In 1907 Einstein's theory clarified the mystery and helped to gain acceptance for the new quantum concepts.

PARAMAGNETISM

7·8 *General calculation of magnetization*

We have considered a simple example of paramagnetism in Sec. 6·3. Here we shall discuss the general case of arbitrary spin.

Consider a system consisting of N noninteracting atoms in a substance at absolute temperature T and placed in an external magnetic field $\mathbf{H}$ pointing along the z direction. Then the magnetic energy of an atom can be written as

$$\epsilon = -\boldsymbol{\mu} \cdot \mathbf{H} \tag{7·8·1}$$

Here $\boldsymbol{\mu}$ is the magnetic moment of the atom. It is proportional to the total angular momentum $\hbar\mathbf{J}$ of the atom and is conventionally written in the form

$$\boldsymbol{\mu} = g\mu_0\mathbf{J} \tag{7·8·2}$$

where μ_0 is a standard unit of magnetic moment (usually the Bohr magneton $\mu_0 \equiv e\hbar/(2mc)$, m being the electron mass) and where g is a number of the order of unity, the so-called g factor of the atom.*

> ***Remark*** Strictly speaking the magnetic field $\mathbf{H}$ used in (7·8·1) is the local magnetic field acting on the atom. It is not quite the same as the external magnetic field since it includes also the magnetic field produced by all the other atoms. Suitable corrections for the difference can be made by standard arguments of electromagnetic theory. The distinction between external and local field becomes increasingly unimportant when the concentration of magnetic atoms is kept small.

By combining (7·8·1) and (7·8·2) one obtains

$$\epsilon = -g\mu_0\mathbf{J} \cdot \mathbf{H} = -g\mu_0 H J_z \tag{7·8·3}$$

since $\mathbf{H}$ points in the z direction. In a quantum-mechanical description the values which J_z can assume are discrete and are given by

$$J_z = m$$

where m can take on all values between $-J$ and $+J$ in integral steps; i.e.,

$$m = -J, -J + 1, -J + 2, \ldots, J - 1, J \tag{7·8·4}$$

Thus there are $2J + 1$ possible values of m corresponding to that many possible projections of the angular momentum vector along the z axis. By virtue of

* In the case of atoms having both electronic spin and orbital angular momentum, g would be the Landé g factor.

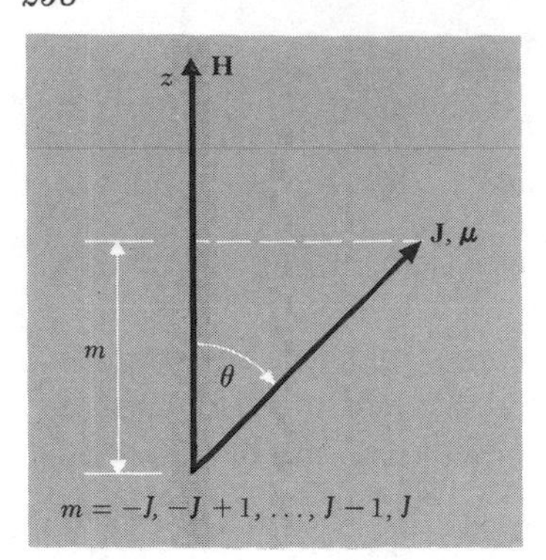

Fig. 7·8·1 Relative orientation of the angular momentum J with respect to H.

(7·8·3), the possible magnetic energies of the atom are then

$$\epsilon_m = -g\mu_0 H m \tag{7·8·5}$$

For example, if $J = \frac{1}{2}$ as would be the case for an atom with a single net electron spin, there are only two possible energies corresponding to $m = \pm\frac{1}{2}$. This was the simple case treated in Sec. 6·3.

The probability P_m that an atom is in a state labeled m is given by

$$P_m \propto e^{-\beta\epsilon_m} = e^{\beta g\mu_0 H m}$$

The z component of its magnetic moment in this state is, by (7·8·2), equal to

$$\mu_z = g\mu_0 m$$

The mean z component of the magnetic moment of an atom is therefore

$$\bar{\mu}_z = \frac{\sum_{m=-J}^{J} e^{\beta g\mu_0 H m}(g\mu_0 m)}{\sum_{m-J}^{J} e^{\beta g\mu_0 H m}} \tag{7·8·6}$$

Here the numerator can conveniently be written as a derivative with respect to the external parameter H, i.e.,

$$\sum_{m=-J}^{J} e^{\beta g\mu_0 H m}(g\mu_0 m) = \frac{1}{\beta}\frac{\partial Z_a}{\partial H}$$

where

$$Z_a \equiv \sum_{m=-J}^{J} e^{\beta g\mu_0 H m} \tag{7·8·7}$$

is the partition function of one atom. Hence (7·8·6) becomes*

▶ $$\bar{\mu}_z = \frac{1}{\beta}\frac{1}{Z_a}\frac{\partial Z_a}{\partial H} = \frac{1}{\beta}\frac{\partial \ln Z_a}{\partial H} \tag{7·8·8}$$

* This expression is valid even if the dependence of the energy levels of an atom on H is more complicated than in (7·8·5). See Problem 11.1.

To calculate Z_a, introduce the abbreviation

$$\eta \equiv \beta g\mu_0 H = \frac{g\mu_0 H}{kT} \tag{7·8·9}$$

which is a dimensionless parameter which measures the ratio of the magnetic energy $g\mu_0 H$, which tends to align the magnetic moment, to the thermal energy kT, which tends to keep it randomly oriented. Thus (7·8·7) becomes

$$Z_a = \sum_{m=-J}^{J} e^{\eta m} = e^{-\eta J} + e^{-\eta(J-1)} + \cdots + e^{\eta J}$$

which is simply a finite geometric series where each term is obtained from the preceding one as a result of multiplication by e^{η}. This can immediately be summed to give

$$Z_a = \frac{e^{-\eta J} - e^{\eta(J+1)}}{1 - e^{\eta}}$$

This can be brought to more symmetrical form by multiplying both numerator and denominator by $e^{-\eta/2}$. Then

$$Z_a = \frac{e^{-\eta(J+\frac{1}{2})} - e^{\eta(J+\frac{1}{2})}}{e^{-\frac{1}{2}\eta} - e^{\frac{1}{2}\eta}}$$

or

▶ $$Z_a = \frac{\sinh (J + \frac{1}{2})\eta}{\sinh \frac{1}{2}\eta} \tag{7·8·10}$$

where we have used the definition of the hyperbolic sine

$$\sinh y \equiv \frac{e^y - e^{-y}}{2} \tag{7·8·11}$$

Thus $$\ln Z_a = \ln \sinh (J + \tfrac{1}{2})\eta - \ln \sinh \tfrac{1}{2}\eta \tag{7·8·12}$$

By (7·8·8) and (7·8·9) one then obtains

$$\bar{\mu}_z = \frac{1}{\beta}\frac{\partial \ln Z_a}{\partial H} = \frac{1}{\beta}\frac{\partial \ln Z_a}{\partial \eta}\frac{\partial \eta}{\partial H} = g\mu_0 \frac{\partial \ln Z_a}{\partial \eta}$$

Hence

$$\bar{\mu}_z = g\mu_0 \left[\frac{(J + \frac{1}{2}) \cosh (J + \frac{1}{2})\eta}{\sinh (J + \frac{1}{2})\eta} - \frac{\frac{1}{2} \cosh \frac{1}{2}\eta}{\sinh \frac{1}{2}\eta} \right]$$

or

▶ $$\bar{\mu}_z = g\mu_0 J B_J(\eta) \tag{7·8·13}$$

where

▶ $$B_J(\eta) \equiv \frac{1}{J}\left[\left(J + \frac{1}{2}\right) \coth \left(J + \frac{1}{2}\right)\eta - \frac{1}{2} \coth \frac{1}{2}\eta \right] \tag{7·8·14}$$

The function $B_J(\eta)$ thus defined is sometimes called the "Brillouin function." Let us investigate its limiting behavior for large and small values of the parameter η.

The hyperbolic cotangent is defined as

$$\coth y \equiv \frac{\cosh y}{\sinh y} = \frac{e^y + e^{-y}}{e^y - e^{-y}} \tag{7·8·15}$$

For $y \gg 1$,
$$e^{-y} \ll e^y \quad \text{and} \quad \coth y = 1 \tag{7·8·16}$$

Conversely, for $y \ll 1$, both e^y and e^{-y} can be expanded in power series. Retaining all terms quadratic in y, the result is

$$\begin{aligned}\coth y &= \frac{1 + \frac{1}{2}y^2 + \cdots}{y + \frac{1}{6}y^3 + \cdots} \\ &= \left(1 + \frac{1}{2}y^2\right)\left[\frac{1}{y}\left(1 + \frac{1}{6}y^2\right)^{-1}\right] \\ &= \frac{1}{y}\left(1 + \frac{1}{2}y^2\right)\left(1 - \frac{1}{6}y^2\right) \\ &= \frac{1}{y}\left(1 + \frac{1}{3}y^2\right)\end{aligned}$$

For $y \ll 1$,
$$\coth y = \frac{1}{y} + \frac{1}{3}y \tag{7·8·17}$$

Applying these results to the function $B_J(\eta)$ defined in (7·8·14) yields

for $\eta \gg 1$,
$$B_J(\eta) = \frac{1}{J}\left[\left(J + \frac{1}{2}\right) - \frac{1}{2}\right] = 1 \tag{7·8·18}$$

In the opposite limit where $\eta \ll 1$,

$$\begin{aligned}B_J(\eta) &= \frac{1}{J}\left\{\left(J + \frac{1}{2}\right)\left[\frac{1}{(J + \frac{1}{2})\eta} + \frac{1}{3}\left(J + \frac{1}{2}\right)\eta\right] - \frac{1}{2}\left[\frac{2}{\eta} + \frac{\eta}{6}\right]\right\} \\ &= \frac{1}{J}\left\{\frac{1}{3}\left(J + \frac{1}{2}\right)^2 \eta - \frac{1}{12}\eta\right\} \\ &= \frac{\eta}{3J}\left\{J^2 + J + \frac{1}{4} - \frac{1}{4}\right\}\end{aligned}$$

For $\eta \ll 1$,
$$B_J(\eta) = \frac{(J + 1)}{3}\eta \tag{7·8·19}$$

Figure 7·8·2 shows how $B_J(\eta)$ depends on η for various values of J.

If there are N_0 atoms per unit volume, the mean magnetic moment per unit volume (or magnetization) becomes by (7·8·13)

▶
$$\bar{M}_z = N_0\bar{\mu}_z = N_0 g\mu_0 J B_J(\eta) \tag{7·8·20}$$

If $\eta \ll 1$, (7·8·19) implies that $\bar{M}_z \propto \eta \propto H/T$. One can write this relation

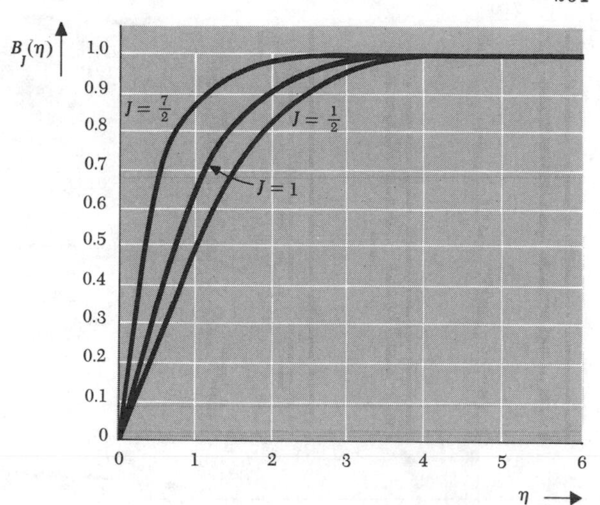

Fig. 7·8·2 Dependence of the Brillouin function $B_J(\eta)$ on its argument η for various values of J.

in the following form:

for $g\mu_0 H/kT \ll 1$, $$\bar{M}_z = \chi H \tag{7·8·21}$$

where the constant of proportionality, i.e., the susceptibility χ, is given by

$$\chi = N_0 \frac{g^2\mu_0{}^2 J(J+1)}{3kT} \tag{7·8·22}$$

Thus $\chi \propto T^{-1}$, a result known as Curie's law. In the other limiting case

when $g\mu_0 H/kT \gg 1$, $$\bar{M}_z \to N_0 g\mu_0 J \tag{7·8·23}$$

One gets then saturation behavior where each atom has the maximum z component of magnetic moment, $g\mu_0 J$, that it can possibly have.

Although the general results (7·8·20) and (7·8·21) are quite important, all the physical ideas are exactly the same as those already discussed in Sec. 6·3 for the special simple case of $J = \frac{1}{2}$. Note that our discussion is equally valid if the total angular momentum $\boldsymbol{J}$ and magnetic moment $\boldsymbol{\mu}$ of the atom are due to unpaired electrons of the atom (e.g., a gadolinium or iron atom); or if the atom has no unpaired electrons and $\boldsymbol{J}$ and $\boldsymbol{\mu}$ are due solely to the nucleus of the atom (e.g., a He^3 atom or fluorine F^- ion). The difference is one of magnitudes. In the first case μ is of the order of a Bohr magneton. But in the second case the magnetic moment is smaller by approximately the ratio of the electron to the nucleon mass; i.e., it is of the order of a nuclear magneton, about 1000 times smaller than the Bohr magneton. Nuclear paramagnetism is thus about 1000 times smaller than electronic paramagnetism. Correspondingly, it requires an absolute temperature about 1000 times smaller to achieve the same extent of preferential nuclear spin orientation along an applied magnetic field as it does to achieve this extent of orientation for an electronic spin (see Problem 6.3).

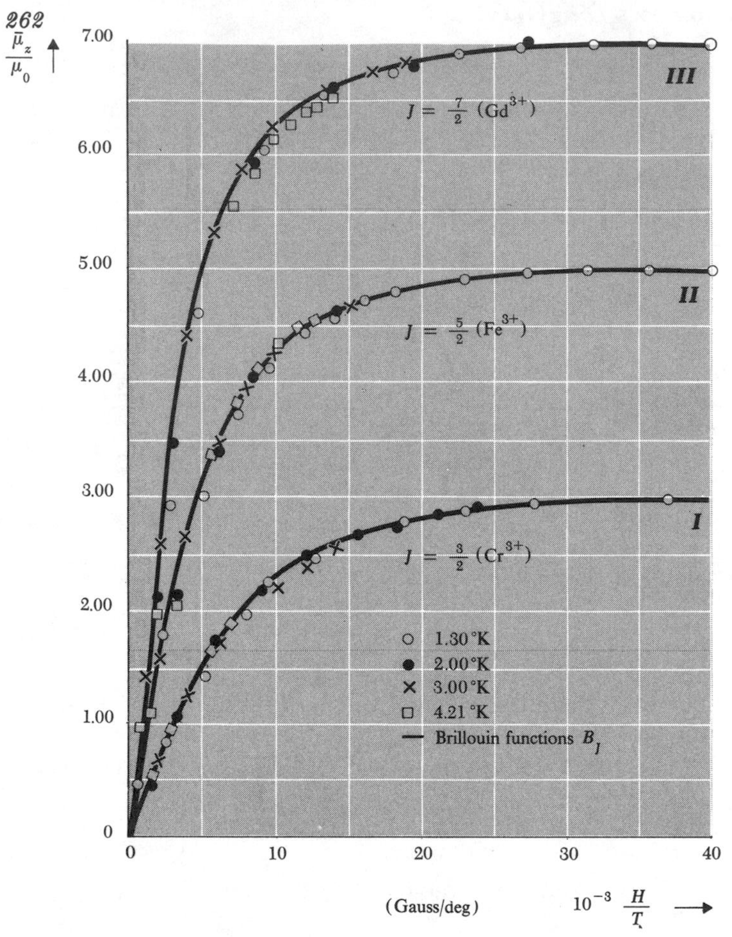

Fig. 7·8·3 Plots of the mean magnetic moment $\bar{\mu}_z$ of an ion (in units of the Bohr magneton μ_0) as a function of H/T. The solid curves are Brillouin functions. The experimental points are those for (I) potassium chromium alum, (II) iron ammonium alum, and (III) gadolinium sulfate octahydrate. In all cases, $J = S$, the total electron spin of the ion, and $g = 2$. Note that at 1.3°K a field of 50,000 gauss is sufficient to produce more than 99.5 percent magnetic saturation (after W. E. Henry, Phys. Rev. vol. 88, p. 561 (1952)).

KINETIC THEORY OF DILUTE GASES IN EQUILIBRIUM

7 · 9 *Maxwell velocity distribution*

Consider a molecule of mass m in a dilute gas. The gas may possibly consist of several different kinds of molecules; the molecule under consideration may also be polyatomic. Let us denote the position of the center of mass of this molecule by $\mathbf{r}$ and the momentum of its center of mass by $\mathbf{p}$. If external force fields (e.g., gravity) are neglected, the energy ϵ of this molecule is equal to

$$\epsilon = \frac{\mathbf{p}^2}{2m} + \epsilon^{(\text{int})} \tag{7·9·1}$$

where the first term on the right is the kinetic energy of the center of mass motion; the second term, which arises only if the molecule is not monatomic, designates the internal energy of rotation and vibration of the atoms with respect to the molecular center of mass. Since the gas is supposed to be sufficiently dilute to be considered ideal, any potential energy of interaction with other molecules is supposed to be negligible; thus ϵ does not depend on $\mathbf{r}$.

The translational degrees of freedom can be treated classically to an excellent approximation if the gas is dilute and the temperature is not too low; the internal degrees of freedom must usually be treated by quantum mechanics. The state of the molecule can be described by specifying that the position of the center of mass of the molecule lies in the range $(\mathbf{r}; d\mathbf{r})$, i.e., in a volume element of magnitude $d^3\mathbf{r} = dx\,dy\,dz$ near the position $\mathbf{r}$; that the momentum of its center of mass lies in the range $(\mathbf{p}; d\mathbf{p})$, i.e., within the momentum space volume $d^3\mathbf{p} \equiv dp_x\,dp_y\,dp_z$ near the momentum $\mathbf{p}$; and that the state of internal motion of the molecule is labeled by some quantum numbers s with corresponding internal energy $\epsilon_s^{(\text{int})}$. This particular molecule is in weak interaction with all the other molecules which act, therefore, as a heat reservoir at the temperature T of the gas. If the gas is sufficiently dilute, it is also permissible to think in classical terms and to focus attention on the particular molecule as a distinguishable entity. Then the molecule satisfies all the conditions of a distinct small system in contact with a heat reservoir and obeys, therefore, the canonical distribution. Hence one obtains for the probability $P_s(\mathbf{r},\mathbf{p})\,d^3\mathbf{r}\,d^3\mathbf{p}$ of finding the molecule with center-of-mass variables in the ranges $(\mathbf{r}; d\mathbf{r})$ and $(\mathbf{p}; d\mathbf{p})$ and with internal state specified by s the result

$$\begin{aligned} P_s(\mathbf{r},\mathbf{p})\,d^3\mathbf{r}\,d^3\mathbf{p} &\propto e^{-\beta[p^2/2m+\epsilon_s^{(\text{int})}]}\,d^3\mathbf{r}\,d^3\mathbf{p} \\ &\propto e^{-\beta p^2/2m}\,e^{-\beta\epsilon_s^{(\text{int})}}\,d^3\mathbf{r}\,d^3\mathbf{p} \end{aligned} \tag{7·9·2}$$

The probability $P(\mathbf{r},\mathbf{p})\,d^3\mathbf{r}\,d^3\mathbf{p}$ of finding the molecule with center-of-mass variables in the ranges $(\mathbf{r}; d\mathbf{r})$ and $(\mathbf{p}; d\mathbf{p})$, irrespective of its internal state, is obtained by summing (7·9·2) over all possible internal states s. The sum over the factor $\exp(-\beta\epsilon_s^{(\text{int})})$ contributes then only a constant of proportionality, so that the result of summing (7·9·2) is simply

$$P(\mathbf{r},\mathbf{p})\,d^3\mathbf{r}\,d^3\mathbf{p} \propto e^{-\beta(p^2/2m)}\,d^3\mathbf{r}\,d^3\mathbf{p} \tag{7·9·3}$$

This is, of course, identical with the result (6·3·11) derived previously under less general conditions.

If one multiplies the probability (7·9·3) by the total number N of molecules of the type under consideration, one obtains the mean number of molecules in this position and momentum range. Let us express the result in terms of the velocity $\mathbf{v} = \mathbf{p}/m$ of the molecule's center of mass. For the type of molecule under consideration, we define

$$\left.\begin{array}{l} f(\mathbf{r},\mathbf{v})\,d^3\mathbf{r}\,d^3\mathbf{v} \equiv \text{the mean number of molecules with center} \\ \text{of mass position between } \mathbf{r} \text{ and } \mathbf{r}+d\mathbf{r}\text{, and velocity between } \mathbf{v} \\ \text{and } \mathbf{v}+d\mathbf{v}. \end{array}\right\} \tag{7·9·4}$$

Then (7 · 9 · 3) gives

$$f(\mathbf{r},\mathbf{v})\, d^3\mathbf{r}\, d^3\mathbf{v} = Ce^{-\beta(mv^2/2)}\, d^3\mathbf{r}\, d^3\mathbf{v} \tag{7·9·5}$$

where C is a constant of proportionality which can be determined by the normalization condition

$$\int_{(\mathbf{r})} \int_{(\mathbf{v})} f(\mathbf{r},\mathbf{v})\, d^3\mathbf{r}\, d^3\mathbf{v} = N \tag{7·9·6}$$

That is, summing over molecules with all possible velocities $\mathbf{v}$ from $-\infty$ to ∞ and with all possible positions $\mathbf{r}$ anywhere in the volume V of the container must yield the *total* number of molecules. Substituting (7 · 9 · 5) in (7 · 9 · 6) thus gives

$$C \int_{(\mathbf{r})} \int_{(\mathbf{v})} e^{-\beta(mv^2/2)}\, d^3\mathbf{v}\, d^3\mathbf{r} = N \tag{7·9·7}$$

Since f does *not* depend on $\mathbf{r}$, the integration over this variable yields simply the volume V. The rest of the integration is similar to that leading to (7 · 2 · 6). Thus (7 · 9 · 7) reduces to

$$CV \left(\int_{-\infty}^{\infty} e^{-\frac{1}{2}\beta m v_x^2}\, dv_x \right)^3 = CV \left(\frac{2\pi}{\beta m} \right)^{\frac{3}{2}} = N$$

or

$$C = n \left(\frac{\beta m}{2\pi} \right)^{\frac{3}{2}}, \qquad n \equiv \frac{N}{V} \tag{7·9·8}$$

where n is the total number of molecules (of this type) per unit volume. Hence (7 · 9 · 5) becomes

$$f(\mathbf{v})\, d^3\mathbf{r}\, d^3\mathbf{v} = n \left(\frac{\beta m}{2\pi} \right)^{\frac{3}{2}} e^{-\frac{1}{2}\beta m v^2}\, d^3\mathbf{r}\, d^3\mathbf{v} \tag{7·9·9}$$

or

$$f(\mathbf{v})\, d^3\mathbf{r}\, d^3\mathbf{v} = n \left(\frac{m}{2\pi kT} \right)^{\frac{3}{2}} e^{-mv^2/2kT}\, d^3\mathbf{r}\, d^3\mathbf{v} \tag{7·9·10}$$

Here we have omitted the variable $\mathbf{r}$ in the argument of f since f does not depend on $\mathbf{r}$. This condition must, of course, be true from symmetry considerations since there is no preferred position in space in the absence of external force fields. Furthermore one sees that f depends only on the magnitude of $\mathbf{v}$; i.e.,

$$f(\mathbf{v}) = f(v) \tag{7·9·11}$$

where $v = |\mathbf{v}|$. Again this is obvious by symmetry, since there is no preferred direction in a situation where the container, and thus also the center of mass of the whole gas, is considered to be at rest.

If (7 · 9 · 10) is divided by the element of volume $d^3\mathbf{r}$, one obtains

$f(\mathbf{v})\, d^3\mathbf{v}$ = the mean number of molecules *per unit volume* with center-of-mass velocity in the range between $\mathbf{v}$ and $\mathbf{v} + d\mathbf{v}$. (7 · 9 · 12)

Equation (7 · 9 · 10) is the Maxwell velocity distribution for a molecule of a dilute gas in thermal equilibrium.

7·10 *Related velocity distributions and mean values*

Distribution of a component of velocity Various other distributions of physical interest follow immediately from (7·9·10). For example, one may be interested in the quantity

$$\left. \begin{array}{l} g(v_x)\, dv_x = \text{the mean number of molecules per unit volume} \\ \text{with } x \text{ component of velocity in the range between } v_x \text{ and} \\ v_x + dv_x, \text{ irrespective of the values of their other velocity} \\ \text{components.} \end{array} \right\} \tag{7·10·1}$$

Clearly, one obtains this by adding up all molecules with x component of velocity in this range; i.e.,

$$g(v_x)\, dv_x = \int_{(v_y)} \int_{(v_z)} f(\mathbf{v})\, d^3\mathbf{v} \tag{7·10·2}$$

where one sums over all possible y and z velocity components of the molecules. By (7·9·10) this becomes

$$\begin{aligned} g(v_x)\, dv_x &= n \left(\frac{m}{2\pi kT}\right)^{\frac{3}{2}} \int_{(v_y)} \int_{(v_z)} e^{-(m/2kT)(v_x^2+v_y^2+v_z^2)}\, dv_x\, dv_x\, dv_z \\ &= n \left(\frac{m}{2\pi kT}\right)^{\frac{3}{2}} e^{-mv_x^2/2kT}\, dv_x \int_{-\infty}^{\infty} e^{-(m/2kT)v_y^2}\, dv_y \int_{-\infty}^{\infty} e^{-(m/2kT)v_z^2}\, dv_z \\ &= n \left(\frac{m}{2\pi kT}\right)^{\frac{3}{2}} e^{-mv_x^2/2kT}\, dv_x \left(\sqrt{\frac{2\pi kT}{m}}\right)^2 \end{aligned}$$

or

$$g(v_x)\, dv_x = n \left(\frac{m}{2\pi kT}\right)^{\frac{1}{2}} e^{-mv_x^2/2kT}\, dv_x \tag{7·10·3}$$

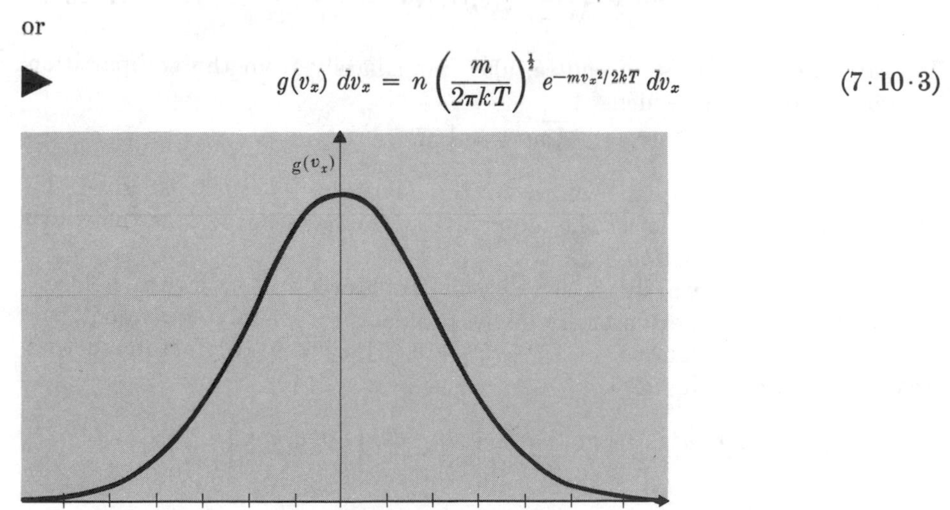

Fig. 7·10·1 ***Maxwellian distribution of a molecular velocity component.***

Of course, this expression is properly normalized so that

$$\int_{-\infty}^{\infty} g(v_x)\, dv_x = n \qquad (7 \cdot 10 \cdot 4)$$

Equation (7 · 10 · 3) shows that each component of velocity is distributed with a Gaussian distribution *symmetric* about the mean value

$$\bar{v}_x = 0 \qquad (7 \cdot 10 \cdot 5)$$

It is physically clear that $\bar{v}_x = 0$ by symmetry, since the x component of velocity of a molecule is as likely to be positive as negative. Mathematically this result follows, since

$$\bar{v}_x = \frac{1}{n} \int_{-\infty}^{\infty} g(v_x)\, v_x\, dv_x$$

Here the integrand is an odd function of v_x (i.e., reverses its sign when v_x reverses sign) because $g(v_x)$ is an even function of v_x (i.e., does not reverse its sign under this operation, since it depends only on v_x^2). Thus contributions to the integral from $+v_x$ and $-v_x$ cancel each other. A similar argument shows immediately that

if k is any odd integer, $$\overline{v_x^k} = 0 \qquad (7 \cdot 10 \cdot 6)$$

Of course, $\overline{v_x^2}$ is intrinsically positive and is, by virtue of (7 · 10 · 5), the dispersion of v_x. By direct integration, using (7 · 10 · 3), or by recalling the properties of the Gaussian distribution already studied in (1 · 6 · 9), it follows that

$$\overline{v_x^2} = \frac{1}{n} \int_{-\infty}^{\infty} g(v_x)\, v_x^2\, dv_x = \frac{kT}{m} \qquad (7 \cdot 10 \cdot 7)$$

The same result follows, of course, also immediately from the equipartition theorem, according to which

$$\overline{\tfrac{1}{2} m v_x^2} = \tfrac{1}{2} kT$$

Thus the root-mean-square width of the Gaussian (7 · 10 · 3) is given by $\Delta^* v_x = \sqrt{kT/m}$. The lower the temperature, the narrower will be the width of the distribution function $g(v_x)$.

Needless to say, exactly the same results hold for v_y and v_z, since all velocity components are, by the symmetry of the problem, completely equivalent.

Note also that, since $v^2 = v_x^2 + v_y^2 + v_z^2$, Eq. (7 · 9 · 10) factors so that it can be written in the form

$$\frac{f(\mathbf{v})\, d^3\mathbf{v}}{n} = \left[\frac{g(v_x)\, dv_x}{n}\right] \left[\frac{g(v_y)\, dv_y}{n}\right] \left[\frac{g(v_z)\, dv_z}{n}\right]$$

This means that the probability that the velocity lies in the range between $\mathbf{v}$ and $\mathbf{v} + d\mathbf{v}$ is just equal to the product of the probabilities that the velocity components lie in their respective ranges. Thus the individual velocity components behave like statistically independent quantities.

Distribution of speed Another quantity of physical interest is

$$\left.\begin{array}{l} F(v)\,dv = \text{the mean number of molecules per unit volume} \\ \text{with a speed } v \equiv |\boldsymbol{v}| \text{ in the range between } v \text{ and } v + dv. \end{array}\right\} \qquad (7\cdot10\cdot8)$$

One clearly obtains this quantity by adding up all molecules with speed in this range irrespective of the *direction* of their velocity. Thus

$$F(v)\,dv = \int' f(\boldsymbol{v})\,d^3\boldsymbol{v}$$

where the integral extends over all velocities, satisfying the condition that

$$v < |\boldsymbol{v}| < v + dv$$

i.e., over all velocity vectors which terminate in velocity space within a spherical shell of inner radius v and outer radius $v + dv$. Since $f(\boldsymbol{v})$ depends only on $|\boldsymbol{v}|$, this integration is just equal to $f(v)$ multiplied by the volume $4\pi v^2\,dv$ of this spherical shell. Thus

$$F(v)\,dv = 4\pi f(v) v^2\,dv \qquad (7\cdot10\cdot9)$$

Using (7 · 9 · 10), this becomes explicitly

$$F(v)\,dv = 4\pi n \left(\frac{m}{2\pi kT}\right)^{\frac{3}{2}} v^2\,e^{-mv^2/2kT}\,dv \qquad (7\cdot10\cdot10)$$

This is the Maxwell distribution of speeds. Note that it has a maximum for the same reason responsible for the maxima encountered in our general discussion of statistical mechanics. As v increases, the exponential factor *decreases*, but the volume of phase space available to the molecule is proportional to v^2 and *increases;* the net result is a gentle maximum. The expression (7 · 10 · 10) is, of course, properly normalized, i.e.,

$$\int_0^\infty F(v)\,dv = n \qquad (7\cdot10\cdot11)$$

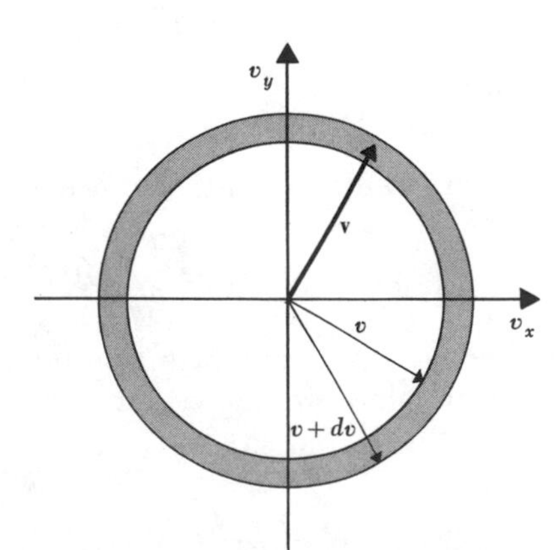

Fig. 7·10·2 Representation (in two dimensions) of the shell in velocity space containing all molecules with velocity v such that $v < |\boldsymbol{v}| < v + dv$.

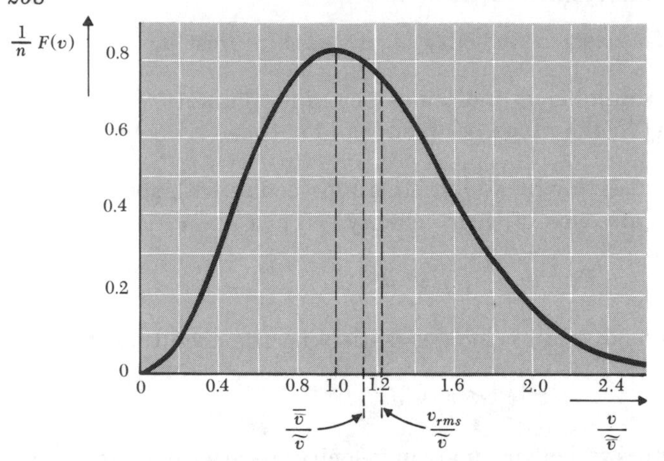

Fig. 7·10·3 Maxwellian distribution of molecular speeds. The speed v is expresed in terms of the speed $\tilde{v} = (2kT/m)^{\frac{1}{2}}$ where F is maximum.

Mean values It is again of interest to calculate some significant mean values. The mean speed is, of course, positive since $v = |\mathbf{v}|$ is intrinsically positive. It is given by

$$\bar{v} = \frac{1}{n}\iiint f(\mathbf{v})v\, d^3\mathbf{v} \tag{7·10·12}$$

where the integration is over all velocities, or equivalently by

$$\bar{v} = \frac{1}{n}\int_0^\infty F(v)v\, dv$$

where the integration is over all speeds. Thus one gets

$$\begin{aligned}\bar{v} &= \frac{1}{n}\int_0^\infty f(v)v \cdot 4\pi v^2\, dv = \frac{4\pi}{n}\int_0^\infty f(v)v^3\, dv \\ &= 4\pi\left(\frac{m}{2\pi kT}\right)^{\frac{3}{2}}\int_0^\infty e^{-mv^2/2kT}\, v^3\, dv \qquad \text{by (7·9·10)} \\ &= 4\pi\left(\frac{m}{2\pi kT}\right)^{\frac{3}{2}} \cdot \frac{1}{2}\left(\frac{m}{2kT}\right)^{-2} \qquad \text{by (A·4·6)}\end{aligned}$$

Hence

$$\bar{v} = \sqrt{\frac{8}{\pi}\frac{kT}{m}} \tag{7·10·13}$$

On the other hand, the mean-square speed is given by

$$\overline{v^2} = \frac{1}{n}\int f(\mathbf{v})v^2\, d^3\mathbf{v} = \frac{4\pi}{n}\int_0^\infty f(v)v^4\, dv \tag{7·10·14}$$

We could again integrate this by using (A·4·6), but we can save ourselves the work since we know that

$$\overline{\tfrac{1}{2}mv^2} = \overline{\tfrac{1}{2}m(v_x^2 + v_y^2 + v_z^2)}$$

Hence by the equipartition theorem (or by using the symmetry argument that

$\overline{v_x^2} = \overline{v_y^2} = \overline{v_z^2}$, so that $\overline{v^2} = 3\overline{v_z^2}$, a quantity already computed in (7·10·7)), we get

$$\tfrac{1}{2}m\overline{v^2} = \tfrac{3}{2}kT$$

or

$$\overline{v^2} = \frac{3kT}{m} \tag{7·10·15}$$

The root-mean-square speed is thus

$$v_{\text{rms}} \equiv \sqrt{\overline{v^2}} = \sqrt{\frac{3kT}{m}} \tag{7·10·16}$$

Finally, one can ask for the most probable speed $\tilde{v}$ of the molecule, i.e., the speed for which $F(v)$ in (7·10·10) is a maximum. This is given by the condition

$$\frac{dF}{dv} = 0$$

i.e.,

$$2v\, e^{-mv^2/2kT} + v^2\left(-\frac{m}{kT}v\right)e^{-mv^2/2kT} = 0$$

or

$$v^2 = \frac{2kT}{m}$$

Hence $\tilde{v}$ is given by

$$\tilde{v} = \sqrt{\frac{2kT}{m}} \tag{7·10·17}$$

All these various speeds are proportional to $(kT/m)^{\frac{1}{2}}$. Thus the molecular speed increases when the temperature is raised; and, for a given temperature, a molecule with larger mass has smaller speed. The various speeds we have calculated are such that their ratios

$$\left.\begin{array}{lc} & v_{\text{rms}} : \bar{v} : \tilde{v} \\ \text{are proportional to} & \sqrt{3} : \sqrt{\frac{8}{\pi}} : \sqrt{2} \\ \text{or to} & 1.224 : 1.128 : 1 \end{array}\right\} \tag{7·10·18}$$

For nitrogen (N_2) gas at room temperature (300°K) one finds by (7·10·16), using $m = 28/(6 \times 10^{23})$ g, that

$$v_{\text{rms}} \approx 5 \times 10^4 \text{ cm/sec} \approx 500 \text{ m/sec} \tag{7·10·19}$$

a number of the order of the velocity of sound in the gas.

7·11 *Number of molecules striking a surface*

It is possible to discuss a number of interesting physical situations by considering the motion of individual molecules in detail. Detailed arguments of this kind constitute the subject matter of what is usually called "kinetic theory." Since we shall for the present restrict ourselves to equilibrium situations, our considerations will be very simple.

Let us focus attention on a dilute gas enclosed in a container and ask the following question: How many molecules per unit time strike a unit area of the wall of this container? This question is very closely related to another question of physical interest: If there is a very small hole in the wall of the container, how many molecules will stream out of this hole per unit time?

Crude calculation To understand the essential features of the situation, it is adequate to adopt a highly simplified approximate point of view. Imagine that the container is a box in the form of a parallelepiped, the area of one end-wall being A. How many molecules per unit time strike this end-wall? Suppose that there are in this gas n molecules per unit volume. Since they all move in random directions, we can say roughly that one-third of them, or $n/3$ molecules per unit volume, have their velocities in a direction predominantly along the z axis (chosen to be normal to the end-wall under consideration as in Fig. 7·11·1). Half of these molecules, i.e., $n/6$ molecules per unit volume, have their velocity in the $+z$ direction so that they will strike the end-wall under consideration. If the mean speed of the molecules is $\bar{v}$, these molecules cover in an infinitesimal time dt a mean distance $\bar{v}\,dt$. Hence all those molecules with velocity $\bar{v}$ in the z direction which lie within a distance $\bar{v}\,dt$ from the end-wall, will, within a time dt, strike the end-wall; those which lie further than a distance $\bar{v}\,dt$ from the end-wall will not. Thus we arrive at the result that [the number of molecules which strike the end-wall of area A in time dt] is equal to [the number of molecules having velocity $\bar{v}$ in the z direction and contained in the cylinder of volume $A\bar{v}\,dt$]; i.e., it is given by

$$\left(\frac{n}{6}\right)(A\bar{v}\,dt) \tag{7·11·1}$$

The total number Φ_0 of molecules which strike unit area of the wall per unit time (i.e., the total molecular "flux") is then given by dividing (7·11·1) by the area A and the time interval dt. Thus

$$\Phi_0 \approx \tfrac{1}{6}n\bar{v} \tag{7·11·2}$$

We emphasize that this result was obtained by a very crude argument in which we did not consider in any detail the velocity distribution of the molecules, either in magnitude or in direction. Nevertheless, arguments of this

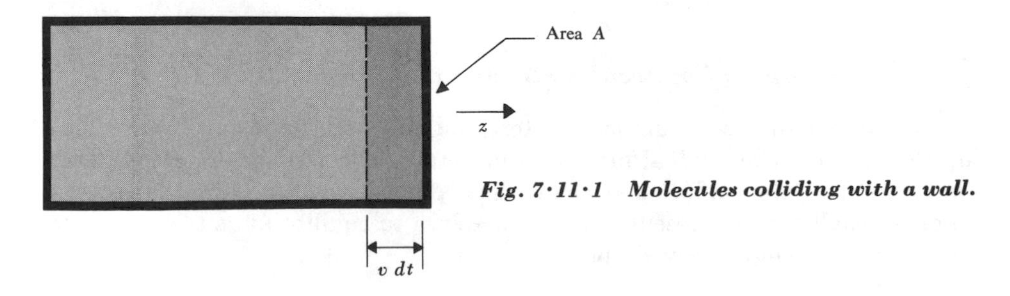

Fig. 7·11·1 Molecules colliding with a wall.

kind are often very useful because they can exhibit the essential features of a phenomenon without the need of exact calculations which are likely to be more laborious. Thus the factor $\frac{1}{6}$ in (7·11·2) is not to be taken very seriously. There ought to be some numerical factor there of the order of $\frac{1}{6}$, but its exact value must certainly depend on the particular way in which various averages are to be computed. (Indeed, the exact calculation will yield a factor $\frac{1}{4}$ instead of $\frac{1}{6}$.) On the other hand, the essential dependence of Φ_0 on n and $\bar{v}$ ought to be correct, i.e., $\Phi_0 \propto n\bar{v}$. Thus (7·11·2) makes the very plausible assertion that Φ_0 is proportionately increased if the concentration of molecules is increased or if their speed is increased.

The dependence of Φ_0 on the temperature T and mean pressure $\bar{p}$ of the gas follows immediately from (7·11·2). The equation of state gives

$$\bar{p} = nkT \qquad \text{or} \qquad n = \frac{\bar{p}}{kT} \tag{7·11·3}$$

Furthermore, by the equipartition theorem,

$$\tfrac{1}{2}m\overline{v^2} = \tfrac{3}{2}kT$$

so that

$$\bar{v} \propto \bar{v}_{\text{rms}} \propto \sqrt{\frac{kT}{m}} \tag{7·11·4}$$

Thus (7·11·2) implies that

$$\Phi_0 \propto \frac{\bar{p}}{\sqrt{mT}} \tag{7·11·5}$$

Exact calculation Consider an element of area dA of the wall of the container. Choose the z axis so as to point along the outward normal of this element of area (see Fig. 7·11·2). Consider first those molecules (in the immediate vicinity of the wall) whose velocity is such that it lies between $\mathbf{v}$ and $\mathbf{v} + d\mathbf{v}$. (That is, the velocity is such that its magnitude lies in the range between v and $v + dv$; its direction, specified by its polar angle θ (with respect to the normal, or z axis) and its azimuthal angle φ, is such that these angles lie between θ and $\theta + d\theta$ and between φ and $\varphi + d\varphi$, respectively.)

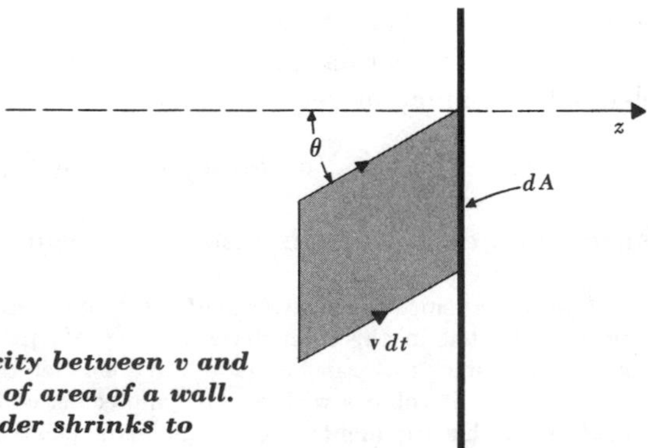

Fig. 7·11·2 Molecules, with velocity between v and v + dv, colliding with an element of area of a wall. (Note that the height of the cylinder shrinks to zero as dt → 0.)

Molecules of this type suffer a displacement $\mathbf{v}\,dt$ in the infinitesimal time interval dt. Hence all molecules of this type which lie within the infinitesimal cylinder of cross-sectional area dA and of a length $v\,dt$ making an angle θ with the z axis will strike the wall within the time interval dt; the molecules lying outside this cylinder will not.* The volume of this cylinder is $dA\,v\,dt\cos\theta$, while the number of molecules per unit volume in this velocity range is $f(\mathbf{v})\,d^3\mathbf{v}$. Hence [the number of molecules of this type which strike the area dA of the wall in time dt] $= [f(\mathbf{v})\,d^3\mathbf{v}][dA\,v\,dt\cos\theta]$. Dividing this by the area dA and the time interval dt, we get for

$\Phi(\mathbf{v})\,d^3\mathbf{v} \equiv$ the number of molecules, with velocity between $\mathbf{v}$ and $\mathbf{v} + d\mathbf{v}$, which strike a unit area of the wall per unit time. (7 · 11 · 6)

the expression

$$\Phi(\mathbf{v})\,d^3\mathbf{v} = d^3\mathbf{v}\,f(\mathbf{v})v\cos\theta \tag{7·11·7}$$

Let $\Phi_0 \equiv$ the *total* number of molecules which strike a unit area of the wall per unit time. (7 · 11 · 8)

This is simply given by summing (7 · 11 · 7) over all possible velocities which a molecule can have and which will cause it to collide with the element of area. This means that we have to sum over all possible speeds $0 < v < \infty$, all possible azimuthal angles $0 < \varphi < 2\pi$, and all angles θ in the range $0 < \theta < \pi/2$. (Those molecules for which $\pi/2 < \theta < \pi$ have their velocity directed away from the wall and hence will *not* collide with it.) In other words, we have to sum over all possible velocities $\mathbf{v}$ subject to the restriction that the velocity component $v_z = v\cos\theta > 0$ (since molecules with $v_z < 0$ will *not* collide with the element of area). Thus we have

$$\Phi_0 = \int_{v_z>0} d^3\mathbf{v}\,f(\mathbf{v})v\cos\theta \tag{7·11·9}$$

The results (7 · 11 · 7) and (7 · 11 · 9) are generally valid even if the gas is not in equilibrium (although f might then also be a function of $\mathbf{r}$ and t). But if we consider a gas in thermal equilibrium, $f(\mathbf{v}) = f(v)$ is only a function of $|\mathbf{v}|$. The element of volume in velocity space can be expressed in spherical coordinates

$$d^3\mathbf{v} = v^2\,dv\,(\sin\theta\,d\theta\,d\varphi)$$

where $\sin\theta\,d\theta\,d\varphi = d\Omega$ is just the element of solid angle. Hence (7 · 11 · 9)

* Note that since the length $v\,dt$ of the cylinder can be considered arbitrarily small, only molecules located in the immediate vicinity of the wall are involved in this argument. Furthermore, since $v\,dt$ can be made much smaller than the mean distance l traveled by a molecule before it collides with another molecule, collisions between molecules need not be considered in this argument; i.e., any molecule located in the cylinder and traveling toward the wall will indeed strike the wall without being deflected by a collision before it gets there.

becomes

$$\Phi_0 = \int_{v_z>0} v^2\, dv \sin\theta\, d\theta\, d\varphi\, f(v)v\cos\theta$$
$$= \int_0^\infty f(v)v^3\, dv \int_0^{\pi/2} \sin\theta\cos\theta\, d\theta \int_0^{2\pi} d\varphi$$

The integration over φ gives 2π, while the integral over θ yields the value $\frac{1}{2}$. Hence

$$\Phi_0 = \pi \int_0^\infty f(v)v^3\, dv \tag{7·11·10}$$

This can be expressed in terms of the mean speed already computed in (7·10·12). Thus

$$\bar{v} = \frac{1}{n}\int d^3\mathbf{v}\, f(v)v = \frac{1}{n}\int_0^\infty \int_0^\pi \int_0^{2\pi} (v^2\, dv \sin\theta\, d\theta\, d\varphi) f(v)v$$

or

$$\bar{v} = \frac{4\pi}{n}\int_0^\infty f(v)v^3\, dv \tag{7·11·11}$$

since the integration over the angles θ and φ is just the total solid angle 4π about a point. Hence (7·11·10) can also be written

▶
$$\Phi_0 = \tfrac{1}{4}n\bar{v} \tag{7·11·12}$$

This rigorous result can be compared with our previous crude estimate in (7·11·2). We see that the latter was off by a factor of only $\frac{2}{3}$.

The mean speed was already computed from the Maxwell distribution in (7·10·13). Combining this with the equation of state (7·11·3), one obtains for (7·11·12)

$$\Phi_0 = \frac{\bar{p}}{\sqrt{2\pi mkT}} \tag{7·11·13}$$

7·12 *Effusion*

If a sufficiently small hole (or slit) is made in the wall of the container, the equilibrium of the gas inside the container is disturbed to a negligible extent. In that case the number of molecules which emerge through the small hole is the same as the number of molecules which would strike the area occupied by the hole if the latter were closed off. The process whereby molecules emerge through such a small hole is called "effusion."

One may ask how small the diameter D of the hole (or the width D of the slit) must be so that there is no appreciable effect on the equilibrium state of the gas. The typical dimension against which D is to be compared is the "mean free path" l, i.e., the mean distance which a molecule in the gas travels before it suffers a collision with another molecule. The concept of mean free path will be discussed in greater detail in Chapter 12. Here it will suffice to make the obvious comment that, at a given temperature, l is inversely proportional to the number of molecules per unit volume. (At room temperature and atmospheric pressure, $l \approx 10^{-5}$ cm in a typical gas.) If $D \ll l$, the hole can be considered as very small. In that case, molecules will now and then emerge from the hole if their velocities happen to be in the right direction. When a few molecules escape through the hole, the remaining molecules in the container are then scarcely affected since l is so large. This is the case of "effusion."

On the other hand, if $D \gg l$, molecules suffer *frequent* collisions with each other within distances of the order of the hole size. When some molecules emerge through this hole (see Fig. 7·12·1), the molecules behind them are in an appreciably different situation. They no longer continue colliding with the molecules on the right which have just escaped through the hole, but they still suffer constant collisions with the molecules on the left. The net result is that the molecules near the hole experience, by virtue of these continuous molecular impacts, a net force to the right which causes them to acquire a drift velocity in the direction toward the hole. The resultant collective motion of all these molecules moving together is then analogous to the flow of water through the hole of a tank. In this case one has not effusion, but "hydrodynamic flow."

Let us consider the situation when the hole is sufficiently small so that molecules emerge through the hole by effusion. If a vacuum is maintained outside the container, the effusing molecules can be collimated further by additional slits so that one is left with a well-defined "molecular beam." Such molecular beams have been extensively used in experimental physics investigations because they provide one with the possibility of studying individual molecules under circumstances where interactions between them are negligible. The number of molecules which have speed in the range between v and $v + dv$ and which emerge per second from a small hole of area A into a solid angle range $d\Omega$ in the forward direction $\theta \approx 0$ is given by (7·11·7) as

$$\begin{aligned} A\Phi(\mathbf{v})\, d^3\mathbf{v} &\propto A[f(v)v \cos \theta](v^2\, dv\, d\Omega) \\ &\propto f(v)v^3\, dv\, d\Omega \propto e^{-mv^2/2kT}v^3\, dv\, d\Omega \end{aligned} \qquad (7\cdot 12\cdot 1)$$

Note that this expression involves the factor v^3, rather than the factor v^2 which occurs in the Maxwellian speed distribution (7·10·10).

Experiments on such a molecular beam can provide a direct test of the Maxwell velocity distribution by checking the prediction (7·12·1). Figure

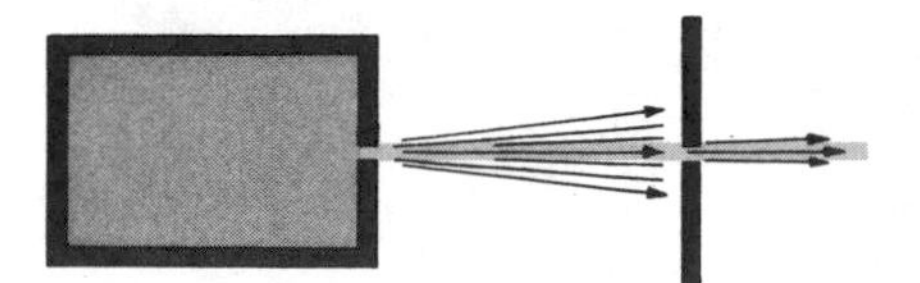

Fig. 7·12·1 Formation of a molecular beam by effusing molecules.

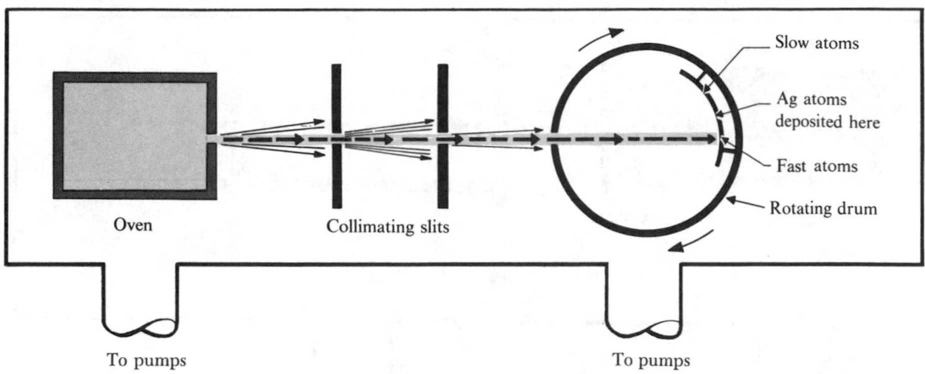

***Fig. 7·12·2** A molecular-beam apparatus for studying the velocity distribution of silver (Ag) atoms. The Ag atoms stick to the drum surface upon impact.*

7 · 12 · 2 shows one experimental arrangement which has been used. Silver (Ag) atoms are produced by evaporation in an oven and emerge through a narrow slit to form a molecular beam. A rotating hollow cylindrical drum, with a slit in it, revolves rapidly about its axis and is located in front of the beam. When molecules enter the slit in the drum, they require different times to reach the opposite side of the drum, a fast molecule requiring less time than a slow one. Since the drum is rotating all the time, the Ag molecules deposited on the opposite inside surface of the drum get spread out on this surface in accordance with their velocity distribution. Thus a measurement of the thickness of the Ag deposit as a function of distance along the drum surface provides a measurement of the molecular velocity distribution.

A more accurate method for determining the velocity distribution involves the use of a velocity selector similar in principle of operation to those used in neutron time-of-flight spectroscopy or for determining the velocity of light (Fizeau wheel). In this method the molecular beam emerges from a hole and is detected by a suitable device at the other end of the apparatus. The velocity selector is placed between the source and the detector and consists, in the simplest case, of a pair of disks mounted on a common axle which can be rotated with known angular velocity. Both disks are identical and have several slots cut along their periphery; thus the rotating disks act as two shutters which are alternately opened and closed. When the disks are properly aligned and not rotating, all molecules can reach the detector by passing through corresponding slots in both disks. But when the disks are rotating, molecules passing through a slot in the first disk can only reach the detector if their velocity is such that [the time of flight required for them to travel to the second disk] is equal to [the time required for the next slot of this disk to rotate to the location of the original slot]. Otherwise they will strike the solid part of the second disk and be stopped. Hence different angular velocities of rotation of the disks allow molecules of different speeds to reach the detector. Measurement of the relative number of molecules arriving there per second then

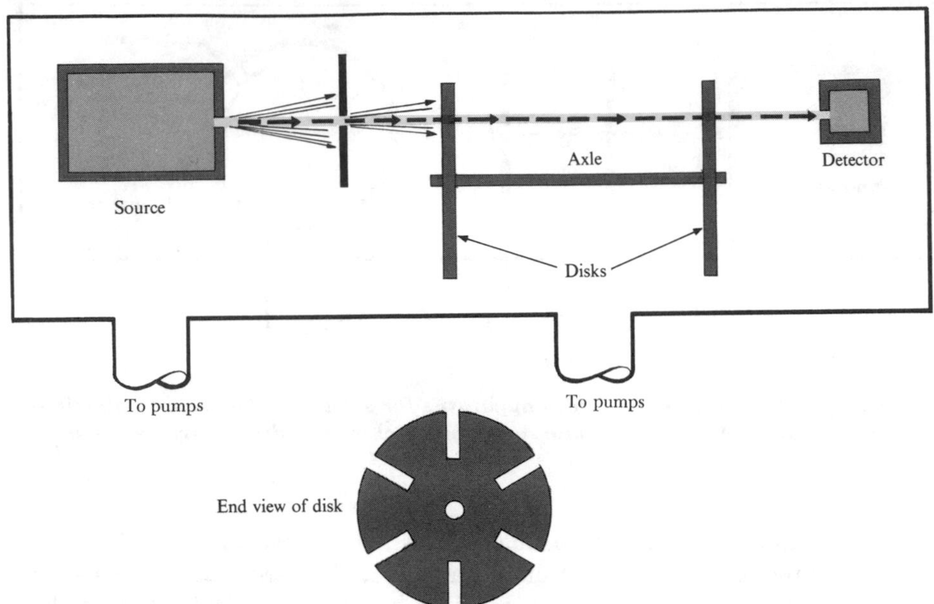

Fig. 7·12·3 Arrangement for studying the molecular velocity distribution by a velocity selector. (***A more effective velocity selector results if more than two similar disks are mounted on the same axle.***)

allows a direct check of the molecular-velocity distribution. The validity of the Maxwellian distribution has been well confirmed by such experiments.*

Equation (7 · 11 · 5) shows that the rate of effusion of a molecule depends on the mass of the molecule, lighter molecules effusing more rapidly than heavier ones. This suggests the application of effusion as a method for the separation of isotopes. Suppose that a container is closed off by a membrane which has very many small holes through which molecules can effuse. If this container is surrounded by a vacuum on the outside and is filled with a gas mixture of two isotopes at some initial time, then the relative concentration of the isotope of larger molecular weight will increase in the container as time goes on. Similarly, the gas pumped off from the surrounding vacuum will be more concentrated in the lighter isotope.†

Another example of interest is illustrated in Fig. 7 · 12 · 4. Here a container is divided into two parts by a partition containing a small hole. The container is filled with gas, but one part of the container is maintained at temperature T_1, the other part at temperature T_2. One may ask the following

* For recent experimental work on velocity distributions see R. C. Miller and P. Kusch' *J. Chem. Phys.*, vol. 25, p. 860 (1956); also P. M. Marcus and J. H. McFee in I. Estermann (ed)., "Recent Research in Molecular Beams," p. 43, Academic Press, New York, 1959.

† The successful large-scale separation by this method of uranium isotopes (in the form of UF_6 gas) was a crucial step in the development of nuclear fission devices (reactors and bombs) and is described in H. de W. Smyth, "Atomic Energy for Military Purposes," chap. 10, Princeton University Press, Princeton, 1947; or *Rev. Mod. Phys.* vol. 17, p. 430 (1945).

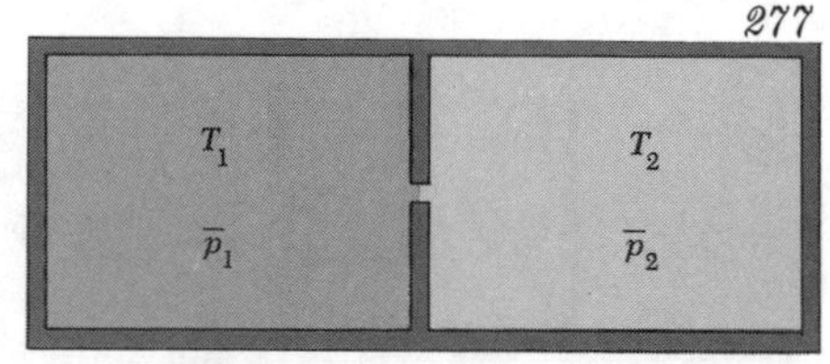

***Fig. 7·12·4** A container divided into two parts by a partition containing a small hole. The gas in the two parts is at different temperatures and pressures.*

question: What is the relation between the mean gas pressures $\bar{p}_1$ and $\bar{p}_2$ in the two parts when the system is in equilibrium, i.e., when a situation is reached where neither $\bar{p}_1$ or $\bar{p}_2$, nor the amount of gas in either part, changes with time? If the linear dimension D of the hole is large ($D \gg l$), then the condition is simply $\bar{p}_2 = \bar{p}_1$; for otherwise the pressure difference would give rise to mass motion of the gas from one side to the other until the pressures on both sides reach equality. But if $D \ll l$, one deals with effusion through the hole rather than with hydrodynamic flow. In this case the equilibrium condition requires that the mass of gas on each side remain constant, i.e., that [the number of molecules which pass per second through the hole from left to right] equals [the number of molecules which pass per second through the hole from right to left]. By (7·11·12) this leads to the simple equality

$$n_1\bar{v}_1 = n_2\bar{v}_2 \tag{7·12·2}$$

By (7·11·5) this condition becomes

$$\frac{\bar{p}_1}{\sqrt{T_1}} = \frac{\bar{p}_2}{\sqrt{T_2}} \tag{7·12·3}$$

Thus the pressures are then not at all equal, but higher gas pressure prevails in the part of the container at higher temperature.

This discussion has practical consequences in experimental work. Suppose, for example, that it is desired to measure the vapor pressure $\bar{p}_v$ (i.e., the pressure of the vapor in equilibrium with the liquid) of liquid helium at 2°K. The experimental arrangement might be as illustrated in Fig. 7·12·5, the mercury manometer at room temperature being used to make the pressure measurement. A small tube of diameter D connects the manometer to the vapor pressure to be measured, and the difference of mercury levels on the two sides of the manometer measures the pressure difference $\bar{p}$. Now at 2°K, $\bar{p}_v$ is still fairly large, i.e., the density of helium vapor is large enough that the mean free path l of molecules in the vapor is much less than the diameter D of the connecting tube. Then the pressure $\bar{p}$ read on the manometer is indeed equal to the vapor pressure $\bar{p}_v$ of interest. But suppose that it is desired to measure $\bar{p}_v$ at lower temperatures, say at 0.5°K. Then $\bar{p}_v$ is small and the density of He vapor is so small that l is comparable to, or large with respect to, the diameter D of the connecting tube. If one assumed that the pressure $\bar{p}$ which one reads on the manometer still equals the vapor pressure $\bar{p}_v$ of interest, one would be fooling oneself very badly. So-called "thermomolecular

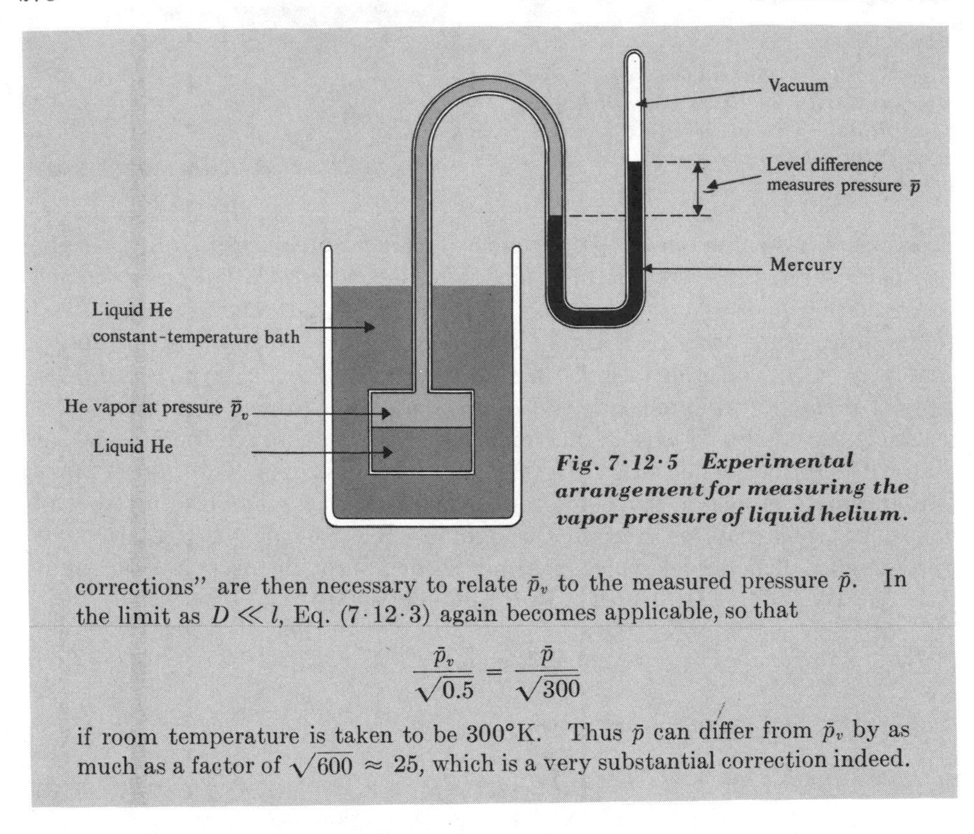

Fig. 7·12·5 Experimental arrangement for measuring the vapor pressure of liquid helium.

corrections" are then necessary to relate $\bar{p}_v$ to the measured pressure $\bar{p}$. In the limit as $D \ll l$, Eq. (7·12·3) again becomes applicable, so that

$$\frac{\bar{p}_v}{\sqrt{0.5}} = \frac{\bar{p}}{\sqrt{300}}$$

if room temperature is taken to be 300°K. Thus $\bar{p}$ can differ from $\bar{p}_v$ by as much as a factor of $\sqrt{600} \approx 25$, which is a very substantial correction indeed.

7 · 13 *Pressure and momentum transfer*

It is of interest to consider from a detailed kinetic point of view how a gas exerts a pressure. The basic mechanism is certainly clear: The mean force exerted on a wall of the container is due to the many collisions of molecules with the wall. Let us examine this mechanism in greater detail. We shall again look at the problem in a highly simplified way before doing the exact calculation.

Crude calculation We can give an argument similar to that used at the beginning of Section 7·11. In Fig. 7·11·1 we again imagine that roughly one-third of the molecules move parallel to the z direction. When such a molecule strikes the right end-wall, its kinetic energy remains unchanged. (This must be true, at least, on the average; otherwise one would not have an equilibrium situation.) The *magnitude* of the momentum of the molecule must then also remain unchanged; i.e., the molecule, approaching the right end-wall with momentum mv in the z direction, must have momentum $-mv$ after it rebounds from the wall. The z component of momentum of the molecule changes then by an amount $\Delta p_z = -2mv$ as a result of the collision with the wall. Correspondingly it follows, by conservation of momentum, that the wall gains in such a collision an amount of momentum $-\Delta p_z = 2mv$. But the mean force exerted on the wall is, by Newton's laws, just equal to the mean rate of change of momentum of the wall. Hence the mean force on the end-wall can be obtained simply by multiplying [the average momentum $2m\bar{v}$ gained by the wall per collision] by [the mean number of collisions $(\frac{1}{6}n\bar{v}A)$ per unit time with the end-wall]. The mean force per unit area, or mean pressure $\bar{p}$ on the wall, is then given by*

$$\bar{p} = \frac{1}{A}(2m\bar{v})\left(\frac{1}{6}n\bar{v}A\right) = \frac{1}{3}nm\bar{v}^2 \tag{7·13·1}$$

Exact calculation Suppose that we wish to calculate the mean force $\boldsymbol{F}$ exerted by the gas on a small element of area dA of the container wall. (See Fig. 7·13·1, where we have chosen the z axis to be normal to the element of area.) Then we must calculate the mean rate of change of momentum of this element of wall, i.e., the mean net momentum delivered to this wall element per unit time by the impinging molecules. If we focus attention on an element of area dA lying inside the gas an infinitesimal distance in front of the wall, then the above calculation is equivalent to finding the mean net molecular momentum which is transported per unit time across this surface from left to right as the molecules cross this surface from both directions.† Let us denote by $\boldsymbol{G}^{(+)}$ the mean molecular momentum crossing this surface dA per unit time from left to right, and by $\boldsymbol{G}^{(-)}$ the mean molecular momentum crossing this surface dA per unit time from right to left. Then one has simply

$$\boldsymbol{F} = \boldsymbol{G}^{(+)} - \boldsymbol{G}^{(-)} \tag{7·13·2}$$

To calculate $\boldsymbol{G}^{(+)}$, consider the element of surface dA in the gas and focus first attention on those molecules with velocity between $\boldsymbol{v}$ and $\boldsymbol{v} + d\boldsymbol{v}$. (See Fig. 7·13·2, which is similar to Fig. 7·11·2.) The mean number of such molecules which cross this area in an infinitesimal time dt is again the mean number of such molecules contained in the cylinder of volume $|dA\ v\ dt\cos\theta|$; i.e., it is equal to $f(\boldsymbol{v})d^3\boldsymbol{v}\,|dA\ v\,dt\cos\theta|$. By multiplying this number by the momentum

* The symbol $\bar{p}$ stands for mean pressure and should not be confused with the momentum variable p.

† Similarly, and quite equivalently, one can consider an element of area anywhere inside the gas and ask for the mean force which the gas on one side exerts on the gas on the other side. Again this is the same as asking what is the net molecular transport of momentum across this area.

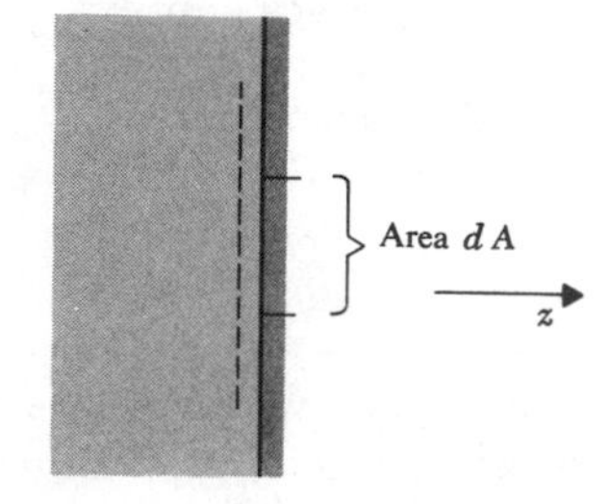

Fig. 7·13·1 ***An element of area dA of the container wall and a surface of area dA lying inside the gas just in front of the wall.***

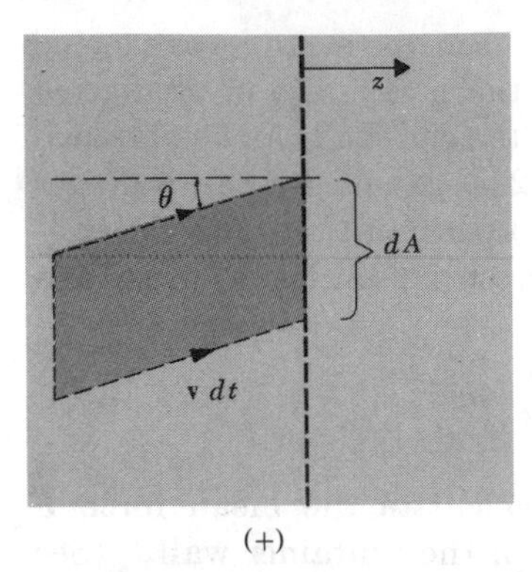

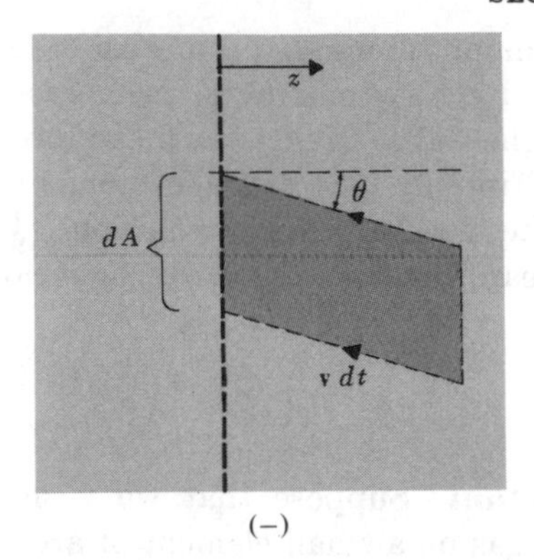

Fig. 7·13·2 Molecules crossing a surface dA in a gas from left to right (+) and from right to left (−). (Note that the height of the cylinders shrinks to zero as dt → 0.)

$m\boldsymbol{v}$ of each such molecule and dividing by the time dt, one obtains the mean momentum transported across the area dA per unit time by molecules with velocity between $\boldsymbol{v}$ and $\boldsymbol{v} + d\boldsymbol{v}$. By summing over all molecules which cross this area from left to right, i.e., over all molecular velocities with $v_z > 0$, one then gets for the total mean molecular momentum $\boldsymbol{G}^{(+)}$ transported across this area from left to right the result

$$\boldsymbol{G}^{(+)} = \int_{v_z>0} f(\boldsymbol{v})\, d^3\boldsymbol{v}\, |dA\, v \cos \theta|(m\boldsymbol{v})$$

or

$$\boldsymbol{G}^{(+)} = dA \int_{v_z>0} d^3\boldsymbol{v}\, f(\boldsymbol{v})\, |v_z|(m\boldsymbol{v}) \tag{7·13·3}$$

where we have put $v_z = v \cos \theta$ and where the integration is over all velocities for which $v_z > 0$. A similar expression gives the total mean molecular momentum $\boldsymbol{G}^{(-)}$ transported across this area from right to left, except that the integration must now be over all molecules for which $v_z < 0$. Thus

$$\boldsymbol{G}^{(-)} = dA \int_{v_z<0} d^3\boldsymbol{v}\, f(\boldsymbol{v})\, |v_z|(m\boldsymbol{v}) \tag{7·13·4}$$

The force (7 · 13 · 2) is then given by the *net* mean momentum transported across the surface, i.e., by subtracting (7 · 13 · 4) from (7 · 13 · 3). But in (7 · 13 · 3), where the integration is only over positive values of v_z, one can simply put $|v_z| = v_z$ in the integrand. In (7 · 13 · 4), where the integration is only over negative values of v_z, one can put $|v_z| = -v_z$ in the integrand. Hence (7 · 13 · 3) gives simply

$$\boldsymbol{F} = \boldsymbol{G}^{(+)} - \boldsymbol{G}^{(-)} = dA \int_{v_z>0} d^3\boldsymbol{v}\, f(\boldsymbol{v}) v_z (m\boldsymbol{v}) + dA \int_{v_z<0} d^3\boldsymbol{v}\, f(\boldsymbol{v}) v_z (m\boldsymbol{v})$$

or

▶
$$\boldsymbol{F} = dA \int d^3\boldsymbol{v}\, f(\boldsymbol{v}) v_z (m\boldsymbol{v}) \tag{7·13·5}$$

where the two integrals have been combined into a single integral over *all* possible velocities. Equation (7 · 13 · 5) is a very general expression and would be valid even if the gas is not in equilibrium, i.e., even if f is completely arbitrary.

If the gas is in equilibrium, then $f(\mathbf{v})$ is only a function of $v \equiv |\mathbf{v}|$. Note first that

$$\bar{F}_x = dA\ m \int d^3\mathbf{v}\ f(\mathbf{v}) v_z v_x = 0 \tag{7·13·6}$$

since the integrand is odd, having opposite signs for $+v_x$ and $-v_x$. Equation (7·13·6) expresses the obvious fact that there can be no mean tangential force on the wall in an equilibrium situation. The mean normal force does, of course, not vanish. Measured per unit area it gives the mean pressure, which is thus, by (7·13·5), equal to

$$\bar{p} = \frac{\bar{F}_z}{dA} = \int d^3\mathbf{v}\ f(\mathbf{v}) m v_z^2$$

or

$$\bar{p} = nm\overline{v_z^2} \tag{7·13·7}$$

Here we have used the definition

$$\overline{v_z^2} \equiv \frac{1}{n} \int d^3\mathbf{v}\ f(\mathbf{v}) v_z^2$$

By symmetry, $\overline{v_x^2} = \overline{v_y^2} = \overline{v_z^2}$ so that

$$\overline{v^2} = \overline{v_x^2 + v_y^2 + v_z^2} = 3\overline{v_z^2}$$

Hence (7·13·7) can be written equivalently as

▶ $$\bar{p} = \tfrac{1}{3} nm\overline{v^2} \tag{7·13·8}$$

This agrees substantially with our crudely derived result (7·13·1) (except that the averaging is now done carefully so that what appears is $\overline{v^2}$ rather than $\bar{v}^2$). Since $\overline{v^2}$ is related to the mean kinetic energy K of a molecule, (7·13·8) implies the general relation

$$\bar{p} = \tfrac{2}{3} n(\tfrac{1}{2} m\overline{v^2}) = \tfrac{2}{3} n\bar{K} \tag{7·13·9}$$

i.e., the mean pressure is just equal to $\frac{2}{3}$ the mean kinetic energy per unit volume of the gas.

Up to now we have not yet made use of the fact that the mean number density of molecules $f(\mathbf{v})\, d^3\mathbf{v}$ is given by the Maxwell velocity distribution.* This information allows us to calculate explicitly $\overline{v^2}$ and is equivalent to using the equipartition theorem result that $\bar{K} = \frac{3}{2} kT$. Then (7·13·9) becomes

$$\bar{p} = nkT \tag{7·13·10}$$

so that one regains the equation of state of a classical ideal gas.

* The expressions (7·13·7) through (7·13·9) are thus equally valid even if f is given by the quantum-mechanical Fermi-Dirac or Bose-Einstein equilibrium distributions to be discussed in Chapter 9.

SUGGESTIONS FOR SUPPLEMENTARY READING

J. F. Lee, F. W. Sears, and D. L. Turcotte: "Statistical Thermodynamics," chaps. 3 and 5, Addison-Wesley Publishing Company, Reading, Mass., 1963.

N. Davidson: "Statistical Mechanics," chaps. 10 and 19, McGraw-Hill Book Company, New York, 1962.

R. D. Present: "Kinetic Theory of Gases," chaps. 2 and 5, McGraw-Hill Book Company, New York, 1958.

PROBLEMS

7.1 Consider a homogeneous mixture of inert monatomic ideal gases at absolute temperature T in a container of volume V. Let there be ν_1 moles of gas 1, ν_2 moles of gas 2, . . . , and ν_k moles of gas k.

(*a*) By considering the classical partition function of this system, derive its equation of state, i.e., find an expression for its total mean pressure $\bar{p}$.

(*b*) How is this total pressure $\bar{p}$ of the gas related to the pressure $\bar{p}_i$ which the ith gas would produce if it alone occupied the entire volume at this temperature?

7.2 An ideal monatomic gas of N particles, each of mass m, is in thermal equilibrium at absolute temperature T. The gas is contained in a cubical box of side L, whose top and bottom sides are parallel to the earth's surface. The effect of the earth's uniform gravitational field on the particles should be considered, the acceleration due to gravity being g.

(*a*) What is the average kinetic energy of a particle?

(*b*) What is the average potential energy of a particle?

7.3 A thermally insulated container is divided by a partition into two compartments, the right compartment having a volume b times as large as the left one. The left compartment contains ν moles of an ideal gas at temperature T and pressure $\bar{p}$. The right compartment also contains ν moles of an ideal gas at the temperature T. The partition is now removed. Calculate

(*a*) the final pressure of the gas mixture in terms of $\bar{p}$;

(*b*) the total change of entropy if the gases are different;

(*c*) the total change of entropy if the gases are identical.

7.4 A thermally insulated container is divided into two parts by a thermally insulated partition. Both parts contain ideal gases which have equal constant heat capacities c_V. One of these parts contains ν_1 moles of gas at a temperature T_1 and pressure $\bar{p}_1$; the other contains ν_2 moles of gas at a temperature T_2 and pressure $\bar{p}_2$. The partition is now removed and the system is allowed to come to equilibrium.

(*a*) Find the final pressure.

(*b*) Find the change ΔS of total entropy if the gases are different.

(*c*) Find ΔS if the gases are identical.

7.5 A rubber band at absolute temperature T is fastened at one end to a peg, and supports from its other end a weight W. Assume as a simple microscopic model of the rubber band that it consists of a linked polymer chain of N segments joined end to end; each segment has length a and can be oriented either parallel or antiparallel to the vertical direction. Find an expression for the resultant mean length $\bar{l}$ of the rubber band as a function of W. (Neglect the kinetic energies or weights of the segments themselves, or any interaction between the segments.)

7.6 Consider a gas which is *not* ideal so that molecules *do* interact with each other. This gas is in thermal equilibrium at the absolute temperature T. Suppose that the translational degrees of freedom of this gas can be treated classically. What is the mean kinetic energy of (center-of-mass) translation of a molecule in this gas?

7.7 Monatomic molecules adsorbed on a surface are free to move on this surface and can be treated as a classical ideal two-dimensional gas. At absolute temperature T, what is the heat capacity per mole of molecules thus adsorbed on a surface of fixed size?

7.8 The electrical resistivity ρ of a metal at room temperature is proportional to the probability that an electron is scattered by the vibrating atoms in the lattice, and this probability is in turn proportional to the mean square amplitude of vibration of these atoms. Assuming classical statistics to be valid in this temperature range, what is the dependence of the electrical resistivity ρ on the absolute temperature T?

7.9 A very sensitive spring balance consists of a quartz spring suspended from a fixed support. The spring constant is α, i.e., the restoring force of the spring is $-\alpha x$ if the spring is stretched by an amount x. The balance is at a temperature T in a location where the acceleration due to gravity is g.

(*a*) If a very small object of mass M is suspended from the spring, what is the mean resultant elongation $\bar{x}$ of the spring?

(*b*) What is the magnitude $\overline{(x - \bar{x})^2}$ of the thermal fluctuations of the object about its equilibrium position?

(*c*) It becomes impracticable to measure the mass of an object when the fluctuations are so large that $[\overline{(x - \bar{x})^2}]^{\frac{1}{2}} = \bar{x}$. What is the minimum mass M which can be measured with this balance?

7.10 A system consists of N very weakly interacting particles at a temperature T sufficiently high so that classical statistical mechanics is applicable. Each particle has mass m and is free to perform one-dimensional oscillations about its equilibrium position. Calculate the heat capacity of this system of particles at this temperature in each of the following cases:

(*a*) The force effective in restoring each particle to its equilibrium position is proportional to its displacement x from this position.

(*b*) The restoring force is proportional to x^3.

7.11 Assume the following highly simplified model for calculating the specific heat of graphite, which has a highly anisotropic crystalline layer structure. Each carbon atom in this structure can be regarded as performing simple harmonic oscillations in three dimensions. The restoring forces in directions parallel to a layer are very large; hence the natural frequencies of oscillations in the x and y directions lying within the plane of a layer are both equal to a value $\omega_{\parallel}$ which is so large that $\hbar\omega_{\parallel} \gg 300k$. On the other hand, the restoring force perpendicular to a layer is quite small; hence the frequency of oscillation $\omega_{\perp}$ of an atom in the z direction perpendicular to a layer is so small that $\hbar\omega_{\perp} \ll 300k$. On the basis of this model, what is the molar specific heat (at constant volume) of graphite at 300°K?

7.12 Consider a solid of compressibility κ. Assume that the atoms in this solid are arranged on a regular cubic lattice, the distance between their nearest neighbors being a. Assume further that a restoring force $-\kappa_0\, \Delta a$ acts on a given atom when it is displaced by a distance Δa from its nearest neighbor.

(*a*) Use simple reasoning to find an approximate relation between the spring constant κ_0 and the compressibility κ of this solid. (Consider the force needed to decrease the length of one edge of a solid parallelepiped by a small amount.)

(*b*) Estimate roughly the order of magnitude of the Einstein temperature Θ_E for copper (atomic weight = 63.5) by assuming that it is a simple cubic structure with density 8.9 g cm^{-3} and compressibility 4.5×10^{-13} cm^2 dyne^{-1}.

7.13 Show that the general expression (7·8·13) for $\bar{\mu}_z$ becomes identical to the simple expression (6·3·3) in the case where $J = \frac{1}{2}$.

7.14 Consider an assembly of N_0 weakly interacting magnetic atoms per unit volume at a temperature T and describe the situation *classically*. Then each magnetic moment $\boldsymbol{\mu}$ can make any arbitrary angle θ with respect to a given direction (call it the z direction). In the absence of a magnetic field, the probability that this angle lies between θ and $\theta + d\theta$ is simply proportional to the solid angle $2\pi \sin\theta \, d\theta$ enclosed in this range. In the presence of a magnetic field H in the z direction, this probability must further be proportional to the Boltzmann factor $e^{-\beta E}$, where E is the magnetic energy of the moment $\boldsymbol{\mu}$ making this angle θ with the z axis. Use this result to calculate the classical expression for the mean magnetic moment $\bar{M}_z$ of these N_0 atoms.

7.15 Consider the expression (7·8·20) for $\bar{M}_z$ in the limit where the spacing between the magnetic energy levels is small compared to kT, i.e., where $\eta \equiv g\mu_0 J/kT \ll 1$. Assume further that the angle θ between $\boldsymbol{J}$ and the z axis is almost continuous, i.e., that J is so large that the possible values of $\cos\theta = m/J$ are very closely spaced; to be specific assume that J is large enough that $J\eta \gg 1$. Show that in this limit the general expression (7·8·20) for $\bar{M}_z$ does approach the classical expression derived in the preceding problem.

7.16 An aqueous solution at room temperature T contains a small concentration of magnetic atoms, each of which has a net spin $\frac{1}{2}$ and a magnetic moment μ. The solution is placed in an external magnetic field H pointing along the z direction. The magnitude of this field is inhomogeneous over the volume of the solution. To be specific, $H = H(z)$ is a monotonic increasing function of z, assuming a value H_1 at the bottom of the solution where $z = z_1$ and a larger value H_2 at the top of the solution where $z = z_2$.

(*a*) Let $n_+(z)\,dz$ denote the mean number of magnetic atoms whose spin points along the z direction and which are located between z and $z + dz$. What is the ratio $n_+(z_2)/n_+(z_1)$?

(*b*) Let $n(z)\,dz$ denote the total mean number of magnetic atoms (of both directions of spin orientation) located between z and $z + dz$. What is the ratio $n(z_2)/n(z_1)$? Is it less than, equal to, or greater than unity?

(*c*) Make use of the fact that $\mu H \ll kT$ to simplify the answers to the preceding questions.

7.17 What fraction of the molecules of a gas have x components of velocity between $-\tilde{v}$ and $+\tilde{v}$, where $\tilde{v}$ is the most probable speed of the molecules? (Suggestion: consult a table of the error function; see Appendix A·5.)

7.18 Use the results of Problem 5.9 to express the velocity of sound in an ideal gas in terms of the most probable speed $\tilde{v}$ of the molecules in the gas and the specific heat ratio $\gamma \equiv c_p/c_V$ of that gas.

In the case of helium (He) gas, what fraction of the molecules have molecular speeds less than the speed of sound in this gas?

7.19 A gas of molecules, each of mass m, is in thermal equilibrium at the absolute temperature T. Denote the velocity of a molecule by $\boldsymbol{v}$, its three cartesian components by v_x, v_y, and v_z, and its speed by v. What are the following mean values:

(a) $\overline{v_x}$ (d) $\overline{v_x^3 v_y}$

(b) $\overline{v_x^2}$ (e) $\overline{(v_x + bv_y)^2}$ where b is a constant

(c) $\overline{v^2 v_x}$ (f) $\overline{v_x^2 v_y^2}$

(If you need to calculate explicitly any integrals in this problem, you are the kind of person who likes to turn cranks but does not think.)

7.20 An ideal monatomic gas is in thermal equilibrium at room temperature T so that the molecular velocity distribution is Maxwellian.

(a) If v denotes the speed of a molecule, calculate $\overline{(1/v)}$. Compare this with $1/\bar{v}$.

(b) Find the mean number of molecules per unit volume whose energy lies in the range between ϵ and $\epsilon + d\epsilon$.

7.21 What is the most probable kinetic energy $\tilde{\epsilon}$ of molecules having a Maxwellian velocity distribution? Is it equal to $\frac{1}{2}m\tilde{v}^2$, where $\tilde{v}$ is the most probable speed of the molecules?

7.22 A gas of atoms, each of mass m, is maintained at the absolute temperature T inside an enclosure. The atoms emit light which passes (in the x direction) through a window of the enclosure and can then be observed as a spectral line in a spectroscope. A stationary atom would emit light at the sharply defined frequency ν_0. But, because of the Doppler effect, the frequency of the light observed from an atom having an x component of velocity v_x is not simply equal to the frequency ν_0, but is approximately given by

$$\nu = \nu_0 \left(1 + \frac{v_x}{c}\right)$$

where c is the velocity of light. As a result, not all of the light arriving at the spectroscope is at the frequency ν_0; instead it is characterized by some intensity distribution $I(\nu)\,d\nu$ specifying the fraction of light intensity lying in the frequency range between ν and $\nu + d\nu$. Calculate

(a) The mean frequency $\bar{\nu}$ of the light observed in the spectroscope.

(b) The root-mean square frequency shift $(\Delta\nu)_{\text{rms}} = [\overline{(\nu - \bar{\nu})^2}]^{\frac{1}{2}}$ (measured from the mean frequency) of the light observed in the spectroscope.

(c) The relative intensity distribution $I(\nu)\,d\nu$ of the light observed in the spectroscope.

7.23 In a molecular beam experiment, the source is a tube containing hydrogen at a pressure $\bar{p}_s = 0.15$ mm of mercury and at a temperature $T = 300°$K. In the tube wall is a slit 20 mm $\times$ 0.025 mm, opening into a highly evacuated region. Opposite the source slit and one meter away from it is a second detector slit parallel to the first and of the same size. This slit is in the wall of a small enclosure in which the pressure $\bar{p}$ can be measured.

(a) How many H_2 molecules leave the source slit per second?

(b) How many H_2 molecules arrive at the detector slit per second?

(c) What is the pressure $\bar{p}_d$ in the detector chamber when a steady state has been reached so that $\bar{p}_d$ is independent of time?

7.24 A thin-walled vessel of volume V, kept at constant temperature, contains a gas which slowly leaks out through a small hole of area A. The outside pressure is low enough that leakage back into the vessel is negligible. Find the time required for the pressure in the vessel to decrease to $1/e$ of its original value. Express your answer in terms of A, V, and the mean molecular speed $\bar{v}$.

7.25 A spherical bulb 10 cm in radius is maintained at room temperature (300°K) except for one square centimeter which is kept at liquid nitrogen temperature (77°K). The bulb contains water vapor originally at a pressure of 0.1 mm of mercury. Assuming that every water molecule striking the cold area condenses and sticks to the surface, estimate the time required for the pressure to decrease to 10^{-6} mm of mercury.

7.26 A vessel is closed off by a porous partition through which gases can pass by effusion and then be pumped off to some collecting chamber. The vessel itself is filled with a dilute gas consisting of two types of molecules which differ because they contain two different atomic isotopes and have correspondingly masses m_1 and m_2. The concentrations of these molecules are c_1 and c_2, respectively, and are maintained constant inside the vessel by constantly replenishing the supply of gas in it by a steady slow flow of fresh gas through the vessel.

(*a*) Let c_1' and c_2' denote the concentrations of the two types of molecules in the collecting chamber. What is the ratio c_2'/c_1'?

(*b*) By using the gas UF_6, one can attempt to separate U^{235} from U^{238}, the first of these isotopes being the one useful in the initiation of nuclear-fission reactions. The molecules in the vessel are then $U^{238}F_6^{19}$ and $U^{235}F_6^{19}$. (The concentrations of these molecules, corresponding to the natural abundance of the two uranium isotopes, are $c_{238} = 99.3$ percent and $c_{235} = 0.7$ percent.) Calculate the corresponding ratio c'_{235}/c'_{238} of the molecules collected after effusion in terms of their original concentration ratio c_{235}/c_{238}.

7.27 A container has as one of its walls a membrane containing many small holes. If the container is filled with gas at some moderate pressure p_0, gas will escape by effusion into the vacuum surrounding the container. It is found that when the container is filled with He gas at room temperature and at pressure p_0, the pressure will have fallen to $\frac{1}{2}p_0$ after one hour.

Suppose that the container is filled at room temperature and at total pressure p_0 with a mixture of helium (He) and neon (Ne), the atomic concentrations of both species being 50 percent (i.e., 50 percent of the atoms are He and 50 percent of them are Ne). What will be the ratio n_{Ne}/n_{He} of the atomic concentrations of Ne to He after one hour? Express your answer in terms of the atomic weights μ_{Ne} of neon and μ_{He} of helium.

7.28 A box of volume V containing an ideal gas of molecular weight μ at temperature T is divided into two equal halves by a partition. Initially the pressure on the left side is $p_1(0)$ and that on the right side is $p_2(0)$. A small hole of area A is now introduced in the partition by opening a valve so that the molecules can effuse through the resulting hole in the (thin) partition.

(*a*) Find the pressure $p_1(t)$ of the gas in the left side of the box as a function of time.

(*b*) Calculate the change of entropy ΔS of the whole gas after the final equilibrium has been reached.

7.29 An enclosure contains a dilute gas at temperature T. Some molecules can escape into a vacuum by effusing through a small hole in one of the walls of the container. Choose the z direction so as to point along the outward normal to the plane of this hole. Let the mass of a molecule be m and the z component of its velocity be denoted by v_z.

(*a*) What is the mean velocity component $\bar{v}_z$ of a molecule inside the container?

(*b*) What is the mean velocity component $\bar{v}_z$ of a molecule which has effused into the vacuum?

7.30 The molecules of a monatomic ideal gas are escaping by effusion through a small hole in a wall of an enclosure maintained at absolute temperature T.

(*a*) By physical reasoning (without actual calculation) do you expect the *mean* kinetic energy $\bar{\epsilon}_0$ of a molecule in the effusing beam to be equal to, greater than, or less than the mean kinetic energy $\bar{\epsilon}_i$ of a molecule within the enclosure?

(*b*) Calculate $\bar{\epsilon}_0$ for a molecule in the effusing beam. Express your answer in terms of $\bar{\epsilon}_i$.

7.31 An enclosure contains gas at a pressure $\bar{p}$ and has in one of its walls a small hole of area A through which gas molecules pass into a vacuum by effusion. In this vacuum, directly in front of the hole at a distance L from it, there is suspended a circular disk of radius R. It is oriented so that the normal to its surface points toward the hole (see figure). Assuming that the molecules in the effusing beam get scattered elastically from this disk, calculate the force exerted on the disk by the molecular beam.

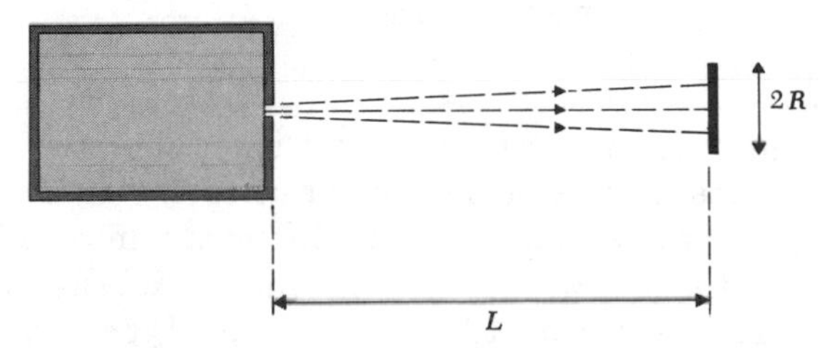

Equilibrium between phases or chemical species

8

THE LAST several chapters have elaborated both the macroscopic and the microscopic aspects of the basic theory of Chapter 3. We are thus well prepared to use this theory to discuss a number of important physical situations. Up to now we have dealt almost exclusively with systems consisting of a single "component" (i.e., of a single type of molecule or chemical species) and a single "phase" (i.e., a single spatially homogeneous state of aggregation). But the situations of greatest interest are often more complicated. For example, one may be interested in a single-component system consisting of several phases (e.g., ice and water in equilibrium, or a liquid and its vapor in equilibrium). Alternatively, one may be interested in a single-phase system consisting of several components (e.g., a gas consisting of several types of molecules which may react chemically with each other). Or, in the case of greatest generality, one may be interested in systems consisting of several components in several phases.

In this chapter we shall show how such more complicated systems can be treated by the methods of statistical thermodynamics. Most of our considerations will be independent of any particular microscopic models and will lead to a number of very general results. These yield much valuable insight into many systems of common occurrence. In addition, they are useful both in establishing important relationships between various macroscopic quantities and in furnishing suitable starting points for detailed microscopic calculations.

GENERAL EQUILIBRIUM CONDITIONS

The following sections will amplify the discussion of Sec. 3·1 with the aim of examining equilibrium conditions for systems in various physical situations.

8 · 1 *Isolated system*

Consider a thermally isolated system A. From our discussion of Sec. 3 · 1, as summarized in the second law of thermodynamics, we know that any spontaneously occurring process is such that the entropy of the system tends to increase. In statistical terms this means that the system tends to approach a situation of larger intrinsic probability. In any such process the spontaneous change of entropy satisfies thus the condition

$$\Delta S \geq 0 \tag{8·1·1}$$

It follows that if a stable equilibrium situation has been attained where no further spontaneous processes (other than ever-present random fluctuations) can take place, then this is a situation where S is maximum; i.e., it is the most probable situation for the system subject to the given constraints. Hence one can make the following statement.

> For a thermally isolated system, the stable equilibrium situation is characterized by the fact that

$$S = \text{maximum} \tag{8·1·2}$$

This means that if one goes away from the situation where $S = S_{\max}$ is maximum, then for very small departures from this equilibrium situation S does not change ($dS = 0$ for an extremum), but for larger departures S must decrease. That is, the change of entropy $\Delta_m S$ measured from a stable equilibrium situation is such that

$$\Delta_m S \equiv S - S_{\max} \leq 0 \tag{8·1·3}$$

Example 1 Let us illustrate the situation schematically (see Fig. 8·1·1). Suppose that the system is characterized by a parameter y (or by several such parameters) which is free to vary. (For example, the system might consist of ice and water, and y might denote the relative concentration of ice.) Then

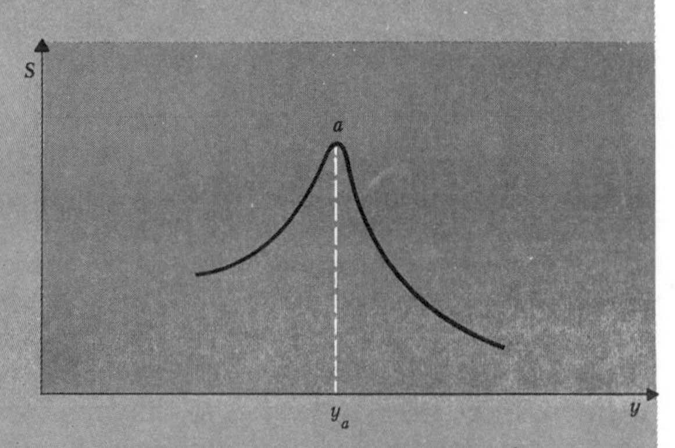

Fig. 8·1·1 Diagram illustrating the dependence of the entropy S on a parameter y.

the point a on the diagram corresponds to a maximum of S, and the stable equilibrium situation of the system corresponds to an adjustment of the parameter y until it attains the value y_a. Here S attains its absolute maximum so that this corresponds to a situation of absolute stability.

Example 2 A more complicated possible situation is illustrated schematically in Fig. 8·1·2. Here the point a corresponds to a local maximum of S. Thus there is no spontaneous process possible whereby y could move away from the value y_a by relatively small amounts. This value of y corresponds then to a situation of relative equilibrium (or metastable equilibrium). On the other hand, it is possible that by some *major* disturbance the parameter y could attain a value in the region near y_b. Then it will tend to approach the value y_b where the entropy S has its absolute maximum. The point b thus represents a situation of absolutely stable equilibrium.

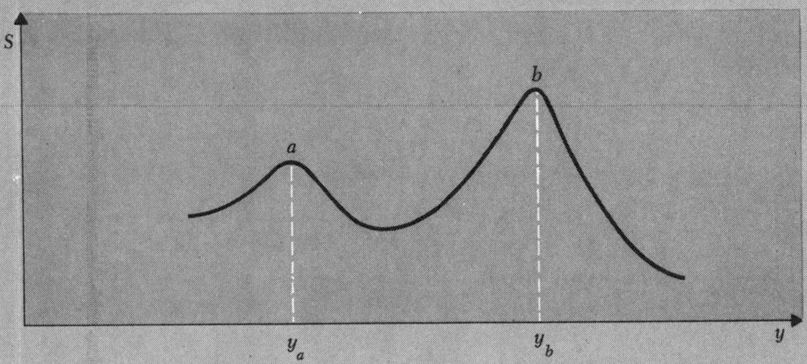

Fig. 8·1·2 *Diagram illustrating the dependence of the entropy S on a parameter y when there exists the possibility of metastable equilibrium.*

In a thermally isolated system of the kind discussed, the first law of thermodynamics implies that

$$Q = 0 = W + \Delta\bar{E}$$

or

$$W = -\Delta\bar{E} \tag{8·1·4}$$

If the external parameters of the system (e.g., its volume) are kept fixed, then no work gets done and

$$\bar{E} = \text{constant} \tag{8·1·5}$$

while S tends to approach its maximum value in accordance with (8·1·2).

The argument leading to (8·1·2) can be phrased in more explicit statistical terms. Suppose that an isolated system is described by a parameter y (or by several such parameters) so that its total energy is constant. Let $\Omega(y)$ denote the number of states accessible to the system when the parameter has a given value between y and $y + \delta y$ (δy being some fixed small interval); the corresponding entropy of the system is then, by definition, $S(y) = k \ln \Omega(y)$. If the parameter y is free to vary, then the fundamental statistical postulate asserts that in an equilibrium situation the probability $P(y)$ of finding the system with the parameter between y and $y + \delta y$ is given by

$$P(y) \propto \Omega(y) = e^{S(y)/k} \tag{8·1·6}$$

Equation (8·1·6) shows explicitly that if y is left free to adjust itself, it will tend to approach a value $\tilde{y}$ where $P(y)$ is maximum, i.e., where $S(y)$ is maximum. In equilibrium the relative probability of the occurrence of a fluctuation where $y \neq \tilde{y}$ is then given by (8·1·6) as

$$\frac{P(y)}{P_{\max}} = e^{\Delta_m S/k} \tag{8·1·7}$$

where $\Delta_m S = S(y) - S_{\max}$.

The relations (8·1·6) or (8·1·7) provide more quantitative statements than the assertion (8·1·2) because they imply not only that the system tends to approach the situation where $S = S_{\max}$, but they also allow one to calculate the probability of occurrence of fluctuations where $S \neq S_{\max}$.

Remark If S depends on a single parameter y, then its maximum occurs for some value $y = \tilde{y}$ determined by the condition

$$\frac{\partial S}{\partial y} = 0$$

Expansion of S about its maximum gives then

$$S(y) = S_{\max} + \frac{1}{2}\left(\frac{\partial^2 S}{\partial y^2}\right)(y - \tilde{y})^2 + \cdots$$

where the second derivative is evaluated at $y = \tilde{y}$ and must be negative to guarantee that S is maximum when $y = \tilde{y}$. Thus one can write $(\partial^2 S/\partial y^2) = -|\partial^2 S/\partial y^2|$ and obtains by (8·1·6) the explicit expression

$$P(y) \propto \exp\left[-\frac{1}{2k}\left|\frac{\partial^2 S}{\partial y^2}\right|(y - \tilde{y})^2\right] \tag{8·1·8}$$

for the probability of fluctuations near the equilibrium situation where $y = \tilde{y}$. The fluctuations are thus described by a Gaussian distribution with a dispersion given by

$$\overline{(y - \tilde{y})^2} = k\left|\frac{\partial^2 S}{\partial y^2}\right|^{-1}$$

8·2 *System in contact with a reservoir at constant temperature*

Knowing the equilibrium conditions for an isolated system, one can readily deduce similar conditions for other situations of physical interest. For example, much experimental work is done under conditions of constant temperature. Thus we should like to investigate the equilibrium conditions for a

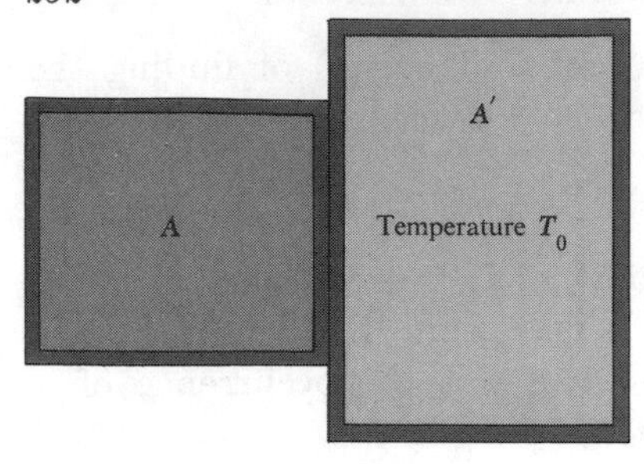

Fig. 8·2·1 A system A in contact with a heat reservoir at temperature T_0.

system A in thermal contact with a heat reservoir A' which is at the constant absolute temperature T_0.

The combined system $A^{(0)}$ consisting of the system A and the heat reservoir A' is an isolated system of the type discussed in Sec. 8·1. The entropy $S^{(0)}$ of $A^{(0)}$ then satisfies the condition (8·1·1) that in any spontaneous process

$$\Delta S^{(0)} \geq 0 \tag{8·2·1}$$

But this condition can readily be written in terms of quantities which refer only to the system A of interest. Thus

$$\Delta S^{(0)} = \Delta S + \Delta S' \tag{8·2·2}$$

where ΔS is the entropy change of A and $\Delta S'$ that of the reservoir A' in the process under consideration. But if A absorbs heat Q from the reservoir A' in this process, then A' absorbs heat $(-Q)$ and suffers a corresponding entropy change

$$\Delta S' = \frac{(-Q)}{T_0}$$

since it remains in internal equilibrium at the constant temperature T_0. Furthermore, the first law gives

$$Q = \Delta\bar{E} + W$$

where $\Delta\bar{E}$ is the internal energy change of A and W the work done by A in the process under consideration. Thus (8·2·2) can be written

$$\Delta S^{(0)} = \Delta S - \frac{Q}{T_0} = \frac{T_0\,\Delta S - (\Delta\bar{E} + W)}{T_0} = \frac{\Delta(T_0 S - \bar{E}) - W}{T_0}$$

or

$$\Delta S^{(0)} = \frac{-\Delta F_0 - W}{T_0} \tag{8·2·3}$$

where we have used the fact that T_0 is a constant and have introduced the definition

$$F_0 \equiv \bar{E} - T_0 S \tag{8·2·4}$$

This reduces to the ordinary Helmholtz free energy $F = \bar{E} - TS$ of the system A if the latter has a temperature T equal to that of the heat reservoir A'. Of course, in the general case when A is not in equilibrium with A', its temperature T is not necessarily equal to T_0.

The total entropy change $\Delta S^{(0)}$ in (8·2·3) is expressed completely in terms of quantities that refer only to the system A of interest. The fundamental condition (8·2·1) then allows us to draw some interesting conclusions. Since T_0 is in all ordinary cases positive, one gets

$$-\Delta F_0 \geq W \tag{8·2·5}$$

This relation implies that the *maximum* work which can be done by a system in contact with a heat reservoir is given by $(-\Delta F_0)$. (This is the reason for the name "free energy" given to F.) The maximum work corresponds, of course, to the equals sign in (8·2·1) and is obtained when the process used is a quasi-static one (so that A is always in equilibrium with A' and $T = T_0$). Equation (8·2·5) should be compared with the rather different relation (8·1·4) which holds for the work done by an *isolated* system.

If the external parameters of the system A (e.g., its volume) are kept fixed, then $W = 0$ and (8·2·5) yields the condition

$$\Delta F_0 \leq 0 \tag{8·2·6}$$

This equation is analogous to Eq. (8·1·1) for an *isolated* system. It implies that if a system in thermal contact with a heat reservoir, its free energy tends to *decrease*. Thus we arrive at the statement:

> If a system, whose external parameters are fixed, is in thermal contact with a heat reservoir, the stable equlibrium situation is characterized by the condition that

▶
$$F_0 = \text{minimum} \tag{8·2·7}$$

This last condition can again be phrased in more explicit statistical terms. Consider the external parameters of A to be fixed so that $W = 0$, and suppose that A is described by some parameter y. The thermodynamic functions of A (e.g., S and $\bar{E}$) have definite values $S(y_1)$ and $\bar{E}(y_1)$ when y has a given value $y = y_1$. If the parameter changes to any other value y, these functions change by corresponding amounts $\Delta S = S(y) - S(y_1)$ and $\Delta \bar{E} = \bar{E}(y) - \bar{E}(y_1) = Q$. The entropy of the heat reservoir A' also changes since it absorbs some heat, and the corresponding change in the total entropy of $A^{(0)}$ is given by (8·2·3) (with $W = 0$) as

$$\Delta S^{(0)} = -\frac{\Delta F_0}{T_0} \tag{8·2·8}$$

But in an equilibrium situation the probability $P(y)$ that the parameter lies between y and $y + \delta y$ is proportional to the number of states $\Omega^{(0)}(y)$ accessible to the total isolated system $A^{(0)}$ when the parameter lies in this range. Thus one has, analogously to (8·1·6),

$$P(y) \propto \Omega^{(0)}(y) = e^{S^{(0)}(y)/k} \tag{8·2·9}$$

But by (8·2·8)

$$S^{(0)}(y) = S^{(0)}(y_1) - \frac{\Delta F_0}{T_0} = S^{(0)}(y_1) - \frac{F_0(y) - F_0(y_1)}{T_0}$$

Since y_1 is just some arbitrary constant value, the corresponding constant terms can be absorbed in the constant of proportionality of (8·2·9) which then becomes

$$P(y) \propto e^{-F_0(y)/kT_0} \tag{8·2·10}$$

This relation shows directly that the most probable situation is one where F_0 is a minimum. Of course, if $\bar{E}$ is a constant independent of y (as it would be for an isolated system), then $F_0 = \bar{E} - T_0S(y)$ and (8·2·10) reduces properly to the relation (8·1·6) which involves only the entropy.

Remark The relation (8·2·10) is, of course, akin to the canonical distribution. According to the latter the desired probability is given by

$$P(y) \propto \sum_r e^{-\beta_0 E_r}, \qquad \beta_0 \equiv (kT_0)^{-1} \tag{8·2·11}$$

where the sum extends over all states r for which the parameter lies between y and $y + \delta y$. If $\Omega(E; y)$ denotes the number of states for which the parameter lies between y and $y + \delta y$ and whose energy lies between E and $E + \delta E$, the relation (8·2·11) becomes

$$P(y) \propto \sum_E \Omega(E; y)\, e^{-\beta_0 E}$$

where the sum is now to be taken over all possible energy intervals. The summand has as a function of E the usual very sharp maximum near some value $\bar{E}(y)$ which depends on y and which is equal to the mean energy of the system for this value of y. Hence only terms near this maximum contribute appreciably to the sum and

$$P(y) \propto \Omega(\bar{E}; y)\, e^{-\beta_0 \bar{E}(y)} = e^{S(y)/k - \beta_0 \bar{E}(y)} = e^{-\beta_0 F_0(y)}$$

8 · 3 *System in contact with a reservoir at constant temperature and pressure*

Another case of physical interest is that where a system A is maintained under conditions of both constant temperature and constant pressure. This is a situation of frequent occurrence in the laboratory, where one may carry out an experiment in a thermostat at, say, atmospheric pressure. A situation such as this implies that the system A is in thermal contact with a heat reservoir A' which is at a constant temperature T_0 and at a constant pressure p_0. The system A can exchange heat with the reservoir A'; but the latter is so large that its temperature T_0 remains unchanged. Similarly, the system A can

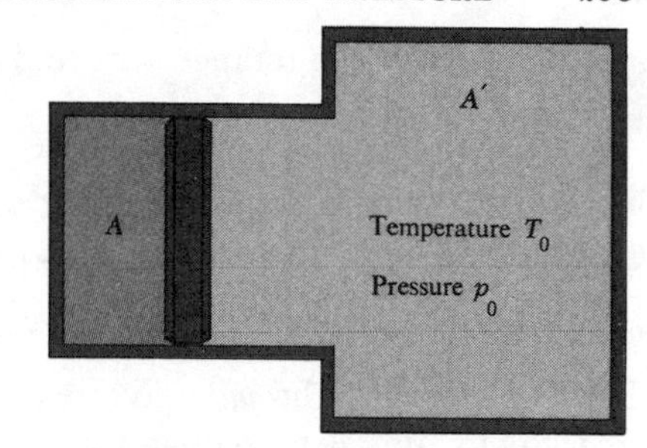

***Fig. 8·3·1** A system A in contact with a reservoir at constant temperature T_0 and constant pressure p_0.*

change its volume V at the expense of the reservoir A', doing work on the reservoir in the process; but again A' is so large that its pressure p_0 remains unaffected by this relatively small volume change.

> ***Remark*** The system A' may be a single reservoir with which A can interact both by heat transfer and pressure work. Alternatively, A' may be a combination of two reservoirs, one at temperature T_0 with which A can interact only by heat transfer, and the other at pressure p_0 with which A can interact only by pressure work.

The analysis of the equilibrium conditions for a system A under these conditions is very similar to that of the last section. Once again the entropy $S^{(0)}$ of the combined system $A^{(0)} = A + A'$ satisfies the condition that in any spontaneous process

$$\Delta S^{(0)} = \Delta S + \Delta S' \geq 0 \tag{8·3·1}$$

If A absorbs heat Q from A' in this process, then $\Delta S' = -Q/T_0$. But now the first law applied to A gives

$$Q = \Delta \bar{E} + p_0 \, \Delta V + W^*$$

where $p_0 \, \Delta V$ is the work done by A against the constant pressure p_0 of the reservoir A' and where W^* denotes any *other* work done by A in the process. (For example, W^* might refer to electric or magnetic work done by A.) Hence one can write

$$\begin{aligned}\Delta S^{(0)} &= \Delta S - \frac{Q}{T_0} = \frac{1}{T_0}[T_0 \, \Delta S - Q] = \frac{1}{T_0}[T_0 \, \Delta S - (\Delta \bar{E} + p_0 \, \Delta V + W^*)] \\ &= \frac{1}{T_0}[\Delta(T_0 S - \bar{E} - p_0 V) - W^*]\end{aligned}$$

or

$$\Delta S^{(0)} = \frac{-\Delta G_0 - W^*}{T_0} \tag{8·3·2}$$

Here we have used the fact that T_0 and p_0 are both constant and have introduced the definition

$$G_0 \equiv \bar{E} - T_0 S + p_0 V \tag{8·3·3}$$

This reduces to the ordinary Gibbs free energy $G = \bar{E} - TS + pV$ for the system A when the temperature and pressure of the latter are equal to those of the reservoir A'.

The total entropy change $\Delta S^{(0)}$ in (8·3·2) is again expressed completely in terms of quantities which refer only to the system A. The fundamental condition (8·3·1) implies then that

$$-\Delta G_0 \geq W^* \tag{8·3·4}$$

This asserts that the *maximum* work (other than work done on the pressure reservoir) which can be done by the system is given by $(-\Delta G_0)$. (This is the reason that G is also called a "free energy.") The maximum work corresponds again to the equals sign in (8·3·1) and corresponds to a quasi-static process.

If all the external parameters of A, *except* its volume, are kept fixed, then $W^* = 0$ and (8·3·4) yields the condition

$$\Delta G_0 \leq 0 \tag{8·3·5}$$

Hence one concludes that

> If a system is in contact with a reservoir at constant temperature and pressure and if its external parameters are fixed so that it can only do work on the pressure reservoir, then the stable equilibrium situation is characterized by the condition that

▶ $$G_0 = \text{minimum} \tag{8·3·6}$$

This last condition can again be phrased in more explicit statistical terms. The probability that a parameter y of the system $A^{(0)}$ assumes a value between y and $y + \delta y$ is given by

$$P(y) \propto e^{S^{(0)}(y)/k} \tag{8·3·7}$$

But with $W^* = 0$, the change in $S^{(0)}$ due to a parameter change away from some standard value y_1 is by (8·3·2)

$$\Delta S^{(0)} = -\frac{\Delta G_0}{T_0} \tag{8·3·8}$$

so that

$$S^{(0)}(y) = S^{(0)}(y_1) - \frac{G_0(y) - G_0(y_1)}{T_0}$$

Since y_1 is just some constant, (8·3·7) leads to the proportionality

▶ $$P(y) \propto e^{-G_0(y)/kT_0} \tag{8·3·9}$$

This again shows explicitly that the most probable situation is one where G_0 is a minimum and allows calculation of the probability of fluctuations about this equilibrium.

8·4 *Stability conditions for a homogeneous substance*

As a simple example of the preceding discussion, consider a one-component system in a single phase (e.g., a simple liquid or solid). Focus attention on

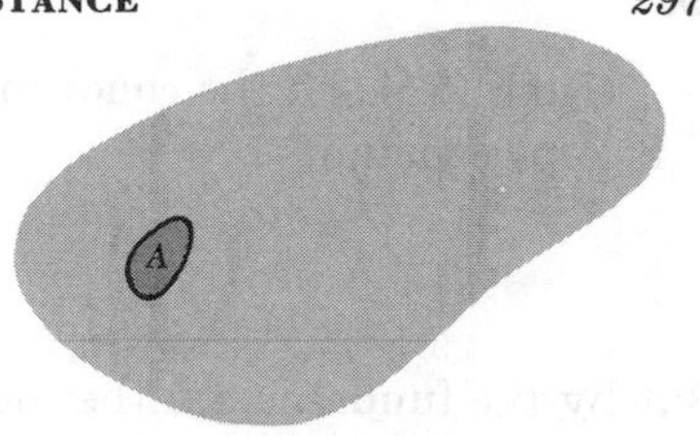

Fig. 8·4·1 A small portion A of a homogeneous substance is singled out for consideration to examine the conditions for stable equilibrium.

some small, but macroscopic, part A of this system where A consists of some fixed number of particles. The rest of the system is then relatively very large and acts like a reservoir at some *constant* temperature T_0 and *constant* pressure p_0. By (8·3·5) the condition for stable equilibrium applied to A is that for this system the function

$$G_0 \equiv \bar{E} - T_0 S + p_0 V = \text{minimum} \tag{8·4·1}$$

Stability against temperature variations Let T and V be the two independent parameters specifying the macrostate of A. Consider first a situation where V is considered fixed but where the temperature parameter T is allowed to vary. Suppose that the minimum of G_0 occurs for $T = \tilde{T}$ when $G_0 = G_{\text{min}}$. Expanding G about its minimum and writing $\Delta T \equiv T - \tilde{T}$, one obtains

$$\Delta_m G_0 = G_0 - G_{\text{min}} = \left(\frac{\partial G_0}{\partial T}\right)_V \Delta T + \frac{1}{2}\left(\frac{\partial^2 G_0}{\partial T^2}\right)_V (\Delta T)^2 + \cdots \tag{8·4·2}$$

Here all derivatives are evaluated at $T = \tilde{T}$. Since G is a minimum, its stationary character at this point implies that $\Delta_m G = 0$ in a first approximation; i.e., the first-order terms proportional to ΔT must vanish so that

$$\left(\frac{\partial G_0}{\partial T}\right)_V = 0 \qquad \text{for } T = \tilde{T} \tag{8·4·3}$$

The fact that G_0 is not only stationary but a *minimum* at $T = \tilde{T}$ requires that in next approximation, when terms in $(\Delta T)^2$ become important, one has

$$\Delta_m G_0 \geq 0$$

or

$$\left(\frac{\partial^2 G_0}{\partial T^2}\right)_V \geq 0 \qquad \text{for } T = \tilde{T} \tag{8·4·4}$$

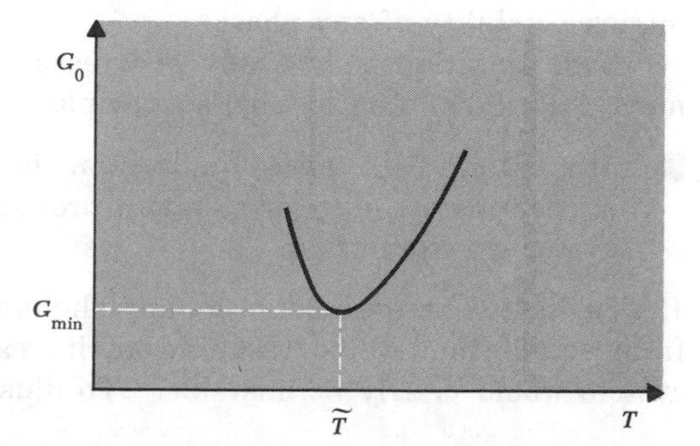

Fig. 8·4·2 Schematic dependence of $G_0(T,V)$ on the temperature T at a fixed volume V.

Using (8 · 4 · 1), the condition (8 · 4 · 3) that G_0 be stationary becomes, when V is kept constant,

$$\left(\frac{\partial G_0}{\partial T}\right)_V = \left(\frac{\partial \bar{E}}{\partial T}\right)_V - T_0 \left(\frac{\partial S}{\partial T}\right)_V = 0 \tag{8·4·5}$$

But by the fundamental thermodynamic relation

$$T\, dS = d\bar{E} + \bar{p}\, dV \tag{8·4·6}$$

it follows that for V constant, $dV = 0$, and

$$T\left(\frac{\partial S}{\partial T}\right)_V = \left(\frac{\partial \bar{E}}{\partial T}\right)_V$$

Thus (8 · 4 · 5) becomes

$$\left(\frac{\partial G}{\partial T}\right)_V = \left(1 - \frac{T_0}{T}\right)\left(\frac{\partial \bar{E}}{\partial T}\right)_V \tag{8·4·7}$$

Setting this equal to zero where $T = \tilde{T}$, one gets simply

$$\tilde{T} = T_0 \tag{8·4·8}$$

Hence we arrive at the obvious conclusion that a necessary condition for equilibrium is that the temperature of the subsystem A is the same as that of the surrounding medium.

Now we proceed to the second-order terms to satisfy the condition (8 · 4 · 4), which guarantees that G_0 is actually a minimum. By (8 · 4 · 7),

$$\left(\frac{\partial^2 G_0}{\partial T^2}\right)_V = \frac{T_0}{T^2}\left(\frac{\partial \bar{E}}{\partial T}\right)_V + \left(1 - \frac{T_0}{T}\right)\left(\frac{\partial^2 \bar{E}}{\partial T^2}\right)_V \geq 0$$

When this is evaluated at the minimum of G, where $T = T_0$ by virtue of (8 · 4 · 8), the second term vanishes and one obtains simply

$$\left(\frac{\partial \bar{E}}{\partial T}\right)_V \geq 0 \tag{8·4·9}$$

But this derivative is just the heat capacity C_V at constant volume. Thus

▶ $$C_V \geq 0 \tag{8·4·10}$$

The condition (8 · 4 · 9) or (8 · 4 · 10) was already derived previously in (3 · 7 · 16) and in (6 · 5 · 8). It is a fundamental condition required to guarantee the intrinsic stability of any phase.

This condition is physically very reasonable. Indeed, the following statement, known as "Le Châtelier's principle," must be true quite generally:

▶ If a system is in *stable* equilibrium, then any spontaneous change of its parameters must bring about processes which tend to restore the system to equilibrium.

If this statement were not true, any slight fluctuation leading to a deviation from equilibrium would result in an increase of this deviation so that the system would clearly be unstable. To illustrate the principle in the present example, suppose that the temperature T of the subsystem A has increased above that of the surroundings A' as a result of a spontaneous fluctuation. Then the process brought into play is heat transfer from the system A at the higher temperature to the surroundings A', and a resulting *decrease* in the energy $\bar{E}$ of A (i.e., $\Delta\bar{E} < 0$). But the stability condition expressed by Le Châtelier's principle requires that this process, induced by the original temperature increase, is such that the temperature is again *decreased* (i.e., $\Delta T < 0$). Hence it follows that $\Delta\bar{E}$ and ΔT must have the same sign, i.e., that $\partial\bar{E}/\partial T > 0$ in agreement with (8·4·9).

Stability against volume fluctuations Suppose that the temperature of the subsystem A is considered fixed at $T = T_0$, but that its volume V is allowed to vary. Then one can write

$$\Delta_m G_0 \equiv G_0 - G_{\min} = \left(\frac{\partial G_0}{\partial V}\right)_T \Delta V + \frac{1}{2}\left(\frac{\partial^2 G_0}{\partial V^2}\right)_T (\Delta V)^2 + \cdots \tag{8·4·11}$$

where $\Delta V \equiv V - \tilde{V}$, and the expansion is about the volume $V = \tilde{V}$ where G_0 is minimum. The condition that G_0 is stationary demands that

$$\left(\frac{\partial G_0}{\partial V}\right)_T = 0 \tag{8·4·12}$$

Using the definition (8·4·1)

$$\left(\frac{\partial G_0}{\partial V}\right)_T = \left(\frac{\partial \bar{E}}{\partial V}\right)_T - T_0\left(\frac{\partial S}{\partial V}\right)_T + p_0$$

But, by virtue of (8·4·6),

$$T\left(\frac{\partial S}{\partial V}\right)_T = \left(\frac{\partial \bar{E}}{\partial V}\right)_T + \bar{p}$$

Hence

$$\left(\frac{\partial G_0}{\partial V}\right)_T = T\left(\frac{\partial S}{\partial V}\right)_T - \bar{p} - T_0\left(\frac{\partial S}{\partial V}\right)_T + p_0$$

or

$$\left(\frac{\partial G_0}{\partial V}\right)_T = -\bar{p} + p_0 \tag{8·4·13}$$

since $T = T_0$. The condition (8·4·12) then implies that at equilibrium, where G is minimum and $V = \tilde{V}$, the corresponding value of the pressure $\bar{p}$ is such that

$$\bar{p} = p_0 \tag{8·4·14}$$

Again this is a rather obvious result which asserts merely that in equilibrium the pressure of the subsystem A must be equal to that of the surrounding medium.

The condition that G is actually a minimum is that $\Delta_m G_0 \geq 0$; or by (8·4·11), that the second derivative of G_0 is positive. By (8·4·13) this yields the condition

$$\left(\frac{\partial^2 G_0}{\partial V^2}\right)_T = -\left(\frac{\partial \bar{p}}{\partial V}\right)_T \geq 0 \tag{8·4·15}$$

In terms of the isothermal compressibility defined by

$$\kappa = -\frac{1}{V}\left(\frac{\partial V}{\partial \bar{p}}\right)_T \tag{8·4·16}$$

the condition (8·4·15) is equivalent to

▶ $$\kappa \geq 0 \tag{8·4·17}$$

The stability condition (8·4·15) is again a physically quite reasonable result consistent with Le Châtelier's principle. Suppose that the volume of the subsystem A has *increased* by an amount ΔV as a result of a fluctuation. The pressure $\bar{p}$ of A must then *decrease* below that of its surroundings (i.e., $\Delta\bar{p} < 0$) to guarantee that the net force exerted on A by its surroundings is of a direction such that it tends to reduce its volume to its former value.

Density fluctuations The preceding considerations permit one also to calculate the fluctuation in the volume V of the small subsystem A. The most probable situation is that where $V = \tilde{V}$ is such that G_0 is a minimum, $G_0(\tilde{V}) = G_{\min}$. Let $\mathcal{P}(V)\,dV$ denote the probability that the volume of A lies between V and $V + dV$. Then one has by (8·3·9)

$$\mathcal{P}(V)\,dV \propto e^{-G_0(V)/kT}\,dV \tag{8·4·18}$$

But when $\Delta V \equiv V - \tilde{V}$ is small, the expansion (8·4·11) is applicable. By virtue of (8·4·12) and (8·4·15) it yields

$$G_0(V) = G_{\min} - \frac{1}{2}\left(\frac{\partial \bar{p}}{\partial V}\right)_T (\Delta V)^2 = G_{\min} + \frac{(\Delta V)^2}{2\tilde{V}\kappa}$$

where we have used the definition (8·4·16) in the last step. Thus (8·4·18) becomes

$$\mathcal{P}(V)\,dV = B\exp\left[-\frac{(V-\tilde{V})^2}{2kT_0\tilde{V}\kappa}\right]dV \tag{8·4·19}$$

where we have absorbed $G_{\min}$ into the proportionality constant B. This constant can, of course, be determined by the normalization requirement that the integral of (8·4·19) over all possible values of the volume V is equal to unity.*

The probability (8·4·19) is simply a Gaussian distribution with a maximum at the volume $V = \tilde{V}$. Thus $\tilde{V}$ is also equal to the mean volume $\bar{V}$ and the general result (1·6·9) implies that (8·4·19) yields a dispersion of the volume equal to

$$\overline{(\Delta V)^2} \equiv \overline{(V-\tilde{V})^2} = kT_0\tilde{V}\kappa \tag{8·4·20}$$

The presence of such volume fluctuations in a small amount of material containing a fixed number N of molecules implies, of course, corresponding fluctuations in the number $n = N/V$ of molecules per unit volume (and thus

* This integral can be extended from $V = -\infty$ to $V = +\infty$, since the integrand (8·4·19) becomes negligible when V differs appreciably from the value $\tilde{V}$ where $\mathcal{P}$ is maximum.

also in the mass density of the substance). The fluctuations in n are centered about the value $\tilde{n} = N/\tilde{V}$, and for relatively small values of $\Delta n \equiv n - \tilde{n}$ one has $\Delta n = -(N/\tilde{V}^2)\,\Delta V = -(\tilde{n}/\tilde{V})\,\Delta V$. Hence (8·4·20) implies for the dispersion in the number density n the result

$$\overline{(\Delta n)^2} = \left(\frac{\tilde{n}}{\tilde{V}}\right)^2 \overline{(\Delta V)^2} = \tilde{n}^2 \left(\frac{kT_0}{\tilde{V}}\,\kappa\right) \tag{8·4·21}$$

Note that this depends on the size of the volume $\tilde{V}$ under consideration.

An interesting case arises when

$$\left(\frac{\partial \bar{p}}{\partial V}\right)_T \to 0 \tag{8·4·22}$$

Then $\kappa \to \infty$ and the density fluctuations become very large.* The conditions of temperature and pressure which are such that $(\partial \bar{p}/\partial V)_T = 0$ define the so-called "critical point" of the substance. The very large density fluctuations at this point lead to a very large scattering of light. As a result a substance, which is ordinarily transparent, will assume a milky white appearance at its critical point (e.g., liquid CO_2 when it approaches its critical point at a temperature of 304°K and pressure of 73 atmospheres). This impressive phenomenon is known as "critical point opalescence."

Remark The result (8·4·19), proved under conditions of constant temperature, can be shown to remain valid even if both V and T are allowed to vary simultaneously (see Problem 8.1). Hence this discussion of density fluctuations is applicable to the actual case of experimental interest.

EQUILIBRIUM BETWEEN PHASES

8·5 *Equilibrium conditions and the Clausius-Clapeyron equation*

Consider a single component system which consists of two phases which we shall denote by 1 and 2. For example, these might be solid and liquid, or liquid and gas. We suppose that the system is in equilibrium with a reservoir at the constant temperature T and constant pressure p so that the system itself has always a temperature T and a mean pressure p. But the system can exist in either of its two possible phases or in a mixture of the two. Let us begin by finding the conditions which must be satisfied so that the two phases can coexist in equilibrium with each other.

In accordance with the discussion of Sec. 8·3, the equilibrium condition is that the Gibbs free energy G of the system is a minimum,

$$G = E - TS + pV = \text{minimum} \tag{8·5·1}$$

* They do not become infinite, since the approximations which allowed us to neglect in (8·4·11) terms beyond $(\Delta V)^2$ are then no longer justified.

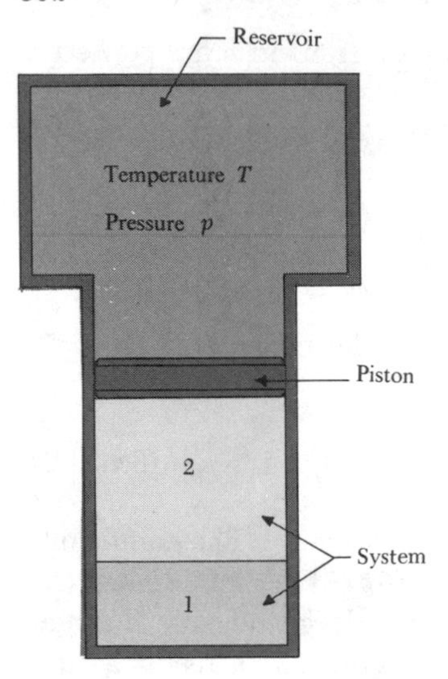

***Fig. 8·5·1** A system consisting of two phases maintained at constant temperature and pressure.*

Let

$$\nu_i = \text{the number of moles of phase } i \text{ present in the system}$$

$$g_i(T,p) = \text{the Gibbs free energy per mole of phase } i \text{ at this temperature } T \text{ and pressure } p$$

Then G can be written

$$G = \nu_1 g_1 + \nu_2 g_2 \qquad (8\cdot5\cdot2)$$

The conservation of matter implies that the total number ν of moles of the substance remain constant, i.e.,

$$\nu_1 + \nu_2 = \nu = \text{constant} \qquad (8\cdot5\cdot3)$$

Thus one can take ν_1 to be the one independent parameter which is free to vary. In equilibrium (8·5·1) requires that G be stationary for changes in ν_1; thus

$$dG = g_1\,d\nu_1 + g_2\,d\nu_2 = 0$$

or

$$(g_1 - g_2)\,d\nu_1 = 0$$

since $d\nu_2 = -d\nu_1$ by (8·5·3). Hence we obtain as a necessary condition for equilibrium that

▶
$$g_1 = g_2 \qquad (8\cdot5\cdot4)$$

Clearly, when this condition is satisfied, the transfer of a mole of substance from one phase to the other does not change G in (8·5·2); hence G is then stationary as required.

Remark One could go further and investigate the conditions that G is actually a *minimum*, but this would not yield anything very interesting other than the conditions that the heat capacity and compressibility of each phase must be positive to guarantee the stability of each phase (as shown in Sec. 8·4).

Let us look at the situation more closely. For given T and p, $g_1(T,p)$ is a well-defined function characteristic of the properties of phase 1; similarly, $g_2(T,p)$ is a well-defined function characteristic of phase 2.

If T and p are such that $g_1 < g_2$, then the minimum value of G in (8·5·2) is achieved if all the ν moles of substance transform into phase 1 so that $G = \nu g_1$. Phase 1 is then the stable one.

If T and p are such that $g_1 > g_2$, then the minimum value of G is achieved if all the substance transforms into phase 2 so that $G = \nu g_2$. Phase 2 is then the stable one.

If T and p are such that $g_1 = g_2$, then the condition (8·5·4) is satisfied and *any* amount ν_1 of phase 1 can coexist in equilibrium with the remaining amount $\nu_2 = \nu - \nu_1$ of phase 2. The value G remains unchanged when ν_1 is varied. The locus of points where T and p are such that the condition (8·5·4) is fulfilled then represents the "phase-equilibrium line" along which the two phases can coexist in equilibrium. This line, along which $g_1 = g_2$, divides the pT plane into two regions: one where $g_1 < g_2$, so that phase 1 is the stable one, and the other where $g_1 > g_2$, so that phase 2 is the stable one.

It is possible to characterize the phase-equilibrium line by a differential equation. In Fig. 8·5·2 consider any point, such as A, which lies *on* the phase-equilibrium line and corresponds to temperature T and pressure p. Then the condition (8·5·4) implies that

$$g_1(T,p) = g_2(T,p) \tag{8·5·5}$$

Consider now a neighboring point, such as B, which also lies on the phase-equilibrium line and corresponds to temperature $T + dT$ and pressure $p + dp$. Then the condition (8·5·4) implies that

$$g_1(T + dT,\, p + dp) = g_2(T + dT,\, p + dp) \tag{8·5·6}$$

Subtracting (8·5·5) from (8·5·6) yields the condition

$$dg_1 = dg_2 \tag{8·5·7}$$

where
$$dg_i = \left(\frac{\partial g_i}{\partial T}\right)_p dT + \left(\frac{\partial g_i}{\partial p}\right)_T dp$$

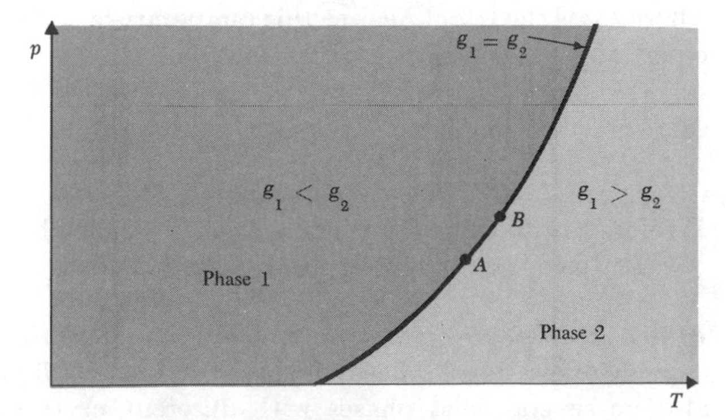

Fig. 8·5·2 Pressure–temperature plot showing the domains of relative stability of two phases and the phase-equilibrium line.

is the change in the molar Gibbs free energy for phase i in going from point A to point B.

But the change dg for each phase can also be obtained by using the fundamental thermodynamic relation

$$d\epsilon = T\,ds - p\,dv$$

expressing the change in mean molar energy ϵ of this phase. Thus

$$dg \equiv d(\epsilon - Ts + pv) = -s\,dT + v\,dp \tag{8·5·8}$$

Hence (8 · 5 · 7) implies that

$$-s_1\,dT + v_1\,dp = -s_2\,dT + v_2\,dp$$

$$(s_2 - s_1)\,dT = (v_2 - v_1)\,dp$$

or

$$\frac{dp}{dT} = \frac{\Delta s}{\Delta v} \tag{8·5·9}$$

where $\Delta s \equiv s_2 - s_1$ and $\Delta v \equiv v_2 - v_1$. This is called the "Clausius-Clapeyron equation." Consider any point on the phase-equilibrium line at a temperature T and corresponding pressure p. Equation (8 · 5 · 9) then relates the slope of the phase-equilibrium line at this point to the entropy change Δs and volume change Δv of the substance in "crossing the line" at this point, i.e., in undergoing a change of phase at this temperature and pressure. (Note that the quantities on the right side of (8 · 5 · 9) do not need to be referred to one mole of the substance; both numerator and denominator can be multiplied by the same number of moles, and dp/dT must obviously be left unchanged.)

Since there is an entropy change associated with the phase transformation, heat must also be absorbed. The "latent heat of transformation" L_{12} is defined as the heat absorbed when a given amount of phase 1 is transformed to phase 2. Since the process takes place at the constant temperature T, the corresponding entropy change is simply

$$\Delta S = S_2 - S_1 = \frac{L_{12}}{T} \tag{8·5·10}$$

where L_{12} is the latent heat at this temperature. Thus the Clausius-Clapeyron equation (8 · 5 · 9) can be written

▶ $$\frac{dp}{dT} = \frac{\Delta S}{\Delta V} = \frac{L_{12}}{T\,\Delta V} \tag{8·5·11}$$

Clearly, if V refers to the molar volume, then L_{12} is the latent heat per mole; if V refers to the volume per gram, then L_{12} is the latent heat per gram.

Let us discuss a few important illustrations.

Phase transformations of a simple substance Simple substances are capable of existing in phases of three types: solid, liquid, and gas. (There may also be several solid phases with different crystal structures.) The phase-equilibrium lines separating these phases appear typically as shown in Fig. 8 · 5 · 3. These lines separate solid from liquid, liquid from gas, and solid from

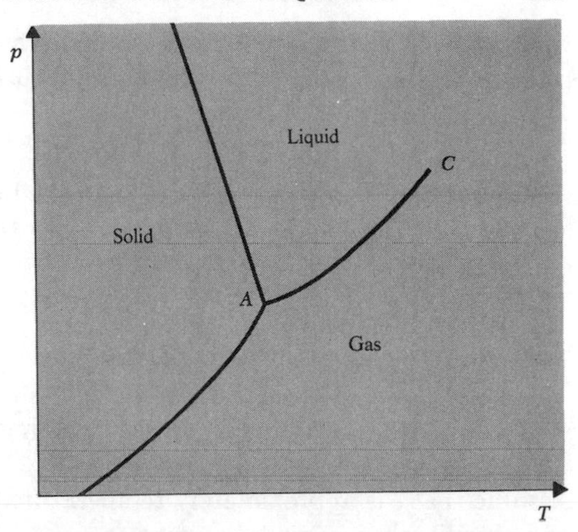

Fig. 8·5·3 Phase diagram for a simple substance. Point A is the triple point, point C the critical point.

gas.* The three lines meet at one common point A, called the "triple point"; at this unique temperature and pressure arbitrary amounts of *all three* phases can therefore coexist in equilibrium with each other. (This is the property which makes the triple point of water so suitable as a readily reproducible temperature standard.) At point C, the so-called "critical point," the liquid–gas equilibrium line ends. The volume change ΔV between liquid and gas has then approached zero; beyond C there is no further phase transformation, since there exists only one "fluid" phase (the very dense gas has become indistinguishable from the liquid).

In going from solid (s) to liquid (l) the entropy of the substance (or degree of disorder) almost always increases.† Thus the corresponding latent heat L_{sl} is positive and heat gets absorbed in the transformation. In most cases the solid expands upon melting, so that $\Delta V > 0$. In that case the Clausius-Clapeyron equation (8·5·11) asserts that the slope of the solid-liquid equilibrium line (i.e., of the melting curve) is positive. There are some substances, like water, which contract upon melting so that $\Delta V < 0$. For these the slope of the melting curve must thus be negative (as drawn in Fig. 8·5·3).

Approximate calculation of the vapor pressure The Clausius-Clapeyron equation can be used to derive an approximate expression for the pressure of a vapor in equilibrium with the liquid (or solid) at a temperature T. This pressure is called the "vapor pressure" of the liquid (or solid) at this temperature. By (8·5·11)

$$\frac{dp}{dT} = \frac{l}{T\,\Delta v} \tag{8·5·12}$$

* The gas phase is sometimes also called the "vapor phase." The transformation from solid to liquid is called "melting," that from liquid to gas is called "vaporization," and that from solid to gas is called "sublimation."

† An exceptional case occurs for solid He^3 in a certain temperature range where the nuclear spins in the solid are randomly oriented while those in the liquid are aligned antiparallel to each other so as to satisfy the quantum-mechanical Fermi-Dirac statistics.

where $l \equiv l_{12}$ is the latent heat per mole and v is the molar volume. Let 1 refer to the liquid (or solid) phase and 2 to the vapor. Then

$$\Delta v = v_2 - v_1 \approx v_2$$

since the vapor is much less dense than the liquid, so that $v_2 \gg v_1$. Let us also assume that the vapor can be adequately treated as an ideal gas, so that its equation of state is simply

$$pv_2 = RT$$

Then $\Delta v = RT/p$ and (8·5·12) becomes

$$\frac{1}{p}\frac{dp}{dT} = \frac{l}{RT^2} \tag{8·5·13}$$

Assume that l is approximately temperature independent. Then (8·5·13) can be immediately integrated to give

$$\ln p = -\frac{l}{RT} + \text{constant}$$

or

$$p = p_0 e^{-l/RT} \tag{8·5·14}$$

where p_0 is some constant. This shows that the vapor pressure p is a very rapidly increasing function of T, the temperature dependence being determined by the magnitude of the latent heat of vaporization.

8·6 *Phase transformations and the equation of state*

Consider a single-component system. Suppose that the equation of state for a mole of this substance

$$p = p(v,T) \tag{8·6·1}$$

is assumed known (by theoretical considerations or from empirical information) for the range of variables where the substance is a gas or a liquid. For example, the equation of state might be the van der Waals equation mentioned in (5·8·13), i.e.,

$$\left(p + \frac{a}{v^2}\right)(v - b) = RT$$

The equation of state (8·6·1) can be represented in a two-dimensional diagram by drawing a set of curves of mean pressure p versus the molar volume v for various values of the temperature T. Such a diagram is illustrated schematically in Fig. 8·6·1.*

Consider the system described by (8·6·1) to be in contact with a reservoir at given temperature T and pressure p. The intensive parameters T and p can then be regarded as the independent variables of the problem. Focus attention

* Upon multiplication by v^2, the van der Waals equation is seen to be cubic in v. Thus it also gives rise to S-shaped curves of the type illustrated in Fig. 8·6·1.

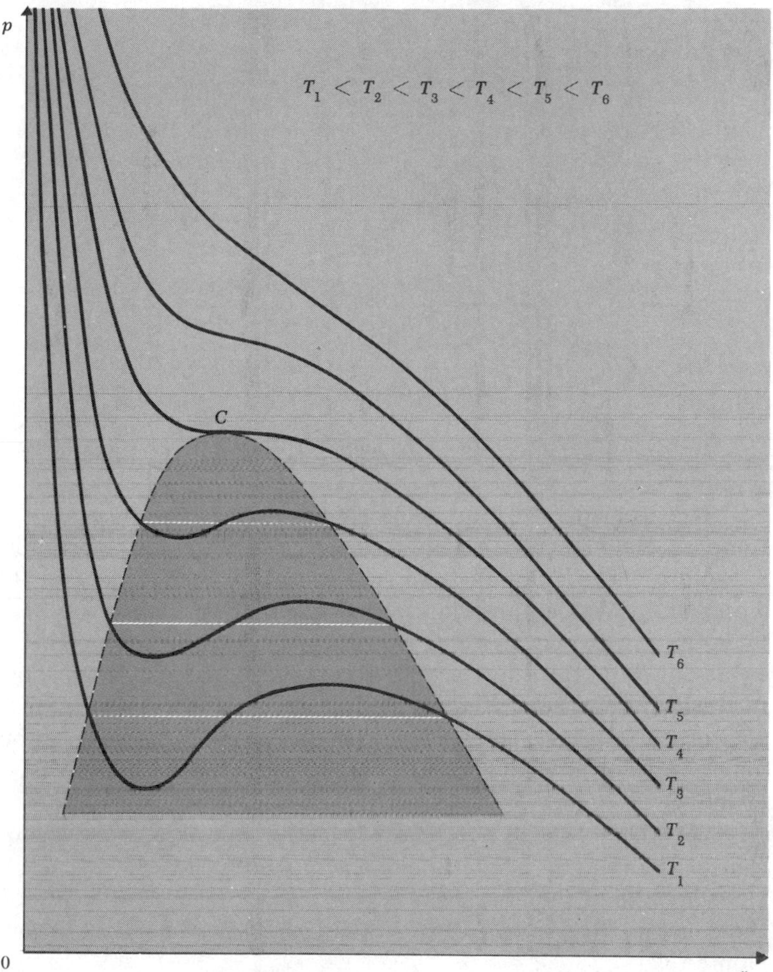

Fig. 8·6·1 ***Schematic diagram showing curves of constant temperature for an equation of state (8·6·1) describing the fluid states of a substance. The point C is the critical point. In the shaded region mixtures of two phases can coexist along the horizontal lines.***

on a particular curve of constant temperature T (or "isotherm") of the equation of state (8·6·1). A curve of this type is illustrated in Fig. 8·6·2 and contains a wealth of information. If at the given temperature T the pressure is sufficiently low so that $p < p_1$, the curve yields, correspondingly, a unique value of v. There exists then a well-defined single phase. Here the slope of the curve $\partial p/\partial v \leq 0$ as is necessary by the stability condition (8·4·15). Also, $|\partial p/\partial v|$ is relatively small, so that the compressibility of this phase is relatively large, as would be the case for a gaseous phase.

If at the given temperature T the pressure is sufficiently high that $p > p_2$, then there exists again a single phase with a unique value of v. The stability condition $\partial p/\partial v \leq 0$ is again satisfied, but $|\partial p/\partial v|$ is relatively large. Hence

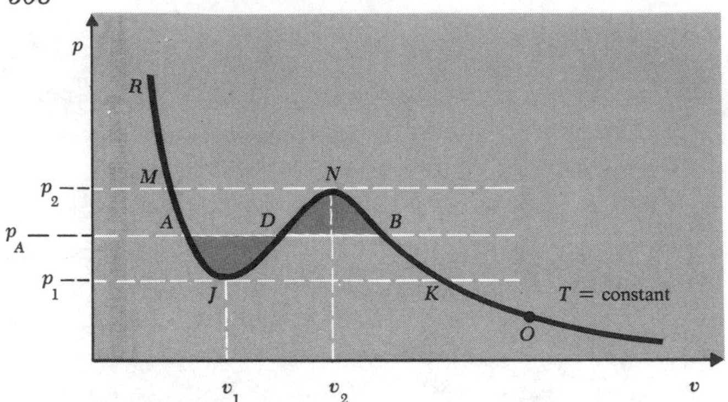

Fig. 8·6·2 Plot of the equation of state (8·6·1) for a particular temperature T.

the compressibility of this phase is relatively small as would be the case for a liquid phase.

Now consider the intermediate pressure range $p_1 < p < p_2$. At the given temperature T there are now, for each pressure p, three possible values of the volume v. The question is which value of v corresponds to the most stable situation. We see immediately that the stability condition $\partial p/\partial v \leq 0$ is violated in the region $v_1 < v < v_2$ where the curve has positive slope. Thus values of v in this range are certainly excluded since they would lead to an intrinsically unstable situation. But this still leaves two possible values of v between which one has to decide on the basis of relative stability. On the diagram, where we have labeled points on the curve by letters, one has to ask which is the more stable situation—where $v = v_A$ or where $v = v_B$. This question then reduces, by virtue of the discussion of Sec. 8·3, to an investigation of the relative magnitude of the molar-free energies $g_A(T,p)$ and $g_B(T,p)$.

Changes of the function $g \equiv \epsilon - Ts + pv$ can readily be computed along the constant temperature curve of Fig. 8·6·2. From the general thermodynamic relation

$$T\,ds = d\epsilon + p\,dv$$

it follows immediately that for a pressure change where T is kept constant,

$$dg = d(\epsilon - Ts + pv) = v\,dp \tag{8·6·2}$$

Differences between g on any point of the curve of Fig. 8·6·2 and some standard point O are then given by

$$g - g_O = \int_{p_O}^{p} v\,dp \tag{8·6·3}$$

The right side represents geometrically the area between the curve and the p axis in the range between p_O and p. Note that, starting from the point O on the curve and performing the integral (8·6·3) along it, the value of this integral first increases until one reaches the point N, then decreases until one reaches the point J, and then increases again as one continues toward the point

M. Hence a curve of $g(T,p)$ versus p along the constant-temperature curve has the appearance shown in Fig. 8·6·3. (The points on this curve are labeled to correspond to those of Fig. 8·6·2.)

From this diagram one can readily see what happens for various values of the pressure. At O only the high-compressibility phase (in our example, the gas) exists. When the pressure is increased to the range $p_1 < p < p_2$, there are three possible values of g. The values of g along the curve $OKXN$ correspond to large values of $v > v_2$ in the region of high compressibility; this corresponds to the gas phase. The values of g along the curve $JXMR$ correspond to small values of $v < v_1$ in the region of low compressibility; this corresponds to the liquid phase. The values of g along the curve JDN correspond to the intrinsically unstable range $v_1 < v < v_2$. If p is only slightly larger than p_1, then Fig. 8·6·3 shows that the gas phase, with volume near v_K, has the lower value of g and is thus the more stable one. This situation prevails until p is increased to the extent that it reaches the value $p = p_X$ corresponding to the point X where the curves KXN and JXM in Fig. 8·6·3 intersect. At this point the free energies g of both gas and liquid become equal. This then is the pressure at which arbitrary proportions of both these phases can coexist in equilibrium with each other. If the pressure is increased beyond p_X, the curve JXM corresponding to the liquid phase yields the lower free energy g so that this phase is the more stable one. At the point X the system shifts, therefore, from the curve $OKXN$ (corresponding to the gas phase) over to the curve $JXMR$ (corresponding to the liquid phase). Thus p_X corresponds to the pressure where the phase transformation from gas to liquid occurs.

Let us look at the phase transformation in greater detail. Assume that in Fig. 8·6·2 the pressure of the phase transformation is $p_A = p_X$. Then A

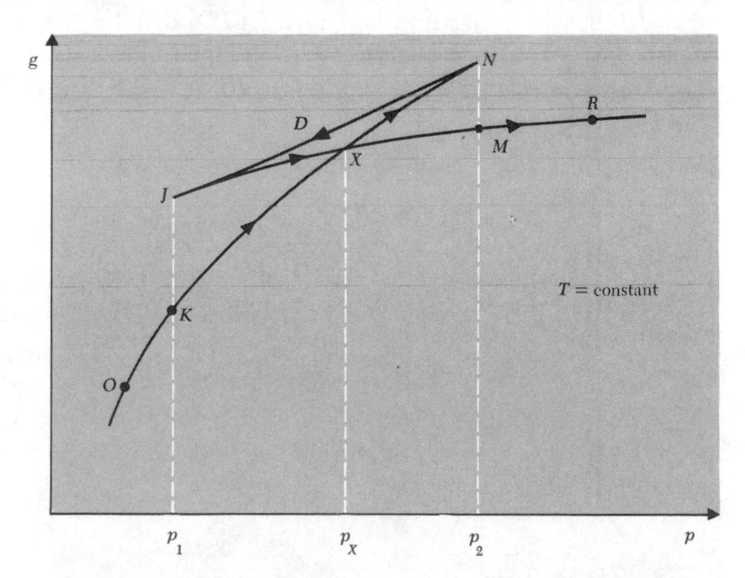

***Fig. 8·6·3** Schematic illustration showing, as a function of pressure p, the behavior of $g(T,p)$ implied by the curve of Fig. 8·6·2.*

and B both correspond to the point X in Fig. 8·6·3. Furthermore v_B is the molar volume of the gas and v_A that of the liquid at the pressure and temperature of the phase transformation. If under these circumstances a fraction ξ of the mole of substance is in the gaseous phase, then the total molar volume v_{tot} is given by

$$v_{tot} = \xi v_B + (1 - \xi) v_A \tag{8·6·4}$$

In the course of the phase transformation the total molar volume changes continuously from the value v_B for the gas to the value v_A for the liquid as the fraction ξ changes continuously from 1 to 0. In this process there will, of course, be a change of entropy and an associated latent heat. In Fig. 8·6·2 the horizontal line BDA along which the phase transformation occurs is characterized by the fact that

$$g_B = g_A \tag{8·6·5}$$

or, by virtue of (8·6·3), by the fact that the integral

$$\int_{BNDJA} v \, dp = 0 \tag{8·6·6}$$

when evaluated along the curve $BNDJA$ in Fig. 8·6·2. This integral can be broken up into several parts giving contributions of different signs:

$$\int_B^N v \, dp + \int_N^D v \, dp + \int_D^J v \, dp + \int_J^A v \, dp = 0$$

$$\left(\int_B^N v \, dp - \int_D^N v \, dp \right) + \left(- \int_J^D v \, dp + \int_J^A v \, dp \right) = 0$$

or

$$\text{area } (DNB) - \text{area } (AJD) = 0$$

where area (DNB) denotes the area enclosed by the straight line DB and the curve DNB, and similarly, where area (AJD) denotes the area enclosed by the straight line AD and the curve AJD. Thus we have the result that the location of the phase-transformation line ADB in Fig. 8·6·2 is determined by the condition that

▶ $$\text{area } (AJD) = \text{area } (DNB) \tag{8·6·7}$$

Remark The molar entropy change Δs and associated latent heat l of the transformation can also be determined from the equation of state. Since T is constant in the transformation so that $dT = 0$, one has simply

$$ds = \left(\frac{\partial s}{\partial v} \right)_T dv = \left(\frac{\partial p}{\partial T} \right)_v dv \tag{8·6·8}$$

where we have used a Maxwell relation in the last step. In going from A to B in Fig. 8·6·2, the corresponding entropy change can then be computed by evaluating the integral

$$\Delta s = s_B - s_A = \int_{AJDNB} \left(\frac{\partial p}{\partial T} \right)_v dv \tag{8·6·9}$$

Consider the given curve for temperature T and a neighboring one for tempera-

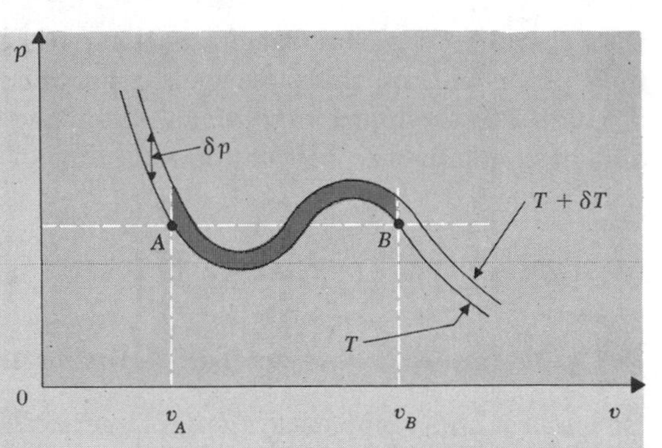

Fig. 8·6·4 Dependence of the pressure p on the molar volume v for two slightly different temperatures.

ture $T + \delta T$. At a given volume v, let δp denote the pressure difference between these two curves. Then (8·6·9) can be written

$$\Delta s = \frac{1}{\delta T} \int_{AJDNB} \delta p \, dv$$

or $\Delta s = \dfrac{1}{\delta T}$ [area between isotherms in the interval $v_A < v < v_B$] (8·6·10)

This area is shown shaded in Fig. 8·6·4.

This area is positive so that $\Delta s > 0$ or $s_B > s_A$, i.e., the entropy of the gas is greater than that of the liquid. The latent heat per mole absorbed in the transformation from liquid to gas is simply given by

$$l = T \Delta s \tag{8·6·11}$$

The change in the molar internal energy of the two phases is given by the fundamental relation $T\,ds = d\epsilon + p\,dv$ as

$$\Delta \epsilon = l - p \Delta v \tag{8·6·12}$$

since T and p are constant. Here $\Delta v = v_B - v_A$ and $\Delta \epsilon = \epsilon_B - \epsilon_A$.

As one goes to higher temperatures the two extremum points where $(\partial p/\partial v)_T = 0$ (indicated in Fig. 8·6·2 by $v = v_1$ and $v = v_2$) move closer together. This also implies that the volume change Δv in the phase transformation decreases. As the temperature is increased further one reaches the situation where these two extremum points v_1 and v_2 just coincide, so that there is no longer a change of sign of $(\partial p/\partial v)$, i.e., the derivative of $(\partial p/\partial v)$ vanishes also. At this point one has then $(\partial^2 p/\partial v^2)_T = 0$ as well as $(\partial p/\partial v)_T = 0$; the point is thus a point of inflection on the pv curve. This unique point is the "critical point" (point C in Fig. 8·6·1) and corresponds to values of T, p, and v which are called the critical temperature, pressure, and volume. There the phase transformation has barely disappeared, the volume change Δv having approached zero.* At still higher temperatures $(\partial p/\partial v) < 0$ everywhere so

* Since $(\partial p/\partial v)_T = 0$ at the critical point, it follows from our previous discussion following (8·4·22) that density fluctuations at this point become very large; i.e., the substance "cannot quite make up its mind" whether to be a liquid or a gas.

that there is no phase transformation. One then deals always with a single fluid phase with no sharp distinction between gas and liquid. As the pressure is raised one then goes *continuously* from the situation of large molar volume v and high compressibility to the situation of small v and low compressibility.

SYSTEMS WITH SEVERAL COMPONENTS; CHEMICAL EQUILIBRIUM

8 · 7 *General relations for a system with several components*

Consider a homogeneous system, of energy E and of volume V, which consists of m different kinds of molecules. Let N_i be the number of molecules of type i. Then the entropy of the system is a function of the following variables:

$$S = S(E, V, N_1, N_2, \ldots, N_m) \tag{8·7·1}$$

These variables can, of course, all change in a general process. For example, the numbers of molecules of each species may change as a result of chemical reactions. In a completely general infinitesimal quasi-static process the entropy change is then given by*

$$dS = \left(\frac{\partial S}{\partial E}\right)_{V,N} dE + \left(\frac{\partial S}{\partial V}\right)_{E,N} dV + \sum_{i=1}^{m} \left(\frac{\partial S}{\partial N_i}\right)_{E,V,N} dN_i \tag{8·7·2}$$

Here the subscript N denotes the fact that all the numbers $\{N_1, \ldots, N_m\}$ are kept constant in taking the partial derivative. In the case of a derivative such as $(\partial S/\partial N_i)$, the subscript N denotes the fact that all the numbers $\{N_1, \ldots, N_{i-1}, N_{i+1}, \ldots, N_m\}$, except the one number N_i with respect to which the derivative is taken, are kept constant in taking the partial derivative.

Equation (8·7·2) is a purely mathematical statement. But in the simple case when all the numbers N_i are kept fixed, the fundamental thermodynamic relation asserts that

$$dS = \frac{đQ}{T} = \frac{dE + p\,dV}{T} \tag{8·7·3}$$

Under these circumstances $dN_i = 0$ for all i in Eq. (8·7·2); comparison of the coefficients of dE and dV in (8·7·2) and (8·7·3) then yields

$$\left(\frac{\partial S}{\partial E}\right)_{V,N} = \frac{1}{T} \qquad \left(\frac{\partial S}{\partial V}\right)_{E,N} = \frac{p}{T} \tag{8·7·4}$$

* The system is, in general, interacting with some other systems. The variables in (8·7·2) and subsequent relations are to be evaluated for the equilibrium (or most probable) situation when the values of these variables are essentially equal to their mean values, $\bar{E}$, $\bar{V}$, $\bar{N}_i$, $\bar{p}$, and so on. We shall, however, omit writing the averaging bars over these symbols.

Let us introduce the abbreviation

$$\mu_j \equiv -T \left(\frac{\partial S}{\partial N_j} \right)_{E,V,N} \tag{8·7·5}$$

The quantity μ_j is called the "chemical potential per molecule" of the jth chemical species and has been defined so that it has the dimensions of energy. Then (8·7·2) can be written in the form

$$dS = \frac{1}{T}\,dE + \frac{p}{T}\,dV - \sum \frac{\mu_i}{T}\,dN_i \tag{8·7·6}$$

or

$$\blacktriangleright \qquad dE = T\,dS - p\,dV + \sum_{i=1}^{m} \mu_i\,dN_i \tag{8·7·7}$$

This is just a generalization of the fundamental relation $dE = T\,dS - p\,dV$ to the case where the numbers of particles are allowed to vary.

Note that the chemical potential μ_j can be written in many forms equivalent to (8·7·5). For example, suppose that all the independent variables other than N_j are kept constant in (8·7·7). Then $dS = dV = 0$, $dN_i = 0$ for $i \neq j$, and (8·7·7) yields the relation

$$\mu_j = \left(\frac{\partial E}{\partial N_j} \right)_{S,V,N} \tag{8·7·8}$$

Alternatively, one can write (8·7·7) in the form

$$d(E - TS) = dF = -S\,dT - p\,dV + \sum_i \mu_i\,dN_i \tag{8·7·9}$$

If all independent variables other than N_j are kept constant, it follows immediately that

$$\mu_j = \left(\frac{\partial F}{\partial N_j} \right)_{T,V,N} \tag{8·7·10}$$

One can also write (8·7·7) in terms of the Gibbs free energy; thus

$$d(E - TS + pV) = dG = -S\,dT + V\,dp + \sum_i \mu_i\,dN_i \tag{8·7·11}$$

Hence one can also write

$$\mu_j = \left(\frac{\partial G}{\partial N_j} \right)_{T,p,N} \tag{8·7·12}$$

If there is *only one* chemical species present, say species j, then

$$G = G(T,p,N_j)$$

But G must be an extensive quantity. Thus, if all the independent extensive parameters are multiplied by a scale factor α, i.e., if N_j is multiplied by α, then G must be multiplied by the same factor α. Thus G must be proportional to N_j and can be written in the form

$$G(T,p,N_j) = N_j g'(T,p)$$

where $g'(T,p)$ does not depend on N_j. Then

$$\mu_j = \left(\frac{\partial G}{\partial N_j}\right)_{T,p} = g'(T,p) \tag{8·7·13}$$

i.e., the chemical potential per molecule is just equal to the Gibbs free energy $g' = G/N_j$ per *molecule.*

When several components are present, then $G = G(T, p, N_1, \ldots, N_m)$ and in general

$$\mu_j = \left(\frac{\partial G}{\partial N_j}\right)_{T,p,N} \neq \frac{G}{N_j}$$

Remark The requirement that extensive quantities scale properly leads, however, to the following general conclusions. Consider, for example, the total energy

$$E = E(S, V, N_1, N_2, \ldots, N_m) \tag{8·7·14}$$

If one increases all extensive variables by the same scale factor, Eq. (8·7·14) must remain valid. That is, if

$$S \to \alpha S, \qquad V \to \alpha V, \qquad N_i \to \alpha N_i$$

then one must also have $E \to \alpha E$. Thus

$$E(\alpha S, \alpha V, \alpha N_1, \ldots, \alpha N_m) = \alpha E(S, V, N_1, \ldots, N_m) \tag{8·7·15}$$

In particular let

$$\alpha = 1 + \gamma$$

where $|\gamma| \ll 1$. Then the left side of (8·7·15) becomes $E(S + \gamma S, V + \gamma V, N_1 + \gamma N_1, \ldots)$ which can be expanded about its value $E(S, V, N_1, \ldots)$, where $\gamma = 0$. Thus (8·7·15) implies the requirement that

$$E + \left(\frac{\partial E}{\partial S}\right)_{V,N} \gamma S + \left(\frac{\partial E}{\partial V}\right)_{S,N} \gamma V + \sum_{i=1}^{m} \left(\frac{\partial E}{\partial N_i}\right)_{S,V,N} \gamma N_i = (1 + \gamma)E$$

or*

$$E = \left(\frac{\partial E}{\partial S}\right)_{V,N} S + \left(\frac{\partial E}{\partial V}\right)_{S,N} V + \sum_{i=1}^{m} \left(\frac{\partial E}{\partial N_i}\right)_{S,V,N} N_i \tag{8·7·16}$$

But the derivatives are precisely given by the respective coefficients of dS, dV, and dN_i in (8·7·7). Hence (8·7·16) is equivalent to the relation

$$E = TS - pV + \sum_i \mu_i N_i \tag{8·7·17}$$

or

▶ $$G = E - TS + pV = \sum_i \mu_i N_i \tag{8·7·18}$$

* This purely mathematical consequence of (8·7·15) is commonly referred to as "Euler's theorem for homogeneous functions."

If only a single kind j of molecule is present, (8·7·18) reduces to the previous relation

$$\mu_j = \frac{G}{N_j}$$

Eq. (8·7·17) implies that

$$dE = T\,dS + S\,dT - p\,dV - V\,dp + \sum_i \mu_i\,dN_i + \sum_i N_i\,d\mu_i$$

But since (8·7·7) must also be valid, one obtains the general result

$$S\,dT - V\,dp + \sum_i N_i\,d\mu_i = 0 \tag{8·7·19}$$

(This is known as the "Gibbs-Duhem relation.")

8·8 *Alternative discussion of equilibrium between phases*

In Sec. 8·5 we treated the problem of the equilibrium between two phases under conditions where the system was considered to be in equilibrium with a reservoir at constant temperature and pressure. It is instructive to treat this problem from a somewhat more general point of view, considering the total system to be isolated. Our discussion will be a straightforward extension of that used at the end of Sec. 3·9.

Consider N molecules forming a substance which consists of two phases denoted by 1 and 2. The whole system is isolated so that its total energy E and its total volume V are both fixed. Let there be N_i molecules of the substance in phase i, and denote the energy of this phase by E_i and its volume by V_i. Then we have the conservation conditions

$$\left.\begin{aligned} E_1 + E_2 &= E = \text{constant} \\ V_1 + V_2 &= V = \text{constant} \\ N_1 + N_2 &= N = \text{constant} \end{aligned}\right\} \tag{8·8·1}$$

The entropy of the whole system (or the total number of states accessible to the whole system) is a function of these parameters. The equilibrium condi-

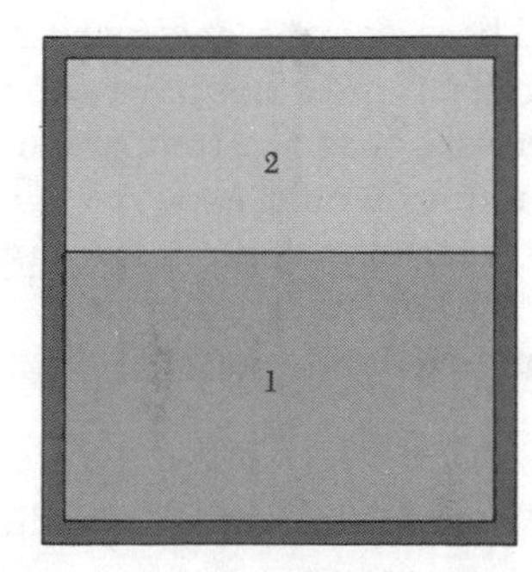

***Fig. 8·8·1** Equilibrium between two phases forming an isolated system of constant total energy and volume.*

tion corresponding to the most probable situation is that the entropy is a maximum, i.e., that

$$S = S(E_1,V_1,N_1;\ E_2,V_2,N_2) = \text{maximum} \tag{8·8·2}$$

But

$$S = S_1(E_1,V_1,N_1) + S_2(E_2,V_2,N_2)$$

where S_i is the entropy of phase i. Thus the maximum condition (8·8·2) yields

$$dS = dS_1 + dS_2 = 0 \tag{8·8·3}$$

subject to the conditions (8·8·1), which become, in differential form,

$$\left.\begin{aligned} dE_1 + dE_2 &= 0 \\ dV_1 + dV_2 &= 0 \\ dN_1 + dN_2 &= 0 \end{aligned}\right\} \tag{8·8·4}$$

By using the relation (8·7·6) for each phase, (8·8·3) gives

$$dS = \left(\frac{1}{T_1}\,dE_1 + \frac{p_1}{T_1}\,dV_1 - \frac{\mu_1}{T_1}\,dN_1\right) + \left(\frac{1}{T_2}\,dE_2 + \frac{p_2}{T_2}\,dV_2 - \frac{\mu_2}{T_2}\,dN_2\right) = 0$$

or

$$dS = \left(\frac{1}{T_1} - \frac{1}{T_2}\right) dE_1 + \left(\frac{p_1}{T_1} - \frac{p_2}{T_2}\right) dV_1 - \left(\frac{\mu_1}{T_1} - \frac{\mu_2}{T_2}\right) dN_1 = 0 \tag{8·8·5}$$

where we have used the conservation conditions (8·8·4). Since (8·8·5) is to be valid for arbitrary variations dE_1, dV_1, dN_1, it follows that the coefficients of all these differentials must separately vanish. Thus one obtains

$$\frac{1}{T_1} - \frac{1}{T_2} = 0$$

$$\frac{p_1}{T_1} - \frac{p_2}{T_2} = 0$$

$$\frac{\mu_1}{T_1} - \frac{\mu_2}{T_2} = 0$$

or

$$\left.\begin{aligned} T_1 &= T_2 \\ p_1 &= p_2 \\ \mu_1 &= \mu_2 \end{aligned}\right\} \tag{8·8·6}$$

These are the necessary conditions for equilibrium between two phases; they reflect the respective conservation equations (8·8·1). The relations (8·8·6) assert that the temperatures and mean pressures of the phases must be equal, as one would expect. The condition that the chemical potentials must also be equal may seem less familiar. But since each phase consists of only a single component, it follows by (8·7·13) that $\mu_i = g_i'$ is the chemical potential per molecule of phase i. The last relation of (8·8·6) is then equivalent to

$$g_1' = g_2' \tag{8·8·7}$$

Thus we regain the condition (8·5·4).

Remark In Sec. 8·5 we used the Gibbs free energy g per *mole;* thus $g_i = N_a g_i'$ where N_a is Avogadro's number. In the same way, it is sometimes useful to define a chemical potential per *mole.* This is given by the relation $(\partial G/\partial \nu_i)$; since $\nu_i = N_i/N_a$, it is N_a times larger than the corresponding chemical potential per molecule $(\partial G/\partial N_i)$.

One could readily extend the arguments of this section to treat the equilibrium between phases containing several components, or to calculate the fluctuations in the number of particles present in each phase.

Finally it is worth noting the microscopic implications of our discussion. One can take the equilibrium condition (8·8·6) or (8·8·7) and write it in the form

$$\mu_1(T,p) = \mu_2(T,p) \tag{8·8·8}$$

where we have expressed the chemical potentials in terms of T and p. Now we know, at least in principle, how to use statistical mechanics to calculate thermodynamic functions such as the entropy S for each phase. We can then go on to calculate for each phase the chemical potential μ (say, by its definition (8·7·5)) and we can express it in terms of T and p. The result is an equation of the form (8·8·8) which can be solved to find p as a function of T. In this way one can, for example, start from first principles to compute the vapor pressure of a substance at a given temperature. The only difficulty in such a calculation is the evaluation of the partition function for each phase. We shall illustrate such a vapor-pressure calculation in the next chapter.

8·9 *General conditions for chemical equilibrium*

Consider a homogeneous system (consisting of a single phase) which contains m different kinds of molecules. Let us designate the chemical symbols of these molecules by $B_1, B_2, \ldots, B_m$. Assume that there exists the possibility of one chemical reaction occurring between these molecules whereby molecules can get transformed into each other. This chemical transformation must be consistent with the conservation of the total number of *atoms* of *each* kind. A properly balanced chemical equation expresses precisely these conservation conditions.

Example Suppose that the system consists of H_2, O_2, and H_2O molecules in the gas phase. There exists the possibility of transformation of molecules into each other through the chemical reaction

$$2H_2 + O_2 \rightleftarrows 2H_2O$$

This chemical equation is properly balanced so that the total number of H atoms is the same on the left and right sides, as is the total number of O atoms.

Let b_i denote the coefficient of B_i in the chemical equation; thus b_i is some small integer designating the number of B_i molecules involved in the chemical transformation. For convenience we adopt the convention that, if the chemical reaction is regarded arbitrarily as proceeding in a given direction, one considers b_i positive for any "product" molecule formed as a result of the chemical reaction, and one considers b_i negative for any "reactant" molecule disappearing as a result of the reaction. For example, the reaction

$$2H_2 + O_2 \rightarrow 2H_2O$$

would be written in standard form as

$$-2H_2 - O_2 + 2H_2O = 0 \qquad (8 \cdot 9 \cdot 1)$$

A general chemical equation can then be written in the form

▶ $$\sum_{i=1}^{m} b_i B_i = 0 \qquad (8 \cdot 9 \cdot 2)$$

Let N_i denote the number of B_i molecules in the system. The numbers N_i can change as a result of the chemical reaction between molecules. But they cannot change in arbitrary ways because the conservation of *atoms* requires that the chemical equation (8·9·2) be satisfied. The changes in the numbers N_i must therefore be proportional to the numbers of molecules appearing in the balanced chemical equation (8·9·2); i.e.,

$$dN_i = \lambda b_i \qquad \text{for all } i \qquad (8 \cdot 9 \cdot 3)$$

where λ is a constant of proportionality. Here $dN_i > 0$ for molecules formed as a result of the reaction, and $dN_i < 0$ for molecules disappearing as a result of the reaction.

Example In the reaction (8·9·1) the number of molecules N_{H_2O}, N_{H_2}, N_{O_2} can only vary in such a way that changes in these numbers are in the proportion

$$dN_{H_2O}:dN_{H_2}:dN_{O_2} = 2:-2:-1$$

Consider now an *equilibrium* situation where molecules confined within an isolated enclosure of volume V can react chemically with each other in accordance with the reaction (8·9·2). Let E denote the total energy of the system. The equilibrium condition then is that

$$S = S(E, V, N_1, \ldots, N_m) = \text{maximum} \qquad (8 \cdot 9 \cdot 4)$$

or

$$dS = 0 \qquad (8 \cdot 9 \cdot 5)$$

Under the assumed conditions of constant V and E, this condition becomes, by virtue of (8·7·6),

$$\sum_{i=1}^{m} \mu_i \, dN_i = 0 \qquad (8 \cdot 9 \cdot 6)$$

Using the fact that the variations dN_i must satisfy the chemical equation (8·9·2), i.e., the restriction (8·9·3), the condition (8·9·6) then becomes simply

$$\sum_{i=1}^{m} b_i\mu_i = 0 \tag{8·9·7}$$

This is the general condition for chemical equilibrium.

Remark If we had assumed that the reaction takes place under conditions of constant temperature T and volume V, then the condition that F is minimum or $dF = 0$ under these circumstances would, by virtue of (8·7·9), again lead to (8·9·6) and (8·9·7). Similarly, for a reaction at constant temperature T and pressure p, the condition that G is minimum or $dG = 0$ would, by virtue of (8·7·11), also lead to (8·9·6) and (8·9·7).

The chemical potentials μ_i are functions of the variables describing the system. For example, $\mu_i = \mu_i\,(E, V, N_1, \ldots, N_m)$, if E and V are chosen as independent variables, or $\mu_i = \mu_i(T, V, N_1, \ldots, N_m)$ if T and V are chosen as independent variables. Hence the condition (8·9·7) implies in an equilibrium situation a definite connection between the mean numbers N_i of molecules of each kind. Since statistical thermodynamics allows one to calculate thermodynamic functions like the entropy S, it also makes it possible to calculate the chemical potentials μ_i and thus to deduce explicitly the connection between the numbers $N_1, \ldots, N_m$ implied by the condition (8·9·7). We shall illustrate such a calculation in the next section.

8·10 *Chemical equilibrium between ideal gases*

Consider that the chemical reaction (8·9·2) can occur between m different types of molecules. Suppose that these form gases which can be considered ideal and that they are confined in a container of volume V at an absolute temperature T. What is the relationship between the mean numbers of the reacting molecules in the equilibrium situation?

The question is readily answered by exploiting the condition (8·9·7) explicity. In other words, suppose that one knows the free energy

$$F = F(T,V,N_1, \ldots, N_m)$$

of this mixture of gases. If at the constant specified temperature T and volume V one imagines a transformation in which, in accordance with the chemical equation (8·9·2), $|b_i|$ of each of the reactant molecules are transformed into b_i of each of the product molecules, then the corresponding small free-energy change ΔF in the reaction is given by

$$\Delta F = \sum_i \left(\frac{\partial F}{\partial N_i}\right)_{T,V,N} b_i = \sum_i \mu_i b_i \tag{8·10·1}$$

Here
$$\mu_i \equiv \left(\frac{\partial F}{\partial N_i}\right)_{T,V,N} \tag{8·10·2}$$

is, in accordance with the relation (8·7·10), just the chemical potential of a molecule of type i and is, like F itself, a function of the variables $T, V, N_1, \ldots, N_m$. In equilibrium the free energy F is a minimum, so that (8·10·1) yields the familiar condition (8·9·7):

$$\Delta F = \sum_i b_i\mu_i = 0 \tag{8·10·3}$$

The remaining task is then merely that of calculating F, and hence the chemical potentials μ_i, of this mixture of gases.

Calculation of the chemical potential We again consider the gases to be at sufficiently high temperature and sufficiently low density that their translational motion can be treated classically.

Let the possible states of the kth molecule in the gas be labeled by s_k and let $\epsilon_k(s_k)$ denote the energy of the molecule in this state. Since there is negligible interaction between the molecules, the total energy of the gas in one of its possible states can always be written as a sum

$$E = \epsilon_1(s_1) + \epsilon_2(s_2) + \epsilon_3(s_3) + \cdots$$

with a number of terms equal to the total number of molecules. Hence the partition function becomes (treating all molecules as distinguishable)

$$Z' = \sum_{s_1,s_2,s_3} e^{-\beta[\epsilon_1(s_1)+\epsilon_2(s_2)+\cdots]}$$

where the summation is over *all* the states of *each* molecule. As usual this factors into the form

$$Z' = \left(\sum_{s_1} e^{-\beta\epsilon(s_1)}\right)\left(\sum_{s_2} e^{-\beta\epsilon_2(s_2)}\right)\cdots \tag{8·10·4}$$

In this product there will be N_i equal factors for all molecules of type i, each of these factors being equal to

$$\zeta_i \equiv \sum_s e^{-\beta\epsilon(s)} \tag{8·10·5}$$

where the sum is over all the states s and corresponding energies of one molecule of type i. Thus (8·10·4) becomes simply

$$Z' = \zeta_1^{N_1}\zeta_2^{N_2}\cdots\zeta_m^{N_m} \tag{8·10·6}$$

Here we have counted as distinct states of the gas all those which differ only by permutations of like molecules. As we saw in Sec. 7·3, it would be erroneous (Gibbs paradox) and inconsistent with the essential indistinguishability of the molecules in quantum mechanics to count these gas states as distinct. To get the correct partition function Z, the expression (8·10·6) must therefore be divided by the $(N_1!\,N_2!\cdots N_m!)$ possible permutations of like molecules among themselves. Thus we get

$$Z = \frac{\zeta_1^{N_1}\zeta_2^{N_2}\cdots\zeta_m^{N_m}}{N_1!N_2!\cdots N_m!} \tag{8·10·7}$$

This can also be written

$$Z = Z_1 Z_2 \cdots Z_m \tag{8·10·8}$$

where

$$Z_i = \frac{\zeta_i^{N_i}}{N_i!} \tag{8·10·9}$$

is the partition function of a gas of N_i molecules occupying the given volume V by itself in the absence of all other gases.

A variety of important results follow from (8·10·8), i.e., from the relation

$$\ln Z = \sum_i \ln Z_i \tag{8·10·10}$$

These results all reflect the fact that the molecules are weakly interacting so that the thermodynamic functions are simply additive. For example, since the mean energy of a system is given quite generally by $\bar{E} = (-\partial \ln Z/\partial\beta)$, it follows from (8·10·10) that

$$\bar{E}(T,V) = \sum_i \bar{E}_i(T,V) \tag{8·10·11}$$

where $\bar{E}_i$ is the mean energy of the ith gas occupying the given volume by itself. Also, since the mean pressure of a system is given quite generally by $\bar{p} = \beta^{-1}(\partial \ln Z/\partial V)$, it follows from (8·10·10) that

$$\bar{p} = \sum_i \bar{p}_i \tag{8·10·12}$$

where $\bar{p}_i$ is the mean pressure that would be exerted by the ith gas if it occupied the given volume V by itself. This quantity $\bar{p}_i$ is called the "partial pressure of the ith gas."

Now we have already derived the equation of state for a single gas; i.e., for gas i occupying the volume V by itself

$$\bar{p}_i = n_i kT, \qquad n_i \equiv \frac{N_i}{V} \tag{8·10·13}$$

Hence (8·10·12) gives us immediately the equation of state for the gas mixture

$$\bar{p} = nkT, \qquad \text{where } n \equiv \sum_{i=1}^{m} n_i \tag{8·10·14}$$

Since the Helmholtz free energy is given quite generally by the relation $F = -kT \ln Z$, it follows from (8·10·10) that

$$F(T,V) = \sum_i F_i(T,V) \tag{8·10·15}$$

where F_i is the free energy of the ith gas by itself. Since $F = \bar{E} - TS$, (8·10·11) and (8·10·14) also establish the additivity of the entropies

$$S(T,V) = \sum_i S_i(T,V) \tag{8·10·16}$$

where S_i is the entropy of the ith gas occupying the given volume by itself.

Let us now proceed to calculate the chemical potential. By (8·10·9),

$$\ln Z_i = N_i \ln \zeta_i - \ln N_i!$$

where $\zeta_i = \zeta_i(T,V)$ is the partition function (8·10·5) for a single molecule and thus does not involve N_i. Using (6·6·9) and (8·10·10), one then obtains

$$F = -kT \ln Z = -kT \sum_i (N_i \ln \zeta_i - \ln N_i!)$$

or

$$F = -kT \sum_i N_i (\ln \zeta_i - \ln N_i + 1) \tag{8·10·17}$$

where we have used Stirling's formula $\ln N! = N \ln N - N$. Since

$$\frac{\partial \ln (N!)}{\partial N} = \ln N$$

(a result already encountered in (1·5·9)), it then follows by (8·10·2) that the chemical potential of the jth kind of molecule is simply given by

$$\mu_j = \left(\frac{\partial F}{\partial N_j}\right)_{T,V,N} = -kT(\ln \zeta_j - \ln N_j)$$

or

► $$\mu_j = -kT \ln \frac{\zeta_j}{N_j} \tag{8·10·18}$$

Law of mass action By (8·10·1), the free-energy change in the reaction is then equal to

$$\Delta F = -kT \sum b_i (\ln \zeta_i - \ln N_i) = \Delta F_0 + kT \sum_i b_i \ln N_i \tag{8·10·19}$$

where

$$\Delta F_0 \equiv -kT \sum_i b_i \ln \zeta_i \tag{8·10·20}$$

is a quantity (the so-called "standard free-energy change of the reaction") which depends only on T and V, but not on the numbers N_i of molecules present. The equilibrium condition (8·10·3) then becomes

$$\Delta F = \Delta F_0 + kT \sum_i b_i \ln N_i = 0$$

$$\sum_i \ln N_i^{b_i} = \ln (N_1^{b_1} N_2^{b_2} \cdots N_m^{b_m}) = -\frac{\Delta F_0}{kT}$$

or

► $$N_1^{b_1} N_2^{b_2} \cdots N_m^{b_m} = K_N(T,V) \tag{8·10·21}$$

where

► $$K_N(T,V) \equiv e^{-\Delta F_0/kT} = \zeta_1^{b_1} \zeta_2^{b_2} \cdots \zeta_m^{b_m} \tag{8·10·22}$$

The quantity K_N is independent of the numbers of molecules present and is called the "equilibrium constant"; it is a function only of T and V through the dependence of the molecular partition functions ζ_i on these quantities.

Equation (8·10·21) is the desired explicit relation between the mean numbers of molecules present in equilibrium; it is called the "law of mass action" and is likely to be familiar from elementary chemistry.

Example Consider the reaction (8·9·1) in the gas phase

$$-2H_2 - O_2 + 2H_2O = 0$$

The law of mass action (8·10·21) then becomes

$$N_{H_2}^{-2} N_{O_2}^{-1} N_{H_2O}^{2} = K_N$$

or

$$\frac{N_{H_2O}^2}{N_{H_2}^2 N_{O_2}} = K_N(T,V)$$

Note that (8·10·22) gives an explicit expression for the equilibrium constant K_N in terms of the partition functions ζ_i for each type of molecule. Hence K_N can be calculated from first principles if the molecules are sufficiently simple so that ζ_i in (8·10·5) can be evaluated from a knowledge of the quantum states of a single molecule. Even when the molecules are more complex, it is still possible to use spectroscopic data to deduce their energy levels and thus to calculate ζ_i and the equilibrium constant K_N.

Remark Note that the factor $N_i!$, present in (8·10·9) to take proper account of the indistinguishability of the molecules, is absolutely essential to the whole theory. If this factor were not present, one would simply have $\ln Z = \Sigma N_i \ln \zeta_i$, so that the chemical potential would become $\mu_j = -kT \ln \zeta_j$, *independent* of N_j. Hence (8·10·3) would give us no relationship between numbers at all! Thus the classical difficulties exemplified by the Gibbs paradox lead to nonsense all down the line.

It is worth noting the simplifying property that $\zeta_i(V,T)$ is, as in (7·2·6), simply proportional to V. Indeed (8·10·5) can be written

$$\zeta_i \propto \int d^3\mathbf{r} \int d^3\mathbf{p}\, e^{-\beta p^2/2m} \sum_s e^{-\beta \epsilon_s^{(\text{int})}}$$

where we have calculated the translational part of the partition function classically. (The remaining sum is over the states of *internal* motion, both vibration and rotation, if the molecule is not monatomic.) Here the center of mass position $\mathbf{r}$ occurs only in the integral $\int d^3\mathbf{r}$ which yields the volume V. Thus one can write

$$\zeta_i(V,T) = V\zeta_i'(T) \tag{8·10·23}$$

where ζ_i' depends only on T. Hence the chemical potential (8·10·18) becomes

$$\mu_j = -kT \ln \frac{\zeta_j'}{n_j} \tag{8·10·24}$$

where $n_j = N_j/V$ is the number of molecules of type j per unit volume.

The fundamental equilibrium condition (8·10·3) can then be written more simply

$$\sum_i b_i \ln n_i = \sum_i b_i \ln \zeta_i'$$

or

$$n_1^{b_1} n_2^{b_2} \cdots n_m^{b_m} = K_n(T) \tag{8·10·25}$$

where

$$K_n(T) = \zeta_1'^{b_1} \zeta_2'^{b_2} \cdots \zeta_m'^{b_m} \tag{8·10·26}$$

and the equilibrium constant $K_n(T)$ depends *only* on the temperature.

By (8·10·22) and (8·10·26) one has the relation

$$K_N(T,V) = V^b K_n(T), \qquad \text{where } b \equiv \sum_{i=1}^{m} b_i \tag{8·10·27}$$

Remark The law of mass action (8·10·25) can be given a detailed kinetic interpretation. Write the reaction (8·9·2) in the form

$$b_1'B_1 + b_2'B_2 + \cdots + b_k'B_k \leftrightarrows b_{k+1}B_{k+1} + \cdots + b_mB_m \tag{8·10·28}$$

where $i = 1, \ldots, k$ refers to the k kinds of reactant molecules (with $b_i' \equiv -b_i > 0$) and where $i = k+1, \ldots, m$ refers to the remaining $(m-k)$ kinds of product molecules. The probability P_+ per unit time that the reaction (8·10·28) occurs from left to right should be proportional to the probability of simultaneous encounter in a given element of volume of b_1' B_1-molecules, b_2' B_2-molecules, . . . , b_k' B_k-molecules. Since molecules in an ideal gas are statistically independent, the probability of molecule i being in the volume element is simply proportional to n_i, and the probability P_+ becomes

$$P_+ = K_+(T) n_1^{b_1'} n_2^{b_2'} \cdots n_k^{b_k'}$$

where $K_+(T)$ is a constant of proportionality which can depend on T. Similarly, the probability P_- per unit time that the reaction (8·10·28) occurs from right to left should be proportional to the probability of simultaneous encounter of b_{k+1} B_{k+1}-molecules, . . . , b_m B_m-molecules, i.e.,

$$P_- = K_-(T) n_{k+1}^{b_{k+1}} n_{k+2}^{b_{k+2}} \cdots n_m^{b_m}$$

In equilibrium one must have

$$P_+ = P_-$$

Hence it follows that

$$\frac{n_{k+1}^{b_{k+1}} n_{k+2}^{b_{k+2}} \cdots n_m^{b_m}}{n_1^{b_1'} n_2^{b_2'} \cdots n_k^{b_k'}} = \frac{K_+(T)}{K_-(T)}$$

which is identical in form with (8·10·25).

Temperature dependence of the equilibrium constant The relation (8·10·22) gives explicitly

$$\ln K_N(T,V) = -\frac{\Delta F_0}{kT} \tag{8·10·29}$$

Hence $$\left(\frac{\partial \ln K_N}{\partial T}\right)_V = -\left(\frac{\partial}{\partial T}\right)_V\left(\frac{\Delta F_0}{kT}\right) = -\left(\frac{\partial}{\partial T}\right)_{V,N}\frac{\Delta F}{kT} \tag{8·10·30}$$

The last expression follows by (8·10·19), since ΔF and ΔF_0 differ only by an expression involving the numbers N_i and since these numbers are supposed to be held constant, as indicated, in the differentiation of $\Delta F/kT$. Hence

$$\left(\frac{\partial \ln K_N}{\partial T}\right)_V = \frac{1}{kT^2}\Delta F - \frac{1}{kT}\left(\frac{\partial \Delta F}{\partial T}\right)_{V,N} \tag{8·10·31}$$

But

$$-\left(\frac{\partial}{\partial T}\right)\Delta F = -\sum_i\left(\frac{\partial}{\partial T}\right)\left(\frac{\partial F}{\partial N_i}\right)b_i = -\sum_i\left(\frac{\partial}{\partial N_i}\right)\left(\frac{\partial F}{\partial T}\right)b_i = \sum\frac{\partial S}{\partial N_i}b_i \equiv \Delta S$$

Here we have used the general relation, implied by (8·7·9), that

$$\left(\frac{\partial F}{\partial T}\right)_{V,N} = -S$$

and have denoted by ΔS the entropy change of the reaction when $|b_i|$ of each of the reactant molecules are transformed into b_i of each of the product molecules. Hence (8·10·31) becomes

$$\left(\frac{\partial \ln K_N}{\partial T}\right)_V = \frac{1}{kT^2}(\Delta F + T\,\Delta S) = \frac{\Delta E}{kT^2} \tag{8·10·32}$$

since $F \equiv E - TS$; thus $\Delta E = \Delta(F + TS)$ is simply the mean energy increase in the reaction. Since the reaction is carried out at constant volume, ΔE is also the heat absorbed in the reaction when $|b_i|$ of each of the reactant molecules are transformed into b_i of each of the product molecules.

Since K_n differs from K_N only by a factor involving the volume V, (8·10·32) implies equivalently that

$$\frac{d \ln K_n}{dT} = \frac{\Delta E}{kT^2} \tag{8·10·33}$$

If $\Delta E > 0$, (8·10·32) asserts that K_N increases as T is increased. This result is again in accord with what would be expected from Le Chatelier's principle. When $E > 0$, heat is absorbed as a result of the reaction. If the temperature T *increases*, *more* molecules must then be produced in order to absorb heat and thus to restore the original temperature. Thus K_N must increase.

SUGGESTIONS FOR SUPPLEMENTARY READING

Stability conditions

H. B. Callen: "Thermodynamics," chap. 8, John Wiley & Sons, Inc., New York, 1960.

Fluctuations of thermodynamic quantities

L. D. Landau and E. M. Lifshitz: "Statistical Physics," secs. 109–111, Addison-Wesley Publishing Company, Reading, Mass., 1959.

Phase transformations

M. W. Zemansky: "Heat and Thermodynamics," 4th ed., chap. 15, McGraw-Hill Book Company, New York, 1957.

W. P. Allis and M. A. Herlin: "Thermodynamics and Statistical Mechanics," secs. 33–36, McGraw-Hill Book Company, New York, 1952.

H. B. Callen: "Thermodynamics," chap. 9, John Wiley & Sons, Inc., New York, 1960.

A. B. Pippard: "The Elements of Classical Thermodynamics," chaps. 8 and 9, Cambridge University Press, Cambridge, 1957. (Includes a good discussion of "higher-order" phase transformations.)

Chemical equilibrium

M. W. Zemansky: "Heat and Thermodynamics," 4th ed., chaps. 17 and 18, McGraw-Hill Book Company, New York, 1957.

H. B. Callen: "Thermodynamics," chap. 12, John Wiley & Sons, Inc., New York, 1960.

T. L. Hill: "An Introduction to Statistical Thermodynamics," chap. 10, Addison-Wesley Publishing Company, Reading, Mass., 1960.

J. F. Lee, F. W. Sears, and D. L. Turcotte: "Statistical Thermodynamics," chap. 13, Addison-Wesley Publishing Company, Reading, Mass., 1963.

PROBLEMS

8.1 In some homogeneous substance at absolute temperature T_0 (e.g., a liquid or gas) focus attention on some small portion of mass M. This small portion is in equilibrium with the rest of the substance; it is large enough to be macroscopic and can be characterized by a volume V and temperature T. Calculate the probability $\mathcal{P}(V,T)\,dV\,dT$ that the volume of this portion lies between V and $V + dV$ *and* that its temperature lies between T and $T + dT$. Express your answer in terms of the compressibility κ of the substance, its density ρ_0, and its specific heat per gram c_V at constant volume.

8.2 The vapor pressure p (in millimeters of mercury) of solid ammonia is given by $\ln p = 23.03 - 3754/T$ and that of liquid ammonia by $\ln p = 19.49 - 3063/T$.

(*a*) What is the temperature of the triple point?

(*b*) What are the latent heats of sublimation and vaporization at the triple point?

(*c*) What is the latent heat of melting at the triple point?

8.3 A simple substance of molecular weight μ has its triple point at the absolute temperature T_0 and pressure p_0. At this point the densities of the solid and liquid are ρ_s and ρ_l, respectively, while the vapor can be approximated by a dilute ideal gas. If at this triple point the slope of the melting curve is $(dp/dT)_m$ and that of the liquid vaporization curve is $(dp/dT)_v$, what is the slope $(dp/dT)_s$ of the sublimation curve of the solid?

8.4 Helium remains a liquid down to absolute zero at atmospheric pressure, but becomes a solid at sufficiently high pressures. The density of the solid is, as usual, greater than that of the liquid. Consider the phase-equilibrium line between the solid and liquid. In the limit as $T \to 0$, is the slope dp/dT of this line positive, zero, or negative?

8.5 Liquid helium boils at a temperature T_0 (4.2°K) when its vapor pressure is equal to $p_0 = 1$ atmosphere. The latent heat of vaporization per mole of the liquid is equal to L and approximately independent of temperature. The liquid is contained within a dewar which serves to insulate it thermally from the room temperature surroundings. Since the insulation is not perfect, an amount of heat $\mathcal{Q}$ per second flows into the liquid and evaporates some of it. (This heat influx $\mathcal{Q}$ is essentially constant, independent of whether the temperature of the liquid is T_0 or less.) In order to reach low temperatures one can reduce the pressure of the He vapor over the liquid by pumping it away with a pump at room temperature T_r. (By the time it reaches the pump, the He vapor has warmed up to room temperature.) The pump has a maximum pumping speed such that it can remove a constant volume $\mathcal{V}$ of gas per second, irrespective of the pressure of the gas. (This is a characteristic feature of ordinary mechanical pumps which simply sweep out a fixed volume of gas per revolution.)

(*a*) Calculate the minimum vapor pressure p_m which this pump can maintain over the surface of the liquid if the heat influx is $\mathcal{Q}$.

(*b*) If the liquid is thus maintained in equilibrium with its vapor at this pressure p_m, calculate its approximate temperature T_m.

8.6 An atomic beam of sodium (Na) atoms is produced by maintaining liquid sodium in an enclosure at some elevated temperature T. Atoms of Na from the vapor above the liquid escape by effusion through a narrow slit in the enclosure and thus give rise to an atomic beam of intensity I. (The intensity I is defined as the number of atoms in the beam which cross unit area per unit time.) The latent heat of vaporization per mole of liquid Na into a vapor of Na atoms is L. To estimate how sensitive the beam intensity is to fluctuations in the temperature of the enclosure, calculate the relative intensity change $I^{-1}(dI/dT)$ in terms of L and the absolute temperature T of the enclosure.

8.7 The molar latent heat of transformation in going from phase 1 to phase 2 at the temperature T and pressure p is l. What is the latent heat of the phase transformation at a slightly different temperature (and corresponding pressure), i.e., what is (dl/dT)? Express your answer in terms of l and the molar specific heat c_p, coefficient of expansion α, and molar volume v of each phase at the original temperature T and pressure p.

8.8 A steel bar of rectangular cross section (height a and width b) is placed on a block of ice with its ends extending a trifle as shown in the figure. A weight of mass m is hung from each end of the bar. The entire system is at 0°C. As a result of the pressure exerted by the bar, the ice melts beneath the bar and refreezes above the bar. Heat is therefore liberated above the bar, conducted through the metal, and then absorbed by the ice beneath the bar. (We assume that this is the most important way in which heat reaches the ice immediately beneath the bar in order to melt it.) Find an approximate expression for the speed with which the bar thus sinks through the ice. The answer should be in terms of the latent heat of fusion l per gram of ice, the densities ρ_i and ρ_w of ice and water respectively, the thermal conductivity κ of steel, the temperature T (0°C) of the ice, the acceleration due to gravity g, the mass m, and the dimensions a, b, and c, where c is the width of the block of ice.

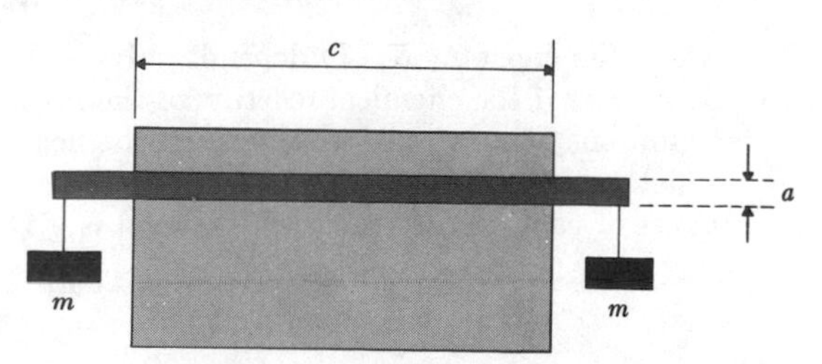

8.9 Careful measurements were made of the vapor pressure of liquid pentane as a function of temperature. The temperature was measured very precisely in terms of the emf of a thermocouple whose reference junction was maintained at the triple point of water. Thus one determined the curve of vapor pressure p versus measured thermocouple emf ϕ. Also measured as a function of ϕ along the vapor pressure curve were the latent heat of vaporization L per gram of liquid pentane, and the volume change ΔV per gram of pentane in going from liquid to vapor. Show that this information is sufficient to calibrate the thermocouple; i.e., write an explicit expression (in terms of an integral) for the absolute temperature T when the measured thermocouple emf is ϕ.

8.10 Consider any substance in equilibrium in the presence of externally applied forces due to gravitational or electromagnetic fields. Focus attention on any two volume elements of this substance, both of fixed size and infinitesimally small on a macroscopic scale. By using the fact that the total entropy of the substance must remain stationary if a small amount of energy or a small number of particles, or both, are transferred from one of these volume elements to the other, show that the temperature T and the chemical potential μ must each have a constant value throughout the substance.

8.11 Consider a classical ideal gas in thermal equilibrium at temperature T in a container of volume V in the presence of a uniform gravitational field. The acceleration due to gravity is g and directed along the $-z$ direction.

(*a*) Calculate the chemical potential μ of an element of volume of such a gas as a function of the pressure p, the temperature T, and the height z.

(*b*) Show that the requirement that μ is constant implies immediately the law of atmospheres which gives the dependence of p on T and z.

8.12 At a fixed temperature $T = 1200°K$, the gases

$$CO_2 + H_2 \rightleftarrows CO + H_2O$$

are in chemical equilibrium in a vessel of volume V. If the volume of this vessel is increased, its temperature being maintained constant, does the relative concentration of CO_2 increase, decrease, or remain the same?

8.13 An experiment on iodine (I) atoms is carried out in a molecular beam apparatus. The beam is obtained by effusion of molecules from a small slit in an oven containing, as a result of thermal dissociation, a mixture of I_2 molecules and I atoms. If the temperature of the oven is kept the same but the total gas pressure within it is doubled, by what factor is the intensity of I atoms in the beam changed?

8.14 Consider the following chemical reaction between ideal gases:

$$\sum_{i=1}^{m} b_i B_i = 0$$

Let the temperature be T, the total pressure be p. Denote the partial pressure of the ith species by p_i. Show that the law of mass action can be put into the form

$$p_1^{b_1} p_2^{b_2} \cdot \cdot \cdot \, p_m^{b_m} = K_p(T)$$

where the constant $K_p(T)$ depends only on T.

8.15 Show that if the chemical reaction of the preceding problem is carried out under conditions of constant total pressure, the heat of reaction per mole (i.e., the heat which must be supplied to transform $|b_i|$ moles of each of the reactants to $|b_i|$ moles of each of the reaction products) is given by the enthalpy change

$$\Delta H = \Sigma b_i h_i$$

where h_i is the enthalpy per mole of the ith gas at the given temperature and pressure.

8.16 Show that

$$\frac{d \ln K_p}{dT} = \frac{\Delta H}{RT^2}$$

where R is the gas constant per mole.

8.17 Suppose that ν_0 moles of H_2O gas are introduced into a container of fixed volume V at a temperature low enough so that virtually all the gas remains undissociated in the form of H_2O vapor. At higher temperatures dissociation can take place according to the reaction

$$2H_2O \rightarrow 2H_2 + O_2$$

Let ξ denote the fraction of H_2O molecules which are dissociated at any temperature T corresponding to a total gas pressure p. Write an equation relating ξ to p and $K_p(T)$.

8.18 In the preceding problem the degree of dissociation ξ at atmospheric pressure and at various temperatures T is experimentally found to have the following values:

T (°K)	ξ
1500	1.97×10^{-4}
1705	1.2×10^{-3}
2155	1.2×10^{-2}

What is the heat required to dissociate one mole of water vapor at 1 atmosphere into O_2 and H_2 at a temperature of 1700°K?

8.19 The partition function for an ideal gas of molecules in a volume V can be written in the form

$$Z = \frac{1}{N!}(V\zeta')^N$$

where $V\zeta'$ is the partition function for a single molecule (involving its kinetic energy, plus internal energy if it is not monatomic) and ζ' depends only on the absolute temperature T.

When these molecules are condensed so as to form a liquid, the crudest approximation consists of treating the liquid as if the molecules still formed a gas of molecules moving independently, provided that (1) each molecule is assumed to have a *constant* potential energy $-\eta$ due to its average interaction with the rest of the molecules; and (2) each molecule is assumed free to move throughout a total volume Nv_0, where v_0 is the (constant) volume available per molecule in the liquid phase.

(*a*) With these assumptions, write down the partition function for a liquid consisting of N_l molecules.

(*b*) Write down the chemical potential μ_g for N_g molecules of the vapor in a volume V_g at the temperature T. Treat it as an ideal gas.

(*c*) Write down the chemical potential μ_l for N_l molecules of liquid at the temperature T.

(*d*) By equating chemical potentials, find an expression relating the vapor pressure to the temperature T where the gas is in equilibrium with the liquid.

(*e*) Calculate the molar entropy difference between gas and liquid in equilibrium at the same temperature and pressure. From this calculate the molar heat of evaporation L. Show that $L = N_A\eta$ if $\eta \gg kT$.

(*f*) The boiling point T_b is that temperature where the vapor pressure is 1 atmosphere. Express the ratio L/RT_b in terms of v_0 and the volume v_g per molecule in the vapor phase at one atmosphere at the temperature T_b.

(*g*) Estimate the order of magnitude of L/RT_b and show that it is a number of the order of 10 for all ordinary liquids. (This result is called "Trouton's rule.")

(*h*) Compare this simple theory with experiment by looking up the densities and molecular weights of some liquids, computing L/T_b, and comparing with the experimental ratio of L/T_b. Data can be found in the "Handbook of Physics and Chemistry" (Chemical Rubber Publishing Company, Cleveland, Ohio). Try nitrogen and benzene, for example.

Quantum statistics of ideal gases

9

THIS CHAPTER will be devoted to a discussion of systems consisting of particles with negligible mutual interaction, i.e., of "ideal gases." But we shall now treat these systems from a completely quantum-mechanical point of view. This will allow us to discuss problems involving gases at low temperatures or high densities and to avoid the problems encountered in Sec. 7·3 in connection with the indistinguishability of the particles. It will also permit us to calculate unique values of entropies, to make absolute calculations of vapor pressures or chemical-equilibrium constants, and to treat distinctly nonclassical gases such as photons or conduction electrons in metals.

MAXWELL - BOLTZMANN, BOSE - EINSTEIN, AND FERMI - DIRAC STATISTICS

9·1 *Identical particles and symmetry requirements*

Consider a gas consisting of N identical structureless particles enclosed within a container of volume V. Let Q_i denote collectively all the coordinates of the ith particle (e.g., its three cartesian position coordinates and its spin coordinate, if any). Let s_i be an index labeling the possible quantum states of this single particle (e.g., each possible value of s_i corresponds to a specification of the three momentum components of the particle and of its direction of spin orientation; we postpone more detailed discussion to Sec. 9·9). The state of the whole gas is then described by the set of quantum numbers

$$\{s_1, s_2, \ldots, s_N\} \tag{9·1·1}$$

which characterize the wave function Ψ of the gas in this state.

$$\Psi = \Psi_{\{s_1, \ldots, s_N\}}(Q_1, Q_2, \ldots, Q_N) \tag{9·1·2}$$

Let us now discuss the various cases of interest.

"Classical" case *(Maxwell-Boltzmann statistics)* In this case the particles are considered to be distinguishable, and any number of particles can be in the same single-particle state s. This "classical" description imposes no symmetry requirements on the wave function when two particles are interchanged. The particles are then said to obey "Maxwell-Boltzmann statistics" (abbreviated "MB statistics"). This description is *not* correct quantum mechanically, but is interesting for purposes of comparison.

Quantum mechanics The quantum-mechanical description is, of course, the one which is actually applicable. But when quantum mechanics is applied to a system of identical particles, it imposes definite symmetry requirements on the wave function (9·1·2) under interchange of any two identical particles. The net result is that one does not obtain a new state of the whole gas by simply interchanging two such particles. When counting the distinct possible states accessible to the whole gas, the particles must thus be considered as intrinsically indistinguishable. In enumerating these possible states of the gas, it does then not matter *which* particle is in *which* particle state, but only *how many* particles there are in each single-particle state s.

The symmetry requirements can be regarded as fundamental quantum-mechanical postulates* and are intimately connected with the spin of the particles. There are two possible cases which may arise: either (*a*) the particles have integral spin or (*b*) the particles have half-integral spin.

a. Particles with integral spin (Bose-Einstein statistics):

This is the case where each particle has a total spin angular momentum (measured in units of $\hbar$) which is integral, i.e., 0, 1, 2, . . . (examples might be He^4 atoms or photons). Then the fundamental quantum-mechanical symmetry requirement is that the total wave function Ψ be *symmetric* (i.e., that it remain unchanged) under interchange of any two particles (i.e., interchange of *both* their spatial and spin coordinates). In symbols,

$$\Psi(\cdots Q_j \cdots Q_i \cdots) = \Psi(\cdots Q_i \cdots Q_j \cdots) \qquad (9 \cdot 1 \cdot 3)$$

(Here we have omitted the subscript $\{s_1, \ldots, s_N\}$ in (9·1·2) for the sake of brevity.) Thus interchange of two particles does *not* lead to a new state of the whole gas. The particles must, therefore, be considered as genuinely indistinguishable in enumerating the distinct states of the gas. Note that there is no restriction on how many particles can be in any one single-particle state s. Particles satisfying the symmetry requirement (9·1·3) are said to obey "Bose-Einstein statistics" (abbreviated "BE statistics") and are sometimes called "bosons."

b. Particles with half-integral spin (Fermi-Dirac statistics):

This is applicable when each particle has a total spin angular momentum (measured in units of $\hbar$) which is half-integral, i.e., $\frac{1}{2}$, $\frac{3}{2}$, . . . (examples might

* These postulates can, however, be derived (as was first done by Pauli) from a much more profound point of view which examines the requirements for a consistent description of the particles in terms of quantum field theory.

be electrons or He^3 atoms). Then the fundamental quantum-mechanical symmetry requirement is that the total wave function Ψ be *antisymmetric* (i.e., that it change sign) under interchange of any two particles. In symbols

$$\Psi(\cdots Q_j \cdots Q_i \cdots) = -\Psi(\cdots Q_i \cdots Q_j \cdots) \qquad (9 \cdot 1 \cdot 4)$$

Once again, interchange of two particles does not lead to a new state of the gas. Hence the particles must again be considered as genuinely indistinguishable in enumerating the distinct states of the gas. But the change of sign in (9·1·4) does imply one additional consequence: Suppose that two particles i and j, both in the *same* single-particle state s, are interchanged. In that case one obviously has

$$\Psi(\cdots Q_j \cdots Q_i \cdots) = \Psi(\cdots Q_i \cdots Q_j \cdots) \qquad (9 \cdot 1 \cdot 5)$$

But since the fundamental symmetry requirement (9·1·4) must also be valid, (9·1·4) and (9·1·5) together imply that

$$\Psi = 0 \qquad \text{when particles } i \text{ and } j \text{ are in the same state } s \qquad (9 \cdot 1 \cdot 6)$$

Thus in the Fermi-Dirac case there exists *no* state of the whole gas for which two or more particles are in the same single-particle state. This is the so-called "Pauli exclusion principle."* In enumerating the distinct states of the gas one must then keep in mind the restriction that there can never be more than one particle in any given single-particle state. Particles satisfying the antisymmetry requirement (9·1·4) are said to obey Fermi-Dirac statistics (abbreviated "FD statistics") and are sometimes called "fermions."

Illustration A very simple example should help to make these general ideas much clearer. Consider a "gas" of only two particles; call them A and B. Assume that each particle can be in one of three possible quantum states, $s = 1, 2, 3$. Let us enumerate the possible states of the whole gas. This is the same as asking in how many distinct ways one can put two particles (A and B) into three single-particle states (labeled 1, 2, 3).

Maxwell-Boltzmann statistics: The particles are to be considered distinguishable. Any number of particles can be in any one state.

1	*2*	*3*
AB	···	···
···	AB	···
···	···	AB
A	B	···
B	A	···
A	···	B
B	···	A
···	A	B
···	B	A

* This principle should be familiar since it applies to the important case of electrons (which have spin $\frac{1}{2}$) and accounts for the periodic table of the elements.

Each of the two particles can be placed in any one of the three states. Hence there exist a total of $3^2 = 9$ possible states for the whole gas.

Bose-Einstein statistics: The particles are to be considered *indistinguishable.* Any number of particles can be in any one state. The indistinguishability implies that $B = A$, so that the three states in the MB case which differed only in interchange of A and B are now no longer to be counted as distinct. The enumeration is then as follows:

1	*2*	*3*
AA	. . .	. . .
. . .	AA	. . .
. . .	. . .	AA
A	A	. . .
A	. . .	A
. . .	A	A

There are now three distinct ways of placing the particles in the same state. There are three distinct ways of placing the particles in different states. Hence there exist a total of $3 + 3 = 6$ possible states for the whole gas.

Fermi-Dirac statistics: The particles are to be considered as *indistinguishable.* No more than one particle can be in any one state. The three states in the BE case where two particles were in the same state must be eliminated in this case. One is thus left with the following enumeration:

1	*2*	*3*
A	A	. . .
A	. . .	A
. . .	A	A

There exist now only a total of 3 possible states for the whole gas.

This example shows one further qualitative feature of interest. Let

$$\xi \equiv \frac{\text{probability that the two particles are found in the same state}}{\text{probability that the two particles are found in different states}}$$

Then we have for the three cases

$$\xi_{\text{MB}} = \tfrac{3}{6} = \tfrac{1}{2}$$
$$\xi_{\text{BE}} = \tfrac{3}{3} = 1$$
$$\xi_{\text{FD}} = \tfrac{0}{3} = 0$$

Thus in the BE case there is a greater relative tendency for particles to bunch together in the same state than in classical statistics. On the other hand, in the FD case there is a greater relative tendency for particles to remain apart in different states than there is in classical statistics.

Discussion in terms of wave functions The same simple example can be discussed equivalently in terms of the possible wave functions for the whole gas. Let

$$\psi_s(Q) = \text{the one-particle wave function for a single particle (with coordinate } Q\text{) in state } s$$

As before, let Ψ be the wave function for the whole gas. Since the particles are noninteracting, Ψ can be written as a simple product of single-particle wave functions, or of proper linear combinations thereof. Let us again discuss the cases in turn.

Maxwell-Boltzmann statistics: There is no particular symmetry requirement on Ψ under particle interchange. Apart from normalization, a complete set of wave functions Ψ for the gas are then the $3 \times 3 = 9$ functions of the form

$$\psi_i(Q_A)\psi_j(Q_B)$$

where $i = 1, 2, 3$ and $j = 1, 2, 3$.

Bose-Einstein statistics: Here Ψ must be symmetric under interchange of the two particles. From the nine wave functions listed above one can construct only six symmetric ones. A complete (unnormalized) set of distinct wave functions are then the three functions of the form

$$\psi_i(Q_A)\psi_i(Q_B)$$

where $i = 1, 2, 3$ and the three functions of the form

$$\psi_i(Q_A)\psi_j(Q_B) + \psi_i(Q_B)\psi_j(Q_A)$$

where $j > i$; again $i = 1, 2, 3$ and $j = 1, 2, 3$.

Fermi-Dirac statistics: Here Ψ must be antisymmetric under interchange of the two particles. From the nine wave functions listed under the MB case one can construct only three antisymmetric ones. A complete (unnormalized) set of distinct wave functions are then the three functions of the form

$$\psi_i(Q_A)\psi_j(Q_B) - \psi_i(Q_B)\psi_j(Q_A)$$

where $j > i$; again $i = 1, 2, 3$ and $j = 1, 2, 3$.

9 · 2 *Formulation of the statistical problem*

We consider a gas of identical particles in a volume V in equilibrium at the temperature T. We shall use the following notation:

Label the possible quantum states of a single particle by r (or s).
Denote the energy of a particle in state r by ϵ_r.
Denote the number of particles in state r by n_r.
Label the possible quantum states of the whole gas by R.

The assumption of negligibly small interaction between the particles allows us to write for the total energy of the gas, when it is in some state R where there are n_1 particles in state $r = 1$, n_2 particles in state $r = 2$, etc., the additive expression

$$E_R = n_1\epsilon_1 + n_2\epsilon_2 + n_3\epsilon_3 + \cdots = \sum_r n_r\epsilon_r \tag{9·2·1}$$

where the sum extends over all the possible states r of a particle. Furthermore, if the total number of particles in the gas is known to be N, one must have

$$\sum_r n_r = N \tag{9·2·2}$$

In order to calculate the thermodynamic functions of the gas (e.g., its entropy), it is necessary to calculate its partition function

$$Z = \sum_R e^{-\beta E_R} = \sum_R e^{-\beta(n_1\epsilon_1 + n_2\epsilon_2 + \cdots)} \tag{9·2·3}$$

Here the sum is over all the possible states R of the whole gas, i.e., essentially over all the various possible values of the numbers $n_1, n_2, n_3, \ldots$.

Since $\exp[-\beta(n_1\epsilon_1 + n_2\epsilon_2 + \cdots)]$ is the relative probability of finding the gas in a particular state where there are n_1 particles in state 1, n_2 particles in state 2, etc., one can write for the mean number of particles in a state s

$$\bar{n}_s = \frac{\sum_R n_s e^{-\beta(n_1\epsilon_1 + n_2\epsilon_2 + \cdots)}}{\sum_R e^{-\beta(n_1\epsilon_1 + n_2\epsilon_2 + \cdots)}} \tag{9·2·4}$$

Hence $$\bar{n}_s = \frac{1}{Z}\sum_R \left(-\frac{1}{\beta}\frac{\partial}{\partial\epsilon_s}\right) e^{-\beta(n_1\epsilon_1 + n_2\epsilon_2 + \cdots)} = -\frac{1}{\beta Z}\frac{\partial Z}{\partial \epsilon_s}$$

or

▶ $$\bar{n}_s = -\frac{1}{\beta}\frac{\partial \ln Z}{\partial \epsilon_s} \tag{9·2·5}$$

Thus the mean number of particles in a given single-particle state s can also be expressed in terms of the partition function Z.

Calculation of the dispersion One can similarly write down an expression for the dispersion of the number of particles in state s. One can use the general relation

$$\overline{(\Delta n_s)^2} = \overline{(n_s - \bar{n}_s)^2} = \overline{n_s^2} - \bar{n}_s^2 \tag{9·2·6}$$

But for $\overline{n_s^2}$ one can write, by definition,

$$\overline{n_s^2} = \frac{\sum_R n_s^2 e^{-\beta(n_1\epsilon_1 + n_2\epsilon_2 + \cdots)}}{\sum_R e^{-\beta(n_1\epsilon_1 + n_2\epsilon_2 + \cdots)}} \tag{9·2·7}$$

Hence $$\overline{n_s^2} = \frac{1}{Z}\sum \left(-\frac{1}{\beta}\frac{\partial}{\partial\epsilon_s}\right)\left(-\frac{1}{\beta}\frac{\partial}{\partial\epsilon_s}\right) e^{-\beta(n_1\epsilon_1 + n_2\epsilon_2 + \cdots)} = \frac{1}{Z}\left(-\frac{1}{\beta}\frac{\partial}{\partial\epsilon_s}\right)^2 Z$$

or
$$\overline{n_s^2} = \frac{1}{\beta^2 Z} \frac{\partial^2 Z}{\partial \epsilon_s^2} \tag{9·2·8}$$

This can be put into a more convenient form involving $\bar{n}_s$ in (9·2·5). Thus

$$\overline{n_s^2} = \frac{1}{\beta^2}\left[\frac{\partial}{\partial \epsilon_s}\left(\frac{1}{Z}\frac{\partial Z}{\partial \epsilon_s}\right) + \frac{1}{Z^2}\left(\frac{\partial Z}{\partial \epsilon_s}\right)^2\right] = \frac{1}{\beta^2}\left[\frac{\partial}{\partial \epsilon_s}\left(\frac{\partial \ln Z}{\partial \epsilon_s}\right) + \beta^2 \bar{n}_s^2\right]$$

Thus (9·2·6) becomes

$$\overline{(\Delta n_s)^2} = \frac{1}{\beta^2} \frac{\partial^2 \ln Z}{\partial \epsilon_s^2} \tag{9·2·9}$$

or by (9·2·5),

▶
$$\overline{(\Delta n_s)^2} = -\frac{1}{\beta} \frac{\partial \bar{n}_s}{\partial \epsilon_s} \tag{9·2·10}$$

Calculation of all physical quantities of interest thus requires simply the evaluation of the partition function (9·2·3). Let us now be very specific about what we mean by the sum over all possible states R of the gas. In accordance with the discussion of Sec. 9·1 we mean the following:

Maxwell-Boltzmann statistics: Here one must sum over all possible numbers of particles in each state, i.e., over all values

$$n_r = 0, 1, 2, 3, \ldots \quad \text{for each } r \tag{9·2·11}$$

subject to the restriction (9·2·2) of a fixed total number of particles

$$\sum_r n_r = N \tag{9·2·12}$$

But the particles have also to be considered as *distinguishable.* Thus any permutation of two particles in different states must be counted as a distinct state of the whole gas even though the numbers $\{n_1, n_2, n_3, \ldots\}$ are left unchanged. This is so because it is not enough to specify how many particles are in each single-particle state, but it is necessary to specify *which* particular particle is in *which* state.

Bose-Einstein and photon statistics: Here the particles are to be considered as indistinguishable, so that mere specification of the numbers $\{n_1, n_2, n_3, \ldots\}$ is enough to specify the state of the gas. Thus it is necessary only to sum over all possible numbers of particles in each single-particle state, i.e., over all possible values

$$n_r = 0, 1, 2, 3, \ldots \quad \text{for each } r \tag{9·2·13}$$

If the total number of particles is fixed, these numbers must satisfy only the restriction (9·2·2)

$$\sum_r n_r = N \tag{9·2·14}$$

A simpler special case is that where there is *no* requirement fixing the total number of particles. This is the case, for example, when one considers the particles to be photons enclosed in a container of volume V, since the photons can be readily absorbed and emitted by the walls. There is then no equation of constraint (9·2·14) to be satisfied, and one obtains the special case of "photon statistics."

Fermi-Dirac statistics: Here the particles are again to be considered as indistinguishable, so that mere specification of the numbers $\{n_1, n_2, n_3, \ldots\}$ is enough to specify the state of the gas. Thus it is necessary only to sum over all possible numbers of particles in each single-particle state, remembering that there can be no more than one particle in any one such state; i.e., one has to sum over the two possible values

$$n_r = 0, 1 \qquad \text{for each } r \tag{9·2·15}$$

If the total number of particles is fixed, these numbers must satisfy only the restriction (9·2·2)

$$\sum_r n_r = N \tag{9·2·16}$$

9 · 3 *The quantum distribution functions*

Before turning to a systematic calculation of the partition functions in the various cases of interest, we shall devote this section to a simple discussion of the essential features of the quantum theory of ideal gases. We begin by noting that there is a profound difference between gases obeying BE statistics and those obeying FD statistics. This difference becomes most striking in the limit as $T \to 0$, when the gas as a whole is in its state of lowest energy.

Consider a gas consisting of a fixed number N of particles, and suppose that the state of lowest energy of a single particle has an energy ϵ_1. (This corresponds to a state where the particle has essentially zero momentum.) In the case of BE statistics, where there is no restriction on how many particles can be placed in any one single-particle state, the lowest energy of the whole gas is then obtained if *all* the N particles of the gas are put into their lowest-lying state of energy ϵ_1 (e.g., all particles are in their state of zero momentum). This then describes the situation at $T = 0$.

But in the case of FD statistics one cannot put more than one particle into any one single-particle state. If one is interested in obtaining the lowest energy of the whole gas, one is then forced to populate single-particle states of increasing energy; i.e., one can start from the state of lowest energy ϵ_1 and must then fill up single-particle states of successively higher energies, one at a time, until all the N particles have been accommodated. The net result is that, even when $T = 0$ and the gas as a whole is in its state of lowest possible energy, there are particles in the gas which have a very high energy compared to ϵ_1; similarly, the gas as a whole has an energy considerably greater than the energy $N\epsilon_1$ which it would have if the particles obeyed BE statistics. The Pauli exclusion principle thus has very pronounced consequences.

Let us now consider the case of arbitrary temperature T and calculate, for the several cases of interest, the mean number of particles $\bar{n}_s$ in a particular state s. We can proceed directly from the expression (9·2·4) for this mean value, i.e.,

$$\bar{n}_s = \frac{\sum\limits_{n_1,n_2,\ldots} n_s\, e^{-\beta(n_1\epsilon_1+n_2\epsilon_2+\cdots+n_s\epsilon_s+\cdots)}}{\sum\limits_{n_1,n_2,\ldots} e^{-\beta(n_1\epsilon_1+n_2\epsilon_2+\cdots+n_s\epsilon_s+\cdots)}} \tag{9·3·1}$$

By summing first over all possible values of n_s, using the multiplicative property of the exponential function, and rearranging the order of summation, (9·3·1) can also be written in the form

$$\bar{n}_s = \frac{\sum\limits_{n_s} n_s\, e^{-\beta n_s\epsilon_s} \sum\limits_{n_1,n_2,\ldots}^{(s)} e^{-\beta(n_1\epsilon_1+n_2\epsilon_2+\cdots)}}{\sum\limits_{n_s} e^{-\beta n_s\epsilon_s} \sum\limits_{n_1,n_2,\ldots}^{(s)} e^{-\beta(n_1\epsilon_1+n_2\epsilon_2+\cdots)}} \tag{9·3·2}$$

Here the last sums in the numerator and denominator omit from consideration the particular state s (this is indicated by the superscript s on the summation symbol).

Photon statistics This is the case of BE statistics with an unspecified total number of particles. In accordance with the discussion of the last section, the numbers $n_1, n_2, \ldots$ assume here all values $n_r = 0, 1, 2, 3, \ldots$ for each r, without any further restriction. The sums $\Sigma^{(s)}$ in the numerator and denominator of (9·3·2) are then identical and cancel. Thus one is left simply with

$$\bar{n}_s = \frac{\sum\limits_{n_s} n_s\, e^{-\beta n_s\epsilon_s}}{\sum\limits_{n_s} e^{-\beta n_s\epsilon_s}} \tag{9·3·3}$$

The rest of the calculation is straightforward. The result (9·3·3) becomes

$$\bar{n}_s = \frac{(-1/\beta)(\partial/\partial\epsilon_s)\Sigma\, e^{-\beta n_s\epsilon_s}}{\Sigma\, e^{-\beta n_s\epsilon_s}} = -\frac{1}{\beta}\frac{\partial}{\partial\epsilon_s}\ln\,(\Sigma\, e^{-\beta n_s\epsilon_s}) \tag{9·3·4}$$

But the last sum is just an infinite geometric series which can be summed. Thus

$$\sum_{n_s=0}^{\infty} e^{-\beta n_s\epsilon_s} = 1 + e^{-\beta\epsilon_s} + e^{-2\beta\epsilon_s} + \cdots = \frac{1}{1-e^{-\beta\epsilon_s}}$$

Hence (9·3·4) gives

$$\bar{n}_s = \frac{1}{\beta}\frac{\partial}{\partial\epsilon_s}\ln\,(1-e^{-\beta\epsilon_s}) = \frac{e^{-\beta\epsilon_s}}{1-e^{-\beta\epsilon_s}}$$

or

$$\bar{n}_s = \frac{1}{e^{\beta\epsilon_s} - 1} \tag{9·3·5}$$

This is called the "Planck distribution."

Fermi-Dirac statistics Let us now turn to cases where the total number N of particles is fixed. This restriction makes the calculation slightly more complicated. We discuss first the case of FD statistics, since it is somewhat simpler. In accordance with the discussion of Sec. 9·2, the sums in (9·3·2) range here over all values of the numbers $n_1, n_2, \ldots$ such that $n_r = 0$ and 1 for each r; but these numbers must always satisfy the restriction

$$\sum_r n_r = N \tag{9·3·6}$$

This restriction implies, for example, that if one particle is in state s, the sum $\Sigma^{(s)}$ in (9·3·2) extends only over the remaining $(N - 1)$ particles which can be put into the states other than s. Let us then introduce for the sum $\Sigma^{(s)}$, extended over all states except s, the convenient abbreviation

$$Z_s(N) \equiv \sum_{n_1,n_2,\ldots}^{(s)} e^{-\beta(n_1\epsilon_1+n_2\epsilon_2+\cdots)} \tag{9·3·7}$$

if N particles are to be distributed over these remaining states, i.e., if

$$\sum_r^{(s)} n_r = N \qquad \text{(state } s \text{ omitted from this sum)}$$

By performing explicitly the sum over $n_s = 0$ and 1, the expression (9·3·2) becomes then

$$\bar{n}_s = \frac{0 + e^{-\beta\epsilon_s}Z_s(N-1)}{Z_s(N) + e^{-\beta\epsilon_s}Z_s(N-1)} \tag{9·3·8}$$

or

$$\bar{n}_s = \frac{1}{[Z_s(N)/Z_s(N-1)]\, e^{\beta\epsilon_s} + 1} \tag{9·3·9}$$

This can be simplified by relating $Z_s(N-1)$ to $Z_s(N)$. Thus one can write quite generally, if $\Delta N \ll N$,

$$\ln Z_s(N - \Delta N) = \ln Z_s(N) - \frac{\partial \ln Z_s}{\partial N}\,\Delta N = \ln Z_s(N) - \alpha_s\,\Delta N$$

or

$$Z_s(N - \Delta N) = Z_s(N)\, e^{-\alpha_s\Delta N} \tag{9·3·10}$$

where

$$\alpha_s \equiv \frac{\partial \ln Z_s}{\partial N} \tag{9·3·11}$$

But since $Z_s(N)$ is a sum over very many states, one expects that the variation of its logarithm with the total number of particles N should be very insensitive as to which particular state s has been omitted from the sum (9·3·7). Let us then introduce the approximation (whose validity can be verified later) that α_s is independent of s, so that one can write simply

$$\alpha_s = \alpha \tag{9·3·12}$$

for all s. The derivative (9·3·11) can then also be expressed approximately in terms of the derivative of the full partition function $Z(N)$ (over *all* states) which occurs in the denominator of (9·3·1) or (9·3·8); i.e.,

$$\alpha = \frac{\partial \ln Z}{\partial N} \tag{9·3·13}$$

Using (9·3·10) with $\Delta N = 1$ and the approximation (9·3·12), the result (9·3·9) becomes then

$$\bar{n}_s = \frac{1}{e^{\alpha+\beta\epsilon_s} + 1} \tag{9·3·14}$$

This is called the "Fermi-Dirac distribution."

The parameter α in (9·3·14) can be determined from the condition (9·3·6), which demands that the mean values must satisfy the relation

$$\sum_r \bar{n}_r = N \tag{9·3·15}$$

or

$$\sum_r \frac{1}{e^{\alpha+\beta\epsilon_r} + 1} = N \tag{9·3·16}$$

Note that since the free energy $F = -kT \ln Z$, the relation (9·3·13) is equivalent to

$$\alpha = -\frac{1}{kT}\frac{\partial F}{\partial N} = -\frac{\mu}{kT} = -\beta\mu \tag{9·3·17}$$

where μ is the chemical potential per particle defined in (8·7·10). The result (9·3·14) can thus also be written in the form

$$\bar{n}_s = \frac{1}{e^{\beta(\epsilon_s-\mu)} + 1} \tag{9·3·18}$$

Note that $\bar{n}_s \to 0$ if ϵ_s becomes large enough. On the other hand, since the denominator in (9·3·14) can never become less than unity no matter how small ϵ_s becomes, it follows that $\bar{n}_s \leq 1$. Hence

$$0 \leq \bar{n}_s \leq 1$$

a relation which reflects properly the requirement imposed by the Pauli exclusion principle.

Remark concerning the validity of the approximation The partition function given by the denominator of (9·3·1) or (9·3·8) is related to $Z_s(N)$ by

$$Z(N) = Z_s(N) + e^{-\beta\epsilon_s}Z_s(N-1) = Z_s(N)(1 + e^{-\alpha-\beta\epsilon_s})$$

or

$$\ln Z = \ln Z_s + \ln (1 + e^{-\alpha-\beta\epsilon_s})$$

where we have used (9·3·10) and (9·3·12). Hence

$$\frac{\partial \ln Z}{\partial N} = \frac{\partial \ln Z_s}{\partial N} - \frac{e^{-\alpha-\beta\epsilon_s}}{1 + e^{-\alpha-\beta\epsilon_s}}\frac{\partial \alpha}{\partial N}$$

or

$$\alpha = \alpha_s - \bar{n}_s \frac{\partial \alpha}{\partial N}$$

The assumption (9·3·12) is then satisfied if

$$\frac{\partial \alpha}{\partial N} \bar{n}_s \ll \alpha \tag{9·3·19}$$

or, for FD statistics where $\bar{n}_s < 1$, if $\partial \alpha / \partial N \ll \alpha$; i.e., if the number of particles N is large enough so that the chemical potential does not change appreciably upon addition of one more particle to the system.

Bose-Einstein statistics The discussion here is very similar to that just given in the case of FD statistics. Here the sums in (9·3·2) range over all values of the numbers $n_1, n_2, \ldots$ such that $n_r = 0, 1, 2, 3, \ldots$ for each r; but the situation differs from the case of photons because these numbers must always satisfy the restriction (9·3·6) of a fixed total number N of particles. Performing explicitly the sum over n_s, Eq. (9·3·2) then becomes

$$\bar{n}_s = \frac{0 + e^{-\beta\epsilon_s} Z_s(N-1) + 2e^{-2\beta\epsilon_s} Z_s(N-2) + \cdots}{Z_s(N) + e^{-\beta\epsilon_s} Z_s(N-1) + e^{-2\beta\epsilon_s} Z_s(N-2) + \cdots} \tag{9·3·20}$$

where $Z_s(N)$ is defined as in (9·3·7). Using (9·3·10) and the approximation (9·3·12), the result (9·3·20) becomes

$$\bar{n}_s = \frac{Z_s(N)[0 + e^{-\beta\epsilon_s} e^{-\alpha} + 2e^{-2\beta\epsilon_s} e^{-2\alpha} + \cdots]}{Z_s(N)[1 + e^{-\beta\epsilon_s} e^{-\alpha} + e^{-2\beta\epsilon_s} e^{-2\alpha} + \cdots]}$$

or

$$\bar{n}_s = \frac{\sum_s n_s e^{-n_s(\alpha + \beta\epsilon_s)}}{\sum_s e^{-n_s(\alpha + \beta\epsilon_s)}} \tag{9·3·21}$$

But this simple expression is similar to (9·3·3), except that $\beta\epsilon_s$ in that expression is replaced by $(\alpha + \beta\epsilon_s)$. The remainder of the calculation is then identical to that leading to (9·3·4) and yields therefore

▶
$$\bar{n}_s = \frac{1}{e^{\alpha + \beta\epsilon_s} - 1} \tag{9·3·22}$$

This is called the "Bose-Einstein distribution." Note that $\bar{n}_s$ can become very large in this case. The parameter α can again be determined by the condition (9·3·15), i.e., by the relation

$$\sum_r \frac{1}{e^{\alpha + \beta\epsilon_r} - 1} = N \tag{9·3·23}$$

It is again related to the chemical potential μ by the relation $\alpha = -\beta\mu$ of (9·3·17), so that (9·3·22) can also be written in the form

$$\bar{n}_s = \frac{1}{e^{\beta(\epsilon_s - \mu)} - 1} \tag{9·3·24}$$

In the case of photons, the sums are to be performed without any restriction as to the total number N of particles, so that $Z(N)$ [or $Z_s(N)$] does not depend on N. Thus $\alpha = 0$ by (9·3·13), and the Bose-Einstein distribution (9·3·22) reduces properly to the special case of the Planck distribution (9·3·5).

Remark In the case of photons (or other particles whose total number is not fixed) ϵ_s denotes the unambiguously defined energy necessary to create one particle in state s (e.g., $\epsilon_s = \hbar\omega_s$ if ω_s in the angular frequency of the photon). Suppose that the energy scale is shifted by an arbitrary constant η, so that the ground state of the photon gas (the situation when $n_1 = n_2 = n_3 = \cdots = 0$) has energy η instead of zero energy. Then the energy of the gas in a particular state becomes $E = \Sigma n_r \epsilon_r + \eta$. But the constant η cancels in (9·3·1), so that the Planck distribution (9·3·5) is properly unaffected.

In the case of ordinary gases with a fixed number N of particles, ϵ_s denotes the energy level of a particle in state s. Suppose that the energy scale is shifted by an arbitrary constant. Then all single-particle energy levels are shifted by the same constant η' and the energy of all states of the whole gas is shifted by the constant $\eta \equiv N\eta'$. Once again this additive constant cancels in (9·3·1); thus the FD and BE distributions (9·3·18) and (9·3·24) are properly unaffected (the chemical potential μ being also shifted by η').

This completes the discussion of the essential features of the quantum statistics of ideal gases. It is, however, worth looking at the various cases in greater detail with the aim of calculating not only the distribution functions $\bar{n}_s$, but also thermodynamic functions (e.g., the entropy) and the magnitude of fluctuations in the number of particles in a given state. We shall, therefore, devote the next few sections to calculate systematically the partition function Z for each case of interest, i.e., to find an explicit expression for Z in terms of the energy levels of a *single* particle. The remainder of the calculation will then involve only the simple problem of finding explicitly the energy levels of a single particle.

9·4 *Maxwell-Boltzmann statistics*

For purposes of comparison, it is instructive to deal first with the strictly classical case of Maxwell-Boltzmann statistics. Here the partition function is

$$Z = \sum_R e^{-\beta(n_1\epsilon_1 + n_2\epsilon_2 + \cdots)} \tag{9·4·1}$$

where the sum is to be evaluated, as described at the end of Sec. 9·2, by summing over all states R of the gas, i.e., by summing over all possible values of the numbers n_r and taking into account the *distinguishability* of the particles. If there is a total of N molecules, there are, for *given* values of $\{n_1, n_2, \ldots\}$,

$$\frac{N!}{n_1!n_2!\cdots}$$

possible ways in which the particles can be put into the given single-particle states, so that there are n_1 particles in state 1, n_2 particles in state 2, etc. By virtue of the distinguishability of the particles, each of these possible arrangements corresponds then to a *distinct* state for the whole gas. Hence (9·4·1) can be written explicitly as

$$Z = \sum_{n_1, n_2, \ldots} \frac{N!}{n_1! n_2! \cdots} e^{-\beta(n_1\epsilon_1 + n_2\epsilon_2 + \cdots)} \tag{9·4·2}$$

where one sums over all values $n_r = 0, 1, 2, \ldots$ for each r, subject to the restriction

$$\sum_r n_r = N \tag{9·4·3}$$

But (9·4·2) can be written

$$Z = \sum_{n_1, n_2, \ldots} \frac{N!}{n_1! n_2! \cdots} (e^{-\beta\epsilon_1})^{n_1}(e^{-\beta\epsilon_2})^{n_2} \cdots$$

which, by virtue of (9·4·3), is just the result of expanding a polynomial. Thus

$$Z = (e^{-\beta\epsilon_1} + e^{-\beta\epsilon_2} + \cdots)^N$$

or

▶ $$\ln Z = N \ln \left(\sum_r e^{-\beta\epsilon_r} \right) \tag{9·4·4}$$

where the argument of the logarithm is simply the partition function for a single particle.

Alternative method One may equally well write the partition function of the whole gas in the form

$$Z = \sum_{r_1, r_2, \ldots} \exp\left[-\beta(\epsilon_{r_1} + \epsilon_{r_2} + \cdots + \epsilon_{r_N})\right] \tag{9·4·5}$$

where the summation is now over all the possible *states* of *each* individual particle. Clearly, this way of summing considers particles as distinguishable and produces distinct terms in the sum when [particle 1 is in state r_1 and particle 2 is in state r_2] and when [particle 2 is in state r_1 and particle 1 is in state r_2]. Now (9·4·5) immediately factors to give

$$Z = \sum_{r_1, r_2, \ldots} \exp(-\beta\epsilon_{r_1}) \exp(-\beta\epsilon_{r_2}) \cdots$$
$$= \left[\sum_{r_1} \exp(-\beta\epsilon_{r_1})\right]\left[\sum_{r_2} \exp(-\beta\epsilon_{r_2})\right] \cdots$$

or

$$Z = \left[\sum_{r_1} \exp(-\beta\epsilon_{r_1})\right]^N \tag{9·4·6}$$

Thus one regains the result (9·4·4).

By applying (9·2·5) to the partition function (9·4·4), one obtains, by differentiating with respect to the one term involving ϵ_s,

$$\bar{n}_s = -\frac{1}{\beta}\frac{\partial \ln Z}{\partial \epsilon_s} = -\frac{1}{\beta} N \frac{-\beta\, e^{-\beta\epsilon_s}}{\sum_r e^{-\beta\epsilon_r}}$$

or

$$\bar{n}_s = N \frac{e^{-\beta\epsilon_s}}{\sum_r e^{-\beta\epsilon_r}} \tag{9·4·7}$$

This is called the "Maxwell-Boltzmann distribution." It is, of course, just the result we encountered previously in a classical approach where we applied the canonical distribution to a single particle.

Calculation of the dispersion By combining the general result (9·2·10) with (9·4·7) one obtains

$$\overline{(\Delta n_s)^2} = -\frac{1}{\beta}\frac{\partial \bar{n}_s}{\partial \epsilon_s} = -\frac{N}{\beta}\left[\frac{-\beta e^{-\beta\epsilon_s}}{\Sigma e^{-\beta\epsilon_r}} - \frac{-\beta e^{-\beta\epsilon_s} e^{-\beta\epsilon_s}}{(\Sigma e^{-\beta\epsilon_r})^2}\right]$$
$$= \bar{n}_s - \frac{\bar{n}_s{}^2}{N}$$

or

$$\overline{(\Delta n_s)^2} = \bar{n}_s\left(1 - \frac{\bar{n}_s}{N}\right) \approx \bar{n}_s \tag{9·4·8}$$

This last step follows since $\bar{n}_s \ll N$ unless the temperature T is exceedingly low. The relative dispersion is then

$$\frac{\overline{(\Delta n_s)^2}}{\bar{n}_s{}^2} = \frac{1}{\bar{n}_s} \tag{9·4·9}$$

9·5 *Photon statistics*

The partition function is given by

$$Z = \sum_R e^{-\beta(n_1\epsilon_1 + n_2\epsilon_2 + \cdots)} \tag{9·5·1}$$

where, in accordance with the discussion at the end of Sec. 9·2, the summation is simply over all values $n_r = 0, 1, 2, 3, \ldots$ for each r, *without* further restriction. Thus (9·5·1) becomes explicitly

$$Z = \sum_{n_1, n_2, \ldots} e^{-\beta n_1\epsilon_1} e^{-\beta n_2\epsilon_2} e^{-\beta n_3\epsilon_3} \cdots$$

or

$$Z = \left(\sum_{n_1=0}^{\infty} e^{-\beta n_1\epsilon_1}\right)\left(\sum_{n_2=0}^{\infty} e^{-\beta n_2\epsilon_2}\right)\left(\sum_{n_3=0}^{\infty} e^{-\beta n_3\epsilon_3}\right) \cdots \tag{9·5·2}$$

But each sum is just an infinite geometric series whose first term is 1 and where the ratio between successive terms is $e^{-\beta\epsilon_r}$. It can thus be immediately summed. Hence (9·5·2) becomes

$$Z = \left(\frac{1}{1 - e^{-\beta\epsilon_1}}\right)\left(\frac{1}{1 - e^{-\beta\epsilon_2}}\right)\left(\frac{1}{1 - e^{-\beta\epsilon_3}}\right) \cdots$$

or

$$\ln Z = -\sum_r \ln\left(1 - e^{-\beta\epsilon_r}\right) \tag{9·5·3}$$

By applying (9·2·5) to (9·5·3), differentiation with respect to the one term involving ϵ_s yields

$$\bar{n}_s = -\frac{1}{\beta}\frac{\partial \ln Z}{\partial \epsilon_s} = \frac{e^{-\beta\epsilon_s}}{1 - e^{-\beta\epsilon_s}}$$

or

$$\bar{n}_s = \frac{1}{e^{\beta\epsilon_s} - 1} \tag{9·5·4}$$

Thus we regain the Planck distribution previously derived in (9·3·5).

Calculation of the dispersion The dispersion in n_s can be calculated by applying (9·2·10) to (9·5·4). Thus

$$\overline{(\Delta n_s)^2} = -\frac{1}{\beta}\frac{\partial \bar{n}_s}{\partial \epsilon_s} = \frac{e^{\beta\epsilon_s}}{(e^{\beta\epsilon_s} - 1)^2}$$

One can use (9·5·4) to write this in terms of $\bar{n}_s$. Thus

$$\overline{(\Delta n_s)^2} = \frac{(e^{\beta\epsilon_s} - 1) + 1}{(e^{\beta\epsilon_2} - 1)^2} = \bar{n}_s + \bar{n}_s^2$$

Hence

$$\overline{(\Delta n_s)^2} = \bar{n}_s(1 + \bar{n}_s) \tag{9·5·5}$$

or

$$\frac{\overline{(\Delta n_s)^2}}{\bar{n}_s^2} = \frac{1}{\bar{n}_s} + 1 \tag{9·5·6}$$

Note that this dispersion is greater than in the MB case of Eq. (9·4·8). In dealing with photons, therefore, the relative dispersion does *not* become arbitrarily small even if $\bar{n}_s \gg 1$.

9·6 *Bose-Einstein statistics*

The partition function is again given by

$$Z = \sum_R e^{-\beta(n_1\epsilon_1 + n_2\epsilon_2 + \cdots)} \tag{9·6·1}$$

where, in accordance with the discussion of Sec. 9·2, the summation is over all values

$$n_r = 0, 1, 2, \ldots \qquad \text{for each } r \tag{9·6·2}$$

Unlike the photon case, however, these numbers must now satisfy the restrictive condition

$$\sum_r n_r = N \tag{9·6·3}$$

where N is the total number of particles in the gas. If it were not for the equation of constraint (9·6·3), the sum (9·6·1) could be easily evaluated just as in the last section. But the condition (9·6·3) introduces a complication.

There are various ways of handling the problem presented by the condition (9·6·3). Let us use an approximation method similar to that described in Sec. 6·8. As a result of (9·6·3), Z depends on the total number N of particles in the system. If the number of particles were N' instead of N, the partition function would have some other value $Z(N')$. Indeed, since there are so many terms in the sum (9·6·1), $Z(N')$ is a very rapidly increasing function of N'. But, by virtue of (9·6·3), we are interested only in the value of Z for $N' = N$. We can, however, exploit the rapidly increasing property of $Z(N')$ by noting that multiplication by the rapidly decreasing function $e^{-\alpha N'}$ produces a function $Z(N')e^{-\alpha N'}$ with a very sharp maximum which can be made to occur at the value $N' = N$ by proper choice of the positive parameter α. A sum of this function over *all* possible numbers N' thus selects only those terms of interest near $N' = N$, i.e.,

$$\sum_{N'} Z(N')\, e^{-\alpha N'} = Z(N)\, e^{-\alpha N}\, \Delta^* N' \tag{9·6·4}$$

where the right side is just the maximum value of the summand multiplied by the width $\Delta^* N'$ of its maximum (where $\Delta^* N' \ll N$).

Let us introduce the abbreviation

$$\mathcal{Z} \equiv \sum_{N'} Z(N') e^{-\alpha N'} \tag{9·6·5}$$

Taking the logarithm of (9·6·4) one then obtains to an *excellent* approximation

$$\ln Z(N) = \alpha N + \ln \mathcal{Z} \tag{9·6·6}$$

where we have neglected the term $\ln (\Delta^* N')$ which is utterly negligible compared to the other terms which are of order N. Here the sum (9·6·5) is easily performed, since it extends over all possible numbers *without* any restriction. (The quantity $\mathcal{Z}$ is called a "grand partition function.")

Let us evaluate $\mathcal{Z}$. By (9·6·1) this becomes

$$\mathcal{Z} = \sum_R e^{-\beta(n_1\epsilon_1 + n_2\epsilon_2 + \cdots)}\, e^{-\alpha(n_1 + n_2 + \cdots)} \tag{9·6·7}$$

where the sum is over all possible numbers (9·6·2) without restriction. By regrouping terms one obtains

$$\begin{aligned}\mathcal{Z} &= \sum_{n_1, n_2, \ldots} e^{-(\alpha+\beta\epsilon_1)n_1 - (\alpha+\beta\epsilon_2)n_2 - \cdots} \\ &= \left(\sum_{n_1=0}^{\infty} e^{-(\alpha+\beta\epsilon_1)n_1}\right)\left(\sum_{n_2=0}^{\infty} e^{-(\alpha+\beta\epsilon_2)n_2}\right)\cdots\end{aligned}$$

This is just a product of simple geometric series. Hence

$$\mathcal{Z} = \left(\frac{1}{1 - e^{-(\alpha+\beta\epsilon_1)}}\right)\left(\frac{1}{1 - e^{-(\alpha+\beta\epsilon_2)}}\right)\cdots$$

or

$$\ln \mathcal{Z} = -\sum_r \ln (1 - e^{-\alpha-\beta\epsilon_r}) \tag{9·6·8}$$

Equation (9·6·6) then yields

$$\ln Z = \alpha N - \sum_r \ln (1 - e^{-\alpha-\beta\epsilon_r}) \tag{9·6·9}$$

Our argument assumed that the parameter α is to be chosen so that the function $Z(N')e^{-\alpha N'}$ has its maximum for $N' = N$, i.e., so that

$$\frac{\partial}{\partial N'} [\ln Z(N') - \alpha N'] = \frac{\partial \ln Z(N)}{\partial N} - \alpha = 0 \tag{9·6·10}$$

Since this condition involves the particular value $N' = N$, α itself must be a function of N. By virtue of (9·6·6), the condition (9·6·10) is equivalent to

$$\left[\alpha + \left(N + \frac{\partial \ln Z}{\partial \alpha}\right)\frac{\partial \alpha}{\partial N}\right] - \alpha = 0$$

or

$$N + \frac{\partial \ln Z}{\partial \alpha} = \frac{\partial \ln Z}{\partial \alpha} = 0 \tag{9·6·11}$$

Using the expression (9·6·9), the relation (9·6·11) which determines α is then

$$N - \sum_r \frac{e^{-\alpha-\beta\epsilon_r}}{1 - e^{-\alpha-\beta\epsilon_r}} = 0$$

or

$$\sum_r \frac{1}{e^{\alpha+\beta\epsilon_r} - 1} = N \tag{9·6·12}$$

By applying (9·2·5) to (9·6·9) one obtains then

$$\bar{n}_s = -\frac{1}{\beta}\frac{\partial \ln Z}{\partial \epsilon_s} = -\frac{1}{\beta}\left[-\frac{\beta e^{-\alpha-\beta\epsilon_s}}{1 - e^{-\alpha-\beta\epsilon_s}} + \frac{\partial \ln Z}{\partial \alpha}\frac{\partial \alpha}{\partial \epsilon_s}\right]$$

The last term takes into account the fact that α is a function of ϵ_s through the relation (9·6·12). But this term vanishes by virtue of (9·6·11). Hence one has simply

$$\bar{n}_s = \frac{1}{e^{\alpha+\beta\epsilon_s} - 1} \tag{9·6·13}$$

Thus one regains the Bose-Einstein distribution already derived previously in (9·3·22). Note that the condition (9·6·12) which determines α is then equivalent to

$$\sum_r \bar{n}_r = N \tag{9·6·14}$$

the obvious requirement needed to satisfy the conservation of particles (9·6·3).

The chemical potential of the gas is given by

$$\mu = \frac{\partial F}{\partial N} = -kT\frac{\partial \ln Z}{\partial N} = -kT\alpha$$

where we have used (9·6·10). Thus the parameter

$$\alpha = -\beta\mu \qquad (9\cdot 6\cdot 15)$$

is directly related to the chemical potential of the gas. In the case of photons, where there is no restriction on the total number of particles, Z is independent of N; then $\alpha = 0$, and all our relations reduce to those of the preceding section.

Calculation of the dispersion By applying (9·2·10) to (9·6·13) one obtains

$$\overline{(\Delta n_s)^2} = -\frac{1}{\beta}\frac{\partial \bar{n}_s}{\partial \epsilon_s} = \frac{1}{\beta}\frac{e^{\alpha+\beta\epsilon_s}}{(e^{\alpha+\beta\epsilon_s}-1)^2}\left(\frac{\partial\alpha}{\partial\epsilon_s}+\beta\right)$$

But

$$\frac{e^{\alpha+\beta\epsilon_s}}{(e^{\alpha+\beta\epsilon_s}-1)^2} = \frac{(e^{\alpha+\beta\epsilon_s}-1)+1}{(e^{\alpha+\beta\epsilon_s}-1)^2} = \bar{n}_s + \bar{n}_s^2$$

Hence

$$\overline{(\Delta n_s)^2} = \bar{n}_s(1+\bar{n}_s)\left(1+\frac{1}{\beta}\frac{\partial\alpha}{\partial\epsilon_s}\right) \approx \bar{n}_s(1+\bar{n}_s) \qquad (9\cdot 6\cdot 16)$$

and

$$\frac{\overline{(\Delta n_s)^2}}{\bar{n}_s^2} \approx \frac{1}{\bar{n}_s} + 1 \qquad (9\cdot 6\cdot 17)$$

where we have neglected the term $\partial\alpha/\partial\epsilon_s$. This term is usually very small, since α is to be determined by (9·6·11) and (unless the temperature $T = (k\beta)^{-1}$ is so low that only a very few terms in the sum have appreciable magnitude) a small change of *one* energy ϵ_s leaves the sum (and hence α) essentially unchanged.

Note that the relation (9·6·17) is exactly the same as that of (9·5·6) for photons. The relative dispersion is again greater than in the MB case of (9·4·9). Thus the relative dispersion does *not* become arbitrarily small even when $\bar{n}_s \gg 1$.

The correction term in (9·6·16) can, of course, be evaluated explicitly by taking (9·6·12), which determines α, and differentiating it with respect to ϵ_s. Thus

$$-\frac{\beta e^{\alpha+\beta\epsilon_s}}{(e^{\alpha+\beta\epsilon_s}-1)^2} - \sum_r \frac{e^{\alpha+\beta\epsilon_r}}{(e^{\alpha+\beta\epsilon_r}-1)^2}\frac{\partial\alpha}{\partial\epsilon_s} = 0$$

or

$$-\beta(\bar{n}_s+\bar{n}_s^2) - \left[\sum_r(\bar{n}_r+\bar{n}_r^2)\right]\frac{\partial\alpha}{\partial\epsilon_s} = 0$$

Hence

$$\frac{\partial\alpha}{\partial\epsilon_s} = -\beta\frac{\bar{n}_s(1+\bar{n}_s)}{\sum_r \bar{n}_r(1+\bar{n}_r)}$$

and

$$\overline{(\Delta n_s)^2} = \bar{n}_s(1+\bar{n}_s)\left[1-\frac{\bar{n}_s(1+\bar{n}_s)}{\sum_r n_r(1+n_r)}\right] \qquad (9\cdot 6\cdot 18)$$

The dispersion is thus slightly smaller than if the last term in the square brackets were neglected. But if one goes to the limit where $T \to 0$, then all the particles tend to be in the one single-particle state $s = 1$ of lowest energy, so that $\bar{n}_1 \approx N$ while $\bar{n}_s \approx 0$ for all other states. The correction term in (9·6·18) is then important since it predicts properly that the fluctuation in the number of particles in the ground state $s = 1$ goes to zero.

9·7 *Fermi-Dirac statistics*

The discussion here is very similar to that for Bose-Einstein statistics. The problem is again to evaluate the partition function (9·6·1). But, in accord with the discussion of Sec. 9·2, the summation is only over the two values

$$n_r = 0 \text{ and } 1 \qquad \text{for each } r \tag{9·7·1}$$

where these numbers must again satisfy the restrictive condition (9·6·3).

The problem can be handled in a manner identical to that used in the last section for BE statistics. The unrestricted sum $\mathcal{Z}$ of (9·6·5) becomes

$$\begin{aligned}\mathcal{Z} &= \sum_{n_1,n_2,n_3} e^{-\beta(n_1\epsilon_1+n_2\epsilon_2+\cdots)-\alpha(n_1+n_2+\cdots)} \\ &= \left(\sum_{n_1=0}^{1} e^{-(\alpha+\beta\epsilon_1)n_1}\right)\left(\sum_{n_2=0}^{1} e^{-(\alpha+\beta\epsilon_2)n_2}\right)\cdots\end{aligned} \tag{9·7·2}$$

Here each sum consists, by virtue of (9·7·1), of only two terms and is thus trivial. Hence

$$\mathcal{Z} = (1 + e^{-\alpha-\beta\epsilon_1})(1 + e^{-\alpha-\beta\epsilon_2})\cdots$$

or

$$\ln \mathcal{Z} = \sum_r \ln (1 + e^{-\alpha-\beta\epsilon_r}) \tag{9·7·3}$$

Hence (9·6·5) becomes

▶
$$\ln Z = \alpha N + \sum_r \ln (1 + e^{-\alpha-\beta\epsilon_r}) \tag{9·7·4}$$

Except for some important sign changes, this expression is of the same form as (9·6·9) for the BE case. The parameter α is again to be determined from the condition (9·6·11). Thus

$$\frac{\partial \ln Z}{\partial \alpha} = N - \sum_r \frac{e^{-\alpha-\beta\epsilon_r}}{1 + e^{-\alpha-\beta\epsilon_r}} = 0$$

or

▶
$$\sum_r \frac{1}{e^{\alpha+\beta\epsilon_r} + 1} = N \tag{9·7·5}$$

By applying (9·2·5) to (9·7·4), one obtains

$$\bar{n}_s = -\frac{1}{\beta}\frac{\partial \ln Z}{\partial \epsilon_s} = \frac{1}{\beta}\frac{\beta e^{-\alpha-\beta\epsilon_s}}{1 + e^{-\alpha-\beta\epsilon_s}}$$

or

▶
$$\bar{n}_s = \frac{1}{e^{\alpha+\beta\epsilon_s} + 1} \tag{9·7·6}$$

Thus one regains the Fermi-Dirac distribution derived previously in (9·3·14). The relation (9·7·5) which is used to determine α is again just the condition (9·6·14) and the parameter α is again related to the chemical potential μ by the relation (9·6·15).

Calculation of the dispersion By applying (9·2·10) to (9·7·6), one obtains

$$\overline{(\Delta n_s)^2} = -\frac{1}{\beta}\frac{\partial \bar{n}_s}{\partial \epsilon_s} = \frac{1}{\beta}\frac{e^{\alpha+\beta\epsilon_s}}{(e^{\alpha+\beta\epsilon_s}+1)^2}\left(\frac{\partial \alpha}{\partial \epsilon_s}+\beta\right)$$

But
$$\frac{e^{\alpha+\beta\epsilon_s}}{(e^{\alpha+\beta\epsilon_s}+1)^2} = \frac{(e^{\alpha+\beta\epsilon_s}+1)-1}{(e^{\alpha+\beta\epsilon_s}+1)^2} = \bar{n}_s - \bar{n}_s^2$$

Hence
$$\overline{(\Delta n_s)^2} = \bar{n}_s(1-\bar{n}_s)\left(1+\frac{1}{\beta}\frac{\partial \alpha}{\partial \epsilon_s}\right) \approx \bar{n}_s(1-\bar{n}_s) \tag{9·7·7}$$

and

▶
$$\frac{\overline{(\Delta n_s)^2}}{\bar{n}_s^2} \approx \frac{1}{\bar{n}_s} - 1 \tag{9·7·8}$$

Note that the relative dispersion is smaller than in the MB case discussed in Eq. (9·4·9). For example, if $\bar{n}_s \to 1$, the maximum value it can attain in accordance with the exclusion principle, then the dispersion vanishes. There is no fluctuation in $\bar{n}_s$ for states which are completely filled.

9·8 *Quantum statistics in the classical limit*

The preceding sections dealing with the quantum statistics of ideal gases can be summarized by the statements that

▶
$$\bar{n}_r = \frac{1}{e^{\alpha+\beta\epsilon_r} \pm 1} \tag{9·8·1}$$

where the upper sign refers to FD and the lower one to BE statistics. If the gas consists of a fixed number N of particles, the parameter α is to be determined by the condition

▶
$$\sum_r \bar{n}_r = \sum_r \frac{1}{e^{\alpha+\beta\epsilon_r} \pm 1} = N \tag{9·8·2}$$

The partition function Z of the gas is given by

▶
$$\ln Z = \alpha N \pm \sum_r \ln(1 \pm e^{-\alpha-\beta\epsilon_r}) \tag{9·8·3}$$

Let us now investigate the magnitude of α in some limiting cases. Consider first the case of a gas at a given temperature when its concentration is made sufficiently low, i.e., when N is made sufficiently small. The relation (9·8·2) can then only be satisfied if each term in the sum over all states is sufficiently small, i.e., if $\bar{n}_r \ll 1$ or $\exp(\alpha+\beta\epsilon_r) \gg 1$ for all states r. Similarly, consider the case of a gas with some fixed number N of particles when its temperature is made sufficiently large, i.e., when β is made sufficiently small. In the sum of (9·8·2) the terms of appreciable magnitude are those for which $\beta\epsilon_r \ll \alpha$; hence it follows that as $\beta \to 0$, an increasing number of terms with large values of ϵ_r contribute substantially to this sum. To prevent this sum from exceeding N, the parameter α must become large enough so that each term is sufficiently small; i.e., it is again necessary that $\exp(\alpha + \beta\epsilon_r) \gg 1$ or $\bar{n}_r \ll 1$ for all states r. Thus one arrives at the conclusion that, if the concentration is made sufficiently low or if the temperature is made sufficiently high, α must become so large that,

$$\text{for all } r, \qquad e^{\alpha+\beta\epsilon_r} \gg 1 \tag{9·8·4}$$

Equivalently this means that the occupation numbers become then small enough so that,

$$\text{for all } r, \qquad \bar{n}_r \ll 1 \tag{9·8·5}$$

We shall call the limit of sufficiently low concentration or sufficiently high temperature where (9·8·4) or (9·8·5) are satisfied the "classical limit."

In this limit it follows by (9·8·4) that for *both* FD and BE statistics (9·8·1) reduces to

$$\bar{n}_r = e^{-\alpha-\beta\epsilon_r} \tag{9·8·6}$$

By virtue of (9·8·2), the parameter α is then determined by the condition

$$\sum_r e^{-\alpha-\beta\epsilon_r} = e^{-\alpha} \sum_r e^{-\beta\epsilon_r} = N$$

or

$$e^{-\alpha} = N \left(\sum_r e^{-\beta\epsilon_r}\right)^{-1} \tag{9·8·7}$$

Thus

$$\bar{n}_r = N \frac{e^{-\beta\epsilon_r}}{\sum_r e^{-\beta\epsilon_r}} \tag{9·8·8}$$

Hence it follows that in the classical limit of sufficiently low density or sufficiently high temperature the quantum distribution laws, whether FD or BE, reduce to the MB distribution.

The present conclusion is in agreement with our discussion of Sec. 7·4, where we estimated more quantitatively just how low the concentration and how high the temperature must be for classical results to be applicable.

Let us now consider the partition function of (9·8·3). In the classical limit, where (9·8·4) is satisfied, one can expand the logarithm in (9·8·3) to get

$$\ln Z = \alpha N \pm \sum_r (\pm e^{-\alpha-\beta\epsilon_r}) = \alpha N + N$$

But by virtue of (9·8·7)

$$\alpha = -\ln N + \ln \left(\sum_r e^{-\beta\epsilon_r}\right)$$

Hence

$$\ln Z = -N \ln N + N + N \ln \left(\sum_r e^{-\beta\epsilon_r}\right) \tag{9·8·9}$$

Note that this does *not* equal the partition function Z_{MB} computed in Eq. (9·4·4) for MB statistics

$$\ln Z_{\text{MB}} = N \ln \left(\sum_r e^{-\beta\epsilon_r}\right) \tag{9·8·10}$$

Indeed

$$\ln Z = \ln Z_{\text{MB}} - (N \ln N - N)$$

Thus

$$\ln Z = \ln Z_{\text{MB}} - \ln N!$$

or

$$Z = \frac{Z_{\text{MB}}}{N!} \tag{9·8·11}$$

where we have used Stirling's formula since N is large. Here the factor $N!$ corresponds simply to the number of possible permutations of the particles, permutations which are physically meaningless when the particles are identical. It was precisely this factor which we had to introduce in an *ad hoc* fashion in Sec. 7·3 to save ourselves from the nonphysical consequences of the Gibbs paradox. What we have done in this section is to justify the whole discussion of Sec. 7·3 as being appropriate for a gas treated properly by quantum mechanics in the limit of sufficiently low concentration or high temperature. The partition function is automatically correctly evaluated by (9·8·9), there is no Gibbs paradox, and everything is consistent.

A gas in the classical limit where (9·8·6) is satisfied is said to be "nondegenerate." On the other hand, if the concentration and temperature are such that the actual FD or BE distribution (9·8·1) must be used, the gas is said to be "degenerate."

IDEAL GAS IN THE CLASSICAL LIMIT

9·9 *Quantum states of a single particle*

Wave function To complete the discussion of the statistical problem it is necessary to enumerate the possible quantum states s and corresponding energies ϵ_s of a single noninteracting particle. Consider this particle to be nonrelativistic and denote its mass by m, its position vector by $\mathbf{r}$, and its momentum by $\mathbf{p}$. Suppose that the particle is confined within a container of volume V within which the particle is subject to no forces. Neglecting for the time being the effect of the bounding walls, the wave function $\Psi(\mathbf{r},t)$ of the particle is then simply described by a plane wave of the form

$$\Psi = Ae^{i(\boldsymbol{\kappa}\cdot\mathbf{r}-\omega t)} = \psi(\mathbf{r})\, e^{-i\omega t} \tag{9·9·1}$$

which propagates in a direction specified by the "wave vector" $\boldsymbol{\kappa}$ and which has some constant amplitude A. Here the energy ϵ of the particle is related to the frequency ω by

$$\epsilon = \hbar\omega \tag{9·9·2}$$

while its momentum is related to its wave vector $\boldsymbol{\kappa}$ by the de Broglie relation

$$\mathbf{p} = \hbar\boldsymbol{\kappa} \tag{9·9·3}$$

Thus one has

$$\epsilon = \frac{\mathbf{p}^2}{2m} = \frac{\hbar^2\boldsymbol{\kappa}^2}{2m} \tag{9·9·4}$$

The basic justification for these statements is, of course, the fact that Ψ must satisfy the Schrödinger equation

$$i\hbar\frac{\partial\Psi}{\partial t} = \mathcal{H}\Psi \tag{9·9·5}$$

Since one can choose the potential energy to be zero inside the container, the Hamiltonian $\mathcal{H}$ reduces there to the kinetic energy alone; i.e.,

$$\mathcal{H} = \frac{1}{2m} \boldsymbol{p}^2 = \frac{1}{2m} \left(\frac{\hbar}{i} \nabla \right)^2 = -\frac{\hbar^2}{2m} \nabla^2$$

where
$$\nabla^2 = \frac{\partial^2}{\partial x^2} + \frac{\partial^2}{\partial y^2} + \frac{\partial^2}{\partial z^2}$$

Putting
$$\Psi = \psi \, e^{-i\omega t} = \psi \, e^{-(i/\hbar)\epsilon t} \tag{9·9·6}$$

where ψ does not depend on time, (9·9·5) reduces to the time-independent Schrödinger equation

$$\mathcal{H}\psi = \epsilon\psi \tag{9·9·7}$$

or
$$\nabla^2\psi + \frac{2m\epsilon}{\hbar^2} \psi = 0 \tag{9·9·8}$$

Equation (9·9·7) shows that ϵ corresponds to the possible values of $\mathcal{H}$ and is thus the energy of the particle. The wave equation (9·9·8) has solutions of the general form

$$\psi = A \, e^{i(\kappa_x x + \kappa_y y + \kappa_z z)} = A \, e^{i\boldsymbol{\kappa}\cdot\boldsymbol{r}} \tag{9·9·9}$$

where $\boldsymbol{\kappa}$ is the constant "wave vector" with components κ_x, κ_y, κ_z. By substitution of (9·9·9) into (9·9·8) one finds that the latter equation is satisfied if

$$-(\kappa_x{}^2 + \kappa_y{}^2 + \kappa_z{}^2) + \frac{2m\epsilon}{\hbar^2} = 0$$

Thus
$$\epsilon = \frac{\hbar^2\boldsymbol{\kappa}^2}{2m} \tag{9·9·10}$$

and ϵ is only a function of the magnitude $\kappa \equiv |\boldsymbol{\kappa}|$ of $\boldsymbol{\kappa}$. Since

$$\boldsymbol{p}\psi = \frac{\hbar}{i} \nabla\psi = \hbar\boldsymbol{\kappa}\psi$$

one obtains then the relations (9·9·3) and (9·9·4).

Up to now we have considered only the translational degrees of freedom. If the particle also has an intrinsic spin angular momentum, the situation is scarcely more complicated; there is then simply a different function ψ for each possible orientation of the particle spin. For example, if the particle has spin $\frac{1}{2}$ (e.g., if it is an electron), then there are two possible wave functions $\psi_\pm$ corresponding to the two possible values $m = \pm\frac{1}{2}$ of the quantum number specifying the orientation of the particle's spin angular momentum.

Boundary conditions and enumeration of states The wave function ψ must satisfy certain boundary conditions. Accordingly, not all possible values of $\boldsymbol{\kappa}$ (or $\boldsymbol{p}$) are allowed, but only certain discrete values. The corresponding energies of the particle are then also quantized by virtue of (9·9·4).

The boundary conditions can be treated in a very general and simple way in the usual situation where the container enclosing the gas of particles is large enough that its smallest linear dimension L is much greater than the de Broglie wavelength $\lambda = 2\pi/|\boldsymbol{\kappa}|$ of the particle under consideration.* It is then physically clear that the detailed properties of the bounding walls of the container (e.g., their shape or the nature of the material of which they are made) must become of negligible significance in describing the behavior of a particle located well within the container.† To make the argument more precise, let us consider any macroscopic volume element which is large compared to λ and which lies well within the container so that it is everywhere removed from the container walls by distances large compared to λ. The actual wave function anywhere within the container can always be written as a superposition of plane waves (9·9·1) with all possible wave vectors $\boldsymbol{\kappa}$. Hence one can regard the volume element under consideration as being traversed by waves of the form (9·9·1) traveling in all possible directions specified by $\boldsymbol{\kappa}$, and with all possible wavelengths related to the magnitude of $\boldsymbol{\kappa}$. Since the container walls are far away (compared to λ), it does not really matter just how each such wave is ultimately reflected from these walls, or which wave gets reflected how many times before it passes again through the volume element under consideration. The number of waves of each kind traversing this volume element should be quite insensitive to any such details which describe what happens near the container walls and should be substantially unaffected if the shape or properties of these walls are modified. Indeed, it is simplest if one imagines these walls moved out to infinity, i.e., if one effectively eliminates the walls altogether. One can then avoid the necessity of treating the problem of reflections at the walls, a problem which is really immaterial in describing the situation in the volume element under consideration. It does not matter whether a given wave enters this volume element after having been reflected somewhere far away, or after coming in from infinity without ever having been reflected at all.

The foregoing comments show that, for purposes of discussing the properties of a gas anywhere but in the immediate vicinity of the container walls, the exact nature of the boundary conditions imposed on each particle should be unimportant. One can therefore formulate the problem in a way which makes these boundary conditions as simple as possible. Let us therefore choose the basic volume V of gas under consideration to be in the shape of a rectangular parallelepiped with edges parallel to the x, y, z axes and with respective edge lengths equal to L_x, L_y, L_z. Thus $V = L_xL_yL_z$. The simplest boundary conditions to impose are such that a traveling wave of the form (9·9·1) is indeed an exact solution of the problem. This requires that the wave (9·9·1) be able to propagate indefinitely without suffering any reflections. In order to make the boundary conditions consistent with this simple situation, one can neglect

* This condition is ordinarily very well satisfied for essentially all molecules of a gas since a typical order of magnitude, already estimated in Sec. 7·4, is $\lambda \approx 1$ Å for an atom of thermal energy at room temperature.

† Note that the fraction of particles near the surface of the container, i.e., within a distance λ of its walls, is of the order of $\lambda L^2/L^3 = \lambda/L$ and is thus ordinarily utterly negligible for a macroscopic container.

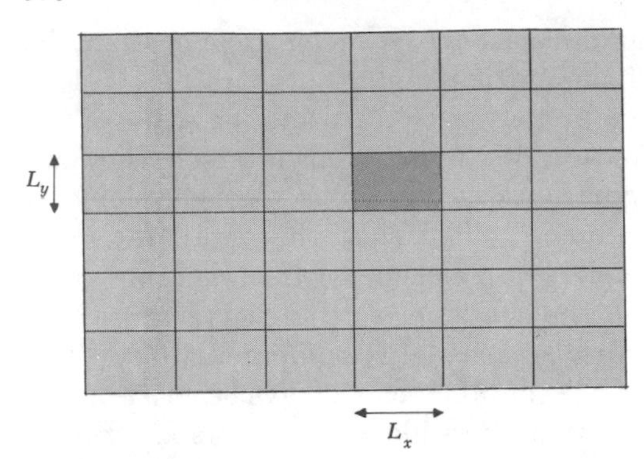

Fig. 9·9·1 The volume under consideration (indicated in darker gray) is here considered embedded in an array of similar volumes extending throughout all space. Wall effects are thus effectively eliminated.

completely the presence of any container walls and can imagine that the volume of gas under consideration is embedded in an infinite set of similar volumes in each of which the physical situation is exactly the same (see Fig. 9·9·1). The wave function must then satisfy the conditions

$$\left.\begin{aligned} \psi(x + L_x, y, z) &= \psi(x,y,z) \\ \psi(x, y + L_y, z) &= \psi(x,y,z) \\ \psi(x, y, z + L_z) &= \psi(x,y,z) \end{aligned}\right\} \qquad (9 \cdot 9 \cdot 11)$$

The requirement that the wave function be the same in any of the parallelepipeds should not affect the physics of interest in the one volume under consideration if its dimensions are large compared to the de Broglie wavelength λ of the particle.

Remark Suppose that the problem were one-dimensional so that a particle moves in the x direction in a container of length L_x. Then one can eliminate the effects of reflections by imagining the container to be bent around in the form of a circle as shown in Fig. 9·9·2. If L_x is very large, the curvature is quite negligible so that the situation inside the container is substantially the same as before. But the advantage is that there are now no container walls to worry about. Hence traveling waves described by (9·9·1) and going around without reflection are perfectly good solutions of the problem. It is only necessary to note that the points x and $x + L_x$ are now coincident; the requirement that the wave function be single-valued implies the condition

$$\psi(x + L_x) = \psi(x) \qquad (9 \cdot 9 \cdot 12)$$

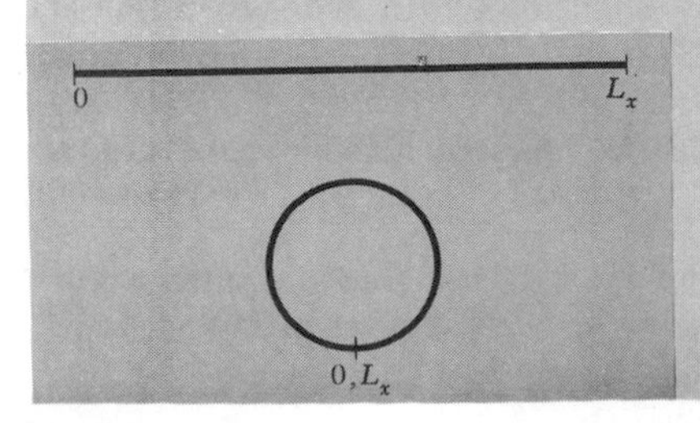

Fig. 9·9·2 A one-dimensional container of length L_x bent into a circle by joining its ends.

This is precisely the analog of (9·9·11) in one dimension. Indeed, one could regard the condition (9·9·11) as resulting from the attempt to eliminate reflections in three dimensions by imagining the original parallelepiped to be bent into a doughnut in four dimensions. (This is, admittedly, difficult to visualize.)

This point of view, which describes the situation in terms of simple traveling waves satisfying the periodic boundary conditions (9·9·11), is very convenient and mathematically exceedingly easy. By virtue of (9·9·1) or (9·9·9)

$$\psi = e^{i\boldsymbol{\kappa}\cdot\mathbf{r}} = e^{i(\kappa_x x + \kappa_y y + \kappa_z z)}$$

To satisfy (9·9·11) one must require that

$$\kappa_x(x + L_x) = \kappa_x x + 2\pi n_x \qquad (n_x \text{ integral})$$

or

$$\left.\begin{aligned} \kappa_x &= \frac{2\pi}{L_x} n_x \\ \text{Similarly,} \qquad \kappa_y &= \frac{2\pi}{L_y} n_y \\ \text{and} \qquad \kappa_z &= \frac{2\pi}{L_z} n_z \end{aligned}\right\} \tag{9·9·13}$$

Here the numbers n_x, n_y, n_z are *any* set of integers—positive, negative, or zero.

The components of $\boldsymbol{\kappa} = \mathbf{p}/\hbar$ are thus quantized in discrete units. Accordingly (9·9·4) yields the possible quantized particle energies

$$\epsilon = \frac{\hbar^2}{2m}(\kappa_x^2 + \kappa_y^2 + \kappa_z^2) = \frac{2\pi^2\hbar^2}{m}\left(\frac{n_x^2}{L_x^2} + \frac{n_y^2}{L_y^2} + \frac{n_z^2}{L_z^2}\right) \tag{9·9·14}$$

Note that for any kind of macroscopic volume where L_x, L_y, L_z are large, the possible values of the wave-vector components given by (9·9·13) are very closely spaced. There are thus very many states of the particle (i.e., very many possible integers n_x) corresponding to any small range $d\kappa_x$ of a wave-vector component. It is easy to do some counting. For given values of κ_y and κ_z, it follows by (9·9·13) that the number Δn_x of possible integers n_x for which κ_x lies in the range between κ_x and $\kappa_x + d\kappa_x$ is equal to

$$\Delta n_x = \frac{L_x}{2\pi} d\kappa_x \tag{9·9·15}$$

The number of translational states $\rho(\boldsymbol{\kappa})\, d^3\boldsymbol{\kappa}$ for which $\boldsymbol{\kappa}$ is such that it lies in the range between $\boldsymbol{\kappa}$ and $\boldsymbol{\kappa} + d\boldsymbol{\kappa}$ (i.e., in the range such that its x component is between κ_x and $\kappa_x + d\kappa_x$, its y component between κ_y and $\kappa_y + d\kappa_y$, and its z component between κ_z and $\kappa_z + d\kappa_z$) is then given by the product of the numbers of possible integers in the three component ranges. Thus

$$\rho d^3\boldsymbol{\kappa} = \Delta n_x\, \Delta n_y\, \Delta n_z = \left(\frac{L_x}{2\pi} d\kappa_x\right)\left(\frac{L_y}{2\pi} d\kappa_y\right)\left(\frac{L_z}{2\pi} d\kappa_z\right) = \frac{L_x L_y L_z}{(2\pi)^3} d\kappa_x\, d\kappa_y\, d\kappa_z$$

or

$$\rho\, d^3\boldsymbol{\kappa} = \frac{V}{(2\pi)^3}\, d^3\boldsymbol{\kappa} \tag{9·9·16}$$

where $d^3\boldsymbol{\kappa} \equiv d\kappa_x\, d\kappa_y\, d\kappa_z$ is the element of volume in "$\boldsymbol{\kappa}$ space." Note that the density of states ρ is independent of $\boldsymbol{\kappa}$ and proportional to the volume V under consideration; i.e., the number of states *per unit volume*, with a wave number $\boldsymbol{\kappa}$ (or momentum $\boldsymbol{p} = \hbar\boldsymbol{\kappa}$) lying in some given range, is a constant independent of the magnitude or shape of the volume.

Remark Note that (9·9·3) yields for the number of translational states $\rho_p\, d^3\boldsymbol{p}$ in the momentum range between $\boldsymbol{p}$ and $\boldsymbol{p} + d\boldsymbol{p}$ the expression

$$\rho_p\, d^3\boldsymbol{p} = \rho\, d^3\boldsymbol{\kappa} = \frac{V}{(2\pi)^3}\frac{d^3\boldsymbol{p}}{\hbar^3} = V\frac{d^3\boldsymbol{p}}{h^3} \tag{9·9·17}$$

where $h = 2\pi\hbar$ is the ordinary Planck's constant. Now $V\, d^3\boldsymbol{p}$ is the volume of the classical six-dimensional phase space occupied by a particle in a box of volume V and with momentum between $\boldsymbol{p}$ and $\boldsymbol{p} + d\boldsymbol{p}$. Thus (9·9·17) shows that subdivision of this phase space into cells of size h^3 yields the correct number of quantum states for the particle.

Various other relations can be deduced from the result (9·9·16). For example, let us find the number of translational states $\rho_\kappa\, d\kappa$ for which $\boldsymbol{\kappa}$ is such that its magnitude $|\boldsymbol{\kappa}|$ lies in the range between κ and $\kappa + d\kappa$. This is obtained by summing (9·9·16) over all values of $\boldsymbol{\kappa}$ in this range, i.e., over the volume in $\boldsymbol{\kappa}$ space of the portion of spherical shell lying between radii κ and $\kappa + d\kappa$. Thus

$$\rho_\kappa\, d\kappa = \frac{V}{(2\pi)^3}\,(4\pi\kappa^2\, d\kappa) = \frac{V}{2\pi^2}\,\kappa^2\, d\kappa \tag{9·9·18}$$

Remark Since ϵ depends only on $\kappa = |\boldsymbol{\kappa}|$, (9·9·18) gives immediately, corresponding to this range of κ, the corresponding number of translational states $\rho_\epsilon\, d\epsilon$ for which the energy of the particle lies between ϵ and $\epsilon + d\epsilon$. From the equality of states one has

$$|\rho_\epsilon\, d\epsilon| = |\rho_\kappa\, d\kappa| = \rho_\kappa\left|\frac{d\kappa}{d\epsilon}\right| d\epsilon = \rho_\kappa\left|\frac{d\epsilon}{d\kappa}\right|^{-1} d\epsilon$$

By (9·9·4) one then obtains

$$\rho_\epsilon\, d\epsilon = \frac{V}{2\pi^2}\,\kappa^2\left|\frac{d\kappa}{d\epsilon}\right| d\epsilon = \frac{V}{4\pi^2}\frac{(2m)^{\frac{3}{2}}}{\hbar^3}\,\epsilon^{\frac{1}{2}}\, d\epsilon \tag{9·9·19}$$

Alternative discussion It is, of course, possible to adopt a slightly more complicated point of view which does take into account explicitly reflections occurring at the walls of the container. Since the exact boundary conditions are immaterial let us, for simplicity, assume that the container is in the shape of a rectangular parallelepiped with walls located at $x = 0$ and $x = L_x$, $y = 0$ and $y = L_y$, and $z = 0$ and $z = L_z$. Let us further assume that these walls are perfectly reflecting, i.e., that the potential energy U of the particle equals $U = 0$ inside the box and $U = \infty$ outside the box. Then the wave function ψ must satisfy the requirement that

$$\psi = 0 \qquad \begin{cases} \text{whenever } x = 0 \text{ or } L_x \\ \qquad\qquad y = 0 \text{ or } L_y \\ \qquad\qquad z = 0 \text{ or } L_z \end{cases} \tag{9·9·20}$$

The *particular* solution $\psi = e^{i\boldsymbol{\kappa}\cdot\mathbf{r}}$ of (9·9·9) represents a traveling wave and does *not* satisfy the boundary conditions (9·9·20). But one can construct suitable linear combinations of (9·9·9) (all of which automatically also satisfy the Schrödinger equation (9·9·8)) which do satisfy the boundary conditions (9·9·20). What this means physically is that in this box with perfectly reflecting parallel walls standing waves are set up which result from the superposition of traveling waves propagating back and forth.* Mathematically, since $e^{i\kappa_x x}$ is a solution of (9·9·8), so is $e^{-i\kappa_x x}$. The combination

$$(e^{i\kappa_x x} - e^{-i\kappa_x x}) \propto \sin \kappa_x x \tag{9·9·21}$$

vanishes properly when $x = 0$. It can also be made to vanish for $x = L_x$, provided one chooses κ_x so that

$$\kappa_x L_x = \pi n_x$$

where n_x is any integer. Here the possible values n_x should be restricted to the positive set

$$n_x = 1, 2, 3, \ldots$$

since a sign reversal of n_x (or κ_x) just turns the function (9·9·20) into

$$\sin(-\kappa_x)x = -\sin \kappa_x x$$

which is not a distinct new wave function. Thus a standing wave solution is specified completely by $|\kappa_x|$.

Forming standing waves analogous to (9·9·21) also for the y and z directions, one obtains the product wave function

$$\psi = A(\sin \kappa_x x)(\sin \kappa_y y)(\sin \kappa_z z) \tag{9·9·22}$$

where A is some constant. This satisfies the Schrödinger equation (9·9·8) and also the boundary conditions (9·9·20) provided that

$$\kappa_x = \frac{\pi}{L_x} n_x, \qquad \kappa_y = \frac{\pi}{L_y} n_y, \qquad \kappa_z = \frac{\pi}{L_z} n_z \tag{9·9·23}$$

* Simple standing waves of the form (9·9·21) would not be set up if the walls of the container were not exactly parallel. Hence our previous discussion in terms of traveling waves criss-crossing the volume in all directions, in a manner insensitive to the precise boundary conditions, affords a more convenient and general point of view.

where n_x, n_y, n_z are any *positive* integers. The possible energies of the particle are then given by

$$\epsilon = \frac{\hbar^2}{2m}\kappa^2 = \frac{\pi^2\hbar^2}{2m}\left(\frac{n_x^2}{L_x^2} + \frac{n_y^2}{L_y^2} + \frac{n_z^2}{L_z^2}\right)$$

For given values of κ_y and κ_z, the number of translational states with κ_x in the range between κ_x and $\kappa_x + d\kappa_x$ is now equal to

$$\Delta n_x = \frac{L_x}{\pi}\, d\kappa_x \qquad (9\cdot 9\cdot 24)$$

The number of translational states with $\boldsymbol{\kappa}$ in the range between $\boldsymbol{\kappa}$ and $\boldsymbol{\kappa} + d\boldsymbol{\kappa}$ is then given by

$$\rho\, d^3\boldsymbol{\kappa} = \Delta n_x\, \Delta n_y\, \Delta n_z = \left(\frac{L_x}{\pi}\, d\kappa_x\right)\left(\frac{L_y}{\pi}\, d\kappa_y\right)\left(\frac{L_z}{\pi}\, d\kappa_z\right)$$

or

$$\rho\, d^3\boldsymbol{\kappa} = \frac{V}{\pi^3}\, d^3\boldsymbol{\kappa} \qquad (9\cdot 9\cdot 25)$$

The number of translational states $\rho_\kappa\, d\kappa$ for which $\boldsymbol{\kappa}$ is such that its *magnitude* lies in the range between κ and $\kappa + d\kappa$ is obtained by summing (9·9·25) over all values of $\boldsymbol{\kappa}$ in this range, i.e., over the volume in $\boldsymbol{\kappa}$ space of the portion of spherical shell lying between radii κ and $\kappa + d\kappa$ *and* located in the first octant where $\kappa_x, \kappa_y, \kappa_z > 0$ so as to satisfy (9·9·23). Thus (9·9·25) yields

$$\rho_\kappa\, d\kappa = \frac{V}{\pi^3}\left(\frac{4\pi\kappa^2\, d\kappa}{8}\right) = \frac{V}{2\pi^2}\,\kappa^2\, d\kappa \qquad (9\cdot 9\cdot 26)$$

This is the *same* result as was obtained in (9·9·18). The reason is simple. By (9·9·24) there are, compared to (9·9·15), twice as many states lying in a given interval $d\kappa_x$, but since only positive values of κ_x are now to be counted, the number of such intervals is decreased by a compensating factor of 2.

By (9·9·26) it also follows that $\rho_\epsilon\, d\epsilon$ is the same as in (9·9·19). This just illustrates the result (which can also be established by rather elaborate general mathematical arguments)* that this density of states should be the same irrespective of the shape of the container or of the exact boundary conditions imposed on its surface, so long as the de Broglie wavelength of the particle is small compared to the dimensions of the container.

9 · 10 *Evaluation of the partition function*

We are now ready to calculate the partition function Z of a monatomic ideal gas in the classical limit of sufficiently low density or sufficiently high temperature. By (9·8·9) one has

$$\ln Z = N(\ln \zeta - \ln N + 1) \qquad (9\cdot 10\cdot 1)$$

where

$$\zeta \equiv \sum_r e^{-\beta\epsilon_r} \qquad (9\cdot 10\cdot 2)$$

* See, for example, R. Courant and D. Hilbert, "Methods of Mathematical Physics," vol. I, pp. 429–445, Interscience Publishers, New York, 1953.

is the sum over all states of a single particle. The expression (9·10·1) is identical to the result (7·3·3); i.e.,

$$Z = \frac{\zeta^N}{N!} \tag{9·10·3}$$

Since we have just enumerated the possible states of a single particle, the sum (9·10·2) is readily evaluated. By (9·9·14)

$$\zeta = \sum_{\kappa_x,\kappa_y,\kappa_z} \exp\left[-\frac{\beta\hbar^2}{2m}(\kappa_x^2 + \kappa_y^2 + \kappa_z^2)\right] \tag{9·10·4}$$

where the sum is over all possible values of κ_x, κ_y, κ_z given by (9·9·13). Since the exponential function factors, ζ becomes the product of three similar sums

$$\zeta = \left(\sum_{\kappa_x} e^{-(\beta\hbar^2/2m)\kappa_x^2}\right)\left(\sum_{\kappa_y} e^{-(\beta\hbar^2/2m)\kappa_y^2}\right)\left(\sum_{\kappa_z} e^{-(\beta\hbar^2/2m)\kappa_z^2}\right) \tag{9·10·5}$$

Successive terms in a sum like that over $\kappa_x = (2\pi/L_x)n_x$ correspond to a very small increment $\Delta\kappa_x = 2\pi/L_x$ in κ_x and differ, therefore, very little from each other; i.e.,

$$\left|\frac{\partial}{\partial\kappa_x}[e^{-(\beta\hbar^2/2m)\kappa_x^2}]\left(\frac{2\pi}{L_x}\right)\right| \ll e^{-(\beta\hbar^2/2m)\kappa_x^2} \tag{9·10·6}$$

Provided that this condition is satisfied, it is an excellent approximation to replace the sums in (9·10·5) by integrals. A small range between κ_x and $\kappa_x + d\kappa_x$ contains then, by (9·9·15), $\Delta n_x = (L_x/2\pi)\,d\kappa_x$ terms which have nearly the same magnitude and can be grouped together. Summing over all possible ranges of κ_x completes the sum. Thus

$$\sum_{\kappa_x=-\infty}^{\infty} e^{-(\beta\hbar^2/2m)\kappa_x^2} \approx \int_{-\infty}^{\infty} e^{-(\beta\hbar^2/2m)\kappa_x^2}\left(\frac{L_x}{2\pi}\,d\kappa_x\right)$$

$$= \frac{L_x}{2\pi}\left(\frac{2\pi m}{\beta\hbar^2}\right)^{\frac{1}{2}} = \frac{L_x}{2\pi\hbar}\left(\frac{2\pi m}{\beta}\right)^{\frac{1}{2}} \qquad \text{by (A·4·2)}$$

Hence (9·10·5) becomes

$$\blacktriangleright \qquad \zeta = \frac{V}{(2\pi\hbar)^3}\left(\frac{2\pi m}{\beta}\right)^{\frac{3}{2}} = \frac{V}{h^3}(2\pi mkT)^{\frac{3}{2}} \tag{9·10·7}$$

Note that this is the same result as that obtained by the classical calculation in (7·2·6), provided that we set the arbitrary parameter h_0 (which measures the size of a cell in classical phase space) equal to Planck's constant h.

It then follows, by (9·10·1), that

$$\ln Z = N\left(\ln\frac{V}{N} - \frac{3}{2}\ln\beta + \frac{3}{2}\ln\frac{2\pi m}{h^2} + 1\right) \tag{9·10·8}$$

Hence

$$\bar{E} = -\frac{\partial \ln Z}{\partial\beta} = \frac{3}{2}\frac{N}{\beta} = \frac{3}{2}NkT \tag{9·10·9}$$

and

$$S = k(\ln Z + \beta\bar{E}) = Nk\left(\ln\frac{V}{N} + \frac{3}{2}\ln T + \sigma_0\right) \tag{9·10·10}$$

where

$$\sigma_0 \equiv \frac{3}{2}\ln\frac{2\pi mk}{h^2} + \frac{5}{2} \tag{9·10·11}$$

These results are exactly the same as those we obtained in (7·3·5), with one important difference. Since we have now treated the problem by quantum mechanics, the constant σ_0 has a definite value in terms of Planck's constant h (unlike the classical case where h_0 was an arbitrary parameter). The fact that the entropy does not involve any arbitrary constants has important physical consequences which we shall discuss in Sec. 9·11. All quantities such as $\bar{E}$ or the mean pressure $\bar{p}$, which depend only on derivatives of S are, of course, the same as those calculated in Sec. 7·2.

Let us verify that the condition (9·10·6) justifying the replacement of the sum over states by an integral is indeed satisfied. This condition requires that

$$\left|\frac{\beta\hbar^2}{m}\kappa_x\frac{2\pi}{L_x}\right| \ll 1 \tag{9·10·12}$$

But the mean value of κ_x can be estimated from (9·10·9) or the equipartition theorem. Thus

$$\frac{\hbar^2\overline{\kappa_x^2}}{2m} = \frac{1}{3}\frac{\hbar^2\overline{\kappa^2}}{2m} = \frac{1}{2}kT$$

or

$$\hbar\bar{\kappa}_x \approx \sqrt{mkT}$$

Hence (9·10·12) becomes

$$\frac{\hbar}{mkT}\sqrt{mkT}\frac{2\pi}{L_x} = \frac{h}{\sqrt{mkT}}\frac{1}{L_x} \ll 1$$

or approximately

$$\bar{\lambda} \ll L_x \tag{9·10·13}$$

where $\bar{\lambda} = h/\bar{p}$ is the mean de Broglie wavelength of the particle.

Thus (9·10·12) demands only that $\bar{\lambda}$ is smaller than the smallest dimension L of the container. On the other hand, we saw in Sec. 7·4 that the requirement for the very applicability of the classical approximation is that $\bar{\lambda}$ be smaller than the mean interparticle separation, i.e.,

$$\bar{\lambda} \ll \frac{L}{N^{\frac{1}{3}}} \tag{9·10·14}$$

which is a much more stringent condition than (9·10·13).

Finally, let us point out what happens if each particle has also an intrinsic spin angular momentum J. The possible orientations of this spin are specified by its projection $m_J = -J, \ -J+1, \ldots, J-1, \ J$. There are then $(2J+1)$ possible states of the same energy associated with each possible translational state of a particle. The net result is that the sum over states ζ simply is multiplied by $(2J+1)$, so that the entropy is increased by the constant $Nk\ln(2J+1)$.

9·11 *Physical implications of the quantum-mechanical enumeration of states*

Although the results of the quantum-mechanical calculation of Z are virtually the same as those of the semiclassical calculation in Sec. 7·3, there are two significant differences:

a. The correct dependence (9·10·1) of $\ln Z$ on N (i.e., the factor $N!$ in (9·10·3)) is an automatic consequence of the theory. Thus the Gibbs paradox does not arise, and $\ln Z$ in (9·10·8) behaves properly like an extensive quantity under simultaneous change of scale of N and V.

b. There are no arbitrary constants occurring in Z or the entropy S derived therefrom; instead Z is a well-defined number involving Planck's constant h.

These differences reflect the fact that we have now unambiguously counted the number of quantum states available to the gas. We should expect this enumeration to be particularly important in cases involving transfer of particles from one phase to another (or from one component to another), since in these cases a calculation of the equilibrium situation must compare the actual number of states available in one phase with that in another (or for one type of molecule with that for another). Mathematically, this is manifested by the properties of the chemical potential

$$\mu = \left(\frac{\partial F}{\partial N}\right)_{V,T} = -kT\left(\frac{\partial \ln Z}{\partial N}\right)_{V,T} \tag{9·11·1}$$

In the last chapter we saw that the chemical potential is the important parameter determining the equilibrium conditions between phases or chemical components. On the other hand, it is clear from (9·11·1) and (9·10·1) that

$$\mu = -kT \ln \frac{\zeta}{N} \tag{9·11·2}$$

does depend on N and the various constants, such as Planck's constant, involved in ζ. Thus the quantum-mechanical calculation of Z in terms of these constants allows one to make predictions completely outside the realm of any theory based on classical statistical mechanics. We shall give two representative illustrations.

Thermal ionization of hydrogen atoms Suppose that H atoms are enclosed in a container of volume V at a high temperature T.* There then exists the possibility of ionization into a hydrogen ion H^+ and an electron e^-. This can be described in terms of the reaction

$$H \rightleftarrows H^+ + e^- \tag{9·11·3}$$

Let ϵ_0 denote the energy necessary to ionize the atom, i.e., its "ionization potential." This means that the ground state of the H atom has an energy $(-\epsilon_0)$ relative to the state where the proton H^+ and the electron e^- are at rest

* We assume that this temperture is high enough that the number of H_2 molecules is negligible, practically all of them being dissociated into H atoms.

separated by an infinite distance from each other.* Viewing (9·11·3) as a chemical equilibrium of the type discussed in Sec. 8·9, we can write it in the standard form

$$-\mathrm{H} + \mathrm{H}^+ + \mathrm{e}^- = 0$$

so that the law of mass action (8·10·21) becomes

$$\frac{N_+N_-}{N_{\mathrm{H}}} = K_N \tag{9·11·4}$$

where

$$K_N = \frac{\zeta_+\zeta_-}{\zeta_{\mathrm{H}}} \tag{9·11·5}$$

Here N denotes the mean number of particles of each kind and the subscripts $+$, $-$, and H refer to the H^+ ion, the electron, and the H atom, respectively.

We are now in a position to calculate the quantities ζ from first principles. It is only necessary to be sure, for the sake of consistency, that all energies in the problem are measured from the same standard state. We shall choose this standard state to be the one where the electron and proton are at rest at infinite separation from each other. Furthermore, we shall assume that the H^+ and e^- concentrations are relatively small. The classical limit is then applicable at these high temperatures, and any coulomb attraction between the separated protons and electrons can be neglected.

Thus one can use (9·10·7) to write for the electron of mass m,

$$\zeta_- = 2\frac{V}{h^3}(2\pi mkT)^{\frac{3}{2}} \tag{9·11·6}$$

Here the factor of 2 is introduced, since the *electron* has spin $\frac{1}{2}$ and, therefore, has two possible spin states for each translational state. Similarly, for the freely moving proton of mass M, one obtains

$$\zeta_+ = 2\frac{V}{h^3}(2\pi MkT)^{\frac{3}{2}} \tag{9·11·7}$$

Here the factor of 2 is introduced, because the *nuclear* spin of the proton is $\frac{1}{2}$, so that there are two possible nuclear spin orientations for each translational state of the proton.

The H atom has a mass $M + m \approx M$, since $m \ll M$. Its *internal* energy measured with respect to our chosen standard state is $(-\epsilon_0)$, since practically all H atoms are in their ground state at the temperature under consideration.† Hence one can write for the H atom

$$\zeta_{\mathrm{H}} = 4\frac{V}{h^3}(2\pi MkT)^{\frac{3}{2}}e^{\epsilon_0/kT} \tag{9·11·8}$$

* From atomic physics we know that $\epsilon_0 = \frac{1}{2}(e^2/a_0)$, where $a_0 = \hbar^2/me^2$ is the Bohr radius. Numerically $\epsilon_0 = 13.6$ electron volts. (This is about three times larger than the energy necessary to dissociate a H_2 molecule.)

† The first excited state has an energy $-\frac{1}{4}\epsilon_0$, so that the relative probability of finding an atom in this state rather than in the ground state is

$$e^{\frac{1}{4}\beta\epsilon_0}/e^{\beta\epsilon_0} = e^{-\frac{3}{4}\beta\epsilon_0}$$

which is very small, even if $T = (k\beta)^{-1} = 10{,}000°\mathrm{K}$.

Here the factor of 4 is introduced, since there are four possible states of the atom for each translation state: two states of possible electron spin orientation, and for each of these, two states of possible nuclear spin orientation.

Combining these various expressions, one obtains, by (9·11·5),

$$K_N = \frac{V}{h^3}(2\pi mkT)^{\frac{3}{2}}\, e^{-\epsilon_0/kT} \tag{9·11·9}$$

which is the desired expression for the equilibrium constant. Note that all the statistical weighting factors due to the existence of spin have cancelled.

What (9·11·4) and (9·11·9) say physically is that the large ionization potential ϵ_0 tends to favor the existence of the H atom since this is the system of lowest energy. On the other hand, many more states become accessible to the system, i.e., its entropy tends to be greater, if one deals with two separate particles. The equilibrium situation represents the compromise between these two tendencies. More generally speaking, the most probable situation is that where the free energy $F = E - TS$ is minimum. At low temperatures where $F \approx E$ this favors the situation of low energy, i.e., the H atom. On the other hand, when T becomes large, F can become small if the entropy S is large, and this favors dissociation.

Suppose that a number N_0 of H atoms are present in the container at some temperature low enough that $N_- = N_+ \approx 0$, and that the temperature is then raised to the value T. Let ξ denote the fraction of atoms dissociated at this temperature, i.e.,

$$\xi \equiv \frac{N_+}{N_0} \tag{9·11·10}$$

But by virtue of (9·11·3)

$$N_+ = N_- = N_0\xi$$

and

$$N_H = N_0 - N_0\xi = N_0(1 - \xi) \approx N_0$$

since $\xi \ll 1$. Then the law of mass action (9·11·4) gives, by (9·11·9),

$$\xi^2 = \left(\frac{V}{N_0}\right)\left(\frac{2\pi mkT}{h^2}\right)^{\frac{3}{2}} e^{-\epsilon_0/kT} \tag{9·11·11}$$

so that the degree of dissociation can be readily calculated. Note that Planck's constant appears quite explicitly in this relation.

Vapor pressure of a solid Consider a solid consisting of monatomic molecules, e.g., solid argon. If it is in equilibrium with its vapor the equilibrium condition is, by (8·8·8),

$$\mu_1 = \mu_2 \tag{9·11·12}$$

where μ_1 is the chemical potential of the vapor and μ_2 that of the solid. Unless the temperature is exceedingly high, the vapor is not too dense and can be treated as an ideal gas. Then for the chemical potential of N_1 atoms of vapor in a volume V_1, (9·11·2) and (9·10·7) give

$$\mu_1 = -kT \ln\left[\frac{V_1}{N_1}\left(\frac{2\pi mkT}{h^2}\right)^{\frac{3}{2}}\right] \tag{9·11·13}$$

Here we have assumed, for simplicity, that the atoms of mass m have no spin degrees of freedom.

Let us now turn to a discussion of the solid. If it consists of N_2 atoms and has a volume V_2, its chemical potential is related to its partition function Z by

$$\mu_2 = \left(\frac{\partial F}{\partial N_2}\right)_{T,V_2} = -kT\left(\frac{\partial \ln Z}{\partial N_2}\right)_{T,V_2} \tag{9·11·14}$$

Although we could try to calculate Z by using a model such as the Einstein model of Sec. 7·7, let us keep the discussion more general and relate Z directly to specific heat information. The mean energy of the solid is related to Z by

$$\bar{E}(T) = -\left(\frac{\partial \ln Z}{\partial \beta}\right)_V = kT^2\left(\frac{\partial \ln Z}{\partial T}\right)_V$$

This can be integrated immediately to give

$$\ln Z(T) - \ln Z(T_0) = \int_{T_0}^{T} \frac{\bar{E}(T')}{kT'^2}\, dT' \tag{9·11·15}$$

Here we shall choose $T_0 \to 0$.

Since the solid is almost incompressible, its volume V_2 is very nearly constant, and its thermodynamic functions are essentially only functions of T. Let us denote by $c(T)$ the specific heat per *atom* of the solid. (It matters little whether it is measured at constant volume or constant pressure, since the solid is nearly incompressible.) Since $(\partial\bar{E}/\partial T)_V = N_2c$, we can express $\bar{E}(T)$ in terms of the specific heat. Thus

$$\bar{E}(T) = -N_2\eta + N_2\int_0^T c(T'')\, dT'' \tag{9·11·16}$$

Here we have put $\bar{E}(0) \equiv -N_2\eta$. This is simply the ground-state energy of the solid measured from the same standard state as that of the vapor, i.e., from the state where all atoms are at rest at very large distances from each other. Thus η is the latent heat of sublimation per atom at $T = 0$.

Finally we note that as $T \to 0$ or $\beta \to \infty$

$$Z = \Sigma\, e^{-\beta E_r} \to \Omega_0\, e^{-\beta(-N_2\eta)}$$

or

$$\ln Z(T_0) = \frac{N_2\eta}{kT_0} \qquad \text{as } T_0 \to 0 \tag{9·11·17}$$

since the number of states Ω_0 accessible to the solid in its ground state is of the order of unity.* (The atoms were assumed to have no spin degrees of freedom which might lead to many states at $T_0 = 0$.) Using (9·11·16) and (9·11·17) in (9·11·15), and putting $T_0 \to 0$, one obtains

$$\ln Z(T) = \frac{N_2\eta}{kT} + N_2\int_0^T \frac{dT'}{kT'^2}\int_0^{T'} c(T'')\, dT'' \tag{9·11·18}$$

Hence (9·11·14) yields

$$\mu_2(T) = -\eta - T\int_0^T \frac{dT'}{T'^2}\int_0^{T'} c(T'')\, dT'' \tag{9·11·19}$$

* That is, the entropy $S = k \ln \Omega$ of the solid vanishes as $T \to 0$, in accordance with the third law.

The equilibrium condition (9·11·12) becomes then

$$\ln\left[\frac{V_1}{N_1}\left(\frac{2\pi mkT}{h^2}\right)^{\frac{3}{2}}\right] = -\frac{\mu_2(T)}{kT} \tag{9·11·20}$$

To find the vapor pressure $\bar{p}$, one needs only to use the ideal gas equation of state $\bar{p}V_1 = N_1kT$ for the vapor. Thus (9·11·20) becomes

$$\ln\left[\frac{kT}{\bar{p}}\left(\frac{2\pi mkT}{h^2}\right)^{\frac{3}{2}}\right] = -\frac{\mu_2}{kT}$$

Hence $$\ln\bar{p} = \ln\left[\frac{(2\pi m)^{\frac{3}{2}}}{h^3}(kT)^{\frac{5}{2}}\right] + \frac{\mu_2}{kT}$$

and $$\bar{p}(T) = \frac{(2\pi m)^{\frac{3}{2}}}{h^3}(kT)^{\frac{5}{2}}\exp\left[-\frac{\eta}{kT} - \frac{1}{k}\int_0^T\frac{dT'}{T'^2}\int_0^{T'}c(T'')\,dT''\right] \tag{9·11·21}$$

This is the desired expression for the vapor pressure. Note that it again involves Planck's constant in an essential way.

The specific heat of the solid can be obtained either from microscopic calculation by some model (e.g., the Einstein model of Sec. 7·7) or from experimental measurements. The double integral in (9·11·21) is seen to be a positive increasing function of T; it converges without difficulty, since $c \to 0$ sufficiently rapidly as $T \to 0$.

Note that if we had tried to compute the vapor pressure by means of the Clapeyron equation in a manner similar to that used at the end of Sec. 8·5, we could not have determined the constant of integration, i.e., all the constants on the right side of (9·11·21).

*9·12 *Partition functions of polyatomic molecules*

Let us sketch briefly how one goes about calculating the partition function for an ideal gas consisting of N *polyatomic* molecules. In the classical limit where the mean de Broglie wavelength $\bar{\lambda}$ associated with the momentum of the center-of-mass motion is small compared to the mean separation of the molecules one has again

$$Z = \frac{\zeta^N}{N!} \tag{9·12·1}$$

Here $$\zeta = \sum_s e^{-\beta\epsilon(s)} \tag{9·12·2}$$

is the partition function for an individual molecule, the summation being over all the quantum states s of the molecule. To a good approximation one can write the Hamiltonian of a molecule in the additive form

$$\mathcal{H} = \mathcal{H}_t + \mathcal{H}_e + \mathcal{H}_r + \mathcal{H}_v \tag{9·12·3}$$

and correspondingly the energy levels of the molecule in the form

$$\epsilon(s) = \epsilon_t(s_t) + \epsilon_e(s_e) + \epsilon_r(s_r) + \epsilon_v(s_v) \tag{9·12·4}$$

$\mathcal{H}_t$ denotes the Hamiltonian describing the translational motion of the center of mass of the molecule; $\epsilon_t(s_t)$ denotes the corresponding translational energy of the translational state labeled s_t.

$\mathcal{H}_e$ denotes the Hamiltonian describing the motion of the electrons about the nuclei assumed in a fixed configuration; $\epsilon_e(s_e)$ denotes the corresponding electronic state labeled s_e.

$\mathcal{H}_r$ denotes the Hamiltonian describing the rotation of the nuclei of the molecule about their center of mass; $\epsilon_r(s_r)$ denotes the corresponding rotational energy of the rotational state labeled s_r.

$\mathcal{H}_v$ denotes the Hamiltonian describing the vibrational motion of the nuclei of the molecule relative to each other; $\epsilon_v(s_v)$ denotes the corresponding vibrational energy of the vibrational state labeled by s_v.

The additivity of (9 · 12 · 4) implies immediately that the partition function ζ factors into a product; i.e.,

$$\zeta = \sum_{s_t, s_e, \ldots} e^{-\beta[\epsilon_t(s_t)+\epsilon_e(s_e)+\epsilon_r(s_r)+\epsilon_v(s_v)]}$$

$$= \Big(\sum_{s_t} e^{-\beta\epsilon_t(s_t)}\Big)\Big(\sum_{s_e} e^{-\beta\epsilon_e(s_e)}\Big)\Big(\sum_{s_r} e^{-\beta\epsilon_r(s_r)}\Big)\Big(\sum_{s_v} e^{-\beta\epsilon_v(s_v)}\Big)$$

or

$$\zeta = \zeta_t\zeta_e\zeta_r\zeta_v \tag{9·12·5}$$

where ζ_t is the partition function for the translational motion of the center of mass, ζ_e is the partition function for electronic motion, etc.

Let us discuss these partition functions specifically for a diatomic molecule with atoms of masses m_1 and m_2.

Translational motion of the center of mass The center of mass moves like a particle of mass $m_1 + m_2$. Thus

$$\mathcal{H}^{(t)} = \frac{\mathbf{p}^2}{2(m_1 + m_2)}$$

where $\mathbf{p}$ denotes the linear momentum of the center of mass. Using the mass $m_1 + m_2$, the translational states are then the same as those discussed in connection with the monatomic gas. Hence the sum over translational states gives by comparison with (9 · 10 · 7)

$$\zeta_t = \frac{V}{h^3}[2\pi(m_1 + m_2)kT]^{\frac{3}{2}} \tag{9·12·6}$$

Electronic motion We turn next to the internal motion of the atoms relative to the center of mass. Consider first the possible electronic states of the molecule. For *fixed* nuclei the electronic ground-state energy ϵ_{e0} can be calculated as a function of the internuclear separation R and yields a curve of the type shown in Fig. 9 · 12 · 1.

The minimum of this curve determines, for the electronic ground state of the molecule, the equilibrium internuclear separation R_0, where $\epsilon_{e0} = -\epsilon_D'$.

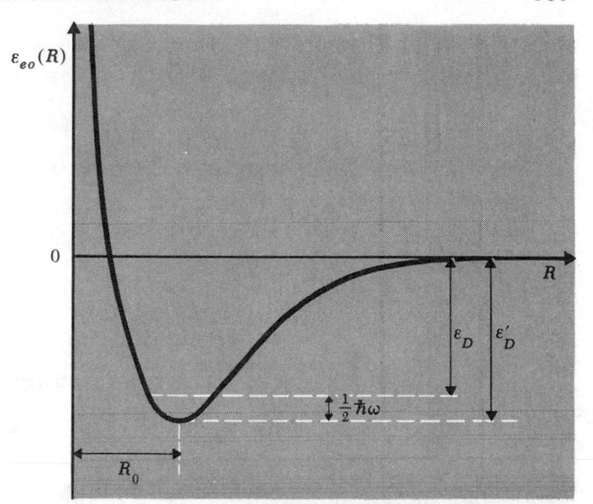

***Fig. 9·12·1** Energy of the electronic ground state $\epsilon_{e0}(R)$ of a diatomic molecule as a function of the internuclear separation R. The dissociation energy is denoted by ϵ_D, the vibrational zero-point energy by $\frac{1}{2}\hbar\omega$.*

This energy is negative when measured with respect to a standard state where the nuclei are at rest at infinite separation from each other. Since the first excited electronic state is, for almost all molecules, higher than the ground state by an energy of the order of a few electron volts, i.e., very large compared to kT, all terms in the electronic partition function other than the one of lowest energy are negligible. (That is, the molecule is with overwhelming probability in its electronic ground state.) Thus one has simply

$$\zeta_e = \Omega_0 \, e^{\beta \epsilon_D'} \tag{9·12·7}$$

where Ω_0 is the degree of degeneracy (if any) of the electronic ground state.

Rotation Consider now the rotation of the molecule. This is in first approximation like the rotation of a rigid dumbbell consisting of two masses m_1 and m_2 separated by the atomic equilibrium distance R_0 in the molecule. The moment

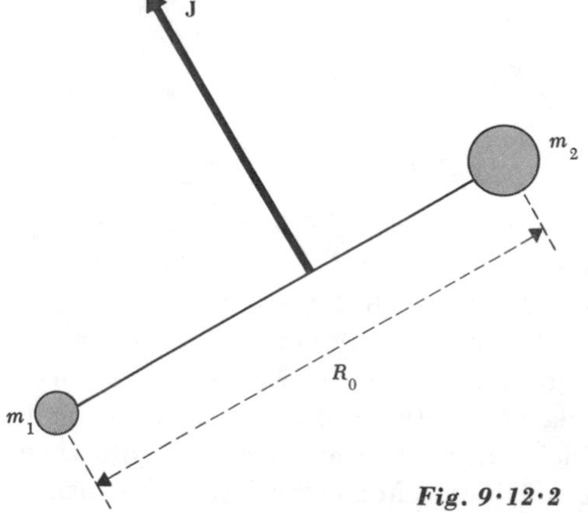

***Fig. 9·12·2** Rotation of a rigid dumbbell molecule.*

of inertia A of the molecule about an axis through the center of mass and perpendicular to the line joining the atomic nuclei is given by

$$A = \tfrac{1}{2}\mu^* R_0^2 \tag{9·12·8}$$

where μ^* is the reduced mass of the two atoms

$$\mu^* = \frac{m_1 m_2}{m_1 + m_2} \tag{9·12·9}$$

If $\hbar \boldsymbol{J}$ denotes the rotational angular momentum of this dumbbell, its classical energy is given by $(\hbar \boldsymbol{J})^2/2A$. Quantum mechanically $\boldsymbol{J}^2$ can assume the possible values $J(J+1)$, where the quantum number $J = 0, 1, 2, 3, \ldots$. Hence the rotational energy levels are given by

$$\epsilon_r = \frac{\hbar^2}{2A} J(J+1) \tag{9·12·10}$$

(Note that a small moment of inertia implies a large spacing of the rotational energy levels.) The vector $\boldsymbol{J}$ can, of course, have several discrete spatial orientations labeled by m_J, its projection along some axis. The possible values of m_J are

$$m_J = -J, -J+1, \ldots (J-1), J$$

so that for each value of J, there are $2J + 1$ possible quantum states of the same energy (9·12·10). The rotational partition function becomes then

$$\zeta_r = \sum_{J=0}^{\infty} (2J+1)\, e^{-(\beta\hbar^2/2A)J(J+1)} \tag{9·12·11}$$

The significant parameter in (9·12·11) is the argument of the exponential, i.e., the ratio of rotational to thermal energy. For low temperature T or small moment of inertia, $\hbar^2/(2AkT) \gg 1$; then practically all molecules are in the very lowest rotational states, and all terms in the sum (9·12·11) beyond the first few are negligible.

Remark Note that in writing down (9·12·11) we have not worried about any angular-momentum component parallel to the axis of the dumbbell. The reason is that the moment of inertia about this axis is very small. Any state with such an angular-momentum component different from zero would, in analogy to (9·12·10), have very high energy compared to kT and can therefore be neglected.

On the other hand, suppose that the temperature T is reasonably large and the moment of inertia is not too small, so that $\hbar^2 J(J+1)(2AkT)^{-1} \ll 1$. (This is the case for many diatomic molecules in which the spacing between the rotational energy levels (9·12·10) is of the order of 10^{-4} ev. Exceptions are molecules such as H_2 below room temperature, because these have such small moments of inertia.) Then the spacing of rotational-energy levels is small compared to kT. This implies that the rotation of the molecule could then also be treated by classical statistical mechanics. Mathematically this means that successive terms in the sum (9·12·11) differ from each other by relatively small amounts so that this sum can be approximated by an integral. Thus one can write, putting $u = J(J+1)$,

$$\zeta_r \approx \int_0^\infty du\, e^{-(\beta\hbar^2/2A)u} = \frac{2A}{\beta\hbar^2}$$

or

$$\zeta_r \approx \frac{2AkT}{\hbar^2} \qquad (9\cdot12\cdot12)$$

If the two nuclei of the molecule are identical, then we must again be concerned about their essential indistinguishability (just as we were concerned about the factor $N!$ in the translational partition function). In the classical limit, where (9·12·12) is valid, the indistinguishability is easily handled. Turning the molecule end-for-end is the same as interchanging the two identical nuclei. We have counted such a turning over by 180° as a distinct state in calculating (9·12·12), and properly so for unlike nuclei. But it is *not* to be counted as a distinguishable state for *like* nuclei; in this case (9·12·12) is too large by a factor of 2. Hence one should generally put

$$\zeta_r = \frac{2AkT}{\hbar^2\sigma} \qquad (9\cdot12\cdot13)$$

where

$$\sigma = \begin{cases} 1 & \text{if the nuclei are unlike} \\ 2 & \text{if they are identical} \end{cases} \qquad (9\cdot12\cdot14)$$

In the case where the quasi-classical treatment of rotation is not applicable (e.g., for H_2 at low temperatures) the situation is more complicated and leads to the involvement of nuclear spins with the rotation in a very intimate way. We shall forego discussing the interesting peculiarities arising in such cases.

Remark In the classical limit where (9·12·13) is applicable

$$\ln \zeta_r = -\ln \beta + \text{constant}$$

Hence the mean energy of rotation is given by

$$\epsilon_r = -\frac{\partial}{\partial\beta}\ln \zeta_r = \frac{1}{\beta} = kT \qquad (9\cdot12\cdot15)$$

This is indeed what one would get from the classical equipartition theorem applied to the two degrees of freedom which represent classical rotation, namely, rotation about the two orthogonal principal axes which are perpendicular to the line joining the two nuclei. (We already mentioned in our last remark that the rotation *about* the line joining the nuclei cannot be treated in the classical limit.)

Vibration Finally, the nuclei are also free to vibrate relative to each other about their equilibrium separation R_0. The potential energy of the nuclei as a function of their separation R is given by the electronic ground-state energy $\epsilon_{e0}(R)$ of Fig. 9·12·1. Near its minimum it can be expanded in the form

$$\epsilon_{e0}(R) = -\epsilon_D' + \tfrac{1}{2}b\xi^2 \qquad (9\cdot12\cdot16)$$

where $$b \equiv \frac{\partial^2 \epsilon_{e0}(R_0)}{\partial R^2} \qquad \text{and} \qquad \xi \equiv R - R_0 \tag{9·12·17}$$

The kinetic energy of vibration of the nuclei relative to their center of mass is given by

$$K = \tfrac{1}{2}\mu^* \dot{R}^2 = \tfrac{1}{2}\mu^* \dot{\xi}^2 \tag{9·12·18}$$

By (9·12·16) and (9·12·18), one would obtain classically simple harmonic motion of angular frequency

$$\omega = \sqrt{\frac{b}{\mu^*}} \tag{9·12·19}$$

Quantum mechanically, (9·12·16) and (9·12·18) yield the Hamiltonian of a simple harmonic oscillator whose possible vibrational energy levels are given by

$$\epsilon_v = \hbar\omega(n + \tfrac{1}{2}) \tag{9·12·20}$$

Here the possible quantum states are labeled n, which can assume all values $n = 0, 1, 2, 3, \ldots$.

Hence the vibrational partition function is

$$\zeta_v = \sum_{n=0}^{\infty} e^{-\beta\hbar\omega(n+\frac{1}{2})} \tag{9·12·21}$$

We already evaluated this simple geometric series in (7·6·10). Thus

$$\zeta_v = \frac{e^{-\frac{1}{2}\beta\hbar\omega}}{1 - e^{-\beta\hbar\omega}} \tag{9·12·22}$$

For most diatomic molecules at ordinary temperatures $\hbar\omega$ is so large (of the order of 0.1 ev) that $\beta\hbar\omega \gg 1$. In that case (9·12·21) reduces to its first term

$$\zeta_v \approx e^{-\frac{1}{2}\beta\hbar\omega}$$

The vibrational degrees of freedom can then certainly *not* be treated classically.

Remark Note that even at $T = 0$ the nuclei still have a zero-point energy $\frac{1}{2}\hbar\omega$ in their lowest vibrational state. Hence ϵ_D' in Fig. 9·12·1 is not equal to the dissociation energy ϵ_D which must be provided at $T = 0$ to dissociate the molecule into two atoms at rest at an infinite distance from each other. Instead, one has (see Fig. 9·12·1)

$$\epsilon_D = \epsilon_D' - \tfrac{1}{2}\hbar\omega \tag{9·12·23}$$

We have now calculated all the essential ingredients necessary for the evaluation of the partition function (9·12·5) for an ideal gas of diatomic molecules. (Some examples of applications will be found in the problems.) If the nuclei of the molecules have spin, then ζ in (9·12·5) must also be multiplied by the possible number of nuclear-spin states. If one deals with molecules consisting of more than two atoms, the decomposition (9·12·4) or (9·12·5) is in general still valid, but the rotational and vibrational partition functions ζ_r and ζ_v become more complicated.

BLACK - BODY RADIATION

9·13 *Electromagnetic radiation in thermal equilibrium inside an enclosure*

Let us consider the electromagnetic radiation (or in quantum-mechanical language, the assembly of photons) which exists in thermal equilibrium inside an enclosure of volume V whose walls are maintained at the absolute temperature T. In this situation photons are continuously absorbed and reemitted by the walls; it is, of course, by virtue of these mechanisms that the radiation inside the container depends on the temperature of the walls. But, as usual, it is not at all necessary to investigate the exact mechanisms which bring about the thermal equilibrium, since the general probability arguments of statistical mechanics suffice to describe the equilibrium situation.

Let us regard the radiation as a collection of photons. These must, of course, be considered as indistinguishable particles. The total number of photons inside the enclosure is not fixed, but depends on the temperature T of the walls. The state s of each photon can be specified, in a manner discussed below, by the magnitude and direction of its momentum and by the direction of polarization of the electric field associated with the photon. The radiation field existing in thermal equilibrium inside the enclosure is completely described if one knows the mean number $\bar{n}_s$ of photons in each possible state. The calculation of this number is precisely the problem already solved in (9·3·5). The result is the Planck distribution

$$\bar{n}_s = \frac{1}{e^{\beta \epsilon_s} - 1} \tag{9·13·1}$$

where ϵ_s is the energy of a photon in state s.

To make this result more concrete we have to consider in greater detail how the state of each photon is specified. Since we are dealing with electromagnetic radiation, the electric field $\boldsymbol{\mathcal{E}}$ (or each component thereof) satisfies the wave equation

$$\nabla^2 \boldsymbol{\mathcal{E}} = \frac{1}{c^2} \frac{\partial^2 \boldsymbol{\mathcal{E}}}{\partial t^2} \tag{9·13·2}$$

This is satisfied by (the real part of) plane wave solutions of the form

$$\boldsymbol{\mathcal{E}} = \mathbf{A}\, e^{i(\boldsymbol{\kappa} \cdot \mathbf{r} - \omega t)} = \boldsymbol{\mathcal{E}}_0(\mathbf{r})\, e^{-i\omega t} \tag{9·13·3}$$

(where $\mathbf{A}$ is any constant), provided that the wave vector $\boldsymbol{\kappa}$ satisfies the condition

$$\kappa = \frac{\omega}{c}, \qquad \kappa \equiv |\boldsymbol{\kappa}| \tag{9·13·4}$$

Remark Note that the spatial part $\boldsymbol{\mathcal{E}}_0(\boldsymbol{r})$ on the right side of (9 · 13 · 3) satisfies the time independent wave equation

$$\nabla^2 \boldsymbol{\mathcal{E}}_0 + \frac{\omega^2}{c^2} \boldsymbol{\mathcal{E}}_0 = 0$$

which is, for each component of $\boldsymbol{\mathcal{E}}_0$, of exactly the same form as the time-independent Schrödinger equation (9 · 9 · 8) for a nonrelativistic particle.

If the electromagnetic wave is regarded as quantized, then the associated photon is described in the familiar way as a relativistic particle of energy ϵ and momentum $\boldsymbol{p}$ given by the familiar relations

$$\left.\begin{aligned} \epsilon &= \hbar\omega \\ \boldsymbol{p} &= \hbar\boldsymbol{\kappa} \end{aligned}\right\} \tag{9·13·5}$$

Thus (9 · 13 · 4) implies that

$$|\boldsymbol{p}| = \frac{\hbar\omega}{c} \tag{9·13·6}$$

Since an electromagnetic wave satisfies the Maxwell equation $\nabla \cdot \boldsymbol{\mathcal{E}} = 0$, it follows by (9 · 13 · 3) that $\boldsymbol{\kappa} \cdot \boldsymbol{\mathcal{E}} = 0$, i.e., that $\boldsymbol{\mathcal{E}}$ is transverse to the direction of propagation determined by the vector $\boldsymbol{\kappa}$. For each $\boldsymbol{\kappa}$, there are thus only two possible components of $\boldsymbol{\mathcal{E}}$, perpendicular to $\boldsymbol{\kappa}$, which can be specified. In terms of photons this means that, for each $\boldsymbol{\kappa}$, there are two possible photons corresponding to the two possible directions of polarization of the electric field $\boldsymbol{\mathcal{E}}$.

As in the case of the particle discussed in Sec. 9 · 9, not all possible values of $\boldsymbol{\kappa}$ are allowed, but only certain discrete values depending on the boundary conditions. Let us again take the enclosure to be in the form of a paralellepiped with edges L_x, L_y, L_z in length. We suppose that the smallest of these lengths, call it L, is so large that $L \gg \lambda$ where $\lambda = 2\pi/\kappa$ is the longest wavelength of significance in the discussion. Then we can again neglect effects occurring near the walls of the container and can describe the situation in terms of simple traveling waves of the form (9 · 13 · 3). To eliminate wall effects it is only necessary to proceed as in Sec. 9 · 9 by imposing the periodic boundary conditions (9 · 9 · 11). The enumeration of possible states is then exactly identical to that in Sec. 9 · 9, the possible values of $\boldsymbol{\kappa}$ being those given by (9 · 9 · 13).

Let $f(\boldsymbol{\kappa})\, d^3\boldsymbol{\kappa}$ = the mean number of photons per unit volume, with *one* specified direction of polarization, whose wave vector lies between $\boldsymbol{\kappa}$ and $\boldsymbol{\kappa} + d\boldsymbol{\kappa}$.

There are, by virtue of (9 · 9 · 16), $(2\pi)^{-3}\, d^3\boldsymbol{\kappa}$ photon states of this kind per *unit* volume. Each of these has an energy $\epsilon = \hbar\omega = \hbar c\kappa$. Since the mean number of photons with *one* definite value of $\boldsymbol{\kappa}$ in this range is given by (9 · 13 · 1),

▶
$$f(\boldsymbol{\kappa})\, d^3\boldsymbol{\kappa} = \frac{1}{e^{\beta\hbar\omega} - 1} \frac{d^3\boldsymbol{\kappa}}{(2\pi)^3} \tag{9·13·7}$$

Obviously $f(\boldsymbol{\kappa})$ is only a function of $|\boldsymbol{\kappa}|$.

Let us find the mean number of photons per unit volume of *both* directions of polarization and with angular frequency in the range between ω and $\omega + d\omega$. This is given by summing (9·13·7) over all the volume of $\boldsymbol{\kappa}$ space contained within the spherical shell of radius $\kappa = \omega/c$ and $\kappa + d\kappa = (\omega + d\omega)/c$ and then multiplying by 2 to include both directions of polarization; i.e., it equals

$$2f(\kappa)(4\pi\kappa^2\,d\kappa) = \frac{8\pi}{(2\pi c)^3}\frac{\omega^2\,d\omega}{e^{\beta\hbar\omega}-1} \tag{9·13·8}$$

Let $\bar{u}(\omega;\,T)\,d\omega$ denote the mean energy per unit volume (i.e., the mean "energy *density*") of photons of both directions of polarization in the frequency range between ω and $\omega + d\omega$. Since each photon of this type has an energy $\hbar\omega$, one obtains

$$\bar{u}(\omega;\,T)\,d\omega = [2f(\kappa)(4\pi\kappa^2\,d\kappa)](\hbar\omega) = \frac{8\pi\hbar}{c^3}f(\kappa)\omega^3\,d\omega \tag{9·13·9}$$

or

$$\bar{u}(\omega;\,T)\,d\omega = \frac{\hbar}{\pi^2c^3}\frac{\omega^3\,d\omega}{e^{\beta\hbar\omega}-1} \tag{9·13·10}$$

Note that the significant dimensionless parameter of the problem is

$$\eta \equiv \beta\hbar\omega = \frac{\hbar\omega}{kT} \tag{9·13·11}$$

the ratio of photon energy to thermal energy. Thus one can write $\bar{u}$ in terms of η as

$$\bar{u}(\omega;\,T)\,d\omega = \frac{\hbar}{\pi^2c^3}\left(\frac{kT}{\hbar}\right)^4\frac{\eta^3\,d\eta}{e^\eta-1} \tag{9·13·12}$$

A plot of $\bar{u}$ as a function of η is shown in Fig. 9·13·1. The curve has a maximum for some value $\eta = \tilde{\eta} \approx 3$. Note a simple scaling property. If at tem-

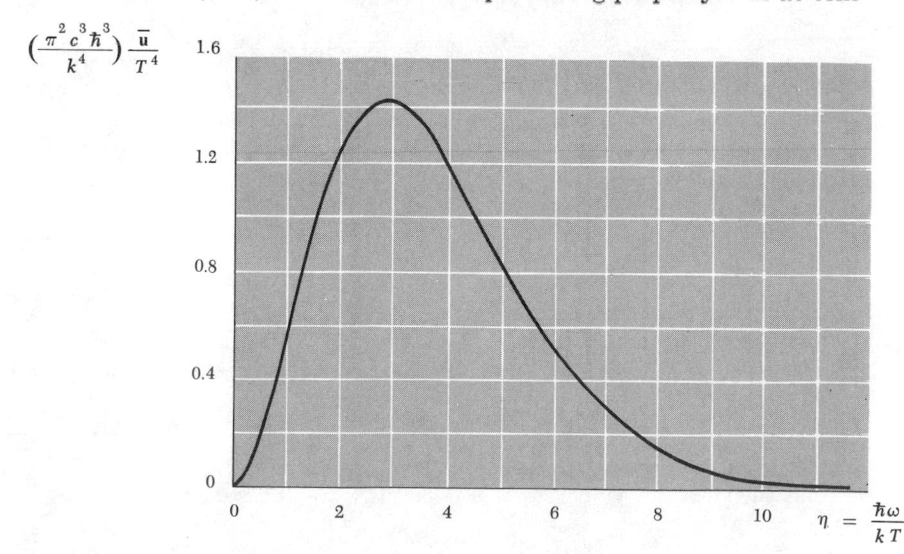

Fig. 9·13·1** **The energy density $\bar{u}(\eta)$ (per unit dimensionless frequency range $d\eta$) as a function of $\eta = \hbar\omega/kT$.

perature T_1 the maximum occurs at the angular frequency $\tilde{\omega}_1$, and at some other temperature T_2 the maximum occurs at $\tilde{\omega}_2$, then one must have

$$\frac{\hbar\tilde{\omega}_1}{kT_1} = \frac{\hbar\tilde{\omega}_2}{kT_2} = \tilde{\eta}$$

or

$$\frac{\tilde{\omega}_1}{T_1} = \frac{\tilde{\omega}_2}{T_2} \tag{9·13·13}$$

This result is known as "Wien's displacement law."

The mean *total* energy density $\bar{u}_0$ in *all* frequencies is given by

$$\bar{u}_0(T) = \int_0^\infty \bar{u}(T;\omega)\,d\omega$$

By (9·13·11) this becomes

$$\bar{u}_0(T) = \frac{\hbar}{\pi^2 c^3}\left(\frac{kT}{\hbar}\right)^4 \int_0^\infty \frac{\eta^3\,d\eta}{e^\eta - 1} \tag{9·13·14}$$

The definite integral here is just some constant. Hence one is left with the interesting result that

▶ $$\bar{u}_0(T) \propto T^4 \tag{9·13·15}$$

This statement is called the Stefan-Boltzmann law.

The integral (9·13·14) can easily be integrated numerically. Although this has no physical importance, it can also be evaluated exactly (see Appendix A·11). The result is

$$\int_0^\infty \frac{\eta^3\,d\eta}{e^\eta - 1} = \frac{\pi^4}{15} \tag{9·13·16}$$

Thus one obtains the explicit expression

$$\bar{u}_0(T) = \frac{\pi^2}{15}\frac{(kT)^4}{(c\hbar)^3} \tag{9·13·17}$$

The proportionality (9·13·15) is an obvious result reflecting the fact that space is three-dimensional. This can be seen by the following simple argument. At a temperature T most of the photons must have an energy of the order of kT or less, i.e., they must be photons with wave vector $\boldsymbol{\kappa}$ of magnitude less than κ' corresponding to an energy

$$\hbar\omega' = \frac{\hbar\kappa'}{c} \approx kT$$

But in three-dimensional space, the number of photon states with $|\boldsymbol{\kappa}|$ less κ' is proportional to the volume in $\boldsymbol{\kappa}$ space contained within a sphere of radius κ'. Hence the total mean number $\bar{N}$ of photons at temperature T must be proportional to

$$\bar{N} \propto \kappa'^3 \propto T^3 \tag{9·13·18}$$

The typical energy of these photons is of the order of kT. Hence it follows that the mean energy density $\bar{u}_0$ satisfies the proportionality

$$\bar{u}_0 \propto \bar{N}(kT) \propto T^4 \tag{9·13·19}$$

Calculation of radiation pressure It is of interest to calculate the mean pressure $\bar{p}$ exerted by the radiation on the walls of the enclosure. The pressure contribution from a photon in state s is given by $-\partial\epsilon_s/\partial V$; hence the mean pressure due to all the photons is*

$$\bar{p} = \sum_s \bar{n}_s \left(-\frac{\partial \epsilon_s}{\partial V}\right) \tag{9·13·20}$$

where $\bar{n}_s$ is given by (9·13·1). To evaluate $-\partial\epsilon_s/\partial V$, consider for simplicity that the enclosure is a cube of edge lengths $L_x = L_y = L_z \equiv L$ so that its volume is $V = L^3$. With the possible values of $\boldsymbol{\kappa}$ given by (9·9·13), one has for a state s specified by the integers n_x, n_y, n_z

$$\epsilon_s = \hbar\omega = \hbar c\kappa = \hbar c(\kappa_x{}^2 + \kappa_y{}^2 + \kappa_z{}^2)^{\frac{1}{2}} = \hbar c\left(\frac{2\pi}{L}\right)(n_x{}^2 + n_y{}^2 + n_z{}^2)^{\frac{1}{2}}$$

or

$$\epsilon_s = CL^{-1} = CV^{-\frac{1}{3}}, \qquad \text{where } C = \text{constant} \tag{9·13·21}$$

Hence

$$\frac{\partial \epsilon_s}{\partial V} = -\frac{1}{3}CV^{-\frac{4}{3}} = -\frac{1}{3}\frac{\epsilon_s}{V} \tag{9·13·22}$$

Thus (9·13·20) becomes

$$\bar{p} = \sum_s \bar{n}_s\left(\frac{1}{3}\frac{\epsilon_s}{V}\right) = \frac{1}{3V}\sum_s \bar{n}_s\epsilon_s = \frac{1}{3V}\bar{E}$$

or

$$\bar{p} = \tfrac{1}{3}\bar{u}_0 \tag{9·13·23}$$

The radiation pressure is thus very simply related to the mean energy density of the radiation.

It is also instructive to calculate the radiation pressure by detailed kinetic arguments similar to those used in Sec. 7·13 for computing the mean pressure exerted by a classical gas of particles. Photons impinging upon an element of area dA of the container wall (normal to the z direction) impart to it in unit time a mean z component of momentum $G_z^{(+)}$. In equilibrium, an equal number of photons leaves the wall and gives rise to an equal momentum flow $-G_z^{(+)}$ in the opposite direction. Hence the net force per unit area, or pressure on the wall, is related to the mean rate of change of momentum by

$$\bar{p} = \frac{1}{dA}[G_z^{(+)} - (-G_z^{(+)})] = \frac{2G_z^{(+)}}{dA}$$

Consider, in Fig. 9·13·2, all photons with wave vector between $\boldsymbol{\kappa}$ and $\boldsymbol{\kappa} + d\boldsymbol{\kappa}$. There are $2f(\boldsymbol{\kappa})\, d^3\boldsymbol{\kappa}$ photons of this kind (of both possible polarizations) per unit volume. Since photons travel with speed c, all photons contained in the cylindrical volume $c\, dt\, dA \cos\theta$ strike the area dA in time dt and carry z component of momentum $\hbar\kappa_z$. The total photon momentum arriving at dA per unit time is then

$$G_z^{(+)} = \frac{1}{dt}\int_{\kappa_z>0} [2f(\boldsymbol{\kappa})\, d^3\boldsymbol{\kappa}](c\, dt\, dA \cos\theta)(\hbar\kappa_z)$$

* The same result could be obtained from the general relation $\bar{p} = \beta^{-1}(\partial \ln Z/\partial V)$ by using the partition function for photons in (9·5·3).

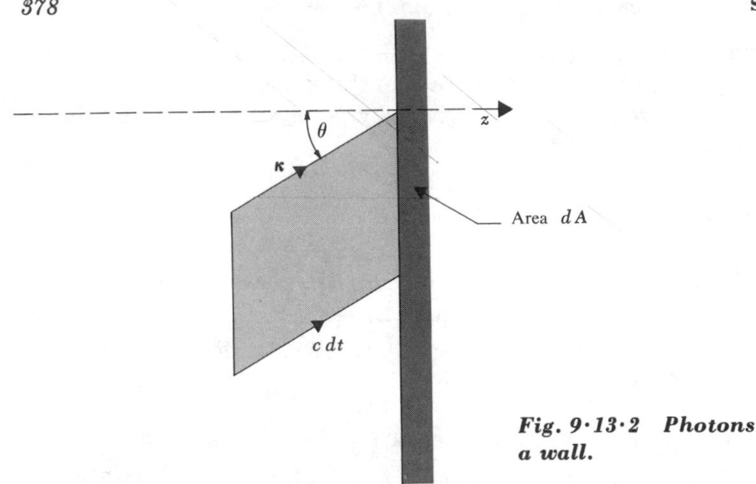

Fig. 9·13·2 Photons impinging upon a wall.

Hence
$$\bar{p} = 2c\hbar \int_{\kappa_z>0} [2f(\boldsymbol{\kappa})\, d^3\boldsymbol{\kappa}] \frac{\kappa_z^2}{\kappa}$$

where we have put $\cos\theta = \kappa_z/\kappa$. But $f(\boldsymbol{\kappa})$ depends only on $|\boldsymbol{\kappa}|$, so that the integrand is an even function of κ_z. Thus one can extend the integration over *all* values of $\boldsymbol{\kappa}$ and write

$$\bar{p} = c\hbar \int [2f(\kappa)\, d^3\boldsymbol{\kappa}] \frac{\kappa_z^2}{\kappa} = \frac{1}{3} c\hbar \int [2f(\kappa)\, d^3\boldsymbol{\kappa}] \frac{(\kappa_x^2 + \kappa_y^2 + \kappa_z^2)}{\kappa}$$

where the last result follows by symmetry, since all directions are equivalent. Since $\kappa_x^2 + \kappa_y^2 + \kappa_z^2 = \kappa^2$, one then obtains

$$\bar{p} = \tfrac{1}{3} \int [2f(\boldsymbol{\kappa})\, d^3\boldsymbol{\kappa}](c\hbar\kappa) = \tfrac{1}{3}\bar{u}_0$$

since $c\hbar\kappa$ is simply the energy of a photon of wave vector $\boldsymbol{\kappa}$.

9 · 14 *Nature of the radiation inside an arbitrary enclosure*

The full generality of the results of the preceding section can be made apparent by a few simple physical arguments. Consider an enclosure which has an arbitrary shape and which may contain several bodies within it. Its walls, which may consist of any material, are maintained at an absolute temperature T. The enclosure thus acts as a heat reservoir. From our general arguments of statistical thermodynamics we know that the equilibrium situation of greatest probability, or entropy, is the one where the radiation, as well as the bodies inside the enclosure, are all characterized by the same temperature T.

The nature of the radiation field existing at this temperature T inside the enclosure can be described in terms of

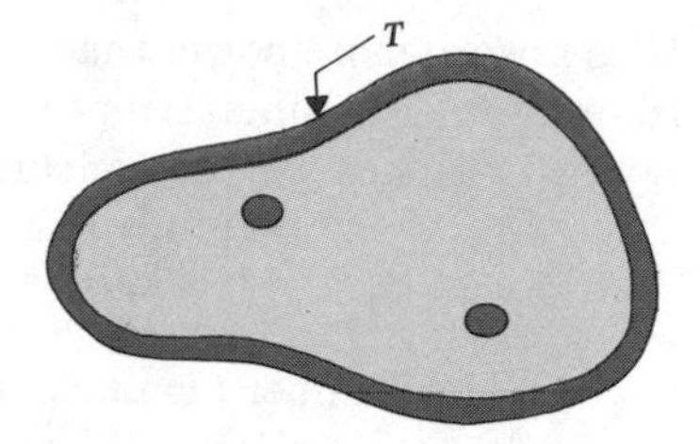

Fig. 9·14·1 Electromagnetic radiation in equilibrium inside an enclosure of arbitrary shape. The radiation must be homogeneous.

$f_\alpha(\boldsymbol{\kappa},\boldsymbol{r})\, d^3\boldsymbol{\kappa}$ = the mean number of photons per unit volume at the point $\boldsymbol{r}$, with wave vector between $\boldsymbol{\kappa}$ and $\boldsymbol{\kappa} + d\boldsymbol{\kappa}$ and with polarization specified by the index α (i.e., by some unit vector $\boldsymbol{b}_\alpha \perp \boldsymbol{\kappa}$)

As usual we assume that the dimensions of the enclosure are large compared to the wavelengths $\lambda = 2\pi\kappa^{-1}$ of interest.

If the enclosure is in equilibrium, one can immediately make several general statements about f.

1. The number f is independent of $\boldsymbol{r}$; i.e., the radiation field is homogeneous.

Argument: Suppose that $f_\alpha(\boldsymbol{\kappa},\boldsymbol{r})$ were different at two positions in the enclosure. Consider what would happen if two *identical* small bodies at temperature T were placed at these positions. (Imagine these to be surrounded by filters which transmit only frequencies in the specified range $\omega = \omega(\boldsymbol{\kappa})$ and which transmit only radiation of the specified direction of polarization α.) Since different amounts of radiation would be incident on the two bodies, they would absorb different amounts of energy per unit time and their temperatures would therefore become different. This would contradict the equilibrium condition of maximum entropy according to which the temperature must be uniform throughout the enclosure. Hence

$$f_\alpha(\boldsymbol{\kappa},\boldsymbol{r}) = f_\alpha(\boldsymbol{\kappa}) \qquad \text{independent of } \boldsymbol{r}$$

2. The number f is independent of the direction of $\boldsymbol{\kappa}$, but depends only on $|\boldsymbol{\kappa}|$; i.e., the radiation field is isotropic.

Argument: Suppose that $f_\alpha(\boldsymbol{\kappa})$ did depend on the direction $\boldsymbol{\kappa}$, e.g., that f is greater if $\boldsymbol{\kappa}$ points north than if it points east. We could again imagine that two identical small bodies at temperature T (and surrounded by the same filters as before) are introduced into the enclosure as shown in Fig. 9·14·2. Then the body on the north side would have more radiation incident on it and

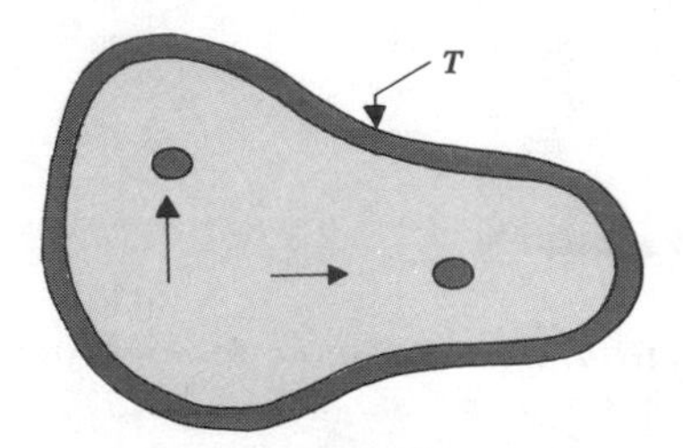

Fig. 9·14·2 The radiation in equilibrium inside the enclosure is isotropic.

thus absorb more power than the body on the east side. This would again lead to a nonpermissible temperature difference being produced between these bodies. Hence we can conclude that

$$f_\alpha(\boldsymbol{\kappa}) = f_\alpha(\kappa), \qquad \text{where } \kappa \equiv |\boldsymbol{\kappa}|$$

3. The number f is independent of the direction of polarization of the radiation, i.e., the radiation field in the enclosure is unpolarized.

Argument: Suppose that $f_\alpha(\kappa)$ did depend on the direction of polarization specified by α. Then we could imagine that two identical small bodies at the temperature T are introduced side by side into the enclosure and are surrounded by filters which transmit different directions of polarization. Hence different amounts of radiation would be incident upon these bodies, and a temperature difference would be developed between them in contradiction to the equilibrium condition. Hence

$$f_1(\kappa) = f_2(\kappa)$$

is independent of the polarization index.

4. The function f does not depend on the shape nor volume of the enclosure, nor on the material of which it is made, nor on the bodies it may contain.

Argument: Consider two different enclosures, both at the temperature T, and suppose that the fractions $f_\alpha^{(1)}(\kappa)$ and $f_\alpha^{(2)}(\kappa)$ describing their radiation fields were different. Imagine that we connect the two enclosures through a small hole (containing a filter which transmits only radiation in a narrow frequency range about $\omega(\kappa)$ and of the specified direction of polarization). This would represent an equilibrium situation if both enclosures are at the same temperature T. But if $f^{(1)} > f^{(2)}$, more radiation per unit time would pass from enclosure 1 into enclosure 2 than in the opposite direction. A temperature difference would then develop between the two enclosures, in contradiction to the equilibrium condition of uniform temperature. Hence one concludes that

$$f_\alpha^{(1)}(\kappa) = f_\alpha^{(2)}(\kappa)$$

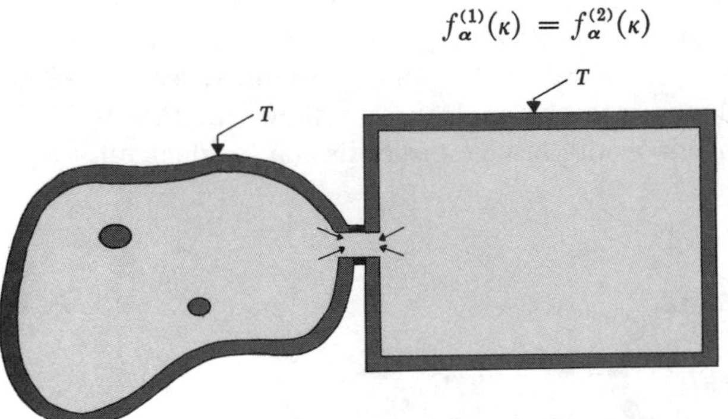

***Fig. 9·14·3** Two different enclosures at the same temperature joined through a small hole.*

Thus we arrive at the result that in thermal equilibrium $f_\alpha(\kappa)$ depends *only* on the temperature T of the enclosure. In particular, it also follows that f is the same for an arbitrary cavity as it is for the rectangular parallellepiped cavity which we used for simplicity in the discussion of Sec. 9·13.

9·15 *Radiation emitted by a body at temperature T*

In the preceding sections we considered the very simple *equilibrium* situation of electromagnetic radiation confined within an enclosure at absolute temperature T. We arrived at a set of interesting results on the basis of very general arguments of statistical mechanics without having to be concerned about the detailed mechanisms of absorption or emission of radiation by the walls. The results of this equilibrium discussion can, however, serve as a basis for treating much more general cases. Consider, for example, a body maintained at some elevated absolute temperature T; as a concrete example, think of the hot filament of a light bulb hanging from the ceiling. We know that this body emits electromagnetic radiation, and we may be interested in how much energy per unit time (or power) $\mathcal{P}_e(\omega)\,d\omega$ this body emits by radiation in the frequency range between ω and $\omega + d\omega$. The situation envisaged here is certainly *not* one of equilibrium; the walls of the room are at a much lower temperature than the light-bulb filament, and there is a continuous transfer of energy by radiation from the hot filament to the colder walls. It might seem, therefore, that we can no longer use the methods of equilibrium statistical mechanics to discuss this problem and that we need to undertake a detailed investigation of the processes whereby the atoms in the body emit radiation. This would indeed be a formidable problem in quantum mechanics and electromagnetic theory! It is possible, however, to circumvent completely such an analysis by reverting to very clever general arguments based on the equilibrium situation. The method of approach consists of imagining the radiating body to *be* in an equilibrium situation inside an enclosure containing radiation at this temperature T, and then investigating the conditions that must prevail so that the equilibrium is indeed maintained. The fundamental argument used here is one of "detailed balance"; i.e., one argues that if the body is to remain in equilibrium, then each process of emission by the body must be balanced by an inverse process of absorption of incident radiation. But the radiation incident on the body in an *equilibrium* situation is easily calculated from the results of the preceding sections dealing with the simple case of an ideal gas of photons. Thus one can immediately find the power emitted in such a process *without* engaging in the vastly more complicated calculation of how a collection of interacting atoms in the body emit radiation. Now that we have outlined the general nature of the arguments to be used, let us make them more precise.

Bodies as emitters and absorbers of radiation Consider an arbitrary body at absolute temperature T. The electromagnetic radiation emitted by

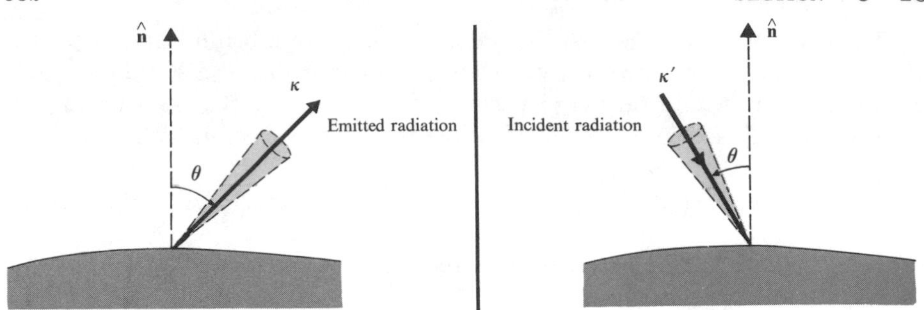

Fig. 9·15·1 Diagram illustrating emission and absorption of radiation by a body.

this body can be described in terms of the energy per unit time, or power, emitted by the body. Specifically one can define its "emissivity" as

> $\mathcal{P}_e(\boldsymbol{\kappa};\,\alpha)\,d\omega\,d\Omega$ = the power, per unit area of the body, emitted with polarization α into a range about $\boldsymbol{\kappa}$ (i.e., into an angular frequency range between ω and $\omega + d\omega$, and into a solid angle $d\Omega$ about the direction $\boldsymbol{\kappa}$)

The emissivity depends on the nature of the body and on its temperature.

Having seen how to describe a body as an emitter of radiation, let us now try to describe it as an absorber of radiation. For this purpose, consider radiation of polarization α and with a wave vector in a small range about $\boldsymbol{\kappa}'$ (i.e., with angular frequency in the range between ω and $\omega + d\omega$ and propagating in a direction lying within the solid-angle range $d\Omega$ about $\boldsymbol{\kappa}'$). Suppose that radiation of this type is *incident* on the body so that power $\mathcal{P}_i(\boldsymbol{\kappa}',\alpha)\,d\omega\,d\Omega$ is incident per unit area of the body. Of this a fraction $a(\boldsymbol{\kappa}',\alpha)$ is absorbed by the body. (By conservation of energy the rest of the incident power is then reflected into various directions if we assume the body to be sufficiently thick that none of the incident radiation is transmitted through it.) The parameter $a(\boldsymbol{\kappa}',\alpha)$ (sometimes called the "absorptivity") is characteristic of the particular body and depends, in general, also on its temperature T. This parameter describes the properties of the body as an absorber of radiation.

The principle of detailed balance We pointed out earlier that it would be quite difficult to calculate directly quantities such as the power $\mathcal{P}_e$ radiated by a body at temperature T. To circumvent this problem, we imagine that the body under consideration is placed inside an enclosure at the same temperature T so as to be in equilibrium with the radiation field existing therein. The characteristics of this radiation field are well known from our previous simple discussion based on equilibrium statistical thermodynamics. Let us now, however, consider more closely the various mechanisms whereby the equilibrium state of the body in this enclosure is actually maintained. Under these circumstances the body emits radiation. On the other hand, radiation is continually incident upon the body which absorbs a certain fraction of it. In the equilibrium situation the energy of the body must remain unchanged. Hence we can conclude that these processes must balance so that

$$\text{Power radiated by body} = \text{power absorbed by body.} \qquad (9 \cdot 15 \cdot 1)$$

We shall, however, want to make statements which are much stronger than this simple condition of over-all energy balance by asserting that the processes which maintain the equilibrium also balance each other in *detail*. For example, it might be conceivable that in one frequency range the body radiates more power than it absorbs, while in another frequency range it radiates less power than it absorbs, in such a way that the over-all energy balance (9 · 15 · 1) is preserved. A simple physical argument shows, however, that this cannot be the case. Imagine that the body is surrounded by a shield (a "filter") which absorbs completely all radiation except that, in one small element of area, it is completely transparent to radiation of one direction of polarization and of one narrow frequency range between ω and $\omega + d\omega$. The presence of this shield cannot affect such intrinsic parameters of the body as its emissivity or absorptivity; nor can it, by the arguments of the preceding section, affect the nature of the radiation in the enclosure. Since the equilibrium situation can equally well exist in the presence of the shield, it follows that the energy balance (9 · 15 · 1) must hold for this particular element of area, direction of polarization, and frequency range. Since any kind of shield could have been used, one thus arrives at the "principle of *detailed* balance," which asserts that in equilibrium the power radiated and absorbed by the body must be equal for *any* particular element of area of the body, for *any* particular direction of polarization, and for *any* frequency range.

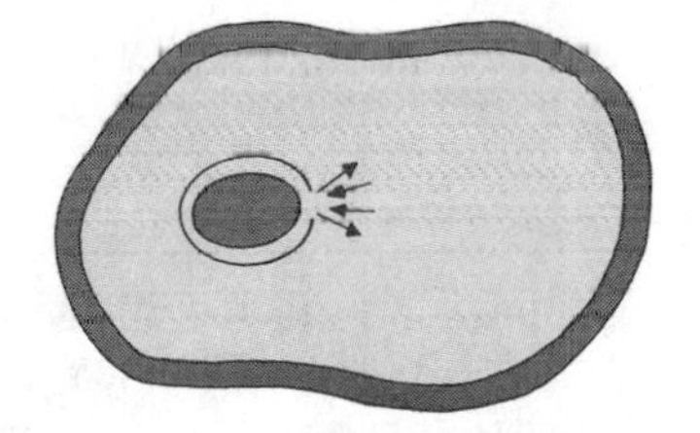

***Fig. 9·15·2** A body located inside an enclosure and surrounded by a shield which is only transparent in one small element of area to radiation of one direction of polarization and of one narrow frequency range.*

Microscopic discussion The principle of detailed balance is a very fundamental result based on considerations more general than those pertaining to ensembles representing systems in thermal equilibrium. The basic justification of the principle rests on the fundamental laws of microscopic physics, e.g., the Schrödinger equation of quantum mechanics and Maxwell's equations of electromagnetic theory. Consider a single isolated system consisting of several weakly interacting parts (e.g., a body and electromagnetic radiation). In the absence of interaction between these parts the system can be in any one of its quantum states labeled by indices r, s, etc. The presence of the interaction causes transitions between these states. From the fundamental microscopic laws one can compute the resulting transition probability w_{rs} per unit time from state r to state s. But these microscopic laws are all invariant under reversal of the time from t to $-t$. Under such a time reversal a state r goes over into a state r^*, etc. (e.g., a state of a particle labeled by its momentum $\mathbf{p} = \hbar\boldsymbol{\kappa}$ goes over into one labeled by momentum $-\mathbf{p}$). If we call the "reverse" transition the one from state s^* to r^*, then the invariance of the microscopic laws under time reversal implies that

$$w_{s^*r^*} = w_{rs} \tag{9·15·2}$$

This expresses the "principle of microscopic reversibility." For example, consider the process of emission of a photon with wave vector $\boldsymbol{\kappa}$. The reverse process obtained by reversing the sign of the time t is the absorption of a photon of wave vector $-\boldsymbol{\kappa}$. The microscopic reversibility (9·15·2) asserts that these two processes occur with equal probability.

Once one knows the transition probability for the occurrence of a process in a single system, one can readily calculate the rate of occurrence of this process when one deals with a statistical ensemble of such systems. Let us consider the process of transition from a set A of states labeled by r to some set B of states labeled by s. Let P_r denote the probability in the ensemble that the system is in state r. Then the probability W_{AB} of occurrence of the process $A \to B$ in the ensemble is given by

$$W_{AB} = \sum_r \sum_s P_r w_{rs} \tag{9·15·3}$$

Here one sums over all the initial states r in the set A from which the system can start out, each of these states being weighted by the probability that the system is found in this state; then one sums this probability over all the possible set B of final states s. Similarly, one can write for the rate of occurrence of the reverse process

$$W_{B^*A^*} = \sum_{s^*} \sum_{r^*} P_{s^*} w_{s^*r^*} \tag{9·15·4}$$

But our fundamental statistical postulate asserts that, in an equilibrium situation, an isolated system is found with equal probability in any accessible state of the ensemble. All the probabilities P_r are then equal to the same value P. Hence one obtains by (9·15·2)

$$W_{B^*A^*} = P \sum_{s^*} \sum_{r^*} w_{s^*r^*} = P \sum_s \sum_r w_{rs}$$

so that

$$W_{B^*A^*} = W_{AB} \tag{9·15·5}$$

This is the principle of detailed balance. In words it asserts that in the statistical ensemble representing a system in equilibrium, the probability of occurrence of *any* process must be equal to the probability of occurrence of the reverse process. By a process we mean transitions from one set of states of the system to another such set of states, the probability of the process being proportional to the number of such transitions occurring per unit time. The reverse process is the one that would result if the sign of the time were reversed, in particular, if all velocities were reversed so that everything went backward in time.

Radiation emitted by a body Let us now apply the principle of detailed balance to a body at temperature T in equilibrium with radiation inside an enclosure at this temperature. On a unit area of this body radiation power $\mathcal{P}_i(\boldsymbol{\kappa},\alpha)$ is incident per unit frequency and solid angle range about the vector $\boldsymbol{\kappa}$; a fraction $a(\boldsymbol{\kappa},\alpha)$ of this is absorbed, the rest being reflected. We know that the reverse process occurs with equal probability. In this process an amount of power $\mathcal{P}_e(-\boldsymbol{\kappa},\alpha)$ is emitted by this area of the body per unit frequency and solid angle range about the direction $-\boldsymbol{\kappa}$. By equating the powers involved in these two processes, one obtains

$$\mathcal{P}_e(-\boldsymbol{\kappa},\alpha) = a(\boldsymbol{\kappa},\alpha)\mathcal{P}_i(\boldsymbol{\kappa},\alpha) \tag{9·15·6}$$

or

$$\frac{\mathcal{P}_e(-\boldsymbol{\kappa},\alpha)}{a(\boldsymbol{\kappa},\alpha)} = \mathcal{P}_i(\boldsymbol{\kappa},\alpha) \tag{9·15·7}$$

Note that on the left side of this last equation there are quantities that depend only on the nature of the particular body and on its temperature. They are parameters that could be calculated from first principles (if we were clever enough to carry through the computation), and they are not affected at all by the fact that the body happens to be located in the radiation field of the enclosure in the particular equilibrium situation which we are envisaging. On the other hand, the incident radiation power $\mathcal{P}_i$ on the right side of (9·15·7) depends only on the temperature of the equilibrium radiation field inside the enclosure and is *independent* of the nature of the body. Hence one can immediately conclude that the ratio on the left side of (9·15·7) can only depend on the temperature. There exists, therefore, a very close connection between the emissivity $\mathcal{P}_e$ and the absorptivity a of a body. *A good emitter of radiation is also a good absorber of radiation, and vice versa.* This is a qualitative statement of "Kirchhoff's law." Note that this statement refers only to properties of the body and is thus generally valid, even if the body is *not* in equilibrium; but we arrived at this conclusion by investigating the conditions which must be fulfilled to make the properties of the body consistent with a possible equilibrium situation.

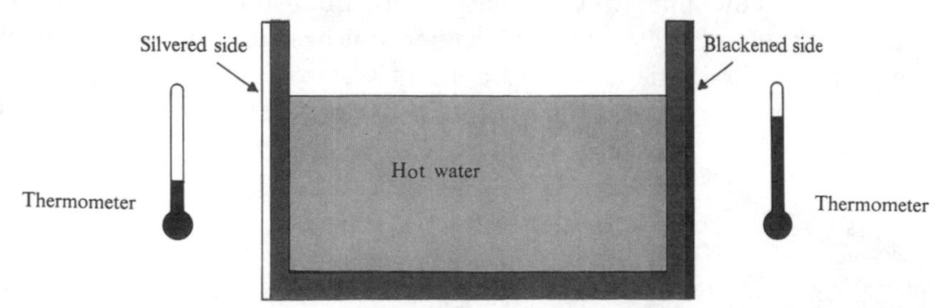

Fig. 9·15·3 A classical experiment illustrating Kirchhoff's law. The container is filled with hot water. Its left side is silvered on the outside so that it is a poor absorber; its right side is blackened so that it is a good absorber. Since the left side is then a poorer emitter of radiation than the right side, the thermometer on the left is found to indicate a lower temperature than the one on the right.

Remark Kirchhoff's law is a reasonable result by virtue of the following microscopic considerations. Focus attention on a pair of energy levels of the body; transitions between these give rise to emission or absorption of radiation at some frequency ω. If transitions between these levels are readily produced (i.e., if the transition probability is large), then the electric field in the incident radiation can readily induce absorption in transitions from the lower to the upper level; but then the thermal agitation can also readily induce emission in transitions from the upper to the lower level.

A particularly simple case arises if $a(\boldsymbol{\kappa},\alpha) = 1$ for all polarizations, frequencies, and directions of the incident radiation. A body having this property is a perfect absorber of radiation and is called a "black body." (The reason for the name is clear, since a body which absorbs all radiation incident on it would *look* black.) For a *black* body (9 · 15 · 6) becomes simply

$$\mathcal{P}_{eb}(-\boldsymbol{\kappa},\alpha) = \mathcal{P}_i(\boldsymbol{\kappa},\alpha) \tag{9·15·8}$$

Substances such as lampblack are reasonable approximations to black bodies, but by no means perfectly so, since they do not absorb all radiation at *all* frequencies. The best approximation to a black body is a small hole in the wall of some enclosure. Consider such a hole. Any radiation incident on this hole from outside gets trapped inside the enclosure with negligible probability of escaping through the hole as a result of several reflections. Thus the hole acts like a perfect absorber of all radiation incident on it, i.e., like a black body. By (9 · 15 · 8) the power emitted by *any* black body has the same characteristics. In particular, the hole cut in the enclosure can serve (and in practice *does* serve) as the prototype of a black body emitter. The emission characteristics of this hole are, of course, particularly easily calculated. The problem here is simply that of "effusion" of photons from the enclosure through the hole (analogous to the effusion of molecules discussed in Sec. 7 · 12).

Let us now put (9 · 15 · 6) into more quantitative form by calculating explicitly the power $\mathcal{P}_i(\boldsymbol{\kappa},\alpha)$ incident per unit area of a body in an enclosure at temperature T. This is readily done in terms of the mean number $f(\kappa)\, d^3\boldsymbol{\kappa}$ of photons per unit volume and of a given polarization, the quantity found in

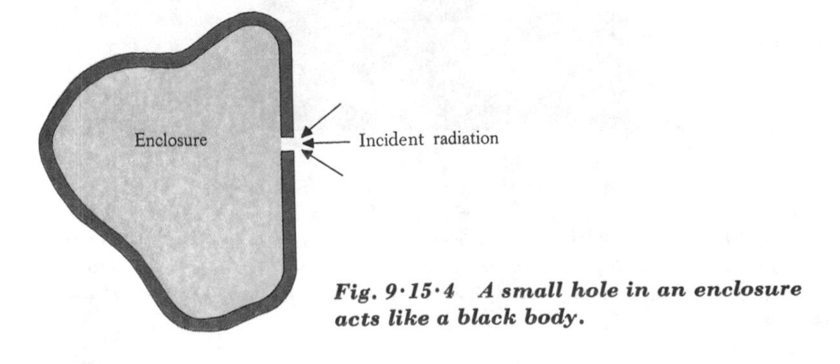

Fig. 9 · 15 · 4 A small hole in an enclosure acts like a black body.

(9·13·7). Referring to Fig. 9·13·2, the familiar argument demonstrates that $(c\,dt\cos\theta)f(\kappa)\,d^3\boldsymbol{\kappa}$ photons of this kind strike a unit area of the body in time dt. Since each photon carries energy $\hbar\omega$ one obtains

$$\mathcal{P}_i(\boldsymbol{\kappa},\alpha)\,d\omega\,d\Omega = (\hbar\omega)(c\cos\theta\,f(\kappa)\,d^3\boldsymbol{\kappa})$$

Expressing the volume element $d^3\boldsymbol{\kappa}$ in spherical coordinates and using the relation $\kappa = \omega/c$, one has

$$d^3\boldsymbol{\kappa} = \kappa^2\,d\kappa\,d\Omega = \frac{\omega^2}{c^3}\,d\omega\,d\Omega$$

Hence

$$\mathcal{P}_i(\boldsymbol{\kappa},\alpha) = \frac{\hbar\omega^3}{c^2}\,f(\kappa)\cos\theta \tag{9·15·9}$$

This is independent of the direction of polarization α, since f does not depend on it, but *does* depend on the angle θ of the direction of incidence with respect to the normal to the surface.

From the detailed balance argument (9·15·6) one thus obtains for the power emitted by a body in the direction $\boldsymbol{\kappa}' = -\boldsymbol{\kappa}$

$$\mathcal{P}_e(\boldsymbol{\kappa}',\alpha)\,d\omega\,d\Omega = a(-\boldsymbol{\kappa}',\alpha)\,\frac{\hbar\omega^3}{c^2}\,f(\kappa)\cos\theta\,d\omega\,d\Omega \tag{9·15·10}$$

If the body absorbs isotropically so that $a(-\boldsymbol{\kappa}',\alpha)$ is independent of the direction of $\boldsymbol{\kappa}'$, this shows that the power emitted is proportional to $\cos\theta$, where θ is the angle between the direction of emission and the normal to the surface. This result is known as "Lambert's law."

Let us now find the total power $\mathcal{P}_e(\omega)\,d\omega$ emitted per unit area into the frequency range between ω and $\omega + d\omega$ for *both* directions of polarization. Then one must integrate (9·15·10) over all possible directions of emission, i.e., over all solid angles in the polar angle range $0 < \theta < \pi/2$ and azimuthal angle range $0 < \varphi < 2\pi$. Then one must multiply by 2 to include both directions of polarization. Assume, for simplicity, that the absorptivity $a = a(\omega)$ is independent of the direction and polarization of the incident radiation. Since $d\Omega = \sin\theta\,d\theta\,d\varphi$, one then obtains

$$\begin{aligned}\mathcal{P}_e(\omega)\,d\omega &= 2\int_\Omega \mathcal{P}_e(\boldsymbol{\kappa}',\alpha)\,d\omega\,d\Omega\\ &= a(\omega)\,\frac{2\hbar\omega^3}{c^2}\,f(\kappa)\,d\omega\left(2\pi\int_0^{\pi/2}\cos\theta\sin\theta\,d\theta\right)\end{aligned}$$

or

$$\mathcal{P}_e(\omega)\,d\omega = a(\omega)\,\frac{2\pi\hbar\omega^3}{c^2}\,f(\kappa)\,d\omega \tag{9·15·11}$$

Here the right side is proportional to $(\hbar\omega)f(\kappa)\,d^3\boldsymbol{\kappa}$, i.e., to the mean radiation energy density $\bar{u}(\omega)\,d\omega$ inside an enclosure. Thus it can be expressed explicitly in terms of $\bar{u}(\omega)$ by (9·13·9) to yield the simple result

▶
$$\mathcal{P}_e(\omega)\,d\omega = a(\omega)[\tfrac{1}{4}c\bar{u}(\omega)\,d\omega] \tag{9·15·12}$$

Using the equilibrium relation (9·13·7) for $f(\kappa)$, or equivalently, the relation

(9·13·10), one thus obtains

$$\mathcal{P}_e(\omega)\,d\omega = a(\omega)\,\frac{\hbar}{4\pi^2c^2}\,\frac{\omega^3\,d\omega}{e^{\beta\hbar\omega}-1} \tag{9·15·13}$$

For a black body $a(\omega) = 1$, and this becomes the famous Planck law for the spectral distribution of black-body radiation. By virtue of (9·15·12), the frequency and temperature dependence of $\mathcal{P}_e(\omega)$ is then the same as that illustrated in Fig. 9·13·1.

The *total* power $\mathcal{P}_e^{(0)}$ emitted per unit area of the body is obtained by integrating (9·15·13) over all frequencies. If $a(\omega)$ has a constant value a in the frequency range where $\mathcal{P}_e(\omega)$ is not negligibly small, then the integral is the same as that encountered in obtaining (9·13·17). Thus one gets simply

$$\mathcal{P}_e^{(0)} = a(\tfrac{1}{4}c\bar{u}_0) = a(\sigma T^4) \tag{9·15·14}$$

where

$$\sigma \equiv \frac{\pi^2}{60}\,\frac{k^4}{c^2\hbar^3} \tag{9·15·15}$$

The relation (9·15·14) is called the Stefan-Boltzmann law, and the coefficient σ is called the Stefan-Boltzmann constant. Its numerical value is

$$\sigma = (5.6697 \pm 0.0029) \times 10^{-5} \text{ erg sec}^{-1}\text{ cm}^{-2}\text{ deg}^{-4} \tag{9·15·16}$$

In the case of a black body, $a = 1$. If one is dealing with radiation in the infrared region, $a \approx 0.98$ for a substance such as lampblack; on the other hand, $a \approx 0.01$ for a metal such as gold with a well-polished surface.

CONDUCTION ELECTRONS IN METALS

9 · 16 *Consequences of the Fermi-Dirac distribution*

It is possible to neglect, to a first approximation, the mutual interaction of the conduction electrons in a metal. These electrons can therefore be treated as an ideal gas. Their concentration in a metal is, however, so high that they cannot be treated at ordinary temperatures by the approximation of classical statistics. This conclusion was made apparent by the numerical example at the end of Sec. 7·4. Hence the appropriate Fermi-Dirac statistics must be used to discuss conduction electrons in a metal.

In (9·3·14) we found for the mean number of particles in state s the FD distribution

$$\bar{n}_s = \frac{1}{e^{\alpha+\beta\epsilon_s}+1} = \frac{1}{e^{\beta(\epsilon_s-\mu)}+1} \tag{9·16·1}$$

Here we have used the definition

$$\mu \equiv -\frac{\alpha}{\beta} = -kT\alpha \tag{9·16·2}$$

The quantity μ is called the "Fermi energy" of the system. (Incidentally, we showed in (9·3·17) that μ is also the chemical potential of the gas). The parameter α or μ is to be determined by the condition that

$$\sum_s \bar{n}_s = \sum_s \frac{1}{e^{\beta(\epsilon_s - \mu)} + 1} = N \tag{9·16·3}$$

where N is the total number of particles in the volume V. By virtue of (9·16·3), μ is then a function of the temperature.

Let us look at the behavior of the "Fermi function"

$$F(\epsilon) \equiv \frac{1}{e^{\beta(\epsilon - \mu)} + 1} \tag{9·16·4}$$

as a function of ϵ, this energy being measured above its lowest possible value $\epsilon = 0$. If μ is such that $\beta\mu \ll 0$, then $e^{\beta(\epsilon-\mu)} \gg 1$ and F reduces to the Maxwell-Boltzmann distribution. In the present case, however, we are interested in the opposite limit where

$$\beta\mu = \frac{\mu}{kT} \gg 1 \tag{9·16·5}$$

In this case, if $\epsilon \ll \mu$, then $\beta(\epsilon - \mu) \ll 0$ so that $F(\epsilon) = 1$. On the other hand, if $\epsilon \gg \mu$, then $\beta(\epsilon - \mu) \gg 0$ so that $F(\epsilon) = e^{\beta(\mu-\epsilon)}$ falls off exponentially like a classical Boltzmann distribution. If $\epsilon = \mu$, then $F = \frac{1}{2}$. The transition

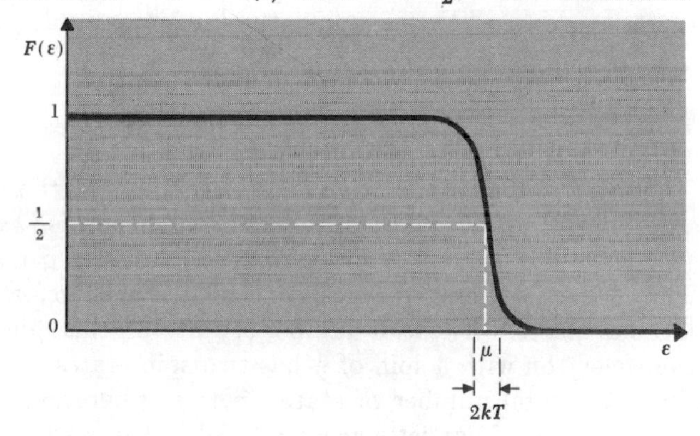

Fig. 9·16·1 The Fermi function at a finite temperature T.

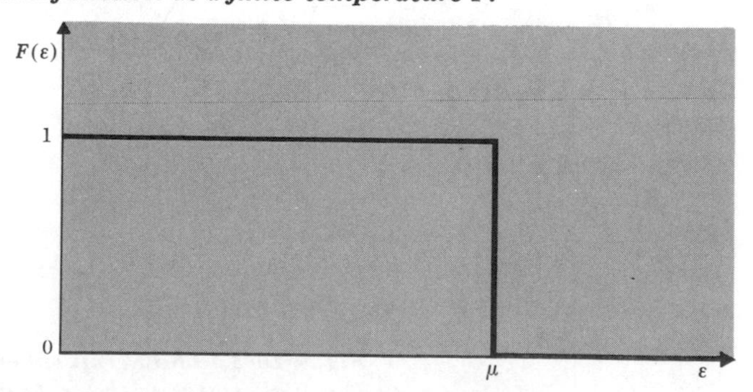

Fig. 9·16·2 The Fermi function at T = 0.

region in which F goes from a value close to 1 to a value close to zero corresponds to an energy interval of the order of kT about $\epsilon = \mu$ (see Fig. 9·16·1). In the limit when $T \to 0$ (or $\beta \to \infty$) this transition region becomes infinitesimally narrow. In this case $F = 1$ for $\epsilon < \mu$ and $F = 0$ for $\epsilon > \mu$, as illustrated in Fig. 9·16·2. This is an obvious result for the absolute-zero situation where the gas is in its ground state of lowest energy. Since the exclusion principle requires that there be no more than one particle per single-particle state, the lowest energy of the gas is obtained by piling all particles into the lowest available unoccupied states until all the particles are accomodated. The last particle thus added to the pile has quite a considerable energy μ, since all the lower energy states have already been used up. Thus one sees that the exclusion principle implies that a FD gas has a large mean energy even at absolute zero.

Let us calculate the Fermi energy $\mu = \mu_0$ of a gas at $T = 0$. The energy of each particle is related to its momentum $\mathbf{p} = \hbar\boldsymbol{\kappa}$ by

$$\epsilon = \frac{p^2}{2m} = \frac{\hbar^2\kappa^2}{2m} \tag{9·16·6}$$

At $T = 0$ all states of lowest energy are filled up to the Fermi energy μ, which corresponds to a "Fermi momentum" of magnitude $p_F = \hbar\kappa_F$ such that

$$\mu_0 = \frac{p_F{}^2}{2m} = \frac{\hbar^2\kappa_F{}^2}{2m} \tag{9·16·7}$$

Thus at $T = 0$ all states with $\kappa < \kappa_F$ are filled, all those with $\kappa > \kappa_F$ are empty. The volume of the sphere of radius κ_F in $\boldsymbol{\kappa}$ space is $(\frac{4}{3}\pi\kappa_F{}^3)$. But, by virtue of (9·9·16), there are $(2\pi)^{-3}V$ translational states per unit volume of $\boldsymbol{\kappa}$ space. The "Fermi sphere" of radius κ_F contains therefore $(2\pi)^{-3}V\,(\frac{4}{3}\pi\kappa_F{}^3)$ translational states. The *total* number of states in this sphere is twice as large, since each electron with a spin of $\frac{1}{2}$ has two spin states for each translational state. Since the total number of states in this sphere must at $T = 0$ be equal the total number of particles accomodated in these states, it follows that

$$2\,\frac{V}{(2\pi)^3}\left(\frac{4}{3}\pi\kappa_F{}^3\right) = N$$

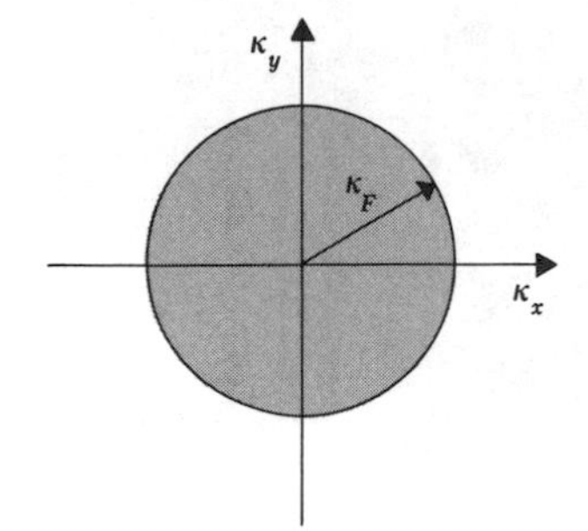

***Fig. 9·16·3** The Fermi sphere in $\boldsymbol{\kappa}$ space. At $T = 0$ all states with $\kappa < \kappa_F$ are completely occupied by particles, those with $\kappa > \kappa_F$ are completely empty.*

or

$$\kappa_F = \left(3\pi^2 \frac{N}{V}\right)^{\frac{1}{3}} \tag{9·16·8}$$

Thus

$$\lambda_F \equiv \frac{2\pi}{\kappa_F} = \frac{2\pi}{(3\pi^2)^{\frac{1}{3}}} \left(\frac{V}{N}\right)^{\frac{1}{3}} \tag{9·16·9}$$

This means that the de Broglie wavelength λ_F corresponding to the Fermi momentum is of the order of the mean interparticle separation $(V/N)^{\frac{1}{3}}$. All particle states with de Broglie wavelength $\lambda = 2\pi\kappa^{-1} > \lambda_F$ are occupied at $T = 0$, all those with $\lambda < \lambda_F$ are empty.

By (9·16·7) one obtains for the Fermi energy at $T = 0$

$$\mu_0 = \frac{\hbar^2}{2m} \kappa_F^2 = \frac{\hbar^2}{2m} \left(3\pi^2 \frac{N}{V}\right)^{\frac{2}{3}} \tag{9·16·10}$$

Numerical estimate Let us calculate the Fermi energy at $T = 0$ of copper, a typical metal. Its density is 9 grams/cm³ and its atomic weight is 63.5. There are then 9/(63.5) = 0.14 moles of Cu per cm³ or, with one conduction electron per atom, $N_a/V = 8.4 \times 10^{22}$ electrons/cm³. Taking the electron mass $m \approx 10^{-27}$ grams, one obtains, by (9·16·10),

$$T_F \equiv \frac{\mu_0}{k} \approx 80{,}000^\circ\text{K} \tag{9·16·11}$$

The quantity T_F is called the "Fermi temperature." For a metal such as Cu it is seen to be much greater than any temperature T of the order of room temperature ($T \approx 300^\circ$K). Hence even at such relatively high temperatures the electron gas is highly degenerate. The Fermi distribution at room temperature has then the appearance of Fig. 9·16·1 where $kT \ll \mu$, and the Fermi energy μ differs only slightly from its value μ_0 at $T = 0$, i.e.,

$$\mu \approx \mu_0$$

Since in a metal there are so many electrons with $\epsilon \ll \mu$, all of which are in completely filled states, these electrons have in many cases very little effect on the macroscopic properties of the metal. Consider, for example, the contribution of the conduction electrons to the specific heat of the metal. The heat capacity C_V at constant volume of these electrons can be computed from a knowledge of their mean energy $\bar{E}(T)$ as a function of T, i.e.,

$$C_V = \left(\frac{\partial \bar{E}}{\partial T}\right)_V \tag{9·16·12}$$

If the electrons obeyed classical MB statistics, so that $F \propto e^{-\beta\epsilon}$ for *all* electrons, then the equipartition theorem would give, classically,

$$\bar{E} = \tfrac{3}{2}NkT \qquad \text{and} \qquad C_V = \tfrac{3}{2}Nk \tag{9·16·13}$$

But in the actual situation where the FD distribution has the form shown in Fig. 9·16·1, the situation is very different. A small change of T does not affect the many electrons in states with $\epsilon \ll \mu$, since all these states are completely filled and remain so when the temperature is changed. The mean energy of these electrons is therefore unaffected by temperature so that these electrons contribute nothing to the heat capacity (9 · 16 · 12). On the other hand, the small number N_{eff} of electrons in the small energy range of order kT near the Fermi energy μ do contribute to the specific heat. In the tail end of this region $F \propto e^{-\beta\epsilon}$ behaves as a MB distribution; hence, in accordance with (9 · 16 · 13), one expects that each electron in this region contributes roughly an amount $\frac{3}{2}k$ to the heat capacity. If $\rho(\epsilon)\,d\epsilon$ is the number of states in the energy range between ϵ and $\epsilon + d\epsilon$, the effective number of electrons in this region is

$$N_{\text{eff}} \approx \rho(\mu)kT \tag{9·16·14}$$

Hence one obtains for the heat capacity

$$C_V \approx N_{\text{eff}}(\tfrac{3}{2}k) \approx \tfrac{3}{2}k^2\rho(\mu)T \tag{9·16·15}$$

More roughly still, one can say that only a fraction kT/μ of the total electrons are in the tail region of the FD distribution so that

$$N_{\text{eff}} \approx \left(\frac{kT}{\mu}\right)N \tag{9·16·16}$$

and

$$C_V \approx \frac{3}{2}Nk\left(\frac{kT}{\mu}\right) = \nu\frac{3}{2}R\left(\frac{T}{T_F}\right) \tag{9·16·17}$$

Since T/T_F is quite small, the molar specific heat c_V of the electrons is very much less than their classical specific heat $\frac{3}{2}R$. This is a welcome result because it accounts for the fact that the molar heat capacity of metals is about the same as that of insulators (at room temperature, approximately $3R$ by virtue of the vibrations of the atoms in the lattice). Before the advent of quantum mechanics the classical theory predicted incorrectly that the presence of conduction electrons should raise the heat capacity of metals by 50 percent (i.e., by $\frac{3}{2}R$) compared to insulators.

Note also that the specific heat (9 · 16 · 17) is not temperature-independent as it would be classically. Using the superscript e to denote the *electronic* specific heat, the molar specific heat is of the form

$$c_V^{(e)} = \gamma T \tag{9·16·18}$$

where γ is a constant of proportionality. At room temperature $c_V^{(e)}$ is completely masked by the much larger specific heat $c_V^{(L)}$ due to lattice vibrations. But at very low temperatures $c_V^{(L)} = AT^3$, where A is a constant of proportionality,* and approaches, therefore, zero much faster than the electronic contribution (9 · 16 · 18) which approaches zero only proportionally to T. Thus it follows that low temperature experiments on metals permit measurements of the magnitude and the temperature dependence of the electronic specific heat. Indeed, the *total* measured specific heat of a metal at low temperatures should

* This result will be derived theoretically in Sec. 10 · 2.

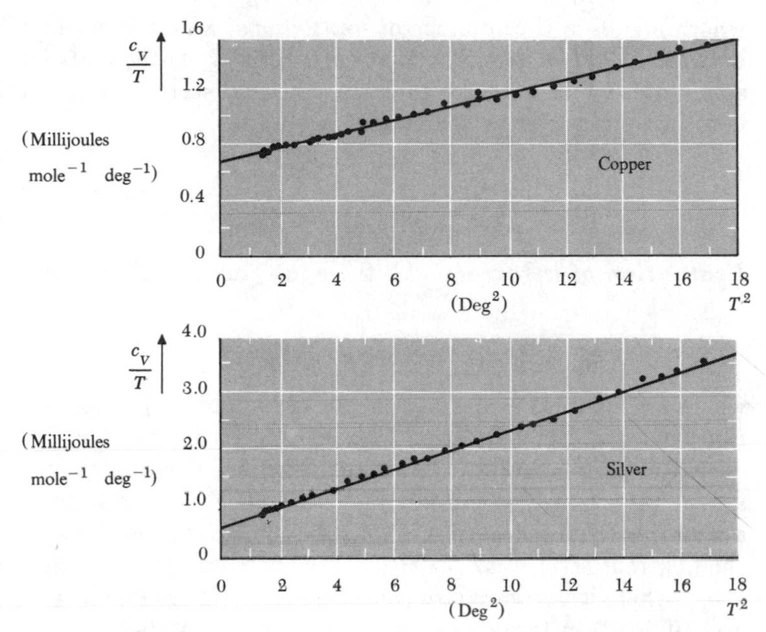

***Fig. 9·16·4** The measured specific heat c_V for copper and silver presented in plots of c_V/T versus T^2 (after Corak, Garfunkel, Satterthwaite, and Wexler, Phys. Rev., vol. 98, p. 1699 (1955)).*

be of the form

$$c_V = c_V^{(e)} + c_V^{(L)} = \gamma T + AT^3 \tag{9·16·19}$$

Hence

$$\frac{c_V}{T} = \gamma + AT^2 \tag{9·16·20}$$

and a plot of c_V/T versus T^2 should yield a straight line whose intercept on the vertical axis gives the coefficient γ. Figure 9·16·4 shows such plots. The fact that good straight lines are obtained experimentally verifies that the temperature dependences predicted by (9·16·19) are indeed correct.

*9·17 *Quantitative calculation of the electronic specific heat*

Let us calculate the specific heat of the electron gas in detail and verify our order-of-magnitude estimates. The mean energy of the electron gas is given by

$$\bar{E} = \sum_r \frac{\epsilon_r}{e^{\beta(\epsilon_r - \mu)} + 1}$$

Since the energy levels of the particles are very closely spaced, the sum can be replaced by an integral. Thus

$$\bar{E} = 2 \int F(\epsilon)\epsilon\, \rho(\epsilon)\, d\epsilon = 2 \int_0^\infty \frac{\epsilon}{e^{\beta(\epsilon - \mu)} + 1}\, \rho(\epsilon)\, d\epsilon \tag{9·17·1}$$

where $\rho(\epsilon)\,d\epsilon$ is the number of *translational* states lying in the energy range between ϵ and $\epsilon + d\epsilon$. The factor of 2 in (9 · 17 · 1) accounts for the two possible spin states which exist for each translational state. Here the Fermi energy μ is to be determined by the condition (9 · 16 · 3), i.e.,

$$2\int F(\epsilon)\rho(\epsilon)\,d\epsilon = 2\int_0^\infty \frac{1}{e^{\beta(\epsilon-\mu)}+1}\rho(\epsilon)\,d\epsilon = N \qquad (9\cdot 17\cdot 2)$$

Evaluation of integrals All these integrals are of the form

$$\int_0^\infty F(\epsilon)\varphi(\epsilon)\,d\epsilon \qquad (9\cdot 17\cdot 3)$$

where $F(\epsilon)$ is the Fermi function (9 · 16 · 4) and $\varphi(\epsilon)$ is some smoothly varying function of ϵ. The function $F(\epsilon)$ has the form shown in Fig. 9 · 16 · 1, i.e., it decreases quite abruptly from 1 to 0 within a narrow range of order kT about $\epsilon = \mu$, but is nearly constant everywhere else. This immediately suggests evaluating the integral (9 · 17 · 3) by an approximation procedure which exploits the fact that $F'(\epsilon) \equiv dF/d\epsilon = 0$ everywhere except in a range of order kT near $\epsilon = \mu$ where it becomes large and negative. Thus one is led to write the integral (9 · 17 · 3) in terms of F' by integrating by parts.

Let
$$\psi(\epsilon) \equiv \int_0^\epsilon \varphi(\epsilon')\,d\epsilon' \qquad (9\cdot 17\cdot 4)$$

Then
$$\int_0^\infty F(\epsilon)\varphi(\epsilon)\,d\epsilon = [F(\epsilon)\psi(\epsilon)]_0^\infty - \int_0^\infty F'(\epsilon)\psi(\epsilon)\,d\epsilon$$

But the integrated term vanishes, since $F(\infty) = 0$, while $\psi(0) = 0$ by (9 · 17 · 4). Hence

$$\int_0^\infty F(\epsilon)\varphi(\epsilon)\,d\epsilon = -\int_0^\infty F'(\epsilon)\psi(\epsilon)\,d\epsilon \qquad (9\cdot 17\cdot 5)$$

Here one has the advantage that, by virtue of the behavior of $F'(\epsilon)$, only the relatively narrow range of order kT about $\epsilon = \mu$ contributes appreciably to the integral. But in this small region the relatively slowly varying function ψ can

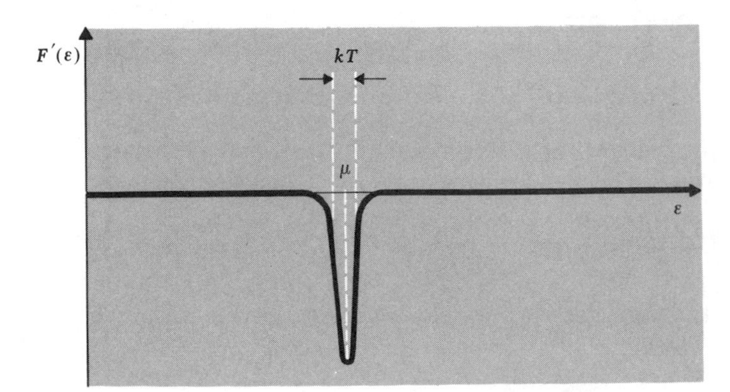

Fig. 9 · 17 · 1 *The derivative $F'(\epsilon)$ of the Fermi function as a function of ϵ.*

be expanded in a power series

$$\psi(\epsilon) = \psi(\mu) + \left[\frac{d\psi}{d\epsilon}\right]_\mu (\epsilon - \mu) + \frac{1}{2}\left[\frac{d^2\psi}{d\epsilon^2}\right]_\mu (\epsilon - \mu)^2 + \cdots$$

$$= \sum_{m=0}^{\infty} \frac{1}{m!}\left[\frac{d^m\psi}{d\epsilon^m}\right]_\mu (\epsilon - \mu)^m$$

where the derivatives are evaluated for $\epsilon = \mu$. Hence (9·17·5) becomes

$$\int_0^\infty F\varphi\, d\epsilon = - \sum_{m=0}^{\infty} \frac{1}{m!}\left[\frac{d^m\psi}{d\epsilon^m}\right]_\mu \int_0^\infty F'(\epsilon)(\epsilon - \mu)^m\, d\epsilon \qquad (9\cdot 17\cdot 6)$$

But $\displaystyle\int_0^\infty F'(\epsilon)(\epsilon - \mu)^m\, d\epsilon = - \int_0^\infty \frac{\beta e^{\beta(\epsilon-\mu)}}{(e^{\beta(\epsilon-\mu)} + 1)^2} (\epsilon - \mu)^m\, d\epsilon$

$$= -\beta^{-m} \int_{-\beta\mu}^{\infty} \frac{e^x}{(e^x + 1)^2} x^m\, dx$$

where

$$x \equiv \beta(\epsilon - \mu) \qquad (9\cdot 17\cdot 7)$$

Since the integrand has a sharp maximum for $\epsilon = \mu$, (i.e., for $x = 0$) and since $\beta\mu \gg 1$, the lower limit can be replaced by $-\infty$ with negligible error. Thus one can write

$$\int_0^\infty F'(\epsilon)(\epsilon - \mu)^m\, d\epsilon = -(kT)^m I_m \qquad (9\cdot 17\cdot 8)$$

where

$$I_m \equiv \int_{-\infty}^{\infty} \frac{e^x}{(e^x + 1)^2} x^m\, dx \qquad (9\cdot 17\cdot 9)$$

Note that

$$\frac{e^x}{(e^x + 1)^2} = \frac{1}{(e^x + 1)(e^{-x} + 1)}$$

is an even function of x. If m is odd, the integrand in (9·17·9) is then an odd function of x so that the integral vanishes; thus

$$I_m = 0 \qquad \text{if } m \text{ is odd} \qquad (9\cdot 17\cdot 10)$$

Also

$$I_0 = \int_{-\infty}^{\infty} \frac{e^x}{(e^x + 1)^2}\, dx = -\left[\frac{1}{e^x + 1}\right]_{-\infty}^{\infty} = 1 \qquad (9\cdot 17\cdot 11)$$

By using (9·17·8), the relation (9·17·6) can then be written in the form

$$\int_0^\infty F\varphi\, d\epsilon = \sum_{m=0}^{\infty} I_m \frac{(kT)^m}{m!}\left[\frac{d^m\psi}{d\epsilon^m}\right]_\mu = \psi(\mu) + I_2 \frac{(kT)^2}{2}\left[\frac{d^2\psi}{d\epsilon^2}\right]_\mu + \cdots \qquad (9\cdot 17\cdot 12)$$

The integral I_2 can readily be evaluated (see Problems 9.26 and 9.27). One finds

$$I_2 = \frac{\pi^2}{3}$$

Hence (9 · 17 · 12) becomes

$$\int_0^\infty F(\epsilon)\varphi(\epsilon)\,d\epsilon = \int_0^\mu \varphi(\epsilon)\,d\epsilon + \frac{\pi^2}{6}(kT)^2\left[\frac{d\varphi}{d\epsilon}\right]_\mu + \cdots \tag{9·17·13}$$

Here the first term on the right is just the result one would obtain for $T \to 0$ corresponding to Fig. 9 · 16 · 2. The second term represents a correction due to the finite width ($\approx kT$) of the region where F decreases from 1 to 0.

Calculation of the specific heat We now apply the general result (9 · 17 · 13) to the evaluation of the mean energy (9 · 17 · 1). Thus one obtains

$$\bar{E} = 2\int_0^\mu \epsilon\rho(\epsilon)\,d\epsilon + \frac{\pi^2}{3}(kT)^2\left[\frac{d}{d\epsilon}(\epsilon\rho)\right]_\mu \tag{9·17·14}$$

Since for the present case, where $kT/\mu \ll 1$, the Fermi energy μ differs only slightly from its value μ_0 at $T = 0$, the derivative in the second small correction term in (9 · 17 · 14) can be evaluated at $\mu = \mu_0$ with negligible error. Furthermore one can write

$$2\int_0^\mu \epsilon\rho(\epsilon)\,d\epsilon = 2\int_0^{\mu_0} \epsilon\rho(\epsilon)\,d\epsilon + 2\int_{\mu_0}^\mu \epsilon\rho(\epsilon)\,d\epsilon = \bar{E}_0 + 2\mu_0\rho(\mu_0)(\mu - \mu_0)$$

since the first integral on the right is by (9 · 17 · 1) just the mean energy $\bar{E}_0$ at $T = 0$. Since

$$\frac{d}{d\epsilon}(\epsilon\rho) = \rho + \epsilon\rho', \qquad \rho' \equiv \frac{d\rho}{d\epsilon}$$

Eq. (9 · 17 · 14) becomes

$$\bar{E} = \bar{E}_0 + 2\mu_0\rho(\mu_0)(\mu - \mu_0) + \frac{\pi^2}{3}(kT)^2[\rho(\mu_0) + \mu_0\rho'(\mu_0)] \tag{9·17·15}$$

Here we still need to know the change $(\mu - \mu_0)$ of the Fermi energy with temperature. Now μ is determined by the condition (9 · 17 · 2) which becomes, by (9 · 17 · 13),

$$2\int_0^\mu \rho(\epsilon)\,d\epsilon + \frac{\pi^2}{3}(kT)^2\rho'(\mu) = N \tag{9·17·16}$$

Here the derivative in the correction term can again be evaluated at μ_0 with negligible error, while

$$2\int_0^\mu \rho(\epsilon)\,d\epsilon = 2\int_0^{\mu_0} \rho(\epsilon)\,d\epsilon + 2\int_{\mu_0}^\mu \rho(\epsilon)\,d\epsilon = N + 2\rho(\mu_0)(\mu - \mu_0)$$

since the first integral on the right side is just the condition (9 · 17 · 2) which determined μ_0 at $T = 0$. Thus (9 · 17 · 16) becomes

$$2\rho(\mu_0)(\mu - \mu_0) + \frac{\pi^2}{3}(kT)^2\rho'(\mu_0) = 0$$

or

$$(\mu - \mu_0) = -\frac{\pi^2}{6}(kT)^2\frac{\rho'(\mu_0)}{\rho(\mu_0)} \tag{9·17·17}$$

Hence Eq. (9·17·15) becomes

$$\bar{E} = \bar{E}_0 - \frac{\pi^2}{3}(kT)^2\mu_0\rho'(\mu_0) + \frac{\pi^2}{3}(kT)^2[\rho(\mu_0) + \mu_0\rho'(\mu_0)]$$

or

$$\bar{E} = \bar{E}_0 + \frac{\pi^2}{3}(kT)^2\rho(\mu_0) \tag{9·17·18}$$

since terms in ρ' cancel. The heat capacity (at constant volume) becomes then

▶
$$C_V = \frac{\partial \bar{E}}{\partial T} = \frac{2\pi^2}{3}k^2\rho(\mu_0)T \tag{9·17·19}$$

This agrees with the simple order of magnitude calculation of Eq. (9·16·15).

The density of states ρ can be written explicitly for the free-electron gas by (9·9·19):

$$\rho(\epsilon)\,d\epsilon = \frac{V}{(2\pi)^3}\left(4\pi\kappa^2\frac{d\kappa}{d\epsilon}\,d\epsilon\right) = \frac{V}{4\pi^2}\frac{(2m)^{\frac{3}{2}}}{\hbar^3}\epsilon^{\frac{1}{2}}\,d\epsilon \tag{9·17·20}$$

But

$$\mu_0 = \frac{\hbar^2}{2m}\left(3\pi^2\frac{N}{V}\right)^{\frac{2}{3}} \quad \text{by (9·16·10)}$$

Hence

$$\rho(\mu_0) = V\frac{m}{2\pi^2\hbar^2}\left(3\pi^2\frac{N}{V}\right)^{\frac{1}{3}} \tag{9·17·21}$$

Equivalently this can be written in terms of N and μ_0 by eliminating the volume V between the last two equations. Thus one obtains

$$\rho(\mu_0) = \left[\frac{m}{2\pi^2\hbar^2}(3\pi^2N)^{\frac{1}{3}}\right]\left[\frac{1}{\mu_0}\frac{\hbar^2}{2m}(3\pi^2N)^{\frac{2}{3}}\right] = \frac{3}{4}\frac{N}{\mu_0} \tag{9·17·22}$$

Hence (9·17·19) gives

$$C_V = \frac{\pi^2}{2}k^2\frac{N}{\mu_0}T = \frac{\pi^2}{2}kN\frac{kT}{\mu_0} \tag{9·17·23}$$

or, per mole,

▶
$$c_V = \frac{3}{2}R\left(\frac{\pi^2}{3}\frac{kT}{\mu_0}\right) \tag{9·17·24}$$

SUGGESTIONS FOR SUPPLEMENTARY READING

D. K. C. MacDonald: "Introductory Statistical Mechanics for Physicists," chap. 3, John Wiley & Sons, Inc., New York, 1963.

C. Kittel: "Elementary Statistical Physics," secs. 19–22, John Wiley & Sons, Inc., New York, 1958.

L. D. Landau and E. M. Lifshitz: "Statistical Physics," chap. 5, Addison-Wesley Publishing Company, Reading, Mass., 1958.

J. E. Mayer and M. G. Mayer: "Statistical Mechanics," chap. 16, John Wiley & Sons, Inc., New York, 1940. (Detailed discussion of degenerate gases.)

T. L. Hill: "An Introduction to Statistical Thermodynamics," chaps. 4, 8, 10, Addison-Wesley Publishing Company, Reading, Mass., 1960. (Ideal gases in the classical limit.)

PROBLEMS

9.1 Consider a system consisting of two particles, each of which can be in any one of three quantum states of respective energies 0, ϵ, and 3ϵ. The system is in contact with a heat reservoir at temperature $T = (k\beta)^{-1}$.

(*a*) Write an expression for the partition function Z if the particles obey classical MB statistics and are considered distinguishable.

(*b*) What is Z if the particles obey BE statistics?

(*c*) What is Z if the particles obey FD statistics?

9.2 (*a*) From a knowledge of the partition function Z derived in the text, write an expression for the entropy S of an ideal FD gas. Express your answer solely in terms of $\bar{n}_r$, the mean number of particles in state r.

(*b*) Write a similar expression for the entropy S of a BE gas.

(*c*) What do these expressions for S become in the classical limit when $\bar{n}_r \ll 1$?

9.3 Calculate the partition function of a monatomic gas in the classical limit by considering the particles enclosed in a rectangular box with perfectly reflecting walls and describing each particle in terms of the wave function ψ of (9·9·22) which vanishes at the walls. Show that the result is identical to that obtained in (9·10·7) by using the description in terms of traveling waves.

9.4 (*a*) An ideal gas of N atoms of mass m is contained in a volume V at absolute temperature T. Calculate the chemical potential μ of this gas. You may use the classical approximation for the partition function, taking into account the indistinguishability of the particles.

(*b*) A gas of N' such weakly interacting particles, adsorbed on a surface of area A on which they are free to move, can form a two-dimensional ideal gas on such a surface. The energy of an adsorbed molecule is then $(\mathbf{p}^2/2m) - \epsilon_0$, where $\mathbf{p}$ denotes its (two-component) momentum vector and ϵ_0 is the binding energy which holds a molecule on the surface. Calculate the chemical potential μ' of this adsorbed ideal gas. The partition function can again be evaluated in the classical approximation.

(*c*) At the temperature T, the equilibrium condition between molecules adsorbed on the surface and molecules in the surrounding three-dimensional gas can be expressed in terms of the respective chemical potentials. Use this condition to find at temperature T the mean number n' of molecules adsorbed per unit area of the surface when the mean pressure of the surrounding gas is $\bar{p}$.

9.5 Consider a nonrelativistic free particle in a cubical container of edge length L and volume $V = L^3$.

(*a*) Each quantum state r of this particle has a corresponding kinetic energy ϵ_r which depends on V. What is $\epsilon_r(V)$?

(*b*) Find the contribution to the gas pressure $p_r = -(\partial\epsilon_r/\partial V)$ of a particle in this state in terms of ϵ_r and V.

(*c*) Use this result to show that the mean pressure $\bar{p}$ of any ideal gas of weakly interacting particles is always related to its mean total kinetic energy $\bar{E}$ by $\bar{p} = \frac{2}{3}\bar{E}/V$, irrespective of whether the gas obeys classical, FD, or BE statistics.

(*d*) Why is this result different from the relation $\bar{p} = \frac{1}{3}\bar{E}/V$ valid for a gas of photons?

(*e*) Calculate the mean pressure $\bar{p}$ by using a semiclassical kinetic theory calculation which computes $\bar{p}$ from the momentum transfer resulting from the molecular impacts with a wall. Show that the result thus obtained is consistent with that derived in (*c*) in all cases.

9.6 Consider an ideal gas of N very weakly interacting ordinary nonrelativistic particles in equilibrium in a volume V at the absolute temperature T. By continuing to reason along lines similar to those used in the preceding problem, establish the following results, irrespective of whether the gas obeys classical, FD, or BE statistics.

(*a*) Calculate the dispersion $\overline{(\Delta p)^2}$ of the gas pressure, and show that it is quite generally directly related to the dispersion $\overline{(\Delta E)^2}$ of the total gas energy.

(*b*) By using the expressions for $\bar{E}$ and $\overline{(\Delta E)^2}$ in terms of the partition function Z, express $\bar{p}$ and $\overline{(\Delta p)^2}$ in terms of Z. Show that

$$\overline{(\Delta p)^2} = \frac{2kT^2}{3V}\frac{\partial \bar{p}}{\partial T}$$

(*c*) In the case of this gas in the classical *limit*, calculate explicitly the fractional dispersion of the pressure $\overline{(\Delta p)^2}/\bar{p}^2$.

***9.7** What is the molar specific heat at constant volume of a diatomic gas at room temperature T_0? Use the fact that for practically all diatomic molecules the spacing between rotational energy levels is small compared to kT_0, while that between their vibrational energy levels is large compared to kT_0.

***9.8** Suppose that N molecules of HD gas are put into a flask and kept at a temperature T until complete equilibrium has been established. The flask will then contain H_2 and D_2 molecules in addition to some mean number n of HD molecules. Calculate the ratio n/N. Express your answer in terms of T, the mass m of a hydrogen atom, the mass M of a deuterium atom, and the (angular) vibrational frequency ω_0 of the HD molecule. You may assume that T is about room temperature, so that the rotational degrees of freedom of the molecules can be treated as classical, while $\hbar\omega_0 \gg kT$, so that all the molecules are essentially in their lowest vibrational state.

9.9 Electromagnetic radiation at temperature T_i fills a cavity of volume V. If the volume of the thermally insulated cavity is expanded quasistatically to a volume $8V$, what is the final temperature T_f? (Neglect the heat capacity of the cavity walls.)

9.10 Apply the thermodynamic relation $T\,dS = d\bar{E} + \bar{p}\,dV$ to a photon gas. Here one can write that $\bar{E} = V\bar{u}$ where $\bar{u}(T)$, the mean energy density of the radiation field, is independent of the volume V. The radiation pressure $\bar{p} = \frac{1}{3}\bar{u}$.

(*a*) Considering S as a function of T and V, express dS in terms of dT and dV. Find $(\partial S/\partial T)_V$ and $(\partial S/\partial V)_T$.

(*b*) Show that the mathematical identity $(\partial^2 S/\partial V \partial T) = (\partial^2 S/\partial T \partial V)$ gives immediately a differential equation for $\bar{u}$ which can be integrated to yield the Stefan-Boltzmann law $\bar{u} \propto T^4$.

9.11 A dielectric solid has an index of refraction n_0 which can be assumed to be constant up to infrared frequencies. Calculate the contribution of the black-body radiation in the solid to its heat capacity at a temperature $T = 300°K$. Compare this result with the classical lattice heat capacity of $3R$ per mole.

9.12 It has been reported that a nuclear fission explosion produces a temperature of the order of 10^6°K. Assuming this to be true over a sphere 10 cm in diameter, calculate approximately

(a) the total rate of electromagnetic radiation from the surface of this sphere;

(b) the radiation flux (power incident per unit area) at a distance of 1 km;

(c) the wavelength corresponding to the maximum in the radiated power spectrum.

9.13 The surface temperature of the sun is T_0 (= 5500°K); its radius is R (= 7×10^{10} cm) while the radius of the earth is r (= 6.37×10^8 cm). The mean distance between the sun and the earth is L (= 1.5×10^{13} cm). In first approximation one can assume that both the sun and the earth absorb all electromagnetic radiation incident upon them.

The earth has reached a steady state so that its mean temperature T does not change in time despite the fact that the earth constantly absorbs and emits radiation.

(a) Find an approximate expression for the temperature T of the earth in terms of the astronomical parameters mentioned above.

(b) Calculate this temperature T numerically.

9.14 What is the total number $\mathfrak{N}$ of molecules that escape per unit time from a unit area of the surface of a liquid which is at temperature T, where its vapor pressure is p? Use detailed balance arguments by considering a situation where the liquid is in thermal equilibrium with its vapor at this temperature and pressure. Treat the vapor as an ideal gas, and assume that molecules striking the surface of the liquid are not appreciably reflected.

Calculate the number $\mathfrak{N}$ of molecules escaping per unit time from unit area of the surface of water in a glass at 25°C. The vapor pressure of water at this temperature is 23.8 mm Hg.

9.15 To measure the vapor pressure of a metal (e.g., nickel) at some elevated temperature T, one can enclose a wire of this metal in an evacuated glass envelope. By passing a current through the wire, one can bring it to the desired temperature T for a time t. (This temperature can be measured by optical means by observing the wavelength distribution of the emitted radiation.) All the molecules of the vapor emitted from the wire and striking the glass envelope condense there, since this envelope is at a much lower temperature. By weighing the wire before and after the experiment, one can measure the small loss of mass ΔM per unit length of this wire. The radius r of the wire can also be measured. (It changes only by a negligible amount during the experiment.) The metal is known to have a molecular weight μ.

Derive an explicit expression showing how the vapor pressure $p(T)$ of the metal at the temperature T can be determined from these experimentally measured quantities. (Assume that when the metal is in equilibrium with its vapor, every molecule of the vapor striking the surface of the metal condenses.)

9.16 An ideal Fermi gas is at rest at absolute zero and has a Fermi energy μ. The mass of each particle is m. If $\mathbf{v}$ denotes the velocity of a molecule, find $\bar{v}_x$ and $\overline{v_x^2}$.

9.17 Consider an ideal gas of N electrons in a volume V at absolute zero.

(a) Calculate the total mean energy $\bar{E}$ of this gas.

(b) Express $\bar{E}$ in terms of the Fermi energy μ.

(c) Show that $\bar{E}$ is properly an extensive quantity, but that for a fixed volume V, $\bar{E}$ is not proportional to the number N of particles in the container. How do you account for this last result despite the fact that there is no interaction potential between the particles?

9.18 Find the relation between the mean pressure $\bar{p}$ and the volume V of an ideal Fermi-Dirac gas at $T = 0$.

(a) Compute this by the general relation $\bar{p} = -(\partial \bar{E}/\partial V)_T$ valid at $T = 0$. Here $\bar{E}$ is the total mean energy of the gas at $T = 0$.

(b) Compute this from the relation $\bar{p} = \frac{2}{3}\bar{E}/V$ derived in Problem 9·5.

(c) Use the result to calculate the approximate pressure exerted by the conduction electrons in copper metal on the solid lattice which confines them within the volume of the metal. Express your answer in atmospheres.

9.19 Use arguments based on the specific heat to answer the following questions without detailed calculations:

(a) By what factor does the entropy of the conduction electrons in a metal change when the temperature is changed from 200 to 400°K?

(b) By what factor does the entropy of the electromagnetic radiation field inside an enclosure change when the temperature is increased from 1000 to 2000°K?

9.20 The atomic weight of sodium (Na) is 23, the density of the metal is 0.95 gm/cm³. There is one conduction electron per atom.

(a) Use an approximate expression for the Fermi energy of the conduction electrons in Na metal to calculate the numerical value of the Fermi temperature $T_F \equiv \mu/k$.

(b) It is desired to cool a sample consisting of 100 cm³ of Na metal from 1°K down to 0.3°K. At these low temperatures the lattice heat capacity of the metal is negligible compared to that due to its conduction electrons. The metal can be cooled by bringing it in thermal contact with liquid He^3 at 0.3°K. If 0.8 joules of heat input are required to evaporate 1 cm³ of liquid He^3, estimate how much of the liquid must be vaporized to cool the Na sample.

9.21 Use qualitative arguments (similar to those used in the text in connection with the discussion of the specific heat of the conduction electrons in a metal) to discuss the paramagnetic susceptibility χ due to the spin magnetic moments of the conduction electrons. In particular,

(a) what is the temperature dependence of χ?

(b) what is the order of magnitude of χ per mole of conduction electrons? By what factor would this answer differ if the electrons obeyed Maxwell-Boltzmann statistics?

9.22 A metal has n conduction electrons per unit volume, each electron having spin $\frac{1}{2}$ and an associated magnetic moment μ_m. The metal is at $T = 0$°K and is placed in a small external magnetic field H. The total energy of the conduction electrons in the presence of the field H must then be as small as possible. Use this fact to find an explicit expression for the paramagnetic susceptibility due to the spin magnetic moments of these conduction electrons.

9.23 The lowest possible energy of a conduction electron in a metal is $-V_0$ below the energy of a free electron at infinity. The conduction electrons have a Fermi energy (or chemical potential) μ. The minimum energy needed to remove an electron from the metal is then $\Phi = V_0 - \mu$ and is called the work function of the metal. The figure illustrates these relations in a diagram of energy versus spatial location of an electron.

Consider an electron gas outside the metal in thermal equilibrium with the electrons in the metal at the temperature T. The density of electrons outside the metal is quite small at all laboratory temperatures where $kT \ll \Phi$. By equating chemical potentials for the electrons outside and inside the metal, find the mean number of electrons per unit volume outside the metal.

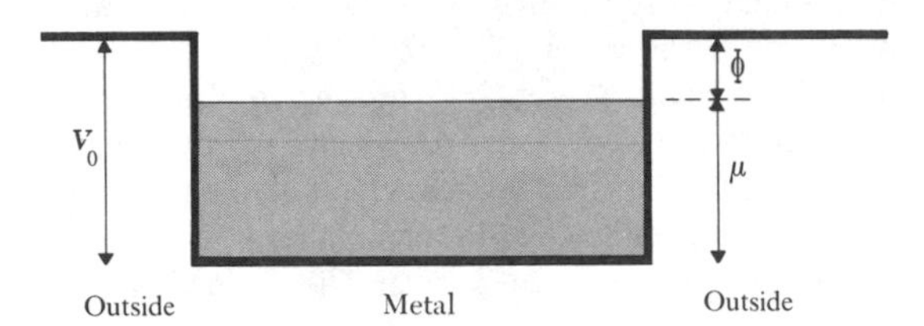

9.24 Calculate the number of electrons emitted per second per unit area from the surface of a metal at temperature T. Hence calculate the resultant electron current density. Use detailed balance arguments by considering the situation when such a metal is in thermal equilibrium with an electron gas outside the metal as discussed in the last problem. Assume that a fraction r of the electrons striking the metal is reflected.

9.25 Write an expression for $f(\boldsymbol{\kappa})\, d^3\boldsymbol{\kappa}$, the mean number of conduction electrons per unit volume which are inside the metal and which have a particular direction of spin orientation and a wave vector $\boldsymbol{\kappa}$ (or momentum $\boldsymbol{p} = \boldsymbol{\kappa}/\hbar$) lying in the range between $\boldsymbol{\kappa}$ and $\boldsymbol{\kappa} + d\boldsymbol{\kappa}$. Use this result and detailed kinetic arguments to calculate how many of the electrons inside the metal strike unit area of its surface per unit time with enough energy to emerge from the metal. Assume that a fraction r of these electrons are reflected back into the metal without emerging from it. Compare your answer with that obtained in the preceding problem.

***9.26** (*a*) Show that the integral I_2, defined in (9·17·9), can by integration by parts, be written in the form

$$I_2 = 4 \int_0^\infty \frac{x\, dx}{e^x + 1}$$

(*b*) By expanding the integrand in powers of e^{-x} and integrating term by term, show that I_2 can be expressed as an infinite series.

(*c*) By summing this series by a method similar to that used in Appendix A·11, show that $I_2 = \pi^2/3$.

***9.27** The calculations involving Fermi-Dirac statistics give rise to the integrals I_m defined in (9·17·9).

(*a*) Show that all these integrals can be obtained if it is possible to evaluate the single integral

$$J(k) \equiv \int_{-\infty}^{\infty} \frac{e^{ikx}\, dx}{(e^x + 1)(e^{-x} + 1)} \tag{1}$$

since power series expansion of $J(k)$ yields the result

$$J(k) = \sum_{n=0}^{\infty} \frac{(ik)^m}{m!} I_m \tag{2}$$

(*b*) Evaluate the integral $J(k)$ by putting $x = z$, where z is a complex variable, and using contour integration. In order to exploit the periodicity of the exponential function e^z, compare the integral along the real axis (where $z = x$) with the integral along the parallel line a distance 2π above the real axis (where $z = x + 2\pi i$). These paths can be distorted into each other if one bypasses the singularity at $z = i\pi$.

(*c*) By expanding $J(k)$ and comparing the coefficients of k^2 on both sides of (2), obtain in particular the value of I_2.

***9.28** Consider a gas of weakly interacting particles obeying BE statistics and maintained at a temperature T. The total number of particles is *not* specified exactly, but only their mean total number $\bar{N}$ is given. The whole gas can then be described in terms of the grand canonical distribution of (6·9·4).

(*a*) Use this distribution to calculate the mean number of particles $\bar{n}_s$ in the single-particle state s. Compare your answer with (9·6·13).

(*b*) Use this distribution to calculate the dispersion $\overline{(\Delta n_s)^2}$. Show that this result agrees with (9·6·17), but does not include the correction terms contained in (9·6·18).

(*c*) Use the grand canonical distribution to calculate $\bar{n}_s$ and $\overline{n_s^2}$ if the particles obey FD statistics.

***9.29** The partition function (9·6·1) subject to the condition (9·6·3) can be written in the form

$$Z = \sum e^{-\beta(n_1\epsilon_1 + n_2\epsilon_2 + \cdots)} \left\{ \frac{1}{2\pi} \int_{-\pi}^{\pi} \exp \left[\left(N - \sum_r n_r \right) (\alpha + i\alpha') \right] d\alpha' \right\} \tag{1}$$

Here the factor in braces has, by (A·7·15), the property that (irrespective of the value assumed by the parameter α) it equals unity when (9·6·3) is satisfied, but vanishes otherwise. Hence the sum in (1) can be performed over *all* possible values $n_r = 0, 1, 2, 3, \ldots$ for each r, without any further restriction.

By using arguments similar to those of method 2 in Sec. 6·8, derive an explicit approximate expression for $\ln Z$ and show how the value of α is to be determined. In this way, rederive the results (9·6·9) and (9·6·12).

10 *Systems of interacting particles*

STATISTICAL MECHANICS has been used repeatedly in the preceding chapters to calculate the equilibrium properties of systems on the basis of a knowledge of their microscopic constituents. Up to now we have, however, restricted ourselves to simple situations where the systems consist of particles which interact with each other to a negligible extent. The calculations are then, of course, quite easy. But most systems in the real world are more complicated than ideal gases and consist of many particles which *do* interact with each other appreciably (e.g., liquids and solids). In principle, the discussion of the equilibrium properties of such a system requires again only a calculation of its partition function. But, although the problem is thus very well defined, it can become both complex and subtle. Indeed, some of the most fascinating and difficult questions explored in present-day research concern systems of interacting particles.

An important and frequently occurring situation arises when one is dealing with a system at a sufficiently low absolute temperature. In this case the probability that the system is in any one of its states is appreciable only if it is a state of low energy. There is therefore no need to examine all the possible quantum states of the system; an investigation of the relatively few quantum states of energy not too far above the ground-state energy of the system is sufficient to discuss the problem. The analysis is thereby greatly facilitated. The general procedure consists in trying to simplify the dynamical problem by introducing new variables in terms of which the low-lying excited states of the system can be described most simply. These excited states represent then particularly simple possible modes of motion ("collective modes") of the system as a whole, rather than of the individual particles. To be specific, one attempts to choose the new variables so that the Hamiltonian of the system, when expressed in terms of these variables, becomes identical in form to the Hamiltonian of an assembly of weakly interacting particles. (These are commonly called "quasi-particles" to avoid confusion with the actual particles constituting the system.) If the *dynamical* problem of reducing the Hamiltonian to this simple form can be solved, then the whole problem becomes equivalent to that of an ideal gas of quasi-particles; the *statistical* problem thus becomes trivial.

Let us mention a few illustrations of the above method of analysis in terms of collective modes. Consider, for example, a solid, i.e., a system where the interaction between particles is sufficiently strong that they become arranged in a lattice of definite crystal structure. If the temperature is not too high, the amplitudes of vibration of the individual atoms are relatively small. The collective modes of motion involving many atoms are then the possible sound waves which can propagate through the solid. When these sound waves are quantized, they exhibit particlelike behavior and act like weakly interacting quasi-particles called "phonons." (The situation is analogous to light waves, which exhibit particlelike behavior when quantized, the quasi-particles being the familiar photons.) Another example might be a ferromagnet at very low temperatures. Here all the spins and associated magnetic moments interact strongly with each other and all point in one particular direction at $T = 0$. Small deviations from perfect alignment constitute low-lying excited states of somewhat higher energy. These deviations can propagate through the ferromagnet like waves (they are called "spin waves") and, when quantized, have particlelike properties (they are then called "magnons"). Other systems, such as liquid helium near $T = 0$, can similarly be described in terms of quasi-particles. These general remarks are admittedly rather vague (although the detailed discussion of solids in Sec. 10·1 will illustrate the method of approach quite specifically in a simple case). They should, however, be sufficient to indicate that it is often possible to reduce a problem of interacting particles near $T = 0$ to a trivial equivalent problem of essentially noninteracting particles.

Another situation which is relatively simple to treat is the opposite limit, where the temperature T of the system is high enough that kT is large compared to the mean energy of interaction of a particle with the other particles of the system. Since the interaction between particles is then relatively small, it can be taken into account by systematic approximation methods (like appropriate power series expansions). These yield the correction terms which describe the extent to which the properties of the system depart from those which would exist if interactions between particles were absent. One instance where such approximation methods are applicable is an ordinary gas which is sufficiently dilute and at a sufficiently high temperature that its behavior does not differ excessively from that of an ideal gas, although it may differ from it significantly. Another illustration is a system of interacting spins at a temperature much higher than that below which ferromagnetism occurs; departures of the susceptibility from Curie's law then do occur, but the correction terms are relatively small.

The two situations described in the preceding paragraphs represent extremes of comparative simplicity. In the limit of sufficiently low temperature the systems are almost completely ordered (e.g., the lattice of a solid is almost rigid, or the spins of a ferromagnet are almost perfectly aligned); the small departures from perfect order can then be discussed fairly readily. In the other limit of sufficiently high temperature the systems are almost completely disordered (e.g., the motions of individual molecules in a gas are almost completely uncorrelated, or the spins of a spin system point in almost random directions); the small departures from perfect randomness can then be readily discussed. It is the intermediate situation which is the most difficult and fascinating, for it raises the question of how a system does make the transition from a disordered to an ordered configuration as its temperature is reduced. The ordering can be expected to take place at some temperature T which is such that kT is of the order of the mean interaction energy of a particle with the other particles in the system. But just how rapidly does the system become ordered when its temperature is lowered? The answer is that the ordering, or at least the onset of ordering, can occur very abruptly at a sharply defined critical temperature T_c. For example, a gas at a fixed pressure turns suddenly into a liquid at a sharply defined temperature; or an assembly of spins becomes suddenly ferromagnetic below a well-defined critical temperature. The reason for the suddenness of these transitions is that the presence of appreciable interactions between particles can lead to "cooperative" behavior involving *all* the particles; i.e., when a few particles become locally ordered, this facilitates the ordering of some more particles further away, with the net result that the order propagates through the entire system (like a row of collapsing dominoes). For example, when a few gas molecules condense to form a liquid, this process helps other molecules to condense; or when a few spins become oriented in the same direction, they produce an effective magnetic field which induces neighboring spins also to line up in this direction. Since such cooperative effects involve correlations between all the particles, they are very difficult to discuss theoretically. In principle it is, of course, true that the exact partition function Z would describe all these phenomena, including the occurrence of abrupt phase transitions whose existence would be manifested by a singularity of the function Z as a function of T at a critical temperature T_c. But the theoretical problem is precisely the circumstance that it is very difficult to calculate Z when the correlated motions of all particles must be taken into account. Only the most trivial problems can be discussed exactly (with great mathematical ingenuity), and it is a challenging task to develop approximation methods which are powerful enough to treat physically important situations. It is characteristic of these methods that they become least satisfactory at a temperature close to T_c where the correlations resulting in cooperative behavior become most significant.

In this chapter we shall discuss only some simple, but important, systems consisting of interacting particles and shall treat them by the simplest methods. We shall discuss a solid as an example of an almost ordered system at comparatively low temperatures; a slightly nonideal gas as an example of an almost random system at comparatively high temperatures; and the case of ferromagnetism as an example of cooperative ordering.

SOLIDS

10 · 1 *Lattice vibrations and normal modes*

Consider a solid consisting of N atoms. Denote the position vector of the ith atom of mass m_i by $\mathbf{r}_i$, its corresponding cartesian coordinates by x_{i1}, x_{i2}, x_{i3}; denote the equilibrium position of this atom by $\mathbf{r}_i^{(0)}$. Each atom is free to vibrate with relatively small amplitude about its equilibrium position. To measure displacements from the equilibrium position, introduce the variable

$$\xi_{i\alpha} \equiv x_{i\alpha} - x_{i\alpha}^{(0)}, \qquad \text{where } \alpha = 1, 2, \text{ or } 3 \tag{10·1·1}$$

The kinetic energy of vibration of the solid is then

$$K = \frac{1}{2} \sum_{i=1}^{N} \sum_{\alpha=1}^{3} m_i \dot{x}_{i\alpha}^2 = \frac{1}{2} \sum_{i=1}^{N} \sum_{\alpha=1}^{3} m_i \dot{\xi}_{i\alpha}^2 \tag{10·1·2}$$

where $\dot{x}_{i\alpha} = \dot{\xi}_{i\alpha}$ denotes the α component of velocity of the ith atom.

The potential energy $V = V(x_{11}, x_{12}, \ldots, x_{N3})$ can be expanded in Taylor's series since the displacements $\xi_{i\alpha}$ are small. Thus one gets

$$V = V_0 + \sum_{i\alpha} \left[\frac{\partial V}{\partial x_{i\alpha}}\right]_0 \xi_{i\alpha} + \frac{1}{2} \sum_{i\alpha,j\gamma} \left[\frac{\partial^2 V}{\partial x_{i\alpha}\, \partial x_{j\gamma}}\right]_0 \xi_{i\alpha}\xi_{j\gamma} + \cdots \tag{10·1·3}$$

Here the sums over i and j are from 1 to N; the sums over α and γ are from 1 to 3. The derivatives are all evaluated at the equilibrium positions of the atoms where $x_{i\alpha} = x_{i\alpha}^{(0)}$ for all i and α. These derivatives are therefore simply constants. In particular, V_0 is merely the potential energy in the equilibrium configuration of the atoms. Since V must be a minimum in this equilibrium configuration, $[\partial V/\partial x_{i\alpha}]_0 = 0$, i.e., the force acting on any atom must then vanish. Introducing as an abbreviation the constants

$$A_{i\alpha,j\gamma} \equiv \left[\frac{\partial^2 V}{\partial x_{i\alpha}\, \partial x_{j\gamma}}\right]_0 \tag{10·1·4}$$

(10·1·3) becomes thus, neglecting terms of order higher than quadratic in ξ,

$$V = V_0 + \frac{1}{2} \sum_{i\alpha,j\gamma} A_{i\alpha,j\gamma}\xi_{i\alpha}\xi_{j\gamma} \tag{10·1·5}$$

Hence the total Hamiltonian, or energy, associated with the vibrations of the atoms in the solid assumes the form

$$\mathcal{H} = V_0 + \frac{1}{2} \sum_{i\alpha} m_i \dot{\xi}_{i\alpha}^2 + \frac{1}{2} \sum_{i\alpha,j\gamma} A_{i\alpha,j\gamma}\xi_{i\alpha}\xi_{j\gamma} \tag{10·1·6}$$

The kinetic energy term is simple since it involves a sum of terms each of which refers only to a single coordinate. The potential energy term is complicated, however, since it involves all possible products of different coordinates. This reflects, of course, just the fact that the atoms interact so that they do not behave like independent particles.

Since the potential energy is quadratic in the coordinates, the problem presented by (10 · 1 · 6) can, however, immediately be reduced to much simpler form by a change of variables which eliminates the cross-product terms in the potential energy (10 · 1 · 5) without destroying the simple form (10 · 1 · 2) of the kinetic energy. (The procedure is analogous to that of rotating the coordinate axes in such a way as to reduce the general equation of an ellipsoid to the simple standard form which involves only the square of each coordinate but no cross-product terms). Indeed, it can readily be shown* that it is always possible to go from the $3N$ old coordinates $\xi_{i\alpha}$ to some new set of $3N$ generalized coordinates q_r by a linear transformation of the form

$$\xi_{i\alpha} = \sum_{r=1}^{3N} B_{i\alpha,r} q_r \tag{10·1·7}$$

such that a proper choice of the coefficients $B_{i\alpha,r}$ transforms $\mathcal{H}$ in (10 · 1 · 6) to the simple form

$$\mathcal{H} = V_0 + \tfrac{1}{2} \sum_{r=1}^{3N} (\dot{q}_r^2 + \omega_r^2 q_r^2) \tag{10·1·8}$$

Here the coefficients ω_r^2 are positive constants, and there are *no* cross terms involving products of two different variables. The new variables q_r are called the "normal coordinates" of the system. In terms of these variables the Hamiltonian (10 · 1 · 8) is simply a sum of $3N$ independent terms each of which refers to only a single variable. Indeed (10 · 1 · 8) is identical in form to the Hamiltonian of $3N$ independent one-dimensional harmonic oscillators, the oscillator of coordinate q_r having an angular ("normal mode") frequency ω_r. The change of variable (10 · 1 · 7) has thus reduced the complicated problem of N interacting atoms to the equivalent problem of $3N$ noninteracting harmonic oscillators. The discussion of the latter problem is, of course, very simple.

To give the quantum-mechanical discussion, consider first the simple one-dimensional harmonic oscillator Hamiltonian

$$\mathcal{H}_r = \tfrac{1}{2}(\dot{q}_r^2 + \omega_r^2 q_r^2) \tag{10·1·9}$$

involving the single variable q_r. The possible quantum states of this oscillator are labeled by the quantum number n_r which can assume the values

$$n_r = 0, 1, 2, 3, 4, \ldots \tag{10·1·10}$$

The corresponding energies are given by

$$\epsilon_r = (n_r + \tfrac{1}{2})\hbar\omega_r \tag{10·1·11}$$

One can then write immediately the solution for the complete Hamiltonian (10 · 1 · 8) of a system of $3N$ independent harmonic oscillators. The quantum state of the whole system is specified by the set of $3N$ quantum numbers $\{n_1, n_2, \ldots, n_{3N}\}$ each of which can assume any of the integral values listed

* See, for example, H. Goldstein, "Classical Mechanics," chap. 10, Addison-Wesley Publishing Company Reading, Mass., 1950; or K. R. Symon, "Mechanics," 2d ed., secs. 12-1–12-3, Addison-Wesley Publishing Company, Reading, Mass., 1960.

in (10·1·10). The corresponding total energy is simply the sum of the one-dimensional oscillator energies, i.e.,

$$E_{n_1,\ldots,n_{3N}} = V_0 + \sum_{r=1}^{3N} (n_r + \tfrac{1}{2})\hbar\omega_r \tag{10·1·12}$$

This can be written in slightly different form as

▶ $$E_{n_1,\ldots,n_{3N}} = -N\eta + \sum_{r=1}^{3N} n_r\hbar\omega_r \tag{10·1·13}$$

where $$-N\eta \equiv V_0 + \tfrac{1}{2}\sum_r \hbar\omega_r \tag{10·1·14}$$

is a constant independent of the quantum numbers n_r. By (10·1·13), $-N\eta$ represents the lowest possible energy of the atoms measured with respect to a standard state where they are at rest at an infinite separation from each other. (Note that this quantity differs from V_0 by the "zero-point energy" $\frac{1}{2}\Sigma\hbar\omega_r$.) Thus η represents the binding energy per atom in the solid at absolute zero.

Remark We have arrived at the result that the state of the system is completely specified by the set of integers $\{n_1, \ldots, n_{3N}\}$ with a corresponding energy given by (10·1·13). Note that this result is exactly the same as if one were dealing with a system of particles each of which can be in any one of the $3N$ states labeled by $r = 1, \ldots, 3N$ with respective energies $\hbar\omega_1$, $\hbar\omega_2, \ldots, \hbar\omega_{3N}$, there being n_1 particles present in state 1, n_2 particles in state 2, . . . , and n_{3N} particles in state $3N$. From this point of view the state of the system would be specified by stating the *numbers* n_r of particles of each type r. Only these *numbers* are important, i.e., there is *no* mention of any distinguishability of particles, and, therefore, it does not matter which particular particle is in which state. Furthermore, any number of particles ($n_r = 0, 1, 2, 3, \ldots$) can be in any one state r, and the total number Σn_r of particles is not fixed in any way. These particles would then obey Bose-Einstein statistics. One can summarize these comments by noting that a quantum-mechanical discussion of the harmonic-oscillator Hamiltonian (10·1·8) gives rise to a specification of the system in terms of integral quantum numbers n_r, and that the results can then be interpreted *as if* one dealt with a system of indistinguishable particles where these integers specify the number of particles in each state. This is, of course, merely an *interpretation* of the results in terms of particles. To emphasize this fact we shall refer to these "particles" as "quasi-particles" to avoid confusion with the real particles of the problem, i.e., the atoms which form the solid. This interpretation in terms of quasi-particles represents a very useful point of view. In the present case the quantization of the lattice vibrations leads to quasi-particles which are called "phonons" (since they are basically quantized sound waves). The situation is completely analogous to the case where one quantizes electromagnetic radiation; the quasi-particles which arise in that problem are called "photons" and have the familiar particlelike properties of light.

The calculation of the partition function with (10·1·13) is immediate. One has

$$Z = \sum_{n_1, n_2, n_3 \ldots} e^{-\beta[-N\eta + n_1\hbar\omega_1 + n_2\hbar\omega_2 + \cdots + n_{3N}\hbar\omega_{3N}]}$$

$$= e^{\beta N\eta} \left(\sum_{n_1=0}^{\infty} e^{-\beta\hbar\omega_1 n_1} \right) \cdots \left(\sum_{n_{3N}=0}^{\infty} e^{-\beta\hbar\omega_{3N} n_{3N}} \right)$$

or

$$Z = e^{\beta N\eta} \left(\frac{1}{1 - e^{-\beta\hbar\omega_1}} \right) \cdots \left(\frac{1}{1 - e^{-\beta\hbar\omega_{3N}}} \right) \qquad (10 \cdot 1 \cdot 15)$$

since this is simply the product of one-dimensional harmonic-oscillator partition functions which are geometric series.

Thus

$$\ln Z = \beta N\eta - \sum_{r=1}^{3N} \ln \left(1 - e^{-\beta\hbar\omega_r}\right) \qquad (10 \cdot 1 \cdot 16)$$

The possible normal mode frequencies ω_r are closely spaced, and it is convenient to define the quantity

$\sigma(\omega)\, d\omega \equiv$ the number of normal modes with angular frequency in the range between ω and $\omega + d\omega$. $\qquad (10 \cdot 1 \cdot 17)$

A curve of $\sigma(\omega)$ versus ω might have the kind of shape shown in Fig. 10·1·1. In terms of the definition (10·1·17), ln Z in (10·1·16) can be expressed as an integral

$$\ln Z = \beta N\eta - \int_0^{\infty} \ln \left(1 - e^{-\beta\hbar\omega}\right) \sigma(\omega)\, d\omega \qquad (10 \cdot 1 \cdot 18)$$

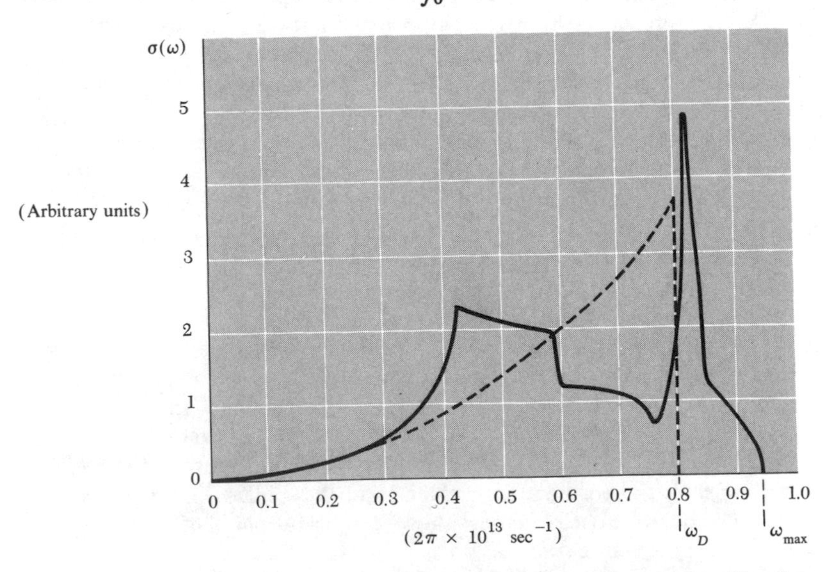

***Fig. 10·1·1** The normal-mode vibrational frequency distribution $\sigma(\omega)$ for aluminum. The solid curve is a measured curve deduced from X-ray scattering measurements at 300°K (after C. B. Walker, Phys. Rev., vol. 103, p. 547 (1956)). The dashed curve represents the Debye approximation of Sec. 10·2 with a value $\Theta_D = 382°K$ deduced from the specific heat.*

Thus the mean energy of the solid becomes

$$\bar{E} = -\frac{\partial \ln Z}{\partial \beta} = -N\eta + \int_0^\infty \frac{\hbar\omega}{e^{\beta\hbar\omega} - 1} \sigma(\omega)\, d\omega \qquad (10 \cdot 1 \cdot 19)$$

Its heat capacity at constant volume is then

$$C_V = \left(\frac{\partial \bar{E}}{\partial T}\right)_V = -k\beta^2 \left(\frac{\partial \bar{E}}{\partial \beta}\right)_V$$

or

▶ $$C_V = k \int_0^\infty \frac{e^{\beta\hbar\omega}}{(e^{\beta\hbar\omega} - 1)^2} (\beta\hbar\omega)^2 \sigma(\omega)\, d\omega \qquad (10 \cdot 1 \cdot 20)$$

The *statistical* problem is thus very simple. The entire complication of the problem revolves about the transformation of the Hamiltonian from (10·1·6) to (10·1·8); i.e., one must solve a mechanics problem to find the actual normal mode frequencies of the solid so as to determine the function $\sigma(\omega)$ for the solid under consideration.

Irrespective of the exact shape of $\sigma(\omega)$, one can make general statements about the high temperature limit. The significant dimensionless parameter occurring in (10·1·18) is $\beta\hbar\omega = \hbar\omega/kT$. Let ω_{max} denote the maximum frequency of the normal mode spectrum, i.e., let ω_{max} be such that

$$\sigma(\omega) = 0 \qquad \text{if } \omega > \omega_{max} \qquad (10 \cdot 1 \cdot 21)$$

If T is large enough so that $\beta\hbar\omega_{max} \ll 1$, then $\beta\hbar\omega \ll 1$ throughout the range of integration of the integral (10·1·20), so that one can write

$$e^{\beta\hbar\omega} = 1 + \beta\hbar\omega + \cdots$$

Thus (10·1·20) becomes simply,

$$\text{for } kT \gg \hbar\omega_{max}, \qquad C_V = k \int_0^\infty \sigma(\omega)\, d\omega = 3Nk \qquad (10 \cdot 1 \cdot 22)$$

since the integral equals simply the total number of normal modes, i.e.,

$$\int_0^\infty \sigma(\omega)\, d\omega = 3N \qquad (10 \cdot 1 \cdot 23)$$

The relation (10·1·22) is simply the Dulong and Petit law already obtained in Sec. 7·7 by applying the equipartition theorem in the classical high-temperature limit.

10·2 *Debye approximation*

The calculation of the number $\sigma(\omega)$ of normal-mode frequencies is a complicated problem. Although fairly good calculations of $\sigma(\omega)$ can be made for solids of simple structure, it is useful to employ less laborious methods to obtain approximate estimates of $\sigma(\omega)$.

Consider a solid consisting of N atoms whose masses are not too dissimilar. The approximation method used by Debye proceeds by neglecting the discreteness of the atoms in the solid and treating the latter as if it were a continuous elastic medium. Each normal mode of vibration of this elastic continuum is characterized by a wavelength λ. Let a denote the mean interatomic separation in the solid. If $\lambda \gg a$, neighboring atoms in the solid are displaced by nearly the same amount. In this case the fact that atoms are at a finite separation a from each other is not very significant, and the normal modes of vibration of the elastic medium are expected to be very nearly the same as those of the actual solid. On the other hand, those modes of vibration of the elastic continuum for which λ becomes comparable to a correspond to markedly different displacements of neighboring atoms. The discrete spacing of atoms is then quite important, and the actual normal modes of vibration of the atoms are consequently quite different from those of the elastic continuum. In short, treating the real solid as though it were an elastic continuum should be a good approximation for modes of long wavelength $\lambda \gg a$. Equivalently, this means that for *low* frequencies ω the density of modes $\sigma_c(\omega)$ for the continuous elastic medium should be very nearly the same as the mode density $\sigma(\omega)$ for the actual solid. For shorter wavelengths, or higher frequencies, $\sigma(\omega)$ and $\sigma_c(\omega)$ differ increasingly. Finally, when $\lambda \lesssim a$, $\sigma(\omega)$ and $\sigma_c(\omega)$ are completely different; indeed the real solid has no modes at such high frequencies (i.e., $\sigma(\omega) = 0$ for $\omega > \omega_{\max}$), while there is no limitation on how short a wavelength or how high a frequency a continuous medium can have.

Let us then investigate the normal modes of vibration of the solid considered as an isotropic elastic continuous medium of volume V. Let $\boldsymbol{u}(\boldsymbol{r},t)$ denote the displacement of a point in this medium from its equilibrium position. (In the limit of long wavelengths, the α component of the atomic displacement $\xi_{i\alpha}$ defined in (10·1·1) is then approximately given by $\xi_{i\alpha}(t) \approx u_\alpha(\boldsymbol{r}_i^{(0)},t)$.) The displacement $\boldsymbol{u}$ must then satisfy a wave equation which describes the propagation through the medium of sound waves traveling with some effective velocity c_s. The analysis of normal modes is then completely analogous to that in Sec. 9·9. A sound wave of wave vector $\boldsymbol{\kappa}$ corresponds to an angular frequency $\omega = c_s\kappa$, and the number of possible wave modes with frequency between ω and $\omega + d\omega$ (corresponding to a magnitude of $\boldsymbol{\kappa}$ between κ and $\kappa + d\kappa$) is given analogously to (9·9·18) by

$$\sigma_c(\omega)\, d\omega = 3\,\frac{V}{(2\pi)^3}\,(4\pi\kappa^2\, d\kappa) = 3\,\frac{V}{2\pi^2 c_s^3}\,\omega^2\, d\omega \tag{10·2·1}$$

where the factor of 3 accounts for the fact that there are three possible directions of polarization of $\boldsymbol{u}$ (one longitudinal one, as well as two transverse ones) for each wave vector $\boldsymbol{\kappa}$.

The analysis can readily be made more detailed. The displacement $\boldsymbol{u}$ can quite generally be written in the form

$$\boldsymbol{u} = \boldsymbol{u}_t + \boldsymbol{u}_l \tag{10·2·2}$$

where

$$\operatorname{div} \boldsymbol{u}_t = 0 \tag{10·2·3}$$

and

$$\operatorname{curl} \boldsymbol{u}_l = 0 \tag{10·2·4}$$

Elasticity theory shows* that the vectors $\boldsymbol{u}_t$ and $\boldsymbol{u}_l$ then satisfy, with different velocities of propagation c_t and c_l, respectively, the wave equations

$$\begin{aligned} \nabla^2 \boldsymbol{u}_t &= \frac{1}{c_t^2} \frac{\partial^2 \boldsymbol{u}_t}{\partial t^2} \\ \nabla^2 \boldsymbol{u}_l &= \frac{1}{c_l^2} \frac{\partial^2 \boldsymbol{u}_l}{\partial t^2} \end{aligned} \tag{10·2·5}$$

Here c_l and c_t can be expressed in terms of the elastic constants of the medium by the relations

$$c_t = \left(\frac{\mu}{\rho}\right)^{\frac{1}{2}}, \qquad c_l = \left(\frac{b + \frac{4}{3}\mu}{\rho}\right)^{\frac{1}{2}} \tag{10·2·6}$$

where ρ is the density of the medium, μ its shear modulus, and b its bulk modulus (the reciprocal of its compressibility).

Equations (10·2·5) have plane wave solutions of the form

$$\boldsymbol{u}_t = \boldsymbol{A}_t\, e^{i(\boldsymbol{\kappa}_t \cdot \boldsymbol{r} - \omega t)}, \qquad |\boldsymbol{\kappa}_t| = \frac{\omega}{c_t} \tag{10·2·7}$$

$$\boldsymbol{u}_l = \boldsymbol{A}_l\, e^{i(\boldsymbol{\kappa}_l \cdot \boldsymbol{r} - \omega t)}, \qquad |\boldsymbol{\kappa}_l| = \frac{\omega}{c_l} \tag{10·2·8}$$

where $\boldsymbol{A}_t$ and $\boldsymbol{A}_l$ are constants. By virtue of (10·2·3)

$$\boldsymbol{\kappa}_t \cdot \boldsymbol{u}_t = 0, \qquad \text{so that } \boldsymbol{u}_t \perp \boldsymbol{\kappa}_t \tag{10·2·9}$$

Hence $\boldsymbol{u}_t$ represents a displacement which is perpendicular to the direction of propagation of the wave, so that $\boldsymbol{u}_t$ in (10·2·7) represents the propagation of a sound wave polarized transversely to the direction of propagation. Similarly, it follows from (10·2·4) that

$$\boldsymbol{\kappa}_l \times \boldsymbol{u}_l = 0, \qquad \text{so that } \boldsymbol{u}_l \| \boldsymbol{\kappa}_l \tag{10·2·10}$$

Hence $\boldsymbol{u}_l$ represents a displacement which is parallel to the direction of propagation of the wave, so that $\boldsymbol{u}_l$ in (10·2·8) represents the propagation of a sound wave polarized longitudinally to the direction of propagation.

By (9·9·18) the number of longitudinal wave modes in the frequency range between ω and $\omega + d\omega$ corresponding to a wave vector of magnitude between κ_l and $\kappa_l + d\kappa_l$ is then

$$\sigma_c^{(l)}(\omega)\, d\omega = \frac{V}{(2\pi)^3} 4\pi \kappa_l^2\, d\kappa_l = \frac{V}{2\pi^2 c_l^3}\, \omega^2\, d\omega \tag{10·2·11}$$

For the number of transverse wave modes in the frequency range between ω and $\omega + d\omega$ one obtains a similar result except for multiplication by a factor of 2, since there are two possible components of the transverse displacement $\boldsymbol{u}_t$. Thus

$$\sigma_c^{(t)}(\omega)\, d\omega = 2 \frac{V}{2\pi^2 c_t^3}\, \omega^2\, d\omega \tag{10·2·12}$$

* See, for example, L. Page, "Introduction to Theoretical Physics," 3d ed., p. 195, D. Van Nostrand Company, Inc. New York, 1953.

Hence the total number of modes of the elastic continuum in this frequency range is given by

$$\sigma_c(\omega)\,d\omega = [\sigma_c^{(l)}(\omega) + \sigma_c^{(t)}(\omega)]\,d\omega = 3\,\frac{V}{2\pi^2 c_s^3}\,\omega^2\,d\omega \qquad (10\cdot2\cdot13)$$

where we have defined an effective sound velocity c_s by the relation

$$\frac{3}{c_s^3} \equiv \frac{1}{c_l^3} + \frac{2}{c_t^3} \qquad (10\cdot2\cdot14)$$

so that c_s reduces simply to the velocity of sound if $c_l = c_t$. Except for the fact that (10·2·14) and (10·2·6) express c_s directly in terms of the measured sound velocities or elastic constants, (10·2·13) is, of course, identical to (10·2·1).

The Debye approximation consists in approximating $\sigma(\omega)$ by $\sigma_c(\omega)$ not only at very low frequencies where these should be nearly the same, but for all the $3N$ lowest-frequency modes of the elastic continuum. Specifically, Debye approximates $\sigma(\omega)$ by the distribution $\sigma_D(\omega)$ defined by

$$\sigma_D(\omega) = \begin{cases} \sigma_c(\omega) & \text{for } \omega < \omega_D \\ 0 & \text{for } \omega > \omega_D \end{cases} \qquad (10\cdot2\cdot15)$$

where the "Debye frequency" ω_D is chosen so that $\sigma_D(\omega)$ yields the correct total number of $3N$ normal modes (just as σ does in (10·1·23)); i.e.,

$$\int_0^\infty \sigma_D(\omega)\,d\omega = \int_0^{\omega_D} \sigma_c(\omega)\,d\omega = 3N \qquad (10\cdot2\cdot16)$$

A plot of $\sigma_D(\omega)$ versus ω is shown in Fig. 10·2·1 and is compared with a real frequency spectrum $\sigma(\omega)$ in Fig. 10·1·1.

Substitution of (10·2·1) into (10·2·16) yields

$$\frac{3V}{2\pi^2 c_s^3}\int_0^{\omega_D} \omega^2\,d\omega = \frac{V}{2\pi^2 c_s^3}\,\omega_D^3 = 3N \qquad (10\cdot2\cdot17)$$

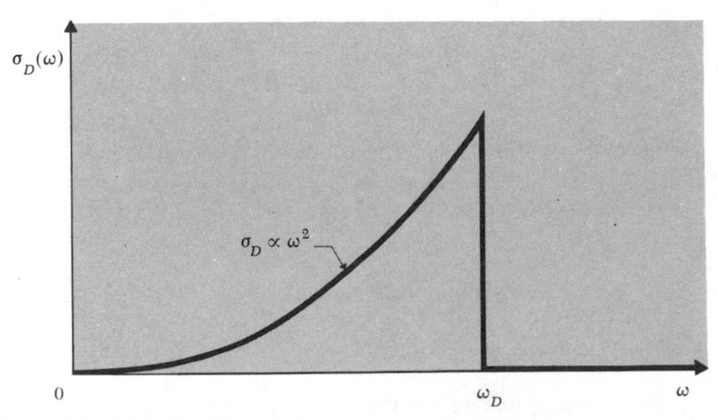

Fig. 10·2·1 The Debye frequency spectrum.

or

$$\omega_D = c_s \left(6\pi^2 \frac{N}{V}\right)^{\frac{1}{3}} \tag{10·2·18}$$

Thus the Debye frequency depends only on the sound velocities in the solid and on the number of atoms per unit volume. It is approximately such that the corresponding wavelength $2\pi c_s/\omega_D$ is of the order of the interatomic spacing $a \approx (V/N)^{\frac{1}{3}}$. Since $c_s \approx 5 \cdot 10^5$ cm/sec and $a \approx 10^{-8}$ cm, $\omega_D \approx 10^{14}$ sec^{-1} is typically a frequency lying in the infrared region of the electromagnetic spectrum.

Using the Debye approximation (10·2·15), the heat capacity (10·1·20) becomes

$$C_V = k \int_0^{\omega_D} \frac{e^{\beta\hbar\omega}(\beta\hbar\omega)^2}{(e^{\beta\hbar\omega} - 1)^2} \frac{3V}{2\pi^2 c_s^3} \omega^2 \, d\omega \tag{10·2·19}$$

In terms of the dimensionless variable $x \equiv \beta\hbar\omega$ this gives

$$C_V = k \frac{3V}{2\pi^2(c_s\beta\hbar)^3} \int_0^{\beta\hbar\omega_D} \frac{e^x}{(e^x - 1)^2} x^4 \, dx \tag{10·2·20}$$

To facilitate comparison with the classical result $C_V = 3Nk$, the volume V can be expressed in terms of N by (10·2·17), giving

$$V = 6\pi^2 N \left(\frac{c_s}{\omega_D}\right)^3 \tag{10·2·21}$$

Thus (10·2·20) can be written in the form

$$C_V = 3Nkf_D(\beta\hbar\omega_D) = 3Nkf_D\left(\frac{\Theta_D}{T}\right) \tag{10·2·22}$$

where we have defined the "Debye function" $f_D(y)$ by

$$f_D(y) \equiv \frac{3}{y^3} \int_0^y \frac{e^x}{(e^x - 1)^2} x^4 \, dx \tag{10·2·23}$$

and have further defined the "Debye temperature" Θ_D to be such that

$$k\Theta_D \equiv \hbar\omega_D \tag{10·2·24}$$

At high temperatures where $T \gg \Theta_D$, $f_D(\Theta_D/T) \to 1$ by the general arguments at the end of Sec. 10·1. Indeed, for small y, one can put $e^x = 1 + x$ in the integrand of (10·2·22) so that

for $y \to 0$,
$$f_D(y) \to \frac{3}{y^3} \int_0^y x^2 \, dx = 1 \tag{10·2·25}$$

The more interesting limiting case is that of very low temperatures. Then $\beta\hbar\omega \gg 1$ for relatively low frequencies where $\omega \ll \omega_D$. Physically this means that only oscillators of *low* frequency ω are then thermally excited to an appreciable extent and contribute to the heat capacity. Mathematically this means that the exponential factors are such that the integrand in (10·1·20) is appreciably only for small values of ω. In this case a knowledge of $\sigma(\omega)$ for

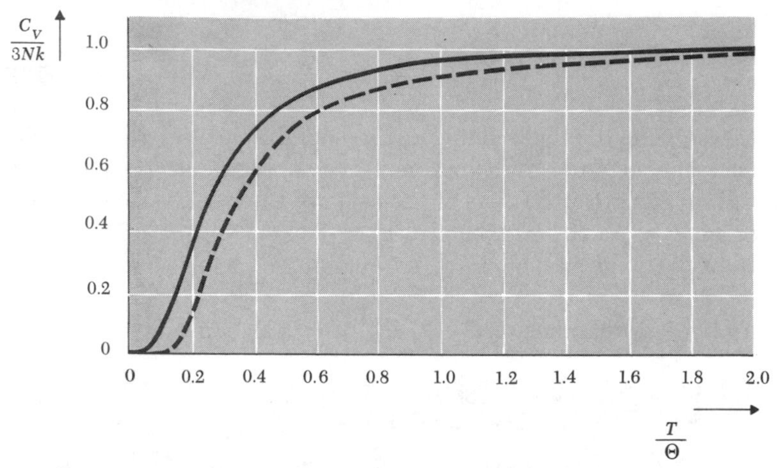

Fig. 10·2·2 Temperature dependence of the heat capacity C_V according to the Debye theory (10·2·22). The dashed curve shows, for comparison, the temperature dependence according to the simple Einstein model of Sec. 7·7, with $\Theta_E = \Theta_D$.

low frequencies ω is sufficient to calculate C_V at low temperatures, and it is precisely in this region that the Debye approximation of replacing the solid by an elastic continuum is best.

In this low-temperature region the upper limit $\beta\hbar\omega_D = \Theta_D/T$ of the integral (10·2·20) can be replaced by ∞ so that this integral becomes simply a constant. Hence it follows immediately that for $T \ll \Theta_D$,

$$C_V \propto \beta^{-3} \propto T^3 \tag{10·2·26}$$

Remark At sufficiently low temperatures normal modes of frequencies as high as ω_D are not excited. Hence it becomes irrelevant that an upper cutoff frequency ω_D exists in this problem and the situation becomes completely analogous to that of electromagnetic waves, where all frequencies are permissible. The remarks made in (9·13·18) in connection with photons are thus equally applicable in this case of "phonons" or normal modes of sound. The mean number of phonons in our three-dimensional problem is thus given by

$$\bar{N} \propto \omega^3 \propto T^3 \tag{10·2·27}$$

and correspondingly their mean energy by

$$\bar{E} \propto \bar{N}(kT) \propto T^4 \tag{10·2·28}$$

Hence it follows that

$$\bar{C}_V = \frac{\partial \bar{E}}{\partial T} \propto T^3 \tag{10·2·29}$$

The T^3 relation of (10·2·26) therefore reflects simply the fact that we are dealing with a three-dimensional solid.

The result (10·2·26) can easily be made more quantitative. Since the upper limit in (10·2·20) can be replaced by ∞, we note that C_V becomes independent of the precise value chosen for the upper cutoff frequency ω_D. The resulting integral was already encountered in our discussion of black-body radiation in (9·13·16); it can be evaluated numerically or can be computed exactly (see Appendix A·11). Thus one finds

$$\int_0^\infty \frac{e^x}{(e^x - 1)^2} x^4 \, dx = 4 \int_0^\infty \frac{x^3}{e^x - 1} \, dx = \frac{4\pi^4}{15} \tag{10·2·30}$$

where the second integral is obtained by integration by parts. Equivalently, this implies that

$$\text{for } y \gg 1, \qquad f_D(y) = \frac{4\pi^4}{5} \frac{1}{y^3} \tag{10·2·31}$$

Hence (10·2·20) becomes

$$C_V = \frac{2\pi^2}{5} Vk \left(\frac{kT}{c_s \hbar}\right)^3$$

Alternatively, this can also be expressed in terms of the quantity ω_D or Θ_D given by (10·2·18) and (10·2·24). Using (10·2·21), or using (10·2·22) and (10·2·23), one obtains

► $$C_V = \frac{12\pi^4}{5} Nk \left(\frac{T}{\Theta_D}\right)^3 \tag{10·2·32}$$

The fact that the specific heat of solids is proportional to T^3 at sufficiently low temperatures is quite well verified experimentally, although it may be necessary to go to temperatures low enough so that $T < 0.02\Theta_D$ (see Figs. 9·16·4 and 10·2·3). The Debye temperature Θ_D can be obtained from such low-temperature specific-heat measurements by comparison with the coefficient of T^3 in (10·2·32). Theoretically, Θ_D should be given by (10·2·18) so that it can be computed from the elastic constants of the solid, either from the known

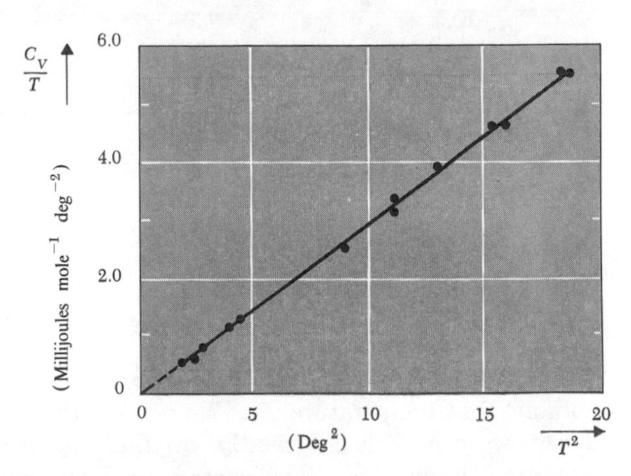

***Fig. 10·2·3** A curve of C_V/T versus T^2 for KCl showing the validity of the T^3 law at low temperatures. (The experimental points are those of P. H. Keesom and N. Pearlman, Phys. Rev., vol. 91, p. 1354 (1953).)*

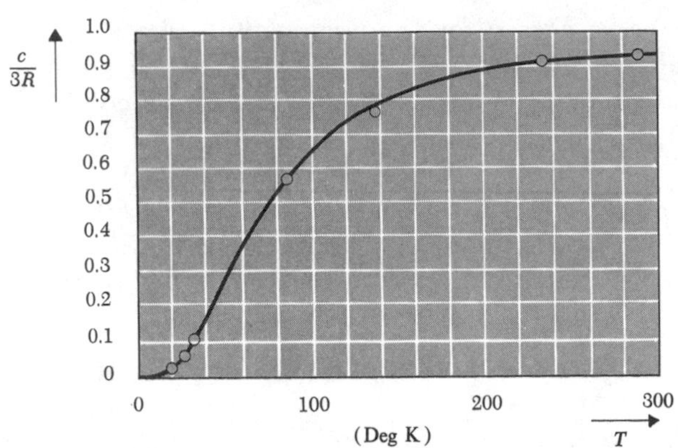

Fig. 10·2·4 The molar specific heat of copper: Debye theory with $\Theta_D = 309°K$ and experimental points. (*After P. Debye, Ann. Physik, vol. 39, p. 789 (1912)*.)

velocities of sound c_l and c_t, or by (10·2·6) from the known elastic moduli. The agreement is fairly good, as shown in Table 10·2·1.

The Debye theory (10·2·22) also provides in many cases a fairly good, although not perfect, representation of the temperature dependence of the specific heat over the entire temperature range. Figure 10·2·4 shows an example taken from Debye's original paper.

Table 10 · 2 · 1 Comparison of Debye temperatures obtained from low-temperature specific-heat measurements and calculated from elastic constants*

Solid	Θ_D *from specific heat* (°K)	Θ_D *from elastic constants* (°K)
NaCl	308	320
KCl	230	246
Ag	225	216
Zn	308	305

* Compilation taken from C. Kittel "Introduction to Solid State Physics," 2d ed., p. 132, John Wiley & Sons, Inc., New York, 1956.

NONIDEAL CLASSICAL GAS

10 · 3 *Calculation of the partition function for low densities*

Consider a monatomic gas of N identical particles of mass m in a container of volume V at temperature T. We assume that T is sufficiently high and the density $n \equiv N/V$ is sufficiently low that the gas can be treated by classical statistical mechanics. The energy, or Hamiltonian, of this system can be written in the form

$$\mathcal{H} = K + U \tag{10·3·1}$$

where

$$K = \frac{1}{2m} \sum_{j=1}^{N} \mathbf{p}_j^2 \tag{10·3·2}$$

is the kinetic energy of the gas and U is the potential energy of interaction between the molecules. We denote the potential energy of interaction between molecules j and k by $u_{jk} \equiv u(R_{jk})$ and assume that it depends only on their relative separation $R_{jk} \equiv |\mathbf{r}_j - \mathbf{r}_k|$. We assume further that U is simply given by the sum of all interactions between pairs of molecules. Thus

$$U = u_{12} + u_{13} + u_{14} + \cdots + u_{23} + u_{24} + \cdots + u_{N-1,N}$$

or

$$U = \sum_{\substack{j=1 \\ j<k}}^{N} \sum_{k=1}^{N} u_{jk} = \frac{1}{2} \sum_{\substack{j=1 \\ j\neq k}}^{N} \sum_{k=1}^{N} u_{jk} \tag{10·3·3}$$

The potential energy of interaction u between a pair of molecules has the general shape shown in Fig. 10·3·1; i.e., it is strongly repulsive when the molecules are very close together and more weakly attractive at larger separations. For simple molecules it is possible to obtain $u(R)$ by quantum-mechanical calculations. A useful semiempirical potential, the so-called "Lennard-Jones potential," is given by

$$u(R) = u_0 \left[\left(\frac{R_0}{R} \right)^{12} - 2 \left(\frac{R_0}{R} \right)^{6} \right] \tag{10·3·4}$$

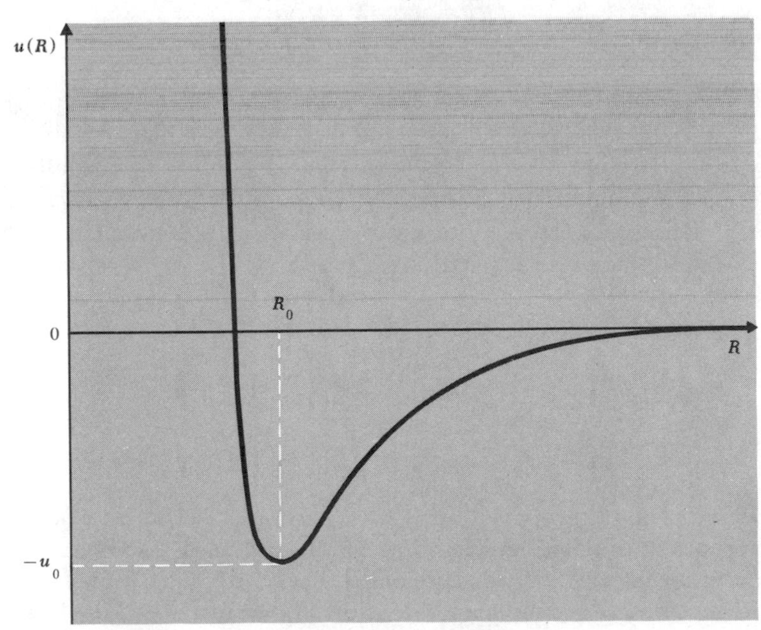

Fig. 10·3·1 Diagram illustrating the potential energy $u(R)$ describing the interaction between two molecules separated by a distance R.

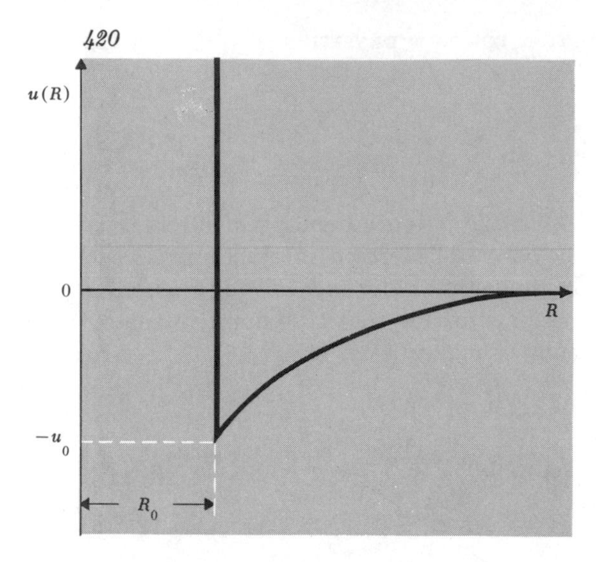

***Fig. 10·3·2** Graph illustrating the potential energy $u(R)$ of Eq. (10·3·5) as a function of R.*

This has the general shape shown in Fig. 10·3·1. The two constant parameters $-u_0$ and R_0 denote, respectively, the minimum value of u and the separation at which this minimum occurs. The fact that $u \propto R^{-6}$ when R is large has theoretical justification (see Problem 10.6).

A different potential which is a less realistic approximation, but mathematically slightly simpler, is of the form

$$u(R) = \begin{cases} \infty & \text{for } R < R_0 \\ -\, u_0 \left(\dfrac{R_0}{R}\right)^s & \text{for } R > R_0 \end{cases} \tag{10·3·5}$$

The fact that $u \to \infty$ for $R < R_0$ means that the minimum possible separation between molecules is R_0, i.e., the molecules act like weakly attracting hard spheres of radius $\frac{1}{2}R_0$. The choice of exponent $s = 6$ is usually most appropriate. This potential is illustrated in Fig. 10·3·2.

To discuss the equilibrium properties of the gas, it is then necessary to calculate the classical partition function

$$\begin{aligned} Z &= \frac{1}{N!} \iint \cdots \int e^{-\beta(K+U)} \frac{d^3\mathbf{p}_1 \cdots d^3\mathbf{p}_N\, d^3\mathbf{r}_1 \cdots d^3\mathbf{r}_N}{h^{3N}} \\ &= \frac{1}{h^{3N}N!} \int \cdots \int e^{-\beta K(p_1, \ldots, p_N)}\, d^3\mathbf{p}_1 \cdots d^3\mathbf{p}_N \\ &\qquad \int \cdots \int e^{-\beta U(r_1, \ldots, r_N)}\, d^3\mathbf{r}_1 \cdots d^3\mathbf{r}_N \end{aligned} \tag{10·3·6}$$

where K and U are given by (10·3·2) and (10·3·3). The factor $N!$ takes into account the indistinguishability of the particles, and we have set $h_0 = h$ in accordance with our discussion of Sec. 9·10. The first integral over the momenta is very simple and identical to that evaluated in (7·2·6) for the ideal gas. Denote the second integral by

$$Z_U \equiv \int \cdots \int e^{-\beta U(\mathbf{r}_1, \ldots, \mathbf{r}_N)}\, d^3\mathbf{r}_1 \cdots d^3\mathbf{r}_N \tag{10·3·7}$$

Then (10·3·6) becomes

$$Z = \frac{1}{N!}\left(\frac{2\pi m}{h^2\beta}\right)^{\frac{3}{2}N} Z_U \qquad (10\cdot3\cdot8)$$

The evaluation of the integral Z_U, where each $\mathbf{r}_j$ ranges over the volume V of the container, is quite complicated because U in (10·3·3) does not involve the coordinates one at a time. The calculation of Z_U constitutes thus the essential difficulty in the discussion of nonideal gases. Of course, Z_U becomes simple in the limiting case when the gas is ideal so that $U \to 0$, or when the temperature is so high that $\beta \to 0$; for then $e^{-\beta U} \to 1$ and $Z_U \to V^N$.

If the gas density n is not too large, it is possible to develop systematic approximation procedures for evaluating the partition function Z_U. These procedures involve essentially a series expansion in successively higher powers of the density n. For low densities the first terms in the expansion are then sufficient to give an accurate description of the gas. Indeed, the first approximation to Z_U can be easily obtained by energy considerations alone. The mean potential energy of the gas is given by

$$\bar{U} = \frac{\int e^{-\beta U} U \, d^3\mathbf{r}_1 \cdots d^3\mathbf{r}_N}{\int e^{-\beta U} \, d^3\mathbf{r}_1 \cdots d^3\mathbf{r}_N} = -\frac{\partial}{\partial\beta} \ln Z_U \qquad (10\cdot3\cdot9)$$

Hence

$$\ln Z_U(\beta) = N \ln V - \int_0^\beta \bar{U}(\beta')\, d\beta' \qquad (10\cdot3\cdot10)$$

since $Z_U(0) = V^N$ for $\beta = 0$. But by Eq. (10·3·3) the mean potential energy of the $\frac{1}{2}N(N-1)$ pairs of molecules is simply

$$\bar{U} = \tfrac{1}{2}N(N-1)\bar{u} \approx \tfrac{1}{2}N^2\bar{u} \qquad (10\cdot3\cdot11)$$

since $N \gg 1$. Furthermore, when the gas is sufficiently dilute, one can assume in first approximation that the motion of any pair of molecules is not correlated appreciably with the motion of the remaining molecules which act, therefore, simply like a heat reservoir at temperature T. The probability that molecule j is located at a position $\mathbf{R} \equiv \mathbf{r}_j - \mathbf{r}_k$ relative to molecule k is then proportional to $e^{-\beta u(R)}\, d^3\mathbf{R}$. Hence the mean potential energy $\bar{u}$ between this pair of molecules is given by

$$\bar{u} = \frac{\int e^{-\beta u} u \, d^3\mathbf{R}}{\int e^{-\beta u}\, d^3\mathbf{R}} = -\frac{\partial}{\partial\beta} \ln \int e^{-\beta u}\, d^3\mathbf{R} \qquad (10\cdot3\cdot12)$$

where the integration is over all possible values of the relative position $\mathbf{R}$, i.e., essentially over the whole volume V of the container. Since $u \approx 0$ and $e^{-\beta u} \approx 1$ practically everywhere except when R is small, it is convenient to write the integral in the form

$$\int e^{-\beta u}\, d^3\mathbf{R} = \int [1 + (e^{-\beta u} - 1)]\, d^3\mathbf{R} = V + I = V\left(1 + \frac{I}{V}\right) \qquad (10\cdot3\cdot13)$$

where

▶ $$I(\beta) \equiv \int (e^{-\beta u} - 1)\, d^3\mathbf{R} = \int_0^\infty (e^{-\beta u} - 1) 4\pi R^2\, dR \qquad (10\cdot3\cdot14)$$

is relatively small so that $I \ll V$. By using (10·3·13) in (10·3·12), one then obtains

$$\bar{u} = -\frac{\partial}{\partial\beta}\left[\ln V + \ln\left(1 + \frac{I}{V}\right)\right] \approx 0 - \frac{\partial}{\partial\beta}\left(\frac{I}{V} + \cdots\right)$$

or

$$\bar{u} = -\frac{1}{V}\frac{\partial I}{\partial\beta} \tag{10·3·15}$$

Thus (10·3·11) becomes

$$\bar{U} = -\frac{1}{2}\frac{N^2}{V}\frac{\partial I}{\partial\beta}$$

Hence (10·3·10) yields simply, since $I = 0$ for $\beta = 0$,

▶
$$\ln Z_U(\beta) = N \ln V + \frac{1}{2}\frac{N^2}{V} I(\beta) \tag{10·3·16}$$

10 · 4 *Equation of state and virial coefficients*

The equation of state can readily be obtained from the partition function by the general relation

$$\bar{p} = \frac{1}{\beta}\frac{\partial \ln Z}{\partial V} = \frac{1}{\beta}\frac{\partial \ln Z_U}{\partial V} \tag{10·4·1}$$

since only Z_U in (10·3·8) involves the volume V. Hence (10·3·16) yields the result

$$\beta\bar{p} = \frac{\bar{p}}{kT} = \frac{N}{V} - \frac{1}{2}\frac{N^2}{V^2} I \tag{10·4·2}$$

This is of the general form

$$\frac{\bar{p}}{kT} = n + B_2(T)n^2 + B_3(T)n^3 + \cdots \tag{10·4·3}$$

where $n \equiv N/V$ is the number of molecules per unit volume. Equation (10·4·3) is an expansion in powers of n which is the so-called "virial expansion" already mentioned in (5·10·12). The coefficients $B_2, B_3, \ldots$ are called the "virial coefficients." For the ideal gas $B_2 = B_3 = \cdots = 0$. If n is not too large, only the first few terms in (10·4·3) are appreciable. We have evaluated the first correction to Z when terms in n^2 become important. Hence we have found the second virial coefficient B_2; by identification with (10·4·2) and reference to (10·3·14) it is simply given by

▶
$$B_2 = -\tfrac{1}{2}I = -2\pi \int_0^\infty (e^{-\beta u} - 1)R^2\, dR \tag{10·4·4}$$

A knowledge of the intermolecular potential u allows therefore an immediate evaluation of the first correction term to the ideal-gas equation of state.

Keeping in mind the general behavior of the intermolecular potential illustrated in Fig. 10·3·1, it is easy to discuss the temperature dependence of B_2. Consider the dependence of the integrand in (10·4·4) as a function of R.

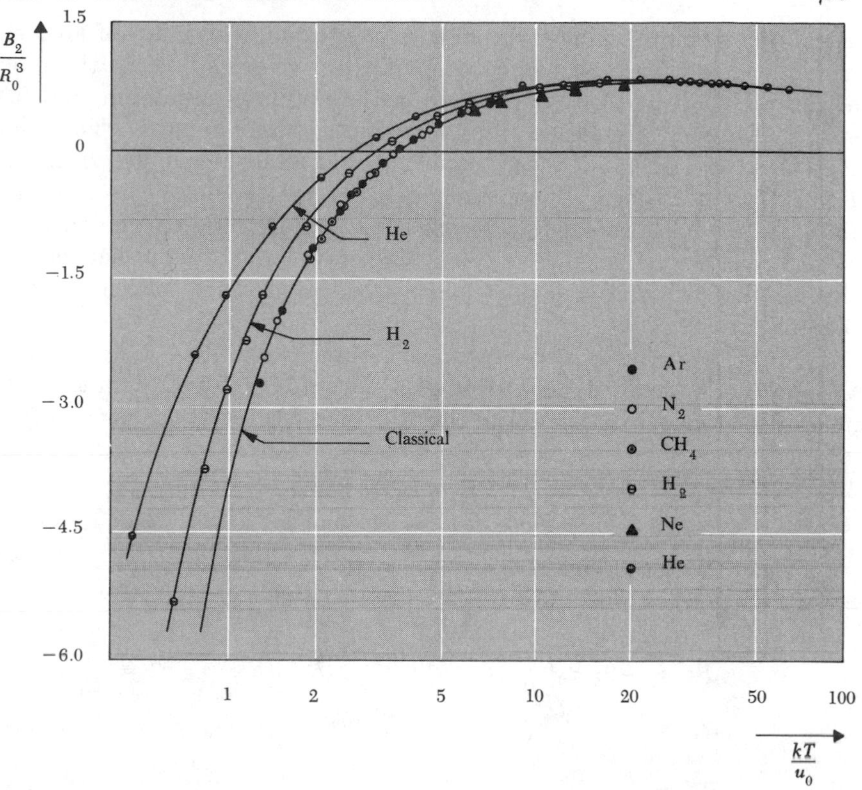

***Fig. 10·4·1** Dimensionless plot showing the dependence of B_2 on the temperature T. The curve labeled "classical" shows the result of the classical calculation using the Lennard-Jones potential (10·3·4). The other two curves indicate, for comparison, calculated curves for He and H_2 gas if quantum-mechanical effects are taken into account. (It is seen that quantum effects for these light gases become important at low temperatures.) The points indicate experimental results for several gases (after J. O. Hirschfelder, C. F. Curtiss, and R. B. Bird, "Molecular Theory of Gases and Liquids," p. 164, John Wiley & Sons, Inc., New York, 1954).*

When R is small, u is large and positive so that $(e^{-\beta u} - 1)$ is negative; in this region the integrand in (10·4·4) thus makes a positive contribution to B_2. But for larger values of R, u is negative so that $(e^{-\beta u} - 1)$ is positive and larger for larger values of β (i.e., for smaller values of T); in this region the integrand thus makes a negative contribution to B_2. At low temperatures this negative contribution is predominant so that B_2 is negative; but at high temperatures it is of minor significance so that B_2 is positive. At some intermediate temperature B_2 must then vanish. The temperature dependence of B_2 exhibits, therefore, the behavior shown in Fig. 10·4·1.

One can understand this behavior in more physical terms. At low temperatures where $kT < u_0$, the molecules are most likely to be in configurations of lowest mutual energy where the intermolecular forces are attractive. This attraction tends to reduce the pressure of the gas below that of an ideal gas, i.e., B_2 is negative. At higher temperatures, where $kT > u_0$, the molecules are scarcely affected by the presence of the potential minimum, and it is the strong repulsive interaction whose influence is predominant. This repulsion increases the pressure of the gas above that of an ideal gas, i.e., B_2 is positive. Finally, when T is very large, the molecular kinetic energy becomes so large that the molecules can also overcome some of their repulsive potential energy to come closer together than at lower temperatures. Thus, at sufficiently high temperatures, there is a slight tendency for the pressure, and hence for B_2, to decrease again.

All these qualitative features are exhibited in Fig. 10·4·1, which illustrates the result of an explicit calculation of B_2 by using (10·4·4) with the Lennard-Jones potential (10·3·4). Note that this potential is of the general form

$$u(R) = u_0\,\phi\left(\frac{R}{R_0}\right) \tag{10·4·5}$$

where u_0 and R_0 are two parameters and ϕ is some function of the relative separation R/R_0. Thus (10·4·4) can be written

$$B_2 = -2\pi R_0{}^3 \int_0^\infty (e^{-\beta u_0 \phi(R')} - 1) R'^2\, dR', \qquad R' \equiv \frac{R}{R_0}$$

or

$$B_2' = -2\pi \int_0^\infty (e^{-\phi(R')/T'} - 1) R'^2\, dR' \tag{10·4·6}$$

where

$$B_2' \equiv \frac{B_2}{R_0{}^3} \qquad \text{and} \qquad T' \equiv \frac{kT}{u_0} = \frac{T}{u_0/k} \tag{10·4·7}$$

Hence the potential (10·4·5) implies that, when expressed in terms of these dimensionless variables, B_2' is the same universal function of T' for all gases. This function is shown in Fig. 10·4·1 for the Lennard-Jones potential (10·3·4). Experimental measurements of the equation of state provide points which can be plotted on such a graph, the question being how good a fit with the theoretical curve can be obtained by using only the two adjustable parameters R_0 and u_0. Figure 10·4·1 shows that rather good agreement can be obtained. The values of R_0 and u_0 which give the best fit then yield information about the intermolecular potential. For example, for argon one thus obtains* $R_0 = 3.82$ Å and $u_0/k = 120$°K.

The van der Waals equation Let us illustrate the calculation of B_2 for a special simple case. Suppose that the potential u can be adequately represented by (10·3·5). Then (10·4·4) becomes

$$B_2 = 2\pi \int_0^{R_0} R^2\, dR - 2\pi \int_{R_0}^\infty (e^{-\beta u} - 1) R^2\, dR \tag{10·4·8}$$

* Values for some other gases can be found summarized in T. L. Hill, "Introduction to Statistical Thermodynamics," p. 484, Addison-Wesley Publishing Company, Reading, Mass., 1960.

Assume that the temperature is high enough that

$$\beta u_0 \ll 1 \tag{10·4·9}$$

Then $e^{-\beta u} \approx 1 - \beta u$ in the second integral and (10·4·8) becomes

$$B_2 = \frac{2\pi}{3} R_0^3 - 2\pi\beta u_0 \int_{R_0}^{\infty} \left(\frac{R_0}{R}\right)^s R^2\, dR$$

or

$$B_2 = \frac{2\pi}{3} R_0^3 \left(1 - \frac{3}{s-3}\frac{u_0}{kT}\right)$$

where we have assumed that $s > 3$ so that the integral converges. Thus B_2 assumes the form

$$B_2 = b' - \frac{a'}{kT} \tag{10·4·10}$$

where

$$b' \equiv \frac{2\pi}{3} R_0^3 \qquad \text{and} \qquad a' \equiv \left(\frac{3}{s-3}\right) b' u_0 \tag{10·4·11}$$

The equation of state (10·4·3) becomes then, neglecting terms of order higher than n^2,

$$\frac{\bar{p}}{kT} = n + \left(b' - \frac{a'}{kT}\right) n^2 \tag{10·4·12}$$

Hence

$$\bar{p} = nkT + (b'kT - a')n^2$$

or

$$\bar{p} + a'n^2 = nkT(1 + b'n) \approx \frac{nkT}{1 - b'n} \tag{10·4·13}$$

In the last step we have made the approximation that

$$b'n \ll 1 \tag{10·4·14}$$

i.e., that the density of the gas is low enough that the mean volume $n^{-1} = V/N$ available per molecule in the container is large compared to the volume of the hard-sphere core of the molecule. Thus (10·4·13) becomes

$$(\bar{p} + a'n^2)\left(\frac{1}{n} - b'\right) = kT \tag{10·4·15}$$

which is essentially the van der Waals equation. One can write this in more familiar form in terms of the molar volume $v = V/\nu$, where ν is the number of moles of gas. Thus

$$n = \frac{N}{V} = \frac{\nu N_a}{V} = \frac{N_a}{v}$$

where N_a is Avogadro's number. Thus (10·4·15) becomes

$$\left(\bar{p} + \frac{a}{v^2}\right)(v - b) = RT \tag{10·4·16}$$

where $R = N_a k$ is the gas constant and

$$a \equiv N_a^2 a' \qquad \text{and} \qquad b \equiv N_a b' \tag{10·4·17}$$

By virtue of (10·4·11), the van der Waals constants a and b are then expressed in terms of the parameters describing the intermolecular potential (10·3·5).

10 · 5 *Alternative derivation of the van der Waals equation*

It is instructive to discuss the problem of the nonideal gas in an alternative way which, though very crude, does not specifically assume that the gas is dilute. One proceeds by focusing attention on a single molecule and approximates the situation by assuming that this molecule moves in an effective potential $U_e(\mathbf{r})$ due to all the other molecules (which are assumed to remain unaffected by the presence of the molecule under consideration). The partition function for the system reduces then to that for a system of N *independent* particles, each with kinetic energy $(2m)^{-1}\mathbf{p}^2$ and potential energy U_e. In this approximation one has classically

$$Z = \frac{1}{N!}\left[\iint e^{-\beta(p^2/2m+U_e)} \frac{d^3\mathbf{p}\, d^3\mathbf{r}}{h^3}\right]^N \tag{10·5·1}$$

where the factor $N!$ again takes into account the indistinguishability of the molecules. The integral over momentum is identical to that evaluated in (7·2·6) for the ideal gas. Thus (10·5·1) becomes

$$Z = \frac{1}{N!}\left(\frac{2\pi m}{h^2\beta}\right)^{\frac{3}{2}N}\left[\int e^{-\beta U_e(r)}\, d^3\mathbf{r}\right]^N \tag{10·5·2}$$

This remaining integral extends over the volume V of the container. To make further progress, we note that there are regions where $U_e \to \infty$ because of the strong repulsion between molecules. Thus the integrand vanishes in these regions of total volume V_x. In the remaining volume $(V - V_x)$, where U_e does not vary too rapidly with the intermolecular separation, we shall replace it by some effective *constant* average value $\bar{U}_e$. Thus (10·5·2) becomes

$$Z = \frac{1}{N!}\left[\left(\frac{2\pi m}{h^2\beta}\right)^{\frac{3}{2}}(V - V_x)\, e^{-\beta\bar{U}_e}\right]^N \tag{10·5·3}$$

It remains to estimate the values of the parameters $\bar{U}_e$ and V_x by some self-consistency requirements. The total mean potential energy of the molecules is given by $N\bar{U}_e$. But since there are $\frac{1}{2}N(N-1) \approx \frac{1}{2}N^2$ pairs of molecules, it follows by (10·3·3) that their total mean potential energy is also given by $\frac{1}{2}N^2\bar{u}$. Equating these expressions one requires that $N\bar{U}_e = \frac{1}{2}N^2\bar{u}$, i.e., that

$$\bar{U}_e = \tfrac{1}{2}N\bar{u} \tag{10·5·4}$$

To estimate the mean potential energy $\bar{u}$ between a pair of molecules, let us assume that the potential (10·3·5) is an adequate representation of reality. Focusing attention on a given molecule j, one can say in the crudest approximation that any other molecule is equally likely to be anywhere in the container if it is at distance R greater than R_0 from molecule j. The probability of its being at a distance between R and $R + dR$ is then $(4\pi R^2\, dR)/V$ so that

$$\bar{u} = \frac{1}{V}\int_{R_0}^{R} u(R)4\pi R^2\, dR = -\frac{4\pi u_0}{V}\int_{R_0}^{R}\left(\frac{R_0}{R}\right)^s R^2\, dR$$

We again assume that $s > 3$, i.e., that $u(R)$ falls off sufficiently rapidly for the integral to converge properly. Then (10·5·4) becomes

$$\bar{U}_e = \frac{1}{2} N\bar{u} = -a' \frac{N}{V} \tag{10·5·5}$$

where

$$a' \equiv \frac{2\pi}{3} R_0^3 \left(\frac{3}{s-3}\right) u_0 \tag{10·5·6}$$

By (10·3·5) the distance of closest approach between molecules is R_0. In each encounter between a pair of molecules there is thus a volume excluded to one molecule by virtue of the presence of the other molecule, this volume being that of a sphere of radius R_0 (see Fig. 10·5·1). Since there are $\frac{1}{2}N(N-1) \approx \frac{1}{2}N^2$ pairs of molecules, the total excluded volume is $\frac{1}{2}N^2(\frac{4}{3}\pi R_0^3)$. But, for the sake of self-consistency, this must be equal to NV_x, since V_x was understood to be the volume excluded per molecule. Thus it follows that

$$V_x = b'N \tag{10·5·7}$$

where

$$b' = \frac{2\pi}{3} R_0^3 = 4\left[\frac{4\pi}{3}\left(\frac{R_0}{2}\right)^3\right] \tag{10·5·8}$$

is just four times the volume of a hard-sphere molecule.

This completes our crude evaluation of the partition function. One can now calculate the equation of state by the general relation (6·5·12). Applied to (10·5·3) this gives

$$\bar{p} = \frac{1}{\beta}\frac{\partial \ln Z}{\partial V} = \frac{1}{\beta}\frac{\partial}{\partial V}\left[N \ln (V - V_x) - N\beta\bar{U}_e\right]$$

Using (10·5·5) and (10·5·7), this becomes

$$\bar{p} = \frac{kTN}{V - b'N} - a'\frac{N^2}{V^2}$$

or

$$\left(\bar{p} + a'\frac{N^2}{V^2}\right)\left(\frac{V}{N} - b'\right) = kT \tag{10·5·9}$$

Thus we regain the van der Waals equation (10·4·15).

The arguments used in this section to derive the equation of state (10·5·9) have been quite crude. They have, however, been more general than those of the preceding section, since they did not assume specifically that the gas is of low density. Hence one expects that, although the van der Waals equation (10·5·9) is only a very approximate equation of state, it retains also some

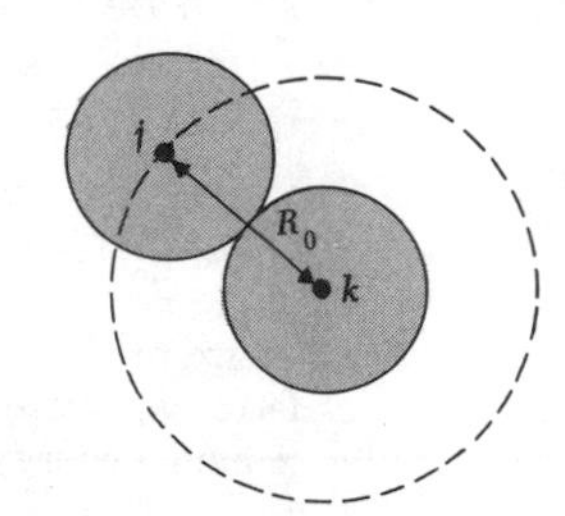

***Fig. 10·5·1** Illustration showing that the presence of a molecule k makes a spherical volume of radius R_0 inaccessible to molecule j. The molecules are considered hard spheres of radius $\frac{1}{2}R_0$.*

rough measure of validity even when used to describe the dense liquid state. It should thus be permissible to use this equation to discuss approximately the occurrence of the gas–liquid phase transformation by the arguments of Sec. 8 · 6.

FERROMAGNETISM

10 · 6 *Interaction between spins*

Consider a solid consisting of N identical atoms arranged in a regular lattice. Each atom has a net electronic spin $\mathbf{S}$ and associated magnetic moment $\boldsymbol{\mu}$. Using a notation similar to that of Sec. 7 · 8, the magnetic moment of an atom is then related to its spin by*

$$\boldsymbol{\mu} = g\mu_0 \mathbf{S} \tag{10·6·1}$$

where μ_0 is the Bohr magneton and the g factor is of order unity. In the presence of an externally applied magnetic field H_0 along the z direction, the Hamiltonian $\mathcal{H}_0$ representing the interaction of the atoms with this field is then

$$\mathcal{H}_0 = -g\mu_0 \sum_{j=1}^{N} \mathbf{S}_j \cdot \mathbf{H}_0 = -g\mu_0 H_0 \sum_{j=1}^{N} S_{jz} \tag{10·6·2}$$

In addition, each atom is also assumed to interact with neighboring atoms. This interaction is not just the magnetic dipole-dipole interaction due to the magnetic field produced by one atom at the position of another one. This interaction is in general much too small to produce ferromagnetism. The predominant interaction is usually the so-called "exchange" interaction. This is a quantum-mechanical consequence of the Pauli exclusion principle. Since electrons cannot occupy the same state, two electrons on neighboring atoms which have parallel spin (i.e., which cannot occupy the same orbital state) cannot come too close to each other in space (i.e., cannot occupy the same orbital state); on the other hand, if these electrons have antiparallel spins, they are already in different states, and there is no exclusion-principle restriction on how close they can come to each other. Since different spatial separations of the electrons give rise to different electrostatic interactions between them, this qualitative discussion shows that the *electrostatic* interaction (which can be of the order of 1 ev and can thus be much larger than any magnetic interaction) between two neighboring atoms does also depend on the relative orientations of their spins. This is the origin of the exchange interaction, which for two atoms j and k can be written in the form

$$\mathcal{H}_{jk} = -2J\mathbf{S}_j \cdot \mathbf{S}_k \tag{10·6·3}$$

Here J is a parameter (depending on the separation between the atoms) which measures the strength of the exchange interaction. If $J > 0$, the interaction

* In Sec. 7 · 8 we used the symbol $\mathbf{J}$ instead of $\mathbf{S}$, but the latter notation is customary in discussions of ferromagnetism; it also avoids confusion with the conventional use of J to designate the exchange energy in (10 · 6 · 3).

energy $\mathcal{H}_{jk}$ is lower when the spins are parallel than when they are antiparallel. The state of lowest energy will then be one which favors *parallel* spin orientation of the atoms, i.e., one which tends to produce ferromagnetism. Note also that, since the exchange interaction depends on the degree to which electrons on the two atoms can overlap so as to occupy approximately the same region in space, J falls off rapidly with increasing separation between atoms; hence the exchange interaction is negligible except when the atoms are sufficiently close to each other. Thus each atom will interact appreciably only with its n nearest neighbor atoms.

Remark Let us show explicitly that the *magnetic* interaction between atoms is far too small to account for ordinary ferromagnetism. Since an atom produces a magnetic field at a distance r of the order of μ_0/r^3, the magnetic interaction of an atom with its n neighboring atoms at a distance r is approximately $(n\mu_0^2/r^3)$. Taking $n = 12$, $\mu_0 \approx 10^{-20}$ ergs gauss^{-1} (the Bohr magneton), and $r = 2 \cdot 10^{-8}$ cm, this gives for the interaction energy 1.5×10^{-16} ergs or, dividing by k, about 1°K. This magnitude of interaction energy might well produce ferromagnetism below 1°K, but certainly not in the region below 1000°K where metallic iron is ferromagnetic!

To simplify the interaction problem, we shall replace (10·6·3) by the simpler functional form

$$\mathcal{H}_{jk} = -2JS_{jz}S_{kz} \tag{10·6·4}$$

This approximate form leaves the essential physical situation intact and avoids the complications introduced by the vector quantities. (The simpler form (10·6·4) is called the "Ising model".)

The Hamiltonian $\mathcal{H}'$ representing the interaction energy between the atoms can then be written in the form

$$\mathcal{H}' = \frac{1}{2}\left(-2J \sum_{j=1}^{N} \sum_{k=1}^{n} S_{jz}S_{kz}\right) \tag{10·6·5}$$

where J is the exchange constant for neighboring atoms and the index k refers to atoms in the nearest neighbor shell surrounding the atom j. (The factor $\frac{1}{2}$ is introduced because the interaction between the same two atoms is counted twice in performing the sums).

The total Hamiltonian of the atoms is then

$$\mathcal{H} = \mathcal{H}_0 + \mathcal{H}' \tag{10·6·6}$$

The problem is to calculate the thermodynamic functions of this system, e.g., its mean magnetic moment $\bar{M}$, as a function of the temperature T and the applied field H_0. The presence of interactions makes this task quite complicated despite the extreme simplicity of (10·6·5). Although the problem has been solved exactly for a two-dimensional array of spins when $H_0 = 0$, the three-dimensional problem is already so difficult that it has up to now defied exact solution. We shall therefore attack the problem by the simplest method of approximation, the molecular-field theory of Pierre Weiss.

10 · 7 *Weiss molecular-field approximation*

Focus attention on a particular atom j, which we shall call the "central atom." The interactions of this atom are described by the Hamiltonian

$$\mathcal{H}_j = -g\mu_0 H_0 S_{jz} - 2J S_{jz} \sum_{k=1}^{n} S_{kz} \tag{10·7·1}$$

The last term represents the interaction of this central atom with its n nearest neighbors. As an approximation we replace the sum over these neighbors by its mean value, i.e., we put

$$2J \overline{\sum_{k=1}^{n} S_{kz}} \equiv g\mu_0 H_m \tag{10·7·2}$$

where H_m is a parameter defined so as to have the dimensions of a magnetic field. It is called the "molecular" or "internal" field and is to be determined in such a way that it leads to a self-consistent solution of the statistical problem. In terms of this parameter (10 · 7 · 1) becomes just

$$\mathcal{H}_j = -g\mu_0 (H_0 + H_m) S_{jz} \tag{10·7·3}$$

The effect of neighboring atoms has thus simply been replaced by an effective magnetic field H_m. The problem presented by (10 · 7 · 3) is just the elementary one of a *single* atom in an external field $(H_0 + H_m)$, a problem discussed in Sec. 7 · 8. The energy levels of the central jth atom are then

$$E_m = -g\mu_0 (H_0 + H_m) m_s, \qquad m_s = -S, (-S+1), \ldots, S \tag{10·7·4}$$

From this one can immediately calculate the mean z component of spin of this atom. One has by (7 · 8 · 13)

$$\overline{S_{jz}} = SB_S(\eta) \tag{10·7·5}$$

where

$$\eta \equiv \beta g\mu_0 (H_0 + H_m), \qquad \beta \equiv (kT)^{-1} \tag{10·7·6}$$

and $B_S(\eta)$ is the Brillouin function for spin S defined in (7 · 8 · 14).

The expression (10 · 7 · 5) involves the unknown parameter H_m. To determine it in a self-consistent way, we note that there is nothing which distinguishes the central jth atom from any of its neighboring atoms. Hence any one of these neighboring atoms might equally well have been considered as the central atom of interest and its mean value of $\bar{S}_z$ must also be given by (10 · 7 · 5). To obtain self-consistency we must then require that (10 · 7 · 2) reduce to

$$2JnSB_S(\eta) = g\mu_0 H_m \tag{10·7·7}$$

Since η is related to H_m by (10 · 7 · 6), the condition (10 · 7 · 7) is an equation which determines H_m and thus completes the solution of the entire problem. Expressing H_m in terms of η, (10·7·7) becomes

$$B_S(\eta) = \frac{kT}{2nJS}\left(\eta - \frac{g\mu_0 H_0}{kT}\right) \tag{10·7·8}$$

which determines η and thus H_m. In particular, in the absence of external field (10·7·8) becomes,

for $H_0 = 0$,
$$B_S(\eta) = \frac{kT}{2nJS}\eta \tag{10·7·9}$$

The solution of the equations (10·7·8) or (10·7·9) can readily be obtained by drawing on the same graph (as shown in Fig. 10·7·1) both the Brillouin function $y = B_s(\eta)$ and the straight line

$$y = \frac{kT}{2nJS}\left(\eta - \frac{g\mu_0 H_0}{kT}\right)$$

and finding the point of intersection $\eta = \eta'$ of these two curves.

Once the molecular field parameter H_m is determined, the total magnetic moment of the sample is of course known. One has by (10·7·5) simply

$$\bar{M} = g\mu_0 \sum_j \overline{S_{jz}} = Ng\mu_0 SB_S(\eta) \tag{10·7·10}$$

Consider now the case when the external field $H_0 = 0$. It is then always true that $\eta = 0$ is a solution of (10·7·9) so that the molecular field H_m vanishes. But there exists also the possibility of a solution where $\eta \neq 0$ so that H_m

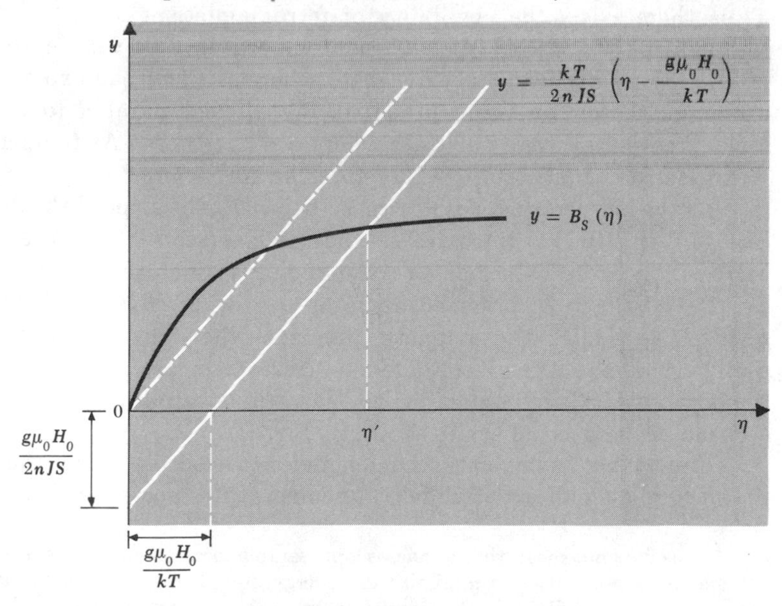

Fig. 10·7·1 Graphical solution of Eq. (10·7·8) determining the molecular field H_m corresponding to the intersection of the curves at $\eta = \eta'$. The dashed straight line corresponds to the case where the external field $H_0 = 0$.

assumes a finite value; correspondingly, there exists then a magnetic moment given by (10·7·10). The presence of such spontaneous magnetization in the absence of an external field is, of course, the distinguishing characteristic of ferromagnetism. To have such a solution where $\eta \neq 0$ it is necessary that the curves in Fig. 10·7·1 intersect at a point $\eta \neq 0$ when both curves start out at the origin. The condition for this to occur is that the initial slope of the curve $y = B_S(\eta)$ is larger than that of the straight line, i.e., that

$$\left[\frac{dB_S}{d\eta}\right]_{\eta=0} > \frac{kT}{2nJS} \tag{10·7·11}$$

But when $\eta \ll 1$, B_S assumes the simple form given by (7·8·19)

$$B_S(\eta) \approx \tfrac{1}{3}(S+1)\eta \tag{10·7·12}$$

Hence (10·7·11) becomes

$$\frac{1}{3}(S+1) > \frac{kT}{2nJS}$$

or

$$T < T_c$$

where

$$kT_c \equiv \frac{2nJS(S+1)}{3} \tag{10·7·13}$$

Thus there exists the possibility of ferromagnetism below a certain critical temperature T_c, called the "Curie temperature," given in terms of J by (10·7·13). This ferromagnetic state where all spins can exploit their mutual exchange energy by being preferentially aligned parallel to each other has lower free energy than the state where $\eta = H_m = 0$. At temperatures below T_c the ferromagnetic state is therefore the stable one.*

As the temperature T is decreased below T_c, the slope of the dashed straight line in Fig. 10·7·1 decreases so that it intersects the curve $y = B_S(\eta)$ at increasingly large values of η corresponding to increasingly large values of y. For $T \to 0$, the intersection occurs for $\eta \to \infty$ where $B_S(\eta) \to 1$; then (10·7·10) gives $\bar{M} \to Ng\mu_0 S$, the magnetic moment when all spins are aligned completely parallel. For all these temperatures one can, of course, use (10·7·10) to compute $\bar{M}(T)$ corresponding to the various values of η. One then obtains a curve of the general shape shown in Fig. 10·7·2.

Finally we investigate the magnetic susceptibility of the solid in the presence of a small external field at temperatures *above* the Curie temperature

* This does not mean that a macroscopic sample in zero external field necessarily has a net magnetic moment. To minimize the energy stored in the magnetic field, the sample tends to become subdivided into many domains, each magnetized along a definite direction, but with these directions differing from one domain to another. (See C. Kittel, "Introduction to Solid State Physics," 2d ed., chap. 15, John Wiley & Sons, Inc., New York, 1956.) Our discussion thus applies to a single domain.

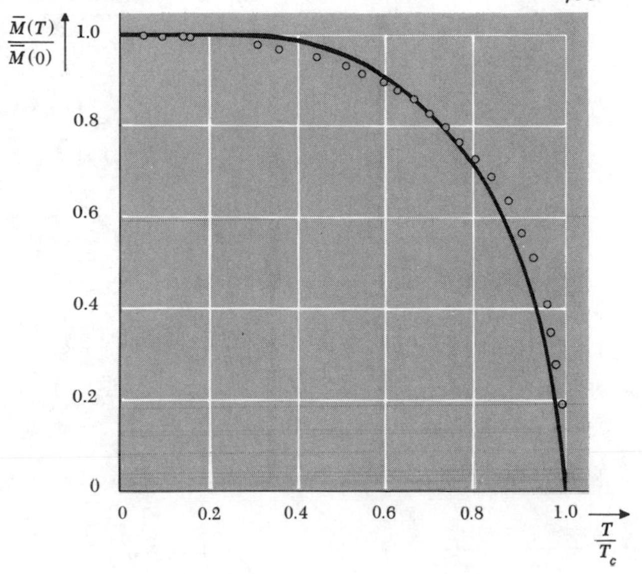

***Fig. 10·7·2** Spontaneous magnetization $\bar{M}$ of a ferromagnet as a function of temperature T in zero external magnetic field. The curve is based on the molecular field theory of (10·7·10) and (10·7·9) with $S = \frac{1}{2}$. The points indicate experimental values for nickel (measured by P. Weiss and R. Forrer, Ann. Phys., vol. 5, p. 153 (1926)).*

(10·7·13). Then we are in a region where η in Fig. 10·7·1 is small. Thus one can use the approximation (10·7·12) to write the general consistency condition (10·7·8) in the form

$$\frac{1}{3}(S+1)\eta = \frac{kT}{2nJS}\left(\eta - \frac{g\mu_0 H_0}{kT}\right)$$

Solving this for η gives, using the quantity kT_c defined in (10·7·13),

$$\eta = \frac{g\mu_0 H_0}{k(T - T_c)} \tag{10·7·14}$$

Thus (10·7·10) yields

$$\bar{M} = \tfrac{1}{3}Ng\mu_0 S(S+1)\eta$$

so that

$$\blacktriangleright \qquad \chi \equiv \frac{\bar{M}}{H_0} = \frac{Ng^2\mu_0{}^2 S(S+1)}{3k(T - T_c)} \tag{10·7·15}$$

is the magnetic susceptibility of N atoms. This is called the Curie-Weiss law. It differs from Curie's law (7·8·22) by the presence of the parameter T_c in the denominator. Thus χ in (10·7·15) becomes infinite when $T \to T_c$, i.e., at the Curie temperature where the substance becomes ferromagnetic.

Experimentally the Curie-Weiss law is well obeyed at temperatures well

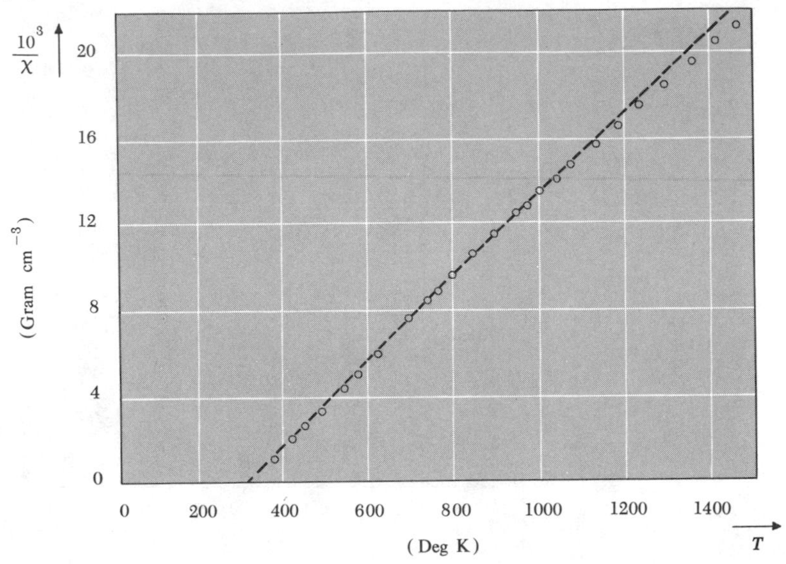

***Fig. 10·7·3** Plot of χ^{-1} versus T per gram of gadolinium metal above its Curie temperature. The curve is (except for some slight departures at high temperatures) a straight line, in accord with what would be expected from the Curie-Weiss law (10·7·15). The intercept of the line with the temperature axis gives T_c = 310°K. The metal becomes ferromagnetic below 289°K. (Experimental data of S. Arajs and R. V. Colvin, J. Appl. Phys., vol. 32 (suppl.), p. 336 (1961).)*

above the Curie temperature. It is, however, not true that the temperature T_c occurring in (10·7·15) is exactly the same as the Curie temperature at which the substance becomes ferromagnetic. Furthermore, the shape of the magnetization curve calculated by the Weiss molecular-field theory in Fig. 10·7·2 is quantitatively not quite correct. One of the most serious discrepancies of the present theory concerns the behavior of the specific heat at the Curie temperature in zero external field. Experimentally, the specific heat has a very sharp discontinuity at that temperature, whereas the theory just discussed predicts a much less abrupt change. The existence of these discrepancies is not surprising in view of the drastic approximations used in this simple theory which replaced all spins by some average effective field and neglected the existence of any correlated fluctuations in the orientations of different spins. The simple theory is, nevertheless, remarkably successful in exhibiting all the main features of ferromagnetism. Needless to say, more refined approximation methods have been devised which improve agreement with experiment considerably.*

* One of the simplest of these, the so-called Bethe-Peierls-Weiss approximation (the Weiss involved here being a different person from Pierre Weiss, who introduced the concept of the molecular field), is a straightforward generalization of the method used in this section. It simply treats a central atom *and* its nearest neighbors exactly and replaces all the *other* atoms by an effective molecular field. See P. R. Weiss, *Phys. Rev.*, vol. 74, p. 1493 (1948).

SUGGESTIONS FOR SUPPLEMENTARY READING

Solids

C. Kittel: "Introduction to Solid State Physics," 2d ed., chaps. 5 and 6, John Wiley & Sons, Inc., New York, 1956.

J. F. Lee, F. W. Sears, and D. L. Turcotte: "Statistical Thermodynamics," chap. 12, Addison-Wesley Publishing Company, Reading, Mass., 1963.

R. Becker: "Theorie der Wärme," chap. 5, Springer-Verlag, Berlin, 1955. (In German.)

M. Blackman: "The Specific Heats of Solids," in "Handbuch der Physik," vol. 7/1, pp. 325–382, Springer-Verlag, Berlin, 1955. (In English.)

(The last two references give more detailed discussions of lattice vibrations.)

Nonideal Gases

T. L. Hill: "An Introduction to Statistical Thermodynamics," chaps. 15 and 16, Addison-Wesley Publishing Company, Reading, Mass., 1960.

———: "Statistical Mechanics," chap. 5, McGraw-Hill Book Company, New York, 1956. (An advanced discussion.)

N. G. Van Kampen: *Physica*, vol. 27, p. 783 (1961). (Simple derivation of the virial expansion.)

R. Brout: *Phys. Rev.*, vol. 115, p. 824 (1959). (An alternative moderately simple derivation of the virial expansion.)

Ferromagnetism

C. Kittel: "Introduction to Solid State Physics," 2d ed., chap. 15, John Wiley & Sons, Inc., New York, 1956. (A simple discussion, including domain formation.)

G. H. Wannier: "Elements of Solid State Theory," chap. 4, Cambridge University Press, 1959. (The Ising model and other cooperative phenomena.)

P. R. Weiss: *Phys. Rev.*, vol. 74, p. 1493 (1948). (Improved approximation method for treatment of ferromagnetism.)

R. Brout: "Phase Transitions," W. A. Benjamin, Inc., New York, 1965. (A general discussion of types of cooperative behavior leading to various kinds of phase transformations.)

PROBLEMS

10.1 For the quantized lattice waves (phonons) discussed in connection with the Debye theory of specific heats, the frequency ω of a propagating wave is related to its wave vector $\boldsymbol{\kappa}$ by $\omega = c\kappa$, where $\kappa = |\boldsymbol{\kappa}|$ and c is the velocity of sound. On the other hand, in a ferromagnetic solid at low temperatures quantized waves of magnetization (spin waves) have their frequency ω related to their wave number κ according to $\omega = A\kappa^2$, where A is a constant. At low temperatures, find the temperature dependence of the heat capacity due to such spin waves.

10.2 Use the Debye approximation to find the following thermodynamic functions of a solid as a function of the absolute temperature T:

(*a*) $\ln Z$, where Z is the partition function

(*b*) the mean energy $\bar{E}$

(*c*) the entropy S

Express your answers in terms of the function

$$D(y) \equiv \frac{3}{y^3} \int_0^y \frac{x^3\,dx}{e^x - 1}$$

and in terms of the Debye temperature $\Theta_D = \hbar\omega_{\max}/k$.

10.3 Evaluate the function $D(y)$ in the limits when $y \gg 1$ and $y \ll 1$. Use these results to express the thermodynamic functions $\ln Z$, $\bar{E}$, and S calculated in the preceding problem in the limiting cases when $T \ll \Theta_D$ and when $T \gg \Theta_D$.

10.4 In the expression for the energy E of (10·1·13) both η and the normal mode frequencies depend, in general, on the volume V of the solid. Use the Debye approximation to find the equation of state of the solid; i.e., find the pressure $\bar{p}$ as a function of V and T. What are the limiting cases valid when $T \ll \Theta_D$ and when $T \gg \Theta_D$? Express your answer in terms of the quantity

$$\gamma \equiv -\frac{V}{\Theta_D}\frac{d\Theta_D}{dV}$$

10.5 Assume that γ is a constant, independent of temperature. (It is called the Grüneisen constant.) Show that the coefficient of thermal expansion α is then related to γ by the relation

$$\alpha = \frac{1}{V}\left(\frac{\partial V}{\partial T}\right)_p = \kappa\left(\frac{\partial p}{\partial T}\right)_V = \kappa\gamma\frac{C_V}{V}$$

where C_V is the heat capacity of the solid and κ is its compressibility.

10.6 As an electron moves in a molecule, there exists at any instant of time a separation of positive and negative charge in the molecule. The latter has, therefore, an electric dipole moment p_1 which varies in time.

(*a*) How does the electric field due to this dipole vary with the distance R from the molecule? (Here R is assumed to be much larger than the molecular diameter.)

(*b*) If another molecule of polarizability α is located at a distance R from the first molecule, how does the dipole moment p_2 induced in this molecule by the first depend on the distance R?

(*c*) How does the interaction energy between p_1 and p_2 depend on R? Show that this is the distance dependence of the "van der Waals potential," i.e., of the long-range part of the Lennard-Jones potential.

10.7 Show that the molar entropy of a monatomic classical nonideal gas at absolute temperature T, and with a density corresponding to n atoms per unit volume, can in first approximation be written in the form

$$S(T,n) = S_0(T,n) + A(T)n$$

where $S_0(T,n)$ is the entropy of the gas at this temperature and density if it were ideal (i.e., if the intermolecular interaction potential $u(R) = 0$). Show that the coefficient $A(T)$ is always negative (as expected since correlations ought to decrease the degree of randomness in the gas) and find an explicit expression for $A(T)$ in terms of u.

10.8 An adsorbed surface layer of area A consists of N atoms which are free to move over the surface and can be treated as a classical two-dimensional gas. The atoms interact with each other according to a potential $u(R)$ which depends only on their mutual separation R. Find the film pressure, i.e., the mean force per unit length, of this gas (up to terms involving the second virial coefficient).

10.9 Consider an assembly of N magnetic atoms in the absence of an external field and described by the Hamiltonian (10·6·5). Treat this problem by the simple Weiss molecular-field approximation.

(*a*) Calculate the behavior of the mean energy of this system in the limiting cases where $T \ll T_c$, where $T \approx T_c$, and where $T \gg T_c$. Here T_c denotes the Curie temperature.

(*b*) Calculate the behavior of the heat capacity in the same three temperature limits.

(*c*) Make a sketch showing the approximate temperature dependence of the heat capacity of this system.

Magnetism and low temperatures

11

MAGNETIC INTERACTIONS are of considerable interest throughout much of physics; they are of particular importance in the study of matter at low temperature and provide also the means for attaining extremely low temperatures. Before leaving the subject of systems in thermal equilibrium, we shall therefore devote some attention to the application of thermodynamic ideas to these topics.

The study of a macroscopic system at very low temperatures provides an opportunity for investigating this system when it is in its ground state and in the quantum states which lie very close to it. The number of such states accessible to the system, or its corresponding entropy, is then quite small. The system exhibits, therefore, much less randomness, or a much greater degree of order, than it would at higher temperatures. The low temperature situation is thus characterized by a fundamental simplicity* and by the possibility that some systems may exhibit in striking fashion a high degree of order on a macroscopic scale. An example of such order is provided by a system of spins all of which, at sufficiently low temperatures, become aligned parallel to each other, thus giving rise to ferromagnetism. A more spectacular example is provided by liquid helium, which remains a liquid down to absolute zero (provided that its pressure is not increased above 25 atmospheres). Below a critical temperature of 2.18°K (the so-called "lambda point") this liquid becomes "superfluid"; it then exhibits completely frictionless flow and can pass through extremely small holes with the greatest of ease. Another set of spectacular examples is provided by many metals (e.g., lead or tin) which become "superconducting" below characteristic sharply defined critical temperatures. The conduction electrons in these metals then exhibit completely frictionless flow with the result that the metals become perfect conductors of electricity (with strictly zero dc electrical resistivity) and manifest striking magnetic properties. We refer the interested reader to the references at the

* There is at least simplicity in principle, since the task of understanding the nature of the ground state of a many-particle system may, at times, be far from trivial.

end of this chapter for more detailed discussions of these remarkable properties. The foregoing comments should, however, be sufficient to indicate why the field of low temperature physics has become a well-developed active field of current research.

It is worth inquiring just how close to its ground state a macroscopic system can be brought in practice, i.e., to how low an absolute temperature it can be cooled. The technique is to insulate the system at low temperatures from its room temperature surroundings by enclosing it in a "dewar." (This is a glass or metal vessel which provides thermal insulation by a double-walled construction; a vacuum maintained between these walls minimizes heat conduction by residual gases and proper polishing of the walls minimizes heat influx due to radiation.)* Helium is the gas which liquefies at the lowest temperature, at 4.2°K under atmospheric pressure. The temperature of the liquid can be readily reduced further to about 1°K simply by pumping away the vapor over the liquid and thus reducing its pressure to the lowest practically feasible value.† Thus it is quite easy by modern techniques to bring any substance to 1°K simply by immersing it in a heat bath consisting of liquid helium. By using liquid He^3, the liquid consisting entirely of the rare isotope He^3 (normally constituting less than 1 part in 10^6 of ordinary helium, which consists almost entirely of He^4), one can apply similar methods to attain fairly readily temperatures down to 0.3°K. Appreciably greater effort and different techniques are necessary in order to work at still lower temperatures. By using a method (to be discussed in Sec. 11·2) which involves the performance of magnetic work by a thermally isolated system of spins, it is feasible to attain temperatures as low as 0.01°K or even 0.001°K. Extensions of this method have even made it possible to reach 10^{-6}°K!

After these general remarks about low-temperature physics and some of its connections with magnetism, we are ready to turn to a specific discussion of magnetic systems. Any subject involving electromagnetism raises immediately the question of choice of units. Since we are discussing problems in physics rather than in electrical engineering, we shall use the units which are currently in most common use in the physics journals of all countries, namely Gaussian cgs units. We recall that in these units all electrical quantities (such as current and voltage) are measured in electrostatic units, while magnetic quantities (such as magnetic field or magnetization) are measured in gauss.

11·1 *Magnetic work*

We consider a system of volume V in an externally applied field $\boldsymbol{H}_a$. The system might, for example, be a sample consisting of a magnetic solid. In

* An ordinary thermos bottle is a familiar example of a dewar.

† The principle of the method should be familiar to any hiker ambitious enough to have cooked out of doors. The boiling point of water on a mountain top is reduced below that at sea level because of the reduced atmospheric pressure.

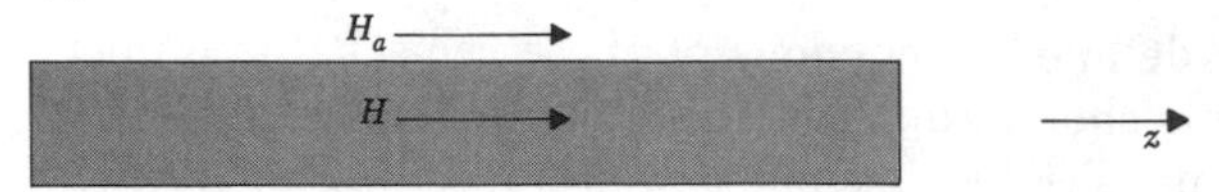

Fig. 11·1·1 A long cylindrical sample in the presence of an externally applied magnetic field H_a. Here $H = H_a$ and $\bar{M}_0 = \chi H$.

order to avoid uninstructive complications and problems of detail which are predominantly in the realm of electromagnetic theory, we shall focus attention on a physically simple situation. We assume that the externally applied field $\mathbf{H}_a$, even when it varies in space, is substantially uniform over the volume of the relatively small sample. We further assume that the sample is in the shape of a cylinder which is very long compared to its cross-sectional dimensions, and that it is always kept oriented parallel to the direction of $\mathbf{H}_a$. Then the mean magnetic moment per unit volume $\bar{\mathbf{M}}_0 = \bar{\mathbf{M}}/V$ is essentially uniform throughout the sample and parallel to $\mathbf{H}_a$. (These properties would also be true for any ellipsoidal sample.) In addition, if $\mathbf{H}$ denotes the magnetic field inside the sample, $\mathbf{H} = \mathbf{H}_a$ by virtue of the boundary condition that tangential components of $\mathbf{H}$ must be continuous. We also recall that quite generally the magnetic induction $\mathbf{B}$ is related to $\mathbf{H}$ by the relation

$$\mathbf{B} = \mathbf{H} + 4\pi\bar{\mathbf{M}}_0 \tag{11·1·1}$$

Outside the sample where $\bar{\mathbf{M}}_0 = 0$, $\mathbf{B} = \mathbf{H}_a$. The magnetic susceptibility χ per unit volume of the sample is defined by the ratio $\chi \equiv \bar{M}_0/H$ so that (11·1·1) can also be written

$$\mathbf{B} = \mu'\mathbf{H} = (1 + 4\pi\chi)\mathbf{H} \tag{11·1·2}$$

where μ' is called the magnetic permeability of the sample.

The starting point for applying macroscopic arguments of statistical thermodynamics to such a magnetic system is again the fundamental relation (3·9·6)

$$đQ = T\,dS = d\bar{E} + đW \tag{11·1·3}$$

valid for any quasi-static process. Here the system is, in general, characterized by two external parameters, the volume V and the applied magnetic field $\mathbf{H}_a$. Hence the total work $đW$ done by the system includes not only the mechanical work $\bar{p}\,dV$ done by the pressure in a volume change dV but also the magnetic work $đW^{(m)}$ associated with changes in $\mathbf{H}_a$. We proceed to derive an expression for this magnetic work.

To keep the geometry simple by making the problem one dimensional, we suppose that the applied magnetic field H_a points in the z direction and that the cylindrical sample is always oriented parallel to this direction. Then the magnetic field H inside the sample (and its magnetic moment $\bar{M}$) also points in the z direction and $H = H_a$. Suppose then that the sample is in a particular state r, where its total magnetic moment is M_r, and that the external magnetic field $H_a = H$ at the position of the sample is changed slowly by a small amount. The work done in this process cannot depend on just how the field is changed.

Let us therefore imagine that the magnitude of the applied field is not quite uniform in space, but that it vanishes at infinity and varies gradually so as to attain the value H_a in the region of experimental interest. The magnetic field then exerts on the sample a force having a component $F_x = M_r(\partial H/\partial x)$ in the x direction (see Fig. 11·1·2). The magnetic field at the position of the sample can now be changed by moving the sample slowly from a position x where $H = H(x)$ to a neighboring position $x + dx$ where $H = H(x + dx)$. In this process one must exert on the sample a force $-F_x$ in the x direction and must do *on* the sample an amount of work $đ\mathfrak{W}_r^{(m)}$ which goes to increase the energy of the sample in this state by an amount dE_r. Thus

$$đ\mathfrak{W}_r^{(m)} = dE_r = (-F_x)\,dx = \left(-M_r\frac{\partial H}{\partial x}\right)dx$$

or*

$$đ\mathfrak{W}_r^{(m)} = dE_r = -M_r\,dH \tag{11·1·4}$$

Thus

$$M_r = -\frac{\partial E_r}{\partial H} \tag{11·1·5}$$

i.e., the magnetic moment is the "generalized force" conjugate to the magnetic field regarded as an external parameter. Taking the statistical average of (11·1·4) over an equilibrium statistical ensemble of similar systems, one then obtains for the macroscopic magnetic work $dW^{(m)}$ done *by* the sample when the field in which it is located changes by an amount dH the result

$$đW^{(m)} = -đ\mathfrak{W}^{(m)} = \bar{M}\,dH \tag{11·1·6}$$

* Note that this expression justifies the familiar result $E_r = -M_rH$ for the energy of a magnetic moment of *fixed* size in an external field H.

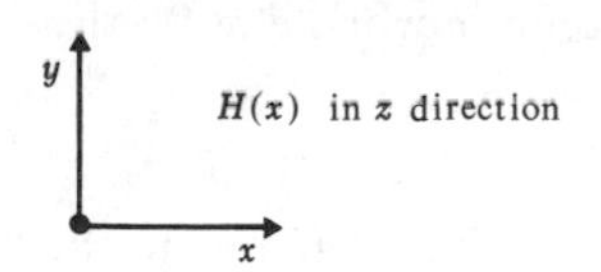

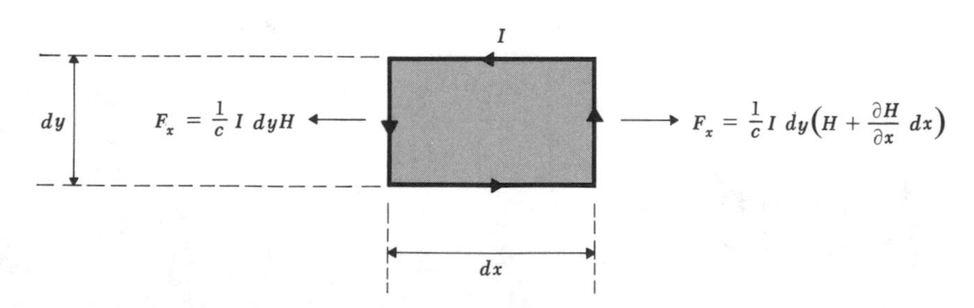

***Fig. 11·1·2** Diagram illustrating the force exerted by a magnetic field H on a magnetic moment represented by a small rectangular current loop. There is a net x component of force given by $F_x = c^{-1}I\,dy\,(\partial H/\partial x)\,dx = M(\partial H/\partial x)$, where I is the current and $M \equiv c^{-1}I(dx\,dy)$ is the magnetic moment of the loop. The force on a large sample can be regarded as due to the superposition of forces on many such infinitesimal moments.*

where $\bar{M}$ is the total mean magnetic moment of the sample. Hence the fundamental thermodynamic relation (11 · 1 · 3) can be written

$$T\,dS = d\bar{E} + \bar{p}\,dV + \bar{M}\,dH \tag{11·1·7}$$

where the last two terms represent the total work done by the sample in a general infinitesimal quasi-static process.

The relation (11 · 1 · 7) can, as usual, be rewritten in a variety of other forms. For example, if it is desired to consider $\bar{M}$ rather than H as an independent variable, one can write $\bar{M}\,dH = d(\bar{M}H) - H\,d\bar{M}$, so that (11 · 1 · 7) becomes

$$T\,dS = d\bar{E}^* + \bar{p}\,dV - H\,d\bar{M} \tag{11·1·8}$$

where $\bar{E}^* \equiv \bar{E} + \bar{M}H$ is the analog of some kind of enthalpy. The thermodynamic consequences of (11 · 1 · 8) or (11 · 1 · 7) are, of course, equivalent; the essential content of these relations is that both $d\bar{E}$ and $d\bar{E}^*$ are exact differentials of well-defined quantities characteristic of the macrostate of the system.

Alternative point of view There is another way in which one can calculate the magnetic work. Imagine that the sample is placed inside a close-fitting solenoid whose length l and area A are then equal to those of the sample so that $lA = V$, the volume of the sample. The solenoid is supposed to consist of N turns of wire and to have negligible electrical resistance. It can be connected to a source of emf (e.g., a battery) as shown in Fig. 11 · 1 · 3. Work must be done by the source of emf on the system consisting of the coil and sample in order to produce the desired magnetic field. The reason is that, in trying to change the magnetic field inside the coil, a counter-emf $\mathcal{V}$ is induced across the coil. The source of emf must then provide an emf $\mathcal{V}$ to overcome this induced emf. If the current in the circuit is I, the magnetic work $đ\mathcal{W}'^{(m)}$ thus done by the source in time dt is

$$đ\mathcal{W}'^{(m)} = \mathcal{V} I\,dt \tag{11·1·9}$$

Let us now express $\mathcal{V}$ and I in terms of the fields B and H inside the solenoid. Since the magnetic flux passing through each turn of the solenoid is BA, the magnitude of the induced emf is given by Faraday's law as

$$\mathcal{V} = \frac{1}{c} N \frac{d}{dt}(AB) \tag{11·1·10}$$

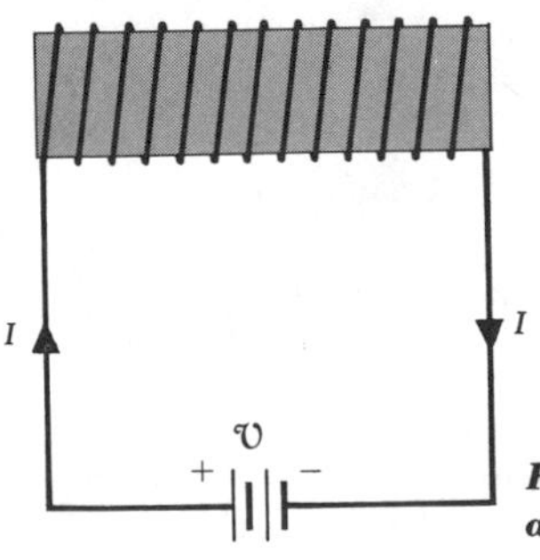

Fig. 11·1·3 ***A magnetic sample placed inside a solenoid.***

where the constant c is the velocity of light (since we use Gaussian units) and where B is expressed in gauss and $\mathcal{V}$ in statvolts. Also, by Ampere's circuital theorem, H inside the solenoid satisfies the relation

$$Hl = \frac{4\pi}{c}(NI) \tag{11·1·11}$$

Hence (11·1·9) becomes

$$đ\mathcal{W}'^{(m)} = \left(\frac{NA}{c}\frac{dB}{dt}\right)\left(\frac{c}{4\pi}\frac{l}{N}H\right)dt = \frac{Al}{4\pi}H\,dB$$

or

$$đ\mathcal{W}'^{(m)} = \frac{V}{4\pi}H\,dB \tag{11·1·12}$$

Using (11·1·1) and $\bar{M} = V\bar{M}_0$ this becomes

$$đ\mathcal{W}'^{(m)} = \frac{V}{4\pi}H(dH + 4\pi\,d\bar{M}_0) = d\left(\frac{VH^2}{8\pi}\right) + H\,d\bar{M} \tag{11·1·13}$$

This expression represents the work necessary to magnetize the sample *and* to establish the magnetic field; i.e., it is the work done *on* the system consisting of the sample *plus* the magnetic field. On the other hand, (11·1·6) represents the work done on the sample in some given magnetic field, i.e., it is the work done on the system consisting of the sample alone. It is, of course, equally legitimate to consider either the sample alone, or the sample plus electromagnetic field, as the system of interest.

The following analogy may serve to clarify the situation. Consider a gas and a spring contained in a cylinder, as shown in Fig. 11·1·4. It is possible to consider the gas as the system of interest. In this case the spring is part of the environment capable of doing work on this system. Alternatively, one can consider the gas plus the spring as the system of interest. The potential energy of the spring is then part of the internal energy of this system.

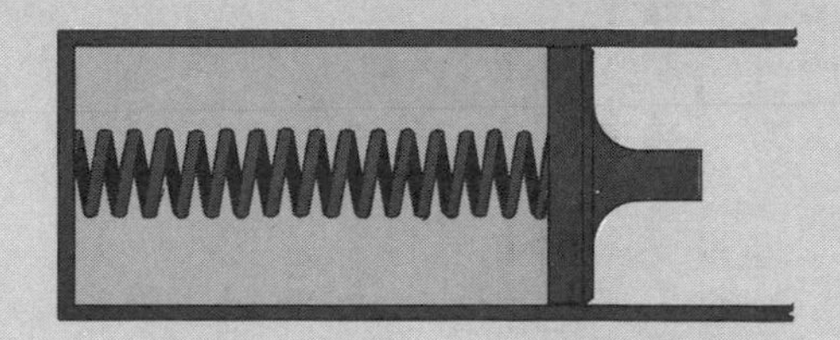

***Fig. 11·1·4** A gas and a spring contained within a cylinder closed by a movable piston.*

If one adopts the point of view that the system of interest consists of sample plus field, the thermodynamic relation (11·1·3) becomes, using for the magnetic work $đW'^{(m)} \equiv -đ\mathcal{W}'^{(m)}$ done *by* this system the expression (11·1·13),

$$T\,dS = d\left(\bar{E}' - \frac{VH^2}{8\pi}\right) + \bar{p}\,dV - H\,d\bar{M} \tag{11·1·14}$$

where $\bar{E}'$ denotes the mean energy of this system. Putting $\bar{E}^* \equiv \bar{E}' - VH^2/8\pi$, this relation is identical with (11·1·8) and thus equivalent to (11·1·7). This shows explicitly that the thermodynamic consequences of our discussion are the same irrespective of which system one chooses to consider.

****Remark*** It is instructive to exhibit in detail the equivalence of the expressions (11·1·6) and (11·1·13) for magnetic work. To show this explicitly, consider the situation illustrated in Fig. 11·1·3. Suppose one starts with $H = 0$ and an unmagnetized sample with $\bar{M} = 0$. What then is the work $\mathcal{W}$ which one must do to reach the final situation where the field is H_0 and the magnetic moment of the sample is $\bar{M}(H_0)$? By using the reasoning leading to (11·1·13), one gets

$$\mathcal{W} = \frac{VH_0^2}{8\pi} + \int_0^{H_0} H\,d\bar{M} \qquad (11\cdot1\cdot15)$$

Let us now look at the problem from a different point of view. Imagine that one starts with $H = 0$, and that the sample with $\bar{M} = 0$ is located at infinity. The final situation can then be brought about in the following steps:

1. Turn on the field H_0 inside the coil.
2. Bring the sample from infinity into the coil, magnetizing it in the process. This requires work for two reasons:
   a. Work must be done, for *fixed* current I_0 in the coil (i.e., for *fixed* H_0), to move the sample into the coil against the forces exerted on it by the field.
   b. Work must be done by the battery to keep the current I_0 constant even though an emf is induced in this coil by the moving magnetized sample which produces a changing flux through the coil.

By (11·1·13) the work done in step (1) is simply

$$\mathcal{W} = \frac{VH_0^2}{8\pi} \qquad (11\cdot1\cdot16)$$

The work done *on* the sample in step (2*a*) is given by (11·1·6) so that

$$\mathcal{W} = -\int_0^{H_0} \bar{M}(H)\,dH \qquad (11\cdot1\cdot17)$$

Finally in step (2*b*), where $H = H_0$ is maintained constant, the work done by the battery is given by (11·1·12)

$$\mathcal{W} = \frac{V}{4\pi} H_0(B_f - B_i) \qquad (11\cdot1\cdot18)$$

where B_i is the initial and B_f the final value of the magnetic induction inside the coil. But when the sample is initially at infinity, $B_i = H_0$; when the sample is finally inside the coil, (11·1·1) yields $B_f = H_0 + 4\pi\bar{M}(H_0)/V$. Hence (11·1·18) becomes

$$\mathcal{W} = \frac{V}{4\pi} H_0 \frac{4\pi\bar{M}(H_0)}{V} = H_0\bar{M}(H_0) \qquad (11\cdot1\cdot19)$$

Adding the three works (11·1·16), (11·1·17) and (11·1·19), one gets then

$$\mathcal{W} = \frac{VH_0^2}{8\pi} - \int_0^{H_0} M(H)\,dH + M(H_0)H_0 \qquad (11\cdot1\cdot20)$$

Integration by parts shows that this is indeed identical to (11·1·15).

11·2 *Magnetic cooling*

Since it is possible to do work on a sample by changing the applied magnetic field, it is also possible to heat or cool a thermally insulated sample by changing a magnetic field. This provides a commonly used method to attain very low temperatures. The nature of this method can be made clearer by comparing it with a more familiar mechanical analogue. Suppose that it is desired to cool a gas by means of mechanical work. One can proceed in the manner illustrated in the top part of Fig. 11·2·1. The gas is initially in thermal contact with a heat bath at temperature T_i, e.g., with a water bath. One can now compress the gas to a volume V_i. In this process work is done on the gas, but it can give off heat to the bath and thus remains at the temperature T_i after equilibrium has been reached. The gas is then thermally insulated (e.g., by removing the water bath) and is allowed to expand quasi-statically to some final volume V_f. In this adiabatic process the gas does work at the expense of its internal energy and, as a result, its temperature falls to some final value T_f less than T_i.

The method of magnetic cooling is very similar and is illustrated in the bottom part of Fig. 11·2·1. The system of interest is a magnetic sample initially in thermal contact with a heat bath at temperature T_i. In practice this heat bath is liquid helium near 1°K, thermal contact of the sample with the bath being established by heat conduction through some helium gas at low pressure. One can now switch on a magnetic field until it attains some value H_i. In this process the sample becomes magnetized and work is done, but the sample can give off heat to the bath and thus remains at the temperature T_i after equilibrium has been reached. The sample is then thermally insulated (e.g., by pumping off the helium gas which provided the thermal contact with the bath) and the magnetic field is reduced quasi-statically to a final value H_f (usually $H_f = 0$).* As a result of this "adiabatic demagnetization" the temperature of the sample falls to some final value T_f less than T_i. In this way temperatures as low as 0.01°K can readily be attained. Indeed, temperatures close to 10^{-6}°K have been achieved by elaboration of this method.

Let us now analyze the method in greater detail in order to understand how the temperature reduction comes about. The first step is an isothermal process: here the system is kept at a constant temperature T_i while it is brought from some macrostate a to some other macrostate b by a change of external parameter. The second step is an adiabatic process: here the system is thermally isolated and is then brought quasi-statically from the macrostate b to a macrostate c by a change of external parameter. The entropy S of the system therefore remains constant in this last step. The whole method is then most conveniently illustrated in a diagram of entropy S versus temperature T. Such a diagram is shown schematically in Fig. 11·2·2 for a paramagnetic sample where the significant external parameter is the magnetic field H. For such a sample the entropy S becomes smaller when the individual atomic

* Internal equilibrium is usually attained rapidly enough that reducing the field to zero in a few seconds is sufficiently slow to be considered quasi-static.

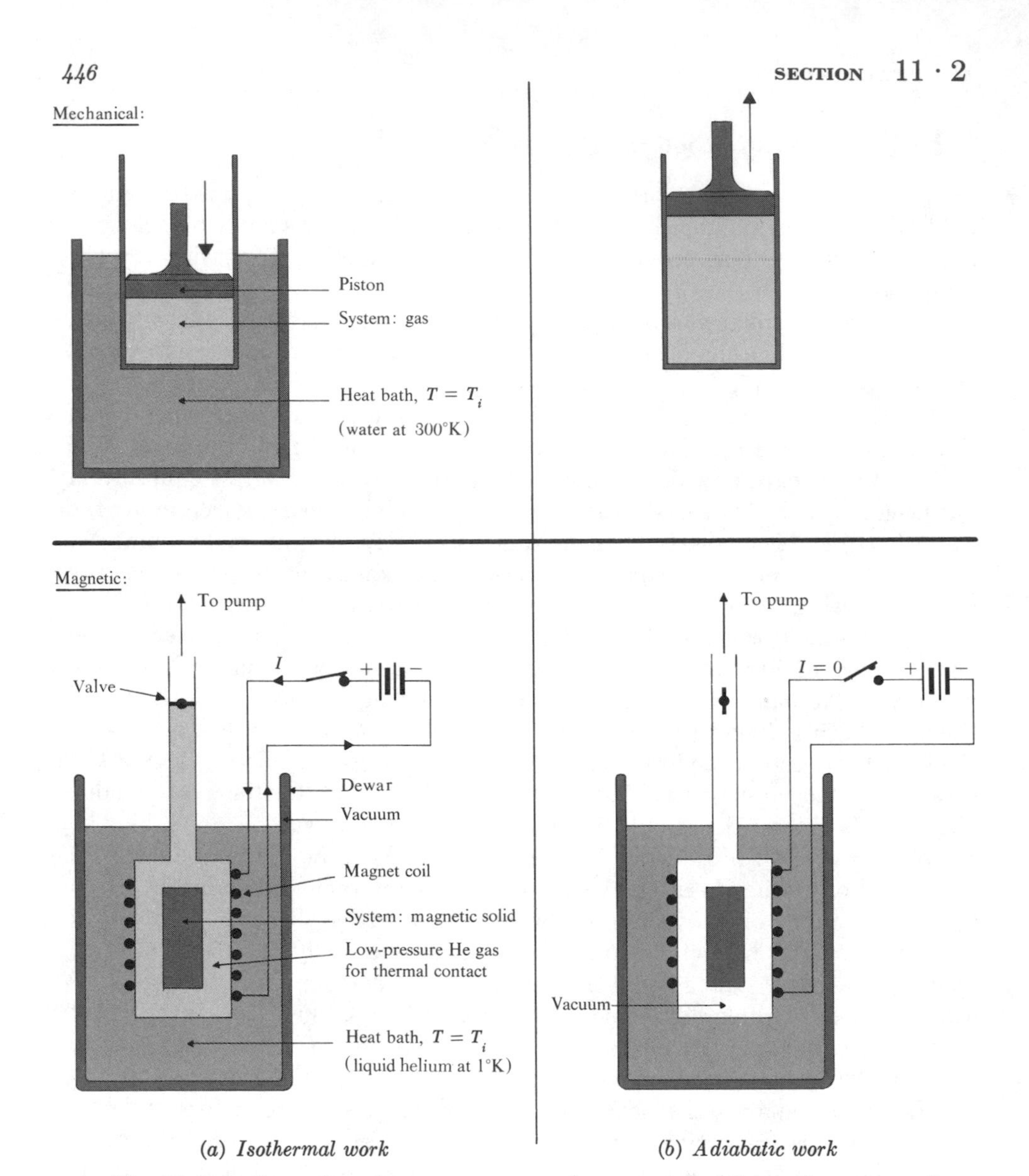

Fig. 11·2·1 Lowering the temperature of a system by the performance of adiabatic work. Top: A gas on which mechanical work is done. Bottom: A magnetic sample on which magnetic work is done.

magnetic moments are more nearly aligned since this is the more ordered situation. Hence S is decreased when the temperature is decreased or when the magnetic field H is increased. The curves of S versus T for various values of H are drawn in Fig. 11·2·2 so as to reflect this situation.*

* At temperatures below 1°K lattice vibrations contribute very little to the heat capacity and entropy of a solid. Practically the entire heat capacity and entropy of a magnetic solid are then due to its magnetic atoms and depend correspondingly on the magnetic field H.

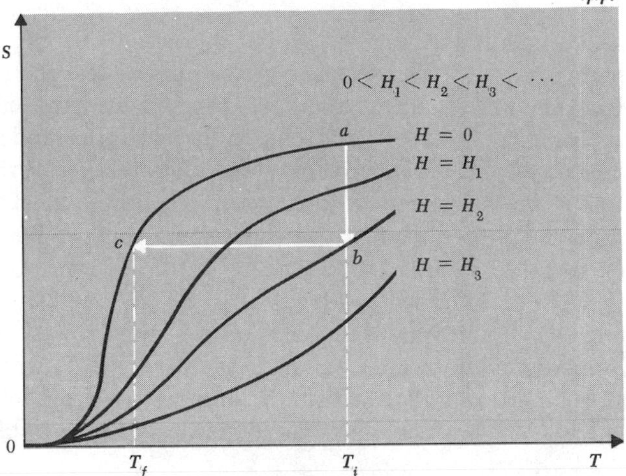

Fig. 11·2·2 Behavior of the entropy S of a paramagnetic sample as a function of T for various values of the magnetic field H. The indicated adiabatic demagnetization process $b \rightarrow c$ is from the initial field $H_i = H_2$ to the final field $H_f = 0$.

The essence of the method is made apparent by this diagram. In the isothermal step $a \rightarrow b$ (field increased from $H = 0$ to $H = H_2$ in Fig. 11·2·2) the external parameter is changed in such a way that the entropy of the system is decreased, i.e., the distribution of such systems over their accessible states becomes less random (see Fig. 11·2·3). In the quasi-static adiabatic step $b \rightarrow c$ (field decreased from $H = H_2$ to $H = 0$ in Fig. 11·2·2) the entropy

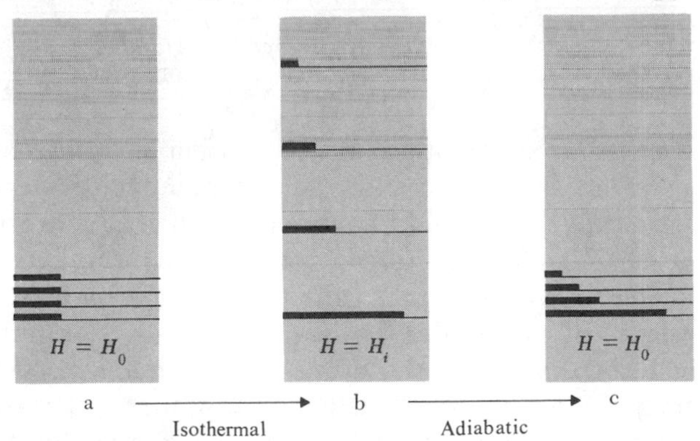

Fig. 11·2·3 Diagram illustrating the effects of isothermal magnetization and adiabatic demagnetization in terms of the energy levels of a single atom (of spin $\frac{3}{2}$). The lengths of the heavy bars indicate the relative numbers of atoms in the respective states. These relative numbers change in the isothermal process, since the Boltzmann factor changes when the energy levels are changed. The relative numbers do not change in the adiabatic process where the entropy remains constant. Since the final population differences are large despite the small energy-level separations, the final temperature must be very low.

must remain unchanged. Hence one sees from Fig. 11·2·2 that the final temperature T_f attained by the system must be less than its initial temperature T_i. In other words, in this adiabatic process the external parameter is changed in such a direction as would tend to increase the randomness of the distribution of the system over its possible states if its temperature were to remain the same. But since the degree of ordering among states must remain unchanged in this process, the temperature of the system must suffer a compensating decrease (see Fig. 11·2·3).

If one knows the entropy $S = S(T,H)$ as a function of T and H, i.e., if one has a quantitative diagram of the type shown in Fig. 11·2·2, then one can immediately determine the final temperature T_f which is reached if one starts at temperature T_i in a field H_i and reduces the field adiabatically to a final value H_f. The constancy of the entropy implies that

$$S(T_f,H_f) = S(T_i,H_i) \tag{11·2·1}$$

and this relation determines the unknown temperature T_f.

Consider first the case where the mutual interaction between the magnetic atoms in the solid can be considered negligible. The only significant parameter of the problem is then the ratio $\mu H/kT$ of the magnetic energy μH of an atom (of magnetic moment μ) compared to the thermal energy kT. The partition function and entropy can then only depend on this ratio. (The analysis of Sec. 7·8 shows this explicitly.) Hence $S = S(H/T)$ is only a function of H/T, and (11·2·1) becomes

$$S\left(\frac{H_f}{T_f}\right) = S\left(\frac{H_i}{T_i}\right)$$

so that

$$\frac{H_f}{T_f} = \frac{H_i}{T_i} \quad \text{or} \quad \frac{T_f}{T_i} = \frac{H_f}{H_i} \tag{11·2·2}$$

This is indeed an approximate relation allowing one to predict the final temperature T_f attained as a result of an adiabatic demagnetization. It does *not* permit one to conclude, however, that demagnetization to $H_f = 0$ allows one to reach $T_f = 0$. The reason is that the assumption of mutually noninteracting atoms breaks down when the temperature becomes too low. More precisely, one needs to consider the fact that, in addition to the externally applied field H_a, there acts on each atom an "internal" or "molecular" field H_m due to neighboring atoms. This field $H_m \approx \mu/r^3$ if it is due to the magnetic dipole-dipole interaction with neighboring magnetic atoms located at a mean distance r from a given atom. (More generally, if the mean energy of interaction between atoms is ϵ_m, one can define H_m by the relation $\mu H_m = \epsilon_m$.) Mutual interaction between atoms is then negligible if T is large enough so that $\mu H_m/kT \ll 1$, but becomes important otherwise. If one still wants to look at the demagnetization approximately in terms of (11·2·2), then H_i in that relation is simply the initially applied field since it is usually much larger than H_m. On the other hand, if in the attempt to reach very low temperatures the applied field is reduced to zero, the final field H_f acting on each atom does not vanish but becomes equal to H_m. Thus one has approximately

$$\frac{T_f}{T_i} \approx \frac{H_m}{H_i} \tag{11·2·3}$$

In order to achieve very low temperatures, it is therefore necessary to use magnetic samples in which the interaction between magnetic atoms is small. It is thus helpful to use as samples solids in which the concentration of magnetic atoms is not too large. Typically, one uses salts containing paramagnetic ions separated by many other atoms. An example is iron ammonium alum ($FeNH_4(SO_4)_2 \cdot 12\ H_2O$) which contains magnetic Fe^{3+} ions; the interactions between these ions are such that, starting at about 1°K in a field $H_i \approx 5000$ gauss, one can attain final temperatures of about 0.09°K. In a salt such as cerium magnesium nitrate ($Ce_2Mg_3(NO_3)_{12} \cdot 24\ H_2O$) the internal interactions affecting the magnetic cerium ions are considerably smaller so that one can attain a final temperature of less than 0.01°K under similar conditions. In attempting to minimize interactions between magnetic atoms one can also try to reduce their concentration by substituting nonmagnetic atoms (such as Zn) for an appreciable fraction of magnetic atoms (such as Fe) in a crystal lattice. Of course, one cannot push this process of dilution too far because then the heat capacity associated with the magnetic atoms becomes too small. If this heat capacity becomes smaller than the small but finite heat capacity associated with lattice vibrations, then it is impossible to reduce appreciably the temperature of the sample (consisting of spins plus lattice).*

The degrees of freedom associated with the possible orientations of the spins of the magnetic atoms constitute a "spin system," which is in thermal interaction with the degrees of freedom of translational motion of all the atoms which constitute the "lattice system." (The interaction between the two systems occurs because the translational motion of the magnetic atoms produces fluctuating magnetic fields which can reorient their magnetic moments and associated spins.) The external magnetic field acts on the spin system alone, but if this field is not changed too rapidly, the lattice and spin systems remain always in equilibrium with each other. Thus the temperature of the lattice is reduced to the same extent as that of the spin system. Furthermore, the total heat capacity of the sample consists of that of the spin system plus that of the lattice.

Thermodynamic analysis The method of adiabatic demagnetization illustrated in Fig. 11·2·2 can readily be discussed in general terms by considerations very similar to those used in discussing the Joule-Thomson effect in Sec. 5·10. The volume V of the solid sample remains essentially constant in these experiments. Only magnetic work is then done and the parameters T and H can be taken to specify completely the macrostate of the sample. The entropy

* It is impossible to cool the water in a swimming pool appreciably by putting it in thermal contact with a golf ball, no matter how cold this golf ball may be.

$S(T,H)$ remains constant during the quasi-static adiabatic demagnetization. A field change dH in this process is then related to the corresponding temperature change dT by the condition

$$dS = \left(\frac{\partial S}{\partial T}\right)_H dT + \left(\frac{\partial S}{\partial H}\right)_T dH = 0 \qquad (11\cdot2\cdot4)$$

or

$$\frac{dT}{dH} = \left(\frac{\partial T}{\partial H}\right)_S = -\frac{\left(\frac{\partial S}{\partial H}\right)_T}{\left(\frac{\partial S}{\partial T}\right)_H} \qquad (11\cdot2\cdot5)$$

But

$$T\left(\frac{\partial S}{\partial T}\right)_H = C_H(T,H) \qquad (11\cdot2\cdot6)$$

is the heat capacity of the sample measured under conditions of constant magnetic field. To simplify the numerator in (11·2·5) one can use the fundamental thermodynamic relation (11·1·7), with $dV = 0$:

$$d\bar{E} = T\,dS - \bar{M}\,dH \qquad (11\cdot2\cdot7)$$

Here the variable pairs (T,S) and $(\bar{M},H)$ occur on the right side, and $(\partial S/\partial H)_T$ is a derivative of a variable from one pair with respect to a variable of the other pair. Hence (11·2·7) suggests immediately a Maxwell relation of the type

$$\left(\frac{\partial S}{\partial H}\right)_T = \left(\frac{\partial \bar{M}}{\partial T}\right)_H \qquad (11\cdot2\cdot8)$$

The rigorous proof follows by writing (11·2·7) in the form

$$dF = -S\,dT - \bar{M}\,dH$$

where the quantity $F \equiv \bar{E} - TS$ can be considered a function of T and H. The relation

$$\frac{\partial^2 F}{\partial H\,\partial T} = \frac{\partial^2 F}{\partial T\,\partial H}$$

then immediately implies (11·2·8).

Writing

$$\bar{M} = V\chi H \qquad (11\cdot2\cdot9)$$

where $\chi = \chi(T,H)$ is the magnetic susceptibility per unit volume of the sample, (11·2·5) becomes then

$$\left(\frac{\partial T}{\partial H}\right)_S = -\frac{VTH}{C_H}\left(\frac{\partial \chi}{\partial T}\right)_H \qquad (11\cdot2\cdot10)$$

Thus a knowledge of $\chi(T,H)$ (i.e., of the magnetic equation of state of the system) and of the heat capacity $C_H(T,H)$ is sufficient to calculate* $(\partial T/\partial H)_S$.

* Note that this derivative is analogous to the Joule-Thomson coefficient $(\partial T/\partial p)_H$ defined in (5·10·8).

Actually, the dependence of C_H on H can also be found from a knowledge of $\chi(T,H)$. Proceeding as in (5·8·5), one obtains

$$\left(\frac{\partial C_H}{\partial H}\right)_T = \left(\frac{\partial}{\partial H}\right)_T \left[T\left(\frac{\partial S}{\partial T}\right)_H\right] = T\frac{\partial^2 S}{\partial H \partial T} = T\frac{\partial^2 S}{\partial T \partial H} = T\left(\frac{\partial}{\partial T}\right)_H \left(\frac{\partial S}{\partial H}\right)_T$$

By (11·2·8) this becomes

$$\left(\frac{\partial C_H}{\partial H}\right)_T = T\left(\frac{\partial^2 \bar{M}}{\partial T^2}\right)_H = VTH\left(\frac{\partial^2 \chi}{\partial T^2}\right)_H \qquad (11\cdot 2\cdot 11)$$

An integration over H for a fixed value of T then yields the relation

$$C_H(T,H) = C_H(T,0) + VT\int_0^H \left(\frac{\partial^2 \chi(T,H')}{\partial T^2}\right)_H H'\,dH' \qquad (11\cdot 2\cdot 12)$$

Hence a knowledge of the heat capacity $C_H(T,0)$ in *zero* magnetic field plus a knowledge of $\chi(T,H)$ is sufficient to compute $C_H(T,H)$ for all fields H and to find the quantity $(\partial T/\partial H)_S$ in (11·2·10).

Example As a very special case, suppose that in a certain range of temperature and magnetic field the susceptibility χ is approximately given by Curie's law

$$\chi = \frac{a}{T}$$

where a is constant. Suppose further that in zero magnetic field the sample has a heat capacity (owing predominantly to interaction between magnetic atoms) which assumes in this temperature range the form

$$C_H(T,0) = \frac{Vb}{T^2}$$

where b is another constant. Then it follows by (11·2·12) that

$$C_H(T,H) = \frac{Vb}{T^2} + VT\int_0^H \left(\frac{2a}{T^3}\right) H'\,dH' = \frac{V}{T^2}(b + aH^2)$$

Furthermore, by virtue of (11·2·8) and (11·2·9),

$$\left(\frac{\partial S}{\partial H}\right)_T = VH\left(\frac{\partial \chi}{\partial T}\right)_H = -\frac{aVH}{T^2}$$

Hence (11·2·4) becomes

$$dS = 0 = \frac{V}{T^3}(b + aH^2)\,dT - \frac{aVH}{T^2}\,dH$$

or

$$\frac{dT}{T} = \frac{aH\,dH}{b + aH^2} = \frac{1}{2}d\,[\ln\,(b + aH^2)]$$

Thus

$$\ln\frac{T_f}{T_i} = \frac{1}{2}\ln\frac{b + aH_f^2}{b + aH_i^2} \qquad \text{by integration}$$

or

$$\frac{T_f}{T_i} = \left(\frac{b + aH_f^2}{b + aH_i^2}\right)^{\frac{1}{2}}$$

11 · 3 *Measurement of very low absolute temperatures*

Measurements of the absolute temperature in the region below 1°K present some difficulties. Since gases have condensed to become liquids or solids, thermometers using ideal gases are not available. Other theoretical relationships (for example, Curie's law according to which $\chi \propto T^{-1}$) may be useful in determining T, but their range of validity is limited. We shall now show how the second law of thermodynamics can be used to establish the absolute temperature scale in this region.

Before proceeding with this discussion, it may be worth pointing out that the actual measurement of absolute temperatures in this range is sufficiently important in physical investigations to warrant considerable effort. First, it should be kept in mind that the range below 1°K is *not* a "small" temperature range. What counts in physical phenomena, and what appears in the Boltzmann factor, is the ratio of kT to significant energies in the systems under consideration. Thus it is temperature *ratios* which tend to be important, and the range from 0.001°K to 1°K can span a variety of physical phenomena comparable to that between 1°K and 1000°K. Second, all theoretical predictions involve the absolute temperature T. Hence no comparison between theory and experiment would be possible if one could not determine the absolute temperature at which a particular experiment is carried out.

It is not difficult to measure *a* temperature. One needs only to proceed as in Sec. 3·5 by choosing an arbitrary macroscopic parameter ϑ of some system as a thermometric parameter while keeping all its other macroscopic parameters fixed. For example, the system used as thermometer might be a paramagnetic solid maintained at a fixed pressure. Its magnetic susceptibility χ can readily be determined, by measuring the inductance of a coil surrounding the solid, and can be used as its thermodynamic paramater ϑ. This parameter ϑ is some unknown function of the absolute temperature T, i.e., $\vartheta = \vartheta(T)$. The problem is how to use a knowledge of ϑ based on such an arbitrary thermometer to obtain the corresponding value of the absolute temperature T.

The second law of thermodynamics is a general relation involving T and provides the starting point of the discussion. The law states that in an infinitesimal quasi-static process the heat absorbed $đQ$ and the corresponding entropy change of a system are related by $dS = đQ/T$. Hence

$$T = \frac{đQ}{dS} \tag{11·3·1}$$

This relation can be applied to the system used as thermometer. Assume that the infinitesimal process envisaged in (11·3·1) is one where the external parameters of the system are kept fixed. To be specific, consider the case where the system used as a thermometer is a paramagnetic solid and where the applied magnetic field H is the one external parameter of significance. Then the infinitesimal process envisaged in (11·3·1) is one where H is kept fixed so that $H = H_0$ (ordinarily $H_0 = 0$). Dividing both the numerator and denominator in (11·3·1) by the change $d\vartheta$ of the thermometric parameter in this infinitesimal process, one obtains

$$T = \frac{(đQ/d\vartheta)_0}{(dS/d\vartheta)_0} \tag{11·3·2}$$

where the subscript 0 indicates that H is to be kept fixed at the value H_0 in evaluating these quantities.

The numerator is readily measured. When the system is in a macrostate characterized by a particular value of ϑ, one keeps the field constant at the value $H = H_0$. One then adds to the system a known small amount of heat $đQ$ (e.g., by sending a current through an electrical-resistor heater embedded in the system, or by knowing the rate of decay and energy release per disintegration of a radioactive source embedded in the system) and measures the resultant increment $d\vartheta$ (e.g., by measuring the change of magnetic susceptibility of the sample). One then calculates the ratio

$$\left(\frac{đQ}{d\vartheta}\right)_0 \equiv C_0(\vartheta) \tag{11·3·3}$$

(a heat capacity with respect to ϑ), which depends, of course, on the particular value of ϑ at which this measurement was performed. One can repeat measurements of this sort for a whole set of temperatures ϑ and construct in this way a curve of $C_0(\vartheta)$ versus ϑ of the type shown in Fig. 11·3·1.

We now turn our attention to the evaluation of the denominator of (11·3·2). The entropy S can be considered a function of ϑ and H so that $S = S(\vartheta,H)$. Assume that $S(\vartheta_i,H)$ is known as a function of H at some temperature $\vartheta = \vartheta_i$. (This may be some high temperature above 1°K where ideal gas thermometry can be used, so that the corresponding absolute temperature T_i is known.) Suppose then that the system, originally at the temperature $\vartheta = \vartheta_i$, is thermally insulated and that the magnetic field is changed quasi-statically from H to the value H_0. In this adiabatic demagnetization the entropy remains unchanged, and at the end of the process the thermometric parameter of the system attains some final value ϑ which can be measured. The final entropy is then given by the relation

$$S(\vartheta,H_0) = S(\vartheta_i,H) \tag{11·3·4}$$

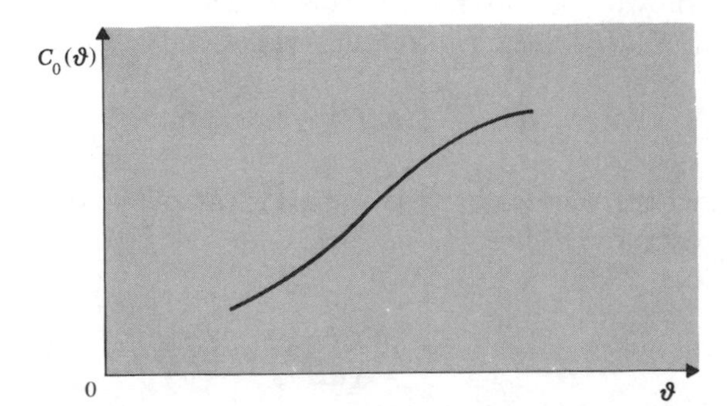

***Fig. 11·3·1** Schematic curve showing $C_0(\vartheta)$ as a function of ϑ.*

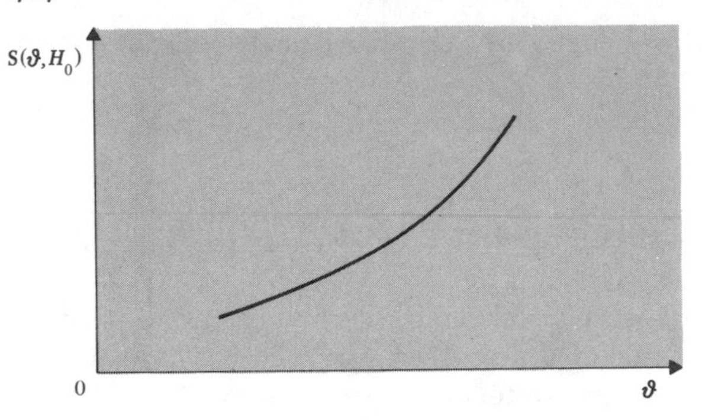

Fig. 11·3·2 Schematic curve showing $S(\vartheta,H_0)$ as a function of ϑ for the given value $H = H_0$.

One can repeat this kind of adiabatic process starting always at $\vartheta = \vartheta_i$, but from a variety of initial fields H. Thus one can reach a variety of final temperatures ϑ in the final field H_0. The corresponding values of $S(\vartheta,H_0)$ are then given by (11·3·4). In this way one can construct a curve of $S(\vartheta,H_0)$ as a function of ϑ for the given value H_0 (see Fig. 11·3·2). From this curve one can then find the slope $(\partial S/\partial\vartheta)_0$ needed in (11·3·2).

Remark Indeed, one can determine this slope quite directly in terms of $S(\vartheta_i,H)$. Suppose that a demagnetization from field H yields a final temperature corresponding to ϑ, while a demagnetization from the slightly different field $H + dH$ yields a final temperature corresponding to $\vartheta + d\vartheta$. Then it follows by (11·3·4) that

$$\left(\frac{\partial S}{\partial\vartheta}\right)_0 = \frac{S(\vartheta + d\vartheta,H_0) - S(\vartheta,H_0)}{d\vartheta} = \frac{S(\vartheta_i,H + dH) - S(\vartheta_i,H)}{d\vartheta}$$

so that

$$\left(\frac{\partial S}{\partial\vartheta}\right)_0 = \frac{\partial S(\vartheta_i,H)}{\partial H}\frac{dH}{d\vartheta} \tag{11·3·5}$$

It is now possible to answer the question how to determine T if ϑ is known. For the given value of ϑ one can find the slope $(\partial S/\partial\vartheta)_0$ of the curve in Fig. 11·3·2. For this value of ϑ one can also read off the corresponding value of $C_0 = (đQ/d\vartheta)_0$ from the curve of Fig. 11·3·1. Hence one can use (11·3·2) to compute the corresponding value of the absolute temperature T.

This discussion assumed that $S(\vartheta_i,H)$ is known as a function of H at the known absolute temperature T_i where $\vartheta = \vartheta_i$. Equivalently, this implies that $(\partial S/\partial H)_\vartheta$ is known for this value ϑ_i; for then

$$S(\vartheta_i,H) = \text{constant} + \int_0^H \frac{\partial S(\vartheta_i,H')}{\partial H'}\,dH' \tag{11·3·6}$$

where the constant is independent of H and is irrelevant in computing the entropy *differences* necessary to calculate the derivative $(\partial S/\partial\vartheta)_0$. But by (11·2·8) one has, for $\vartheta = \vartheta_i$,

$$\left(\frac{\partial S}{\partial H}\right)_\vartheta = \left(\frac{\partial \bar{M}}{\partial T}\right)_H = VH\left(\frac{\partial \chi}{\partial T}\right)_H \tag{11·3·7}$$

so that a knowledge of the susceptibility χ near the relatively high absolute temperature T_i (where Curie's law is usually well obeyed) is sufficient to provide the necessary knowledge of $S(\vartheta_i,H)$ in (11·3·6).

11·4 *Superconductivity*

We already mentioned that many metals become "superconducting" when cooled to sufficiently low temperatures. When such a metal is cooled below a sharply defined temperature T_c, which depends on the external magnetic field $\boldsymbol{H}$ in which the metal is located, the dc electrical resistance of the metal falls abruptly to zero. At the same time currents are set up in the metal in such a way that the magnetic induction $\boldsymbol{B}$ vanishes inside the metal* irrespective of the applied field $\mathbf{H}$. This superconducting state of the metal exists as long as the temperature is sufficiently low and the applied field is sufficiently small; otherwise the metal is normal. The situation is illustrated in Fig. 11·4·1.

To give a concrete example, lead becomes superconducting in zero applied field at a critical temperature $T_c = 7.2°K$. The critical field necessary to destroy its superconductivity in the limit where $T \rightarrow 0$ is $H_c = 800$ gauss. The absence of resistance can be impressively demonstrated by setting up a current in a ring of superconducting metal and then removing any batteries or other power sources. One finds that even after waiting a year, the current is still flowing with no measurable decrease in its magnitude!

Although superconductivity was first discovered by Kammerlingh Onnes

* We restrict our discussion to the so-called "soft" superconductors or "superconductors of the first kind." The situation is somewhat more complicated in "superconductors of the second kind" which have aroused much interest recently in connection with the production of very high magnetic fields.

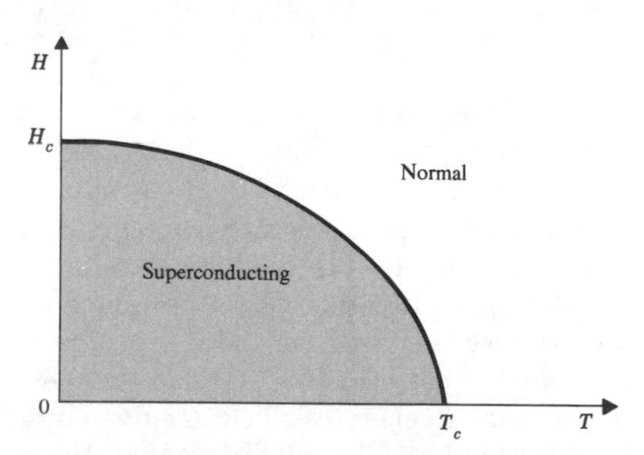

Fig. 11·4·1 Diagram showing the domain of temperature T and magnetic field H in which a metal is superconducting. The curve separating the domain where the metal is superconducting from that where it is normal is the critical-temperature curve, which determines the critical temperatures and corresponding critical fields at which the transition of the metal from the normal to the superconducting states (or vice versa) occurs.

in 1911, it was only in 1957 (after years of immense efforts) that a successful microscopic theory of this remarkable phenomenon was finally proposed. The essence of the microscopic explanation is that a very weak interaction between the conduction electrons of a metal can, at sufficiently low temperatures, lead to quantum states characterized by a highly correlated motion of all these electrons. The theory is complicated precisely because it must give a proper quantum-mechanical description of the correlated motion of very many particles obeying Fermi-Dirac statistics. The reader is referred for further details to the references at the end of this chapter. In the present section we shall only want to make use of the fundamental property that

$$\mathbf{B} = 0 \qquad \text{inside a superconducting metal} \tag{11·4·1}$$

Both the normal and the superconducting states of the metal can be treated as well-defined thermodynamic macrostates of the metal. The transition from one to the other is completely analogous to a phase transformation like that between a liquid and a solid. Thus the superconducting transition can be readily discussed by similar thermodynamic reasoning.

Consider a metal in the form of a long cylinder placed parallel to an externally applied field H as shown in Fig. 11·1·1. When the metal is normal, its magnetic moment $\bar{M} = 0$ to a very good approximation, since its susceptibility χ is very small (of the order of 10^{-5}). On the other hand, when the metal is superconducting, the property (11·4·1) implies that the sample has a large magnetic moment. Indeed, since H inside the metal is the same as the applied field, it follows by (11·1·1) that inside the superconducting metal

$$B = H + 4\pi \frac{\bar{M}}{V} = 0 \tag{11·4·2}$$

Hence the total magnetic moment $\bar{M}_s$ of the superconducting sample is

$$\bar{M}_s = -\frac{V}{4\pi} H \tag{11·4·3}$$

There is another more direct way of obtaining this result. We adopt a microscopic point of view so that there is no distinction between B and H. Then $B = H$ is produced by *all* currents, both those that are macroscopically applied and those microscopic currents that are responsible for the magnetization of the sample. In the presence of an externally applied field H a circulating current I must then be set up (along the periphery of the superconducting metal) of such a direction and magnitude that it cancels this applied field so as to make the net field $B = 0$ inside the metal. By Ampere's circuital theorem this current I generates a field H_I given by $H_I l = (4\pi/c)I$, where l is the length of the sample. The condition that the total field inside the sample must vanish becomes then

$$H + H_I = H + \frac{4\pi}{cl} I = 0$$

The current flowing along the periphery of the sample has thus a magnitude

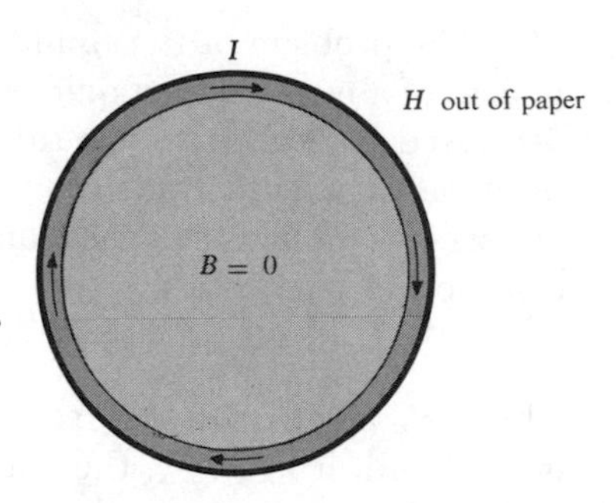

Fig. 11·4·2 End view of a cylindrical superconducting sample in an applied field H. A circulating current is set up in a very thin (about 5 × 10⁻⁶ cm thick) peripheral layer of the cylinder and produces a field −H inside the cylinder so as to reduce B to zero inside the metal.

$$I = -\frac{cl}{4\pi} H \tag{11·4·4}$$

If the sample has a cross-sectional area A, this current produces a magnetic moment

$$\bar{M}_s \equiv \frac{1}{c} IA = \frac{1}{c}\left(-\frac{cl}{4\pi} H\right) A = -\frac{V}{4\pi} H \tag{11·4·5}$$

where $V = Al$ is the volume of the sample. Thus we regain the result (11·4·3).

Work must be done in setting up this current I or magnetic moment $\bar{M}_s$. Since the volume V of the sample remains unchanged to excellent approximation, the fundamental thermodynamic relation (11·1·7) applied to the sample becomes then simply

$$đQ = T\,dS = d\bar{E} + \bar{M}\,dH \tag{11·4·6}$$

Since the temperature T (rather than the entropy S) is the independent variable at our disposal, we rewrite (11·4·6) in the form

$$d(TS) - S\,dT = d\bar{E} + \bar{M}\,dH$$

or

$$dF = -S\,dT - \bar{M}\,dH \tag{11·4·7}$$

where

$$F \equiv \bar{E} - TS \tag{11·4·8}$$

is a free energy. The condition of increasing entropy for an isolated system implies then, analogously to (8·2·3), that

$$\Delta S - \frac{Q}{T} \geq 0 \qquad \text{or} \qquad Q - T\,\Delta S \leq 0 \tag{11·4·9}$$

when the sample absorbs heat Q from a heat reservoir at the constant temperature T. But, if the magnetic field H is kept constant, it follows by (11·4·6) that the heat Q absorbed by the system is simply $Q = \Delta\bar{E}$. Hence (11·4·9) shows that, when the system is kept at a constant temperature T in a constant magnetic field H,

$$\Delta F \leq 0 \tag{11·4·10}$$

i.e., the condition for stable equilibrium is the usual one that the free energy F of (11·4·8) must satisfy the condition

$$F = \text{minimum} \tag{11·4·11}$$

The problem of the equilibrium between the normal and superconducting phases of a metal is then quite analogous to the ordinary phase transformations discussed in Sec. 8·5. To satisfy the minimum property (11·4·11), F must be stationary if one transforms the metal from the normal to the superconducting state. Hence the condition of equilibrium between the normal and superconducting phases at a given temperature T and magnetic field H is

$$F_n = F_s \tag{11·4·12}$$

where F_n and F_s are the respective free energies of the two phases. On the other hand, if $F_n < F_s$, the normal phase of the metal is the more stable one; if $F_s < F_n$, the superconducting phase is the more stable one.

The condition (11·4·12) is satisfied everywhere along the phase transformation (i.e., critical temperature) curve of Fig. 11·4·1. Proceeding as in the derivation of the Clausius-Clapeyron equation (8·5·9), the relation (11·4·12) applied to a particular point (T,H) on the critical temperature curve yields

$$F_n(T,H) = F_s(T,H)$$

At a neighboring point on this critical temperature curve one has similarly

$$F_n(T + dT,\, H + dH) = F_s(T + dT,\, H + dH)$$

Subtracting these two relations, one gets

$$dF_n = dF_s \tag{11·4·13}$$

Here dF, whether in the normal or superconducting phases, is given by (11·4·7). Hence (11·4·13) becomes

$$-S_n\, dT - \bar{M}_n\, dH = -S_s\, dT - \bar{M}_s\, dH$$

where the subscripts n and s denote functions in the normal and superconducting phases respectively. Thus

$$(S_n - S_s)\, dT = -(\bar{M}_n - \bar{M}_s)\, dH$$

Since $\bar{M}_n \approx 0$, while $\bar{M}_s$ is given by (11·4·3), this becomes

▶ $$S_n - S_s = -\frac{V}{4\pi} H \frac{dH}{dT} \tag{11·4·14}$$

This is the analog of the Clausius-Clapeyron equation (8·5·9); it relates the entropy difference to the slope of the transition temperature curve.

Since $dH/dT \leq 0$, it follows that $S_n \geq S_s$. Thus the superconducting state of the metal has lower entropy (i.e., is more ordered) than the normal state. Associated with this entropy difference, there is a latent heat

$$L = T(S_n - S_s) \tag{11·4·15}$$

which must be absorbed by the metal in the transformation from the superconducting to the normal state. Note that $S_n - S_s = 0$ when the transition takes place in zero external field so that $H = 0$. In that case there is *no* latent heat associated with the transition. When $T \to 0$, the third law of thermodynamics requires that $S_n - S_s \to 0$. The latent heat must then also vanish as $T \to 0$. In addition, Eq. (11·4·14) implies then that $(dH/dT) \to 0$ as $T \to 0$, i.e., the transition curve in Fig. 11·4·1 must start out at $T = 0$ with zero slope. All these conclusions are in agreement with experiment.

SUGGESTIONS FOR SUPPLEMENTARY READING

Magnetic work

C. Kittel: "Elementary Statistical Physics," sec. 18, John Wiley & Sons, Inc., New York, 1958.

H. B. Callen: "Thermodynamics," chap. 14 and sec. 4·10, John Wiley & Sons, Inc., New York, 1960.

General references on low-temperature physics

D. K. C. MacDonald: "Near Zero," Anchor Books, New York, 1961. (A very elementary survey.)

M. W. Zemansky: "Heat and Thermodynamics," 4th ed., chap. 16, McGraw-Hill Book Company, New York, 1957.

K. Mendelssohn: "Cryophysics," Interscience Publishers, New York, 1960.

L. C. Jackson: "Low-temperature Physics," 5th ed., John Wiley & Sons, Inc., New York, 1962.

C. T. Lane: "Superfluid Physics," McGraw-Hill Book Company, New York, 1962. (Superfluid helium and superconductivity.)

E. M. Lifshitz: "Superfluidity," *Scientific American*, June, 1958.

F. Reif: "Superfluidity and Quasi-particles," *Scientific American*, November, 1960.

(The last two articles give simple accounts of superfluid liquid helium.)

Magnetic cooling

F. E. Simon, N. Kurti, J. F. Allen, and K. Mendelssohn: "Low-temperature Physics," Pergamon Press, London, 1952. (The second chapter, by Kurti, presents a good elementary discussion of adiabatic demagnetization.)

N. Kurti: *Physics Today*, vol. 13, pp. 26–29, October, 1960. (Popular account describing the attainment of temperatures close to 10^{-6}°K.)

Superconductivity

E. A. Lynton: "Superconductivity," John Wiley & Sons, Inc., New York, 1962. (A good modern introduction.)

J. E. Kunzler and M. Tanenbaum: "Superconducting Magnets," *Scientific American*, June, 1962.

T. A. Buchhold: "Applications of Superconductivity," *Scientific American*, March, 1960.

(The last two articles describe practical applications of superconductivity.)

PROBLEMS

11.1 Suppose that the energy of a system in state r is given by $E_r(H)$ when the system is in a magnetic field H. Then its magnetic moment M_r in this state is quite generally given by (11·1·5), so that $M_r = -\partial E_r/\partial H$. Use this result to show that when the system is in equilibrium at the temperature T, its mean magnetic moment is given by $\bar{M} = \beta^{-1}\,\partial \ln Z/\partial H$, where Z is the partition function of this system.

11.2 The magnetic susceptibility per unit volume of a magnetic solid is given by $\chi = A/(T - \theta)$ where A and θ are constants independent of magnetic field. How much does the entropy per unit volume of this solid change if, at the temperature T, the magnetic field is increased from $H = 0$ to $H = H_0$?

11.3 The magnetic susceptibility per *mole* of a substance containing interacting magnetic atoms is given by the Curie-Weiss law, $\chi = A/(T - \theta)$, where A and θ are constants independent of temperature and of magnetic field. The parameter θ depends, however, on the pressure p according to the relation $\theta = \theta_0(1 + \alpha p)$ where θ_0 and α are constants. Calculate the change in molar volume of this substance when, at a fixed temperature and pressure, the magnetic field is increased from $H = 0$ to the value $H = H_0$.

11.4 (*a*) Show that for a fixed temperature, the entropy of a metal is independent of magnetic field in both its superconducting and its normal states. (The magnetic susceptibility in the normal state is negligibly small.)

(*b*) Given the critical field curve $H = H(T)$ for a superconductor, find a general expression for the difference $(C_s - C_n)$ between the heat capacities of the metal in the superconducting and normal states at the same temperature T.

(*c*) What is the answer to part (*b*) at the transition temperature $T = T_c$?

11.5 At these low temperatures the temperature dependence of the heat capacity C_n in the normal state is to a good approximation given by

$$C_n = aT + bT^3$$

where a and b are constants. In the superconducting state the heat capacity $C_s \to 0$ as $T \to 0$ more rapidly than T; i.e., $C_s/T \to 0$ as $T \to 0$. Assume that the critical field curve has the parabolic shape $H = H_c[1 - (T/T_c)^2]$. What is the temperature dependence of C_s?

11.6 Consider a metal in zero magnetic field and at atmospheric pressure. The heat capacity of the metal in the normal state is $C_n = \gamma T$; in the superconducting state it is approximately $C_s = \alpha T^3$. Here γ and α are constants; T is the absolute temperature.

(*a*) Express the constant α in terms of γ and the critical temperature T_c.

(*b*) Find the difference between the internal energy of the metal in the normal and superconducting states at $T = 0$. Express the answer in terms of γ and T_c.

(Remember that the entropy of the normal and superconducting states is the same both at $T = 0$ and at $T = T_c$.)

11.7 The heat capacity C_n of a normal metal at low temperatures T is given by $C_n = \gamma T$, where γ is a constant. But, if the metal is superconducting below the transition temperature T_c, then its heat capacity C_s in the superconducting state in the temperature range $0 < T < T_c$ is approximately given by the relation $C_s = \alpha T^3$, where α is some constant. The entropies S_n and S_s of the metal in the normal and superconducting states are equal at the transition temperature $T = T_c$; it also follows by the third law that $S_n = S_s$ as $T \to 0$. Use the above information to find the relation between C_s and C_n at the transition temperature T_c.

12

Elementary kinetic theory of transport processes

IN THE preceding chapters our concern has been almost exclusively with equilibrium situations. General statistical arguments were quite sufficient to treat problems of this sort, and there was no need to investigate in detail the interaction processes which bring about the equilibrium. Many problems of great physical interest deal, however, with nonequilibrium situations.

Consider, for example, the case where the two ends of a copper rod are maintained at different temperatures. This is *not* an equilibrium situation, since the entire bar would then be at the same temperature. Instead, energy in the form of heat flows through the bar from the high- to the low-temperature end, the rate of this energy transfer being measured by the "thermal conductivity" of the copper bar. A calculation of the coefficient of thermal conductivity thus requires a more detailed consideration of the nonequilibrium processes whereby energy is transported from one end of the bar to the other. Calculations of this sort can become quite complicated, even in the rather simple case of ideal gases which we shall treat in these next chapters. It is therefore very valuable to develop simple approximate methods which yield physical insight into basic mechanisms, which elucidate the main features of phenomena in a semiquantitative way, and which can be extended to the discussion of more complicated cases where more rigorous methods might become hopelessly complex. Indeed, it is very often found that simple approximate calculations of this sort lead to the correct dependence of all significant parameters, like temperature and pressure, and to numerical values which differ by no more than 50 percent from the results of rigorous calculations based on the solution of complicated integrodifferential equations. In this chapter, therefore, we shall begin by discussing some of the simplest approximate methods for dealing with nonequilibrium processes. Although we shall treat the case of dilute gases, the same methods are useful in more advanced work, e.g., in discussing transport processes in solids in terms of "dilute gases" of electrons, "phonons" (quantized sound waves with particlelike properties), or "magnons" (quantized waves of magnetization).

In a gas, molecules interact with each other through collisions. If such a gas is initially not in an equilibrium situation, these collisions are also responsible for bringing about the ultimate equilibrium situation where a Maxwell-Boltzmann velocity distribution prevails. We shall discuss the case of a gas which is *dilute*. The problem is then relatively simple because of the following features:

a. Each molecule spends a relatively large fraction of its time at distances far from other molecules so that it does not interact with them. In short, the time *between* collisions is much greater than the time involved *in* a collision.

b. The probability of *more* than two molecules coming close enough to each other at any time so as to interact with each other *simultaneously* is negligibly small compared to the probability of only two molecules coming sufficiently close to another to interact. In short, triple collisions occur very rarely compared to two-particle collisions. Thus the analysis of collisions can be reduced to the relatively simply mechanical problem of only *two* interacting particles.

c. The mean de Broglie wavelength of molecules is small compared to the mean separation between molecules. The behavior of a molecule between collisions can then be described adequately by the motion of a wave packet or classical particle trajectory, even though a quantum-mechanical calculation may be necessary to derive the scattering cross section describing a collision between two molecules.

Finally, it is worth adding a very general comment about the distinction between equilibrium and steady-state situations. An *isolated* system is said to be in equilibrium when none of its parameters depends on the time. It is, however, also possible to have a nonequilibrium situation where a system A, which is *not* isolated, is maintained in such a way that all of its parameters are time-independent. The system A is then said to be in a "steady state," but this situation is not one of equilibrium, since the combined isolated system $A^{(0)}$ consisting of A and its surroundings is not in equilibrium, i.e., since the parameters of A' vary in time.

Example Consider a copper rod A connecting two heat reservoirs B_1 and B_2 at different temperatures T_1 and T_2. A steady-state situation would be one where the temperatures T_1 and T_2 at the two ends of the rod are maintained constant and where one has waited a sufficiently long time so that the local temperature in each macroscopically small region of the rod has attained a constant value. If B_1 and B_2 are sufficiently large, the temperatures T_1 and T_2 will vary very slowly, despite the transfer of heat from one reservoir to the other; but B_1 and B_2 are certainly not in equilibrium, and their temperatures will gradually change and reach equality if one waits long enough to attain the final equilibrium situation. Similarly, if one constantly does work on one reservoir and uses a refrigerator on the other reservoir to keep their respective temperatures strictly constant, the environment of A is most certainly not in equilibrium.

12 · 1 *Collision time*

Consider a molecule with velocity $\mathbf{v}$. Let

$$\left.\begin{array}{l} P(t) = \text{the probability that such a molecule survives a time} \\ t \text{ without suffering a collision.} \end{array}\right\} \qquad (12\cdot 1\cdot 1)$$

Of course $P(0) = 1$, since a molecule has no chance of colliding in a time $t \to 0$, i.e., it certainly manages to survive for a vanishingly short time. On the other hand, $P(t)$ decreases as the time t increases, since a molecule is constantly exposed to the danger of suffering a collision; hence its probability of surviving a time t without suffering such a fate decreases as time goes on. Finally, $P(t) \to 0$ at $t \to \infty$. (The situation is similar to one very familiar to all of us; being constantly exposed to the vicissitudes of disease and accident, each one of us must die sooner or later.) The net result is that a plot of $P(t)$ versus t must have the shape indicated in Fig. 12·1·1.

To describe the collisions, let

$$\left.\begin{array}{l} w\,dt = \text{the probability that a molecule suffers a collision} \\ \text{between time } t \text{ and } t + dt. \end{array}\right\} \qquad (12\cdot 1\cdot 2)$$

The quantity w is thus the probability per unit time that a molecule suffers a collision, or the "collision rate." We shall assume that the probability w is *independent* of the past history of the molecule; i.e., it does not matter when the molecule suffered its last collision. In general w may, however, depend on the speed v of the particular molecule under consideration, so that $w = w(v)$.

Knowing the collision probability w, it is possible to calculate the survival probability $P(t)$. This can be done by noting that [the probability that a molecule survives a time $t + dt$ without suffering a collision] must be equal to [the probability that this molecule survives a time t without suffering a collision] multiplied by [the probability that it does not suffer a collision in the

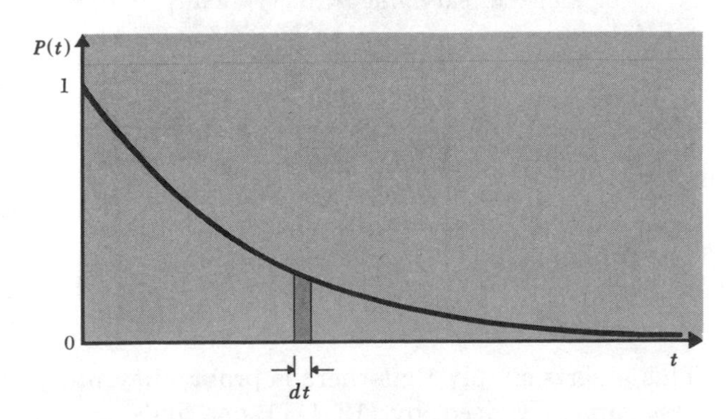

***Fig. 12·1·1** Probability $P(t)$ of surviving a time t without suffering a collision. (The shaded area represents the probability $\mathcal{P}(t)\,dt$ of suffering a collision in the time interval between t and $t + dt$ after surviving a time t without collisions.)*

subsequent time interval between t and $t + dt$]. In symbols, this statement becomes

$$P(t + dt) = P(t)(1 - w\,dt) \tag{12·1·3}$$

Hence

$$P(t) + \frac{dP}{dt}dt = P(t) - P(t)w\,dt$$

or

$$\frac{1}{P}\frac{dP}{dt} = -w \tag{12·1·4}$$

Between collisions (i.e., in a time or the order of w^{-1}) the speed v of a molecule does not change at all; or, if the molecule is subject to external forces due to gravity or electromagnetic fields, its speed changes usually only by a relatively small amount in the short time w^{-1}. Hence the probability w, even if it is a function of v, can ordinarily be considered essentially a constant independent of time. The integration of (12·1·4) is then immediate and gives

$$\ln P = -wt + \text{constant}$$

or

$$P = C\,e^{-wt}$$

Here the constant of integration C can be determined by the condition that $P(0) = 1$. Thus one obtains $C = 1$ and

▶
$$P(t) = e^{-wt} \tag{12·1·5}$$

Multiplication of (12·1·1) by (12·1·2) gives then

$$\left.\begin{array}{l}\mathcal{P}(t)\,dt = \text{the probability that a molecule, after surviving}\\ \text{without collisions for a time } t\text{, suffers a collision in the time}\\ \text{interval between } t \text{ and } t + dt.\end{array}\right\} \tag{12·1·6}$$

Thus

▶
$$\mathcal{P}(t)\,dt = e^{-wt}w\,dt \tag{12·1·7}$$

Remark One could also find $\mathcal{P}(t)\,dt$ by arguing that it must be equal to [the probability of surviving a time t] minus [the probability of surviving a time $t + dt$].

Thus

$$\mathcal{P}(t)\,dt = P(t) - P(t + dt) = -\frac{dP}{dt}dt$$

By using (12·1·5), this yields again the relation (12·1·7).

The probability (12·1·7) should be properly normalized in the sense that

$$\int_0^\infty \mathcal{P}(t)\,dt = 1 \tag{12·1·8}$$

This asserts simply that there is probability unity that a molecule collides at *some* time. Indeed, by (12·1·7), one finds

$$\int_0^\infty e^{-wt}w\,dt = \int_0^\infty e^{-y}\,dy = 1$$

so that the normalization condition (12·1·8) is verified.

Let $\tau \equiv \bar{t}$ be the mean time between collisions. This is also called the "collision time" or "relaxation time" of the molecule. By (12·1·7) one can write

$$\tau \equiv \bar{t} = \int_0^\infty \mathcal{P}(t)\,dt\,t$$
$$= \int_0^\infty e^{-wt}w\,dt\,t$$
$$= \frac{1}{w}\int_0^\infty e^{-y}y\,dy = \frac{1}{w}$$

since the integral is of the type evaluated in (A·3·3). Thus

$$\tau = \frac{1}{w} \tag{12·1·9}$$

and (12·1·7) can equally well be written in the form

$$\blacktriangleright \qquad \mathcal{P}(t)\,dt = e^{-t/\tau}\frac{dt}{\tau} \tag{12·1·10}$$

Since in general $w = w(v)$, τ may depend on the speed v of the molecule. The mean distance traveled by such a molecule between collisions is called the "mean free path" l of the molecule. One has thus

$$l(v) = v\,\tau(v) \tag{12·1·11}$$

A gas of molecules can then conveniently be characterized by the average collision time, or the average mean free path, of the molecules traveling with a mean speed $\bar{v}$.

Remark If w depends on the speed v and the latter *does* change appreciably in times of the order of w^{-1}, then w on the right side of (12·1·4) becomes a function of the time t. The integration would then give, instead of (12·1·7),

$$\mathcal{P}(t)\,dt = \left\{\exp\left[-\int_0^t w(t')\,dt'\right]\right\}w(t)\,dt \tag{12·1·12}$$

Remarks on the similarity to a game of chance The problem of molecular collisions formulated in the preceding paragraphs is similar to a simple game of chance. The molecule in danger of a collision is analogous to a man who keeps on throwing a die in a game where he has to pay $100 whenever the throw of the die results in the "fatal event" of a 6 landing uppermost. (The game of Russian roulette described in Problem 1.5, might be a more bloody analogue.) Let

p = the probability that the fatal event occurs in a given trial.

This probability p is assumed to be independent of the past history of occurrence of fatal events. Then

$q \equiv 1 - p$ = the probability that the fatal event does *not* occur in a given trial.

The probability P_n of surviving n trials without a fatal event is thus given by

$$P_n = (1 - p)^n \tag{12·1·13}$$

The probability $\mathcal{P}_n$ of surviving $(n - 1)$ trials without a fatal event and then suffering the fatal event at the nth trial is then

$$\mathcal{P}_n = (1 - p)^{n-1}p = q^{n-1}p \tag{12·1·14}$$

This probability is properly normalized so that the probability of suffering the fatal event at *some* time is unity, i.e.,

$$\sum_{n=1}^{\infty} \mathcal{P}_n = 1 \tag{12·1·15}$$

This can be verified by using (12·1·14); thus

$$\sum_{n=1}^{\infty} \mathcal{P}_n = \sum_{n=1}^{\infty} q^{n-1}p = p(1 + q + q^2 + \cdots)$$

By summing the geometric series, we obtain properly

$$\sum_{n=1}^{\infty} \mathcal{P}_n = \frac{p}{1-q} = \frac{p}{p} = 1$$

The mean number of trials is given by

$$\bar{n} = \sum_{n=1}^{\infty} \mathcal{P}_n n = \sum_{n=1}^{\infty} q^{n-1}pn = \frac{p}{q}\sum_{n=1}^{\infty} q^n n \tag{12·1·16}$$

By considering q as an arbitrary parameter one can write

$$\begin{aligned} \sum_{n=1}^{\infty} q^n n &= q \frac{\partial}{\partial q} \sum_{n=1}^{\infty} q^n \\ &= q \frac{\partial}{\partial q}\left(\frac{q}{1-q}\right) \\ &= q\,\frac{(1-q)+q}{(1-q)^2} = \frac{q}{(1-q)^2} \end{aligned}$$

Hence (12·1·16) becomes, putting $q = 1 - p$,

$$\bar{n} = \frac{p}{(1-q)^2} = \frac{p}{p^2} = \frac{1}{p} \tag{12·1·17}$$

All these results are analogous to those obtained in discussing a molecule. To make the correspondence exact, consider time to be divided into fixed infinitesimal intervals of magnitude dt. Each such interval represents a "trial" for the molecule in its game of chance. The fatal event is, of course, the suffering of a collision. In terms of the collision rate w of (12·1·2), the probability p is then given by

$$p = w\,dt \tag{12·1·18}$$

Furthermore, the number of trials experienced by a molecule in a time t is given by

$$n = \frac{t}{dt} \tag{12·1·19}$$

Note that as $dt \to 0$, $p \to 0$ and $n \to \infty$ in such a way that

$$pn = wt \tag{12·1·20}$$

By (12·1·13) the survival probability is then given by

$$P(t) = (1 - p)^n$$

Since $p \ll 1$, this can be approximated by writing

$$\ln P = n \ln (1 - p) \approx -np$$

Hence

$$P(t) = e^{-np} = e^{-wt} \tag{12·1·21}$$

where we have used (12·1·20). Thus we regain (12·1·5).

Similarly, one obtains by (12·1·14)

$$\mathcal{P}(t)\, dt = (1 - p)^{n-1}p = e^{-wt}w\, dt \tag{12·1·22}$$

which agrees with (12·1·7). Finally (12·1·19) gives

$$\bar{t} = \bar{n}\, dt$$

which becomes, by (12·1·17),

$$\tau = \frac{1}{p}\, dt = \frac{dt}{w\, dt} = \frac{1}{w} \tag{12·1·23}$$

and agrees thus with (12·1·9).

12·2 *Collision time and scattering cross section*

Scattering cross section An encounter (or collision) between two particles is described in terms of a "scattering cross section" which can be computed by the laws of mechanics if the interaction potential between the particles is known. Consider two particles of respective masses m_1 and m_2. Denote their respective position vectors by $\mathbf{r}_1$ and $\mathbf{r}_2$, and their respective velocities by $\mathbf{v}_1$ and $\mathbf{v}_2$. View the situation from a frame of reference fixed with respect to particle 2; the motion of particle 1 relative to 2 is then described by the relative position vector $\mathbf{R} = \mathbf{r}_1 - \mathbf{r}_2$, and the relative velocity $\mathbf{V} = \mathbf{v}_1 - \mathbf{v}_2$. In this frame of reference where the "target" particle 2 is at rest, consider that there is a uniform flux of $\mathfrak{F}_1$ type 1 particles per unit area per unit time incident with relative velocity $\mathbf{V}$ on the target particle 2. As a result of the scattering process a number $d\mathfrak{N}$ of particles of type 1 will emerge per unit time at large distances from the target particle with final velocity in the range between $\mathbf{V}'$ and $\mathbf{V}' + d\mathbf{V}'$. This defines a small solid angle range $d\Omega'$ about the direction $\hat{\mathbf{V}}' \equiv \mathbf{V}'/|\mathbf{V}'|$ of the scattered beam. (If the collision process is elastic so that

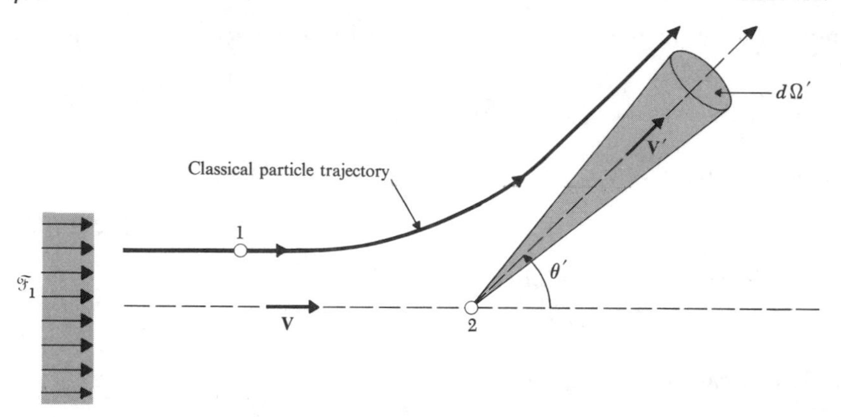

Fig. 12·2·1 Scattering process viewed from the frame of reference where the target particle 2 is at rest.

energy is conserved, $|\boldsymbol{V}'| = |\boldsymbol{V}|$.) This number $d\mathfrak{N}$ is proportional to the incident flux $\mathfrak{F}_1$ and to the solid angle $d\Omega'$. One can then write

$$d\mathfrak{N} = \mathfrak{F}_1 \sigma \, d\Omega' \tag{12·2·1}$$

where the factor of proportionality σ is called the "differential scattering cross section." It depends in general on the magnitude V of the relative velocity of the incident particle and on the particular direction $\hat{\boldsymbol{V}}'$ (specified by the polar angle θ' and azimuthal angle φ') of the scattered beam relative to the incident direction $\boldsymbol{V}$. This differential scattering cross section $\sigma = \sigma(V; \hat{\boldsymbol{V}}')$ can be computed by classical or quantum mechanics if the interaction potential between the particles is known. Note that σ has the dimensions of an area, since the flux $\mathfrak{F}_1$ is expressed per unit area.

The *total* number $\mathfrak{N}$ of particles scattered per unit time in *all* directions is obtained by integrating (12·2·1) over all solid angles. Thus

$$\mathfrak{N} = \int_{\Omega'} \mathfrak{F}_1 \sigma \, d\Omega' \equiv \mathfrak{F}_1 \sigma_0 \tag{12·2·2}$$

where

$$\sigma_0(V) = \int_{\Omega'} \sigma(V; \hat{\boldsymbol{V}}') \, d\Omega' \tag{12·2·3}$$

is called the "total scattering cross section." In general σ_0 depends on the relative speed V of the incident particles.

The calculation of scattering cross sections for various types of forces between particles is a problem discussed in courses on mechanics. Let us here recall only briefly the very simple result obtained in classical mechanics for the total scattering cross section between two "hard spheres" of respective radii a_1 and a_2. (This means that the interaction potential $V(R)$ between the particles is a function of the distance R between their centers such that $V(R) = 0$ when $R > (a_1 + a_2)$ and $V(R) \to \infty$ when $R < (a_1 + a_2)$.) The relative initial motion of the two spheres is indicated in Fig. 12·2·2. Note that scattering takes place only if the distance b (called the "impact parameter") is such that $b < (a_1 + a_2)$. Hence, out of an incident flux of $\mathfrak{F}_1$ particles per

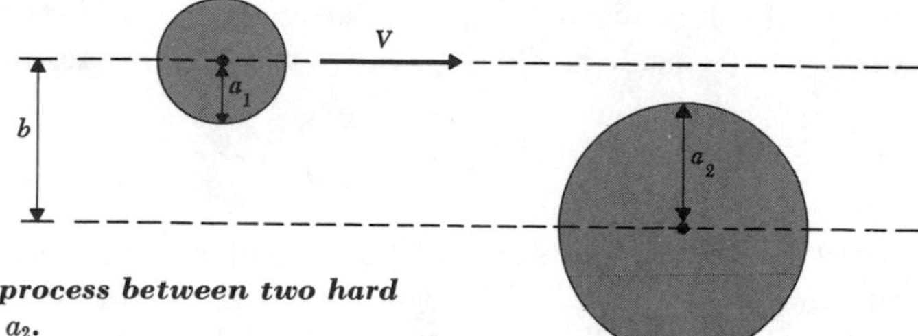

Fig. 12·2·2 Collision process between two hard spheres of radii a_1 and a_2.

unit area per unit time, only that fraction of particles incident on the circular area $\pi(a_1 + a_2)^2$ is scattered. By the definition (12·2·2) one thus obtains for the total scattering cross section between two hard spheres

$$\sigma_0 = \frac{\mathfrak{N}}{\mathfrak{F}_1} = \pi(a_1 + a_2)^2 \tag{12·2·4}$$

If the two particles are identical, this reduces simply to

$$\sigma_0 = \pi\, d^2 \tag{12·2·5}$$

where $d = 2a$ is the diameter of the spherical particle.

Relation between collision time and scattering cross section If the scattering cross section σ for collisions between molecules is known, one can readily find the probability τ^{-1} per unit time that a given molecule in a gas suffers a collision. We shall give the argument in simplified fashion without being too careful about the rigorous way of taking various averages.

Consider a gas consisting of only a single kind of molecule. Denote the mean number of molecules per unit volume by n. Let $\bar{v}$ be the mean speed of these molecules, $\bar{V}$ their mean *relative* speed, and σ_0 their mean total scattering cross section at this speed. Focus attention on the particular type of molecules (say those of velocity near $\mathbf{v}_1$) whose collision rate τ^{-1} we wish to calculate, and let n_1 denote the number of such molecules per unit volume. Consider now how this type of molecule (call it type 1) is scattered by all the molecules in an

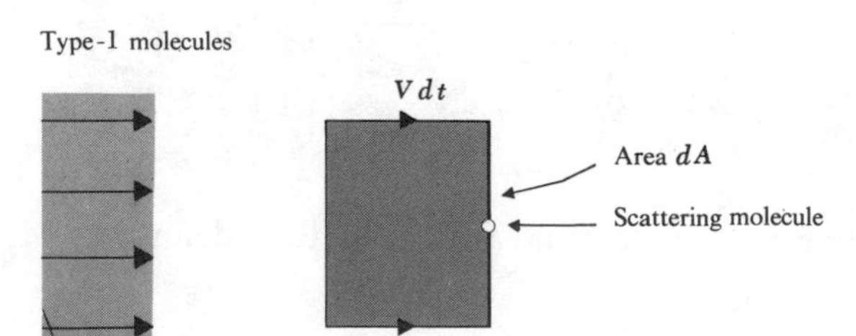

Fig. 12·2·3 If there are n_1 molecules per unit volume with relative velocity near V, all of these contained in the volume ($V\,dt\,dA$) collide with the area dA in time dt and thus constitute a flux n_1V incident upon the scattering molecule.

element of volume $d^3\mathbf{r}$ of the gas. The relative flux of type 1 molecules incident on any *one* molecule in $d^3\mathbf{r}$ is given by the familiar argument as

$$\mathfrak{F}_1 = \frac{n_1(\bar{V}\,dt\,dA)}{dt\,dA} = n_1\bar{V} \tag{12·2·6}$$

By (12·2·2) a number $n_1\bar{V}\sigma_0$ of these incident molecules is then scattered per unit time in all possible directions by this *one* target molecule. The total number of type 1 molecules scattered by *all* the molecules in $d^3\mathbf{r}$ is then given by

$$(n_1\bar{V}\sigma_0)(n\,d^3\mathbf{r})$$

Dividing this by the number $n_1\,d^3\mathbf{r}$ of type 1 molecules in the element of volume under consideration, one then obtains the collision probability $w = \tau^{-1}$ per unit time for one molecule of this type. Hence

▶ $$\tau^{-1} = \bar{V}\sigma_0 n \tag{12·2·7}$$

The collision probability is thus enhanced by a large density of molecules, a large molecular speed, and a large scattering cross section.

****Remark*** This calculation can readily be made rigorous by careful averaging. Let $f(\mathbf{v})\,d^3\mathbf{v}$ be the mean number of molecules per unit volume with velocity in the range between $\mathbf{v}$ and $\mathbf{v} + d\mathbf{v}$. (In the equilibrium situation envisaged here this is just the Maxwell velocity distribution.) We wish to calculate the collision probability $\tau^{-1}(\mathbf{v}_1)$ for a molecule with velocity between $\mathbf{v}_1$ and $\mathbf{v}_1 + d\mathbf{v}_1$. The relative flux of these molecules with respect to a molecule of velocity $\mathbf{v}$ is given by $[f(\mathbf{v}_1)\,d^3\mathbf{v}_1 V]$ where $V = |\mathbf{v}_1 - \mathbf{v}|$. Multiplying this by the differential scattering cross section $\sigma(V;\hat{\mathbf{V}}')$, and integrating over all solid angles $d\Omega'$ corresponding to the various directions of scattering $\hat{\mathbf{V}}'$, gives then the total number of molecules scattered by *one* molecule in the volume $d^3\mathbf{r}$. One then has to integrate over all the scattering molecules in $d^3\mathbf{r}$, and to divide by the number of molecules with velocity near $\mathbf{v}_1$ in this volume. Thus

$$\tau^{-1}(\mathbf{v}_1) = \frac{\int_{\mathbf{v}}\int_{\Omega'} [f(\mathbf{v}_1)\,d^3\mathbf{v}_1 V]\sigma(V;\hat{\mathbf{V}}')\,d\Omega'[f(\mathbf{v})\,d^3\mathbf{v}\,d^3\mathbf{r}]}{f(\mathbf{v}_1)\,d^3\mathbf{v}_1\,d^3\mathbf{r}}$$

or

$$\tau^{-1}(\mathbf{v}_1) = \int_{\mathbf{v}}\int_{\Omega'} V\sigma(V;\hat{\mathbf{V}}')f(\mathbf{v})\,d\Omega'\,d^3\mathbf{v} \tag{12·2·8}$$

Equation (12·2·7) yields for the average mean free path l defined in (12·1·11) the result

$$l = \tau\bar{v} = \frac{\bar{v}}{\bar{V}}\frac{1}{n\sigma_0} \tag{12·2·9}$$

Here the ratio $(\bar{v}/\bar{V})$ of mean speed to mean relative speed is close to unity. Actually $\bar{V}$ is somewhat larger than $\bar{v}$. The following simple argument makes this clear. Consider two different molecules with velocities $\mathbf{v}_1$ and $\mathbf{v}_2$. Their relative velocity is then given by

$$\mathbf{V} = \mathbf{v}_1 - \mathbf{v}_2$$

Hence

$$\mathbf{V}^2 = \mathbf{v}_1^2 + \mathbf{v}_2^2 - 2\mathbf{v}_1 \cdot \mathbf{v}_2 \tag{12·2·10}$$

If one takes the average of both sides of this equation, $\overline{\mathbf{v}_1 \cdot \mathbf{v}_2} = 0$, since the cosine of the angle between $\mathbf{v}_1$ and $\mathbf{v}_2$ is as likely to be positive as negative for molecules moving in random directions. Thus (12·2·10) becomes

$$\overline{V^2} = \overline{v_1^2} + \overline{v_2^2}$$

Neglecting the distinction between root-mean-square and mean values, this can be written

$$\bar{V} \approx \sqrt{\bar{v}_1^2 + \bar{v}_2^2} \tag{12·2·11}$$

When the molecules in the gas are all identical, $\bar{v}_1 = \bar{v}_2$ and (12·2·11) becomes

$$\bar{V} \approx \sqrt{2}\,\bar{v} \tag{12·2·12}$$

Then (12·2·9) becomes

▶
$$l \approx \frac{1}{\sqrt{2}\,n\sigma_0} \tag{12·2·13}$$

Although this fact is not too interesting, it may be remarked parenthetically that a suitable average of (12·2·8) over the Maxwell velocity distribution yields for hard-sphere collisions precisely the result (12·2·13).

It is of interest to estimate the mean free path for a typical gas at room temperature ($\approx 300°$K) and atmospheric pressure (10^6 dynes cm^{-2}). The number density n can be calculated from the equation of state. Thus

$$n = \frac{\bar{p}}{kT} = \frac{10^6}{(1.4 \times 10^{-16})(300)} = 2.4 \times 10^{19} \text{ molecules/cm}^3$$

A typical molecular diameter d might be 2×10^{-8} cm. Hence (12·2·5) gives $\sigma_0 \approx \pi(2 \times 10^{-8})^2 \approx 12 \times 10^{-16}$ cm^2 and (12·2·13) yields the estimate

$$l \approx 3 \times 10^{-5} \text{ cm} \tag{12·2·14}$$

Thus

$$l \gg d \tag{12·2·15}$$

so that our approximations based on relatively infrequent encounters between molecules are justified. If the gas is nitrogen, the mean speed of a N_2 molecule is, by (7·10·19), of the order of $\bar{v} \approx 5 \times 10^4$ cm/sec. Its mean time τ between collisions is then roughly $\tau \approx l/\bar{v} \approx 6 \times 10^{-10}$ sec. Thus its collision rate is $\tau^{-1} \approx 2 \times 10^9$ sec^{-1}, which is a frequency in the microwave region of the electromagnetic spectrum.

12·3 *Viscosity*

Definition of the coefficient of viscosity Consider a fluid (liquid or gas). Imagine in this fluid some plane with its normal pointing along the z direction.

Fig. 12·3·1 A plane z = constant in a fluid. The fluid below the plane exerts a force $\mathbf{P}_z$ on the fluid above it.

Then the fluid below this plane (i.e., on the side of smaller z) exerts a mean force $\mathbf{P}_z$ per unit area (or "stress") on the fluid above the plane. Conversely, it follows by Newton's third law that the fluid above the plane exerts a stress $-\mathbf{P}_z$ on the fluid below the plane.

The mean force per unit area normal to the plane, i.e., the z component of $\mathbf{P}_z$, measures just the mean pressure $\bar{p}$ in the fluid; to be precise, $P_{zz} = \bar{p}$. When the fluid is in equilibrium, so that it is at rest or moving with *uniform* velocity throughout, then there is no mean component of stress *parallel* to the plane. Thus $P_{zx} = 0$. Note that the quantity P_{zx} is labeled by two indices, the first of them designating the orientation of the plane and the second the component of the force exerted across this plane.*

Consider now a nonequilibrium situation where the liquid does *not* move with the same velocity throughout. To be specific, consider the case where the fluid has a constant mean velocity u_x in the x direction, the magnitude of u_x depending on z so that $u_x = u_x(z)$. This kind of situation could be produced if the fluid is enclosed between two plates a distance L apart, the plate at $z = 0$ being stationary and the plate at $z = L$ moving in the x direction with constant velocity u_0. The layers of fluid adjacent to the plates assume, to a good approximation, the velocities of the plates, so that there is no relative velocity of slip between the fluid and the plates. Layers of fluid between the plates have then different mean velocities u_x varying in magnitude between 0 and u_0. In this case the fluid exerts a tangential force on the moving plate, tending to slow it down so as to restore the equilibrium situation.

More generally, any layer of fluid below a plane z = constant exerts a tangential stress P_{zx} on the fluid above it. We already saw that $P_{zx} = 0$ in the equilibrium situation where $u_x(z)$ does *not* depend on z. In the present nonequilibrium case where $\partial u_x/\partial z \neq 0$ one expects, therefore, that P_{zx} should

* The quantity $P_{\alpha\gamma}$ (where α and γ can denote x, y, or z) is called the "pressure tensor."

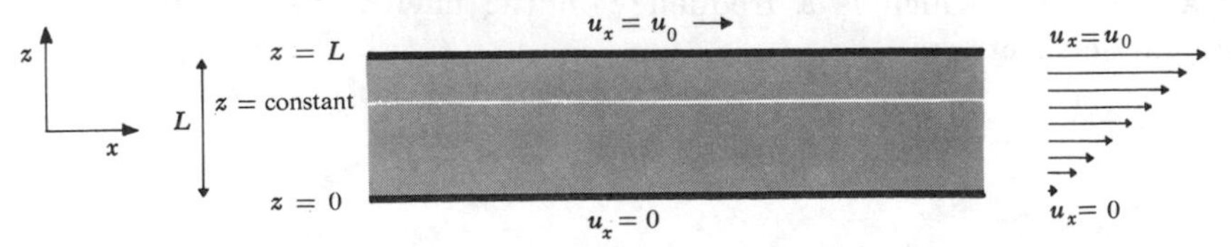

Fig. 12·3·2 A fluid contained between two plates. The lower plate is at rest while the upper one moves with velocity u_0; there thus exists a velocity gradient ($\partial u_x/\partial z$) in the fluid.

be some function of derivatives of u_x with respect to z such that it vanishes when u_x is independent of z. But if $\partial u_x/\partial z$ is assumed to be relatively small, the leading term in a power-series expansion of this function should be an adequate approximation, i.e., one expects a linear relation of the form

$$P_{zx} = -\eta \frac{\partial u_x}{\partial z} \tag{12·3·1}$$

Here the constant of proportionality η is called the "coefficient of viscosity." If u_x increases with increasing z, then the fluid below the plane tends to slow down the fluid above the plane and thus exerts on it a force in the $-x$ direction. That is, if $\partial u_x/\partial z > 0$, then $P_{zx} < 0$. Hence the minus sign was introduced explicitly in (12·3·1) so as to make the coefficient η a positive quantity. The cgs unit of η is, by (12·3·1), that of gm cm^{-1} sec^{-1}. (It is also commonly called a "poise" in honor of the physicist Poiseuille.) The proportionality implied by (12·3·1) between the stress P_{zx} and the velocity gradient $\partial u_x/\partial z$ is experimentally well satisfied by liquids and gases if the velocity gradient is not too large.

Note the various forces which act in the x direction in Fig. 12·3·2. The fluid below the plane z = constant exerts a force P_{zx} per unit area on the fluid above it. Since the fluid between this plane and the plate at $z = L$ moves without acceleration, this plate must exert a force $-P_{zx}$ per unit area on the fluid adjacent to it. By Newton's third law, the fluid must then also exert on the plate at $z = L$ a force per unit area $+P_{zx}$, given by (12·3·1).

Calculation of the coefficient of viscosity for a dilute gas In the simple case of a dilute gas, the coefficient of viscosity can be calculated fairly readily on the basis of microscopic considerations of kinetic theory. Suppose that the mean fluid velocity component u_x (which is assumed to be very small compared to the mean thermal speed of the molecules) is a function of z. How does the stress P_{zx} come about in this situation? The qualitative reason is that in Fig. 12·3·2 molecules above the plane z = constant have a somewhat larger x component of momentum than molecules below this plane. As molecules cross back and forth across this plane they carry this x component of momentum with them. Hence the gas below the plane gains momentum in the x direction from the molecules coming from above the plane; conversely, the gas above the plane loses momentum in the x direction by virtue of the molecules arriving from below the plane. Since the rate of change of momentum of a system is, by Newton's second law, equal to the force acting on the system, it follows that the gas above the plane is acted on by a force due to the gas below the plane. More precisely,

P_{zx} = mean increase, per unit time and per unit area of the plane, of the x component of momentum of the gas above the plane due to the net transport of momentum by molecules crossing this plane. (12·3·2)

Remark An analogy may make this mechanism of viscosity by momentum transfer clearer. Suppose two railroad trains move side by side along parallel tracks, the speed of one train being greater than that of the other. Imagine that workers on each train constantly pick up sandbags from their train and throw them onto the other train. Then there is a transfer of momentum between the trains so that the slower train tends to be accelerated and the faster train to be decelerated.

Let us now give an approximate simple calculation of the coefficient of viscosity. If there are n molecules per unit volume, roughly one-third of them have velocities along the z direction. Half of these, or $\frac{1}{6}n$ molecules per unit volume, have mean velocity $\bar{v}$ in the $+z$ direction; the other half have a mean velocity $\bar{v}$ in the $-z$ direction. On the average there are, therefore, $(\frac{1}{6}n\bar{v})$ molecules which in unit time cross a unit area of the plane z = constant from below; similarly, there are $(\frac{1}{6}n\bar{v})$ molecules which in unit time cross a unit area of this plane from above. But molecules which cross the plane from below have, on the average, experienced their last collision at a distance l (l = mean free path) below the plane. Since the mean velocity $u_x = u_x(z)$ is a function of z, the molecules at this position $(z - l)$ had on the average a mean x component of velocity $u_x(z - l)$. Thus each such molecule of mass m transports across the plane a mean x component of momentum $[mu_x(z - l)]$. Hence one concludes that

The mean x component of momentum transported per unit time per unit area across the plane in the *up*ward direction = $(\frac{1}{6}n\bar{v})[mu_x(z - l)]$. (12·3·3)

Similarly, the mean x component of momentum transported per unit time per unit area across the plane in the *down*ward direction = $(\frac{1}{6}n\bar{v})[mu_x(z + l)]$. (12·3·4)

By subtracting (12·3·4) from (12·3·3) one obtains the *net* molecular transport of x component of momentum per unit time per unit area from below to above the plane, i.e., the stress P_{zx} described in (12·3·2). Thus

$$P_{zx} = (\tfrac{1}{6}n\bar{v})[mu_x(z - l)] - (\tfrac{1}{6}n\bar{v})[mu_x(z + l)]$$

or

$$P_{zx} = \tfrac{1}{6}n\bar{v}m[u_x(z - l) - u_x(z + l)] \tag{12·3·5}$$

Here $u_x(z)$ can be expanded in Taylor's series and higher-order terms can be neglected, since the velocity gradient $\partial u_x/\partial z$ is assumed to be small (i.e., small

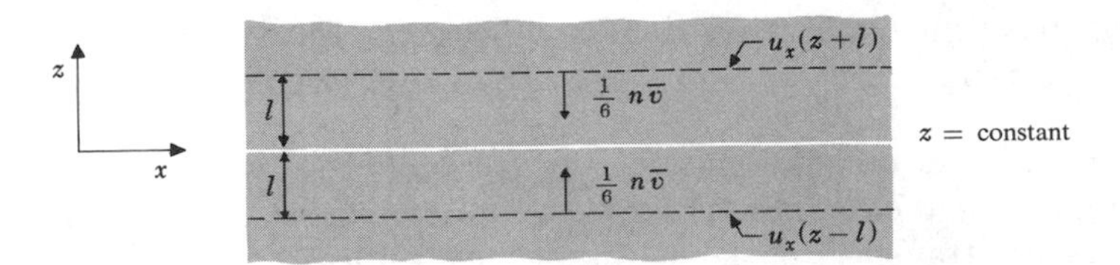

Fig. 12·3·3 Momentum transport by molecules crossing a plane.

enough that the mean velocity u_x does not vary appreciably over a distance of the order of l). Thus

$$u_x(z+l) = u_x(z) + \frac{\partial u_x}{\partial z} l \cdots$$

$$u_x(z-l) = u_x(z) - \frac{\partial u_x}{\partial z} l \cdots$$

Hence
$$P_{zx} = \frac{1}{6} n\bar{v}m \left(-2 \frac{\partial u_x}{\partial z} l \right) \equiv -\eta \frac{\partial u_x}{\partial z} \tag{12·3·6}$$

where

▶
$$\eta = \tfrac{1}{3} n\bar{v}ml \tag{12·3·7}$$

Thus P_{zx} is indeed proportional to the velocity gradient $\partial u_x/\partial z$ (as expected by (12·3·1)), and (12·3·7) provides an explicit approximate expression for the viscosity coefficient η in terms of the microscopic parameters characterizing the molecules of the gas.

Our calculation has been very simplified and careless about the exact way various quantities ought to be averaged. Hence the factor $\frac{1}{3}$ in (12·3·7) is not to be trusted too much; the constant of proportionality given by a more careful calculation might be somewhat different. On the other hand, the essential dependence of η on the parameters n, $\bar{v}$, m, and l ought to be correct.

Discussion The result (12·3·7) leads to some interesting predictions. By (12·2·13)

$$l \approx \frac{1}{\sqrt{2}\, n\sigma_0} \tag{12·3·8}$$

Thus the factor n cancels in (12·3·7), and one obtains

$$\eta = \frac{1}{3\sqrt{2}} \frac{m}{\sigma_0} \bar{v} \tag{12·3·9}$$

But the mean molecular speed, given by (7·10·13) as*

$$\bar{v} = \sqrt{\frac{8}{\pi} \frac{kT}{m}} \tag{12·3·10}$$

depends only on the temperature but not on the gas density n. Hence (12·3·9) is *independent* of the gas density n, or equivalently, of the gas pressure $\bar{p} = nkT$.

This is a remarkable result. It asserts that in the situation illustrated in Fig. 12·3·2, the viscous retarding force exerted by the gas on the moving upper plate is the same whether the pressure of the gas between the two plates is, for example, equal to 1 mm Hg or is increased to 1000 mm Hg. At first sight such a conclusion seems strange, since a naïve intuition might lead one

* In the approximate calculations of this chapter one could just as well replace the mean speed $\bar{v}$ by the rms speed $\sqrt{3kT/m}$ obtained from the equipartition theorem.

to expect that the tangential force transmitted by the gas should be proportional to the number of gas molecules present. The paradox is resolved by noting that if the number of gas molecules is doubled, there are indeed twice as many molecules available to transport momentum from one plate to the other; but the mean free path of each molecule is then also halved, so that it can transport this momentum only half as effectively. Thus the net rate of momentum transfer is left unchanged. The fact that the viscosity η of a gas at a given temperature is independent of its density was first derived by Maxwell in 1860 and was confirmed by him experimentally.

It is clear, however, that this result cannot hold over an arbitrarily large density range of the gas. Indeed, we made two assumptions in deriving the relation (12·3·7):

1. We assumed that the gas is sufficiently dilute that there is negligible probability that *more* than two molecules come simultaneously so close together as to interact appreciably among themselves. Thus we were allowed to consider only two-particle collisions. This assumption is justified if the density n of the gas is sufficiently low so that

$$l \gg d \tag{12·3·11}$$

where $d \approx \sqrt{\sigma_0}$ is a measure of the molecular diameter.

2. On the other hand, we assumed that the gas is dense enough that the molecules collide predominantly with other molecules rather than with the walls of the container. This assumption implies that n is sufficiently large that

$$l \ll L \tag{12·3·12}$$

where L is a measure of the smallest linear dimension of the containing vessel (e.g., L is the spacing between the plates in Fig. 12·3·2).

If the gas is made so dilute that the condition (12·3·12) is violated, then the viscosity η must decrease, since in the limiting case when $n \to 0$ (perfect vacuum) the tangential force on the moving plate in Fig. 12·3·2 must clearly go to zero. (Indeed, in this limit the mean free path l in (12·3·7) must approach the container dimension L.) Note, however, that the range of densities where both (12·3·11) and (12·3·12) are simultaneously satisfied is quite large, because $L \gg d$ in usual macroscopic experiments. Thus the coefficient of viscosity η of a gas is independent of its pressure over a very considerable range.

Remark The foregoing comments can be made somewhat more quantitative. The *total* probability per unit time τ_0^{-1} that a molecule suffers a collision can be written as the sum of two probabilities

$$\tau_0^{-1} = \tau^{-1} + \tau_w^{-1} \tag{12·3·13}$$

where τ^{-1} is the probability per unit time that the molecule collides with other molecules and τ_w^{-1} is the probability per unit time that the molecule collides with the walls of the container. But by (12·2·7)

$$\tau^{-1} = \bar{V} n \sigma_0 = \frac{\bar{v}}{l} \tag{12·3·14}$$

where $l \approx (\sqrt{2}\, n\sigma_0)^{-1}$ is the mean free path due to collisions with other molecules. On the other hand, the mean time τ_w required to cross the smallest dimension L of the container is of the order of $L/\bar{v}$ so that

$$\tau_w^{-1} \approx \frac{\bar{v}}{L} \tag{12·3·15}$$

Hence (12·3·13) gives for the resultant mean free path $l_0 \equiv \tau_0 \bar{v}$ the relation

$$\frac{1}{l_0} = \frac{1}{l} + \frac{1}{L} \approx \sqrt{2}\, n\sigma_0 + \frac{1}{L} \tag{12·3·16}$$

It is this value of l_0 which should be used in (12·3·7) if approximate account is to be taken of collisions with the walls. When the density n becomes sufficiently small, $l_0 \to L$ and $\eta \propto n$. It should be pointed out, however, that in the very low density limit where $l \gg L$ (a gas satisfying this condition is called a "Knudsen gas") the concept of coefficient of viscosity tends to lose its meaning. The reason is that molecular collisions with the walls are then of overwhelming importance so that geometrical factors involving the shape of the container have to be considered in detail.

Let us now discuss the temperature dependence of η. If the scattering of molecules is similar to that of hard spheres, then the cross section σ_0 is by (12·2·5) a number independent of T. Then it follows by (12·3·9) that the temperature dependence of η is the same as that of $\bar{v}$; i.e., for hard-sphere scattering,

$$\eta \propto T^{\frac{1}{2}} \tag{12·3·17}$$

More generally, $\sigma_0 = \sigma_0(\bar{V})$ depends on the mean relative speed of the molecules. Since $\bar{V} \propto T^{\frac{1}{2}}$, σ_0 becomes then also temperature dependent. The result is that η tends to vary with temperature more rapidly than in (12·3·17), somewhat more like $T^{0.7}$. This can be qualitatively understood since there exists not only a repulsive but also a longer-range attractive interaction between the molecules. This latter interaction tends to increase the scattering probability of a molecule and becomes more effective at low temperatures, where the molecules have low velocities and are thus more readily deflected. Hence the scattering cross section σ_0 tends to *decrease* with increasing temperature. As T increases, the viscosity $\eta \propto T^{\frac{1}{2}}/\sigma_0$ tends, therefore, to increase with temperature more rapidly than $T^{\frac{1}{2}}$.

Note that the viscosity of a gas *increases* as the temperature is raised. This behavior is quite different from that of the viscosity of liquids, which generally *decreases* rapidly with increasing temperature. The reason is that in a liquid, where molecules are close together, momentum transfer across a plane occurs by direct forces between molecules on adjacent sides of the plane as well as by virtue of their motion across this plane.

Finally we estimate the magnitude of η. Since the mean pressure of the gas is, by (7·13·1), approximately given by $\bar{p} = \frac{1}{3}nm\bar{v}^2$, the expression (12·3·7) can be written as

$$\eta = \frac{\bar{p}}{\bar{v}}\, l = \frac{\bar{p}}{\bar{v}/l} \qquad (12\cdot3\cdot18)$$

In words this says that the coefficient of viscosity is of such a magnitude that it would give rise to a stress equal to the gas pressure in the presence of a velocity gradient equal to the mean molecular speed divided by the mean free path. For air at atmospheric pressure (10^6 dynes cm^{-2}) and room temperature (300°K) one has approximately $\bar{v} \approx 5 \times 10^4$ cm sec^{-1} and $l \approx 3 \times 10^{-5}$ cm· hence (12·3·18) gives as an order-of-magnitude estimate

$$\eta \approx 10^6/(1.7 \times 10^9) = 6 \times 10^{-4} \text{ gm cm}^{-1} \text{ sec}^{-1}$$

The measured value of η for N_2 gas at this temperature is 1.78×10^{-4} gm cm^{-1} sec^{-1}.

For purposes of later comparisons with more exact calculations, we can combine (12·3·9) and (12·3·10) to obtain the following explicit expression for η obtained by our simple theory:

$$\eta = \frac{2}{3\sqrt{\pi}} \frac{\sqrt{mkT}}{\sigma_0} = 0.377 \frac{\sqrt{mkT}}{\sigma_0} \qquad (12\cdot3\cdot19)$$

12 · 4 *Thermal conductivity*

Definition of the coefficient of thermal conductivity Consider a substance in which the temperature is *not* uniform throughout. In particular, imagine that the temperature T is a function of the z coordinate so that $T = T(z)$. Then the substance is certainly not in a state of equilibrium. As a result, energy in the form of heat flows from the region of higher to that of lower temperature. Let

$$\left.\begin{array}{l}\mathcal{Q}_z = \text{the heat crossing unit area of a plane (in the } z \text{ direction} \\ \text{normal to the plane) per unit time.}\end{array}\right\} \qquad (12\cdot4\cdot1)$$

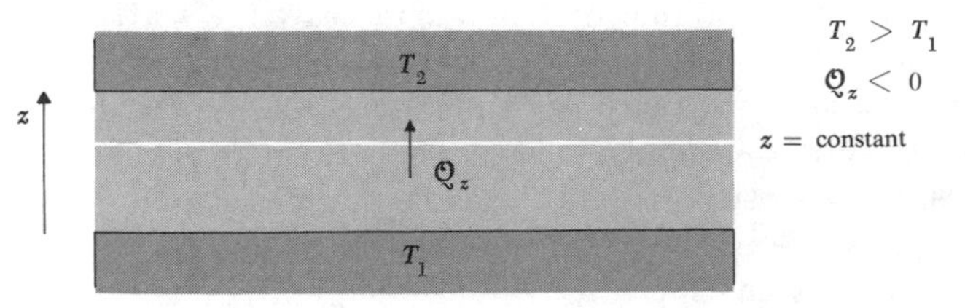

Fig. 12·4·1 A substance in thermal contact with two heat reservoirs at constant temperatures T_1 and T_2. If $T_2 > T_1$, heat flows in the $-z$ direction from the region of higher to that of lower temperature.

The quantity $\mathcal{Q}_z$ is called the "heat flux" in the z direction. If the temperature were uniform, $\mathcal{Q}_z = 0$. If it is not uniform, one expects that $\mathcal{Q}_z$ should to good approximation be proportional to the temperature gradient $\partial T/\partial z$ if the latter is not too large. Thus one can write

$$\mathcal{Q}_z = -\kappa \frac{\partial T}{\partial z} \tag{12·4·2}$$

The constant of proportionality κ is called the "coefficient of thermal conductivity" of the substance. Since heat flows from the region of higher to that of lower temperature, $\mathcal{Q}_z < 0$ if $\partial T/\partial z > 0$. The minus sign was introduced explicitly in (12·4·2) so as to make κ a positive quantity. The relation (12·4·2) is found to be well obeyed in practically all gases, liquids, and isotropic solids.

Calculation of the coefficient of thermal conductivity for a dilute gas
In the simple case of a dilute gas the coefficient of thermal conductivity can be readily calculated by simple microscopic arguments similar to those used in discussing the viscosity of a gas. Consider a plane z = constant in the gas where $T = T(z)$. The mechanism of heat transport is due to the fact that molecules cross this plane from above and below. But if $\partial T/\partial z > 0$, a molecule coming from above has a mean energy $\bar{\epsilon}(T)$ which is larger than that of a molecule coming from below. Thus there results a net transport of energy from the region above the plane to that below it. More quantitatively, there are again roughly $\frac{1}{6}n\bar{v}$ molecules which in unit time cross unit area of this plane from below and an equal number of molecules which cross it from above.* Here n is the mean number of molecules per unit volume at the plane z = constant, and $\bar{v}$ is their mean speed. Now molecules which cross this plane from below have, on the average, experienced their last collision at a distance l (l = mean free path) below the plane. But the mean energy $\bar{\epsilon}$ of a molecule is a function of T and, since $T = T(z)$ in the present case, consequently a function of z so that $\bar{\epsilon} = \bar{\epsilon}(z)$. Hence the molecules crossing the plane from below carry with them a mean energy $\bar{\epsilon}(z - l)$ corresponding to the

* Since the thermal conductivity of a gas is measured under steady-state conditions where there is no convective motion of the gas, the number of molecules crossing unit area of any plane per second from one side must always be equal to the number of molecules crossing the plane in the opposite direction. It is therefore unnecessary in this simple discussion to worry about the fact that the temperature gradient causes n and $\bar{v}$ to be slightly different above and below the plane. (Such questions can be investigated more carefully by the methods of the next chapters.)

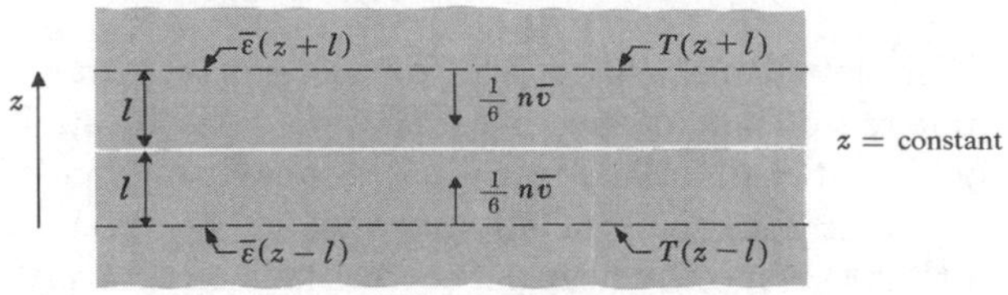

$\bar{\epsilon}(z+l)$ $T(z+l)$ z l $\frac{1}{6}n\bar{v}$ z = constant l $\frac{1}{6}n\bar{v}$ $\bar{\epsilon}(z-l)$ $T(z-l)$

Fig. 12·4·2 Energy transport by molecules crossing a plane.

mean energy assumed by them at their last collision at the position $(z - l)$. Thus one obtains

$$\left.\begin{array}{l}\text{Mean energy transported per unit time per unit area across} \\ \text{the plane from below } = \frac{1}{6}n\bar{v}\bar{\epsilon}(z - l).\end{array}\right\} \tag{12·4·3}$$

Similarly, in considering molecules coming from above the plane where they suffered their last collision at $(z + l)$, one obtains

$$\left.\begin{array}{l}\text{Mean energy transported per unit time per unit area across} \\ \text{the plane from above } = \frac{1}{6}n\bar{v}\bar{\epsilon}(z + l).\end{array}\right\} \tag{12·4·4}$$

By subtracting (12·4·4) from (12·4·3) one then obtains the *net* flux of energy $\mathcal{Q}_z$ crossing the plane from below in the $+z$ direction

$$\begin{aligned}\mathcal{Q}_z &= \tfrac{1}{6}n\bar{v}\{\bar{\epsilon}(z - l) - \bar{\epsilon}(z + l)\} \\ &= \frac{1}{6}n\bar{v}\left\{\left[\bar{\epsilon}(z) - l\frac{\partial\bar{\epsilon}}{\partial z}\right] - \left[\bar{\epsilon}(z) + l\frac{\partial\bar{\epsilon}}{\partial z}\right]\right\}\end{aligned}$$

or

$$\mathcal{Q}_z = \frac{1}{6}n\bar{v}\left(-2l\frac{\partial\bar{\epsilon}}{\partial z}\right) = -\frac{1}{3}n\bar{v}l\frac{\partial\bar{\epsilon}}{\partial T}\frac{\partial T}{\partial z} \tag{12·4·5}$$

since $\bar{\epsilon}$ depends on z through the temperature T. Let us introduce the abbreviation

$$c \equiv \frac{\partial\bar{\epsilon}}{\partial T} \tag{12·4·6}$$

which is the specific heat per *molecule*. Then (12·4·5) becomes

$$\mathcal{Q}_z = -\kappa\frac{\partial T}{\partial z} \tag{12·4·7}$$

where

▶ $$\kappa = \tfrac{1}{3}n\bar{v}cl \tag{12·4·8}$$

The relation (12·4·7) shows that $\mathcal{Q}_z$ is indeed proportional to the temperature gradient (as expected by (12·4·2)), and (12·4·8) provides an explicit expression for the thermal conductivity κ of the gas in terms of fundamental molecular quantities.

Once again the precise factor $\frac{1}{3}$ in (12·4·8) is not to be trusted too much in this simplified calculation, but the dependence on the other parameters ought to be correct. Since $l \propto n^{-1}$, the density n again cancels; i.e., using (12·3·8), the thermal conductivity (12·4·8) becomes

$$\kappa = \frac{1}{3\sqrt{2}}\frac{c}{\sigma_0}\bar{v} \tag{12·4·9}$$

which is *independent* of the pressure of the gas. This result is due to the same reasons mentioned in connection with the similar property of the viscosity coefficient η and is again valid in a density range where $d \ll l \ll L$.

Note that for a monatomic gas the equipartition theorem gives $\bar{\epsilon} = \frac{3}{2}kT$ so that the specific heat per molecule is simply given by $c = \frac{3}{2}k$.

Since $\bar{v} \propto T^{\frac{1}{2}}$ and since c is usually temperature independent, (12·4·9) gives for hard-sphere interaction between molecules

$$\kappa \propto T^{\frac{1}{2}} \tag{12·4·10}$$

More generally, σ_0 also tends to vary with the temperature in the manner discussed in the last section in connection with the viscosity. As a result, κ increases again somewhat more rapidly with increasing temperature than is indicated by (12·4·10).

An estimate of the order of magnitude of κ for a gas at room temperature can readily be obtained by substituting typical numbers into (12·4·8). A representative value is the measured thermal conductivity of argon at 273°K, namely $\kappa = 1.65 \times 10^{-4}$ watts cm^{-1} deg^{-1}.

By using the result (12·3·10) for $\bar{v}$, the approximate expression (12·4·9) for the thermal conductivity becomes explicitly

$$\kappa = \frac{2}{3\sqrt{\pi}} \frac{c}{\sigma_0} \sqrt{\frac{kT}{m}} \tag{12·4·11}$$

Finally, comparison between the expressions (12·4·8) for the thermal conductivity κ and (12·3·7) for the viscosity η shows that these are quite similar in form. Indeed, one obtains for their ratio the relation

$$\frac{\kappa}{\eta} = \frac{c}{m} \tag{12·4·12}$$

Equivalently, multiplying both numerator and denominator by Avogadro's number N_a,

$$\frac{\kappa}{\eta} = \frac{c_V}{\mu} \tag{12·4·13}$$

where $c_V = N_a c$ is the molar specific heat of the gas at constant volume and where $\mu = N_a m$ is its molecular weight. Thus there exists a very simple relation between the two transport coefficients κ and η, a relation which can readily be checked experimentally. One finds that the ratio $(\kappa/\eta)(c/m)^{-1}$ lies somewhere in the range between 1.3 and 2.5 instead of being unity as predicted by (12·4·12). In view of the very simplified nature of the arguments leading to these expressions for η and κ, there is greater justification for being pleased by the extent of agreement with experiment than there is cause for surprise at the discrepancy. Indeed, part of the latter is readily explained by the mere fact that our calculation did not take into account effects due to the distribution of molecular velocities. Thus faster molecules cross a given plane more frequently than slower ones. In the case of thermal conductivity these faster molecules also transport more kinetic energy; but in the case of viscosity they do not carry any greater mean x component of momentum. Thus the ratio κ/η should be increased to a value larger than that given by (12·4·12).

Application to nonclassical gases It is worth pointing out that the simple considerations of this section are applicable to a much wider class of physical situations. Consider, for example, the thermal conductivity of a metal. Heat in such a metal is predominantly transported by the conduction electrons. The latter would travel through a perfect periodic crystal lattice without being scattered (since the electrons have wave properties in a quantum description). They do, however, get scattered because every metal contains some impurities or other lattice imperfections, and because at a finite temperature the lattice vibrates (i.e., the perfect periodicity of the lattice is then disturbed by thermally excited sound waves, or phonons, traveling through the lattice).

In order to apply (12·4·8) to the conduction electrons which form a highly degenerate Fermi-Dirac gas, we note first that only those electrons lying within a range of the order of kT around the Fermi energy μ, i.e., only the fraction kT/μ of electrons which contribute to the electronic specific heat (an amount $\frac{3}{2}k$ per electron) contribute to the thermal conductivity κ. Hence the product nc in (12·4·8) involves only these effective electrons; thus it becomes approximately $n(kT/\mu)(\frac{3}{2}k)$, i.e., it is proportional to T. All these electrons move nearly with the Fermi velocity v_F; thus $\bar{v} \approx v_F$ in (12·4·8), and this is essentially temperature independent. If the temperature is low enough, the number n_p of thermally excited phonons per unit volume becomes sufficiently small compared to the number n_i of impurities per unit volume that impurity scattering of the electrons is predominant. But since n_i is a fixed number independent of T, the electron mean free path $l \propto n_i^{-1}$ is independent of T (assuming the electron-impurity scattering cross section to be essentially constant). Hence (12·4·8) predicts that, for impurity scattering,

$$\kappa_i \propto T \tag{12·4·14}$$

This proportionality is experimentally found to be well satisfied for metals (and alloys, which are, of course, very impure) at sufficiently low temperatures.

At higher temperatures, scattering by phonons becomes predominant. If the temperature is still sufficiently low that all thermally excited phonons (or lattice vibrations) have large wavelengths compared to the interatomic spacing (i.e., if T is still appreciably less than the Debye temperature of Sec. 10·2), then the problem is quite analogous to that of photons, and the mean number of phonons per unit volume $n_p \propto T^3$ (see (10·2·27)). Hence the electron mean free path due to collision with phonons is given by $l \propto n_p^{-1} \propto T^{-3}$ (assuming an essentially constant electron-phonon scattering cross section). The temperature dependence of κ in (12·4·8) then becomes, for phonon scattering,

$$\kappa_p \propto T\left(\frac{1}{T^3}\right) \propto \frac{1}{T^2} \tag{12·4·15}$$

More generally, the electrons are scattered independently by both impurities and phonons at the same time. Hence the thermal resistivities (i.e., the reciprocals of the respective conductivities) due to these processes simply add. The resultant thermal conductivity κ must then be given by a relation of the form

$$\frac{1}{\kappa} = \frac{1}{\kappa_i} + \frac{1}{\kappa_p} = \frac{a}{T} + bT^2 \tag{12·4·16}$$

where κ_i and κ_p are given by (12·4·14) and (12·4·15), and where a and b are two constants. The temperature dependence (12·4·16), with its characteristic maximum as a function of T, is experimentally well verified.

Let us finally consider the thermal conductivity of an insulating solid at low temperatures. Since there are no conduction electrons, the thermal conductivity is low and is entirely due to heat transport by lattice vibrations, i.e., by phonons. In order to apply (12·4·8) to these phonons, we note that $n_p \propto T^3$ if T is sufficiently low. The speed $\bar{v}$ of a phonon is the velocity of sound, which is essentially temperature-independent. The mean energy $\bar{\epsilon}$ of a phonon is of the order of kT, so that $c = \partial\bar{\epsilon}/\partial T$ is of the order k and temperature-independent. If T is sufficiently low, the mean free path of a phonon is essentially limited by scattering from the boundaries of the specimen; thus l is of the order of the specimen dimensions and therefore temperature-independent. Hence one obtains for an insulator at low temperature simply

$$\kappa \propto T^3 \tag{12·4·17}$$

This temperature dependence is experimentally found to be approximately correct.

12 · 5 *Self-diffusion*

Definition of the coefficient of self-diffusion Consider a substance consisting of similar molecules, but assume that a certain number of these molecules are labeled in some way. For example, some of the molecules might be labeled by the fact that their nuclei are radioactive. Let n_1 be the mean number of labeled molecules per unit volume. In an equilibrium situation the labeled molecules would be distributed uniformly throughout the available volume, so that n_1 is independent of position. Now suppose that their distribution is not uniform, so that n_1 *does* depend on position, e.g., $n_1 = n_1(z)$, even though the *total* mean number n of molecules per unit volume remains constant. (This constancy guarantees that there is no mass motion of the whole substance.) This is not an equilibrium situation and thus there will be a motion of labeled molecules tending to increase the entropy, i.e., tending to make the concentration n_1 more nearly uniform. Let the flux of labeled molecules be denoted by $\boldsymbol{J}$, i.e., let

J_z = the mean number of labeled molecules crossing unit area of a plane (in the z direction normal to the plane) per unit time. (12·5·1)

If n_1 were uniform, $J_z = 0$. If n_1 is not uniform one expects that J_z should to good approximation be proportional to the concentration gradient of labeled molecules. Thus one can write

$$J_z = -D\frac{\partial n_1}{\partial z} \tag{12·5·2}$$

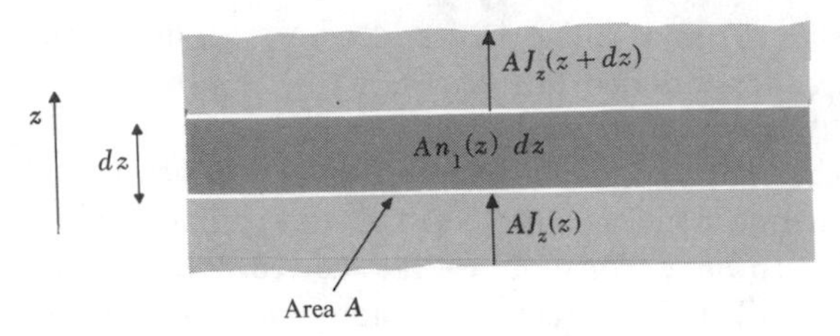

***Fig. 12·5·1** Diagram illustrating the conservation of the number of molecules during diffusion.*

The constant of proportionality D is called the "coefficient of self-diffusion" of the substance. If $\partial n_1/\partial z > 0$, the flow of labeled particles is in the $-z$ direction so as to equalize the concentration, i.e., $J_z < 0$. Hence the minus sign was introduced explicitly in (12·5·2) to make D a positive quantity. The relation (12·5·2) is found to describe quite adequately the self-diffusion* of molecules in gases, liquids, or isotropic solids.

It is useful to point out that the quantity n_1 satisfies, by virtue of the relation (12·5·2), a simple differential equation. Consider a one-dimensional problem where $n_1(z,t)$ is the mean number of labeled molecules per unit volume located at time t near the position z. Focus attention on a slab of substance of thickness dz and of area A. Since the total number of labeled molecules is conserved, one can make the statement that the [increase per unit time in the number of labeled molecules contained within the slab] must be equal to [the number of labeled molecules entering the slab per unit time through its surface at z] minus [the number of labeled molecules leaving the slab per unit time through its surface at $z + dz$]. In symbols,

$$\frac{\partial}{\partial t}(n_1 A\,dz) = AJ_z(z) - AJ_z(z+dz)$$

or

$$\frac{\partial n_1}{\partial t}\,dz = J_z(z) - \left[J_z(z) + \frac{\partial J_z}{\partial z}\,dz\right]$$

Thus

$$\frac{\partial n_1}{\partial t} = -\frac{\partial J_z}{\partial z} \tag{12·5·3}$$

This equation expresses just the conservation of the number of labeled molecules. Using the relation (12·5·2), this becomes

$$\blacktriangleright \qquad \frac{\partial n_1}{\partial t} = D\,\frac{\partial^2 n_1}{\partial z^2} \tag{12·5·4}$$

This is the desired partial differential equation, the "diffusion equation," satisfied by $n_1(z,t)$.

Calculation of the coefficient of self-diffusion for a dilute gas In the simple case of a dilute gas, the coefficient of self-diffusion can readily be calculated by mean-free-path arguments similar to those used in the last two sections. Consider a plane $z =$ constant in the gas. The mean number of

* One speaks of *self*-diffusion if the diffusing molecules are, except for being labeled, identical to the remaining molecules of the substance. The more general and complicated situation would be that of *mutual* diffusion where the molecules are *unlike*, e.g., the diffusion of He molecules in argon gas.

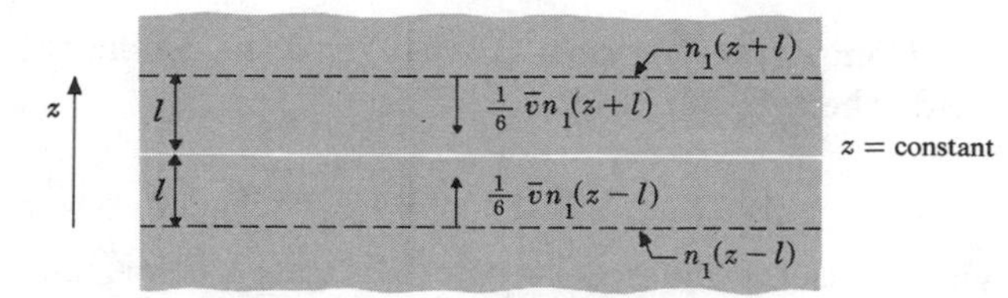

Fig. 12·5·2 Transport of labeled molecules across a plane.

labeled molecules which in unit time cross a unit area of this plane from below is equal to $\frac{1}{6}\bar{v}n_1(z-l)$; the mean number of labeled molecules which in unit time cross a unit area of this plane from above is $\frac{1}{6}\bar{v}n_1(z+l)$. Hence one obtains for the net flux of labeled molecules crossing the plane from below in the $+z$ direction

$$J_z = \tfrac{1}{6}\bar{v}n_1(z-l) - \tfrac{1}{6}\bar{v}n_1(z+l)$$
$$= \frac{1}{6}\bar{v}[n_1(z-l) - n_1(z+l)] = \frac{1}{6}\bar{v}\left(-2\frac{\partial n_1}{\partial z}l\right)$$

or
$$J_z = -D\frac{\partial n_1}{\partial z} \tag{12·5·5}$$

where

$$D = \tfrac{1}{3}\bar{v}l \tag{12·5·6}$$

Thus (12·5·5) shows explicitly that J_z is proportional to the concentration gradient (in accordance with the general relation (12·5·2)), and (12·5·6) gives an approximate expression for the coefficient of self-diffusion in terms of fundamental molecular quantities.

To express D in more explicit form one need only use the relations

$$l = \frac{1}{\sqrt{2}\,n\sigma_0} = \frac{1}{\sqrt{2}\,\sigma_0}\frac{kT}{\bar{p}} \tag{12·5·7}$$

and
$$\bar{v} = \sqrt{\frac{8}{\pi}\frac{kT}{m}} \tag{12·5·8}$$

Thus
$$D = \frac{2}{3\sqrt{\pi}}\frac{1}{\bar{p}\sigma_0}\sqrt{\frac{(kT)^3}{m}} \tag{12·5·9}$$

Hence the coefficient of self-diffusion D *does* depend on the pressure of the gas. At a fixed temperature T,

$$D \propto \frac{1}{n} \propto \frac{1}{\bar{p}} \tag{12·5·10}$$

Also at a fixed pressure,

$$D \propto T^{\frac{3}{2}} \tag{12·5·11}$$

if the scattering is like that between hard spheres so that σ_0 is a constant independent of T.

By virtue of (12·5·6), the order of magnitude of D at room temperature and atmospheric pressure is $\frac{1}{3}\bar{v}l \approx \frac{1}{3}(5 \times 10^4)(3 \times 10^{-5}) \approx 0.5$ cm² sec⁻¹. The experimentally measured value for N_2 gas at 273°K and 1 atmosphere pressure is 0.185 cm² sec⁻¹.

Comparison between (12·5·6) and the coefficient of viscosity η in (12·3·7) yields the relation

$$\frac{D}{\eta} = \frac{1}{nm} = \frac{1}{\rho} \tag{12·5·12}$$

where ρ is the mass density of the gas. Experimentally, one finds that the ratio $(D\rho/\eta)$ lies in the range between 1.3 and 1.5 instead of being unity as predicted by (12·5·12). In view of the crude nature of our simple calculations this extent of agreement between theory and experiment can be regarded as quite satisfactory.

Diffusion regarded as a random walk problem It is possible to look upon the diffusion problem as a random walk executed by the labeled molecule. Assume that successive displacements suffered by the molecule between collisions are statistically independent and denote by ζ_i the z component of the ith displacement of the molecule. If the molecule starts at $z = 0$, the z component of its position vector after a total of N displacements is then given by

$$z = \sum_{i=1}^{N} \zeta_i \tag{12·5·13}$$

We calculate mean values as in Sec. 1·9. By virtue of the random direction of each displacement, $\overline{\zeta_i} = 0$ so that $\bar{z} = 0$. On the other hand, one obtains for the dispersion

$$\overline{z^2} = \sum_i \overline{\zeta_i^2} + \sum_i \sum_{\substack{j \\ i \neq j}} \overline{\zeta_i \zeta_j} \tag{12·5·14}$$

Now, by virtue of the statistical independence, $\overline{\zeta_i \zeta_j} = \overline{\zeta_i}\,\overline{\zeta_j} = 0$ so that (12·5·14) reduces simply to

$$\overline{z^2} = N\overline{\zeta^2} \tag{12·5·15}$$

The mean-square displacement $\overline{\zeta^2}$ per step can readily be computed. The z component of this displacement in time t is $\zeta = v_z t$. Hence

$$\overline{\zeta^2} = \overline{v_z^2}\,\overline{t^2}$$

But, by symmetry, $\overline{v_z^2} = \frac{1}{3}\overline{v^2}$. Furthermore, one has by (12·1·10)

$$\overline{t^2} = \int_0^\infty e^{-t/\tau} \frac{dt}{\tau} \cdot t^2 = \tau^2 \int_0^\infty e^{-u}\, u^2\, du = 2\tau^2$$

Hence

$$\zeta^2 = \tfrac{2}{3}\overline{v^2}\tau^2 \tag{12·5·16}$$

Since each displacement between collisions requires a mean time τ, the total number N of displacements suffered in a time t is equal to t/τ. Hence (12·5·15) yields for the mean-square z component of displacement of a molecule in time t the result

► $$\overline{z^2(t)} = (\tfrac{2}{3}\overline{v^2}\tau)\, t \tag{12·5·17}$$

On the other hand, one can also calculate the mean-square displacement $\overline{z^2(t)}$ by purely macroscopic reasoning based on the diffusion equation (12·5·4).

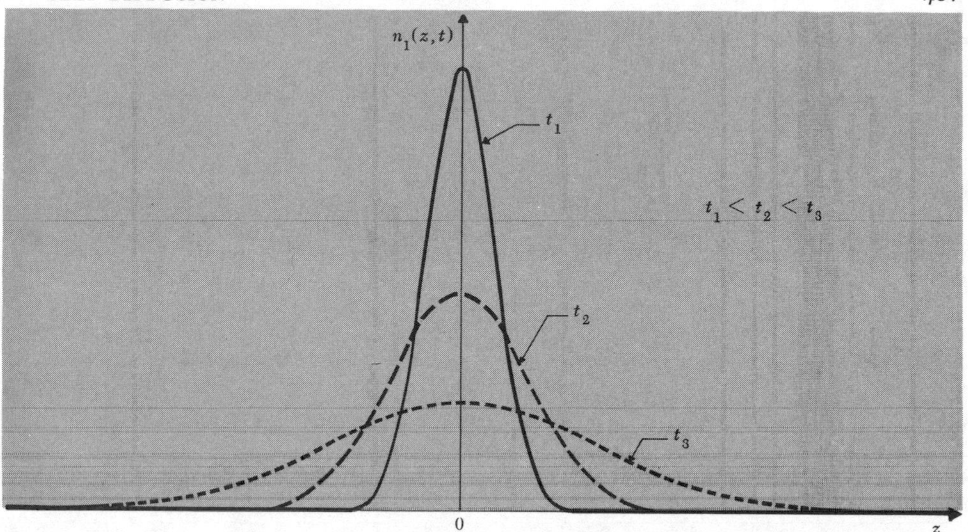

***Fig. 12·5·3** The number density $n_1(z,t)$ as a function of z at various times t after molecules are introduced at time $t = 0$ near the plane $z = 0$. The areas under all the curves are the same and equal to the total number N_1 of labeled molecules.*

Imagine that a total of N_1 labeled molecules per unit area are introduced at time $t = 0$ in an infinitesimally thick slab near $z = 0$. The molecules then proceed to diffuse (see Fig. 12·5·3). Conservation of the total number of labeled molecules requires that

$$\int_{-\infty}^{\infty} n_1(z,t)\, dz = N_1 \tag{12·5·18}$$

at all times. By definition one also has

$$\overline{z^2(t)} = \frac{1}{N_1} \int_{-\infty}^{\infty} z^2 n_1(z,t)\, dz \tag{12·5·19}$$

To find how $\overline{z^2}$ depends on t, multiply the diffusion equation (12·5·4) by z^2 and integrate over z. This yields

$$\int_{-\infty}^{\infty} z^2 \frac{\partial n_1}{\partial t}\, dz = D \int_{-\infty}^{\infty} z^2 \frac{\partial^2 n_1}{\partial z^2}\, dz \tag{12·5·20}$$

The left side gives, by (12·5·19),

$$\int_{-\infty}^{\infty} z^2 \frac{\partial n_1}{\partial t}\, dz = \frac{\partial}{\partial t} \int_{-\infty}^{\infty} z^2 n_1\, dz = N_1 \frac{\partial}{\partial t} \overline{(z^2)}$$

The right side of (12·5·20) can be simplified by successive integrations by part

$$\begin{aligned}\int_{-\infty}^{\infty} z^2 \frac{\partial^2 n_1}{\partial z^2}\, dz &= \left[z^2 \frac{\partial n_1}{\partial z} \right]_{-\infty}^{\infty} - 2 \int_{-\infty}^{\infty} z \frac{\partial n_1}{\partial z}\, dz \\ &= 0 - 2[z n_1]_{-\infty}^{\infty} + 2 \int_{-\infty}^{\infty} n_1\, dz \\ &= 0 + 2N_1\end{aligned}$$

since n_1 and $(\partial n_1/\partial z) \to 0$ as $|z| \to \infty$. Thus (12·5·20) becomes

$$\frac{\partial}{\partial t}\,\overline{(z^2)} = 2D \qquad (12\cdot5\cdot21)$$

or

$$\overline{z^2} = 2Dt \qquad (12\cdot5\cdot22)$$

where the constant of integration has been set equal to zero, since $\overline{z^2} = 0$ for $t = 0$ by virtue of the initial condition that all molecules start at $z = 0$.

By comparing (12·5·22) with the random walk result (12·5·17) one obtains

$$D = \tfrac{1}{3}\overline{v^2}\tau \qquad (12\cdot5\cdot23)$$

or

$$D = \tfrac{1}{3}\bar{v}l \qquad (12\cdot5\cdot24)$$

if one neglects the distinction between $\overline{v^2}$ and $\bar{v}^2$ and sets $\bar{v}\tau = l$. Thus one regains the result (12·5·6).

12 · 6 *Electrical conductivity*

Consider a system (liquid, solid, or gas) containing charged particles which are free to move. If a small uniform electric field $\mathcal{E}$ is applied in the z direction, a nonequilibrium situation results in which an electric current density j_z is set up in this direction. By definition

j_z = the mean electric charge crossing a unit area (perpendicular to the z direction) per unit time. (12·6·1)

If the electric field $\mathcal{E}$ is sufficiently small, one expects that

$$j_z = \sigma_{el}\,\mathcal{E} \qquad (12\cdot6\cdot2)$$

where the constant of proportionality σ_{el} is called the "electrical conductivity" of the system. The relation (12·6·2) is called "Ohm's law."

Consider now a dilute gas of particles having mass m and charge e and interacting with some other system of particles by which they can get scattered with collision time τ. A particularly simple case would be that of a relatively small number of ions (or electrons) in a gas where these ions are predominantly scattered by collisions with the neutral gas molecules.

Remark Another example would be that of the conduction electrons in a metal where the electrons are scattered by the atoms of the solid. This case involves, however, some subtleties by virtue of the Fermi-Dirac statistics obeyed by the electrons. It will be considered more specifically in Chapter 13.

When an electric field $\mathcal{E}$ is applied in the z direction, it gives rise to a mean z component of velocity $\bar{v}_z$ of the charged particles. The mean number of such particles crossing a unit area (perpendicular to the z direction) per unit time is then given by $n\bar{v}_z$ if n is the mean number of charged particles per unit volume. Since each particle carries a charge e, one thus obtains

$$j_z = ne\bar{v}_z \tag{12·6·3}$$

It only remains to calculate $\bar{v}_z$. Let us measure time from the instant $t = 0$ immediately after the particle's last collision. The equation of motion of the particle between this collision and the next one is

$$m\frac{dv_z}{dt} = e\mathcal{E}$$

Hence

$$v_z = \frac{e\mathcal{E}}{m}t + v_z(0) \tag{12·6·4}$$

In order to calculate the mean value $\bar{v}_z$, we assume that as a result of each collision the particle is, at least on the average, restored to thermal equilibrium; its velocity $\mathbf{v}(0)$ has then random direction and $\bar{v}_z(0) = 0$ irrespective of the particle's past history before that collision.* Taking the mean value of (12·6·4) over all possible times t between collisions, as given by the probability (12·1·10), we then obtain

$$\bar{v}_z = \frac{e\mathcal{E}}{m}\bar{t} = \frac{e}{m}\int_0^\infty e^{-t/\tau}\frac{dt}{\tau}t = \frac{e\mathcal{E}}{m}\tau \tag{12·6·5}$$

that is, we have just used the familiar result that the mean time $\bar{t}$ between collisions is equal to τ. We have also treated the collision probability τ^{-1} per unit time as a constant, even though it may depend on the particle speed. This is justified because the electric field $\mathcal{E}$ was assumed to be sufficiently small that the increment in the particle's speed produced by $\mathcal{E}$ between successive collisions is negligibly small compared to the thermal speed of the particles.

By using (12·6·5) in (12·6·3), one obtains

$$j_z = \frac{ne^2}{m}\tau\mathcal{E} \tag{12·6·6}$$

or

$$j_z = \sigma_{el}\,\mathcal{E} \tag{12·6·7}$$

where

► $$\sigma_{el} = \frac{ne^2}{m}\tau \tag{12·6·8}$$

Thus j_z is indeed proportional to $\mathcal{E}$ (as expected by (12·6·2)) and (12·6·8) provides an explicit expression for the electrical conductivity σ_{el} in terms of microscopic parameters of the gas.

If the conductivity is due to a relatively small number of ions in a gas, the collisions limiting the mean free path are predominantly those between

* One can expect this to be a very good approximation if the charged particle suffers collisions with particles of much larger mass. Otherwise the charged particle retains after each collision some memory of the z component of velocity it had before that collision. For the time being we neglect corrections due to such "persistence-of-velocity" effects.

ions and neutral gas molecules.* Suppose that the total scattering cross section of an ion by a molecule is σ_{im} and that there are, per unit volume, n_1 molecules of mass $m_1 \gg m$. The thermal speed of the ions is then much greater than that of the molecules, and the mean relative speed of an ion-molecule encounter is simply the mean ion speed $\bar{v}$. Thus the collision rate of an ion is approximately equal to

$$\tau^{-1} \approx n_1 \bar{v} \sigma_{\text{im}} = n_1 \left(\frac{8}{\pi}\frac{kT}{m}\right)^{\frac{1}{2}} \sigma_{\text{im}}$$

and

$$\sigma_{\text{el}} \approx \sqrt{\frac{\pi}{8}} \frac{ne^2}{n_1 \sigma_{\text{im}} \sqrt{mkT}} \tag{12·6·9}$$

SUGGESTIONS FOR SUPPLEMENTARY READING

R. D. Present: "Kinetic Theory of Gases," chap. 3, McGraw-Hill Book Company, New York, 1958. (See also secs. 8·1–8·2 on collision dynamics.)

J. F. Lee, F. W. Sears, and D. L. Turcotte: "Statistical Thermodynamics," chap. 4, Addison-Wesley Publishing Company, Reading, Mass., 1963.

C. Kittel: "Introduction to Solid State Physics," 2d ed., pp. 138–153, John Wiley & Sons, Inc., New York, 1956. (Applications to the thermal conductivity of solids.)

PROBLEMS

12.1 A large number of throws are made with a single die.

(*a*) What is the mean number of throws between the appearances of a six?

At any stage of the process what is the mean number of throws

(*b*) before the next appearance of a six;

(*c*) since the last appearance of a six?

12.2 Let l denote the mean free path of a molecule in a gas. Suppose that such a molecule has just suffered a collision. What is the mean distance

(*a*) it travels before it suffers the next collision;

(*b*) it has traveled since it suffered the last collision?

(*c*) What is the mean distance traveled by the molecule between two successive collisions?

12.3 An ion of mass m and electric charge e is moving in a dilute gas of molecules with which it collides. The mean time between collisions suffered by the ion is τ. Suppose that a uniform electric field $\mathcal{E}$ is applied in the x direction.

(*a*) What is the mean distance $\bar{x}$ (in the direction of $\mathcal{E}$) which the ion travels between collisions if it starts out with zero x component of velocity after each collision?

(*b*) In what fraction of cases does the ion travel a distance x less than $\bar{x}$?

12.4 Calculate the differential scattering cross section σ for the scattering of a hard

* Actually, even if ion-ion collisions occurred frequently, they would not affect the electrical conductivity, since the colliding ions would, effectively, simply exchange roles in carrying the electric current. This will be shown in greater detail in Sec. 14·6.

sphere of radius a_1 by a stationary hard sphere of radius a_2. How does the result depend on the angle of scattering θ'? (Use classical mechanics.)

12.5 The total scattering cross section for an electron-air molecule collision is about 10^{-15} cm^2. At what gas pressure will 90 percent of the electrons emitted from a cathode reach an anode 20 cm away? (Assume that any electron scattered out of the beam does not reach the anode; i.e., neglect multiple scattering.)

12.6 Estimate the magnitude of the coefficient of viscosity η of argon gas at 25°C and 1 atmosphere pressure. To estimate the dimension of an argon atom, consider the atoms as hard spheres which, in the solid at low temperatures, form a close-packed structure having a density of 1.65 gm/cm^3. The atomic weight of Ar is 39.9. Compare your estimate with the experimentally observed value of $\eta = 2.27 \times 10^{-4}$ gm cm^{-1} sec^{-1}.

12.7 In the Millikan oil-drop experiment, the terminal velocity with which the oil drop falls is inversely proportional to the viscosity of the air. If the temperature of the air increases, does the terminal velocity of the drop increase, decrease, or remain the same? What happens when the atmospheric pressure increases?

12.8 It is desired to measure the coefficient of viscosity η of air at room temperature, since this parameter is essential for determining the electronic charge by Millikan's oil-drop experiment. It is proposed to perform the measurement in a viscometer consisting of a stationary inner cylinder (of radius R and length L) supported by a torsion fiber, and an outer cylinder (of slightly larger inner radius $R + \delta$) rotating slowly with angular velocity ω. The narrow annular region of thickness $\delta (\delta \ll R)$ is filled with the gas under consideration and one measures the torque G on the inner cylinder.

(*a*) Find the torque G in terms of η and the parameters of this experimental apparatus.

(*b*) To determine what quartz fiber is needed, estimate the magnitude of the viscosity of air from first principles, and use this result to estimate the magnitude of the torque which has to be measured in an apparatus of this kind. Take as dimensions $R = 2$ cm, $\delta = 0.1$ cm, $L = 15$ cm, and $\omega = 2\pi$ radians/second.

12.9 Suppose that the molecules of a gas interact with each other through a radial force F which depends on the intermolecular separation R according to $F = CR^{-s}$, where s is some positive integer and C is a constant.

(*a*) Use arguments of dimensional analysis to show how the total scattering cross section σ_0 of the molecules depends on their relative speed V. Assume a classical calculation so that σ_0 can only depend on V, the molecular mass m, and the force constant C.

(*b*) How does the coefficient of viscosity η of this gas depend on the absolute temperature T?

12.10 A fluid of viscosity η flows through a tube of length L and radius a as a result of a pressure difference, the pressure being p_1 at one end and p_2 at the other end of the tube. Write the conditions necessary to ensure that a cylinder of fluid of radius r moves without acceleration under the influence of the pressure difference and the shearing force due to the viscosity of the liquid. Hence derive an expression for the mass $\dot{M}$ of fluid flowing per second through the tube for the following two cases:

(*a*) The fluid is an incompressible liquid of density ρ.

(*b*) The fluid is an ideal gas of molecular weight μ.

(These results are Poiseuille's flow formulas.) Assume that the layer of fluid in contact with the walls of the tube is at rest. Note also that the same mass of fluid must cross any cross-sectional area of the tube per unit time.

12.11 Consider a general situation where the temperature T of a substance is a function of the time t and the spatial coordinate z. The density of the substance is ρ, its specific heat per unit mass is c, and its thermal conductivity is κ. By macroscopic reasoning similar to that used in deriving the diffusion equation (12·5·4), obtain the general partial differential equation which must be satisfied by the temperature $T(z,t)$.

12.12 A long cylindrical wire of radius a and electrical resistance R per unit length is stretched along the axis of a long cylindrical container of radius b. This container is maintained at a fixed temperature T_0 and is filled with a gas having a thermal conductivity κ. Calculate the temperature difference ΔT between the wire and the container walls when a small constant electrical current I is passed through the wire, and show that a measurement of ΔT provides a means for determining the thermal conductivity of the gas. Assume that a steady-state condition has been reached so that the temperature T at any point has become independent of time. (Suggestion: Consider the condition which must be satisfied by any cylindrical shell of the gas contained between radius r and radius $r + dr$.)

12.13 Consider a cylindrical dewar vessel (e.g., thermos bottle) of the usual double-walled construction. The outer diameter of the inner wall is 10 cm, the inner diameter of the outer wall is 10.6 cm. The dewar contains a mixture of ice and water; the outside of the dewar is at room temperature, i.e., at about 25°C.

(*a*) If the space between the two walls of the dewar contains He gas at atmospheric pressure, calculate approximately the heat influx (in watts per cm height of the dewar) due to heat conduction by the gas. (A reasonable estimate for the radius of a helium atom is about 10^{-8} cm.)

(*b*) Estimate to what value (in mm Hg) the pressure of the gas between the walls must be reduced before the heat influx due to conduction is reduced below the value calculated in part (*a*) by a factor of 10.

12.14 The coefficient of viscosity of He gas at $T = 273°\text{K}$ and 1 atmosphere is η_1, that of Ar gas is η_2. The atomic weights of these gases are μ_1 and μ_2, respectively.

(*a*) What is the ratio σ_2/σ_1 of the Ar-Ar atom total scattering cross section σ_2 as compared to that of the He-He atom total scattering cross section σ_1?

(*b*) What is the ratio κ_2/κ_1 of the thermal conductivity κ_2 of argon gas compared to the thermal conductivity κ_1 of He gas when $T = 273°\text{K}$?

(*c*) What is the ratio D_2/D_1 of the diffusion coefficients of these gases when $T = 273°\text{K}$?

(*d*) The atomic weights of He and A are respectively $\mu_1 = 4$ and $\mu_2 = 40$. The measured viscosities at 273°K are respectively $\eta_1 = 1.87 \times 10^{-4}$ and $\eta_2 = 2.105 \times 10^{-4}$ gm cm^{-1} sec^{-1}. Use this information to calculate approximate values of the cross sections σ_1 and σ_2.

(*e*) If the atoms are considered to scatter like hard spheres, estimate the diameter d_1 of a He atom and the diameter d_2 of an Ar atom.

12.15 It is desired to do an experiment on an isotopic mixture of N_2 gas. For this purpose one takes a spherical storage vessel, 1 meter in diameter, containing N_2^{14} gas at room temperature and atmospheric pressure, and introduces through a valve at one side of the container a small amount of N_2^{15} gas. In the absence of any convection in the gas, make a *rough* estimate of how long one has to wait before one can be reasonably sure that the N_2^{14} and N_2^{15} molecules are uniformly mixed throughout the container.

12.16 A satellite, in the form of a cube of edge length L, moves through outer space with a velocity V parallel to one of its edges. The surrounding gas consists of molecules of mass m at a temperature T, the number n of molecules per unit volume being very small, so that the mean free path of the molecules is much larger than L. Assuming that collisions of the molecules with the satellite are elastic, calculate the mean retarding force exerted on the satellite by the interplanetary gas. You can assume that V is small compared to the mean speed of the gas molecules. If the mass of the satellite is M and it is not subject to external forces, after how long a time will the velocity of the satellite be reduced to half its original value?

13 Transport theory using the relaxation-time approximation

IN THE preceding chapter we presented a highly simplified discussion of nonequilibrium transport processes in dilute gases. Although this discussion was valuable and illuminated the main physical processes at play, it was quite crude; e.g., no attempt whatever was made to take into account the distribution of molecular velocities. In this chapter we shall treat the theory of transport phenomena from a somewhat more sophisticated point of view. Our main aim will be to gain insight into how the velocity distribution of the molecules is actually modified when nonequilibrium conditions prevail. We shall then use this information to investigate the resulting transport processes. Up to a point we shall thus set up the problem quite correctly, taking proper account of the distribution of molecular velocities and acquiring a better understanding of the significant parameters involved. Nevertheless, we shall still use some fairly drastic approximations in order to avoid the complications involved in a detailed analysis of molecular collisions. The resulting theoretical approach is thus still fairly simple and very useful in practice, especially in more complicated problems. On the one hand it has the advantage, compared to the most elementary theory, that it provides a systematic way of formulating a problem and that it isolates clearly the essential assumptions in a calculation; on the other hand, it is often tractable even when the more rigorous theory proves excessively difficult. In the next chapter we shall then show how this theoretical approach can be extended so as to yield a more nearly exact theory.

13 · 1 *Transport processes and distribution functions*

The general theory of transport processes is based on the following observation. Suppose that one knows for a given situation, which is in general *not* an equilibrium situation, the *actual* molecular distribution function $f(\mathbf{r},\mathbf{v},t)$. As usual, this is defined so that

$$f(\boldsymbol{r},\boldsymbol{v},t)\,d^3\boldsymbol{r}\,d^3\boldsymbol{v} \equiv \left.\begin{array}{l}\text{the mean number of molecules whose center of mass at time } t \text{ is located between } \boldsymbol{r} \text{ and } \boldsymbol{r}+d\boldsymbol{r} \text{ and}\\ \text{has a velocity between } \boldsymbol{v} \text{ and } \boldsymbol{v}+d\boldsymbol{v}.\end{array}\right\} \tag{13·1·1}$$

The function $f(\boldsymbol{r},\boldsymbol{v},t)$ provides a complete description of the macroscopic state of the dilute gas (neglecting possible nonequilibrium perturbations of the internal degrees of freedom of the molecules) and should therefore permit calculation of all quantities of physical interest, e.g., of viscosity coefficients or thermal conductivities. Hence any transport problem can be solved by attempting to calculate the actual distribution function $f(\boldsymbol{r},\boldsymbol{v},t)$ for the physical situation of interest.

To elaborate these comments, let us point out a few general relations involving the distribution function $f(\boldsymbol{r},\boldsymbol{v},t)$. Let

$$n(\boldsymbol{r},t)\,d^3\boldsymbol{r} \equiv \left.\begin{array}{l}\text{the mean number of molecules (irrespective of}\\ \text{velocity) which at time } t \text{ are located between } \boldsymbol{r} \text{ and } \boldsymbol{r}+d\boldsymbol{r}.\end{array}\right\} \tag{13·1·2}$$

Then one has by the definition (13·1·1)

$$n(\boldsymbol{r},t) = \int d^3\boldsymbol{v} f(\boldsymbol{r},\boldsymbol{v},t) \tag{13·1·3}$$

where the integration is over all possible velocities. Furthermore, let $\chi(\boldsymbol{r},\boldsymbol{v},t)$ be any function that denotes a property of a molecule located at time t near $\boldsymbol{r}$ with a velocity near $\boldsymbol{v}$. For example, χ might denote the energy ϵ of the molecule; or it might denote a vector quantity like the momentum $\boldsymbol{p}$ of the molecule. The mean value of χ at time t at the position $\boldsymbol{r}$ will be denoted interchangeably either by a bar or by angular brackets and is defined by

$$\langle\chi(\boldsymbol{r},t)\rangle \equiv \bar{\chi}(\boldsymbol{r},t) \equiv \frac{1}{n(\boldsymbol{r},t)}\int d^3\boldsymbol{v}\, f(\boldsymbol{r},\boldsymbol{v},t)\chi(\boldsymbol{r},\boldsymbol{v},t) \tag{13·1·4}$$

In particular, the mean velocity $\boldsymbol{u}(\boldsymbol{r},t)$ of a molecule at position $\boldsymbol{r}$ and time t is defined by

$$\boldsymbol{u}(\boldsymbol{r},t) \equiv \langle\boldsymbol{v}(\boldsymbol{r},t)\rangle = \frac{1}{n(\boldsymbol{r},t)}\int d^3\boldsymbol{v}\, f(\boldsymbol{r},\boldsymbol{v},t)\boldsymbol{v} \tag{13·1·5}$$

This velocity $\boldsymbol{u}(\boldsymbol{r},t)$ describes the mean velocity of flow of the gas at a given point. (This is just the "hydrodynamic velocity" of the fluid described by macroscopic hydrodynamics.) It is useful to measure the velocity $\boldsymbol{v}$ of a molecule with respect to this mean velocity. We shall therefore define the "peculiar velocity" $\boldsymbol{U}$ of a molecule by the relation

$$\boldsymbol{U} \equiv \boldsymbol{v} - \boldsymbol{u} \tag{13·1·6}$$

Thus it follows, by the definition (13·1·5), that

$$\langle\boldsymbol{U}\rangle = \langle\boldsymbol{v}\rangle - \boldsymbol{u} = 0 \tag{13·1·7}$$

In considering transport phenomena, one is interested in calculating the fluxes of various quantities. Consider at time t and at the point $\boldsymbol{r}$ an infinitesimal element of area dA whose normal is denoted by the unit vector $\hat{\boldsymbol{n}}$. This element of area divides the gas into two regions, a (+) region on the side

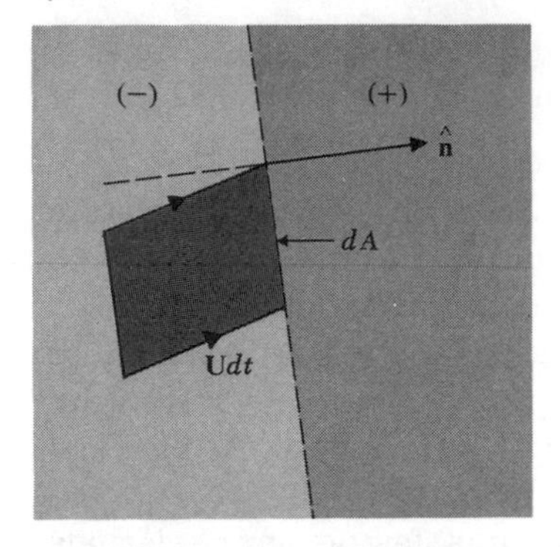

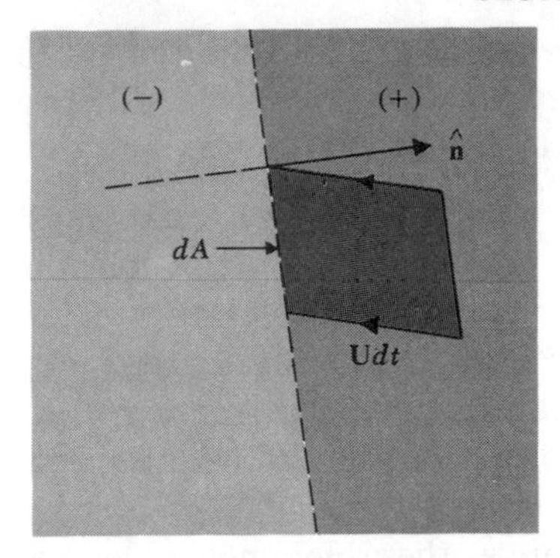

Fig. 13·1·1 The element of area dA with normal n̂ divides the gas into a (+) and (−) region and moves with velocity u. The figure illustrates molecules crossing the element of area in time dt from the (−) to the (+) side (left diagram), and from the (+) to the (−) side (right diagram).

toward which $\hat{\boldsymbol{n}}$ points and a (−) region on the other side. If the *mean* molecular velocity $\boldsymbol{u}(\boldsymbol{r},t)$ does not vanish, we imagine that the element of area moves along with the fluid, i.e., that it moves without change of orientation with the velocity $\boldsymbol{u}(\boldsymbol{r},t)$. Because of their random velocities $\boldsymbol{U}$ about the mean velocity $\boldsymbol{u}$, molecules move back and forth across this element of area. Each such molecule carries the property χ with it. One then defines $\mathfrak{F}_n(\boldsymbol{r},t)$, the $\hat{\boldsymbol{n}}$ component of the flux of χ across this element of area, by the statement that

$$\left.\begin{array}{l}\mathfrak{F}_n(\boldsymbol{r},t) \equiv \text{the net amount of } \chi \text{ which is transported per unit} \\ \text{time per unit area of an element of area (oriented with its} \\ \text{normal along } \hat{\boldsymbol{n}}) \text{ from its } (-) \text{ to its } (+) \text{ side.}\end{array}\right\} \qquad (13\cdot1\cdot8)$$

Referring to Fig. 13·1·1, it is easy to calculate this flux by familiar methods. Consider molecules at time t located near $\boldsymbol{r}$ with a velocity near $\boldsymbol{v}$. Their velocity relative to dA is given by $\boldsymbol{U} = \boldsymbol{v} - \boldsymbol{u}$. If $U_n \equiv \hat{\boldsymbol{n}} \cdot \boldsymbol{U} > 0$, these molecules will cross the element of area from the (−) to the (+) side. The number of such molecules crossing dA in the infinitesimal time dt is the number of such molecules contained in the infinitesimal cylinder of area dA and of length $|\boldsymbol{U}\,dt|$ i.e., of corresponding height $|\hat{\boldsymbol{n}} \cdot \boldsymbol{U}\,dt|$. Since its volume is $|\hat{\boldsymbol{n}} \cdot \boldsymbol{U}\,dt\,dA|$, it contains thus $f(\boldsymbol{r},\boldsymbol{v},t)\,d^3\boldsymbol{v}\,|\hat{\boldsymbol{n}} \cdot \boldsymbol{U}\,dt\,dA|$ such molecules. Since each such molecule carries the property $\chi(\boldsymbol{r},\boldsymbol{v},t)$, the *total* amount of χ carried by the molecules across the area dA in time dt from the (−) to the (+) side is given by

$$\int_{\hat{\boldsymbol{n}} \cdot \boldsymbol{U}>0} f(\boldsymbol{r},\boldsymbol{v},t)\,d^3\boldsymbol{v}\,|\hat{\boldsymbol{n}} \cdot \boldsymbol{U}\,dt\,dA|\,\chi(\boldsymbol{r},\boldsymbol{v},t) \qquad (13\cdot1\cdot9)$$

where the integration is over all velocities $\boldsymbol{v}$ for which $\hat{\boldsymbol{n}} \cdot \boldsymbol{U} > 0$.

The analysis for molecules crossing dA from the (+) to the (−) side is similar. These are the molecules for which $U_n \equiv \hat{\boldsymbol{n}} \cdot \boldsymbol{U} < 0$. The total amount of χ carried by these molecules across the area dA in time dt from the (+) to the (−) side is then given by the corresponding integral

$$\int_{\hat{\boldsymbol{n}} \cdot \boldsymbol{U}<0} f(\boldsymbol{r},\boldsymbol{v},t)\,d^3\boldsymbol{v}\,|\hat{\boldsymbol{n}} \cdot \boldsymbol{U}\,dt\,dA|\,\chi(\boldsymbol{r},\boldsymbol{v},t) \qquad (13\cdot1\cdot10)$$

To obtain the *net* flux of χ from the $(-)$ to the $(+)$ side one needs then only subtract (13·1·10) from (13·1·9) and to divide by $dt\,dA$. Thus

$$\mathfrak{F}_n(\mathbf{r},t) = \int_{\hat{\mathbf{n}}\cdot\mathbf{U}>0} d^3\mathbf{v}\, f|\hat{\mathbf{n}}\cdot\mathbf{U}|\,\chi - \int_{\hat{\mathbf{n}}\cdot\mathbf{U}<0} d^3\mathbf{v}\, f|\hat{\mathbf{n}}\cdot\mathbf{U}|\,\chi \qquad (13\cdot1\cdot11)$$

In the first integral $|\hat{\mathbf{n}}\cdot\mathbf{U}| = \hat{\mathbf{n}}\cdot\mathbf{U}$ since $\hat{\mathbf{n}}\cdot\mathbf{U}$ is positive. But in the second integral $|\hat{\mathbf{n}}\cdot\mathbf{U}| = -\hat{\mathbf{n}}\cdot\mathbf{U}$ since $\hat{\mathbf{n}}\cdot\mathbf{U}$ is negative. Hence the integrand in the second integral assumes a net positive sign, and the two integrals combine to give a single integral over *all* possible velocities:

$$\mathfrak{F}_n(\mathbf{r},t) = \int d^3\mathbf{v}\; f\hat{\mathbf{n}}\cdot\mathbf{U}\chi \qquad (13\cdot1\cdot12)$$

In terms of the definition (13·1·4) this can be written

$$\mathfrak{F}_n(\mathbf{r},t) = n\,\langle\hat{\mathbf{n}}\cdot\mathbf{U}\chi\rangle \qquad (13\cdot1\cdot13)$$

Thus $\mathfrak{F}_n$ can be regarded as the $\hat{\mathbf{n}}$ component of a flux *vector* $\boldsymbol{\mathfrak{F}}$ such that

$$\mathfrak{F}_n = \hat{\mathbf{n}}\cdot\boldsymbol{\mathfrak{F}} \qquad (13\cdot1\cdot14)$$

where

$$\boldsymbol{\mathfrak{F}} = n\langle\mathbf{U}\chi\rangle \qquad (13\cdot1\cdot15)$$

Examples In calculating the viscosity of a gas (as we did in Sec. 12·3), one is interested in finding $P_{z\alpha}$, the α component of the mean stress exerted, on a unit area of surface with normal oriented along the z axis, by the fluid below this surface on the fluid above this surface. The corresponding rate of change of momentum is given by the net flux of α component of molecular momentum transported from below to above the surface. The quantity transported is thus $\chi = mv_\alpha$, while $\hat{\mathbf{n}}\cdot\mathbf{U} = U_z$. Hence this stress, or momentum flux, is by (13·1·13)

$$P_{z\alpha} = nm\langle U_z v_\alpha\rangle \qquad (13\cdot1\cdot16)$$

This can also be written in the form

$$P_{z\alpha} = nm\langle U_z(u_\alpha + U_\alpha)\rangle = nm[u_\alpha\langle U_z\rangle + \langle U_z U_\alpha\rangle]$$

or

$$P_{z\alpha} = nm\langle U_z U_\alpha\rangle \qquad (13\cdot1\cdot17)$$

since $u_\alpha(\mathbf{r},t)$ does not depend on $\mathbf{v}$ and $\langle U_z\rangle = 0$.

For instance, in the physical situation illustrated in Fig. 12·3·2, one has $u_x \neq 0$ and $u_z = 0$. Then $U_z = v_z$, and (13·1·16) becomes simply

$$P_{z\alpha} = nm\langle v_z v_\alpha\rangle = m\int d^3\mathbf{v}\, f v_z v_\alpha \qquad (13\cdot1\cdot18)$$

The mean pressure $\bar{p}$ is just the stress exerted normal to the surface, i.e., in the z direction. Hence one has

$$\bar{p} = P_{zz} = nm\langle v_z^2\rangle \qquad (13\cdot1\cdot19)$$

This result agrees with that of Eq. (7·13·7) and is, of course, always a positive quantity. In calculating $P_{z\alpha}$ in (13·1·18) one must use the actual function $f(\mathbf{r},\mathbf{v},t)$ for the nonequilibrium situation. If f were simply the equilibrium distribution function for a gas at rest, f would depend only on $|\mathbf{v}|$. In calculating P_{zx}, the integrand in (13·1·18) would then be an odd function of v_z and v_x so that the integral over all velocities would vanish by symmetry. The existence of a shearing stress $P_{zx} \neq 0$ in the situation of Fig. 12·3·2 is due precisely to the fact that f is *different* from the Maxwellian equilibrium distribution.

Similarly, one can write down the flux of other molecular quantities. For example, consider a monatomic gas at rest so that $\boldsymbol{u} = 0$. The entire energy of a molecule is then its kinetic energy $\frac{1}{2}mv^2$, and the heat flux (or energy flux) $\mathcal{Q}_z$ in the z direction is by (13·1·13) simply

$$\mathcal{Q}_z = n\langle v_z(\tfrac{1}{2}mv^2)\rangle = \tfrac{1}{2}nm\langle v_z v^2\rangle \tag{13·1·20}$$

13 · 2 *Boltzmann equation in the absence of collisions*

In order to find the distribution function $f(\boldsymbol{r},\boldsymbol{v},t)$, we should like to know what relations this function must satisfy. Let us assume that each molecule of mass m is subject to an external force $\boldsymbol{F}(\boldsymbol{r},t)$ which may be due to gravity or electric fields. (For simplicity we assume that $\boldsymbol{F}$ does *not* depend on the velocity $\boldsymbol{v}$ of the molecule. We thus exclude magnetic forces from the present discussion.) We begin by considering the particularly simple situation when interactions between molecules (i.e., collisions) can be completely neglected. What statements can one make about $f(\boldsymbol{r},\boldsymbol{v},t)$ under these circumstances?

Consider the molecules which at time t have positions and velocities in the range $d^3\boldsymbol{r}\,d^3\boldsymbol{v}$ near $\boldsymbol{r}$ and $\boldsymbol{v}$, respectively. At an infinitesimally later time $t' = t + dt$ these molecules will, as a result of their motion under the influence of the force $\boldsymbol{F}$, have positions and velocities in the range $d^3\boldsymbol{r}'\,d^3\boldsymbol{v}'$ near $\boldsymbol{r}'$ and $\boldsymbol{v}'$, respectively. Here

$$\boldsymbol{r}' = \boldsymbol{r} + \dot{\boldsymbol{r}}\,dt = \boldsymbol{r} + \boldsymbol{v}\,dt \tag{13·2·1}$$

and

$$\boldsymbol{v}' = \boldsymbol{v} + \dot{\boldsymbol{v}}\,dt = \boldsymbol{v} + \frac{1}{m}\boldsymbol{F}\,dt \tag{13·2·2}$$

The situation is illustrated schematically in Fig. 13·2·1. In the absence of collisions this is all that happens; i.e., all molecules in the range $d^3\boldsymbol{r}\,d^3\boldsymbol{v}$ near $\boldsymbol{r}$

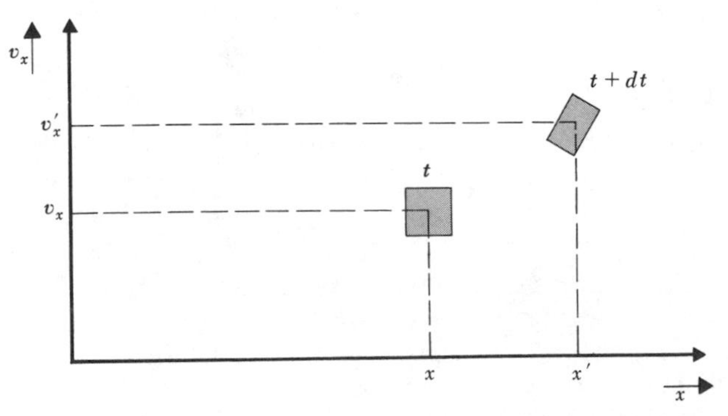

Fig. 13·2·1 ***Figure illustrating the motion of a particle in one dimension in a two-dimensional phase space specified by the particle position x and its velocity v_x.***

and $\boldsymbol{v}$ will, after the time interval dt, be found in the new range $d^3\boldsymbol{r}'\,d^3\boldsymbol{v}'$ near $\boldsymbol{r}'$ and $\boldsymbol{v}'$. In symbols,

$$f(\boldsymbol{r}',\boldsymbol{v}',t')\,d^3\boldsymbol{r}'\,d^3\boldsymbol{v}' = f(\boldsymbol{r},\boldsymbol{v},t)\,d^3\boldsymbol{r}\,d^3\boldsymbol{v} \tag{13·2·3}$$

Remark The element of volume $d^3\boldsymbol{r}\,d^3\boldsymbol{v}$ in the six-dimensional $(\boldsymbol{r},\boldsymbol{v})$ phase space may become distorted in shape as a result of the motion. But its new volume is simply related to the old one by the relation

$$d^3\boldsymbol{r}'\,d^3\boldsymbol{v}' = |J|\,d^3\boldsymbol{r}\,d^3\boldsymbol{v} \tag{13·2·4}$$

where J is the Jacobian of the transformation (13·2·1) and (13·2·2) from the old variables $\boldsymbol{r}$, $\boldsymbol{v}$ to the new variables $\boldsymbol{r}'$, $\boldsymbol{v}'$. The partial derivatives appearing in J are (for the various components $\alpha, \gamma = 1, 2, 3$)

$$\frac{\partial x_\alpha'}{\partial x_\gamma} = \delta_{\alpha\gamma}; \qquad \frac{\partial x_\alpha'}{\partial v_\gamma} = \delta_{\alpha\gamma}\,dt \qquad \text{by (13·2·1)}$$

$$\frac{\partial v_\alpha'}{\partial x_\gamma} = \frac{1}{m}\frac{\partial F_\alpha}{\partial x_\gamma}\,dt; \qquad \frac{\partial v_\alpha'}{\partial v_\gamma} = \delta_{\alpha\gamma} \qquad \text{by (13·2·2)}$$

where we have used the fact that $\boldsymbol{F}$ is independent of $\boldsymbol{v}$. Hence J is given by

$$J = \frac{\partial(x',y',z',v_x',v_y',v_z')}{\partial(x,y,z,v_x,v_y,v_z)} = \begin{vmatrix} 1 & 0 & 0 & dt & 0 & 0 \\ 0 & 1 & 0 & 0 & dt & 0 \\ 0 & 0 & 1 & 0 & 0 & dt \\ \frac{1}{m}\frac{\partial F_x}{\partial x}\,dt & \cdots & \cdots & 1 & 0 & 0 \\ \cdots & \cdots & \cdots & 0 & 1 & 0 \\ \cdots & \cdots & \cdots & 0 & 0 & 1 \end{vmatrix}$$

where all nine terms in the lower left corner of the determinant are proportional to dt. Hence

$$J = 1 + \mathcal{O}(dt^2)$$

so that $J = 1$ is correct up to and including first-order terms in the infinitesimal time interval dt. Hence it follows by (13·2·4) that

$$d^3\boldsymbol{r}'\,d^3\boldsymbol{v}' = d^3\boldsymbol{r}\,d^3\boldsymbol{v} \tag{13·2·5}$$

By (13·2·5) the relation (13·2·3) becomes simply*

$$f(\boldsymbol{r}',\boldsymbol{v}',t') = f(\boldsymbol{r},\boldsymbol{v},t) \tag{13·2·6}$$

or

$$f(\boldsymbol{r} + \dot{\boldsymbol{r}}\,dt,\ \boldsymbol{v} + \dot{\boldsymbol{v}}\,dt,\ t + dt) - f(\boldsymbol{r},\boldsymbol{v},t) = 0$$

By expressing this in terms of partial derivatives, one obtains

$$\left[\left(\frac{\partial f}{\partial x}\dot{x} + \frac{\partial f}{\partial y}\dot{y} + \frac{\partial f}{\partial z}\dot{z}\right) + \left(\frac{\partial f}{\partial v_x}\dot{v}_x + \frac{\partial f}{\partial v_y}\dot{v}_y + \frac{\partial f}{\partial v_z}\dot{v}_z\right) + \frac{\partial f}{\partial t}\right]dt = 0$$

* Since J in (13·2·4) differs from 1 only by terms of second order in dt, it follows that (13·2·5) and (13·2·6) hold actually for *all* times and not merely for infinitesimal times.

More compactly, this can be written

$$Df = 0 \tag{13·2·7}$$

where
$$Df \equiv \frac{\partial f}{\partial t} + \dot{\boldsymbol{r}} \cdot \frac{\partial f}{\partial \boldsymbol{r}} + \dot{\boldsymbol{v}} \cdot \frac{\partial f}{\partial \boldsymbol{v}} = \frac{\partial f}{\partial t} + \boldsymbol{v} \cdot \frac{\partial f}{\partial \boldsymbol{r}} + \frac{\boldsymbol{F}}{m} \cdot \frac{\partial f}{\partial \boldsymbol{v}} \tag{13·2·8}$$

Here $\partial f/\partial \boldsymbol{r}$ denotes the gradient with respect to $\boldsymbol{r}$, i.e., the vector with components $\partial f/\partial x$, $\partial f/\partial y$, $\partial f/\partial z$. Similarly, $\partial f/\partial \boldsymbol{v}$ denotes the gradient with respect to $\boldsymbol{v}$, i.e., the vector with components $\partial f/\partial v_x$, $\partial f/\partial v_y$, $\partial f/\partial v_z$.

Equation (13·2·7) is a linear partial differential equation satisfied by f. The relations (13·2·6) or (13·2·7) assert that f remains unchanged if one moves along with the molecules in phase space. (This is a special case of Liouville's theorem proved in Appendix A·13.) Equation (13·2·7) is the Boltzmann equation without collisions. (In plasma physics it is sometimes called the "Vlasov equation.")

***Alternative derivation** Instead of following the motion of an element of volume of phase space as we did in Fig. (13·2·1), one can focus attention on a *fixed* element of volume. Thus, consider a *given* range of molecular positions between $\boldsymbol{r}$ and $\boldsymbol{r} + d\boldsymbol{r}$ and of velocities between $\boldsymbol{v}$ and $\boldsymbol{v} + d\boldsymbol{v}$ (see Fig. 13·2·2). The number of molecules in this element of volume $d^3\boldsymbol{r}\, d^3\boldsymbol{v}$ of phase space changes as the positions and velocities of the molecules change. The increase in the number of molecules in this range in time dt is given by $(\partial f/\partial t)\, d^3\boldsymbol{r}\, d^3\boldsymbol{v}\, dt$. This change is due to the number of molecules entering and leaving this range $d^3\boldsymbol{r}\, d^3\boldsymbol{v}$ as a result of their motion.

In the absence of collisions, the molecular positions and velocities change simply in accordance with (13·2·1) and (13·2·2). The number of molecules entering the "volume" $d^3\boldsymbol{r}\, d^3\boldsymbol{v}$ in time dt through the "face" x = constant is then just the number contained in the volume $(\dot{x}\, dt)\, dy\, dz\, dv_x\, dv_y\, dv_z$, i.e., equal to $f(\dot{x}\, dt)\, dy\, dz\, dv_x\, dv_y\, dv_z$. The number leaving through the "face"

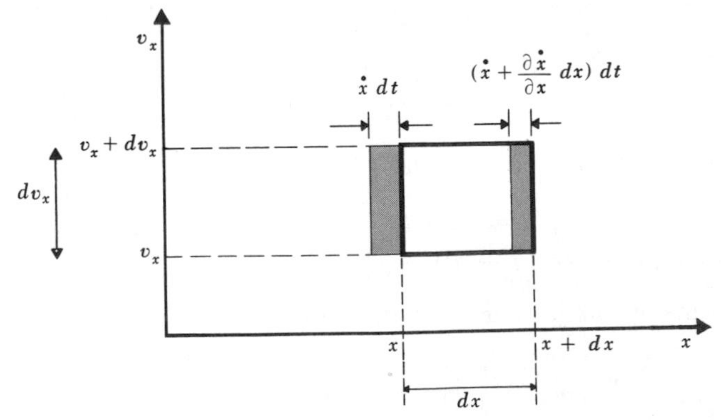

***Fig. 13·2·2** Figure illustrating a fixed element of volume of phase space for a particle moving in one dimension and specified by its position x and velocity v_x.*

$x + dx =$ constant is given by a similar expression except that both f and $\dot{x}$ must be evaluated at $x + dx$ instead of at x. Hence the net number of molecules entering the range $d^3\boldsymbol{r}\, d^3\boldsymbol{v}$ in time dt through the faces x and $x + dx$ is

$$f\dot{x}\, dt\, dy\, dz\, dv_x\, dv_y\, dv_z - \left[f\dot{x} + \frac{\partial}{\partial x}(f\dot{x})\, dx \right] dt\, dy\, dz\, dv_x\, dv_y\, dv_z = -\frac{\partial}{\partial x}(f\dot{x})\, dt\, d^3\boldsymbol{r}\, d^3\boldsymbol{v}$$

Summing similar contributions for molecules entering through the faces labeled by y, z, v_x, v_y, and v_z, one obtains

$$\frac{\partial f}{\partial t}\, dt\, d^3\boldsymbol{r}\, d^3\boldsymbol{v} = -\left[\frac{\partial}{\partial x}(f\dot{x}) + \frac{\partial}{\partial y}(f\dot{y}) + \frac{\partial}{\partial z}(f\dot{z}) + \frac{\partial}{\partial v_x}(f\dot{v}_x) + \frac{\partial}{\partial v_y}(f\dot{v}_y) + \frac{\partial}{\partial v_z}(f\dot{v}_z) \right] dt\, d^3\boldsymbol{r}\, d^3\boldsymbol{v}$$

Thus
$$\frac{\partial f}{\partial t} + \sum_{\alpha=1}^{3} \left[\frac{\partial f}{\partial x_\alpha}\dot{x}_\alpha + \frac{\partial f}{\partial v_\alpha}\dot{v}_\alpha \right] + \sum_{\alpha=1}^{3} f \left[\frac{\partial \dot{x}_\alpha}{\partial x_\alpha} + \frac{\partial \dot{v}_\alpha}{\partial v_\alpha} \right] = 0 \qquad (13\cdot2\cdot9)$$

where
$$x_1 = x,\ x_2 = y,\ x_3 = z$$
and
$$v_1 = v_x,\ v_2 = v_y,\ v_3 = v_z$$
But
$$\frac{\partial \dot{x}_\alpha}{\partial x_\alpha} = \frac{\partial v_\alpha}{\partial x_\alpha} = 0$$

since the variables $\boldsymbol{v}$ and $\boldsymbol{r}$ are independent, and

$$\frac{\partial \dot{v}_\alpha}{\partial v_\alpha} = \frac{1}{m}\frac{\partial F_\alpha}{\partial v_\alpha} = 0$$

since $\boldsymbol{F}$ does not depend on velocity. Hence the last bracket in (13·2·9) vanishes so that this equation reduces to

$$\frac{\partial f}{\partial t} + \dot{\boldsymbol{r}} \cdot \frac{\partial f}{\partial \boldsymbol{r}} + \dot{\boldsymbol{v}} \cdot \frac{\partial f}{\partial \boldsymbol{v}} = 0$$

which is identical to (13·2·7).

***Remark** This last derivation is, of course, similar to that used in the derivation of Liouville's theorem in Appendix A·13. This also suggests how one would generalize the discussion to include magnetic forces which *are* velocity dependent. Since the force could still be derived from a Hamiltonian $\mathcal{H}$, one could introduce a distribution function $f'(\boldsymbol{r},\boldsymbol{p},t)$ involving the momentum $\boldsymbol{p}$ instead of the velocity $\boldsymbol{v}$. By Liouville's theorem this would satisfy the equation

$$\frac{\partial f'}{\partial t} + \dot{\boldsymbol{r}} \cdot \frac{\partial f'}{\partial \boldsymbol{r}} + \dot{\boldsymbol{p}} \cdot \frac{\partial f'}{\partial \boldsymbol{p}} = 0$$

(Here $\boldsymbol{p} = m\boldsymbol{v} + (e/c)\boldsymbol{A}$ if the magnetic field is given in terms of a vector potential $\boldsymbol{A}$.)

13 · 3 *Path integral formulation*

The investigation of transport processes requires a consideration of the effects of molecular collisions. We are interested in finding $f(\mathbf{r},\mathbf{v},t)$. Suppose that there were no collisions. A molecule with position $\mathbf{r}$ and velocity $\mathbf{v}$ at time t would then at any earlier time $t_0 = t - t'$ have a position $\mathbf{r}_0$ and velocity $\mathbf{v}_0$ which could be calculated from the equations of motion for the particle under the influence of the force $\mathbf{F}$. Both $\mathbf{r}_0$ and $\mathbf{v}_0$ depend thus on the elapsed time t' and on the values of $\mathbf{r}$ and $\mathbf{v}$ at the time t. In symbols,

$$\left.\begin{aligned} \mathbf{r}_0 &\equiv \mathbf{r}(t_0) = \mathbf{r}_0(t;\,\mathbf{r},\mathbf{v}) \\ \mathbf{v}_0 &\equiv \mathbf{v}(t_0) = \mathbf{v}_0(t;\,\mathbf{r},\mathbf{v}) \end{aligned}\right\} \tag{13·3·1}$$

In the absence of collisions, $Df = 0$, so that f remains unchanged in any finite time interval, i.e.,

$$f(\mathbf{r},\mathbf{v},t) = f(\mathbf{r}_0,\mathbf{v}_0,t_0) \tag{13·3·2}$$

Let us now consider the effects of collisions. We assume, as in Sec. 12 · 1, that the probability that such a molecule suffered a collision in the time interval between $t_0 = t - t'$ and $t_0 - dt' = t_0 - (t' + dt')$, and then continued to move *without* further collisions during the time t', is given by

$$\frac{dt'}{\tau}\, e^{-t'/\tau} \tag{13·3·3}$$

Let us suppose that the number of molecules near $\mathbf{r}_0$ and $\mathbf{v}_0$ which have just suffered a collision at time t_0 is given by $f^{(0)}(\mathbf{r}_0,\mathbf{v}_0,t_0)\, d^3\mathbf{r}_0\, d^3\mathbf{v}_0$. To avoid a detailed analysis of collision processes, we must introduce some assumption concerning the functional form of $f^{(0)}$. We shall thus assume that the effect of collisions is simply to restore the distribution function to one describing equilibrium conditions existing *locally* near the particular point at the particular time. For example, if the molecules obey Maxwell-Boltzmann statistics, we assume that $f^{(0)}$ has the form.

$$f^{(0)}(\mathbf{r}_0,\mathbf{v}_0,t_0) = n\left(\frac{m\beta}{2\pi}\right)^{\frac{3}{2}} e^{-\frac{1}{2}\beta m(\mathbf{v}_0-\mathbf{u})^2}$$

where the parameters n, β, and $\mathbf{u}$ may be functions of $\mathbf{r}_0$ and t_0 (though *not* of $\mathbf{v}_0$) and represent respectively the local number density, temperature parameter, and mean velocity of the molecules.* It follows, by (13 · 3 · 2), that if there were no collisions, all the $f^{(0)}(\mathbf{r}_0,\mathbf{v}_0,t_0)\, d^3\mathbf{r}_0\, d^3\mathbf{v}_0$ molecules near $\mathbf{r}_0$ and $\mathbf{v}_0$ would arrive near $\mathbf{r}$ and $\mathbf{v}$ at time t. Because of collisions only a fraction (13 · 3 · 3) of these manage to survive so as to arrive near $\mathbf{r}$ and $\mathbf{v}$ at time t. Hence it follows that the number of molecules near $\mathbf{r}$ and $\mathbf{v}$ at time t can be regarded as

* If externally applied forces or temperature gradients are not too large, collisions may readily be effective in bringing about equilibrium conditions within any macroscopically small volume element. But the existence of such *local* equilibria does *not* imply that the various volume elements are in equilibrium with each other; i.e., that there exists *global* equilibrium throughout the whole gas.

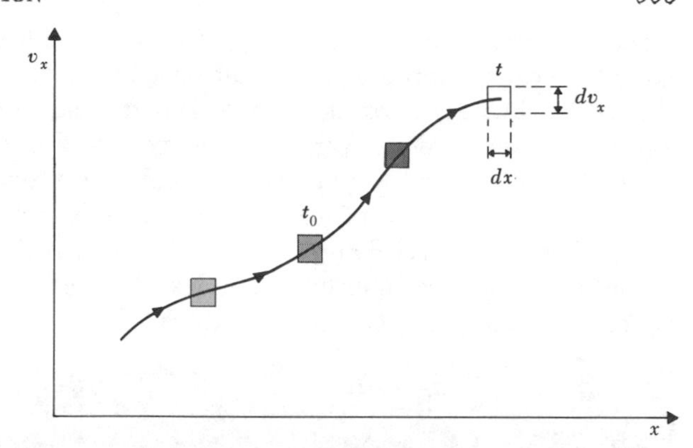

***Fig. 13·3·1** The physical content of Eq. (13·3·4), shown schematically. A particular element $dx\, dv_x$ of phase space is shown at time t. The curve indicates the trajectory in phase space of a particle which moves solely under the influence of the external forces so as to arrive in $dx\, dv_x$ at time t. The other variously shaded elements of volume in this phase space indicate those particles which have just been thrown onto the trajectory as a result of collisions and which succeed in staying on this trajectory without further collisions until time t.*

consisting of the sum total of all molecules which suffered collisions at all possible earlier times $t - t'$ and then moved without further collisions so as to reach the region near $\mathbf{r}$ and $\mathbf{v}$ at time t. In symbols,

$$f(\mathbf{r},\mathbf{v},t) = \int_0^\infty f^{(0)}(\mathbf{r}_0, \mathbf{v}_0, t - t')\, e^{-t'/\tau} \frac{dt'}{\tau} \tag{13·3·4}$$

Here the integration is over all earlier times t', and $\mathbf{r}_0$ and $\mathbf{v}_0$ are, by (13·3·1), also functions of t'. Note that the integrand becomes negligibly small when $t' \gg \tau$. Hence it is sufficient to solve the equation of motion (13·3·1) relating $\mathbf{r}_0$ and $\mathbf{v}_0$ to $\mathbf{r}$ and $\mathbf{v}$ for only relatively short time intervals t' of the order of τ.

Remark In writing (13·3·3) or (13·3·4) we have assumed that the velocity change (due to possible external forces) of a molecule during a time τ is sufficiently small that changes in the collision probability τ^{-1} *between* collisions can be neglected. Then $\tau(v_0) \approx \tau(v)$ and one can simply put $\tau = \tau(v)$ in (13·3·4). This is usually an excellent approximation. If this is not true, τ^{-1} itself becomes a function of time since $v = v(t)$. Then (13·3·3) would have to be replaced by (12·1·12), and (13·3·4) would become

$$f(\mathbf{r},\mathbf{v},t) = \int_0^\infty f^{(0)}(\mathbf{r}_0, \mathbf{v}_0, t - t') \exp\left(-\int_{t-t'}^{t} \frac{ds}{\tau(s)}\right) \frac{dt'}{\tau(t-t')}$$

Equation (13·3·4) is the desired "kinetic" or "path integral" formulation of the transport problem. It allows one to compute f from a knowledge of the collision time $\tau(v)$ and from an assumed function $f^{(0)}$ describing the local equilibrium distribution immediately after a collision. This is, of course, *not* an exact prescription for calculating f. It assumes the existence of a collision time $\tau(v)$ which can be approximately calculated from the scattering cross section by (12·2·8). Furthermore, it assumes that the same distribution $f^{(0)}$ is always restored after collisions; hence it does not take into account "persistence-of-velocity" effects, i.e., the fact that a molecule after a collision has a velocity that is not independent of its velocity just before the collision. Nevertheless, (13·3·4) is a very useful approximation.* It has the great virtue of simplicity so that it can be used to discuss complex physical situations without encountering prohibitive mathematical difficulties.

**Remark* Some care must be taken to verify that the function f calculated by (13·3·4) is not inconsistent with the distribution $f^{(0)}$ assumed immediately after a collision. For example, as a result of a collision, the velocity $\mathbf{v}$ of a molecule can change by arbitrarily large amounts in an infinitesimal time, but its position cannot change instantaneously. Hence the number $n(\mathbf{r},t)$ of molecules per unit volume must be the same immediately before or after a collision; i.e., one must require that

$$\int d^3\mathbf{v}\, f(\mathbf{r},\mathbf{v},t) = \int d^3\mathbf{v}\, f^{(0)}(\mathbf{r},\mathbf{v},t)$$

Finally, it is worth pointing out that (13·3·4) can be written in a convenient alternative form. Let us write $f^{(0)}[t']$ to denote the complete time dependence of $f^{(0)}$, including that contained in $\mathbf{r}_0$ and $\mathbf{v}_0$. Then (13·3·4) can be integrated by parts to give

$$f(\mathbf{r},\mathbf{v},t) = -\int_0^\infty f^{(0)}[t']\, d(e^{-t'/\tau}) = -[f^{(0)}[t']\, e^{-t'/\tau}]_0^\infty + \int_0^\infty \frac{df^{(0)}[t']}{dt'}\, e^{-t'/\tau}\, dt'$$

The first integrated term on the right becomes $f^{(0)}\,[0] = f^{(0)}(\mathbf{r},\mathbf{v},t)$. Hence we obtain

$$\Delta f \equiv f(\mathbf{r},\mathbf{v},t) - f^{(0)}(\mathbf{r},\mathbf{v},t) = \int_0^\infty \frac{df^{(0)}[t']}{dt'}\, e^{-t'/\tau}\, dt' \qquad (13\cdot3\cdot5)$$

This has the advantage that it yields directly the (usually small) departure of f from the distribution function $f^{(0)}$.

13 · 4 *Example: calculation of electrical conductivity*

We illustrate the path-integral method by treating two particular situations of physical interest. The first of these is the case of electrical conductivity already discussed by the simplified arguments of Sec. 12·6. We again consider a gas of particles of mass m and charge e in the presence of a constant uniform

* This path-integral formulation has become popular in recent years, particularly in solid-state physics. See R. G. Chambers, *Proc. Phys. Soc. London,* vol. 65A, p. 458 (1952); also V. Heine, *Phys. Rev.* vol. 107, p. 431 (1957); Cohen, Harrison, and Harrison, *Phys. Rev.* vol. 117, p. 937 (1960).

external electric field $\mathcal{E}$ in the z direction. The particles might be ions in a gas which are scattered by the molecules of the gas; or they might be electrons in a metal which are scattered by the atoms of the lattice.

We assume that collisions lead to a local equilibrium distribution of the form

$$f^{(0)}(\boldsymbol{r},\boldsymbol{v},t) = g(\epsilon), \qquad \epsilon = \tfrac{1}{2}mv^2 \tag{13·4·1}$$

Here $g(\epsilon)$ for the ions in a gas is just the equilibrium Maxwell-Boltzmann distribution, i.e.,

$$g(\epsilon) = n\left(\frac{m\beta}{2\pi}\right)^{\frac{3}{2}} e^{-\beta\epsilon} \tag{13·4·2}$$

where n is the number of particles per unit volume.

Consider a particle with position $\boldsymbol{r}$ and velocity $\boldsymbol{v}$ at time t. One can find its position $\boldsymbol{r}(t_0)$ and velocity $\boldsymbol{v}(t_0)$ (and thus also its energy $\epsilon(t_0)$) at any other time t_0 from the equations of motion. These are in this case

$$\begin{aligned} \frac{dv_x}{dt_0} &= \frac{dv_y}{dt_0} = 0 \\ m\frac{dv_z}{dt_0} &= e\mathcal{E} \end{aligned} \tag{13·4·3}$$

To calculate f by (13·3·5) we note that, with $t' = t - t_0$ and using (13·4·3),

$$\frac{df^{(0)}}{dt'} = -\frac{df^{(0)}}{dt_0} = -\frac{\partial g}{\partial v_z}\frac{dv_z}{dt_0} = -\frac{\partial g}{\partial v_z}\frac{e\mathcal{E}}{m}$$

Thus

$$\frac{df^{(0)}}{dt'} = -\frac{e\mathcal{E}}{m}\frac{\partial g}{\partial v_z} = -e\mathcal{E}v_z\frac{dg}{d\epsilon} \tag{13·4·4}$$

since $\partial\epsilon/\partial v_z = mv_z$. We assume that $\mathcal{E}$ is sufficiently small that v_z does not change appreciably in a time of the order of τ; i.e., we assume that $(dv_z/dt)\tau = (e\mathcal{E}/m)\tau \ll \bar{v}$, where $\bar{v}$ is the mean thermal velocity of the particles. Then $dg/d\epsilon$ and v_z in the integral of (13·3·5) can to good approximation be evaluated at the time t instead of at the earlier time $t_0 = t - t'$. Thus (13·3·5) becomes

$$\Delta f = -e\mathcal{E}v_z\frac{dg}{d\epsilon}\int_0^\infty e^{-t'/\tau}\,dt'$$

or

▶ $$f(\boldsymbol{r},\boldsymbol{v},t) = g(\epsilon) - e\mathcal{E}\tau v_z\frac{dg}{d\epsilon} \tag{13·4·5}$$

The current density j_n in the $\hat{\boldsymbol{n}}$ direction is the flux of charge through an element of area directed along $\hat{\boldsymbol{n}}$. Thus

$$j_n = e\int d^3\boldsymbol{v}\, fv_n \tag{13·4·6}$$

Note that, since both τ and g depend only on $|\boldsymbol{v}|$, the integral $\int d^3\boldsymbol{v}\, gv_n$ has an odd integrand for every velocity component v_n so that it vanishes. This is, of course, an obvious result, since $\boldsymbol{j}$ must be zero in the equilibrium situation in the absence of an electric field.

By symmetry $\boldsymbol{j}$ must be in the direction of the electric field $\mathcal{E}$; thus only j_z does not vanish. By (13·4·5) and (13·4·6) one obtains

$$\sigma_{\text{el}} \equiv \frac{j_z}{\mathcal{E}} = -e^2 \int d^3\boldsymbol{v} \frac{dg}{d\epsilon} \tau v_z^2 \tag{13·4·7}$$

The ratio $j_z/\mathcal{E} = \sigma_{\text{el}}$ is, by definition, the electrical conductivity of the particles. As one would expect for the case of sufficiently low electric fields, (13·4·7) shows that $j_z \propto \mathcal{E}$.

When g is the Maxwell-Boltzmann distribution of (13·4·2), as it would be for ions or for (sufficiently dilute) electrons in a gas,

$$\frac{dg}{d\epsilon} = -\beta g \tag{13·4·8}$$

Then (13·4·7) becomes

$$\sigma_{\text{el}} = \beta e^2 \int d^3\boldsymbol{v}\, g\tau v_z^2 \tag{13·4·9}$$

To evaluate the integral one requires a knowledge of τ as a function of v. This could be calculated by (12·2·8), but the result is rather complicated. If one makes the approximation of replacing $\tau(v)$ by some constant mean value $\bar{\tau}$ and takes it outside the integral, one obtains

$$\sigma_{\text{el}} \approx \beta e^2 \bar{\tau} \int d^3\boldsymbol{v}\, g v_z^2 = \beta e^2 \bar{\tau}(n\overline{v_z^2})$$

Here the average is calculated with the equilibrium function g so that the equipartition result $\frac{1}{2}m\overline{v_z^2} = \frac{1}{2}kT$ applies. Thus

$$\sigma_{\text{el}} = \frac{ne^2}{m} \bar{\tau} \tag{13·4·10}$$

which agrees with our previously derived result (12·6·8).

If it is desired to apply (13·4·7) to the calculation of the conductivity of electrons in a metal, then $g(\epsilon)$ is given by the Fermi distribution so that $g \propto (e^{\beta(\epsilon-\mu)} + 1)^{-1}$. Since this gas is highly degenerate, $dg/d\epsilon$ differs from zero only when $\epsilon \approx \mu$, the Fermi energy (see Sec. 9·16). Hence (13·4·7) shows that *only* the electrons with energy close to the Fermi energy contribute to the conductivity of the metal. Correspondingly, only the value of $\tau = \tau_F$ for electrons near the Fermi energy is needed to compute the integral. Thus (13·4·7) becomes

$$\sigma_{\text{el}} = -e^2 \tau_F \int d^3\boldsymbol{v} \frac{dg}{d\epsilon} v_z^2 \tag{13·4·11}$$

This integral can, however, again be expressed in terms of the *total* number n of electrons per unit volume. Since $\partial g/\partial v_z = (dg/d\epsilon)(\partial\epsilon/\partial v_z) = mv_z(dg/d\epsilon)$,

$$\int_{-\infty}^{\infty} dv_z \frac{dg}{d\epsilon} v_z^2 = \int_{-\infty}^{\infty} dv_z \frac{\partial g}{\partial v_z} \frac{v_z}{m} = \frac{1}{m}[gv_z]_{-\infty}^{\infty} - \frac{1}{m}\int_{-\infty}^{\infty} dv_z\, g$$

where we have performed an integration by parts. The first term on the right vanishes since $g = 0$ for $v_z = \pm\infty$. Hence (13·4·11) becomes

$$\sigma_{\text{el}} = -e^2\tau_F\left(-\frac{1}{m}\int d^3\boldsymbol{v}\, g\right) = \frac{ne^2}{m}\tau_F \tag{13·4·12}$$

since the integral is simply equal to n. The relation (13·4·12) is similar to (13·4·10). But here we have not introduced any approximations concerning the velocity dependence of τ, since only the value of τ for electrons near the Fermi energy is of relevance. Note, however, that n in (13·4·12) is the *total* number of electrons per unit volume.

13·5 *Example: calculation of viscosity*

We treat again the physical situation illustrated in Fig. 12·3·2, where the mean velocity component $u_x(g)$ of the gas does not vanish and depends on z. We assume that collisions tend to produce a local equilibrium distribution *relative* to the gas moving with a mean velocity u_x at the location of the collision. Thus

$$f^{(0)}(\mathbf{r},\mathbf{v},t) = g[v_x - u_x(z), v_y,v_z] = g(U_x,U_y,U_z) \tag{13·5·1}$$

where

$$U_x = v_x - u_x(z), \qquad U_y = v_y, \qquad U_z = v_z \tag{13·5·2}$$

and g is simply the Maxwell distribution

$$g(U_x,U_y,U_z) = g(U) = n\left(\frac{m\beta}{2\pi}\right)^{\frac{3}{2}} e^{-\frac{1}{2}\beta m U^2} \tag{13·5·3}$$

Since there are no external forces acting on a molecule between collisions, the molecular velocity remains constant between collisions. Consider such a molecule which at time t is located at z and has a velocity component v_z. Then as a function of the time t_0 this molecule moves simply so that

$$\frac{dz(t_0)}{dt_0} = v_z(t_0) = v_z \tag{13·5·4}$$

By (13·5·1) we note that $f^{(0)}[t']$ depends on the time $t' = t - t_0$ only through the time dependence of z. Thus

$$\frac{df^{(0)}}{dt'} = -\frac{df^{(0)}}{dt_0} = -\frac{\partial g}{\partial U_x}\frac{\partial U_x}{\partial t_0} = -\frac{\partial g}{\partial U_x}\left(-\frac{\partial u_x}{\partial z}\right)\frac{dz(t_0)}{dt_0}$$

or by (13·5·4)

$$\frac{df^{(0)}}{dt'} = \frac{\partial g}{\partial U_x}\frac{\partial u_x}{\partial z} v_z \tag{13·5·5}$$

This expression is independent of t' so that (13·3·5) becomes simply

▶
$$f = f^{(0)} + \frac{\partial u_x}{\partial z}\frac{\partial g}{\partial U_x} v_z\tau \tag{13·5·6}$$

We now calculate the stress compoent P_{zx}. By (13·1·17) this is

$$P_{zx} = m\int d^3\mathbf{v}\, fU_zU_x \tag{13·5·7}$$

Since $f^{(0)} = g(U)$ is only a function of $|\mathbf{U}|$, the integral $\int d^3\mathbf{v}\, f^{(0)}U_zU_x = 0$ by symmetry, since the integrand is an odd function of U_z and of U_x. Hence (13·5·7) becomes simply, since $v_z = U_z$ by (13·5·2),

$$P_{zx} = -\eta \frac{\partial u_x}{\partial z} \tag{13·5·8}$$

where

$$\eta = -m \int d^3U \frac{\partial g}{\partial U_x} U_z^2 U_x \tau \tag{13·5·9}$$

The coefficient η in (13·5·8) is, by definition, the coefficient of viscosity, and (13·5·9) represents the expression derived for it on the basis of the present formulation of the theory.

If we again make the approximation of replacing $\tau(v)$ by a suitable mean value $\bar{\tau}$ so that it can be taken outside the integral, further simplifications become possible. Thus one gets

$$\eta = -m\bar{\tau} \iint dU_y \, dU_z \, U_z^2 \int_{-\infty}^{\infty} dU_x \frac{\partial g}{\partial U_x} U_x$$

Integrating by parts

$$\int_{-\infty}^{\infty} dU_x \frac{\partial g}{\partial U_x} U_x = [gU_x]_{-\infty}^{\infty} - \int_{-\infty}^{\infty} dU_x \, g = 0 - \int_{-\infty}^{\infty} dU_x \, g$$

Hence

$$\eta = m\bar{\tau} \int d^3U \, g U_z^2 = m\bar{\tau} n \overline{U_z^2} \tag{13·5·10}$$

where the mean value is calculated for the Maxwell distribution (13·5·3) so that the equipartition result $\overline{\frac{1}{2}mU_z^2} = \frac{1}{2}kT$ applies. Hence

$$\eta = nkT\bar{\tau} \tag{13·5·11}$$

By symmetry one can also write $\overline{U_z^2} = \frac{1}{3}\overline{U^2}$ so that (13·5·11) becomes

$$\eta = \tfrac{1}{3} nm\bar{\tau} \overline{U^2} \tag{13·5·12}$$

If one puts approximately $\overline{U^2} \approx \bar{U}^2$, then $\tau\bar{U} = l$, the mean free path. If one considers further the case where u_x is small so that $\bar{U} \approx \bar{v}$, then (13·5·12) reduces to $\frac{1}{3}nml\bar{v}$, the expression (12·3·7) derived by simple mean-free-path arguments.

13 · 6 *Boltzmann differential equation formulation*

It is possible to formulate the problem of calculating $f(\mathbf{r},\mathbf{v},t)$ by an alternative approach which is equivalent to the path-integral method of Sec. 13·3. Proceeding as in Sec. 13·2, focus attention on the molecules which at time t have position and velocity in a range $d^3\mathbf{r}\, d^3\mathbf{v}$ near $\mathbf{r}$ and $\mathbf{v}$. Consider then the situation at an *infinitesimally* later time $t + dt$. If there were no collisions, these particles would simply move, under the influence of the external force, to a position near $\boldsymbol{r}' = \boldsymbol{r} + \dot{\boldsymbol{r}}\,dt$ and to a velocity near $\boldsymbol{v}' = \boldsymbol{v} + \dot{\boldsymbol{v}}\,dt$ so that Eq. (13·2·3) would be valid. But if collisions also take place, the number of molecules in the range $d^3\boldsymbol{r}\,d^3\boldsymbol{v}$ can also change by virtue of collisions. The reason is that, as a result of collisions, molecules originally with positions and velocities *not* in this range $d^3\boldsymbol{r}\,d^3\boldsymbol{v}$ can be scattered *into* this range; conversely, molecules originally *in* this range can be scattered *out* of it. Let $D_Cf\,d^3\boldsymbol{r}\,d^3\boldsymbol{v}$ denote the net increase per unit time in the number of molecules in the range $d^3\boldsymbol{r}\,d^3\boldsymbol{v}$ as a result of such collisions. Then one can assert, instead of (13·2·3), that [the number of molecules which at the time $t + dt$ are in the range near $(\boldsymbol{r} + \dot{\boldsymbol{r}}\,dt)$ and $(\boldsymbol{v} + \dot{\boldsymbol{v}}\,dt)$] must be equal to [the number of molecules which were at time t in the range near $\boldsymbol{r}$ and $\boldsymbol{v}$ and thus moved to $(\boldsymbol{r} + \dot{\boldsymbol{r}}\,dt)$ and $(\boldsymbol{v} + \dot{\boldsymbol{v}}\,dt)$ as a result of the external force] *plus* [the net change in the number of molecules in this range caused by collisions in the time interval dt]. In symbols,

$$f(\boldsymbol{r} + \dot{\boldsymbol{r}}\,dt,\ \boldsymbol{v} + \dot{\boldsymbol{v}}\,dt,\ t + dt)\,d^3\boldsymbol{r}'\,d^3\boldsymbol{v}' = f(\boldsymbol{r},\boldsymbol{v},t)\,d^3\boldsymbol{r}\,d^3\boldsymbol{v} + D_Cf\,d^3\boldsymbol{r}\,d^3\boldsymbol{v}\,dt$$

Using (13·2·5), this becomes with the definition (13·2·8)

$$\blacktriangleright \qquad Df = D_Cf \qquad (13\cdot 6\cdot 1)$$

This is called the "Boltzmann Equation."

The compact form (13·6·1) hides a complicated equation. In particular, one can write down an explicit expression for D_Cf in terms of integrals involving f and describing properly the rate at which molecules enter and leave the range $d^3\boldsymbol{r}\,d^3\boldsymbol{v}$ as a result of collisions. The Boltzmann equation (13·6·1) is then an "integrodifferential" equation, i.e., *both* partial derivatives of f *and* integrals over f enter in this equation for the unknown function f. This equation is the basis of the more exact transport theory to be discussed in the next chapter.

For the present we want to avoid excessive complications by means of suitable approximations for D_Cf. Let us then simply *assume* that the effect of collisions is always to restore a *local* equilibrium situation described by the distribution function $f^{(0)}(\boldsymbol{r},\boldsymbol{v},t)$. Let us further *assume* that, if the molecular distribution is disturbed from the local equilibrium so that the actual distribution $f(\boldsymbol{r},\boldsymbol{v},t)$ is different from $f^{(0)}$, then the effect of the collisions is simply to restore f to the local equilibrium value $f^{(0)}$ exponentially with a relaxation time τ_0 which is of the order of the time between molecular collisions. In symbols this assumption for D_Cf reads

$$D_Cf = -\frac{f - f^{(0)}}{\tau_0} \qquad (13\cdot 6\cdot 2)$$

According to (13·6·2), $D_Cf = 0$ if $f = f^{(0)}$; also for fixed $\boldsymbol{r}$ and $\boldsymbol{v}$, f changes as a result of collisions according to $f(t) = f^{(0)}(t) + [f(0) - f^{(0)}(0)]\exp(-t/\tau_0)$. With the assumption (13·6·2), Eq. (13·6·1) becomes

$$\blacktriangleright \qquad Df \equiv \frac{\partial f}{\partial t} + \boldsymbol{v}\cdot\frac{\partial f}{\partial \boldsymbol{r}} + \frac{\boldsymbol{F}}{m}\cdot\frac{\partial f}{\partial \boldsymbol{v}} = -\frac{f - f^{(0)}}{\tau_0} \qquad (13\cdot 6\cdot 3)$$

This is simply a linear partial differential equation for f and is the Boltzmann equation with the relaxation-time assumption.

13 · 7 *Equivalence of the two formulations*

We now want to show that the special Boltzmann equation (13·6·3) is completely equivalent to the path integral (13·3·4). Indeed, the latter integral is just an integrated form of the partial differential equation (13·6·3). Thus it is quite immaterial whether a given transport problem is discussed in terms of the path-integral method of Sec. 13·3 or by the Boltzmann equation method of Sec. 13·6. For more complicated problems the path-integral method is usually easier, since it provides already an expression in integrated form so that no partial differential equation needs to be solved.

To show the equivalence, we consider $f(\mathbf{r},\mathbf{v},t)$ as given by (13·3·4) and ask what differential equation f satisfies. Remember that in (13·3·4) the variables $\mathbf{r}$ and $\mathbf{v}$ are related to the corresponding variables $\mathbf{r}_0$ and $\mathbf{v}_0$ via the equations of motion (13·3·1). That is, $\mathbf{r}_0$ and $\mathbf{v}_0$ are the values assumed by the position and velocity of a molecule at time $t_0 = t - t'$ if the molecule has position and velocity $\mathbf{r}$ and $\mathbf{v}$ at time t. Or conversely, $\mathbf{r}$ and $\mathbf{v}$ are the values assumed by the position and velocity at time $t = t_0 + t'$ if the corresponding values are $\mathbf{r}_0$ and $\mathbf{v}_0$ at time t_0. Consider then a molecule with *given* values of $\mathbf{r}_0$ and $\mathbf{v}_0$ at time $t_0 = t - t'$. The equations of motion (13·3·1) under the influence of the external force then imply a connection such that at time t the molecule has position $\mathbf{r}$ and velocity $\mathbf{v}$, and that at the time $t + dt$ it has position $\mathbf{r} + \dot{\mathbf{r}}\, dt$ and velocity $\mathbf{v} + \dot{\mathbf{v}}\, dt$. Now by (13·3·4)

$$f(\mathbf{r},\mathbf{v},t) = \int_0^\infty f^{(0)}(\mathbf{r}_0, \mathbf{v}_0, t - t')\, e^{-t'/\tau}\, \frac{dt'}{\tau} \tag{13·7·1}$$

Similarly, one must have

$$f(\mathbf{r} + \dot{\mathbf{r}}\, dt, \mathbf{v} + \dot{\mathbf{v}}\, dt, t + dt) = \int_0^\infty f^{(0)}(\mathbf{r}_0, \mathbf{v}_0, t + dt - t')\, e^{-t'/\tau}\, \frac{dt'}{\tau} \tag{13·7·2}$$

Subtracting (13·7·1) from (13·7·2), one obtains

$$f(\mathbf{r} + \dot{\mathbf{r}}\, dt, \mathbf{v} + \dot{\mathbf{v}}\, dt, t + dt) - f(\mathbf{r},\mathbf{v},t) = \int_0^\infty \frac{\partial f^{(0)}(\mathbf{r}_0, \mathbf{v}_0, t - t')}{\partial t}\, dt\, e^{-t'/\tau}\, \frac{dt'}{\tau} \tag{13·7·3}$$

Since we are considering a situation with fixed given values of $\mathbf{r}_0$ and $\mathbf{v}_0$, we can write

$$\frac{\partial f^{(0)}(\mathbf{r}_0, \mathbf{v}_0, t - t')}{\partial t} = - \frac{\partial f^{(0)}(\mathbf{r}_0, \mathbf{v}_0, t - t')}{\partial t'}$$

Dividing both sides of (13·7·3) by dt then gives

$$Df = -\frac{1}{\tau} \int_0^\infty \frac{\partial f^{(0)}(\mathbf{r}_0, \mathbf{v}_0, t - t')}{\partial t'}\, e^{-t'/\tau}\, dt' \tag{13·7·4}$$

Here the left side is just the quantity Df defined in (13·2·8). Integrating (13·7·4) by parts, one gets

$$Df = -\frac{1}{\tau}\left[f^{(0)}(\boldsymbol{r}_0, \boldsymbol{v}_0, t - t')\, e^{-t'/\tau}\right]_0^\infty - \frac{1}{\tau}\int_0^\infty f^{(0)}(\boldsymbol{r}_0, \boldsymbol{v}_0, t - t')\, e^{-t'/\tau}\frac{dt'}{\tau}$$

$$= -\frac{1}{\tau}[0 - f^{(0)}(\boldsymbol{r},\boldsymbol{v},t)] - \frac{1}{\tau}f(\boldsymbol{r},\boldsymbol{v},t)$$

since the remaining integral is just the one occurring in (13·7·1). Hence one obtains

$$Df = \frac{1}{\tau}(f^{(0)} - f) \tag{13·7·5}$$

which is identical with the Boltzmann equation (13·6·3), provided that the relaxation time τ_0 introduced there is identified with the ordinary mean time τ between molecular collisions. We shall therefore henceforth write $\tau_0 = \tau$ in the Boltzmann equation (13·6·3).

13·8 *Examples of the Boltzmann equation method*

In order to apply the Boltzmann equation (13·6·3) to the discussion of transport problems, further approximations are usually needed to achieve a solution of this partial differential equation. A frequently occurring situation is that where the departures from the equilibrium situation are relatively small. The approximation procedures exploit this fact.

Suppose that the situation is one of complete equilibrium. Then $f = f^{(0)}$ is the actual equilibrium distribution which ought also, therefore, to satisfy the Boltzmann equation (13·6·3). Indeed this is the case. The collision term $D_c f \propto (f - f^{(0)})$ on the right side vanishes then. Furthermore, the left side Df of (13·6·3) also vanishes as a special case of Liouville's theorem.

Remark For example, for a gas at rest one knows by statistical mechanics that $f^{(0)} = f^{(0)}(\epsilon)$ is only a function of the energy ϵ of the molecule. Specifically, $\epsilon = \frac{1}{2}mv^2 + \mathcal{V}(\boldsymbol{r})$ where $\mathcal{V}(\boldsymbol{r})$ is the potential energy from which the force is derived so that $\boldsymbol{F} = -(\partial\mathcal{V}/\partial\boldsymbol{r})$. Then $\partial f^{(0)}/\partial t = 0$. Also the term

$$\boldsymbol{v}\cdot\frac{\partial f^{(0)}}{\partial \boldsymbol{r}} = \boldsymbol{v}\cdot\frac{\partial f^{(0)}}{\partial \epsilon}\frac{\partial \epsilon}{\partial \boldsymbol{r}} = -\frac{\partial f^{(0)}}{\partial \epsilon}\boldsymbol{v}\cdot\boldsymbol{F}$$

just cancels the term $\dfrac{\boldsymbol{F}}{m}\cdot\dfrac{\partial f^{(0)}}{\partial \boldsymbol{v}} = \dfrac{\boldsymbol{F}}{m}\cdot\dfrac{\partial f^{(0)}}{\partial \epsilon}\dfrac{\partial \epsilon}{\partial \boldsymbol{v}} = \dfrac{\partial f^{(0)}}{\partial \epsilon}\dfrac{\boldsymbol{F}}{m}\cdot m\boldsymbol{v}$

Thus $Df^{(0)} = 0$ so that the left side of (13·6·3) vanishes.

Consider now a situation slightly removed from equilibrium. Then one can write $f = f^{(0)} + f^{(1)}$, where $f^{(1)} \ll f^{(0)}$. The collision term on the right side of (13·6·3) becomes just $-f^{(1)}/\tau$. The left side is then also small and can be evaluated approximately by neglecting terms in $f^{(1)}$. This perturbation procedure is best illustrated by some examples. We shall, therefore, apply the Boltzmann equation method to the same two cases already discussed by the path-integral method in Secs. 13·4 and 13·5.

Electrical conductivity The physical situation is the one described in Sec. 13·4. In the absence of an external electric field $\mathcal{E}$, the distribution function is given by

$$f^{(0)} = g(\epsilon), \qquad \epsilon = \tfrac{1}{2}mv^2 \tag{13·8·1}$$

where $g(\epsilon)$ is the MB distribution (13·4·2) in the case of ions or the FD distribution in the case of electrons in a metal. If a spatially uniform time-independent electric field $\mathcal{E}$ is applied in the z direction, one expects that the new distribution function $f(\mathbf{r},\mathbf{v},t)$ will still be independent of $\mathbf{r}$ and t. Then the Boltzmann equation (13·6·3) becomes simply, since $\mathbf{F} = e\boldsymbol{\mathcal{E}}$ has only a z component,

$$\frac{e\mathcal{E}}{m}\frac{\partial f}{\partial v_z} = -\frac{f - f^{(0)}}{\tau} \tag{13·8·2}$$

Let us assume that $\mathcal{E}$ is quite small. Then one expects that f differs only slightly from $f^{(0)} = g$. Thus we put

$$f = g + f^{(1)}, \qquad \text{where } f^{(1)} \ll g \tag{13·8·3}$$

Then (13·8·2) becomes in first approximation

$$\frac{e\mathcal{E}}{m}\frac{\partial g}{\partial v_z} = -\frac{f^{(1)}}{\tau} \tag{13·8·4}$$

Here we have neglected on the left side the term involving $f^{(1)}$, since it is of the order of the product of the *two* small quantities $\mathcal{E}$ and $f^{(1)}$. Thus (13·8·4) becomes

$$f^{(1)} = f - g = -\frac{e\mathcal{E}\tau}{m}\frac{\partial g}{\partial v_z} = -e\mathcal{E}\tau v_z\frac{dg}{d\epsilon} \tag{13·8·5}$$

which is identical with (13·4·5). The rest of the discussion proceeds then as in Sec. 13·4.

By (13·8·3), the above approximations are valid if $f^{(1)} \ll g$. Since $dg/d\epsilon = -\beta g$, this condition becomes by (13·8·5) $e\mathcal{E}\tau v_z\beta = e\mathcal{E}(\tau v_z)/kT \ll 1$. This means that the electric field $\mathcal{E}$ must be sufficiently small that the energy acquired by the particle from the field $\mathcal{E}$ in a mean free path $v_z\tau$ is appreciably less than its mean thermal energy.

Viscosity The physical situation here is the one discussed in Sec. 13·5. In the *absence* of a mean velocity gradient, the equilibrium distribution for a fluid moving with *constant* mean velocity u_x in the x direction is simply

$$f^{(0)} = g(v_x - u_x, v_y, v_z) = g(U_x, U_y, U_z), \tag{13·8·6}$$

the Maxwellian distribution (13·5·3) relative to the moving fluid. This satisfies the Boltzmann equation (13·6·3). If there is a mean velocity gradient in the gas so that u_x in (13·8·6) is a function of z such that $\partial u_x/\partial z \neq 0$, then (13·8·6) no longer satisfies the equation (13·6·3). Since the situation is time-independent, f does not depend on t. But f will depend on z, since this is the direction of the velocity gradient. Here there is no external force, i.e., $\boldsymbol{F} = 0$, so that the Boltzmann equation (13·6·3) becomes simply

$$v_z \frac{\partial f}{\partial z} = -\frac{f - f^{(0)}}{\tau} \tag{13·8·7}$$

Again we assume that $\partial u_x/\partial z$ is sufficiently small, and hence $\partial f/\partial z$ is sufficiently small, that f differs from $f^{(0)}$ only by a small amount. Putting again

$$f = f^{(0)} + f^{(1)}, \qquad \text{with } f^{(1)} \ll f^{(0)} \tag{13·8·8}$$

(13·8·7) becomes then

$$v_z \frac{\partial f^{(0)}}{\partial z} = -\frac{f^{(1)}}{\tau} \tag{13·8·9}$$

where we have neglected terms in $f^{(1)}$ on the left side. Thus

$$f^{(1)} = f - f^{(0)} = -\tau v_z \frac{\partial f^{(0)}}{\partial z} = \tau v_z \frac{\partial g}{\partial U_x} \frac{\partial u_x}{\partial z} \tag{13·8·10}$$

which agrees again with our previous expression (13·5·6).

SUGGESTIONS FOR SUPPLEMENTARY READING

C. Kittel: "Elementary Statistical Physics," secs. 40, 41, 43, John Wiley & Sons, Inc., New York, 1958.

S. Chapman and T. G. Cowling: "The Mathematical Theory of Non-uniform Gases," 2d ed., chap. 2, Cambridge University Press, Cambridge, 1952. (Discussion of distribution functions and fluxes.)

PROBLEMS

13.1 Since the relaxation time τ is only a function of $|\boldsymbol{v}|$ (or of $\epsilon = \frac{1}{2}mv^2$), show by performing the integrations over angles in (13·4·7) that the expression for the electrical conductivity can be written in the form

$$\sigma_{\text{el}} = -\frac{1}{3} e^2 \int \frac{dg}{d\epsilon} \tau v^2 \, d^3\boldsymbol{v}$$

or

$$\sigma_{\text{el}} = -\frac{4\pi}{3} e^2 \int_0^\infty \frac{dg}{d\epsilon} \tau v^4 \, dv$$

where $g = g(\epsilon) = g(\frac{1}{2}mv^2)$ is the equilibrium distribution function.

13.2 Suppose that the particles in the preceding problem obey Maxwell-Boltzmann statistics. Show that their electrical conductivity can be written in the convenient form

$$\sigma_{el} = \frac{ne^2}{m} \langle\tau\rangle_\sigma$$

where $\langle\tau\rangle_\sigma$ is a suitably weighted average of $\tau(v)$ over the velocity distribution and is defined by

$$\langle\tau\rangle_\sigma = \frac{8}{3\sqrt{\pi}} \int_0^\infty ds\, e^{-s^2} s^4 \tau(\tilde{v}s)$$

Here $\tilde{v} \equiv (2kT/m)^{\frac{1}{2}}$ is the most probable speed of the particle in equilibrium and $s \equiv v/\tilde{v}$ is a dimensionless variable expressing molecular speeds in terms of this most probable speed. The average $\langle\tau\rangle_\sigma$ has been defined in such a way that it reduces to τ when this quantity is independent of v.

13.3 Show that the expression (13·5·9) for the viscosity can be written in the form

$$\eta = -\frac{m^2}{15} \int d^3U \frac{dg}{d\epsilon} \tau U^4$$

where $g = g(\epsilon) = g(\frac{1}{2}mU^2)$ is the equilibrium distribution function.

13.4 If the molecules obey Maxwell-Boltzmann statistics, show that their coefficient of viscosity can be written in the form

$$\eta = nkT\langle\tau\rangle_\eta$$

where $\langle\tau\rangle_\eta$ is a mean collision time defined by

$$\langle\tau\rangle_\eta = \frac{16}{15\sqrt{\pi}} \int_0^\infty ds\, e^{-s^2} s^6 \tau(\tilde{v}s)$$

and where the notation is similar to that used in Problem 13.2. If $\tau(v)$ is independent of v, $\langle\tau\rangle_\eta$ again reduces to τ.

13.5 Assume that $\langle\tau\rangle_\eta$ in the last problem (or in (13·5·11)) can be approximated by a constant value $\tau = l/\bar{v} = (\sqrt{2}\, n\sigma_0\bar{v})^{-1}$ where σ_0 is the constant total scattering cross section between rigid-sphere molecules. Calculate η in terms of σ_0, T, and the molecular mass m. Compare the result with the simplest mean-free-path calculation (12·3·19) and with the result of the rigorous calculation (14·8·33).

13.6 A dilute gas of monatomic molecules of mass m is enclosed in a container and maintained in the presence of a small temperature gradient $\partial T/\partial z$ in the z direction. It is desired to find an expression for the thermal conductivity of this gas at the temperature T. If Maxwell-Boltzmann statistics are applicable, find, in first approximation, the distribution function f of the molecules in the presence of this temperature gradient by using the path-integral method. Assume that the molecular collision time τ is independent of molecular speed.

Suggestion: Assume that the distribution function immediately after collision is of the local equilibrium form

$$g = n\left(\frac{\beta m}{2\pi}\right)^{\frac{3}{2}} \exp\left(-\tfrac{1}{2}\beta m v^2\right)$$

where the temperature parameter $\beta = (kT)^{-1}$ and the local density n are both functions of z. Since the experimental conditions are such that no macroscopic mass motion of the gas is allowed, the condition that $\overline{v_z} = 0$ yields a relation between n and β. What does this relation say about the pressure in the gas?

13.7 Find the molecular distribution function in the preceding problem by solving the Boltzmann equation.

13.8 Use the results of the preceding problems to calculate the thermal conductivity of the gas. Show that it can be written in the form

$$\kappa = \frac{5}{2} \frac{nk^2T}{m} \tau$$

where τ is the constant collision time.

13.9 Compare the result obtained in the calculation of the preceding problem with the simplest mean-free-path calculation $\kappa = \frac{1}{3}nc\bar{v}l$ of (12·4·8) for a monatomic gas. Assume a constant relaxation time $\tau \equiv l/\bar{v}$.

13.10 Assuming that the collision time τ can be taken as a constant, use the results of Problems 13.4 and 13.8 to find the ratio κ/η of thermal conductivity to viscosity. Compare this with the value of this ratio obtained by the simplest mean-free-path calculations. Does the more refined calculation improve the agreement of this ratio with experiment?

13.11 The conduction electrons of a metal are primarily responsible for the thermal conductivity of the metal. Calculate the thermal conductivity κ due to these conduction electrons, remembering that these form a highly degenerate Fermi-Dirac gas. In setting up the calculation (either from the point of view of the path integral or the Boltzmann equation method), remember that the thermal conductivity is measured under open-circuit conditions where no electric current flows through the metal. Hence in the presence of the thermal gradient, a slight redistribution of the conduction electrons must occur so as to set up in the metal an electric field $\mathcal{E}$ of just the right amount to reduce the mean drift velocity of the electrons to zero. Express your final answer for κ in terms of the temperature T of the metal, the total number n of conduction electrons per unit volume, the electron mass m, and the collision time τ_F of the electrons having an energy near the Fermi energy.

13.12 Use the result of the preceding problem and that of Eq. (13·4·12) to compute the ratio κ/σ_{el} of the thermal conductivity κ to the electrical conductivity σ_{el} of a metal. Show that this ratio depends only on the temperature T and on the fundamental constants e and k, but is independent of the mass of the electrons, their number density, or their collision times in the particular metal. (This result is known as the Wiedemann-Franz law.)

Calculate the value of this ratio numerically at 0°C ($T = 273$°K). Compare this value with the experimentally measured ratio for the following metals: silver, gold, copper, lead, platinum, tin, tungsten, and zinc. (The experimentally measured values κ and σ_{el} for these metals can be found, for example, in the "Handbook of Physics and Chemistry" (Chemical Rubber Publishing Company) or in the "American Institute of Physics Handbook," 2d ed. (McGraw-Hill Book Company, 1963).)

Near-exact formulation of transport theory

14

OUR TREATMENT of transport processes in the preceding chapter left much to be desired. We assumed the existence of a relaxation time τ and had only approximate means of calculating this quantity. More important, we did not treat the effects of collisions in a detailed way. Thus we neglected correlations between molecular velocities before and after a collision, i.e., persistence of velocity effects. We shall now formulate the problem in a more rigorous and satisfactory way *without* using the concept of a relaxation time. The procedure will be to write down an equation for the distribution function $f(\mathbf{r},\mathbf{v},t)$ directly in terms of the scattering cross section σ for binary collisions between the molecules. The solution of this equation provides, in principle, a solution of the physical problem. Since the equation is quite complicated, the task of solving it is not easy, and approximation methods must again be used. Nevertheless, despite increased complexity, there is an important advantage in formulating the problem in this way. The reason is that the point of departure of the theory is an equation which is fairly rigorous. Hence general theorems can be proved and systematic approximation procedures developed. On the other hand, if one starts from the simpler formulations of the preceding chapter, it is difficult to estimate the errors committed and to know how to correct for certain effects (like persistence of velocities) in a systematic way.

14 · 1 *Description of two-particle collisions*

We begin our discussion by considering in detail collisions between two molecules. Throughout this chapter we shall assume that if the molecules are not monatomic, their states of internal motion (e.g., rotation or vibration) are unaffected by the collisions. Thus the two molecules under consideration can be treated as simple particles with respective masses m_1 and m_2, position vectors $\mathbf{r}_1$ and $\mathbf{r}_2$, and velocities $\mathbf{v}_1$ and $\mathbf{v}_2$. The interaction between these particles depends then in some way on their relative positions and velocities. (If the particles also have spins, we assume for the sake of simplicity that their interaction does *not* depend on their spins.)

The collision problem can be much simplified by an appropriate change of variables. Conservation of the total momentum implies the relation

$$m_1\mathbf{v}_1 + m_2\mathbf{v}_2 = \mathbf{P} = \text{constant} \tag{14·1·1}$$

Thus the velocities $\mathbf{v}_1(t)$ and $\mathbf{v}_2(t)$ are not independent, but must always satisfy the relation (14·1·1). The other quantity of physical interest is the relative velocity

$$\mathbf{v}_1 - \mathbf{v}_2 \equiv \mathbf{V} \tag{14·1·2}$$

One can then use (14·1·1) and (14·1·2) to express $\mathbf{v}_1$ and $\mathbf{v}_2$ in terms of $\mathbf{P}$ and $\mathbf{V}$. Thus

$$\left.\begin{aligned}(m_1 + m_2)\mathbf{v}_1 &= \mathbf{P} + m_2\mathbf{V}\\ (m_1 + m_2)\mathbf{v}_2 &= \mathbf{P} - m_1\mathbf{V}\end{aligned}\right\}$$

or

$$\left.\begin{aligned}\mathbf{v}_1 &= \mathbf{c} + \frac{\mu}{m_1}\mathbf{V}\\ \mathbf{v}_2 &= \mathbf{c} - \frac{\mu}{m_2}\mathbf{V}\end{aligned}\right\} \tag{14·1·3}$$

where

$$\mathbf{c} \equiv \frac{\mathbf{P}}{m_1 + m_2} = \frac{m_1\mathbf{v}_1 + m_2\mathbf{v}_2}{m_1 + m_2} \tag{14·1·4}$$

is the time-independent velocity of the center of mass; i.e.,

$$\mathbf{c} = \frac{d\mathbf{r}_c}{dt}$$

where

$$\mathbf{r}_c \equiv \frac{m_1\mathbf{r}_1 + m_2\mathbf{r}_2}{m_1 + m_2} \tag{14·1·5}$$

is the position vector of the center of mass. In addition, we have introduced the quantity

$$\mu \equiv \frac{m_1 m_2}{m_1 + m_2} \tag{14·1·6}$$

which is called the "reduced mass" of the particles.

The total kinetic energy K of the particles becomes by (14·1·3)

$$K = \tfrac{1}{2}m_1\mathbf{v}_1{}^2 + \tfrac{1}{2}m_2\mathbf{v}_2{}^2 = \tfrac{1}{2}(m_1 + m_2)\mathbf{c}^2 + \tfrac{1}{2}\mu\mathbf{V}^2 \tag{14·1·7}$$

Consider now a collision process. Denote the velocities of the two particles *before* they interact with each other in the collision process by $\mathbf{v}_1$ and $\mathbf{v}_2$; denote their respective velocities *after* the collision by $\mathbf{v}_1'$ and $\mathbf{v}_2'$. In terms of the new variables the situation is described in a particularly simple way. The center-of-mass velocity $\mathbf{c}$ remains unchanged as a consequence of conservation of momentum. The relative velocity changes from the value $\mathbf{V}$ before the collision to the value $\mathbf{V}'$ after the collision. We assume the collisions to be elastic so that the internal energies of the molecules remain unchanged. Then the kinetic energy K remains unchanged in a collision, and it follows, by (14·1·7), that $\mathbf{V}^2$ also remains unchanged so that $|\mathbf{V}'| = |\mathbf{V}|$. Hence the only

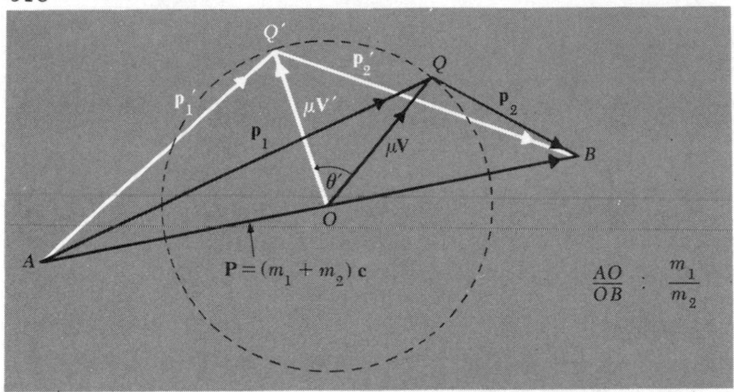

***Fig. 14·1·1** Geometrical construction based on (14·1·8) and illustrating an elastic collision process. Consider the original momenta $p_1 = m_1 v$ and $p_2 = m_2 v_2$ represented by the vectors $\overrightarrow{AQ}$ and $\overrightarrow{QB}$, respectively. Their vector sum yields the total momentum $P = (m_1 + m_2)c = \overrightarrow{AB}$, which remains unchanged in time. Divide this vector $\overrightarrow{AB}$ in the ratio $m_1 : m_2$ to locate the point O; then $\overrightarrow{AO} = m_1 c$ and $\overrightarrow{OB} = m_2 c$. Hence the vector $\overrightarrow{OQ}$ represents μV. Draw a sphere about O of radius OQ. Then the final relative velocity must be such that the vector $\mu V' = \overrightarrow{OQ'}$ terminates somewhere on this same sphere (although not necessarily in the same plane ABQ). The final momenta p_1' and p_2' are simply given by the vectors $\overrightarrow{AQ'}$ and $\overrightarrow{Q'B}$. Their directions with respect to the original vectors p_1 and p_2 are then immediately apparent from the diagram.*

effect of a collision is that $\boldsymbol{V}$ changes its direction without changing its magnitude. The collision process can thus be described by merely specifying the polar angle θ' and azimuthal angle φ' of the final relative velocity $\boldsymbol{V}'$ with respect to the relative velocity $\boldsymbol{V}$ before the collision.

It is simplest to visualize the relationship between the velocities before and after the collision by considering the corresponding molecular momenta $\boldsymbol{p}_1, \boldsymbol{p}_2$ before and $\boldsymbol{p}_1', \boldsymbol{p}_2'$ after the collision. By (14·1·3) one has at all times

$$\left.\begin{aligned} \boldsymbol{p}_1 &= m_1\boldsymbol{c} + \mu\boldsymbol{V} \\ \boldsymbol{p}_2 &= m_2\boldsymbol{c} - \mu\boldsymbol{V} \end{aligned}\right\} \tag{14·1·8}$$

The geometrical relationships are illustrated in Fig. 14·1·1.

One can correspondingly visualize (classically) the positions of the particles at all times. In addition to the center-of-mass position vector $\boldsymbol{r}_c$ of (14·1·5) we introduce, corresponding to (14·1·2), the relative position vector

$$\boldsymbol{r}_1 - \boldsymbol{r}_2 \equiv \boldsymbol{R} \tag{14·1·9}$$

Then, analogously to (14·1·3), one has

$$\left.\begin{aligned} \boldsymbol{r}_1^* &\equiv \boldsymbol{r}_1 - \boldsymbol{r}_c = \frac{\mu}{m_1}\boldsymbol{R} \\ \boldsymbol{r}_2^* &\equiv \boldsymbol{r}_2 - \boldsymbol{r}_c = -\frac{\mu}{m_2}\boldsymbol{R} \end{aligned}\right\} \tag{14·1·10}$$

Thus the position $\boldsymbol{r}_c$ of the center of mass moves with the constant velocity $\boldsymbol{c}$ of

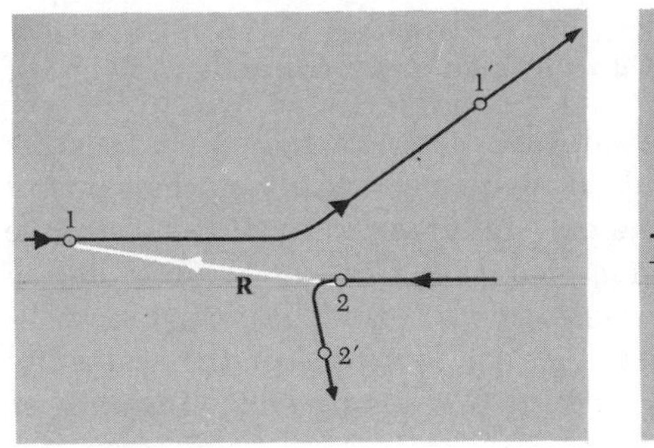

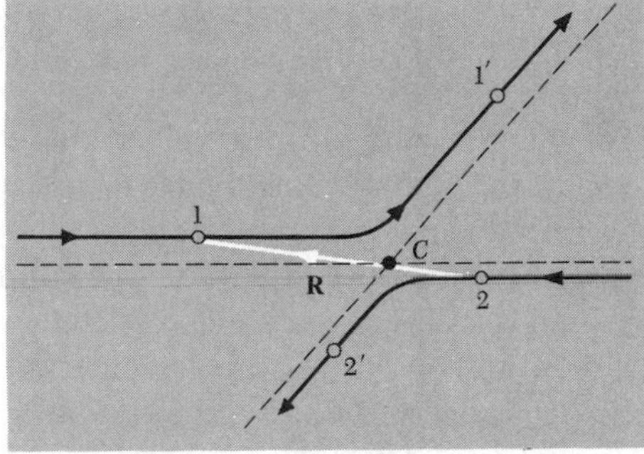

Laboratory system *Center-of-mass system*

Fig. 14·1·2 Classical trajectories for two colliding particles illustrated in the laboratory system and in the reference frame moving with their center of mass C.

(14·1·4). In the frame of reference which moves with the center of mass, the collision process is described very simply: The position vectors $\mathbf{r}_1^*$ and $\mathbf{r}_2^*$ of the particles relative to the center of mass are, by (14·1·10), at all times oppositely directed, and their magnitudes have a fixed ratio so that

$$m_1\mathbf{r}_1^* = -m_2\mathbf{r}_2^*$$

The vector $\mathbf{R}$ joining the particles passes always through their center of mass. If the force exerted on molecule 1 by molecule 2 is denoted by $\mathbf{F}_{12}$, it follows by (14·1·8) that

$$\frac{d\mathbf{p}_1}{dt} = \mu\frac{d^2\mathbf{R}}{dt^2} = \mathbf{F}_{12} \tag{14·1·11}$$

Hence the motion $\mathbf{R}(t)$ of molecule 1 relative to molecule 2 is the same as if it had a mass μ and were acted on by the force $\mathbf{F}_{12}$. The discussion of the two-particle problem is thus reduced to the solution of a simple one-particle problem. With respect to particle 2 the scattering process appears as shown in Fig. 14·1·3.

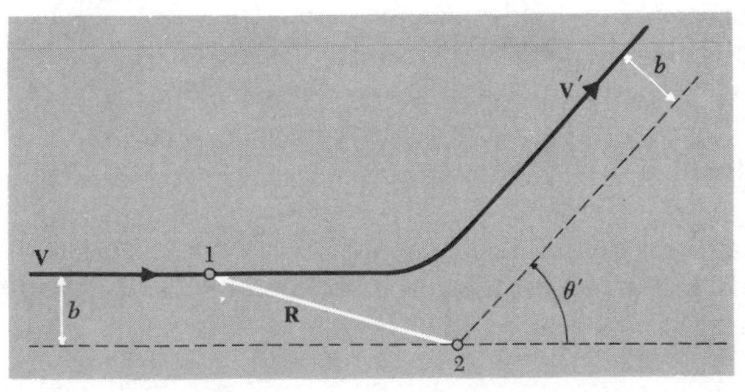

Fig. 14·1·3 Scattering process relative to molecule 2 regarded as fixed. The polar scattering angle is denoted by θ′. Classically, various scattering angles θ′ correspond to different values of the impact parameter b, the distance of closest approach of the molecules if there were no interaction between them.

14 · 2 *Scattering cross sections and symmetry properties*

Molecules having initially velocities $\mathbf{v}_1$ and $\mathbf{v}_2$ can get scattered in their relative motion through various angles θ' and φ' (depending classically on the value of the initial impact parameter b). If the only information available are these initial velocities $\mathbf{v}_1$ and $\mathbf{v}_2$ (and quantum mechanically this is *all* the information one can possibly have, since simultaneous determination of the impact parameter b would be impossible in principle), then the outcome of the scattering process must be described in statistical terms. This can be done in terms of the quantity σ' defined so that

$\sigma'(\mathbf{v}_1,\mathbf{v}_2 \to \mathbf{v}_1',\mathbf{v}_2')\, d^3\mathbf{v}_1'\, d^3\mathbf{v}_2' \equiv$ the number of molecules per unit time (per unit flux of type 1 molecules incident with relative velocity $\mathbf{V}$ upon a type 2 molecule) emerging after scattering with respective final velocities between $\mathbf{v}_1'$ and $\mathbf{v}_1' + d\mathbf{v}_1'$ and between $\mathbf{v}_2'$ and $\mathbf{v}_2' + d\mathbf{v}_2'$. (14 · 2 · 1)

Analogously to (14 · 1 · 3), one has

$$\mathbf{v}_1' = \mathbf{c}' + \frac{\mu}{m_1}\mathbf{V}' \qquad \text{and} \qquad \mathbf{v}_2' = \mathbf{c}' - \frac{\mu}{m_2}\mathbf{V}' \tag{14·2·2}$$

where, by virtue of conservation of momentum and energy, $\mathbf{c}' = \mathbf{c}$ and $|\mathbf{V}'| = |\mathbf{V}|$. Thus σ' must vanish unless $\mathbf{v}_1'$ and $\mathbf{v}_2'$ are such that these conditions are satisfied. Indeed, in terms of the variables $\mathbf{c}$ and $\mathbf{V}$, the scattering process is completely described in terms of the equivalent one-body problem of relative motion of Fig. 14 · 1 · 3, where $\mathbf{V}'$ is specified completely in terms of the polar and azimuthal angles θ' and φ' with respect to $\mathbf{V}$. Hence one can define a simpler but less symmetrical quantity, the differential scattering cross section σ already introduced in (12 · 2 · 1), by the statement that

$\sigma(\mathbf{V}')\, d\Omega' \equiv$ the number of molecules per unit time (per unit flux of type 1 molecules incident with relative velocity $\mathbf{V}$ upon a type 2 molecule) emerging after scattering with final relative velocity $\mathbf{V}'$ with a direction in the solid-angle range $d\Omega'$ about the angles θ' and φ'. (14 · 2 · 3)

Here σ depends in general on the relative speed $|\mathbf{V}| = |\mathbf{V}'|$ and on the angles θ' and φ'; i.e., it depends on the magnitude and direction of $\mathbf{V}'$. By the definitions (14 · 2 · 3) and (14 · 2 · 2), σ is related to σ' by

$$\sigma(\mathbf{V}')\, d\Omega' = \int_{\mathbf{c}'} \int_{V'} \sigma'(\mathbf{v}_1,\mathbf{v}_2 \to \mathbf{v}_1',\mathbf{v}_2')\, d^3\mathbf{v}_1'\, d^3\mathbf{v}_2' \tag{14·2·4}$$

where the integration is over all values of $\mathbf{c}'$ and of $|\mathbf{V}'|$. (The integration is, of course, trivial, since $\sigma' = 0$ unless $\mathbf{c}' = \mathbf{c}$ and $|\mathbf{V}'| = |\mathbf{V}|$.)

Remark It is useful to express the velocity range $d^3\mathbf{v}_1\, d^3\mathbf{v}_2$ in terms of the variables $\mathbf{c}$ and $\mathbf{V}$. One has

$$d^3\mathbf{v}_1\, d^3\mathbf{v}_2 = |J'|\, d^3\mathbf{c}\, d^3\mathbf{V} \tag{14·2·5}$$

where J' is the Jacobian of the transformation (14·1·3). But

$$dv_{1x}\, dv_{2x} = \frac{\partial(v_{1x},v_{2x})}{\partial(c_x,V_x)}\, dc_x\, dV_x = \begin{vmatrix} 1 & \dfrac{\mu}{m_1} \\ 1 & -\dfrac{\mu}{m_2} \end{vmatrix} dc_x\, dV_x$$

$$= -\mu\left(\frac{1}{m_2} + \frac{1}{m_1}\right) dc_x\, dV_x = -dc_x\, dV_x$$

where we have used (14·1·6). The transformation (14·2·5) is just the absolute value of a product of three such terms, corresponding to x, y, and z components. Hence one gets simply

$$d^3\mathbf{v}_1\, d^3\mathbf{v}_2 = d^3\mathbf{c}\, d^3\mathbf{V} \tag{14·2·6}$$

Similarly, one has

$$d^3\mathbf{v}_1'\, d^3\mathbf{v}_2' = d^3\mathbf{c}'\, d^3\mathbf{V}' \tag{14·2·7}$$

Now $\mathbf{c}' = \mathbf{c}$; furthermore $\mathbf{V}'$ and $\mathbf{V}$ differ only in direction but not in magnitude, and since volume elements remain unchanged under simple rotations of coordinates, $d^3\mathbf{V}' = d^3\mathbf{V}$. Hence (14·2·6) and (14·2·7) also imply

$$d^3\mathbf{v}_1'\, d^3\mathbf{v}_2' = d^3\mathbf{v}_1\, d^3\mathbf{v}_2 \tag{14·2·8}$$

The probability σ' has various useful symmetry properties which imply connections between a given collision process and related processes. The interactions between the molecules are basically electromagnetic in origin. The equations of motion must therefore have the following very general properties:

1. The equations of motion must be invariant under reversal of the sign of the time from $t \to -t$. Under such a time reversal, which implies of course a corresponding reversal of all the velocities, one obtains the "reverse" collision in which the particles simply retrace their paths in time. Thus one must have the following relation between scattering probabilities

$$\sigma'(\mathbf{v}_1,\mathbf{v}_2 \to \mathbf{v}_1',\mathbf{v}_2')\, d^3\mathbf{v}_1'\, d^3\mathbf{v}_2' = \sigma'(-\mathbf{v}_1',-\mathbf{v}_2' \to -\mathbf{v}_1,-\mathbf{v}_2)\, d^3\mathbf{v}_1\, d^3\mathbf{v}_2$$

or by (14·2·8)

$$\sigma'(\mathbf{v}_1,\mathbf{v}_2 \to \mathbf{v}_1',\mathbf{v}_2') = \sigma'(-\mathbf{v}_1',-\mathbf{v}_2' \to -\mathbf{v}_1,-\mathbf{v}_2) \tag{14·2·9}$$

****Remark*** If the particles had spin and the interaction between them were spin-dependent, then time reversal would also necessitate simultaneous reversal of all spins. Equation (14·2·9) would therefore not be valid as it stands, but would still hold if averaged over all possible directions of the initial and final spins.

2. The equations of motion must be invariant under the transformation which reverses the sign of all spatial coordinates so that $\boldsymbol{r} \to -\boldsymbol{r}$. Under such a "space inversion," the signs of all velocities also change, but the time order does not. Thus one must have

$$\sigma'(\boldsymbol{v}_1,\boldsymbol{v}_2 \to \boldsymbol{v}_1',\boldsymbol{v}_2') = \sigma'(-\boldsymbol{v}_1,-\boldsymbol{v}_2 \to -\boldsymbol{v}_1',-\boldsymbol{v}_2') \qquad (14 \cdot 2 \cdot 10)$$

It is of particular interest to consider the so-called "inverse" collision which, by definition, is obtained from the original collision by interchanging the initial and final states. Whereas in the original collision the particles collide with velocities $\boldsymbol{v}_1$ and $\boldsymbol{v}_2$ and emerge with velocities $\boldsymbol{v}_1'$ and $\boldsymbol{v}_2'$, in the inverse collision precisely the opposite takes place; i.e., the particles collide with velocities $\boldsymbol{v}_1'$ and $\boldsymbol{v}_2'$ and emerge with velocities $\boldsymbol{v}_1$ and $\boldsymbol{v}_2$ (see Figs. 14·2·1 and 14·2·2). The inverse collision can be obtained from the original collision by considering the operation of time reversal followed by the operation of space inversion which changes the sign of all spatial coordinates. Successive application of (14·2·9) and (14·2·10) shows that the collision probabilities for the original and inverse collisions are also equal; i.e., by applying the operation of space inversion to the right side of (14·2·9), one obtains by (14·2·10)

$$\sigma'(\boldsymbol{v}_1,\boldsymbol{v}_2 \to \boldsymbol{v}_1',\boldsymbol{v}_2') = \sigma'(\boldsymbol{v}_1',\boldsymbol{v}_2' \to \boldsymbol{v}_1,\boldsymbol{v}_2) \qquad (14 \cdot 2 \cdot 11)$$

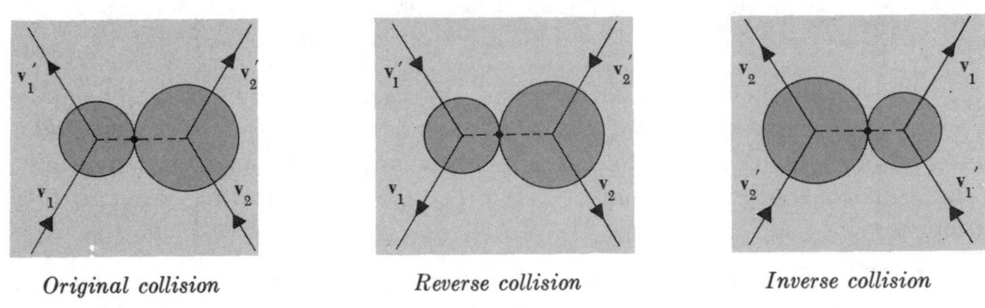

Fig. 14·2·1 ***Figure illustrating related collisions between hard spheres. The scattering cross sections are the same for all these collisions.***

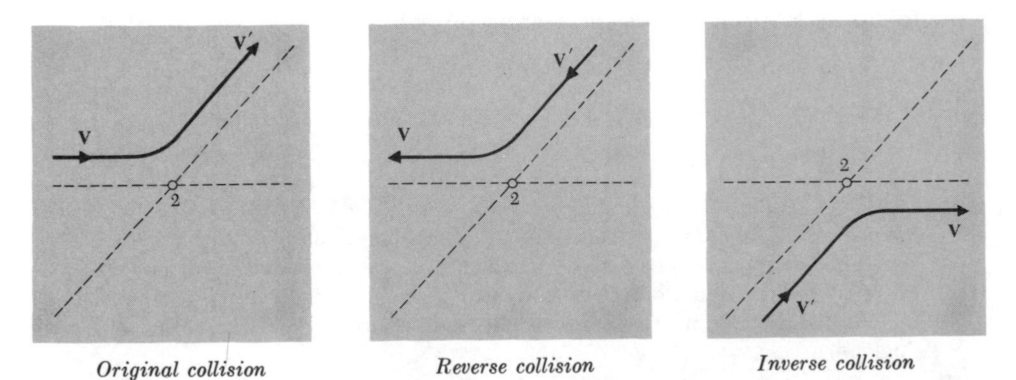

Fig. 14·2·2 ***Figure illustrating classical relative orbits for related collisions.***

14·3 *Derivation of the Boltzmann equation*

We are now in a position to make use of our knowledge of molecular collisions to derive an explicit expression for D_cf in the Boltzmann equation (13·6·1),

$$Df = D_cf \tag{14·3·1}$$

In order to calculate D_cf, the rate of change of f caused by collisions, we shall make the following assumptions:

a. The gas is sufficiently dilute that only two-particle collisions need be taken into account.

b. Any possible effects of the external force $\mathbf{F}$ on the magnitude of the collision cross section can be ignored.

c. The distribution function $f(\mathbf{r},\mathbf{v},t)$ does not vary appreciably during a time interval of the order of the duration of a molecular collision, nor does it vary appreciably over a spatial distance of the order of the range of intermolecular forces.

d. When considering a collision between two molecules one can neglect possible correlations between their initial velocities prior to the collision. This fundamental approximation in the theory is called the assumption of "molecular chaos." It is justified when the gas density is sufficiently low. Then the mean free path l is much greater than the range of intermolecular forces, and two molecules originate before their encounter at a relative separation which is of the order of l and thus sufficiently large that a correlation between their initial velocities is unlikely.

Focus attention on molecules located in the volume element $d^3\mathbf{r}$ located between $\mathbf{r}$ and $\mathbf{r} + d\mathbf{r}$, and consider the collisions which occur there in the time between t and $t + dt$. (Here $d\mathbf{r}$ is taken to be large compared to the range of intermolecular forces and dt to be large compared to the duration of a collision. Nevertheless, they can be considered infinitesimally small with respect to variations in f by virtue of the assumption (*c*).) We are interested in calculating how collisions cause a net change $D_cf(\mathbf{r},\mathbf{v},t)\, d^3\mathbf{r}\, d^3\mathbf{v}\, dt$ in the number of such molecules with velocity between $\mathbf{v}$ and $\mathbf{v} + d\mathbf{v}$. First, the molecules in $d^3\mathbf{r}$ can be thrown *out* of this velocity range by virtue of collisions with other molecules; we denote the resulting *decrease* in time dt of the number of such molecules by $D_c^{(-)}f(\mathbf{r},\mathbf{v},t)\, d^3\mathbf{r}\, d^3\mathbf{v}\, dt$. Second, molecules in $d^3\mathbf{r}$ whose velocity is originally not in the range between $\mathbf{v}$ and $\mathbf{v} + d\mathbf{v}$ can be thrown *into* this velocity range by virtue of collisions with other molecules; we denote the resulting *increase* in time dt of the number of molecules thus scattered into this velocity range by $D_c^{(+)}f(\mathbf{r},\mathbf{v},t)\, d^3\mathbf{r}\, d^3\mathbf{v}\, dt$. Hence one can write

$$D_cf = -D_c^{(-)}f + D_c^{(+)}f \tag{14·3·2}$$

To calculate $D_c^{(-)}f$, we consider in the volume element $d^3\mathbf{r}$ molecules with velocity near $\mathbf{v}$ (call these A molecules) which are scattered out of this velocity range by virtue of collisions with other molecules (call them A_1 molecules) which are in the same volume element $d^3\mathbf{r}$ and which have some velocity $\mathbf{v}_1$.

The probability of occurrence of such a collision where an A molecule changes its velocity from $\mathbf{v}$ to one near $\mathbf{v}'$, while an A_1 molecules changes its velocity from $\mathbf{v}_1$ to one near $\mathbf{v}_1'$ is by (14·2·1) described in terms of the scattering probability $\sigma'(\mathbf{v},\mathbf{v}_1 \to \mathbf{v}',\mathbf{v}_1')\, d^3\mathbf{v}'\, d^3\mathbf{v}_1'$. To obtain the *total* collision-induced decrease $D_C^{(-)}f\, d^3\mathbf{r}\, d^3\mathbf{v}\, dt$ in time dt of the number of molecules located in $d^3\mathbf{r}$ with velocity between $\mathbf{v}$ and $\mathbf{v} + d\mathbf{v}$ one must first multiply $\sigma'\, d^3\mathbf{v}'\, d^3\mathbf{v}_1'$ by the relative flux $|\mathbf{v} - \mathbf{v}_1| f(\mathbf{r},\mathbf{v},t)\, d^3\mathbf{v}$ of A molecules incident upon an A_1 molecule and must then multiply this by the number of A_1 molecules $f(\mathbf{r},\mathbf{v}_1,t)\, d^3\mathbf{r}\, d^3\mathbf{v}_1$ which can do such scattering. Then one has to sum the result over all possible initial velocities $\mathbf{v}_1$ of the A_1 molecules with which A can collide and over all possible final velocities $\mathbf{v}'$ and $\mathbf{v}_1'$ of the scattered A and A_1 molecules. Thus one obtains:

$$D_C^{(-)}f(\mathbf{r},\mathbf{v},t)\, d^3\mathbf{r}\, d^3\mathbf{v}\, dt = \int_{\mathbf{v}_1'} \int_{\mathbf{v}'} \int_{\mathbf{v}_1} [|\mathbf{v} - \mathbf{v}_1| f(\mathbf{r},\mathbf{v},t)\, d^3\mathbf{v}][f(\mathbf{r},\mathbf{v}_1,t)\, d^3\mathbf{r}\, d^3\mathbf{v}_1][\sigma'(\mathbf{v},\mathbf{v}_1 \to \mathbf{v}',\mathbf{v}_1')\, d^3\mathbf{v}'\, d^3\mathbf{v}_1'] \quad (14 \cdot 3 \cdot 3)$$

Here we have used the fundamental assumption (*d*) of molecular chaos in writing for the probability of simultaneous presence in $d^3\mathbf{r}$ of molecules with respective velocities near $\mathbf{v}$ and $\mathbf{v}_1$ an expression proportional to the simple product

$$f(\mathbf{r},\mathbf{v},t)\, d^3\mathbf{v} \cdot f(\mathbf{r},\mathbf{v}_1,t)\, d^3\mathbf{v}_1$$

which assumes the absence of any correlations between the initial velocities $\mathbf{v}$ and $\mathbf{v}_1$, so that these are statistically independent.

We now turn to the calculation of $D_C^{(+)}f$. Considering again the same volume element $d^3\mathbf{r}$, we ask how many molecules will end up after collisions with a velocity in the range between $\mathbf{v}$ and $\mathbf{v} + d\mathbf{v}$. But this involves precisely a consideration of what we called "inverse collisions" in Sec. 14·2. Namely, we should like to consider all molecules in $d^3\mathbf{r}$ with arbitrary initial velocities $\mathbf{v}'$ and $\mathbf{v}_1'$ which are such that, after collision, one molecule acquires a velocity in the range of interest between $\mathbf{v}$ and $\mathbf{v} + d\mathbf{v}$, while the other acquires some velocity between $\mathbf{v}_1$ and $\mathbf{v}_1 + d\mathbf{v}_1$. This scattering process is described by the scattering probability $\sigma'(\mathbf{v}',\mathbf{v}_1' \to \mathbf{v},\mathbf{v}_1)$. The relative flux of molecules with initial velocity near $\mathbf{v}'$ is $|\mathbf{v}' - \mathbf{v}_1'| f(\mathbf{r},\mathbf{v}',t)\, d^3\mathbf{v}'$, and these molecules get scattered by the $f(\mathbf{r},\mathbf{v}_1',t)\, d^3\mathbf{r}\, d^3\mathbf{v}_1'$ molecules with velocity near $\mathbf{v}_1'$. Hence one can write for the total increase in time dt of the number of molecules located in $d^3\mathbf{r}$ with velocity between $\mathbf{v}$ and $\mathbf{v} + d\mathbf{v}$ the expression

$$D_C^{(+)}f(\mathbf{r},\mathbf{v},t)\, d^3\mathbf{r}\, d^3\mathbf{v}\, dt = \int_{\mathbf{v}_1} \int_{\mathbf{v}_1'} \int_{\mathbf{v}'} [|\mathbf{v}' - \mathbf{v}_1'| f(\mathbf{r},\mathbf{v}',t)\, d^3\mathbf{v}'][f(\mathbf{r},\mathbf{v}_1',t)\, d^3\mathbf{r}\, d^3\mathbf{v}_1'][\sigma'(\mathbf{v}',\mathbf{v}_1' \to \mathbf{v},\mathbf{v}_1)\, d^3\mathbf{v}\, d^3\mathbf{v}_1] \quad (14 \cdot 3 \cdot 4)$$

where the integrations are over all the possible initial velocities $\mathbf{v}'$ and $\mathbf{v}_1'$ of the molecules, and over all possible final velocities $\mathbf{v}_1$ of the other molecule whose velocity does not end up in the range of interest near $\mathbf{v}$.

By (14·3·2), D_Cf is then obtained by subtracting (14·3·3) from (14·3·4). Note the following simplifying features. By (14·2·11) the probabilities for inverse collisions are equal so that

$$\sigma'(\mathbf{v}',\mathbf{v}_1' \to \mathbf{v},\mathbf{v}_1) = \sigma'(\mathbf{v},\mathbf{v}_1 \to \mathbf{v}',\mathbf{v}_1')$$

Furthermore, we can introduce the relative velocities

$$\mathbf{V} \equiv \mathbf{v} - \mathbf{v}_1, \qquad \mathbf{V}' \equiv \mathbf{v}' - \mathbf{v}_1' \tag{14·3·5}$$

Then the conservation of energy for elastic collisions implies that

$$|\mathbf{V}'| = |\mathbf{V}| \equiv V$$

In order to save writing, it is also convenient to introduce the abbreviations

$$\begin{aligned} f &\equiv f(\mathbf{r},\mathbf{v},t), \qquad & f_1 &\equiv f(\mathbf{r},\mathbf{v}_1,t) \\ f' &\equiv f(\mathbf{r},\mathbf{v}',t), \qquad & f_1' &\equiv f(\mathbf{r},\mathbf{v}_1',t) \end{aligned} \tag{14·3·6}$$

Then (14·3·2) becomes

$$D_c f = \int_{\mathbf{v}_1} \int_{\mathbf{v}_1'} \int_{\mathbf{v}'} (f'f_1' - ff_1) V\sigma'(\mathbf{v},\mathbf{v}_1 \to \mathbf{v}',\mathbf{v}_1')\, d^3\mathbf{v}_1\, d^3\mathbf{v}'\, d\mathbf{v}_1' \tag{14·3·7}$$

One can use (14·2·4) to express this result in terms of $\mathbf{V}'$ and the solid-angle range $d\Omega'$ about this vector. Using (13·2·8), the Boltzmann equation (14·3·1) for $f(\mathbf{r},\mathbf{v},t)$ can then be written in the explicit form

$$\blacktriangleright \qquad \frac{\partial f}{\partial t} + \mathbf{v} \cdot \frac{\partial f}{\partial \mathbf{r}} + \frac{\mathbf{F}}{m} \cdot \frac{\partial f}{\partial \mathbf{v}} = \int_{\mathbf{v}_1} \int_{\Omega'} (f'f_1' - ff_1) V\sigma\, d\Omega'\, d^3\mathbf{v}_1 \tag{14·3·8}$$

where $\sigma = \sigma(\mathbf{V}')$.

14 · 4 *Equation of change for mean values*

Consider any function $\chi(\mathbf{r},\mathbf{v},t)$ which describes a property of a molecule that has a position $\mathbf{r}$ and a velocity $\mathbf{v}$ at time t. As in (13·1·4), the mean value of χ is defined by

$$\langle \chi(\mathbf{r},t) \rangle \equiv \frac{1}{n(\mathbf{r},t)} \int d^3\mathbf{v}\, f(\mathbf{r},\mathbf{v},t)\chi(\mathbf{r},\mathbf{v},t) \tag{14·4·1}$$

where $n(\mathbf{r},t)$ is the mean number of molecules per unit volume. We should like to derive an equation which describes how $\langle\chi\rangle$ varies as a function of t and $\mathbf{r}$. This can be done in two ways, either by analyzing the situation directly from the beginning, or by starting from the Boltzmann equation (14·3·8). Since both approaches are instructive, we shall illustrate them in turn.

Direct analysis Consider the fixed volume element $d^3\mathbf{r}$, located between $\mathbf{r}$ and $\mathbf{r} + d\mathbf{r}$, which contains $n(\mathbf{r},t)\, d^3\mathbf{r}$ molecules. In a time interval between t and $t + dt$ the total mean value $\langle n\, d^3\mathbf{r}\, \chi\rangle$ of the quantity χ for all molecules in $d^3\mathbf{r}$ increases by an amount

$$\frac{\partial}{\partial t} \langle n\chi\rangle\, d^3\mathbf{r}\, dt = A_{\text{int}} + A_{\text{flux}} + A_{\text{col}} \tag{14·4·2}$$

Note that n can always be taken outside the averaging brackets, since it does not depend on $\mathbf{v}$. The quantities A represent various contributions to be described presently.

1. There is an intrinsic increase A_{int} in the total mean value of χ due to the fact that the quantity $\chi(\boldsymbol{r},\boldsymbol{v},t)$ for each molecule in $d^3\boldsymbol{r}$ changes. In time dt, each molecule changes position by $d\boldsymbol{r} = \boldsymbol{v}\,dt$ and velocity by $d\boldsymbol{v} = (\boldsymbol{F}/m)\,dt$; hence the corresponding change in χ is given by

$$\frac{\partial \chi}{\partial t}\,dt + \frac{\partial \chi}{\partial x_\alpha}\,v_\alpha\,dt + \frac{\partial \chi}{\partial v_\alpha}\frac{F_\alpha}{m}\,dt$$

Here x_α and v_α denote the respective cartesian components of the vectors $\boldsymbol{r}$ and $\boldsymbol{v}$, and we have adopted the "summation convention" that a summation from 1 to 3 is implied whenever a Greek subscript occurs twice. Hence the intrinsic increase in the mean value of χ in $d^3\boldsymbol{r}$ is given by

$$A_{\text{int}} = \langle n\,d^3\boldsymbol{r}\;D\chi\;dt\rangle = n\,d^3\boldsymbol{r}\,dt\,\langle D\chi\rangle \tag{14·4·3}$$

where

$$D\chi \equiv \frac{\partial \chi}{\partial t} + v_\alpha\frac{\partial \chi}{\partial x_\alpha} + \frac{F_\alpha}{m}\frac{\partial \chi}{\partial v_\alpha} = \frac{\partial \chi}{\partial t} + \boldsymbol{v}\cdot\frac{\partial \chi}{\partial \boldsymbol{r}} + \frac{\boldsymbol{F}}{m}\cdot\frac{\partial \chi}{\partial \boldsymbol{v}} \tag{14·4·4}$$

2. There is an increase A_{flux} in the total mean value of χ in $d^3\boldsymbol{r}$ due to the net flux of molecules which enter the volume element $d^3\boldsymbol{r}$ in time dt. By an argument similar to that in Sec. 13·2, the increase in the mean value of χ caused by molecules entering the element of volume in time dt through the face x_1 is the mean value contained in the volume $v_1\,dt\,dx_2\,dx_3$, i.e., $\langle n\chi[v_1\,dt\,dx_2\,dx_3]\rangle$. The decrease in the mean value of χ caused by molecules leaving through the face $x_1 + dx_1$ is correspondingly given by

$$\langle n\chi v_1\,dt\,dx_2\,dx_3\rangle + \frac{\partial}{\partial x_1}\langle n\chi v_1\,dt\,dx_2\,dx_3\rangle\,dx_1$$

By subtracting these two expressions one obtains for the net increase in the mean value of χ due to molecules entering and leaving through these two faces the contribution

$$-\frac{\partial}{\partial x_1}\langle n\chi v_1\,dt\,d^3\boldsymbol{r}\rangle$$

Adding contributions from all other faces one gets thus

$$A_{\text{flux}} = -\frac{\partial}{\partial x_\alpha}\langle n v_\alpha \chi\rangle\,dt\,d^3\boldsymbol{r} \tag{14·4·5}$$

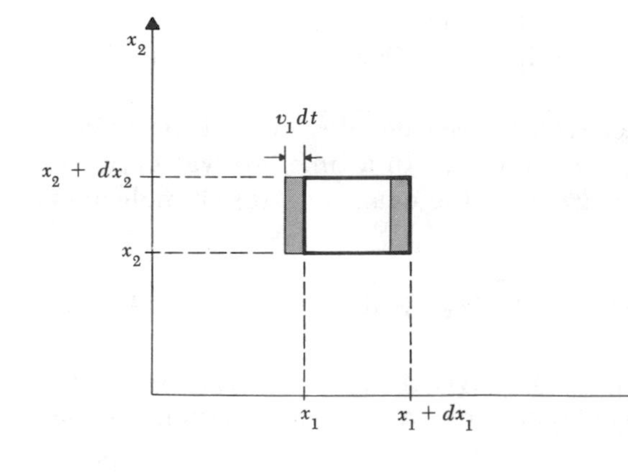

***Fig. 14·4·1** Figure illustrating a two-dimensional projection of the volume element $d^3\boldsymbol{r}$.*

3. Finally, there is an increase A_{col} in the total mean value of χ in $d^3\boldsymbol{r}$ because of collisions between molecules in this volume element. In such a collision where a molecule with velocity $\boldsymbol{v}$ collides with another one of velocity $\boldsymbol{v}_1$, and the molecules then emerge with respective final velocities $\boldsymbol{v}'$ and $\boldsymbol{v}_1'$, the change in χ is equal to

$$\Delta\chi = \chi' + \chi_1' - \chi - \chi_1 \tag{14·4·6}$$

where

$$\begin{aligned} \chi &\equiv \chi(\boldsymbol{r},\boldsymbol{v},t), & \chi_1 &\equiv \chi(\boldsymbol{r},\boldsymbol{v}_1,t) \\ \chi' &\equiv \chi(\boldsymbol{r},\boldsymbol{v}',t), & \chi_1' &\equiv \chi(\boldsymbol{r},\boldsymbol{v}_1',t) \end{aligned} \tag{14·4·7}$$

The number of such collisions is again given by

$$[|\boldsymbol{v} - \boldsymbol{v}_1| f(\boldsymbol{r},\boldsymbol{v},t)\, d^3\boldsymbol{v}][\sigma'\, d^3\boldsymbol{v}'\, d^3\boldsymbol{v}_1'][f(\boldsymbol{r},\boldsymbol{v}_1,t)\, d^3\boldsymbol{r}\, d^3\boldsymbol{v}_1]$$

where we have used the assumption of molecular chaos. Thus A_{col} is obtained by multiplying this number by $\Delta\chi$ and summing over all possible initial velocities $\boldsymbol{v}$ and $\boldsymbol{v}_1$ of the molecules and over all their possible final velocities. This result must then be divided by 2, since in the above sums over molecular velocities each colliding pair of molecules is counted twice. Thus one has

$$A_{col} = \tfrac{1}{2}\, d^3\boldsymbol{r}\, dt \int\!\!\int\!\!\int\!\!\int d^3\boldsymbol{v}\, d^3\boldsymbol{v}_1\, d^3\boldsymbol{v}'\, d^3\boldsymbol{v}_1' f f_1 V \sigma'\, \Delta\chi \tag{14·4·8}$$

By (14·4·3), (14·4·5) and (14·4·8), Eq. (14·4·2) then becomes

$$\blacktriangleright \qquad \frac{\partial}{\partial t}\langle n\chi\rangle = n\langle D\chi\rangle - \frac{\partial}{\partial x_\alpha}\langle n v_\alpha \chi\rangle + \mathcal{C}(\chi) \tag{14·4·9}$$

where $\mathcal{C}(\chi)$ denotes the rate of change of χ per unit volume due to collisions; by (14·4·8) and (14·2·4) it can be written

$$\blacktriangleright \qquad \mathcal{C}(\chi) = \frac{A_{col}}{d^3\boldsymbol{r}\, dt} = \frac{1}{2}\int\!\!\int\!\!\int d^3\boldsymbol{v}\, d^3\boldsymbol{v}_1\, d\Omega'\, f f_1 V \sigma\, \Delta\chi \tag{14·4·10}$$

Note again that n can always be taken outside the averaging angular brackets since it does not depend on $\boldsymbol{v}$. Equation (14·4·9) is sometimes called "Enskog's equation of change."

Analysis based on the Boltzmann equation To find the equation satisfied by $\langle\chi\rangle$ defined in (14·4·1), we multiply both sides of the Boltzmann equation (14·3·8) by χ and then integrate over all velocities $\boldsymbol{v}$. Thus we get

$$\int d^3\boldsymbol{v}\, Df\, \chi = \int d^3\boldsymbol{v}\, D_c f\, \chi \tag{14·4·11}$$

where
$$\int d^3\boldsymbol{v}\, Df\,\chi \equiv \int d^3\boldsymbol{v}\, \frac{\partial f}{\partial t}\chi + \int d^3\boldsymbol{v}\, \boldsymbol{v}\cdot\frac{\partial f}{\partial \boldsymbol{r}}\chi + \int d^3\boldsymbol{v}\, \frac{\boldsymbol{F}}{m}\cdot\frac{\partial f}{\partial \boldsymbol{v}}\chi \tag{14·4·12}$$

and where
$$\mathcal{C}(\chi) \equiv \int d^3\boldsymbol{v}\, D_c f\,\chi = \int\!\!\int\!\!\int d^3\boldsymbol{v}\, d^3\boldsymbol{v}_1\, d\Omega'\, (f'f_1' - f f_1) V\sigma\chi \tag{14·4·13}$$

Let us now transform the integrals in (14·4·12) into quantities which are averages, i.e., into integrals which involve f itself rather than its derivatives. Thus we have

$$\int d^3\boldsymbol{v}\, \frac{\partial f}{\partial t}\chi = \int d^3\boldsymbol{v} \left[\frac{\partial}{\partial t}(f\chi) - f\frac{\partial \chi}{\partial t}\right] = \frac{\partial}{\partial t}\int d^3\boldsymbol{v}\, f\chi - \int d^3\boldsymbol{v}\, f\frac{\partial \chi}{\partial t}$$

since the order of differentiation with respect to t and integration over $\boldsymbol{v}$ can be interchanged. Hence

$$\int d^3\boldsymbol{v}\, \frac{\partial f}{\partial t}\, \chi = \frac{\partial}{\partial t}\, (n\langle\chi\rangle) - n \left\langle \frac{\partial \chi}{\partial t} \right\rangle \tag{14·4·14}$$

The second integral in (14·4·12) can be similarly rewritten. To avoid confusion, we express the vector quantities in terms of their cartesian components denoted by Greek subscripts and again use the summation convention. Keeping in mind that $\boldsymbol{r}$, $\boldsymbol{v}$, t are to be considered as *independent* variables, one has

$$\begin{aligned} \int d^3\boldsymbol{v}\, \boldsymbol{v} \cdot \frac{\partial f}{\partial \boldsymbol{r}}\, \chi &= \int d^3\boldsymbol{v}\, v_\alpha \frac{\partial f}{\partial x_\alpha}\, \chi \\ &= \int d^3\boldsymbol{v} \left[\frac{\partial}{\partial x_\alpha}\, (v_\alpha f \chi) - v_\alpha f \frac{\partial \chi}{\partial x_\alpha} \right] \\ &= \frac{\partial}{\partial x_\alpha} \int d^3\boldsymbol{v}\, f v_\alpha \chi - \int d^3\boldsymbol{v}\, f v_\alpha \frac{\partial \chi}{\partial x_\alpha} \end{aligned}$$

or

$$\int d^3\boldsymbol{v}\, \boldsymbol{v} \cdot \frac{\partial f}{\partial \boldsymbol{r}}\, \chi = \frac{\partial}{\partial x_\alpha}\, (n\langle v_\alpha \chi\rangle) - n \left\langle v_\alpha \frac{\partial \chi}{\partial x_\alpha} \right\rangle \tag{14·4·15}$$

Finally, since we assumed the force $\boldsymbol{F}$ to be independent of velocity, one gets

$$\begin{aligned} \int d^3\boldsymbol{v}\, \frac{\boldsymbol{F}}{m} \cdot \frac{\partial f}{\partial \boldsymbol{v}}\, \chi &= \int d^3\boldsymbol{v}\, \frac{F_\alpha}{m} \frac{\partial f}{\partial v_\alpha}\, \chi \\ &= \int d^3\boldsymbol{v} \left[\frac{\partial}{\partial v_\alpha} \left(\frac{F_\alpha}{m} f \chi \right) - \frac{F_\alpha}{m} f \frac{\partial \chi}{\partial v_\alpha} \right] \\ &= \left[\frac{F_\alpha}{m} f \chi \right]_{v_\alpha = -\infty}^{v_\alpha = +\infty} - \int d^3\boldsymbol{v}\, \frac{F_\alpha}{m} f \frac{\partial \chi}{\partial v_\alpha} \end{aligned}$$

Since $f \to 0$ as $|v_\alpha| \to \infty$, this becomes simply

$$\int d^3\boldsymbol{v}\, \frac{\boldsymbol{F}}{m} \cdot \frac{\partial f}{\partial \boldsymbol{v}}\, \chi = -\frac{F_\alpha}{m}\, n \left\langle \frac{\partial \chi}{\partial v_\alpha} \right\rangle \tag{14·4·16}$$

Hence (14·4·12) is obtained by adding the expressions (14·4·14) through (14·4·16). The result is

▶
$$\int d^3\boldsymbol{v}\, Df\, \chi = \frac{\partial}{\partial t}\, (n\langle\chi\rangle) + \frac{\partial}{\partial x_\alpha}\, (n\langle v_\alpha \chi\rangle) - n\langle D\chi\rangle \tag{14·4·17}$$

where $D\chi$ is defined in (14·4·4).

We now turn to the evaluation of the collision term (14·4·13). This is by (14·2·4) most symmetrically written in the form

$$\mathcal{C}(\chi) = \iiiint d^3\boldsymbol{v}\, d^3\boldsymbol{v}_1\, d^3\boldsymbol{v}'\, d^3\boldsymbol{v}_1'\, (f'f_1' - ff_1) V\sigma'(\boldsymbol{v},\boldsymbol{v}_1 \to \boldsymbol{v}',\boldsymbol{v}_1')\chi(\boldsymbol{r},\boldsymbol{v},t) \tag{14·4·18}$$

The high symmetry of this expression can be exploited by interchanging $\boldsymbol{v}$ and $\boldsymbol{v}_1$ as well as $\boldsymbol{v}'$ and $\boldsymbol{v}_1'$. This leaves σ' unchanged so that one obtains

$$\mathcal{C}(\chi) = \iiiint d^3\boldsymbol{v}\, d^3\boldsymbol{v}_1\, d^3\boldsymbol{v}'\, d^3\boldsymbol{v}_1'\, (f'f_1' - ff_1) V\sigma'\chi(\boldsymbol{r},\boldsymbol{v}_1,t) \tag{14·4·19}$$

Adding (14·4·18) and (14·4·19) then yields

$$\mathcal{C}(\chi) = \tfrac{1}{2}\iiiint d^3\boldsymbol{v}\, d^3\boldsymbol{v}_1\, d^3\boldsymbol{v}'\, d^3\boldsymbol{v}_1'\, (f'f_1' - ff_1)V\sigma'[\chi + \chi_1] \quad (14\cdot 4\cdot 20)$$

with χ and χ_1 defined in (14·4·7).

But one can exploit a further symmetry by interchanging $\boldsymbol{v}$ and $\boldsymbol{v}'$, as well as $\boldsymbol{v}_1$ and $\boldsymbol{v}_1'$. This leads to the inverse collision which leaves σ' also unchanged. Thus one can write

$$\iiiint d^3\boldsymbol{v}\, d^3\boldsymbol{v}_1\, d^3\boldsymbol{v}'\, d^3\boldsymbol{v}_1'\, f'f_1'V\sigma\, [\chi + \chi_1] = \iiiint d^3\boldsymbol{v}\, d^3\boldsymbol{v}_1\, d^3\boldsymbol{v}'\, d^3\boldsymbol{v}_1'\, ff_1V\sigma'[\chi' + \chi_1'] \quad (14\cdot 4\cdot 21)$$

where χ' and χ_1' are defined in (14·4·7).

Substituting (14·4·21) in (14·4·20) then yields

$$\mathcal{C}(\chi) = \tfrac{1}{2}\iiiint d^3\boldsymbol{v}\, d^3\boldsymbol{v}_1\, d^3\boldsymbol{v}'\, d^3\boldsymbol{v}_1'\, ff_1V\sigma'\, \Delta\chi$$

or

▶ $$\mathcal{C}(\chi) = \tfrac{1}{2}\iiint d^3\boldsymbol{v}\, d^3\boldsymbol{v}_1\, d\Omega'\, ff_1V\sigma\, \Delta\chi \quad (14\cdot 4\cdot 22)$$

where $\Delta\chi \equiv \chi' + \chi_1' - \chi - \chi_1$ is the total change in the quantity χ in the collision between two molecules. Substitution of (14·4·17) and (14·4·22) into (14·4·11) leads then again to Eq. (14·4·9).

14·5 *Conservation equations and hydrodynamics*

The equation of change (14·4·9) becomes particularly simple if χ refers to a quantity which is conserved in collisions between molecules so that $\Delta\chi = 0$. Then $\mathcal{C}(\chi) = 0$ and Eq. (14·4·9) reduces simply to*

$$\frac{\partial}{\partial t}\langle n\chi\rangle + \frac{\partial}{\partial x_\alpha}\langle nv_\alpha\chi\rangle = n\langle D\chi\rangle \quad (14\cdot 5\cdot 1)$$

The fundamental quantities which are conserved in a collision are, first, any constant, in particular the mass m of a molecule. Furthermore, each component of the total momentum of the colliding molecules is conserved. Finally, assuming the internal energies of all molecules to remain unchanged in collisions, the total kinetic energy of the colliding molecules is also conserved. These conservation laws lead then to five corresponding cases where $\Delta\chi = 0$ in (14·4·6). These are

conservation of mass	$\chi = m$		(14·5·2)
conservation of momentum	$\chi = mv_\gamma,$	$\gamma = 1, 2, 3$	(14·5·3)
conservation of energy	$\chi = \frac{1}{2}mv^2$		(14·5·4)

One then obtains by (14·5·1) five corresponding conservation laws satisfied by the gas.

* This equation and all subsequent considerations of this section are very general. They depend only on the conservation laws, not on the assumption of molecular chaos and the consequent special form (14.4.10) of $\mathcal{C}(\chi)$ involving ff_1 as a simple product.

Conservation of mass Putting $\chi = m$, Eq. (14·5·1) leads immediately to

$$\frac{\partial}{\partial t}(nm) + \frac{\partial}{\partial x_\alpha}\langle nmv_\alpha\rangle = 0 \tag{14·5·5}$$

Here n is independent of $\boldsymbol{v}$ and can be taken outside the angular averaging brackets. Also by (13·1·5), $\langle \boldsymbol{v}\rangle = \boldsymbol{u}$, the mean velocity of the gas. Furthermore, the mass density of the gas, i.e., its mass per unit volume, is given by

$$\rho(\boldsymbol{r},t) = mn(\boldsymbol{r},t) \tag{14·5·6}$$

Hence (14·5·5) becomes simply

$$\blacktriangleright \qquad \frac{\partial \rho}{\partial t} + \frac{\partial}{\partial x_\alpha}(\rho u_\alpha) = 0 \tag{14·5·7}$$

or using the vector notation of the divergence,

$$\frac{\partial \rho}{\partial t} + \nabla \cdot (\rho \boldsymbol{u}) = 0 \tag{14·5·8}$$

This is the so-called "equation of continuity" of hydrodynamics. It expresses the macroscopic condition necessary to guarantee the conservation of mass.

Conservation of momentum By (14·5·3) we put $\chi = mv_\gamma$ so that Eq. (14·5·1) becomes

$$\frac{\partial}{\partial t}\langle nmv_\gamma\rangle + \frac{\partial}{\partial x_\alpha}\langle nmv_\alpha v_\gamma\rangle = n\langle m\, Dv_\gamma\rangle \tag{14·5·9}$$

By the definition (14·4·4),

$$Dv_\gamma = \frac{F_\alpha}{m}\frac{\partial v_\gamma}{\partial v_\alpha} = \frac{F_\alpha}{m}\delta_{\gamma\alpha} = \frac{F_\gamma}{m}$$

Hence (14·5·9) becomes, using (14·5·7),

$$\frac{\partial}{\partial t}(\rho u_\gamma) + \frac{\partial}{\partial x_\alpha}(\rho\langle v_\alpha v_\gamma\rangle) = \rho F_\gamma' \tag{14·5·10}$$

where

$$\boldsymbol{F}' \equiv \frac{\boldsymbol{F}}{m} \tag{14·5·11}$$

is the force per unit mass of the fluid.

The second term in (14·5·10) is usefully expressed in terms of $\boldsymbol{u}$ and the peculiar velocity $\boldsymbol{U}$. By (13·1·6),

$$\boldsymbol{v} = \boldsymbol{u} + \boldsymbol{U}$$

and $\quad \langle v_\alpha v_\gamma\rangle = \langle (u_\alpha + U_\alpha)(u_\gamma + U_\gamma)\rangle = \langle u_\alpha u_\gamma + U_\alpha U_\gamma + u_\alpha U_\gamma + U_\alpha u_\gamma\rangle$

or

$$\langle v_\alpha v_\gamma\rangle = u_\alpha u_\gamma + \langle U_\alpha U_\gamma\rangle \tag{14·5·12}$$

since

$$\langle u_\alpha U_\gamma\rangle = u_\alpha\langle U_\gamma\rangle = 0$$

Furthermore, we define the "pressure tensor" $P_{\alpha\gamma}$ by

$$P_{\alpha\gamma} \equiv \rho\langle U_\alpha U_\gamma\rangle, \qquad P_{\gamma\alpha} = P_{\alpha\gamma} \tag{14·5·13}$$

This definition agrees with that of (13·1·7). By (14·5·12) and (14·5·13), Eq. (14·5·10) then becomes

$$\frac{\partial}{\partial t}(\rho u_\gamma) + \frac{\partial}{\partial x_\alpha}(\rho u_\alpha u_\gamma) = -\frac{\partial P_{\alpha\gamma}}{\partial x_\alpha} + \rho F_\gamma' \tag{14·5·14}$$

This is the Euler equation of motion of macroscopic hydrodynamics. It can be put into more transparent form by rewriting the left side of (14·5·14) as

$$u_\gamma \frac{\partial \rho}{\partial t} + \rho \frac{\partial u_\gamma}{\partial t} + u_\gamma \frac{\partial}{\partial x_\alpha}(\rho u_\alpha) + \rho u_\alpha \frac{\partial u_\gamma}{\partial x_\alpha}$$
$$= u_\gamma \left[\frac{\partial \rho}{\partial t} + \frac{\partial}{\partial x_\alpha}(\rho u_\alpha)\right] + \rho \left[\frac{\partial u_\gamma}{\partial t} + u_\alpha \frac{\partial u_\gamma}{\partial x_\alpha}\right] = 0 + \rho \frac{du_\gamma}{dt}$$

Here the first square bracket vanishes by the equation of continuity (14·5·7); furthermore, we have defined the "substantial derivative" of any function $\phi(\mathbf{r},t)$ by

$$\frac{d\phi}{dt} \equiv \frac{\partial \phi}{\partial t} + u_\alpha \frac{\partial \phi}{\partial x_\alpha} \tag{14·5·15}$$

This measures the rate of change of the function ϕ if one considers oneself moving along with the mean velocity $\boldsymbol{u}$ of the fluid. Hence (14·5·14) becomes

$$\blacktriangleright \qquad \rho \frac{du_\gamma}{dt} = -\frac{\partial P_{\alpha\gamma}}{\partial x_\alpha} + \rho F_\gamma' \tag{14·5·16}$$

This expresses physically the fact that the rate of change of mean momentum of any element of fluid is due to the stress forces (including the ordinary pressure) of the surrounding fluid, as well as to the external forces acting on the fluid.

We could, similarly, go on to use (14·5·4) to derive the hydrodynamic equation for energy conservation, but we shall not do this here.

The conservation equations (14·5·7) and (14·5·16) are rigorous consequences of the Boltzmann equation (14·3·8). Nevertheless it is clear that, in order to obtain from them practical hydrodynamic equations, one must find explicit expressions for quantities such as the pressure tensor $P_{\alpha\gamma}$. Of course, (14·5·13) provides a prescription for calculating this quantity in terms of molecular quantities, but this requires finding the actual distribution function f which is a solution of the Boltzmann equation (14·3·8). Hydrodynamic equations can thus be obtained to various orders of approximation. Details are discussed in the references.

14·6 *Example: simple discussion of electrical conductivity*

Before turning to detailed applications of the theory of this chapter to the solution of problems of physical interest, we shall show how the present formulation of the theory can also be useful in discussing situations in less rigorous terms. As an example, we shall give a semiquantitative treatment of electrical conductivity which, although almost as simple as the elementary arguments of Chapter 12, will bring out several new features of physical significance.

Consider the case of ions of mass m and charge e which move in a neutral gas consisting of molecules of mass m_1. Let the number of ions per unit volume be n, the number of neutral molecules per unit volume be n_1. The temperature is T, and a small uniform electric field $\mathcal{E}$ is applied in the z direction. We should like to find the electrical conductivity σ_{el} of the ions.

This situation was already discussed in Secs. 12·6, 13·4, and 13·8. Here we shall consider the collision processes in somewhat greater detail. We are interested in finding the electrical current density of the ions

$$j_z = enu_z \tag{14·6·1}$$

Since $\mathcal{E}$ depends on neither $\mathbf{r}$ nor t, it follows that neither n nor the mean ion velocity $\mathbf{u}$ depends on position $\mathbf{r}$ or on the time t once a steady-state situation has been reached. One can immediately write the equation for the mean momentum balance for the ions contained in a unit volume by using the equation of change (14·4·9). The direct physical argument is that the [rate of change of mean momentum of these ions] must be equal to [the mean external force exerted on these ions by the electric field] plus [the mean rate of momentum gain of these ions due to collisions]. In symbols

$$nm \frac{\partial u_z}{\partial t} = ne\mathcal{E} + \mathcal{C}(mv_z)$$

In the steady state $\partial u_z/\partial t = 0$ so that this condition becomes simply

$$ne\mathcal{E} + \mathcal{C}(mv_z) = 0 \tag{14·6·2}$$

To calculate the mean rate of ion-momentum gain caused by collisions, we note first that, when two ions collide, their total momentum is conserved. Hence there is *no* mean change of ion momentum caused by collisions between ions. Thus $\mathcal{C}(mv_z)$ is due entirely to momentum changes suffered by ions in collisions with neutral molecules.

The mean number of such ion-molecule collisions per unit time is approximately given by

$$\tau^{-1} = \bar{V}\sigma_{\text{im}}n_1 \tag{14·6·3}$$

where $\bar{V}$ is the mean relative speed between an ion and a molecule and σ_{im} is the total scattering cross section for scattering of an ion by a molecule. Here one can put approximately, as in (12·2·11),

$$V^2 \approx \overline{V^2} = \overline{v^2} + \overline{v_1{}^2} = 3kT\left(\frac{1}{m} + \frac{1}{m_1}\right) = \frac{3kT}{\mu} \tag{14·6·4}$$

where we have used the equipartition theorem and introduced the reduced mass

$$\mu \equiv \frac{mm_1}{m + m_1} \tag{14·6·5}$$

We calculate next the mean momentum gain $\langle \boldsymbol{p} \rangle$ of an ion in an ion-molecule collision. By (14·1·3) we can write the ion velocity $\boldsymbol{v}$ in terms of the velocity $\boldsymbol{V}$ of an ion relative to the molecule with which it collides and the velocity $\boldsymbol{c}$ of their center of mass. Thus $\boldsymbol{p} = m\boldsymbol{v} = m\boldsymbol{c} + \mu\boldsymbol{V}$. The momentum change of the ion in this collision is then

$$\Delta \boldsymbol{p} = m(\boldsymbol{v}' - \boldsymbol{v}) = \mu(\boldsymbol{V}' - \boldsymbol{V}) = \mu[(\cos\theta' - 1)\boldsymbol{V} + \boldsymbol{V}_\perp'] \qquad (14 \cdot 6 \cdot 6)$$

Here we have resolved $\boldsymbol{V}'$ into components parallel and perpendicular to $\boldsymbol{V}$. The relative velocity $\boldsymbol{V}'$ after collision is such that $|\boldsymbol{V}'| = |\boldsymbol{V}|$ and that it makes an angle θ' with respect to $\boldsymbol{V}$. On the average, $\boldsymbol{V}'$ will have no components perpendicular to $\boldsymbol{V}$ so that $\langle \boldsymbol{V}_\perp' \rangle = 0$. Furthermore, if the collision is like that between hard spheres, all scattering angles θ' are equally probable so that $\cos\theta' = 0$ on the average. (See Problem 12.4.) Hence on the average (14·6·6) gives, for the mean momentum gain of an ion per collision,

$$\langle \Delta \boldsymbol{p} \rangle = -\mu \langle \boldsymbol{V} \rangle = -\mu \langle \boldsymbol{v} - \boldsymbol{v}_1 \rangle$$

or

$$\langle \Delta \boldsymbol{p} \rangle = -\mu \boldsymbol{u} \qquad (14 \cdot 6 \cdot 7)$$

if we assume that the neutral molecules are at rest with respect to the container walls so that their mean velocity $\boldsymbol{u}_1 = 0$.

It is of interest to compare $\langle \Delta \boldsymbol{p} \rangle$ with the mean momentum $m\boldsymbol{u}$ of the ions. (In the present problem, $\boldsymbol{u}$ has, of course, only a nonvanishing component in the z direction.) Then (14·6·7) can be written

$$\langle \Delta \boldsymbol{p} \rangle \equiv -\xi m \boldsymbol{u}, \qquad \xi \equiv \frac{\mu}{m} = \frac{m_1}{m + m_1} \qquad (14 \cdot 6 \cdot 8)$$

where ξ denotes the fractional mean momentum loss of an ion per collision. This shows that if $m \ll m_1$, then $\xi \approx 1$; the ion loses then, on the average, practically all its forward momentum in each collision with the much heavier molecule. On the other hand, if $m \gg m_1$, then $\xi \approx m_1/m$; the ion loses then, on the average, only a relatively small fraction m_1/m of its forward momentum in each collision with the much lighter molecule. In the latter case, collisions with molecules are, of course, not very effective in reducing the electrical conductivity of the ions. The factor ξ of (14·6·8) shows that an ion after a collision may have a velocity which depends strongly on its velocity before the collision, particularly if $m \gg m_1$. Hence this factor takes into account the persistence of velocity effects which we ignored in the preceding two chapters.

The mean rate of collision-induced momentum gain of an ion can then be computed by multiplying the mean number τ^{-1} of ion-molecule collisions per unit time by the mean momentum gain $\langle \Delta \boldsymbol{p} \rangle$ per collision. By (14·6·8) the momentum balance (14·6·2) becomes simply

$$e\mathcal{E} - \tau^{-1}(\xi m u_z) = 0 \qquad (14 \cdot 6 \cdot 9)$$

Hence

$$u_z = \frac{e}{m} \frac{\tau}{\xi} \mathcal{E}$$

By (14·6·1) the electrical conductivity is then

$$\sigma_{\text{el}} \equiv \frac{j_z}{\mathcal{E}} = \frac{ne^2}{m}\frac{\tau}{\xi]} \tag{14·6·10}$$

This differs from the previous expressions (12·6·8) or (13·4·10) by the factor ξ which takes into account the persistence-of-velocity effects. Using (14·6·3) and (14·6·8) one gets explicitly

$$\sigma_{\text{el}} = \frac{ne^2}{m}\left[\sqrt{\frac{3kT}{\mu}}\,\sigma_{\text{im}}n_1\right]^{-1}\left(\frac{m}{\mu}\right)$$

or

$$\sigma_{\text{el}} = \frac{ne^2}{n_1\sigma_{\text{im}}}\frac{1}{\sqrt{3\mu kT}} \tag{14·6·11}$$

Note that this depends only on the *reduced* mass of the ion and molecule. If $m \ll m_1$ so that persistence-of-velocity effects are negligible, then $\mu = m$ and (14·6·11) reduces essentially to (12·6·9). But in the opposite limit, where $m \gg m_1$, one gets $\mu = m_1$, and σ_{el} becomes independent of the mass of the ion.

The relation (14·6·11) exhibits the correct dependence on the various parameters of the problem. In particular, it takes into account persistence-of-velocity effects and shows properly that ion-ion collisions have no appreciable effect on the electrical conductivity. A more careful evaluation of the momentum-balance equation would lead to numerically more accurate results. We leave this as an exercise in one of the problems at the end of the chapter.

14 · 7 *Approximation methods for solving the Boltzmann equation*

To apply the transport theory developed in this chapter to a quantitative discussion of situations of physical interest, it is necessary to find approximate solutions of the Boltzmann equation

$$Df = D_C f \tag{14·7·1}$$

written out explicitly in (14·3·8). Our aim will not be to find the most exact solutions accessible by means of elaborate approximation procedures, but to show how results of good accuracy can be obtained by relatively simple methods.

To find the distribution function $f(\mathbf{r},\mathbf{v},t)$ which satisfies (14·7·1), we assume again that we are dealing with a physical situation which is not too far removed from equilibrium conditions. Then one expects that $f(\mathbf{r},\mathbf{v},t)$ does not differ too much from a Maxwell distribution $f^{(0)}(\mathbf{r},\mathbf{v},t)$, which describes *local* equilibrium conditions near a particular place and time; i.e.,

$$f^{(0)}(\mathbf{r},\mathbf{v},t) = n\left(\frac{m\beta}{2\pi}\right)^{\frac{3}{2}} e^{-\frac{1}{2}\beta m(\mathbf{v}-\mathbf{u})^2} \tag{14·7·2}$$

where n, β, and $\mathbf{u}$ may all be slowly varying functions of $\mathbf{r}$ and t, but do not depend on $\mathbf{v}$. Since the dependence of $f^{(0)}$ on $\mathbf{v}$ is thus the same as that for a genuine equilibrium distribution, and since the collision term in the Boltzmann equation involves only integrations over velocities, it follows that $f^{(0)}$ has the same property as a genuine equilibrium distribution of remaining unchanged under the influence of collisions; i.e.,

$$D_C f^{(0)} = 0 \tag{14·7·3}$$

Remark This can be readily verified by showing that, for any $\mathbf{r}$ and t,

$$f^{(0)}(\mathbf{v})f^{(0)}(\mathbf{v}_1) = f^{(0)}(\mathbf{v}')f^{(0)}(\mathbf{v}_1') \tag{14·7·4}$$

so that the integrand on the right side of (14·3·8) vanishes. Equivalently, it is necessary to show that

$$\ln f^{(0)}(\mathbf{v}) + \ln f^{(0)}(\mathbf{v}_1) = \ln f^{(0)}(\mathbf{v}') + \ln f^{(0)}(\mathbf{v}_1')$$

or by (14·7·2) that

$$\tfrac{1}{2}m(\mathbf{v} - \mathbf{u})^2 + \tfrac{1}{2}m(\mathbf{v}_1 - \mathbf{u})^2 = \tfrac{1}{2}m(\mathbf{v}' - \mathbf{u})^2 + \tfrac{1}{2}m(\mathbf{v}_1' - \mathbf{u})^2 \tag{14·7·5}$$

But the left side of (14·7·5) can be written simply as

$$(\tfrac{1}{2}m\mathbf{v}^2 + \tfrac{1}{2}m\mathbf{v}_1^2) - (m\mathbf{v} + m\mathbf{v}_1) \cdot \mathbf{u} + m\mathbf{u}^2$$

which involves, besides constants, only the total kinetic energy and the total momentum of the particles before collision. Since the right side of (14·7·5) is of the same form, except for referring to velocities after the collision, the conditions of conservation of total kinetic energy and of total momentum in a collision imply immediately the validity of (14·7·5) and hence of (14·7·3).

Of course, $f^{(0)}$ does *not* in general reduce the left side of the Boltzmann equation (14·7·1) to zero. That is, $Df^{(0)} \neq 0$, unless n, β, and $\mathbf{u}$ are independent of $\mathbf{r}$ and t. Only then would $f^{(0)}$ be a genuine, rather than merely a local, equilibrium distribution function satisfying the Boltzmann equation (14·7·1).

To exploit the assumption that the situation is not far removed from equilibrium, one can write f in the form

$$f = f^{(0)}(1 + \Phi), \qquad \text{where } \Phi \ll 1 \tag{14·7·6}$$

In the Boltzmann equation (14·7·1), the contribution of the correction term $f^{(0)}\Phi$ to the left side can then be neglected compared to the contribution of $f^{(0)}$. Thus

$$Df \approx Df^{(0)} \tag{14·7·7}$$

The right side of (14·7·1) is given by (14·3·8) as

$$D_C f = \iint d^3\mathbf{v}_1\, d\Omega'\, (f'f_1' - ff_1)V\sigma \tag{14·7·8}$$

Using (14·7·6), one has

$$ff_1 = f^{(0)}f_1^{(0)}(1 + \Phi + \Phi_1)$$

where we have neglected the small quadratic term $\Phi\Phi_1$. Similarly, one has

$$f'f_1' = f^{(0)}f_1^{(0)}(1 + \Phi' + \Phi_1')$$

where we have used (14·7·4) to put $f^{(0)\prime}f_1^{(0)\prime} = f^{(0)}f_1^{(0)}$.

Substituting these relations into (14·7·8), and using the fact that terms involving only $f^{(0)}$'s lead to a vanishing contribution to the integral (since

$Dcf^{(0)} = 0$ by (14·7·3)), one obtains

$$D_C f = \mathfrak{L}\Phi \tag{14·7·9}$$

where

$$\mathfrak{L}\Phi \equiv \iint d^3v\, d\Omega'\, f^{(0)}f_1^{(0)} V\sigma\, \Delta\Phi \tag{14·7·10}$$

with

$$\Delta\Phi = \Phi' + \Phi_1' - \Phi - \Phi_1 \tag{14·7·11}$$

In terms of these abbreviations the Boltzmann equation (14·7·1) is then, by (14·7·7) and (14·7·9), reduced to the approximate form

$$Df^{(0)} = \mathfrak{L}\Phi \tag{14·7·12}$$

The functional form of $f^{(0)}$ is known by (14·7·2); hence the left side of (14·7·12) is a known function. The unknown function Φ appears in (14·7·12) only in the *integrand* of the right side. Finding the functional form of Φ which satisfies the integral equation (14·7·12) is still not a trivial task. On the other hand, this equation is linear in Φ and very much simpler than the original Boltzmann equation (14·7·1).

Remark We can impose one physical requirement which helps to place some restriction on the possible form of Φ. Let us require that the actual function $f(\mathbf{r},\mathbf{v},t)$ be such that the quantities $n(\mathbf{r},t)$, $\mathbf{u}(\mathbf{r},t)$, and $\beta(\mathbf{r},t) \equiv (kT)^{-1}$ in (14·7·2) preserve their usual meaning of denoting respectively the mean number of particles per unit volume, their mean velocity, and their mean thermal kinetic energy. In more mathematical terms this means that we require the following relations, true in the equilibrium situation when n, $\mathbf{u}$, and T are independent of $\mathbf{r}$ and t, to remain valid even if these parameters do depend on $\mathbf{r}$ and t:

$$\left.\begin{aligned} \int d^3v\, f &= n(\mathbf{r},t) \\ \frac{1}{n}\int d^3v\, f\mathbf{v} &= \mathbf{u}\,(\mathbf{r},t) \\ \frac{1}{n}\int d^3v\, f\left[\frac{1}{2}m(\mathbf{v}-\mathbf{u})^2\right] &= kT(\mathbf{r},t) \end{aligned}\right\} \tag{14·7·13}$$

By (14·7·6), the first term $f^{(0)}$ by itself satisfies all these relations. Hence it follows that, to satisfy (14·7·13), the function Φ must be such that

$$\left.\begin{aligned} \int d^3v\, f^{(0)}\Phi &= 0 \\ \int d^3v\, f^{(0)}\Phi\mathbf{v} &= 0 \\ \int d^3v\, f^{(0)}\Phi(\mathbf{v}-\mathbf{u})^2 &= 0 \end{aligned}\right\} \tag{14·7·14}$$

To determine the function Φ which satisfies (14·7·12), one can assume that it has some reasonable functional form which depends on q parameters $A_1, A_2, A_3, \ldots, A_q$. For example, one could assume that Φ is a function of $\mathbf{U} \equiv \mathbf{v} - \mathbf{u}$ of the form

$$\Phi = \sum_{\lambda=1}^{3} a_\lambda U_\lambda + \sum_{\lambda,\mu=1}^{3} a_{\lambda\mu} U_\lambda U_\mu + \cdots \tag{14·7·15}$$

where the coefficients a_λ and $a_{\lambda\mu}$ are parameters. If one takes this assumed form of Φ and substitutes it into (14·7·12), one will in general find that $\mathcal{L}\Phi$ is not the same function of $\boldsymbol{v}$ as the left-hand side $Df^{(0)}$. Hence no choice of the parameters $A_1, \ldots, A_q$ will really fulfill the requirement that (14·7·12) be satisfied for all values of $\boldsymbol{v}$. Nevertheless, one's guess as to the functional form of Φ may not be too bad, provided that one makes an optimum choice of the parameters $A_1, \ldots, A_q$. One systematic way of making this choice is to replace the task of satisfying Eq. (14·7·12) by the weaker requirement of satisfying it only in an average sort of way. For example, if $\Psi(\boldsymbol{v})$ is any function of $\boldsymbol{v}$, then it follows by (14·7·12) that

$$\int d^3\boldsymbol{v}\, \Psi\, Df^{(0)} = \int d^3\boldsymbol{v}\, \Psi\, \mathcal{L}\Phi \qquad (14\cdot7\cdot16)$$

Since one has integrated over $\boldsymbol{v}$, both sides of this equation are independent of $\boldsymbol{v}$. Although (14·7·12) implies (14·7·16), the converse is, of course, *not* true; i.e., if Φ satisfies (14·7·16) for a given choice of the function Ψ, it does *not* necessarily satisfy the original equation (14·7·12). Only if (14·7·16) were satisfied for *all* possible functions Ψ, could one conclude that (14·7·16) is equivalent to (14·7·12). Nevertheless, if one chooses some set of q functions $\Psi_1, \ldots, \Psi_q$ and tries to satisfy the resulting q equations (14·7·16) for all of these, then one obtains q simple *algebraic* equations for the q unknown parameters $A_1 \ldots A_q$. One would then expect that this choice of the A's would give a reasonably good solution of (14·7·12). Of course, the larger the number q of parameters at one's disposal (and the corresponding number of Ψ's used to obtain independent algebraic equations (14·7·16)), the better can one expect the solution to approximate the real solution of (14·7·12). If one assumes the functional form (14·7·15) for Φ, then the various "test functions" Ψ are most conveniently chosen to be functions of the form U_λ, $U_\lambda U_\mu$, ... The method just described is then called the "method of moments."

Note that (14·7·16) is equivalent to (14·4·11); thus the condition (14·7·16) implies physically the requirement that the *mean* value $\langle\Psi\rangle$ satisfies the correct equation of change (14·4·9).

Solution by use of a variational principle A particularly powerful method of solving (14·7·12) is provided by the use of a suitable variational principle. Note that the integral in (14·7·12) is linear in Φ; i.e., for any two functions Φ and Ψ

$$\mathcal{L}(\Phi + \Psi) = \mathcal{L}\Phi + \mathcal{L}\Psi \qquad (14\cdot7\cdot17)$$

Note also that the integral in (14·7·16) has a lovely degree of symmetry. Using (14·7·10) and (14·2·4) it can be written in the symmetric form

$$\int d^3\boldsymbol{v}\, \Psi\, \mathcal{L}\Phi = \iiiint d^3\boldsymbol{v}\, d^3\boldsymbol{v}_1\, d^3\boldsymbol{v}'\, d^3\boldsymbol{v}_1'\, f^{(0)}f_1^{(0)}V\sigma'(\boldsymbol{v},\boldsymbol{v}_1 \to \boldsymbol{v}',\boldsymbol{v}_1')\Psi\, \Delta\Phi \qquad (14\cdot7\cdot18)$$

One can now proceed as we did in transforming the expression (14·4·18) into (14·4·22). First interchange $\boldsymbol{v}$ and $\boldsymbol{v}_1$, as well as $\boldsymbol{v}'$ and $\boldsymbol{v}_1'$. Then σ' as well as $\Delta\Phi$ are unchanged. Thus (14·7·8) becomes (we omit the volume elements in the integrals for the sake of brevity)

$$\int \Psi\, \mathcal{L}\Phi = \iiiint f^{(0)} f_1^{(0)} V\sigma' \Psi_1\, \Delta\Phi$$

Adding this to (14·7·18) then gives

$$\int \Psi\, \mathcal{L}\Phi = \tfrac{1}{2} \iiiint f^{(0)} f_1^{(0)} V\sigma' [\Psi + \Psi_1]\, \Delta\Phi \qquad (14 \cdot 7 \cdot 19)$$

Interchanging $\mathbf{v}$ and $\mathbf{v}'$ as well as $\mathbf{v}_1$ and $\mathbf{v}_1'$ (i.e., passing to the inverse collision) leaves σ' unchanged and changes the sign of $\Delta\Phi$. Also by (14·7·4), $f^{(0)\prime} f_1^{(0)\prime} = f^{(0)} f_1^{(0)}$. Hence (14·7·19) becomes

$$\int \Psi\, \mathcal{L}\Phi = -\tfrac{1}{2} \iiiint f^{(0)} f_1^{(0)} V\sigma' [\Psi' + \Psi_1']\, \Delta\Phi$$

Adding this to (14·7·19) yields the symmetrical form

$$\int \Psi\, \mathcal{L}\Phi = -\tfrac{1}{4} \iiiint f^{(0)} f_1^{(0)} V\sigma'\, \Delta\Psi\, \Delta\Phi \qquad (14 \cdot 7 \cdot 20)$$

where $\Delta\Psi = \Psi' + \Psi_1' - \Psi - \Psi_1$.

The result (14·7·20) implies two important properties. First it yields the symmetry relation

▶ $$\int \Psi\, \mathcal{L}\Phi = \int \Phi\, \mathcal{L}\Psi \qquad (14 \cdot 7 \cdot 21)$$

Second, if $\Psi = \Phi$, then the integrand in (14·7·20) is positive so that

▶ $$\int \Phi\, \mathcal{L}\Phi \leq 0 \qquad (14 \cdot 7 \cdot 22)$$

Let us now compare the actual solution Φ of (14·7·12) with *any* other function $\Phi' = \Phi + \delta\Phi$. If we multiply both sides of (14·7·12) by Φ, we obtain the expressions $\int \Phi\, \mathcal{L}\Phi$ and $\int \Phi\, Df^{(0)}$. Using instead the function Φ' and exploiting (14·7·21), these expressions become, respectively,

$$\int \Phi'\, \mathcal{L}\Phi' = \int (\Phi + \delta\Phi)\, \mathcal{L}(\Phi + \delta\Phi) = \int \Phi\, \mathcal{L}\Phi + 2\int \delta\Phi\, \mathcal{L}\Phi + \int \delta\Phi\, \mathcal{L}\, \delta\Phi \qquad (14 \cdot 7 \cdot 23)$$

and

$$\int \Phi'\, Df^{(0)} = \int (\Phi + \delta\Phi)\, Df^{(0)} = \int \Phi\, Df^{(0)} + \int \delta\Phi\, Df^{(0)} \qquad (14 \cdot 7 \cdot 24)$$

Hence, if one considers the expression

▶ $$M \equiv \int d^3\mathbf{v}\, \Phi\, \mathcal{L}\Phi - 2 \int d^3\mathbf{v}\, \Phi\, Df^{(0)} = \int d^3\mathbf{v}\, \Phi(\mathcal{L}\Phi - 2\, Df^{(0)}) \qquad (14 \cdot 7 \cdot 25)$$

and the corresponding expression M' calculated with the function Φ', one finds by (14·7·23) and (14·7·24) that

$$M' - M = \int \delta\Phi\, \mathcal{L}\, \delta\Phi \leq 0 \qquad (14 \cdot 7 \cdot 26)$$

where we have deliberately defined M so that, by virtue of (14·7·12), terms linear in $\delta\Phi$ vanish in (14·7·26). Hence M does not change for small changes $\delta\Phi$ in the function Φ, but is always negative for larger values of $\delta\Phi$. Thus we arrive at the "variational principle":

> The expression M assumes its maximum value for that particular function Φ which is a solution of the Boltzmann equation (14·7·12).

This provides one with a *very* powerful method of finding approximate solutions of the Boltzmann equation (14·7·12). If one assumes a functional form for Φ which contains several parameters, then the optimum choice of these parameters which provides one with the best solution to (14·7·12) is that which makes the expression M a maximum. One can also quite systematically find increasingly better approximations to the real solution of (14·7·12) by successively modifying one's first guess of the function Φ in such a way as to make M greater and greater.

As the simplest example, assume that Φ is of the form $\Phi = A\phi$, where ϕ is some function of $\boldsymbol{v}$ and A is a parameter. Then (14·7·25) gives

$$M = A^2 \int \phi \mathfrak{L}\phi - 2A \int \phi\, Df^{(0)}$$

The optimum choice of A is obtained by maximizing M, i.e., by putting

$$\frac{\partial M}{\partial A} = 0 = 2A \int \phi \mathfrak{L}\phi - 2 \int \phi\, Df^{(0)}$$

so that

$$A = \frac{\int \phi\, Df^{(0)}}{\int \phi \mathfrak{L}\phi} \tag{14·7·27}$$

14·8 *Example: calculation of the coefficient of viscosity*

Consider again the physical situation illustrated in Fig. 12·3·2 where the gas has a mean velocity $u_x(z)$ in the x direction and a gradient $\partial u_x/\partial z \neq 0$. We are interested in calculating the shearing stress P_{zx}.

This situation is time-independent and only u_x depends on z. Hence the *local* equilibrium distribution is simply that already used in (13·8·6); i.e.,

$$f^{(0)}(\boldsymbol{r},\boldsymbol{v},t) = n\left(\frac{m\beta}{2\pi}\right)^{\frac{3}{2}} e^{-\frac{1}{2}\beta m\{[v_x - u_x(z)]^2 + v_y^2 + v_z^2\}}$$

where β and n are just constants. More compactly one can write this in the form

$$f^{(0)}(\boldsymbol{r},\boldsymbol{v},t) = g(U) \tag{14·8·1}$$

where

$$U_x(z) \equiv v_x - u_x(z), \qquad U_y = v_y, \qquad U_z = v_z \tag{14·8·2}$$

and

$$g(U) = n\left(\frac{m\beta}{2\pi}\right)^{\frac{3}{2}} e^{-\frac{1}{2}\beta m U^2} \tag{14·8·3}$$

Since there are no external forces so that $\boldsymbol{F} = 0$, one has simply

$$Df^{(0)} = v_z \frac{\partial f^{(0)}}{\partial z} = v_z \frac{\partial g}{\partial U_x}\left(-\frac{\partial u_x}{\partial z}\right) = \beta m g(U) U_z U_x \frac{\partial u_x}{\partial z}$$

where we have put $v_z = U_z$ by (14·8·2). Hence the Boltzmann equation (14·7·12) becomes, using (14·8·3),

$$\left(\beta m \frac{\partial u_x}{\partial z}\right) g(U) U_z U_x = \iint d^3\boldsymbol{U}_1\, d\Omega'\, g(U) g(U_1) V\sigma\, \Delta\Phi \tag{14·8·4}$$

It remains to determine how Φ depends on $\boldsymbol{v}$, or equivalently on $\boldsymbol{U}$, so that it yields upon integration the left side of (14·8·4). This left side transforms under rotation of coordinate axes like the product of the vector components U_zU_x. The right side must, therefore, behave similarly. Thus we are led to expect that Φ has the form

$$\Phi = AU_zU_x \tag{14·8·5}$$

where A, in general, might be some function of $|\boldsymbol{U}|$. We shall, however, assume that to a first approximation A is simply independent of $|\boldsymbol{U}|$. To determine the value of this one parameter A which gives the best solution, we proceed by the method of (14·7·16). We shall multiply both sides of (14·8·4) by the test function U_zU_x, integrate over $\boldsymbol{v}$ (or equivalently $\boldsymbol{U}$), and try to satisfy the resulting equation by proper choice of A. This means that we are trying to satisfy, instead of (14·8·4), the equation

$$\beta m \frac{\partial u_x}{\partial z} \int d^3\boldsymbol{U}\, g(U) U_z^2 U_x^2 = A \iiint d^3\boldsymbol{U}\, d^3\boldsymbol{U}_1\, d\Omega'\, g(U) g(U_1) V\sigma U_z U_x\, \Delta(U_zU_x) \tag{14·8·6}$$

(Note that this choice of A is exactly equivalent to that given by the variational principle in (14·7·27).)

The integral on the left is by (14·8·6) simply

$$\int d^3\boldsymbol{U}\, g(U) U_z^2 U_x^2 = n \left(\frac{m\beta}{2\pi}\right)^{\frac{3}{2}} \int_{-\infty}^{\infty} e^{-\frac{1}{2}\beta m U_y^2}\, dU_y \int_{-\infty}^{\infty} e^{-\frac{1}{2}\beta m U_z^2}\, U_z^2\, dU_z \int_{-\infty}^{\infty} e^{-\frac{1}{2}\beta m U_x^2}\, U_x^2\, dU_x$$

or

$$\int d^3\boldsymbol{U}\, g(U) U_z^2 U_x^2 = n\overline{U_z^2}\;\overline{U_x^2} = n\left(\frac{kT}{m}\right)^2 \tag{14·8·7}$$

where the bars denote equilibrium values calculated with the equilibrium function g; these can therefore be obtained by the equipartition theorem without need of explicit calculation.

In the integral on the right side of (14·8·6) we shall first integrate over all angles of scattering, i.e., over all $d\Omega'$, and then integrate over $\boldsymbol{U}$ and $\boldsymbol{U}_1$. Thus we write this integral in the form

$$I \equiv \iint d^3\boldsymbol{U}\, d^3\boldsymbol{U}_1\, g(U) g(U_1) U_z U_x\, J(\boldsymbol{U}, \boldsymbol{U}_1) \tag{14·8·8}$$

where

$$J(\boldsymbol{U}, \boldsymbol{U}_1) \equiv \int d\Omega'\, V\sigma\, \Delta(U_zU_x) \tag{14·8·9}$$

Using (14·8·7), Eq. (14·8·6) then becomes simply

$$\blacktriangleright \qquad n\left(\frac{kT}{m}\right) \frac{\partial u_x}{\partial z} = AI \tag{14·8·10}$$

Before evaluating I, note that the function Φ of (14·8·5) satisfies, by symmetry, the conditions (14·7·14). Note also that the pressure tensor P_{zx} can be immediately calculated from (14·8·5). By (13·1·17)

$$\begin{aligned} P_{zx} &= m \int d^3\boldsymbol{v}\, f U_z U_x = m \int d^3\boldsymbol{U}\, g(U)(1 + \Phi) U_z U_x \\ &= 0 + m \int d^3\boldsymbol{U}\, g\Phi U_z U_x = Am \int d^3\boldsymbol{U}\, g U_z^2 U_x^2 \end{aligned}$$

where we have used the fact that the integral involving $g(U)$ alone vanishes by symmetry. Since the last integral is just that of (14·8·7), one obtains

$$P_{zx} = nm\left(\frac{kT}{m}\right)^2 A \tag{14·8·11}$$

Thus the calculation is completed once the integral I is evaluated, since A is then known by (14·8·10).

Evaluation of the collision integral I We first evaluate the integral J of (14·8·9). Since the integrand involves V directly as well as in the cross section $\sigma(V)$, we express $\mathbf{U}$ in terms of the relative velocity $\mathbf{V}$ and the center-of-mass velocity $\mathbf{c}$. Thus

$$\mathbf{V} = \mathbf{v} - \mathbf{v}_1 = \mathbf{U} - \mathbf{U}_1 \tag{14·8·12}$$

Also, since all molecules have the same mass m,

$$\mathbf{c} = \frac{m\mathbf{v} + m\mathbf{v}_1}{2m} = \frac{1}{2}(\mathbf{v} + \mathbf{v}_1)$$

or

$$\mathbf{C} \equiv \mathbf{c} - \mathbf{u} = \tfrac{1}{2}(\mathbf{U} + \mathbf{U}_1) \tag{14·8·13}$$

where $\mathbf{C}$ is the center-of-mass velocity measured relative to the mean velocity $\mathbf{u}$ of the gas. By (14·8·12) and (14·8·13) one finds, analogously to (14·1·3),

$$\mathbf{U} = \mathbf{C} + \tfrac{1}{2}\mathbf{V}, \qquad \mathbf{U}_1 = \mathbf{C} - \tfrac{1}{2}\mathbf{V} \tag{14·8·14}$$

After the collision one has similarly

$$\mathbf{U}' = \mathbf{C} + \tfrac{1}{2}\mathbf{V}', \qquad \mathbf{U}_1 = \mathbf{C} - \tfrac{1}{2}\mathbf{V}' \tag{14·8·15}$$

since $\mathbf{C}$ remains constant. Now, by (14·7·11),

$$\Delta(U_zU_x) = U_z'U_x' + U_{1z}'U_{1x}' - U_zU_x - U_{1z}U_{1x}$$

Using (14·8·14), one obtains

$$\begin{aligned} U_zU_x + U_{1z}U_{1x} &= (C_z + \tfrac{1}{2}V_z)(C_x + \tfrac{1}{2}V_x) + (C_z - \tfrac{1}{2}V_z)(C_x - \tfrac{1}{2}V_x) \\ &= 2C_zC_x + \tfrac{1}{2}V_zV_x \end{aligned}$$

Hence

$$\Delta(U_zU_x) = \tfrac{1}{2}(V_z'V_x' - V_zV_x) \tag{14·8·16}$$

Thus (14·8·9) becomes

$$J = \tfrac{1}{2}\int_0^{\pi}\int_0^{2\pi} \sin\theta'\,d\theta'\,d\varphi'\,V\sigma[V_z'V_x' - V_zV_x] \tag{14·8·17}$$

where θ' and φ' are the polar and azimuthal angles of $\mathbf{V}'$ measured with respect to $\mathbf{V}$ as polar axis. This vector $\mathbf{V}$ is, of course, fixed in the integration (14·8·17) which is over the various directions of $\mathbf{V}'$. Imagine that $\mathbf{V}$ is taken to be along the ζ axis of a cartesian coordinate system ξ, η, ζ. Without loss of generality one can then choose the ξ axis, from which the angle φ' is measured, to lie in the $(\mathbf{V},\hat{\mathbf{z}})$ plane. Here $\hat{\boldsymbol{\xi}}, \hat{\boldsymbol{\eta}}, \hat{\boldsymbol{\zeta}}$ denote unit vectors directed along the ξ, η, ζ axes; similarly $\hat{\mathbf{x}}, \hat{\mathbf{y}}, \hat{\mathbf{z}}$ denote unit vectors directed along the labora-

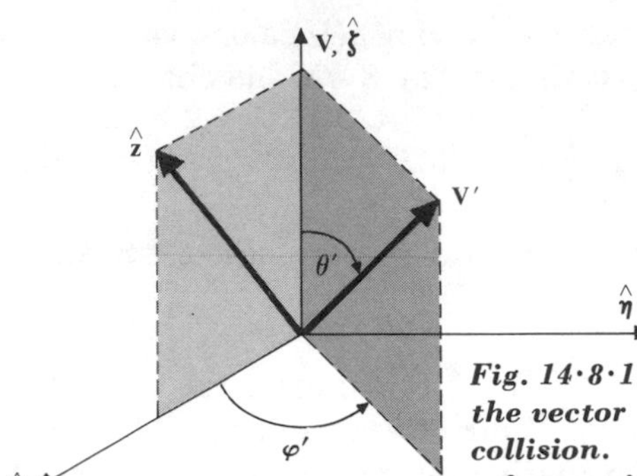

Fig. 14·8·1 Geometrical relationships between the vector V before and the vector V′ after a collision. The $\hat{z}$ axis of the laboratory frame of reference is also shown.

tory x, y, z axes. The geometrical relationships are illustrated in Fig. 14·8·1. Since $|\mathbf{V}'| = |\mathbf{V}|$, one can write

$$\mathbf{V}' = V \cos \theta' \hat{\boldsymbol{\zeta}} + V \sin \theta' \cos \varphi' \hat{\boldsymbol{\xi}} + V \sin \theta' \sin \varphi' \hat{\boldsymbol{\eta}}$$

Remembering that $V\hat{\boldsymbol{\zeta}} = \mathbf{V}$ and that $\hat{\boldsymbol{\eta}} \perp \hat{\mathbf{z}}$, one then gets

$$\begin{aligned} V_z' &= \mathbf{V}' \cdot \hat{\mathbf{z}} = V_z \cos \theta' + V \sin \theta' \cos \varphi' \, \hat{\boldsymbol{\xi}} \cdot \hat{\mathbf{z}} \\ V_x' &= \mathbf{V}' \cdot \hat{\mathbf{x}} = V_x \cos \theta' + V \sin \theta' \cos \varphi' \, \hat{\boldsymbol{\xi}} \cdot \hat{\mathbf{x}} + V \sin \theta' \cos \varphi' \, \hat{\boldsymbol{\xi}} \cdot \hat{\boldsymbol{\eta}} \end{aligned} \qquad (14\cdot8\cdot18)$$

We shall assume that the forces between molecules depend only on the relative distance between them. Then the differential scattering cross section σ is independent of the azimuthal angle φ'; i.e., $\sigma = \sigma(V; \theta')$. The integral (14·8·17) is then much simplified. Consider first the integration over φ' from 0 to 2π. Since $\int \sin \varphi' \, d\varphi' = \int \cos \varphi' \, d\varphi' = \int \sin \varphi' \cos \varphi' \, d\varphi' = 0$, while $\int \cos^2 \varphi' \, d\varphi' = \int \sin^2 \varphi' \, d\varphi' = \pi$, use of (14·8·18) in (14·8·17) shows that all cross terms cancel. Thus (14·8·18) yields

$$\int_0^{2\pi} d\varphi' \, (V_z'V_x' - V_zV_x) = 2\pi V_zV_x \cos^2 \theta' + \pi V^2 \sin^2 \theta' (\hat{\boldsymbol{\xi}} \cdot \hat{\mathbf{z}})(\hat{\boldsymbol{\xi}} \cdot \hat{\mathbf{x}}) - 2\pi V_zV_x \qquad (14\cdot8\cdot19)$$

To eliminate the vector $\hat{\boldsymbol{\xi}}$, we note that the condition $\hat{\mathbf{z}} \perp \hat{\mathbf{x}}$ can be expressed in terms of ξ, η, ζ components (i.e., in terms of direction cosines) as

$$\hat{\mathbf{z}} \cdot \hat{\mathbf{x}} = 0 = (\hat{\boldsymbol{\xi}} \cdot \hat{\mathbf{z}})(\hat{\boldsymbol{\xi}} \cdot \hat{\mathbf{x}}) + (\hat{\boldsymbol{\eta}} \cdot \hat{\mathbf{z}})(\hat{\boldsymbol{\eta}} \cdot \hat{\mathbf{x}}) + (\hat{\boldsymbol{\zeta}} \cdot \hat{\mathbf{z}})(\hat{\boldsymbol{\zeta}} \cdot \hat{\mathbf{x}})$$

Since $\hat{\boldsymbol{\eta}} \perp \hat{\mathbf{z}}$, this implies that

$$V^2(\hat{\boldsymbol{\xi}} \cdot \hat{\mathbf{z}})(\hat{\boldsymbol{\xi}} \cdot \hat{\mathbf{x}}) = -V^2(\hat{\boldsymbol{\zeta}} \cdot \hat{\mathbf{z}})(\hat{\boldsymbol{\zeta}} \cdot \hat{\mathbf{x}}) = -V_zV_x$$

Hence (14·8·19) becomes

$$\int_0^{2\pi} d\varphi' \, (V_z'V_x' - V_zV_x) = \pi V_zV_x(2 \cos^2 \theta' - \sin^2 \theta' - 2) = -3\pi V_zV_x \sin^2 \theta'$$

and (14·8·17) can be written in the form

▶ $$J = -\tfrac{3}{4} V_zV_xV\sigma_\eta(V) \qquad (14\cdot8\cdot20)$$

where

$$\sigma_\eta(V) \equiv 2\pi \int_0^\pi \sigma(V,\theta') \sin^3\theta' \, d\theta' \tag{14·8·21}$$

can be interpreted as the total effective scattering cross section which enters the viscosity calculation.

One can now return to the integral (14·8·8), which becomes

$$I = -\tfrac{3}{4}\iint d^3\boldsymbol{U}\, d^3\boldsymbol{U}_1\, g(U)g(U_1)\, U_z U_x V_z V_x V \sigma_\eta(V)$$

or by (14·8·3)

$$I = -\frac{3}{4} n^2 \left(\frac{m\beta}{2\pi}\right)^3 \iint_{-\infty}^{\infty} d^3\boldsymbol{U}\, d^3\boldsymbol{U}_1\, e^{-\frac{1}{2}\beta m(U^2+U_1^2)} U_z U_x V_z V_x V \sigma_\eta(V) \tag{14·8·22}$$

This integral depends again on the relative velocity $\boldsymbol{V}$. Hence we again make the transformation (14·8·14) to the variables $\boldsymbol{V}$ and $\boldsymbol{C}$. Analogously to (14·2·6), one has $d^3\boldsymbol{U}\, d^3\boldsymbol{U}_1 = d^3\boldsymbol{C}\, d^3\boldsymbol{V}$. Then (14·8·22) becomes

$$\begin{aligned} I &= -\frac{3}{4} n^2 \left(\frac{m\beta}{2\pi}\right)^3 \iint_{-\infty}^{\infty} d^3\boldsymbol{C}\, d^3\boldsymbol{V}\, e^{-\beta m(C^2+\frac{1}{4}V^2)} \\ &\qquad \left(C_z C_x + \frac{1}{4} V_z V_x + \frac{1}{2} C_z V_x + \frac{1}{2} C_x V_z\right) V_z V_x V \sigma_\eta \\ &= -\frac{3}{4} n^2 \left(\frac{m\beta}{2\pi}\right)^3 \int_{-\infty}^{\infty} d^3\boldsymbol{C}\, e^{-\beta m C^2} \int_{-\infty}^{\infty} d^3\boldsymbol{V}\, e^{-\frac{1}{4}\beta m V^2} \left(C_z C_x V_z V_x + \frac{1}{4} V_z^2 V_x^2 \right. \\ &\qquad \left. + \frac{1}{2} C_z V_z V_x^2 + \frac{1}{2} C_x V_z^2 V_x\right) V \sigma_\eta(V) \end{aligned}$$

By symmetry it is clear that three of the terms in the parentheses are odd functions of V_z or V_x and lead therefore, by symmetry, to a vanishing result upon integration over $\boldsymbol{V}$. Thus one is left with

$$I = -\frac{3}{16} n^2 \left(\frac{m\beta}{2\pi}\right)^3 \int_{-\infty}^{\infty} d^3\boldsymbol{C}\, e^{-\beta m C^2} \int_{-\infty}^{\infty} d^3\boldsymbol{V}\, e^{-\frac{1}{4}\beta m V^2} V_z^2 V_x^2 V \sigma_\eta(V) \tag{14·8·23}$$

By Appendix A·4 one has simply

$$\int_{-\infty}^{\infty} d^3\boldsymbol{C}\, e^{-\beta m C^2} = 4\pi \int_0^\infty e^{-\beta m C^2} C^2\, dC = \left(\frac{\pi}{m\beta}\right)^{\frac{3}{2}} \tag{14·8·24}$$

Expressing $\boldsymbol{V}$ in terms of spherical coordinates with $\hat{\boldsymbol{z}}$ as polar axis, one has $V_z = V\cos\theta$, $V_x = V\sin\theta\cos\varphi$, and $d^3\boldsymbol{V} = V^2\, dV \sin\theta\, d\theta\, d\varphi$. Thus one gets

$$\begin{aligned} \int d^3\boldsymbol{V}\, e^{-\frac{1}{4}\beta m V^2} V_z^2 V_x^2 \sigma_\eta(V) &= \int_0^\infty dV\, e^{-\frac{1}{4}\beta m V^2} V^7 \sigma_\eta(V) \int_0^\pi d\theta \sin^3\theta \cos^2\theta \\ &\qquad \int_0^{2\pi} d\varphi \cos^2\varphi \\ &= \left[\left(\frac{2}{\sqrt{m\beta}}\right)^8 \int_0^\infty ds\, e^{-s^2} s^7 \sigma_\eta\left(\frac{2s}{\sqrt{m\beta}}\right)\right] \left[\frac{4}{15}\right] [\pi] \end{aligned} \tag{14·8·25}$$

where we have introduced the dimensionless variable

$$s = \frac{1}{2}\sqrt{m\beta}\,V = \frac{1}{\sqrt{2}}\frac{V}{\tilde{v}}$$

which physically expresses the relative speed V in terms of the most probable speed $\tilde{v}$ of the gas in equilibrium. If σ_η did not depend on the relative speed of the molecules, then the integral in (14·8·25) would, by Appendix A·4, be equal to $3\sigma_\eta$. More generally, let us define a mean effective cross section by

$$\bar{\sigma}_\eta(T) \equiv \frac{1}{3}\int_0^\infty ds\, e^{-s^2}\sigma_\eta\left(\frac{2s}{\sqrt{m\beta}}\right) \tag{14·8·26}$$

Then (14·8·25) becomes

$$\int d^3V\, e^{-\frac{1}{4}\beta m V^2} V_z^2 V_x^2 V \sigma_\eta(V) = \frac{1024\pi}{5}\frac{\bar{\sigma}_\eta}{(m\beta)^4} \tag{14·8·27}$$

Substituting (14·8·27) and (14·8·24) into (14·8·23) yields then

$$I = -\frac{24}{5\sqrt{\pi}}\,n^2\left(\frac{kT}{m}\right)^{\frac{5}{2}}\bar{\sigma}_\eta \tag{14·8·28}$$

Evaluation of the coefficient of viscosity Using (14·8·28), the parameter A is now given by (14·8·10) as

$$A = -\frac{5\sqrt{\pi}}{24}\frac{1}{n\bar{\sigma}_\eta}\left(\frac{m}{kT}\right)^{\frac{3}{2}}\frac{\partial u_x}{\partial z} \tag{14·8·29}$$

Then one obtains by (14·8·11)

$$P_{zx} = -\eta\frac{\partial u_x}{\partial z}$$

where

$$\eta = \frac{5\sqrt{\pi}}{24}\frac{\sqrt{mkT}}{\bar{\sigma}_\eta} \tag{14·8·30}$$

This completes the calculation. It is, of course, apparent by (14·8·30) that η depends only on the temperature, but not on the pressure.

Note that it is the cross section σ_η of (14·8·21), rather than the total cross section $\sigma_0 = 2\pi\int_0^{2\pi}\sigma\sin\theta'\,d\theta'$, which enters this viscosity calculation. The extra $\sin^2\theta'$ factor in the integrand implies that scattering angles near 90° are weighted most heavily. The physical reason is clear; scattering through such angles is most effective in cutting down the rate of momentum transfer responsible for the viscosity.

Example Let us evaluate (14·8·30) for the special case where the molecules can be considered hard spheres of radius a. Then the total scattering cross section is

$$\sigma_0 = \pi(2a)^2 = 4\pi a^2 \tag{14·8·31}$$

The differential scattering cross section for hard spheres does not depend on the scattering angle θ' (see Problem 12.4). Thus

$$\sigma = \frac{\sigma_0}{4\pi} = a^2$$

Then one finds, by (14·8·21),

$$\sigma_\eta = 2\pi\left(\frac{\sigma_0}{4\pi}\right)\int_0^\pi \sin^3\theta'\,d\theta' = \frac{1}{2}\sigma_0\left(\frac{4}{3}\right) = \frac{2}{3}\sigma_0$$

Since this is independent of V, (14·8·26) gives simply

$$\bar{\sigma}_\eta = \sigma_\eta = \tfrac{2}{3}\sigma_0 \tag{14·8·32}$$

Thus (14·8·30) becomes

$$\eta = \frac{5\sqrt{\pi}}{16}\frac{\sqrt{mkT}}{\sigma_0} = 0.553\,\frac{\sqrt{mkT}}{\sigma_0} \tag{14·8·33}$$

This is actually a very good value for the hard-sphere model. More refined approximations would only increase this result by 1.6 percent.

It is of interest to note that (14·8·33) is larger than the result of the most elementary mean-free-path calculation (12·3·19) by a factor of 1.47.

SUGGESTIONS FOR SUPPLEMENTARY READING

Relatively elementary discussions

R. D. Present: "Kinetic Theory of Gases," chaps. 8 and 11, McGraw-Hill Book Company, New York, 1958.

E. A. Guggenheim: "Elements of the Kinetic Theory of Gases," Pergamon Press, New York, 1960.

A. Sommerfeld: "Thermodynamics and Statistical Mechanics," chap. 5, Academic Press, New York, 1956.

E. M. Kennard: "Kinetic Theory of Gases," chap. 4, McGraw-Hill Book Company, New York, 1938.

More advanced books

K. Huang: "Statistical Mechanics," chaps. 3, 5, and 6, John Wiley & Sons, Inc., New York, 1963.

K. M. Watson, J. W. Bond, and J. A. Welch: "Atomic Theory of Gas Dynamics," Addison-Wesley Publishing Company, Reading, Mass., 1965.

S. Chapman and T. G. Cowling: "The Mathematical Theory of Non-uniform Gases," 2d ed., Cambridge University Press, Cambridge, 1952.

J. O. Hirschfelder, C. F. Curtiss, and R. B. Byrd, "Molecular Theory of Gases and Liquids," chaps. 7 and 8, John Wiley & Sons, Inc., New York, 1954.

PROBLEMS

14.1 **Consider the physical situation envisaged in Sec. 14·6 where there are n ions per unit volume, each having charge e and mass m. These ions are free to move in a gas containing n_1 neutral molecules of mass m_1 in unit volume. Assume, for**

the sake of simplicity, that the scattering of an ion by a molecule can be approximated by the scattering of hard spheres of total cross section σ_{im}. The neutral gas has zero drift velocity, since it is at rest with respect to the container. By carrying out the momentum balance argument of Sec. 14·6 exactly, derive an expression for the electrical conductivity of the ions in this gas.

14.2 Consider again the physical situation described in Problem 13.6 where a monatomic dilute gas at temperature T is enclosed in a container and is maintained in the presence of a small temperature gradient in the z direction. The molecular mass is m, and the differential scattering cross section describing collisions between molecules is $\sigma(V,\theta')$.

(*a*) Obtain an approximate form of the molecular distribution function. Refer to the suggestion in Problem 13·6 and note again that the physical situation of no mass motion requires that $\bar{v}_z = 0$.

(*b*) Find an approximate solution of the Boltzmann equation and use this result to find the coefficient of thermal conductivity of this gas. Express your answer in terms of T, m, and the effective total cross section $\bar{\sigma}_\eta(T)$ defined in (14·8·26).

(*c*) If the molecules can be considered hard spheres of total scattering cross section σ_0, calculate the thermal conductivity of this gas in terms of σ_0.

14.3 By comparing the general expressions derived in the case of a monatomic dilute gas for its thermal conductivity κ in Problem 14.2 and for its viscosity coefficient η in (14·8·30), show that the ratio κ/η is a constant independent of the nature of the interactions between the molecules.

(*a*) Find the numerical value of this ratio.

(*b*) Compare this value with the value which would be computed on the basis of the simplest mean-free-path arguments.

(*c*) Compare the value of this ratio predicted by the exact theory with experimental values obtained for several monatomic gases. Appended are a few values of atomic weights μ, viscosity η (in gm cm^{-1} sec^{-1}), and thermal conductivity κ (in ergs cm^{-1} sec^{-1} deg^{-1}) at $T = 373°K$

Gas	μ	η	κ
Neon	20.18	3.65×10^{-4}	5.67×10^3
Argon	39.95	2.70×10^{-4}	2.12×10^3
Xenon	131.3	2.81×10^{-4}	0.702×10^3

14.4 The general expression (6·6·24) for the entropy of a system suggests that the quantity H defined in terms of the distribution function $f(\mathbf{r},\mathbf{v},t)$ by

$$H \equiv \int d^3\mathbf{v}\, f \ln f$$

is related to the entropy of the gas. By using for f the equilibrium Maxwell velocity distribution verify that $H = -S/k$, where S is the entropy per unit volume of a monatomic ideal gas.

14.5 Use the definition

$$H \equiv \int d^3\mathbf{v}\, f \ln f$$

to obtain a general expression for the time derivative dH/dt. Make use of the Boltzmann equation satisfied by f, and exploit the symmetry of the resulting expression in a manner similar to that employed in evaluating the collision term at the end of Sec. 14·4.

(*a*) In this manner show that

$$\frac{dH}{dt} = -\frac{1}{4}\iiint d^3v\, d^3v_1\, d\Omega'\, V\sigma(\ln f'f_1' - \ln ff_1)(f'f_1' - ff_1)$$

(*b*) Since for any y and x,

$$(\ln y - \ln x)(y - x) \geq 0$$

(the equal sign being valid only when $y = x$), show that $dH/dt \leq 0$ and that the equals sign holds if, and only if, $f'f_1' = ff_1$. This is the so-called "Boltzmann H theorem," which proves that the quantity H always tends to decrease (i.e., the generalized entropy defined as $-H/k$ tends to increase).

(*c*) Since in equilibrium it must be true that $dH/dt = 0$, show that when equilibrium has been reached $f'f_1' = ff_1$.

14.6 The equilibrium condition $f'f_1' = ff_1$ is equivalent to

$$\ln f' + \ln f_1' = \ln f + \ln f_1$$

i.e., the sum of quantities before a collision must be equal to the sum of the corresponding quantities after the collision. The only quantities thus conserved are, besides a constant, the three momentum components of a molecule and its kinetic energy. Thus the equilibrium condition can only be satisfied by an expression of the form

$$\ln f = A + B_x mv_x + B_y mv_y + B_z mv_z + C(\tfrac{1}{2}mv^2)$$

where the coefficients A, B_x, B_y, B_z, and C are constants. Show that this implies that f must, therefore, be the Maxwellian velocity distribution (for a gas whose mean velocity does not necessarily vanish).

Irreversible processes and fluctuations

15

IN THIS final chapter we shall attempt to make some general statements about systems which are not necessarily in equilibrium. Thus we shall examine how equilibrium is approached and how rapidly it is approached. We shall also want to gain an understanding of the frictional effects (like those due to viscous forces or to electrical resistance) which lead to dissipation of energy in many systems of interest. Finally, we shall investigate the fluctuations exhibited by certain parameters of systems in thermal *equilibrium*. Although these various questions may at first sight seem unrelated, our discussion will show that they are actually very intimately connected.

TRANSITION PROBABILITIES AND MASTER EQUATION

15·1 *Isolated system*

Consider an isolated system A. Let its Hamiltonian (or energy) be

$$\mathcal{H}_0 = \mathcal{H} + \mathcal{H}_i \qquad (15 \cdot 1 \cdot 1)$$

where $\mathcal{H}$ is the main part of the Hamiltonian and $\mathcal{H}_i \ll \mathcal{H}$ is a small additional part describing some weak interactions not included in $\mathcal{H}$. (For example, in the case of a dilute gas, $\mathcal{H}$ might contain all the kinetic energy terms of the molecules, while $\mathcal{H}_i$ might describe the small interactions between the molecules.) Let the quantum states of $\mathcal{H}$ be denoted by r and their corresponding energy levels by E_r. If $\mathcal{H}_i = 0$, these states would be quantum states of the total Hamiltonian so that the system A would remain in any such state indefinitely. The presence of the additional interaction $\mathcal{H}_i$ makes this no longer true, since $\mathcal{H}_i$ is capable of inducing transitions between the various unperturbed states r. If $\mathcal{H}_i$ is small, if there is a nearly continuous distribution of accessible energy levels,* and if one considers time intervals which are not too small, then there exists a well-defined transition probability W_{rs} per unit time from the unperturbed state r to the unperturbed state s of system A. By conservation of energy W_{rs} is such that

$$\text{if } E_r \neq E_s, \qquad W_{rs} = 0 \tag{15·1·2}$$

Furthermore, there is a symmetry property relating this transition to the inverse transition* from state s to state r,

$$W_{sr} = W_{rs} \tag{15·1·3}$$

Remark This follows from the fact that by quantum mechanics

$$W_{rs} \propto |\langle s|\mathcal{H}_i|r\rangle|^2 = \langle s|\mathcal{H}_i|r\rangle^*\langle s|\mathcal{H}_i|r\rangle$$

where $\langle s|\mathcal{H}_i|r\rangle$ is the matrix element of $\mathcal{H}_i$ between state r and state s. Since $\mathcal{H}_i$ is hermitian, so that $\langle s|\mathcal{H}_i|r\rangle = \langle r|\mathcal{H}_i|s\rangle^*$, the relation (15·1·3) follows immediately. Note also that this symmetry property relating *inverse* transitions is not quite the same as the symmetry property between *reverse* transitions discussed in (9·15·2).

Let $P_r(t)$ denote the probability that system A is found in state r at time t. Then P_r tends to increase with time because systems in other states make transitions to the given state r, and it tends to decrease because systems in this state r make transitions to other states s. The time dependence of P_r can thus be described by the equation

$$\frac{dP_r}{dt} = \sum_s P_s W_{sr} - \sum_s P_r W_{rs} \tag{15·1·4}$$

or

$$\frac{dP_r}{dt} = \sum_s (P_s W_{sr} - P_r W_{rs}) \tag{15·1·5}$$

If there are N states, one can write N such equations in the corresponding number of unknowns P_r. Hence a knowledge of the transition probabilities W_{rs} allows one to compute the probabilities P_r as a function of time.

Equation (15·1·5) is called the "master equation." Note that all terms in it are real and that the time t enters linearly in the first derivative. Hence the master equation does *not* remain invariant as the sign of the time t is reversed from t to $-t$. This equation describes, therefore, the *irreversible* behavior of a system. It is thus quite unlike the detailed microscopic equations of motion, e.g., the Schrödinger equation, which provide a description which

* This can be the case because the energy levels are very closely spaced, or because one considers an ensemble of systems not all of which have exactly the same energy levels. (An example of the latter situation might be an assembly of nuclei located in slightly different local magnetic fields.)

is invariant under time-reversal.* Although the master equation is related to the Schrödinger equation, the precise approximations which lead to the derivation of the master equation are subtle and involve the interesting question of how one passes from reversible microscopic equations to a description which exhibits irreversibility.† We shall discuss this question in somewhat greater detail in Sec. 15·7. In the present context the approximations involved are those, already mentioned, which lead to the existence of well-defined transition probabilities per unit time and those implied by a discussion which concerns itself only with the *probability* P_r that the system is in a given state r. (The complete quantum-mechanical specification of the state of the system at all times involves the complex *probability amplitude* $a_r(t)$, where $P_r = |a_r|^2$. The present discussion thus involves approximations which disregard all information contained in the phases of the amplitudes a_r.)

If the isolated system A is in equilibrium, the fundamental statistical postulate of equal a priori probabilities asserts that, for all r and s,

$$P_r = P_s \tag{15·1·6}$$

By the symmetry property (15·1·3) the right side of (15·1·5) gives then

$$P_sW_{sr} - P_rW_{rs} = P_s(W_{sr} - W_{rs}) = 0 \tag{15·1·7}$$

so that (15·1·5) yields $dP_r/dt = 0$ for all r. Thus Eq. (15·1·5) does then describe correctly an equilibrium situation.

The relation (15·1·7) expresses a condition of detailed balance according to which the rate of occurrence of *any* transition equals the corresponding rate of occurrence of the *inverse* transition; i.e., states of the same energy (for which $W_{rs} \neq 0$) are connected by the relation

$$P_rW_{rs} = P_sW_{sr} \tag{15·1·8}$$

This is certainly a sufficient condition to guarantee that $dP_r/dt = 0$ for all r; it is not, however, a necessary condition. As we have just shown, the condition (15·1·8) is satisfied in equilibrium. It may, however, be possible to encounter nonequilibrium steady-state situations, defined by the fact that $dP_r/dt = 0$ for all r, in which the condition (15·1·8) is *not* satisfied.‡

Note that if it is *assumed* that the condition of detailed balance holds in an equilibrium situation where $P_r = P_s$, then one can immediately conclude from (15·1·8) that $W_{sr} = W_{rs}$. Since the transition probabilities W are dynamical quantities which do not depend on whether the system is in equilibrium or not, it then follows that the relation $W_{sr} = W_{rs}$ must be quite generally valid even if A is not in equilibrium. Thus one arrives again at the result (15·1·3).

* For a discussion of time-reversal invariance, see R. C. Tolman, "The Principles of Statistical Mechanics," secs. 37 and 95, Oxford University Press, Oxford, 1938.

† For a discussion of such derivations and further references, see R. W. Zwanzig in University of Colorado "Lectures in Theoretical Physics," vol. 3, pp. 106–141, Interscience Publishers, New York, 1961.

‡ See, for example, M. J. Klein, *Phys. Rev.* vol. 97, p. 1446 (1955).

15·2 *System in contact with a heat reservoir*

Consider a system A in thermal contact with a much larger system A'. The total Hamiltonian of the combined system $A^{(0)} = A + A'$ is

$$\mathcal{H}^{(0)} = \mathcal{H} + \mathcal{H}' + \mathcal{H}_i$$

where $\mathcal{H}$ is the Hamiltonian of A, $\mathcal{H}'$ is that of the heat reservoir A', and $\mathcal{H}_i$ is very small and describes the weak interaction between A and A'. In the absence of interaction when $\mathcal{H}_i = 0$, denote the energy of A in state r by E_r, and the energy of A' in state r' by $E'_{r'}$. The presence of the interaction $\mathcal{H}_i$ induces transitions between these states and is responsible for bringing about equilibrium between A and A'.

Let P_r be the probability that A is in state r, and $P'_{r'}$ the corresponding probability that A' is in state r'. The state of the combined system $A^{(0)}$ is described by the pair of numbers r and r'; the probability of $A^{(0)}$ being in this state is $P^{(0)}_{rr'} = P_r P'_{r'}$. In this combined system the interaction $\mathcal{H}_i$ causes transitions between states. Under assumptions similar to those of the last subsection, there exists a well-defined transition probability $W^{(0)}(rr' \to ss')$ per unit time from state rr' to state ss'. By conservation of energy,

$$\text{if } E_r + E'_{r'} \neq E_s + E'_{s'}, \qquad W^{(0)}(rr' \to ss') = 0 \tag{15·2·1}$$

In addition, (15·1·3) yields for the isolated system $A^{(0)}$ the symmetry condition

$$W^{(0)}(ss' \to rr') = W^{(0)}(rr' \to ss') \tag{15·2·2}$$

Let us first recall the relations valid in equilibrium. Then P_r is proportional to the total number of states available to $A^{(0)}$ when A is known to be in the given state r; thus, by the reasoning of Sec. 6·2,

$$P_r \propto 1 \cdot \Omega'(E^{(0)} - E_r) \propto e^{-\beta E_r} \tag{15·2·3}$$

where $\Omega'(E')$ is the number of states available to A' when its energy is E', $E^{(0)}$ is the constant total energy of $A^{(0)}$, and $\beta \equiv \partial \ln \Omega'/\partial E'$ (evaluated for $E' = E^{(0)}$) is the temperature parameter of the reservoir A'.

One can also obtain the canonical distribution (15·2·3) for system A by assuming that the condition of detailed balance prevails in equilibrium and that the interaction between A and A' causes transitions between all states of A. Then one can write

$$P_r P'_{r'} W^{(0)}(rr' \to ss') = P_s P'_{s'} W^{(0)}(ss' \to rr') \tag{15·2·4}$$

where $W^{(0)} \neq 0$. Using the equality (15·2·2) this becomes

$$\frac{P_r}{P_s} = \frac{P'_{s'}}{P'_{r'}} \tag{15·2·5}$$

But A' itself satisfies the canonical distribution, since the equilibrium of A' is undisturbed when it is placed in contact with a heat reservoir at temperature β. Hence $P'_{r'} \propto \exp(-\beta E'_{r'})$ and (15·2·5) becomes

$$\frac{P_r}{P_s} = \frac{e^{-\beta E'_{s'}}}{e^{-\beta E'_{r'}}} = e^{-\beta(E'_{s'} - E'_{r'})} \tag{15·2·6}$$

Application of the conservation of energy to (15·2·4) requires that

$$E_r + E'_{r'} = E_s + E'_{s'} \qquad \text{or} \qquad E'_{s'} - E'_{r'} = -(E_s - E_r)$$

Hence (15·2·6) becomes

$$\frac{P_r}{P_s} = e^{-\beta(E_r - E_s)} = \frac{e^{-\beta E_r}}{e^{-\beta E_s}} \tag{15·2·7}$$

and one regains the canonical distribution (15·2·3) for system A.

Suppose now that we consider a general *nonequilibrium* situation, but assume that the heat reservoir A' is so large that, irrespective of what A does, A' always remains in internal equilibrium. Thus, no matter what total energy A' may have at any one time, A' is always supposed to be distributed according to a canonical distribution with the constant temperature parameter β. Under these circumstances, let us ask for the net transition probability W_{rs} per unit time of the subsystem A from its state r to its state s. It should be clear that $W_{rs} \neq W_{sr}$; Fig. 15·2·1 illustrates how this comes about.

The transition probability W_{rs} can be obtained by multiplying the transition probability $W^{(0)}(rr' \to ss')$ for the combined system $A^{(0)}$ by the probability $P'_{r'}$ that A' is in the particular state r' and then summing over all possible initial states r' in which A' can be found and all the possible final states s' in which it can end up. Thus

$$W_{rs} = \sum_{r's'} P_{r'} W^{(0)}(rr' \to ss') = C \sum_{r's'} e^{-\beta E'_{r'}} W^{(0)}(rr' \to ss') \tag{15·2·8}$$

since $P'_{r'} = C \exp(-\beta E'_{r'})$, where C is some constant. Similarly,

$$W_{sr} = C \sum_{r's'} e^{-\beta E'_{s'}} W^{(0)}(ss' \to rr') \tag{15·2·9}$$

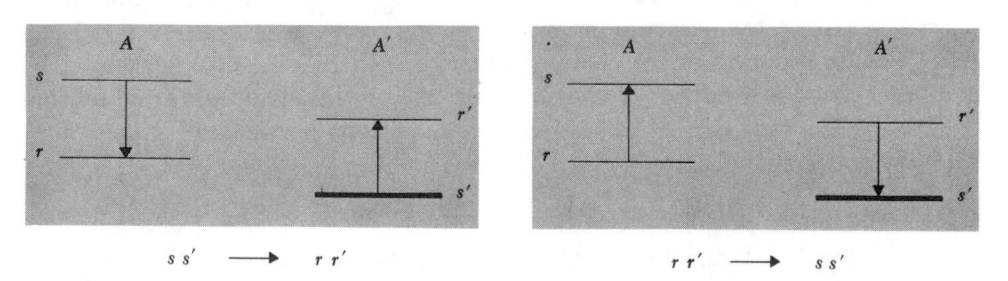

***Fig. 15·2·1** Diagram illustrating a system A in thermal contact with a heat reservoir A'. The lines indicate the energy levels for states r and s of A, and states r' and s' of A'. Since A' remains always in thermal equilibrium, A' is more likely to be in the lower state s' (marked by the heavy line) than in the upper state r'. As a result, transitions of the type $ss' \to rr'$, illustrated in the left diagram, occur more frequently than those of type $rr' \to ss'$, illustrated on the right. Hence $W_{sr} > W_{rs}$. These transitions tend thus to produce the thermal equilibrium distribution where A is more likely to be in state r than in state s.*

But the energy conservation (15·2·1) implies that $E'_{s'} = E'_{r'} + E_r - E_s$. Using the symmetry relation (15·2·2), the expression (15·2·9) becomes then

$$W_{sr} = C \sum_{r's'} e^{-\beta E_{r'}} e^{-\beta(E_r - E_s)} W^{(0)}(rr' \to ss') = e^{-\beta(E_r - E_s)} W_{rs}$$

Thus

$$\frac{W_{sr}}{W_{rs}} = e^{-\beta(E_r - E_s)} = \frac{e^{-\beta E_r}}{e^{-\beta E_s}} \tag{15·2·10}$$

In the ordinary case where $\beta > 0$, $W_{sr} > W_{rs}$ if $E_s > E_r$. In the system A transitions to states of lower energy are thus more probable than transitions in the opposite direction. This is, of course, what is required to bring about the thermal equilibrium distribution in A. Indeed, if it is *assumed* that the condition of detailed balance prevails at equilibrium, then

$$P_r W_{rs} = P_s W_{sr} \tag{15·2·11}$$

The canonical distribution (15·2·7) then implies immediately the relation (15·2·10) between transition probabilities.

It is convenient to make the relationship (15·2·10) apparent by introducing a quantity $\lambda_{rs} = \lambda_{sr}$ defined by

$$e^{-\beta E_s} W_{sr} = e^{-\beta E_r} W_{rs} \equiv \lambda_{rs} \equiv \lambda_{sr}$$

Then one can write

$$W_{sr} = e^{\beta E_s} \lambda_{sr}, \qquad W_{rs} = e^{\beta E_r} \lambda_{rs} \tag{15·2·12}$$

and (15·2·10) follows automatically since $\lambda_{sr} = \lambda_{rs}$.

The probabilities for system A satisfy again the master equation (15·1·5)

$$\frac{dP_r}{dt} = \sum_s (P_s W_{sr} - P_r W_{rs}) = \sum_s \lambda_{sr}(P_s\, e^{\beta E_s} - P_r\, e^{\beta E_r}) \tag{15·2·13}$$

Once again, if equilibrium prevails so that the canonical distribution (15·2·3) holds for system A, then the right side of (15·2·13) vanishes and $dP_r/dt = 0$ for all r, as it should.

15·3 *Magnetic resonance*

An instructive and important example of the ideas developed in the preceding sections is that of magnetic resonance. Consider a substance containing N noninteracting nuclei (or electrons) of spin $\frac{1}{2}$ and magnetic moment μ. If the substance is placed in an applied magnetic field H, each spin can point either "up" (i.e., parallel to H) or "down." We denote the corresponding states by $+$ and $-$, respectively. The two possible energies of each nucleus are then

$$\epsilon_{\pm} = \mp \mu H \tag{15·3·1}$$

Let n_+ be the mean number of spins pointing up and n_- be the mean number of spins pointing down. Clearly, $n_+ + n_- = N$.

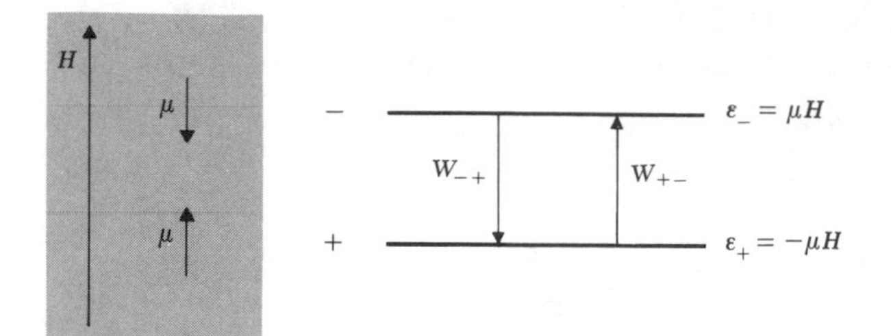

Fig. 15·3·1 Energy levels of a nucleus of spin ½ in an external magnetic field H. The magnetic moment μ is assumed to be positive in this diagram.

The total Hamiltonian of the system can be written as

$$\mathcal{H}^{(0)} = \mathcal{H}_n + \mathcal{H}_L + \mathcal{H}_i$$

Here $\mathcal{H}_n$ is the Hamiltonian expressing the interaction between the nuclear moments and the external field H; $\mathcal{H}_L$ is the Hamiltonian describing the "lattice," i.e., all non–spin degrees of freedom of the nuclei and all other atoms in the substance. The Hamiltonian $\mathcal{H}_i$ describes the interaction between the spins of the nuclei and the lattice, and causes transitions between the possible spin states of the nuclei. (For example, the magnetic moment of a *moving* nucleus produces a fluctuating magnetic field at the positions of other nuclei and this field causes transitions.) Let W_{+-} be the transition probability per unit time that a nucleus flips its spin from "up" to "down" as a result of interaction with the lattice. The lattice itself is a large system which can be regarded as being always very close to internal equilibrium at the absolute temperature $T = (k\beta)^{-1}$. Thus (15·2·10) allows one to write the relationship

$$\frac{W_{-+}}{W_{+-}} = \frac{e^{-\beta\epsilon_+}}{e^{-\beta\epsilon_-}} = e^{\beta(\epsilon_- - \epsilon_+)} \tag{15·3·2}$$

Now (15·3·1) gives $\beta(\epsilon_- - \epsilon_+) = 2\beta\mu H$. For nuclei the magnetic moment $\mu \approx 5 \cdot 10^{-24}$ ergs/gauss so that in laboratory fields of the order of $H = 10^4$ gauss

$$\beta\mu H = \frac{\mu H}{kT} \approx \frac{5 \cdot 10^{-4}}{T} \ll 1$$

for all but *exceedingly* low temperatures. Even for electronic moments, which are about 1000 times larger, this inequality is almost always well satisfied. By expanding the exponential, the relation (15·3·2) can then be written in the form

$$\begin{aligned} W_{+-} &\equiv W \\ \text{and} \qquad W_{-+} &= W(1 + 2\beta\mu H), \qquad \text{where } \beta\mu H \ll 1 \end{aligned} \tag{15·3·3}$$

Finally, there may also be present an externally applied alternating magnetic field of angular frequency ω. If $\hbar\omega \approx \epsilon_- - \epsilon_+ = 2\mu H$, this field will induce transitions between the spin states of a nucleus. (If $H \approx 10^4$ gauss, ω is typically a radio frequency (rf) of the order of 10^8 sec^{-1}.) Let w_{+-} be the transition probability per unit time for the "up" to "down" transition induced by this rf field. Then one again has the symmetry property (15·1·3)

$$w_{+-} = w_{-+} \equiv w \tag{15·3·4}$$

Here $w = w(\omega)$ is only appreciable if ω satisfies the resonance condition $\hbar\omega \approx 2\mu H$.

The master equations for $n_+(t)$ and $n_-(t)$ then become

$$\left.\begin{aligned} \frac{dn_+}{dt} &= n_-(W_{-+} + w) - n_+(W_{+-} + w) \\ \frac{dn_-}{dt} &= n_+(W_{+-} + w) - n_-(W_{-+} + w) \end{aligned}\right\} \qquad (15 \cdot 3 \cdot 5)$$

By subtracting the second equation from the first, one obtains

$$\frac{d}{dt}(n_+ - n_-) = -2n_+(W_{+-} + w) + 2n_-(W_{-+} + w) \qquad (15 \cdot 3 \cdot 6)$$

Introducing the population difference

$$n \equiv n_+ - n_- \qquad (15 \cdot 3 \cdot 7)$$

and using (15·3·3), the relation (15·3·6) becomes

$$\frac{dn}{dt} = -2(W + w)n + 2\beta\mu HWN \qquad (15 \cdot 3 \cdot 8)$$

Here we have put $4\beta\mu HWn_- = 4\beta\mu HW\,(\frac{1}{2}\,N - n) \approx 2\beta\mu HWN$, since one always has $n \ll N$ in the temperature range of interest.

Let us now investigate various cases of interest. Consider first the equilibrium situation in the absence of an applied rf field, i.e., with $w = 0$. Then $dn/dt = 0$, and (15·3·8) yields for the *equilibrium* excess number of spins

$$n_0 = N\beta\mu H \qquad (15 \cdot 3 \cdot 9)$$

This is, of course, the result which follows from the canonical distribution according to which one has *in equilibrium*

$$n_\pm = N\,\frac{e^{\pm\beta\mu H}}{e^{\beta\mu H} + e^{-\beta\mu H}} \approx N\,\frac{1 \pm \beta\mu H}{2} = \frac{1}{2}\,N(1 \pm \beta\mu H)$$

so that $n_0 = n_+ - n_-$ assumes the value (15·3·9)

Hence (15·3·8) can be written in the form

► $$\frac{dn}{dt} = -2W(n - n_0) - 2wn \qquad (15 \cdot 3 \cdot 10)$$

In the absence of a rf field when $w = 0$, this yields upon integration

$$n(t) = n_0 + [n(0) - n_0]\,e^{-2Wt} \qquad (15 \cdot 3 \cdot 11)$$

where $n(0)$ is the population difference at the initial time $t = 0$. Thus $n(t)$ approaches its equilibrium value n_0 exponentially with a characteristic "relaxation time" $(2W)^{-1}$. Obviously, the larger the interaction W of the spins with the lattice heat reservoir, the shorter the relaxation time.

Suppose now that the interaction of the spins with the lattice is very weak,

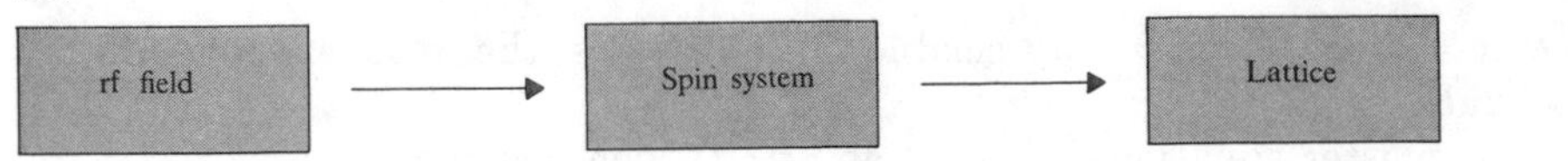

Fig. 15·3·2 Net energy flow in a steady-state resonance absorption experiment.

so that $W \approx 0$, and that a rf magnetic field is applied. Then (15·3·10) becomes

$$\frac{dn}{dt} = -2wn$$

so that

$$n(t) = n(0)\, e^{-2wt} \tag{15·3·12}$$

The population difference then becomes exponentially zero within a characteristic time $(2w)^{-1}$ which is inversely proportional to the strength of the interaction with the applied rf field. The reason for this is quite clear. Since there are more nuclei in the lower state of Fig. 15·3·1 than in the upper state, the transitions induced by the rf field result in a net absorption of power equal to $(n_+ - n_-)w(2\mu H)$. The nuclear spin system is essentially isolated since $W \approx 0$; hence the energy absorbed from the rf field goes to increase the energy and hence the "spin temperature T_s" of the spins until T_s approaches infinity. At this point $n_+ = n_-$, the population difference $n = 0$, and there is no further absorption of rf power. The spin system is then said to be "saturated."

If one *does* want to observe absorption of rf power in a steady state situation, there must be adequate contact between the spin system and the lattice heat reservoir. In that case the power absorbed by the spin system is in turn given off to the lattice (which, being a system of very large heat capacity, experiences only an infinitesimal increase in temperature). In the steady state $dn/dt = 0$ so that (15·3·10) becomes

$$W(n - n_0) = -wn$$

or

$$n = \frac{n_0}{1 + (w/W)} \tag{15·3·13}$$

Thus n is less than the equilibrium value n_0 by an amount which depends on the ratio w/W of the transition probabilities. If $w \ll W$, $n \to n_0$; but if $w \gg W$, $n \to 0$ and one gets saturation.

15 · 4 *Dynamic nuclear polarization; Overhauser effect*

Nonequilibrium methods for achieving nuclear polarization (the so-called "Overhauser effect" being one such method) provide a particularly illuminating illustration of the ideas presented in the last few sections. Consider a substance containing both nuclei of spin $\frac{1}{2}$ and magnetic moment μ_n, and also unpaired electrons of spin $\frac{1}{2}$ and magnetic moment μ_e ($\mu_e < 0$). The substance is placed in an external magnetic field H pointing in the z direction. Suppose that the principal interaction of a nucleus is with an electron through "hyperfine interaction," i.e., through the magnetic field produced by the electron at the position of the nucleus. (This interaction is described by a Hamiltonian of the form $\mathcal{H}_{ne} = a\boldsymbol{I} \cdot \boldsymbol{S}$ where $\boldsymbol{I}$ is the nuclear and $\boldsymbol{S}$ the electronic spin angular momentum.) Since no z component of external torque acts on the system consisting of nucleus and electron (i.e., since the total Hamiltonian of this system is invariant under rotations about the z direction), the *total* z component of angular momentum $(I_z + S_z)$ of this system is a constant of the motion. Thus the transitions induced by the interaction between a nucleus and an electron must always be such that whenever the nucleus flips its spin from "up" to "down," the electron must flip its spin from "down" to "up," and vice versa. We shall denote the transition probability per unit time due to this interaction by $W_{ne}(+\,\boldsymbol{-} \to -\,\boldsymbol{+})$ where $+$ and $-$ indicate up and down orientations of the nucleus n, and $\boldsymbol{+}$ and $\boldsymbol{-}$ up and down orientations of the electron e.

The nuclei interact then predominantly with the electron spins, which in turn interact appreciably with the lattice heat reservoir. It is through this chain of interactions that the nuclear spins attain the thermal equilibrium situation corresponding to the lattice temperature $T = (k\beta)^{-1}$. Let n_+ and n_- denote the mean number of nuclear "up" and "down" spins; let N_+ and N_- denote the mean number of electron "up" and "down" spins. In thermal equilibrium one then obtains, for the nuclei of energy $\epsilon_\pm = \mp \mu_n H$,

$$\frac{n_+}{n_-} = \frac{e^{\beta\mu_n H}}{e^{-\beta\mu_n H}} = e^{2\beta\mu_n H} \tag{15·4·1}$$

Also for the electrons of energy $E_\pm = \mp \mu_e H$,

$$\frac{N_+}{N_-} = \frac{e^{\beta\mu_e H}}{e^{-\beta\mu_e H}} = e^{2\beta\mu_e H} \tag{15·4·2}$$

The degree of polarization of the nuclei and electrons can be measured by the respective ratios defined by

$$\xi_n \equiv \frac{n_+ - n_-}{n_+ + n_-} \quad \text{and} \quad \xi_e \equiv \frac{N_+ - N_-}{N_+ + N_-} \tag{15·4·3}$$

Each ξ lies in the range $-1 \leq \xi \leq 1$. Since $|\mu_e| \approx 1000|\mu_n|$, it is clear by (15·4·1) and (15·4·2) that $\xi_n \ll \xi_e$. Even if one goes to such high applied fields H and low temperatures T that the electrons are appreciably polarized, the degree of nuclear polarization is thus still very small, in particular much too small to do nuclear physics experiments on polarized nuclei.

From the point of view of the transition probabilities one can regard the combined system $(n + e)$ of nucleus and electron as being in thermal contact with the lattice heat reservoir. It then follows by (15·2·10) that the transition probabilities must satisfy the relation

$$\frac{W_{ne}(+\,\boldsymbol{-} \to -\,\boldsymbol{+})}{W_{ne}(-\,\boldsymbol{+} \to +\,\boldsymbol{-})} = e^{-\beta(\epsilon_- + E_+ - \epsilon_+ - E_-)} = e^{-2\beta(\mu_n - \mu_e)H} \tag{15·4·4}$$

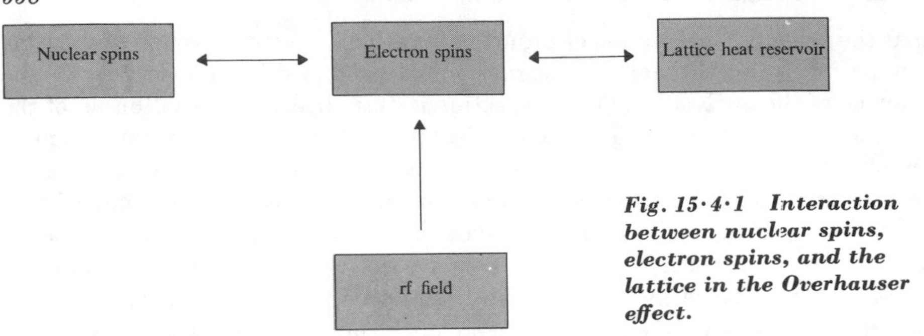

Fig. 15·4·1 Interaction between nuclear spins, electron spins, and the lattice in the Overhauser effect.

If one assumes the condition of detailed balance to be valid in the equilibrium situation,

$$n_+N_-W_{ne}(+\,-\rightarrow -\,+) = n_-N_+W_{ne}(-\,+\rightarrow +\,-) \qquad (15\cdot4\cdot5)$$

By (15·4·4) this condition becomes

$$\frac{n_+}{n_-}\frac{N_-}{N_+} = e^{2\beta(\mu_n-\mu_e)H} \qquad (15\cdot4\cdot6)$$

a result consistent with the equilibrium relations (15·4·1) and (15·4·2).

Imagine, however, that an rf field is applied at the electron-spin resonance frequency and suppose that it is strong enough to saturate the electron-spin system so that $N_+ = N_-$. If one assumes that the detailed balance condition (15·4·5) remains valid in this steady-state situation, (15·4·6) becomes

$$\blacktriangleright \qquad \frac{n_+}{n_-} = e^{2\beta(\mu_n-\mu_e)H} \approx e^{-2\beta\mu_e H} \qquad (15\cdot4\cdot7)$$

since $|\mu_e| \gg |\mu_n|$. Thus the nuclei are now polarized by an amount as large as though they were endowed with the very much larger *electronic* magnetic moment! This remarkable result is called the "Overhauser effect." It allows the achievement of appreciable nuclear polarizations in a nonequilibrium, but steady-state, situation.

Alternative discussion A more general and illuminating discussion of the effect can be given by focusing attention on the lattice heat reservoir. Consider the situation $(+\,-)$ where the nucleus points up and the electron points down; suppose that the reservoir then has the energy E'. The probability $P(+\,-)$ of occurrence of this situation is then proportional to the number of states available to the combined system, i.e., to the number of states Ω' available to the reservoir when the nucleus and electron are in this one state. Thus

$$P(+\,-) \propto \Omega'(E')$$

Compare this now with the situation $(-\,-)$, where the nucleus n points down and the electron e points down as before. If in going to this new situation the reservoir has gained energy $\Delta E'$, then one has similarly

$$P(-\,-) \propto \Omega'(E' + \Delta E') = \Omega'(E')\, e^{\beta\,\Delta E'}$$

where we have expanded $\ln \Omega'$ about E' in the last term and where

$$\beta \equiv \frac{\partial \ln \Omega'}{\partial E'}$$

is the temperature parameter of the reservoir. Hence

$$\frac{P(+\,-)}{P(-\,-)} = e^{-\beta \Delta E'} \tag{15·4·8}$$

or, equivalently,

$$\frac{n_+N_-}{n_-N_-} = \frac{n_+}{n_-} = e^{-\beta \Delta E'} \tag{15·4·9}$$

We now want to calculate $\Delta E'$. According to the interactions illustrated in Fig. 15·4·1, the situation ($-\,-$) must be visualized as arising from the preceding situation ($+\,-$) in two stages, as shown in Fig. 15·4·2. (*a*) The first stage involves a mutual spin flip of n and e caused by the mutual interaction of nucleus and electron. (*b*) The second stage involves interaction between the electron and the lattice (or the rf field) whereby the electron spin alone is flipped. Let E_{ne} = denote the total energy of the system $(n + e)$ consisting of nucleus and electron. Then the change ΔE_{ne} of this energy in the two stages is

Stage *a*: $$\Delta_a E_{ne} = (\epsilon_- - \epsilon_+) + (E_+ - E_-) = 2\mu_n H - 2\mu_e H = 2(\mu_n - \mu_e)H \tag{15·4·10}$$

Stage *b*: $$\Delta_b E_{ne} = (E_- - E_+) = 2\mu_e H \tag{15·4·11}$$

By addition, the total change in energy of the system $(n + e)$ is then

$$\Delta E_{ne} = 2\mu_n H \tag{15·4·12}$$

Consider first the case of thermal equilibrium in the absence of an rf field. Conservation of energy requires then that the total energy change (15·4·12) is provided by the lattice heat reservoir so that

$$\Delta E' = -\Delta E_{ne} = -2\mu_n H \tag{15·4·13}$$

Hence it follows, by (15·4·9), that

$$\frac{n_+}{n_-} = e^{2\mu_n H} \tag{15·4·14}$$

in agreement with (15·4·1).

Consider now, by contrast, the steady-state situation where the applied rf

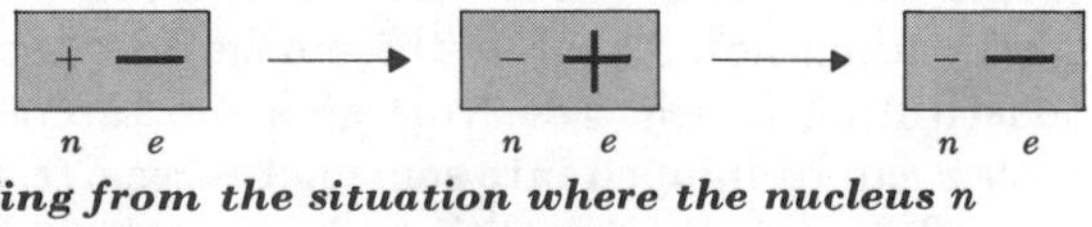

+ — → – + → – —

n e n e n e

Fig. 15·4·2 Interactions leading from the situation where the nucleus n points up and the electron e points down to the situation where n points down and e points down as before.

field keeps the electron-spin system saturated. Since the interaction of the electron-spin system with the rf field is then much stronger than its interaction with the lattice, the electron-spin flip of stage b is in this case achieved by interaction with the rf field; the requisite energy for this process is thus provided by this externally applied field and *not* by the lattice. The total energy supplied by the heat reservoir is then only that for stage a. Thus

$$\Delta E' = -\Delta_a E_{ne} = -2(\mu_n - \mu_e)H \tag{15·4·15}$$

Hence it follows by (15·4·9) that

$$\frac{n_+}{n_-} = e^{2\beta(\mu_n - \mu_e)H} \approx e^{-2\beta\mu_e H} \tag{15·4·16}$$

Thus we obtain again the Overhauser effect predicted in (15·4·7).

SIMPLE DISCUSSION OF BROWNIAN MOTION

15 · 5 *Langevin equation*

A sufficiently small macroscopic particle immersed in a liquid exhibits a random type of motion. This phenomenon, already mentioned in Sec. 7·6, is called "Brownian motion" and reveals very clearly the statistical fluctuations which occur in a system in thermal equilibrium.

There are a variety of important situations which are basically similar. Examples are the random motion of the mirror mounted on the suspension fiber of a sensitive galvanometer, or the fluctuating current existing in an electric resistor. Thus Brownian motion can serve as a prototype problem whose analysis provides considerable insight into the mechanisms responsible for the existence of fluctuations and "dissipation of energy." This problem is also of great practical interest because such fluctuations constitute a background of "noise" which imposes limitations on the possible accuracy of delicate physical measurements.

For the sake of simplicity we shall treat the problem of Brownian motion in *one* dimension. We consider thus a particle of mass m whose center-of-mass coordinate at time t is designated by $x(t)$ and whose corresponding velocity is $v \equiv dx/dt$. This particle is immersed in a liquid at the absolute temperature T. It would be a hopelessly complex task to describe in detail the interaction of the center-of-mass coordinate x with all the many degrees of freedom other than x (i.e., those describing the internal motion of the atoms in the macroscopic particle, as well as those describing the motion of the molecules in the surrounding liquid). But these other degrees of freedom can be regarded as constituting a heat reservoir at some temperature T, and their interaction with x can be lumped into some net force $F(t)$ effective in determining the time dependence of x. In addition, the particle may also interact with some external systems, such as gravity or electromagnetic fields, through a force denoted by

$\mathfrak{F}(t)$. The velocity v of the particle may, in general, be appreciably different from its mean value in equilibrium.*

Focusing attention on the center-of-mass coordinate x, Newton's second law of motion can then be written in the form

$$m\frac{dv}{dt} = \mathfrak{F}(t) + F(t) \tag{15·5·1}$$

Here very little is known about the force $F(t)$ which describes the interaction of x with the many other degrees of freedom of the system. Basically, $F(t)$ must depend on the positions of very many atoms which are in constant motion. Thus $F(t)$ is some rapidly fluctuating function of the time t and varies in a highly irregular fashion. Indeed, one cannot specify the precise functional dependence of F on t. To make progress, one has to formulate the problem in statistical terms. One must, therefore, envisage an ensemble of very many similarly prepared systems, each of them consisting of a particle and the surrounding medium. For each of these the force $F(t)$ is some random function of t (see Fig. 15·5·1). One can then attempt to make statistical statements about this ensemble.

For example, looking at a given time t_1, one can ask for the probability $P(F_1,t_1)\,dF_1$ that the force F in the ensemble assumes at this time a value between F_1 and $F_1 + dF_1$. Or, picking two specific times t_1 and t_2, one can ask for the joint probability $P(F_1t_1; F_2t_2)\,dF_1\,dF_2$ that at time t_1 the force lies between F_1 and $F_1 + dF_1$, *and* that at time t_2 it lies between F_2 and $F_2 + dF_2$. Similarly, one can compute various averages over the ensemble. Thus the mean value of F at time t_1 is

$$\bar{F}(t_1) \equiv \frac{1}{N}\sum_{k=1}^{N} F^{(k)}(t_1)$$

where the sum is over all the N systems, labeled by k, contained in the ensemble.

The following descriptive comments can be made about $F(t)$. The rate at which $F(t)$ varies can be characterized by some "correlation time" τ^* which measures roughly the mean time between two successive maxima (or minima) of the fluctuating function $F(t)$. This time τ^* is quite small on a *macro*scopic scale. (It ought to be roughly of the order of a mean intermolecular separation divided by a mean molecular velocity, e.g., about 10^{-13} sec if $F(t)$ describes

* Indeed, the following question is of interest. Suppose that $\mathfrak{F} = 0$ and that at some initial time the particle has a velocity v different from its thermal mean value $\bar{v} = 0$. (This could be the result of a spontaneous fluctuation; alternatively, this velocity might have been produced by the application of some external force which is then switched off.) How rapidly does the particle's velocity then approach the mean value corresponding to thermal equilibrium?

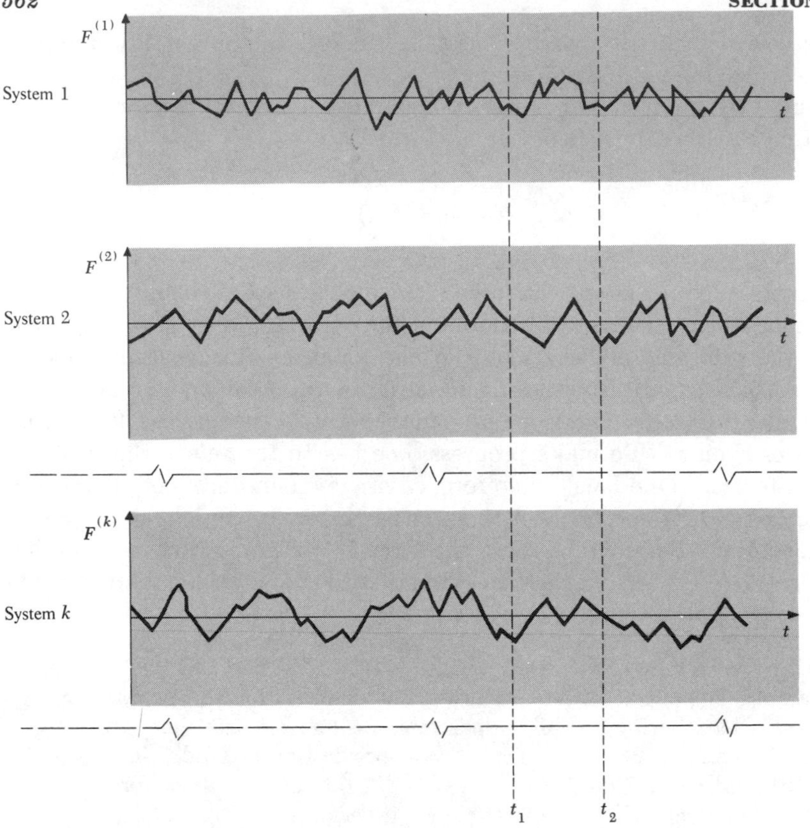

Fig. 15·5·1 Ensemble of systems illustrating the behavior of the fluctuating force $F(t)$ acting on a stationary particle. Here $F^{(k)}(t)$ is the force in the kth system of the ensemble as a function of the time t.

interactions with molecules of a typical liquid.) Furthermore, if one contemplates a situation where the particle is imagined clamped so as to be stationary, there is no preferred direction in space; then $F(t)$ must be as often positive as negative so that the ensemble average $\bar{F}(t)$ vanishes.

Equation (15·5·1) holds for each member of the ensemble, and our aim is to deduce from it statistical statements about v. Since $F(t)$ is a rapidly fluctuating function of time, it follows by (15·5·1) that v also fluctuates in time. But, superimposed upon these fluctuations, the time dependence of v may also exhibit a more slowly varying trend. For example, one can focus attention on the ensemble average $\bar{v}$ of the velocity, which is a much more slowly varying function of the time than v itself, and write

$$v = \bar{v} + v' \tag{15·5·2}$$

where v' denotes the part of v which fluctuates rapidly (although less rapidly than $F(t)$, since the mass m is appreciable) and whose mean value vanishes. The slowly varying part $\bar{v}$ is of crucial importance (even if it is small) because it is of primary significance in determining the behavior of the particle over long periods of time. To investigate its time dependence, let us integrate (15·5·1) over some time interval τ which is small on a *macro*scopic scale, but large in the sense that $\tau \gg \tau^*$. Then one gets

$$m[v(t + \tau) - v(t)] = \mathfrak{F}(t)\tau + \int_t^{t+\tau} F(t')\,dt' \tag{15·5·3}$$

where we have assumed that the external force $\mathfrak{F}$ is varying slowly enough that it changes by a negligible amount during a time τ. The last integral in (15·5·3) ought to be very small since $F(t)$ changes sign many times in the time τ. Hence one might expect that any slowly varying part of v should be due only to the external force $\mathfrak{F}$; i.e., one might be tempted to write

$$m\frac{d\bar{v}}{dt} = \mathfrak{F} \tag{15·5·4}$$

But this order of approximation is too crude to describe the physical situation. Indeed, the interaction with the environment expressed by $F(t)$ must be such that it always tends to restore the particle to the equilibrium situation. Suppose, for example, that the external force $\mathfrak{F} = 0$. The interaction expressed by F must then be such that, if $\bar{v} \neq 0$ at some initial time, it causes $\bar{v}$ to approach gradually its ultimate equilibrium value $\bar{v} = 0$. But (15·5·4) fails to predict this kind of trend of $\bar{v}$ toward its equilibrium value. The reason is that we were too careless in treating the effects of F in (15·5·3). Thus we did not consider the fact that the interaction force F must actually be affected by the motion of the particle in such a way that F itself also contains a slowly varying part $\bar{F}$ tending to restore the particle to equilibrium. Hence we shall write, analogously to (15·5·2),

$$F = \bar{F} + F' \tag{15·5·5}$$

where F' is the rapidly fluctuating part of F whose average value vanishes. The slowly varying part $\bar{F}$ must be some function of $\bar{v}$ which is such that $\bar{F}(\bar{v}) = 0$ in equilibrium when $\bar{v} = 0$. If $\bar{v}$ is not too large, $\bar{F}(\bar{v})$ can be expanded in a power series in $\bar{v}$ whose first nonvanishing term must then be linear in $\bar{v}$. Thus $\bar{F}$ must have the general form

$$\bar{F} = -\alpha\bar{v} \tag{15·5·6}$$

where α is some positive constant (called the "friction constant") and where the minus sign indicates explicitly that the force $\bar{F}$ acts in such a direction that it tends to reduce $\bar{v}$ to zero as time increases. Our present arguments do not permit us to make any statements about the actual magnitude of α. We can, however, surmise that α must in some way be expressible in terms of F itself, since the frictional restoring force is also caused by the interactions described by $F(t)$.

In the general case the slowly varying part of (15 · 5 · 1) becomes then

$$m \frac{d\bar{v}}{dt} = \mathfrak{F} + \bar{F} = \mathfrak{F} - \alpha \bar{v} \tag{15·5·7}$$

If one includes the rapidly fluctuating parts v' and F' of (15 · 5 · 2) and (15 · 5 · 5), Eq. (15 · 5 · 1) can be written

▶
$$m \frac{dv}{dt} = \mathfrak{F} - \alpha v + F'(t) \tag{15·5·8}$$

where we have put $\alpha\bar{v} \approx \alpha v$ with negligible error (since the rapidly fluctuating contribution $\alpha v'$ can be neglected compared to the predominant fluctuating term $F'(t)$). Equation (15 · 5 · 8) is called the "Langevin equation." It differs from the original equation (15 · 5 · 1) by explicitly decomposing the force $F(t)$ into a slowly varying part $-\alpha v$ and into a fluctuating part $F'(t)$ which is "purely random," i.e., such that its mean value always vanishes irrespective of the velocity or position of the particle. The Langevin equation (15 · 5 · 8) describes in this way the behavior of the particle at all later times if its initial conditions are specified.

Since the Langevin equation contains the frictional force $-\alpha v$, it implies the existence of processes whereby the energy associated with the coordinate x of the particle is dissipated in the course of time to the other degrees of freedom (e.g., to the molecules of the liquid surrounding the particle). This is, of course, in accord with our macroscopic experience where frictional forces are commonplace. Nevertheless, one deals here with an example of an interesting and conceptually difficult general problem. Consider a system A in contact with some large system B. The *micro*scopic equations governing the motion of the combined system $(A + B)$ do *not* involve any frictional forces. The total energy is conserved, and the motion is reversible. (That is, if the sign of the time t were reversed, the equations of motion would be essentially unchanged and all particles would (classically) retrace their paths in time.) But if one focuses attention on A, its interaction with the heat reservoir B can be adequately described by equations of motion involving frictional forces. There is thus dissipation of energy from A to the heat reservoir B and the motion of A is not reversible. The question is to understand in detail how this situation comes about, what conditions must be satisfied for this description to be approximately valid, and how the modified equations of motion for A are derivable from the microscopic equations. A discussion of such questions in abstract generality is beyond the scope of this book. We shall, however, return in Sec. 15 · 7 to a more detailed investigation of the approximations which lead to the irreversible Langevin equation (15 · 5 · 8).

Finally, it is useful to point out the analogous electrical problem of an electrical conductor of self-inductance L carrying a current I. Let the applied electromotive force (emf) be $\mathcal{V}(t)$. The current I carried by the electrons is affected by the interactions of these electrons with the atoms of the conductor. The net effect of these interactions on the current I can be represented by some effective fluctuating emf $V(t)$. The latter can be decomposed into a slowly varying part $-RI$ (where R is some positive constant) and into a rapidly fluctuating part $V'(t)$ whose mean value vanishes. The analogue of the Langevin equation (15·5·8) then becomes

$$L\frac{dI}{dt} = \mathcal{V} - RI + V'(t) \tag{15·5·9}$$

The friction constant R is here simply the electrical resistance of the conductor.

15·6 *Calculation of the mean-square displacement*

Let us assume the validity of Langevin's equation as an adequate phenomenological description of Brownian motion and illustrate how it can be applied to the calculation of quantities of physical interest. In the absence of external forces (15·5·8) becomes

$$m\frac{dv}{dt} = -\alpha v + F'(t) \tag{15·6·1}$$

One can go further and calculate the frictional force by applying purely macroscopic hydrodynamic reasoning to the motion of a macroscopic spherical particle of radius a moving with velocity v through a liquid of viscosity η. This calculation yields a frictional force $-\alpha v$ where

$$\alpha = 6\pi\eta a \tag{15·6·2}$$

This is known as "Stokes's law."*

Consider the situation of thermal equilibrium. Clearly the mean displacement $\bar{x}$ of the particle vanishes (i.e., $\bar{x} = 0$) by symmetry, since there is no preferred direction in space. To calculate the magnitude of the fluctuations, we now use (15·6·1) to calculate the mean-square displacement $\langle x^2\rangle = \overline{x^2}$ of the particle in a time interval t. (We shall indicate ensemble averages interchangeably by bars or angular brackets.) Equation (15·6·1) contains the quantities $v = \dot{x}$ and $dv/dt = d\dot{x}/dt$. Multiplying both sides of (15·6·1) by x, one thus gets

$$mx\frac{d\dot{x}}{dt} = m\left[\frac{d}{dt}(x\dot{x}) - \dot{x}^2\right] = -\alpha x\dot{x} + xF'(t) \tag{15·6·3}$$

One can now take the ensemble average of both sides of (15·6·3). As pointed out in connection with the Langevin equation (15·5·8), the mean value of the fluctuating force F' always vanishes, irrespective of the value of v or x. Hence $\langle xF'\rangle = \langle x\rangle\langle F'\rangle = 0$. Furthermore, the equipartition theorem yields $\frac{1}{2}m\langle\dot{x}^2\rangle = \frac{1}{2}kT$. Thus (15·6·3) becomes

$$m\left\langle\frac{d}{dt}(x\dot{x})\right\rangle = m\frac{d}{dt}\langle x\dot{x}\rangle = kT - \alpha\langle x\dot{x}\rangle \tag{15·6·4}$$

* See Problem 15.1. For the rigorous hydrodynamic derivation see, for example, L. Page, "Introduction to Theoretical Physics," 3d ed., p. 286, D. Van Nostrand Company, Princeton, N.J., 1952; or G. Joos, "Theoretical Physics," 3d ed., p. 218, Hafner Publishing Company, N.Y., 1958.

Remark The operations of taking a time derivative and taking an ensemble average commute. Indeed, suppose that a quantity y assumes at time t the value $y^{(k)}(t)$ in the kth system of an ensemble consisting of N systems. Then

$$\frac{d}{dt}\langle y\rangle = \frac{d}{dt}\left(\frac{1}{N}\sum_{k=1}^{N} y^{(k)}(t)\right) = \frac{1}{N}\sum_{k=1}^{N}\frac{dy^{(k)}}{dt} = \left\langle\frac{dy}{dt}\right\rangle$$

since one can interchange the order of differentiation and summation.

The relation (15 · 6 · 4) is a simple differential equation which can immediately be solved for the quantity $\langle x\dot{x}\rangle = \frac{1}{2}(d\langle x^2\rangle/dt)$. Thus one obtains

$$\langle x\dot{x}\rangle = Ce^{-\gamma t} + \frac{kT}{\alpha} \tag{15·6·5}$$

where C is a constant of integration. Here we have introduced the definition

$$\gamma \equiv \frac{\alpha}{m} \tag{15·6·6}$$

so that γ^{-1} denotes a characteristic time constant of the system. Assuming that each particle in the ensemble starts out at $t = 0$ at the position $x = 0$, so that x measures the displacement from the initial position, the constant C in (15 · 6 · 5) must be such that $0 = C + kT/\alpha$. Hence (15 · 6 · 5) becomes

$$\langle x\dot{x}\rangle = \frac{1}{2}\frac{d}{dt}\langle x^2\rangle = \frac{kT}{\alpha}(1 - e^{-\gamma t}) \tag{15·6·7}$$

Integrating once more one obtains the final result

$$\blacktriangleright \qquad \langle x^2\rangle = \frac{2kT}{\alpha}[t - \gamma^{-1}(1 - e^{-\gamma t})] \tag{15·6·8}$$

Note two interesting limiting cases. If $t < \gamma^{-1}$, then

$$e^{-\gamma t} = 1 - \gamma t + \tfrac{1}{2}\gamma^2 t^2 - \cdots$$

Thus

for $t \ll \gamma^{-1}$,
$$\langle x^2\rangle = \frac{kT}{m}t^2 \tag{15·6·9}$$

The particle then behaves during a short initial time interval as though it were a free particle moving with the constant thermal velocity $v = (kT/m)^{\frac{1}{2}}$.

On the other hand, if $t \gg \gamma^{-1}$, $e^{-\gamma t} \to 0$. Thus (15 · 6 · 8) becomes simply

for $t \gg \gamma^{-1}$,
$$\langle x^2\rangle = \frac{2kT}{\alpha}t \tag{15·6·10}$$

The particle then behaves like a diffusing particle executing a random walk so that $\langle x^2\rangle \propto t$. Indeed, since the diffusion equation leads by (12 · 5 · 22) to the relation $\langle x^2\rangle = 2Dt$, comparison with (15 · 6 · 10) shows the corresponding diffusion coefficient to be given by

$$D = \frac{kT}{\alpha} \tag{15·6·11}$$

By using (15·6·2), the relation (15·6·10) yields the explicit result

$$\langle x^2 \rangle = \frac{kT}{3\pi\eta a} t \tag{15·6·12}$$

Observations of particles executing Brownian motion allowed Perrin (ca. 1910) to measure $\langle x^2 \rangle$ experimentally. Indeed, knowing the size and density of the particles as well as the viscosity of the medium, Perrin was able to deduce from these observations a reasonably good value for Boltzmann's constant k and thus, with a knowledge of the gas constant R, the value of Avogadro's number.

A problem intimately related to the one discussed in this section is that of the behavior of the particle in an external force field. If the particle carries an electric charge e and is placed in a uniform electric field $\mathcal{E}$, the Langevin equation (15·5·8) becomes

$$m \frac{dv}{dt} = e\mathcal{E} - \alpha v + F'(t)$$

Taking the mean values of both sides and considering the steady-state situation where $d\bar{v}/dt = 0$, this yields

$$e\mathcal{E} - \alpha\bar{v} = 0$$

This shows that $\bar{v} \propto \mathcal{E}$. The "mobility" $\mu \equiv \bar{v}/\mathcal{E}$ is then given by

$$\mu \equiv \frac{\bar{v}}{\mathcal{E}} = \frac{e}{\alpha} \tag{15·6·13}$$

Thus the mobility μ and the diffusion coefficient D in (15·6·11) are both expressible in terms of α. There exists therefore an intimate connection between these two coefficients, namely,

▶
$$\frac{\mu}{D} = \frac{e}{kT} \tag{15·6·14}$$

This is known as the "Einstein relation."

DETAILED ANALYSIS OF BROWNIAN MOTION

15·7 *Relation between dissipation and the fluctuating force*

In order to gain a better understanding of the frictional force, we shall now return to Eq. (15·5·1) and attempt to analyze it in somewhat greater detail. Let us again consider a time interval τ which *macro*scopically is very small, but which is large on a *micro*scopic scale so that

$$\tau \gg \tau^*$$

where τ^* is a correlation time which is of the order of the mean period of fluctuations of the force $F(t)$. (This time τ^* measures also the relaxation time required for the degrees of freedom responsible for the force F to come to internal equilibrium when disturbed by a sudden small change of x.) We again assume that the external force $\mathfrak{F}$ is slowly varying, and we shall be interested in finding the slowly varying part of the velocity v. Here we call quantities slowly varying if they change by negligible amounts in a time interval τ.

Focus attention on an ensemble of similarly prepared systems each of which satisfies Eq. (15 · 5 · 1). By considering this equation in its integrated form (15 · 5 · 3) and taking the ensemble average of both sides, one obtains

$$m\langle v(t+\tau) - v(t)\rangle = \mathfrak{F}(t)\tau + \int_t^{t+\tau} \langle F(t')\rangle \, dt' \qquad (15 \cdot 7 \cdot 1)$$

If one neglected completely any effect of the particle's motion on the force F exerted on it by the environment, the mean value $\langle F\rangle$ would be the same as the mean value $\langle F\rangle_0 = 0$ that prevails in a static equilibrium situation where the particle is somehow clamped so as to be stationary with respect to its environment. This is, of course, not the actual situation. Indeed, as was pointed out in Sec. 15 · 5, an order of approximation that would simply put $\langle F\rangle = \langle F\rangle_0 = 0$ in (15 · 7 · 1) would be inadequate, since it would not yield a slowly varying velocity tending to restore the particle to thermal equilibrium. One therefore needs to estimate how $\langle F\rangle$ is affected as the velocity v of the particle changes.

To make an approximate analysis of the situation, we can proceed as follows. Let us focus attention on the small system A described by x and the other degrees of freedom with which this coordinate interacts through the force F (e.g., the particle itself and the molecules of the liquid in the immediate vicinity of the particle). All the many other degrees of freedom (e.g., the bulk of the liquid) then constitute a large heat bath B. The temperature $T \equiv (k\beta)^{-1}$ of this heat bath is essentially constant, irrespective of any small changes in its energy. For a given value of v, considered as a parameter, the possible states of A can be labeled r; in such a state, the force F assumes some value F_r.

Suppose that at some arbitrary time t the particle has a velocity $v(t)$. In first approximation, the system A at this time can be assumed to be in an equilibrium situation where $\langle F\rangle = 0$ and where the probability of A being in state r is denoted by $W_r^{(0)}$. In the next approximation one must, however, investigate how $\langle F\rangle$ is affected by the motion of the particle. Consider then the situation at a slightly later time $t' = t + \tau'$ when the particle has a velocity $v(t+\tau')$. The motion of the particle affects its environment and, if τ' is sufficiently short, the mean force $\langle F(t')\rangle$ depends on the situation at the earlier time t. Indeed, as the particle's velocity changes, the internal equilibrium of the environment is disturbed. But after a time of the order of τ^* (i.e., of the order of the time between molecular collisions) the interactions between molecules will have reestablished equilibrium conditions consistent with the new value of the parameter $v = v(t+\tau')$. This means that the heat bath B will again be found with equal likelihood in any one of its Ω accessible states. Suppose then that in a time interval $\tau' > \tau^*$ the velocity of the particle changes by $\Delta v(\tau')$ and that correspondingly the energy of B changes from E' to $E' + \Delta E'(\tau')$. The number of states accessible to B changes accordingly from $\Omega(E')$ to $\Omega(E' + \Delta E')$. Since in a situation of equilibrium the probability of occurrence of a given state r of A is proportional to the corresponding number of states accessible to the heat bath B, it is then possible to compare the probability of occurrence of the *same* configuration r at the times t and $t + \tau'$. If the latter probability is denoted by $W_r(t + \tau')$, one has simply

$$\frac{W_r(t + \tau')}{W_r^{(0)}} = \frac{\Omega(E' + \Delta E')}{\Omega(E')} = e^{\beta\,\Delta E'} \tag{15·7·2}$$

where $\beta \equiv (\partial \ln \Omega/\partial E')$ is the temperature parameter of the heat bath B. Physically, this means that *the likelihood that system A is found in a given state at a somewhat later time is increased if more energy becomes available to the heat reservoir.* Thus

$$W_r(t + \tau') = W_r^{(0)}\, e^{\beta\,\Delta E'} \approx W_r^{(0)}(1 + \beta\,\Delta E') \tag{15·7·3}$$

At the slightly later time $t' = t + \tau'$ the mean value of F is then given by

$$\langle F\rangle = \sum_r W_r(t + \tau')F_r = \sum_r W_r^{(0)}(1 + \beta\,\Delta E')F_r = \langle(1 + \beta\,\Delta E')F\rangle_0$$

where the last mean value is to be computed with the equilibrium probability $W_r^{(0)}$. Since $\langle F\rangle_0 = 0$, this gives

► $$\langle F\rangle = \beta\langle F\,\Delta E'\rangle_0 \tag{15·7·4}$$

which, in general, does *not* vanish.

These approximate considerations can now be used in (15·7·1), where $\tau \gg \tau^*$. The integral in that expression extends then over a time interval sufficiently long that $\tau' = t' - t \gg \tau^*$ over practically the whole range of integration, making it possible to use the approximation (15·7·4) in the integrand.

The energy increase of B in the time $t' - t$ is simply the negative of the work done by the force F on the particle. Thus

$$\Delta E' = -\int_t^{t'} v(t'')F(t'')\,dt'' \approx -v(t)\int_t^{t'} F(t'')\,dt'' \tag{15·7·5}$$

where we have made the approximation, consistent with (15·7·1), that $v(t)$ does not vary appreciably over times of the order of τ. Hence one can use (15·7·4) to write for the integrand in (15·7·1) the expression

$$\langle F(t')\rangle = -\beta\left\langle F(t')v(t)\int_t^{t'} F(t'')\,dt''\right\rangle_0 = -\beta\bar{v}(t)\int_t^{t'} dt''\,\langle F(t')F(t'')\rangle_0 \tag{15·7·6}$$

Here we have first averaged separately over $v(t)$, since it varies much more slowly than $F(t)$. It is useful to write (15·7·6) in terms of the physically significant time difference

$$s \equiv t'' - t' \tag{15·7·7}$$

Equation (15 · 7 · 1) then becomes

$$m\langle v(t+\tau) - v(t)\rangle = \mathfrak{F}(t)\tau - \beta\bar{v}(t)\int_t^{t+\tau} dt' \int_{t-t'}^{0} ds\, \langle F(t')F(t'+s)\rangle_0 \qquad (15\cdot 7\cdot 8)$$

The last term on the right is slowly varying and leads to "dissipation," i.e., to the fact that in the absence of external forces, when $\mathfrak{F} = 0$, the mean velocity $\bar{v}$ goes to zero with increasing time.

15 · 8 *Correlation functions and the friction constant*

The ensemble average which occurs in (15 · 7 · 8), i.e.,

$$K(s) \equiv \langle F(t')F(t'')\rangle_0 = \langle F(t')F(t'+s)\rangle_0 \qquad (15\cdot 8\cdot 1)$$

is called the "correlation function" of the function $F(t)$. The ensemble average is here taken in the equilibrium situation where the distribution of systems in the ensemble is independent of the absolute value of the time. Hence this average is independent of the time t' and depends only on the time *difference* s.

Correlation functions occur quite frequently in statistical physics and have several general interesting properties. In the following we can consider $F(t)$ to be any random function of t and shall, for convenience, write t instead of t' and drop the subscript from the averaging brackets. Then one has

$$K(0) = \langle F(t)F(t)\rangle = \langle F^2(t)\rangle > 0 \qquad (15\cdot 8\cdot 2)$$

Thus $K(0)$ is equal to the mean square value of the fluctuating function F, i.e., to its dispersion if $\langle F\rangle = 0$. (Of course $\langle F^2(t)\rangle$ is independent of the value of the time t in this equilibrium situation.)

If s becomes sufficiently large, then $F(t)$ and $F(t+s)$ must become uncorrelated, i.e., the probability that F assumes a certain value at time $t+s$ must be independent of the value it assumed at the much earlier time t. Thus one has

$$\text{for } s \to \infty, \qquad K(s) \to \langle F(t)\rangle\langle F(t+s)\rangle$$

$$\text{i.e., for } s \to \infty, \qquad K(s) \to 0 \quad \text{if } \langle F\rangle = 0 \qquad (15\cdot 8\cdot 3)$$

Quite generally one can also show that

$$|K(s)| \le K(0) \qquad (15\cdot 8\cdot 4)$$

This follows from the obvious relation that

$$\langle [F(t) \pm F(t+s)]^2\rangle \ge 0$$

Thus

$$\langle F^2(t) + F^2(t+s) \pm 2F(t)F(t+s)\rangle \ge 0$$

or

$$\langle F^2(t)\rangle + \langle F^2(t+s)\rangle \pm 2\langle F(t)F(t+s)\rangle \ge 0$$

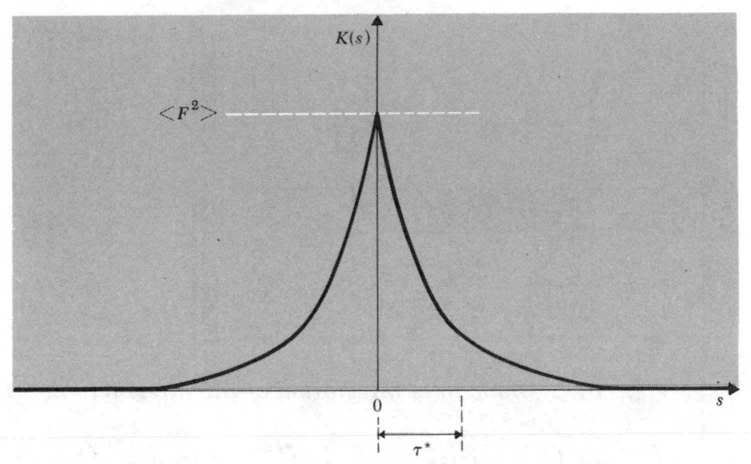

Fig. 15·8·1 Diagram illustrating the correlation function $K(s) \equiv \langle F(t)F(t+s)\rangle$ ***of a random function*** $F(t)$.

Since in the equilibrium ensemble $\langle F^2(t+s)\rangle = \langle F^2(t)\rangle = K(0)$, this becomes

$$K(0) \pm K(s) \geq 0$$

or

$$-K(0) \leq K(s) \leq K(0)$$

which is the result (15·8·4).

Finally, since in the equilibrium ensemble $K(s)$ defined in (15·8·1) is independent of the time t, it follows that for any other time t_1

$$K(s) \equiv \langle F(t)F(t+s)\rangle = \langle F(t_1)F(t_1+s)\rangle$$

Putting $t_1 = t - s$, this becomes

$$K(s) = \langle F(t)F(t+s)\rangle = \langle F(t-s)F(t)\rangle = \langle F(t)F(t-s)\rangle$$

or

$$K(s) = K(-s) \tag{15·8·5}$$

Thus a plot of the correlation function K as a function of s has a symmetric shape of the type shown in Fig. 15·8·1. In our case values assumed by the force $F(t)$ become uncorrelated over times of the order of τ^*; hence $K(s) \to 0$ when $s \gg \tau^*$. It is clear from this discussion that the correlation function $K(s)$ contains an appreciable amount of information about the statistical properties of the random variable $F(t)$.

We now return to the evaluation of the integral in (15·7·8). The domain of integration is shown in Fig. 15·8·2 in the t's plane. (Note that, since the integrand is just the correlation function $K(s)$ of the force F and differs from zero only in the small region where $|s| \lesssim \tau^* \ll \tau$, the integral is proportional to $\tau^*\tau$ rather than to the area τ^2 of the complete domain of integration. Thus the integral is proportional to the *first* power of τ.) Since $K(s)$ is independent of t', the integration over t' can be performed first quite easily. Changing the order of integration in the double integral and reading off the limits from Fig. 15·8·2, one then gets

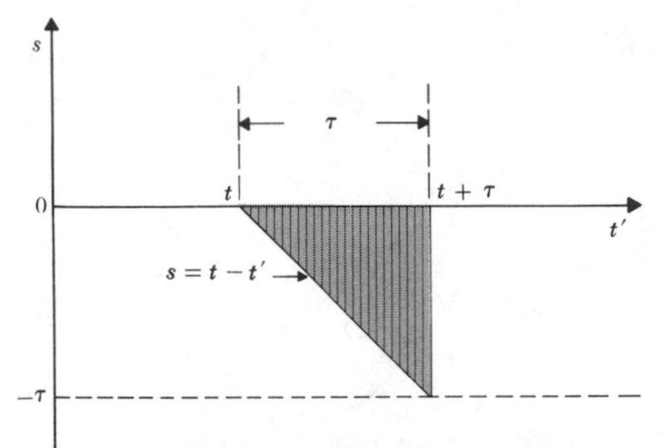

Fig. 15·8·2 Domain of integration of the integral in Eq. (15·7·8).

$$\int_t^{t+\tau} dt' \int_{t-t'}^{0} ds\, K(s) = \int_{-\tau}^{0} ds \int_{t-s}^{t+\tau} dt'\, K(s) = \int_{-\tau}^{0} ds\, K(s)(\tau + s)$$

Since $\tau \gg \tau^*$, while $K(s) \to 0$ when $|s| \gg \tau^*$, one can neglect s compared to τ in the entire range where the integrand is appreciable; furthermore, the lower limit in the last integral can be replaced by $-\infty$ with negligible error. Hence one obtains simply

$$\int_t^{t+\tau} dt' \int_{t-t'}^{0} ds\, K(s) \approx \tau \int_{-\infty}^{0} ds\, K(s) = \tfrac{1}{2}\tau \int_{-\infty}^{\infty} ds\, K(s) \qquad (15\cdot 8\cdot 6)$$

where we have used the symmetry property (15·8·5) in the last step. Thus (15·7·8) can be written

$$m\langle v(t+\tau) - v(t)\rangle = \mathfrak{F}(t)\tau - \alpha\bar{v}(t)\tau \qquad (15\cdot 8\cdot 7)$$

where the constant α is given explicitly by

▶ $$\alpha \equiv \frac{1}{2kT} \int_{-\infty}^{\infty} \langle F(0)F(s)\rangle_0\, ds \qquad (15\cdot 8\cdot 8)$$

The last term in (15·8·7) is thus seen to be proportional to τ in the same way as the term involving the external force $\mathfrak{F}$. Since we assumed that $\bar{v} \equiv \langle v\rangle$ varies only slowly over time intervals of the order of τ, the left side of (15·8·7) is relatively small. Hence one can express (15·8·7) in terms of the ("coarse-grained") time derivative

$$\frac{d\bar{v}}{dt} \equiv \frac{\langle v(t+\tau)\rangle - \langle v(t)\rangle}{\tau} = \frac{\langle v(t+\tau) - v(t)\rangle}{\tau}$$

Thus

▶ $$m\frac{d\bar{v}}{dt} = \mathfrak{F} - \alpha\bar{v} \qquad (15\cdot 8\cdot 9)$$

When $\mathfrak{F} = 0$, this yields upon integration

$$\bar{v}(t) = \bar{v}(0)\, e^{-\gamma t}, \qquad \gamma \equiv \frac{\alpha}{m} \qquad (15\cdot 8\cdot 10)$$

so that $\bar{v}$ approaches its equilibrium value $\bar{v} = 0$ with a time constant γ^{-1}. Since we assumed that $\bar{v}$ is slowly varying, so that $(d\bar{v}/dt)\tau \ll \bar{v}$, this assumption demands, by (15·8·10), that $\gamma\tau \ll 1$ or, by (15·7·1), that

$$\gamma^{-1} \gg \tau^* \tag{15·8·11}$$

The discussion leading to (15·8·9) shows that frictional forces arise in the dynamic description of a system when the surroundings with which it interacts come to internal equilibrium very quickly compared to the smallest time scale of interest in the description of this system. The result (15·8·9) agrees with the previously discussed phenomenological equation (15·5·7) and thus leads again to a description of fluctuations in terms of the Langevin equation (15·5·8). But our present analysis has provided us with greater insight into how the frictional force $-\alpha\bar{v}$ arises from the fluctuating force. In particular, the relation (15·8·8), which is sometimes called the "fluctuation-dissipation theorem," provides us with an explicit expression for the friction constant α in terms of the correlation function of the fluctuating force $F(t)$ in the equilibrium situation. It is plausible that this kind of connection should exist. Indeed, suppose that the interaction of the system (e.g., of the particle) with its surrounding heat reservoir is strong. Then the force $F(t)$ producing fluctuations in the system must be large. But, if the interaction with the heat reservoir is strong, the system will also attain equilibrium with the surrounding heat reservoir quite quickly if it is not in equilibrium initially; i.e., the parameter α describing the rate of approach to equilibrium must then also be large.

Remark on the analogous electrical problem It was pointed out at the end of Sec. 15·5 that the problem of an electrical conductor of self-inductance L and carrying a current I is analogous to the Brownian-motion problem. Indeed, if $\mathcal{V}$ denotes an applied emf and $V(t)$ the effective fluctuating emf representing the interaction of the conduction electrons with all the other degrees of freedom, the current I satisfies the equation

$$L\frac{dI}{dt} = \mathcal{V}(t) + V(t) \tag{15·8·12}$$

This is analogous to (15·5·1) with $L = m$, $I = v$, and $V(t) = F(t)$. If $\mathcal{V}$ is a relatively slowly varying function of time, an analysis similar to that leading to (15·8·9) then gives the equation

$$L\frac{d\bar{I}}{dt} = \mathcal{V} - R\bar{I} \tag{15·8·13}$$

This equation is, of course, simply the conventional electric circuit equation for a conductor of inductance L and electrical resistance R. But the present analysis has examined more closely how the frictional effects represented by this resistance come about and provides, through the analogue of (15·8·8), the following explicit expression for R in terms of the fluctuating emf:

$$R = \frac{1}{2kT}\int_{-\infty}^{\infty} \langle V(0)V(s)\rangle_0\, ds \tag{15·8·14}$$

*15 · 9 *Calculation of the mean-square velocity increment*

The expression (15 · 8 · 8) for the friction constant α in terms of the correlation function of F suggests a connection between α and fluctuations in the velocity v which are caused by this force F. Consider the particle in equilibrium with the surrounding medium in the absence of external forces. Then

$$m \frac{dv}{dt} = F(t)$$

and

$$\Delta v(\tau) \equiv m[v(t + \tau) - v(t)] = \int_t^{t+\tau} F(t')\, dt' \tag{15·9·1}$$

Let us calculate the quantity $\langle[\Delta v(\tau)]^2\rangle$. By (15 · 9 · 1) this ensemble average is given by

$$m^2\langle[\Delta v(\tau)]^2\rangle = \left\langle \int_t^{t+\tau} dt'\, F(t') \cdot \int_t^{t+\tau} dt''\, F(t'') \right\rangle$$

$$= \int_t^{t+\tau} dt' \int_t^{t+\tau} dt''\, \langle F(t')F(t'')\rangle \tag{15·9·2}$$

$$= \int_t^{t+\tau} dt' \int_{t-t'}^{t+\tau-t'} ds\, \langle F(t')F(t' + s)\rangle \tag{15·9·3}$$

where $s = t'' - t'$. Here the ensemble average of the force F can be calculated as if the particle were stationary, since corrections to this approximation would already involve terms containing more than two factors F and would thus be negligible compared to the leading term (i.e., they would go to zero faster than linearly in the small quantity τ).

The integral in (15 · 9 · 2) looks, except for the limits, very much like the integral in (15 · 7 · 8) and can be evaluated similarly. The integrand in (15 · 9 · 3) is the correlation function $K(s)$, and the domain of integration in the t's plane is that shown in Fig. 15 · 9 · 1. Once again one can exploit the fact that the integrand is independent of t' by changing the order of integration so as to

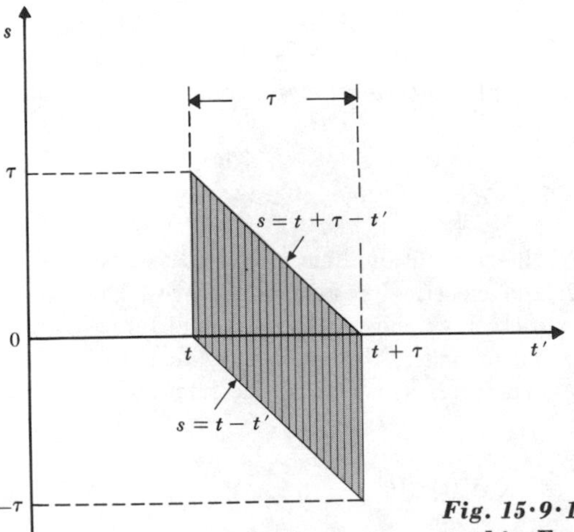

***Fig. 15·9·1** Domain of integration of the integral in Eq. (15·9·3).*

integrate first over t'. Thus one gets, with limits determined from Fig. 15·9·1,

$$\int_t^{t+\tau} dt' \int_{t-t'}^{t+\tau-t'} ds\, K(s) = \int_0^\tau ds \int_t^{t+\tau-s} dt'\, K(s) + \int_{-\tau}^0 ds \int_{t-s}^{t+\tau} dt'\, K(s)$$
$$= \int_0^\tau ds\, K(s)(\tau - s) + \int_{-\tau}^0 ds\, K(s)(\tau + s)$$

Since $\tau \gg \tau^*$, while $K(s) \to 0$ when $|s| \gg \tau^*$, the term s can again be neglected compared to τ in the integrand; also τ in the limit of integration can be replaced by ∞ with negligible error. Thus (15·9·3) gives

$$m^2\langle[\Delta v(\tau)]^2\rangle = \tau \int_0^\infty ds\, K(s) + \tau \int_{-\infty}^0 ds\, K(s) = \tau \int_{-\infty}^\infty ds\, K(s)$$

or

$$\langle[\Delta v(\tau)]^2\rangle = \frac{\tau}{m^2} \int_{-\infty}^\infty \langle F(0)F(s)\rangle_0\, ds \tag{15·9·4}$$

Hence $\langle[\Delta v]^2\rangle$ is directly proportional to τ. Comparison with (15·8·8) shows that this can be expressed in terms of the friction constant α. Thus

▶ $$\langle[\Delta v(\tau)]^2\rangle = \frac{\tau}{m^2}(2kT\alpha) = \frac{2kT}{m}\gamma\tau \tag{15·9·5}$$

where $\gamma \equiv \alpha/m$.

In (15·8·7) we found that, for $\mathfrak{F} = 0$,

▶ $$\langle\Delta v(\tau)\rangle = -\frac{\alpha}{m}\bar{v}\tau = -\bar{v}\gamma\tau \tag{15·9·6}$$

Hence (15·9·5) implies the relation

$$\langle[\Delta v(\tau)]^2\rangle = -\frac{2kT}{m}\frac{\langle\Delta v(\tau)\rangle}{\bar{v}} \tag{15·9·7}$$

*15·10 *Velocity correlation function and mean-square displacement*

As an example of the use of correlation functions, let us compute the mean-square displacement $\langle x^2(t)\rangle$ of the particle in time t by a more direct method than that used in Sec. 15·6. We consider the equilibrium situation in the absence of external forces. Let $x(0) = 0$ at $t = 0$. Since $\dot{x} = v$, one has

$$x(t) = \int_0^t v(t')\, dt' \tag{15·10·1}$$

Thus one obtains

$$\langle x^2(t)\rangle = \left\langle \int_0^t v(t')\, dt' \int_0^t v(t'')\, dt'' \right\rangle = \int_0^t dt' \int_0^t dt''\, \langle v(t')v(t'')\rangle \tag{15·10·2}$$

In this time-independent problem the velocity correlation function K_v can depend only on the time difference $s = t'' - t'$, so that

$$\langle v(t')v(t'')\rangle = \langle v(0)v(s)\rangle \equiv K_v(s) \tag{15·10·3}$$

The integral (15 · 10 · 2) is essentially the same as the one occurring in (15 · 9 · 2). With a change of limits similar to that in Fig. 15 · 9 · 1 one then obtains

$$\begin{aligned}\langle x^2(t)\rangle &= \int_0^t ds \int_0^{t-s} dt'\, K_v(s) + \int_{-t}^0 ds \int_{-s}^t dt'\, K_v(s) \\ &= \int_0^t ds\, K_v(s)(t-s) + \int_{-t}^0 ds\, K_v(s)(t+s)\end{aligned}$$

If we put $s \to -s$ in the second integral and use the symmetry property $K_v(-s) = K_v(s)$ of (15 · 8 · 5), this becomes simply

$$\blacktriangleright \qquad \langle x^2(t)\rangle = 2\int_0^t ds\,(t-s)\langle v(0)v(s)\rangle \qquad (15\cdot 10\cdot 4)$$

A calculation of $\langle x^2(t)\rangle$ then requires only a knowledge of the velocity correlation function (15 · 10 · 3).

Since our information about v relates its derivative to F, it is easiest to find a differential relation for the desired correlation function K_v. One approach consists in starting from the Langevin equation (15 · 5 · 8)

$$m\frac{dv}{dt} = -\alpha v + F'(t) \qquad (15\cdot 10\cdot 5)$$

Then
$$v(s+\tau) - v(s) = -\gamma v(s)\tau + \frac{1}{m}\int_s^{s+\tau} F'(t')\,dt' \qquad (15\cdot 10\cdot 6)$$

where $\gamma \equiv \alpha/m$ and $\tau \gg \tau^*$ is macroscopically small. Multiplying both sides by $v(0)$ and taking the ensemble average gives

$$\begin{aligned}&\langle v(0)v(s+\tau)\rangle - \langle v(0)v(s)\rangle \\ &\qquad = -\gamma\langle v(0)v(s)\rangle\tau + \frac{1}{m}\left\langle v(0)\int_s^{s+\tau} F'(t')\,dt'\right\rangle \qquad (15\cdot 10\cdot 7)\end{aligned}$$

In taking the ensemble average, the fluctuating force F' can here be treated as purely random and independent of v, since any dependence on the motion of the particle has already been explicitly included in the friction term αv of (15 · 10 · 5). Thus one has simply

$$\left\langle v(0)\int_s^{s+\tau} F'(t')\,dt'\right\rangle \approx \langle v(0)\rangle\left\langle\int_s^{s+\tau} F'(t')\,dt'\right\rangle = 0$$

Equation (15 · 10 · 7), which is valid for $\tau > 0$, becomes then to a good approximation

$$\frac{\langle v(0)v(s+\tau)\rangle - \langle v(0)v(s)\rangle}{\tau} = \frac{d}{ds}\langle v(0)v(s)\rangle = -\gamma\langle v(0)v(s)\rangle \qquad (15\cdot 10\cdot 8)$$

By integration this gives for $s > 0$

$$\langle v(0)v(s)\rangle = \langle v^2(0)\rangle\, e^{-\gamma s}$$

By virtue of the symmetry property $\langle v(0)v(-s)\rangle = \langle v(0)v(s)\rangle$, this yields for *all* values of s the expression

$$\blacktriangleright \qquad \langle v(0)v(s)\rangle = \langle v^2(0)\rangle\, e^{-\gamma|s|} = \frac{kT}{m}\, e^{-\gamma|s|} \qquad (15\cdot 10\cdot 9)$$

where we have used the equipartition theorem result $\langle \frac{1}{2}mv^2(0)\rangle = \frac{1}{2}kT$. This correlation function has the functional dependence illustrated in Fig. 15·8·1. Note that the values assumed by the velocity are only correlated over a time interval of the order of γ^{-1}. This time interval is shorter the larger the magnitude of the friction constant.

By using (15·10·9) in the relation (15·10·4) one obtains

$$\langle x^2(t)\rangle = \frac{2kT}{m}\int_0^t ds(t-s)\, e^{-\gamma s} \tag{15·10·10}$$

If $t \gg \gamma^{-1}$, $e^{-\gamma s} \to 0$ when $s \gg \gamma^{-1}$; i.e., $e^{-\gamma s} \to 0$ for values of s that are much less than t. Hence one can put $t - s \approx t$ in the integrand and put $t \to \infty$ in the upper limit of (15·10·10) with negligible error. Thus (15·10·10) becomes

for $t \gg \gamma^{-1}$,
$$\langle x^2(t)\rangle = \frac{2kT}{m}\int_0^\infty ds\, t\, e^{-\gamma s} = \frac{2kT}{m\gamma}t \tag{15·10·11}$$

More generally, the evaluation of (15·10·12) for arbitrary γ yields

$$\langle x^2(t)\rangle = \frac{2kT}{m}\left(t + \frac{\partial}{\partial\gamma}\right)\int_0^t ds\, e^{-\gamma s}$$

or

$$\langle x^2(t)\rangle = \frac{2kT}{m\gamma}\left[t - \frac{1}{\gamma}(1 - e^{-\gamma t})\right] \tag{15·10·12}$$

which agrees with (15·6·8).

CALCULATION OF PROBABILITY DISTRIBUTIONS

*15·11 *The Fokker-Planck equation*

Consider the Brownian-motion problem in the absence of external forces. Instead of inquiring simply how the mean value of the velocity v changes with time t, let us now ask the more detailed question concerning the time dependence of the probability $P(v,t)\,dv$ that the particle's velocity at time t lies between v and $v + dv$. One expects that this probability does not depend on the entire past history of the particle, but that it is determined if one knows that $v = v_0$ at some earlier time t_0. (This assumption characterizes what is known as a "Markoff process.") Hence one can write P more explicitly as a conditional probability which depends on v_0 and t_0 as parameters; i.e.,

$$P\,dv = P(vt|v_0t_0)\,dv \tag{15·11·1}$$

is the probability that the velocity is between v and $v + dv$ at time t if it is known that the velocity is v_0 at the earlier time t_0. Since nothing in this problem depends on the origin from which time is measured, P can actually only depend on the time difference $s = t - t_0$. Thus one can simply write

$$P(vt|v_0t_0)\,dv = P(v,s|v_0)\,dv \tag{15·11·2}$$

to denote the probability that, if the velocity has the value v_0 at any one time, it assumes a value between v and $v + dv$ after a time s. If $s \to 0$, one knows that $v = v_0$. Thus

$$\text{for } s \to 0, \qquad P(v,s|v_0) \to \delta(v - v_0) \tag{15·11·3}$$

where the Dirac δ function on the right side vanishes unless $v = v_0$.

On the other hand, if $s \to \infty$, the particle must come to equilibrium with the surrounding medium at temperature T, irrespective of its past history. Hence P then becomes independent of v_0 and also independent of the time; i.e., it reduces to the canonical distribution. Thus

$$\text{for } s \to \infty, \qquad P(v,s|v_0)\, dv \to \left(\frac{m\beta}{2\pi}\right)^{\frac{1}{2}} e^{-\frac{1}{2}\beta m v^2}\, dv \tag{15·11·4}$$

One can readily write down a general condition that must be satisfied by the probability $P(v,s|v_0)$. In any small time interval τ, the [increase in the probability that the particle has a velocity between v and $v + dv$] must be equal to the [decrease in this probability owing to the fact that the particle, originally with velocity between v and $v + dv$, has a probability $P(v_1,\tau|v)\, dv_1$ of changing its velocity to *any* other value between v_1 and $v_1 + dv_1$] plus the [increase in this probability owing to the fact that the particle, originally with a velocity in *any* range between v_1 and $v_1 + dv_1$, has a probability $P(v,\tau_1|v_1)\, dv$ of changing its velocity to a value in the original range between v and $v + dv$]. In symbols this condition becomes

$$\frac{\partial P}{\partial s}\, dv\, \tau = -\int_{v_1} P(v,s|v_0)\, dv \cdot P(v_1,\tau|v)\, dv_1 + \int_{v_1} P(v_1,s|v_0)\, dv_1 \cdot P(v,\tau|v_1)\, dv \tag{15·11·5}$$

where the integrals extend over all possible velocities v_1. In the first integral $P(v,s|v_0)$ does not depend on v_1, while

$$\int_{v_1} P(v_1,\tau|v)\, dv_1 = 1 \tag{15·11·6}$$

since $P(v_1,\tau|v)$ is a properly normalized probability. Putting $v_1 \equiv v - \xi$, Eq. (15 · 11 · 5) then becomes

$$\frac{\partial P}{\partial s}\tau = -P(v,s|v_0) + \int_{-\infty}^{\infty} P(v - \xi,\, s|v_0) P(v,\, \tau|v - \xi)\, d\xi \tag{15·11·7}$$

Note that Eq. (15 · 11 · 5) is equivalent to the general master equation (15 · 1 · 5) and is also similar to the Boltzmann equation (14 · 3 · 8). In the latter equation a molecule can change its velocity abruptly by a very large amount as a result of a collision with another molecule. But in the present case the velocity v of the *macroscopic* particle can only change by a small amount during the small time interval τ. Hence one can assert that the probability $P(v,\, \tau|v - \xi)$ can only be appreciable when $|\xi| = |v - v_1|$ is sufficiently small. A knowledge of the integrand of (15 · 11 · 7) for small values of ξ should then be sufficient for evaluating the integral. Hence it should be permissible to expand this integrand in a Taylor's series in powers of ξ about the value

$P(v,s|v_0)P(v + \xi, \tau|v)$ and to retain only the lowest-order terms. This expansion becomes

$$P(v - \xi, s|v_0)P(v, \tau|v - \xi) = \sum_{n=0}^{\infty} \frac{(-\xi)^n}{n!} \frac{\partial^n}{\partial v^n} [P(v,s|v_0)P(v + \xi, \tau|v)]$$

Hence (15·11·7) becomes

$$\frac{\partial P}{\partial s}\tau = -P(v,s|v_0) + \sum_{n=0}^{\infty} \frac{(-1)^n}{n!} \frac{\partial^n}{\partial v^n} \left[P(v,s|v_0) \int_{-\infty}^{\infty} d\xi\, \xi^n P(v + \xi, \tau|v) \right] \tag{15·11·8}$$

The term $n = 0$ in the sum is simply $P(v,s|v_0)$ by virtue of the normalization condition (15·11·6). As for the other terms, it is convenient to introduce the abbreviation

$$M_n \equiv \frac{1}{\tau} \int_{-\infty}^{\infty} d\xi\, \xi^n P(v + \xi, \tau|v) = \frac{\langle [\Delta v(\tau)]^n \rangle}{\tau} \tag{15·11·9}$$

where $\langle [\Delta v(\tau)]^n \rangle = \langle [v(\tau) - v(0)]^n \rangle$ is the nth moment of the velocity increment in time τ. Equation (15·11·8) then becomes

$$\frac{\partial P(v,s|v_0)}{\partial s} = \sum_{n=1}^{\infty} \frac{(-1)^n}{n!} \frac{\partial^n}{\partial v^n} [M_n P(v,s|v_0)] \tag{15·11·10}$$

But when $\tau \to 0$, $\langle (\Delta v)^n \rangle \to 0$ more rapidly than τ itself if $n > 2$ (see Problem 15.11). When τ is *macro*scopically infinitesimal (although $\tau \gg \tau^*$) the terms involving M_n with $n > 2$ can therefore be neglected in (15·11·10). Hence this equation for $P(v,s|v_0)$ reduces to

► $$\frac{\partial P}{\partial s} = -\frac{\partial}{\partial v}(M_1 P) + \frac{1}{2}\frac{\partial^2}{\partial v^2}(M_2 P) \tag{15·11·11}$$

This is the so-called "Fokker-Planck equation." It is a partial differential equation for P and involves as coefficients only the two moments M_1 and M_2 of P evaluated for a macroscopically infinitesimal time interval. These moments have already been calculated for the Brownian motion problem in sections 15·8 and 15·9. Alternatively they can also be easily deduced from the Langevin equation (see Problem 15.11). The relations (15·9·5) and (15·9·6) give

$$\left.\begin{aligned} M_1 &= \frac{1}{\tau}\langle \Delta v(\tau) \rangle = -\gamma v \\ \text{and} \qquad M_2 &= \frac{1}{\tau}\langle [\Delta v(\tau)]^2 \rangle = \frac{2kT}{m}\gamma \end{aligned}\right\} \tag{15·11·12}$$

Hence (15·11·11) becomes

$$\frac{\partial P}{\partial s} = \gamma \frac{\partial}{\partial v}(vP) + \gamma \frac{kT}{m}\frac{\partial^2 P}{\partial v^2} \tag{15·11·13}$$

or

► $$\frac{\partial P}{\partial s} = \gamma P + \gamma v \frac{\partial P}{\partial v} + \gamma \frac{kT}{m}\frac{\partial^2 P}{\partial v^2} \tag{15·11·14}$$

Alternative method The condition satisfied by the probability $P(v,t|v_0,t_0)$ can also be formulated quite generally to express the fact that P provides a complete probability description for the problem. Thus consider any time t_1, with $t_0 < t_1 < t$, and denote the velocity at this time by v_1. The particle must have arrived at the final velocity v at time t by passing through any of its possible velocities v_1 at the intermediate time t_1. Now the probability that a particle starting at time t_0 with velocity v_0 attains at time t_1 a velocity between v_1 and $v_1 + dv_1$, and then ends up at time t with a velocity between v and $v + dv$, is given by $P(vt|v_1t_1)\,dv \cdot P(v_1t_1|v_0t_0)\,dv_1$. Hence one must have the general relation

$$P(vt|v_0t_0)\,dv = \int_{-\infty}^{\infty} P(vt|v_1t_1)\,dv \cdot P(v_1t_1|v_0t_0)\,dv_1 \qquad (15\cdot11\cdot15)$$

where the integration is over all possible values of the intermediate velocity v_1. Let us introduce the time differences

$$s \equiv t_1 - t_0 \qquad \text{and} \qquad \tau \equiv t - t_1$$

Then $t - t_0 = s + \tau$, and (15·11·15) can be written in terms of the simpler notation of (15·11·2) as

$$P(v, s + \tau|v_0) = \int_{-\infty}^{\infty} P(v,\tau|v_1)P(v_1,s|v_0)\,dv_1 \qquad (15\cdot11\cdot16)$$

This is an integral equation (called the "Smoluchowski equation"), which must be satisfied by the probability P.

To convert (15·11·16) into a differential equation, it is only necessary to consider τ to be small. Putting $v_1 \equiv v - \xi$, (15·11·16) then becomes

$$P(v,s|v_0) + \frac{\partial P}{\partial s}\tau = \int_{-\infty}^{\infty} P(v,\tau|v - \xi)P(v - \xi, s|v_0)\,d\xi \qquad (15\cdot11\cdot17)$$

where $P(v,\tau|v - \xi)$ is only appreciable when ξ is small. Equation (15·11·17) is identical with (15·11·7) so that one regains the Fokker-Planck equation (15·11·11).

*15 · 12 *Solution of the Fokker-Planck equation*

To find $P(v,s|v_0)$, it is only necessary to solve the Fokker-Planck equation (15·11·14), subject to the initial condition (15·11·3) which requires that

$$\text{for } s \to 0, \qquad P(v,s|v_0) \to \delta(v - v_0) \qquad (15\cdot12\cdot1)$$

We outline briefly the method of solution. Consider first the simpler equation obtained by putting $\partial^2P/\partial v^2 = 0$ in (15·11·14). Then one gets

$$\frac{\partial P}{\partial s} - \gamma v \frac{\partial P}{\partial v} = \gamma P \qquad (15\cdot12\cdot2)$$

This looks similar to the perfect differential expression

$$\frac{\partial P}{\partial s}\,ds + \frac{\partial P}{\partial v}\,dv = dP$$

Multiplying (15·12·2) by the integrating factor $\lambda(v,s)$, the identification would be complete if

$$ds = \lambda, \qquad dv = -\lambda\gamma v, \qquad dP = \lambda\gamma P$$

i.e., if

$$\frac{dv}{ds} = -\gamma v \qquad \text{and} \qquad \frac{dP}{ds} = \gamma P$$

or

$$v = u\,e^{-\gamma s} \qquad \text{and} \qquad P = Q\,e^{\gamma s}$$

where u and Q are constants.

This result suggests using the method of variation of parameters by considering Q and u as variables and trying to simplify the original equation (15·11·14) by a substitution of the form

$$P(v,s) = e^{\gamma s}Q(u,s), \qquad \text{where } u \equiv ve^{\gamma s} \tag{15·12·3}$$

Then one calculates

$$\frac{\partial P}{\partial v} = e^{2\gamma s}\frac{\partial Q}{\partial u}$$

$$\frac{\partial^2 P}{\partial v^2} = e^{3\gamma s}\frac{\partial^2 Q}{\partial u^2}$$

Also $$\frac{\partial P}{\partial s} = \gamma\,e^{\gamma s}Q + e^{\gamma s}\left[\frac{\partial Q}{\partial s} + \frac{\partial Q}{\partial u}\left(\frac{\partial u}{\partial s}\right)_v\right] = \gamma\,e^{\gamma s}Q + e^{\gamma s}\left[\frac{\partial Q}{\partial s} + \gamma u\frac{\partial Q}{\partial u}\right]$$

Thus (15·11·14) reduces to

$$\frac{\partial Q}{\partial s} = \gamma\frac{kT}{m}\,e^{2\gamma s}\,\frac{\partial^2 Q}{\partial u^2} \tag{15·12·4}$$

To eliminate the factor $e^{2\gamma s}$, one can introduce a new time scale θ defined so that $dt \equiv e^{-2\gamma s}\,d\theta$, or

$$\theta \equiv \frac{1}{2\gamma}\,(e^{2\gamma s} - 1) \tag{15·12·5}$$

Then (15·12·5) becomes simply

$$\frac{\partial Q}{\partial \theta} = C\,\frac{\partial^2 Q}{\partial u^2}, \qquad \text{with } C \equiv \gamma\,\frac{kT}{m} \tag{15·12·6}$$

This is the standard diffusion equation with a solution

$$Q = (4\pi C\theta)^{-\frac{1}{2}}\,e^{-(u-u_0)^2/4C\theta} \tag{15·12·7}$$

which satisfies the condition that $Q \to \delta(u - u_0)$ as $\theta \to 0$. In terms of the original variables one obtains then

$$\blacktriangleright \qquad P(v,s|v_0) = \left[\frac{m}{2\pi kT(1 - e^{-2\gamma s})}\right]^{\frac{1}{2}} \exp\left[-\frac{m(v - v_0\,e^{-\gamma s})^2}{2kT(1 - e^{-2\gamma s})}\right] \tag{15·12·8}$$

Note that for $s \to \infty$, this approaches properly the Maxwell distribution (15·11·4). Note also that at any arbitrary time (15·12·8) is a Gaussian distribution with a mean value $\overline{v(s)} = v_0 e^{-\gamma s}$; this result is in accord with (15·8·10).

FOURIER ANALYSIS OF RANDOM FUNCTIONS

15 · 13 *Fourier analysis*

When dealing with a random function of time $y(t)$, it is very often convenient to consider its frequency components obtained by Fourier analysis. This has the advantage that it is commonly easier to focus attention on the amplitudes and phases of simple sinusoidally varying functions than on the very complicated variation in time of the random function $y(t)$. In addition, suppose that $y(t)$ is a quantity which is used as the input to a linear system (e.g., an electrical voltage which is used as input to an electrical circuit involving resistors, inductors, and capacitors); then it is quite easy to discuss what happens to each frequency component passing through the system, but it would be very difficult to analyze the situation without resolving $y(t)$ into Fourier components.

The quantity $y(t)$ has statistical properties described by a representative ensemble similar to that illustrated in Fig. 15·5·1 for the function $F(t)$. We want to represent $y(t)$ within the very large time interval $-\Theta < t < \Theta$ in terms of a superposition of sinusoidally varying functions of time. (We can ultimately go to the limit where $\Theta \to \infty$.) To avoid possible convergence difficulties, we shall then try to find the Fourier representation of the modified function $y_\Theta(t)$ which is identical with $y(t)$ in the entire domain $-\Theta < t < \Theta$, but which vanishes otherwise. Thus $y_\Theta(t)$ is defined by

$$y_\Theta(t) \equiv \begin{cases} y(t) & \text{for } -\Theta < t < \Theta \\ 0 & \text{otherwise} \end{cases} \tag{15·13·1}$$

We make use of the result (A·7·14) that the complex exponential function satisfies the ("orthogonality") property

$$\frac{1}{2\pi} \int_{-\infty}^{\infty} e^{i\omega(t-t')}\, d\omega = \delta(t - t') \tag{15·13·2}$$

where $\delta(t - t')$ is the Dirac δ function. *For any one system of the ensemble*, we can then write the function $y_\Theta(t)$ in the form

$$\begin{aligned} y_\Theta(t) &= \int_{-\infty}^{\infty} dt'\, \delta(t - t') y_\Theta(t') \\ &= \frac{1}{2\pi} \int_{-\infty}^{\infty} dt' \int_{-\infty}^{\infty} d\omega\, e^{i\omega(t-t')} y_\Theta(t') \end{aligned}$$

or

▶ $$y_\Theta(t) = \int_{-\infty}^{\infty} C(\omega)\, e^{i\omega t}\, d\omega \tag{15·13·3}$$

where $$C(\omega) = \frac{1}{2\pi} \int_{-\infty}^{\infty} y_\Theta(t')\, e^{-i\omega t'}\, dt'$$

or

$$\blacktriangleright \qquad C(\omega) = \frac{1}{2\pi} \int_{-\Theta}^{\Theta} y(t')\, e^{-i\omega t'}\, dt' \tag{15·13·4}$$

and where we have used the definition (15·13·1) in the last step. The relation (15·13·3) is the desired "Fourier integral" representation of the function $y_\Theta(t)$ in terms of a superposition of complex exponential functions of different frequencies ω; the coefficient $C(\omega)$ is then determined by (15·13·4). Since $y(t)$ is real, its complex conjugate satisfies the relation

$$y^*(t) = y(t) \tag{15·13·5}$$

Hence it follows by (15·13·4) that

$$C^*(\omega) = C(-\omega) \tag{15·13·6}$$

In any one system k of the ensemble, the function $y^{(k)}(t)$ can thus be represented by its corresponding Fourier coefficient $C^{(k)}(\omega)$ given by (15·13·4).

15·14 *Ensemble and time averages*

There are two types of averages that are of interest. The first of these is the ordinary statistical average of y *at a given time* over all systems of the ensemble. This *ensemble* average, which we denote interchangeably by $\bar{y}$ or $\langle y \rangle$, is defined by

$$\overline{y(t)} \equiv \langle y(t) \rangle \equiv \frac{1}{N} \sum_{k=1}^{N} y^{(k)}(t) \tag{15·14·1}$$

where $y^{(k)}(t)$ is the value assumed by $y(t)$ in the kth system of the ensemble and where N is the very large total number of systems in the ensemble.

The second average of interest is the average of y for a *given system of the ensemble* over some very large time interval 2Θ (where $\Theta \to \infty$). We shall denote this *time* average by $\{y\}$ and define it for the kth system of the ensemble by

$$\{y^{(k)}(t)\} \equiv \frac{1}{2\Theta} \int_{-\Theta}^{\Theta} y^{(k)}(t + t')\, dt' \tag{15·14·2}$$

(In more pictorial terms illustrated by Fig. 15·5·1, the ensemble average is taken vertically for a given t, while the time average is taken horizontally for a given k.)

Note that the operations of taking a time average and taking an ensemble average commute. Indeed

$$\langle\{y^{(k)}(t)\}\rangle = \frac{1}{N} \sum_{k=1}^{N} \left[\frac{1}{2\Theta} \int_{-\Theta}^{\Theta} y^{(k)}(t + t')\, dt'\right]$$

$$= \frac{1}{2\Theta} \int_{-\Theta}^{\Theta} \left[\frac{1}{N} \sum_{k=1}^{N} y^{(k)}(t + t')\right] dt' = \frac{1}{2\Theta} \int_{-\Theta}^{\Theta} \langle y(t + t')\rangle\, dt'$$

or

$$\langle\{y^{(k)}(t)\}\rangle = \{\langle y(t)\rangle\} \tag{15·14·3}$$

Consider now a situation which is "stationary" with respect to y. This means that there is no preferred origin in time for the statistical description of y, i.e., the same ensemble ensues when all member functions $y^{(k)}(t)$ of the ensemble are shifted by arbitrary amounts in time. (In an equilibrium situation this would, of course, be true for all statistical quantities.) For such stationary ensembles there is an intimate connection between ensemble and time averages if one assumes that (with the possible exception of a negligible number of exceptional systems in the ensemble) the function $y^{(k)}(t)$ for each system of the ensemble will in the course of a sufficiently long time pass through all the values accessible to it. (This is called the "ergodic" assumption.) One can then imagine that one takes, for example, the kth system of the ensemble and subdivides the time scale into very long sections (or intervals) of magnitude 2Θ, as shown in Fig. 15 · 14 · 1. Since Θ is very large, the behavior of $y^{(k)}(t)$ in each such section will then be independent of its behavior in any other section. Some large number of M such sections should then constitute as good a representative ensemble of the statistical behavior of y as the original ensemble. Hence the time average should be equivalent to the ensemble average.

More precisely, in such a stationary ensemble the time average of y taken over some very long time interval Θ must be independent of the time t. Furthermore, the ergodic assumption implies that the time average must be the

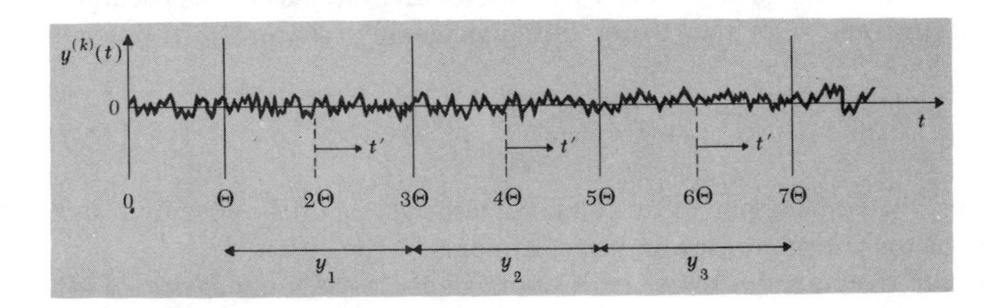

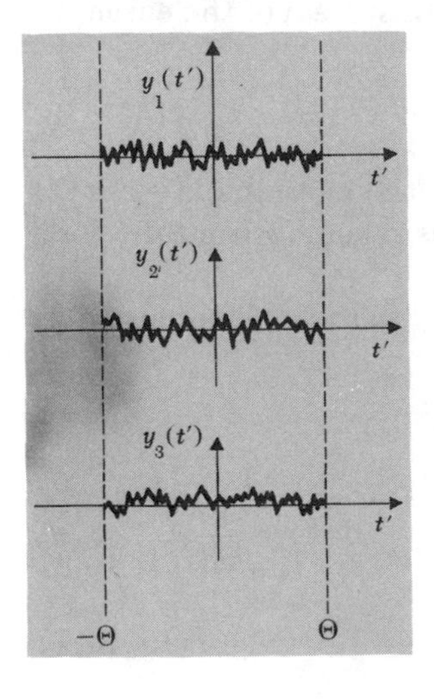

***Fig. 15·14·1** The time dependence of $y^{(k)}(t)$ in the kth member of a stationary ergodic ensemble. The time scale is shown broken up into sections of very long duration 2Θ. These sections are shown rearranged vertically in the bottom part of the figure to form another representative ensemble equivalent to the original one. (Here $y_j(t') = y^{(k)}(2j\Theta + t')$, with $-\Theta < t' < \Theta$.)*

same for essentially all systems of the ensemble. Thus

$$\{y^{(k)}(t)\} = \{y\} \qquad \text{independent of } k \tag{15·14·4}$$

Similarly, it must be true that in such a stationary ensemble the ensemble average of y must be independent of time. Thus

$$\langle y(t)\rangle = \langle y\rangle \qquad \text{independent of } t \tag{15·14·5}$$

The general relation (15·14·3) leads then immediately to an interesting conclusion. By taking the ensemble average of (15·14·4), one gets simply

$$\langle\{y^{(k)}(t)\}\rangle = \{y\}$$

Furthermore, taking the time average of (15·4·5) gives simply

$$\{\langle y(t)\rangle\} = \langle y\rangle$$

Hence (15·14·3) yields, *for a stationary ergodic ensemble*, the important result

$$\{y\} = \langle y\rangle \tag{15·14·6}$$

15·15 *Wiener-Khintchine relations*

Consider the random function $y(t)$, which is stationary. Its correlation function is then, by definition,

$$K(s) \equiv \langle y(t)y(t+s)\rangle \tag{15·15·1}$$

and is independent of t since $y(t)$ is stationary. Note that $K(0) = \langle y^2\rangle$ gives the dispersion of y if $\langle y\rangle = 0$.

The correlation function of y can, like any other function of time and like y itself, also be expressed as a Fourier integral. Analogously to (15·13·3) one can then write

▶ $$K(s) = \int_{-\infty}^{\infty} J(\omega)\, e^{i\omega s}\, d\omega \tag{15·15·2}$$

where the coefficient $J(\omega)$ is called the "spectral density" of y. From (15·15·2) it follows that $J(\omega)$ can conversely be expressed in terms of $K(s)$. One thus obtains, analogously to (15·13·4),

▶ $$J(\omega) = \frac{1}{2\pi}\int_{-\infty}^{\infty} K(s)\, e^{-i\omega s}\, ds \tag{15·15·3}$$

Remark This follows explicitly from (15·15·2) by multiplying both sides of that relation by $e^{-i\omega' s}$ and then integrating over s. Thus

$$\begin{aligned}\int_{-\infty}^{\infty} ds\, K(s)\, e^{-i\omega' s} &= \int_{-\infty}^{\infty} ds \int_{-\infty}^{\infty} d\omega\, J(\omega)\, e^{i(\omega-\omega')s} \\ &= 2\pi \int_{-\infty}^{\infty} d\omega\, J(\omega)\, \delta(\omega-\omega') \\ &= J(\omega')\end{aligned}$$

which is identical to (15·15·3) if one puts $\omega' = \omega$.

The correlation function $K(s)$ is real and satisfies the symmetry property (15·8·5). Thus

$$K^*(s) = K(s) \qquad \text{and} \qquad K(-s) = K(s) \tag{15·15·4}$$

Hence it follows by (15·15·3) that $J(\omega)$ is also real and satisfies similar symmetry properties, i.e.,

$$J^*(\omega) = J(\omega) \qquad \text{and} \qquad J(-\omega) = J(\omega) \tag{15·15·5}$$

Remark The proofs are immediate since, by virtue of (15·15·3) and (15·15·4),

$$J^*(\omega) = \frac{1}{2\pi}\int_{-\infty}^{\infty} K(s)\, e^{i\omega s}\, ds = \frac{1}{2\pi}\int_{-\infty}^{\infty} K(s)\, e^{-i\omega s}\, ds = J(\omega)$$

and $$J(-\omega) = \frac{1}{2\pi}\int_{-\infty}^{\infty} K(s)\, e^{i\omega s}\, ds = \frac{1}{2\pi}\int_{-\infty}^{\infty} K(s)\, e^{-i\omega s}\, ds = J(\omega)$$

where we have changed the variable of integration from s to $-s$ in the second set of integrals.

Note that (15·15·2) implies the particularly important result that

$$\langle y^2 \rangle = K(0) = \int_{-\infty}^{\infty} J(\omega)\, d\omega = \int_0^{\infty} J_+(\omega)\, d\omega \tag{15·15·6}$$

where $$J_+(\omega) \equiv 2J(\omega) \tag{15·15·7}$$

is the spectral density for positive frequencies.

The Fourier integrals (15·15·2) and (15·15·3) are known as the Wiener-Khintchine relations. They can also be written in explicitly real form by putting $e^{\pm i\omega s} = (\cos \omega s \pm i \sin \omega s)$ and noting that, by virtue of (15·15·4) and (15·15·5), the part of the integrand involving $\sin \omega s$ is odd and leads to a vanishing integral. Thus (15·15·2) and (15·15·3) become

$$K(s) = \int_{-\infty}^{\infty} J(\omega) \cos \omega s\, d\omega = 2\int_0^{\infty} J(\omega) \cos \omega s\, d\omega \tag{15·15·8}$$

$$J(\omega) = \frac{1}{2\pi}\int_{-\infty}^{\infty} K(s) \cos \omega s\, ds = \frac{1}{\pi}\int_0^{\infty} K(s) \cos \omega s\, ds \tag{15·15·9}$$

It is of interest to express $K(s)$ and $J(\omega)$ directly in terms of the Fourier coefficients $C(\omega)$ of the original random function $y(t)$. If $y(t)$ is stationary and ergodic, $K(s)$ is independent of time, and the ensemble average can be replaced by a time average over any system of the ensemble. Hence (15·15·1) can be written

$$K(s) = \langle y(0)y(s) \rangle = \{y(0)y(s)\}$$

or $$K(s) = \frac{1}{2\Theta}\int_{-\Theta}^{\Theta} dt'\, y(t')y(s + t')$$

where we have used the definition (15·14·2). By replacing $y(t')$ by the modified function $y_\Theta(t')$ of (15·13·1), this becomes

$$K(s) = \frac{1}{2\Theta}\int_{-\infty}^{\infty} dt'\, y_\Theta(t')y_\Theta(s+t') \qquad (15\cdot 15\cdot 10)$$

(Replacement of the function $y(s+t')$ by $y_\Theta(s+t')$ in the integrand is permissible since it causes an error only of the order of s/Θ, which is negligible in the assumed limit $\Theta \to \infty$). By using the Fourier expansion (15·13·3), the last expression can then be written

$$\begin{aligned} K(s) &= \frac{1}{2\Theta}\int_{-\infty}^{\infty} dt' \int_{-\infty}^{\infty} d\omega'\, C(\omega')\, e^{i\omega' t'} \int_{-\infty}^{\infty} d\omega\, C(\omega)\, e^{i\omega(s+t')} \\ &= \frac{1}{2\Theta}\int_{-\infty}^{\infty} d\omega' \int_{-\infty}^{\infty} d\omega\, C(\omega')C(\omega)\, e^{i\omega s} \int_{-\infty}^{\infty} dt'\, e^{i(\omega'+\omega)t'} \\ &= \frac{1}{2\Theta}\int_{-\infty}^{\infty} d\omega \int_{-\infty}^{\infty} d\omega'\, C(\omega')C(\omega)\, e^{i\omega s}[2\pi\, \delta(\omega'+\omega)] \\ &= \frac{\pi}{\Theta}\int_{-\infty}^{\infty} d\omega\, C(-\omega)C(\omega)\, e^{i\omega s} \\ &= \frac{\pi}{\Theta}\int_{-\infty}^{\infty} d\omega\, |C(\omega)|^2\, e^{i\omega s} \qquad \text{by (15·13·6)} \end{aligned}$$

or

$$K(s) = \int_{-\infty}^{\infty} J(\omega)\, e^{i\omega s}\, ds$$

where

$$J(\omega) = \frac{\pi}{\Theta}|C(\omega)|^2 \qquad (15\cdot 15\cdot 11)$$

Hence one also obtains

$$\langle y^2\rangle = K(0) = \frac{\pi}{\Theta}\int_{-\infty}^{\infty} |C(\omega)|^2\, d\omega \qquad (15\cdot 15\cdot 12)$$

The relation (15·15·11) provides an expression for the spectral density $J(\omega)$ in terms of the Fourier coefficient $C(\omega)$ of any system in the ensemble. It also shows explicitly that $J(\omega)$ can never be negative.

15·16 *Nyquist's theorem*

Suppose that an electrical resistor R is connected across the input terminals of a linear amplifier which is tuned so as to pass (angular) frequencies in the range between ω_1 and ω_2. The fluctuating current $I(t)$ due to the random thermal motion of electrons in the resistor gives rise to a random output signal (or "noise") in the amplifier. The interactions responsible for this random current can be represented by an effective fluctuating emf $V(t)$ in the resistor. If this emf $V(t)$ is expressed in terms of Fourier components, then one can write

$$\langle V^2\rangle = \int_0^{\infty} J_+(\omega)\, d\omega \qquad (15\cdot 16\cdot 1)$$

where $J_+(\omega)$ is the spectral density of the emf $V(t)$. Since $J_+(\omega)$ is intrinsically nonnegative, it provides an appropriate measure of the magnitude of the random noise input $V(t)$. The part of this input which contributes to the amplifier noise output is then given by the integral of $J_+(\omega)$ between the limits ω_1 and ω_2. To assess the importance of thermal noise in electrical measurements it is therefore necessary to know the spectral density of the emf $V(t)$.

Our general discussion of Sec. 15 · 8 has already shown that there exists a close connection between the fluctuating emf $V(t)$ and the resistance R of an electrical conductor maintained at the absolute temperature T. The explicit relation was obtained in (15 · 8 · 14) in terms of the correlation function of V in an equilibrium situation. Thus

$$R = \frac{1}{2kT} \int_{-\infty}^{\infty} \langle V(0)V(s)\rangle_0 \, ds \tag{15·16·2}$$

By virtue of (15 · 15 · 3), the right side is immediately expressible in terms of the spectral density $J(\omega)$ of V, i.e.,

$$R = \frac{1}{2kT} [2\pi J(0)] = \frac{\pi}{kT} J(0) \tag{15·16·3}$$

or

$$J_+(0) \equiv 2J(0) = \frac{2}{\pi} kTR \tag{15·16·4}$$

But the correlation time τ^* of the fluctuating voltage is very short (roughly of the order of the time an electron travels between collisions, i.e., roughly 10^{-14} sec.). Thus $K(s) \equiv \langle V(0)V(s)\rangle_0 = 0$ if $|s| \gg \tau^*$, and this correlation function is very sharply peaked near $s = 0$. Correspondingly, it follows from the Fourier integral (15 · 15 · 3) that in the domain where $K(s)$ does *not* vanish, $e^{-i\omega s} = 1$ so long as ω is small enough that $\omega\tau^* \lesssim 1$. For all values of ω smaller than this, the integral therefore has the same value; i.e.,

$$\text{for } |\omega| \ll \frac{1}{\tau^*}, \qquad J(\omega) = J(0) \tag{15·16·5}$$

Thus a sharply peaked correlation function in the time domain implies a correspondingly very broad spectral density in the frequency domain. (This is analogous to the Heisenberg uncertainty principle $\Delta\omega\,\Delta t \gtrsim 1$ in quantum

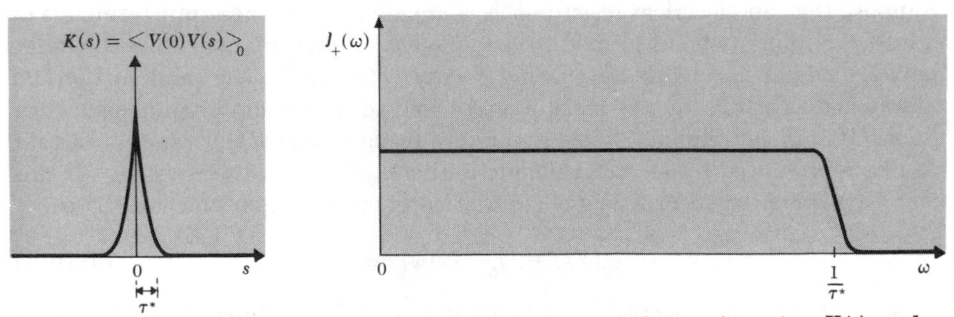

***Fig. 15 · 16 · 1** Schematic diagram showing the correlation function $K(s)$ and spectral density $J_+(\omega)$ of the fluctuating emf $V(t)$.*

mechanics). Using (15·16·4), one can then conclude that,

$$\blacktriangleright \quad \text{for } \omega \ll \frac{1}{\tau^*}, \qquad J_+(\omega) = \frac{2}{\pi} kTR \tag{15·16·6}$$

This important equation relating the spectral density $J_+(\omega)$ of the fluctuating voltage to the resistance R is known as "Nyquist's theorem." It is a special case of the general connection existing between fluctuations and dissipation in physical systems. (For example, in the Brownian-motion problem, there is a similar relation between the spectral density of the fluctuating force $F(t)$ and the friction constant α.)

Note that, by (15·16·5), $J_+(\omega)$ is independent of ω until one approaches frequencies of the order of 10^{14} cps, far beyond the microwave range. A fluctuating quantity such as V which has such a frequency-independent spectral density is in customary jargon said to give rise to "white noise" ("white," because just as in visible white light, all frequencies are equally represented). Note also that J_+ is proportional to R and to the absolute temperature T of the resistor. Thermal noise is thus decreased by reducing the temperature of the resistor.

15·17 *Nyquist's theorem and equilibrium conditions*

Nyquist's theorem is such an important general result that it is worth some further discussion. In particular, it is of interest to verify that the theorem is consistent with the conditions that must prevail in any equilibrium situation. Indeed, it is possible to conceive of *any* convenient equilibrium situation involving an electrical resistor and to derive Nyquist's theorem by requiring that the conditions for equilibrium are properly satisfied.

As a specific illustration, consider the simple circuit illustrated in Fig. 15·17·1 consisting of a resistor R in series with an inductance L and a capacitor C. The whole system is supposed to be in thermal equilibrium at the temperature T. Fluctuations in the current I can again be regarded as due to some effective random emf in the resistor. Focus attention on a particular (complex) Fourier component $V_0(\omega)\, e^{i\omega t}$ of this emf. The corresponding current $I_0(\omega)\, e^{i\omega t}$ at this frequency is then found from the usual circuit equation

$$L\frac{dI}{dt} + RI + \frac{1}{C}\int I\, dt = V(t)$$

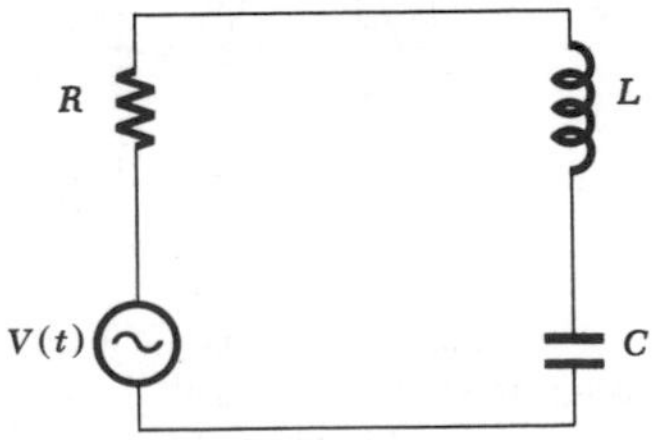

R L V(t) C

Fig. 15·17·1 An electric circuit consisting of a resistance R connected in series with an inductance L and a capacitor C.

This gives, for the particular frequency ω,

$$I_0(\omega) = \frac{V_0(\omega)}{Z(\omega)}, \qquad \text{where } Z(\omega) = R + i\left(\omega L - \frac{1}{\omega C}\right) \tag{15·17·1}$$

is the impedance of the circuit.

General statistical arguments applied to this circuit in thermal equilibrium require, however, that the mean energy stored in the inductance be given by

$$\langle \tfrac{1}{2}LI^2 \rangle = \tfrac{1}{2}kT \tag{15·17·2}$$

The circuit equation (15 · 17 · 1) must therefore be consistent with this result.

It is not immediately obvious that (15·17·2) follows immediately from the classical equipartition theorem.* One can, however, regard the current I as a macroscopic parameter of the system and consider the free energy F of the circuit as a function of I. Then the probability $P(I)\,dI$ that this current has a value between I and $I + dI$ when the circuit is in thermal equilibrium with a heat reservoir at temperature T is, by (8·2·10), proportional to $\exp(-\Delta F/kT)$. Here ΔF is its free energy measured from the situation where $I = 0$. But a bodily motion of all the charges, setting up a current I but leaving their relative state of motion unchanged, should have negligible effect on their entropy S. Thus $\Delta S = 0$ and $\Delta F = \Delta E - T\,\Delta S = \Delta E$ so that

$$P(I)\,dI \propto e^{-\Delta E/kT}\,dI \propto e^{-\frac{1}{2}LI^2/kT}\,dI \tag{15·17·3}$$

since the energy change is that associated with the inductance, i.e., $\Delta E = \frac{1}{2}LI^2$. The functional form (15·17·3) implies that

$$\left\langle \frac{1}{2}LI^2 \right\rangle = \frac{\int_{-\infty}^{\infty} P(I)\,dI\,I^2}{\int_{-\infty}^{\infty} P(I)\,dI} = \frac{1}{2}kT$$

is just given by the equipartition theorem result. A similar argument shows that, if the voltage across a capacitor C is V_C, the mean energy $\langle \frac{1}{2}CV_C^2 \rangle$ stored in this capacitor in thermal equilibrium ought also to be equal to $\frac{1}{2}kT$.

Returning now to the circuit of Fig. 15 · 17 · 1, the mean energy stored in the inductance L can be expressed in terms of Fourier components by (15 · 15 · 12). Thus

$$\left\langle \frac{1}{2}LI^2 \right\rangle = \frac{1}{2}L\frac{\pi}{\Theta}\int_{-\infty}^{\infty} |I_0(\omega)|^2\,d\omega = \frac{1}{2}L\frac{\pi}{\Theta}\int_{-\infty}^{\infty} \frac{|V_0(\omega)|^2}{|Z(\omega)|^2}\,d\omega$$
$$= \frac{1}{2}L\int_{-\infty}^{\infty} \frac{J(\omega)}{|Z(\omega)|^2}\,d\omega = \frac{1}{2}L\int_{0}^{\infty} \frac{J_+(\omega)}{|Z(\omega)|^2}\,d\omega$$

* In order to satisfy the conditions of applicability of this theorem, it would be necessary to show that it is possible to write the total energy of all the particles in the circuit in the form $\frac{1}{2}LI^2$ plus other terms which involve generalized coordinates and momenta of all these particles but *not* I.

where we have used the definition (15·15·11) of the spectral density of the emf $V(t)$

$$J_+(\omega) \equiv 2J(\omega) \equiv \frac{2\pi}{\Theta} |V_0(\omega)|^2 \tag{15·17·4}$$

Using (15·17·1), the condition that (15·17·2) be satisfied in thermal equilibrium requires then that the fluctuating voltage have a spectral density J_+ such that

$$\int_0^\infty \frac{J_+(\omega)\, d\omega}{R^2 + (\omega L - 1/\omega C)^2} = \frac{kT}{L} \tag{15·17·5}$$

Let us write this in the form

$$\frac{1}{R^2} \int_0^\infty \frac{J_+(\omega)\, d\omega}{1 + (L/\omega R)^2(\omega^2 - \omega_0{}^2)^2} = \frac{kT}{L}, \qquad \text{where } \omega_0 \equiv (LC)^{-\frac{1}{2}} \tag{15·17·6}$$

If L is made very large the integrand exhibits a very sharp maximum at the frequency $\omega = \omega_0$. (That is, the circuit is then very sharply turned.) One can then put $J_+(\omega) = J_+(\omega_0)$ outside the integral. Also one can put $\omega^2 - \omega_0{}^2 = (\omega + \omega_0)(\omega - \omega_0) \approx 2\omega_0(\omega - \omega_0)$ in the integrand and replace $L/\omega R$ by $L/\omega_0 R$. The integral of (15·17·6) then becomes

$$\frac{J_+(\omega_0)}{R^2} \int_0^\infty \frac{d\omega}{1 + (2L/R)^2(\omega - \omega_0)^2} = \frac{J_+(\omega_0)}{R^2} \int_{-\infty}^\infty \frac{d\eta}{1 + (2L/R)^2\eta^2} = \frac{J_+(\omega_0)}{R^2} \left[\pi \left(\frac{R}{2L}\right)\right]$$

where we have put $\eta \equiv \omega - \omega_0$ and extended the limits from $-\infty$ to $+\infty$ with negligible error, since the integrand is negligible for $\eta \neq 0$. Thus (15·17·6) yields the result

$$J_+(\omega_0) = \frac{2}{\pi} kTR \tag{15·17·7}$$

Since ω_0 can be *any* frequency obtained by proper choice of C, we regain in (15·17·7) Nyquist's theorem. We have also shown that it is consistent with the mean thermal energy (15·17·2) stored in an inductance.

As another illustration of an equilibrium argument, consider the circuit of Fig. 15·17·2, which is in thermal equilibrium at temperature T. Here $Z'(\omega) = R'(\omega) + iX'(\omega)$ is any arbitrary impedance, with resistive and reactive parts which are in general frequency-dependent. The fluctuating thermal

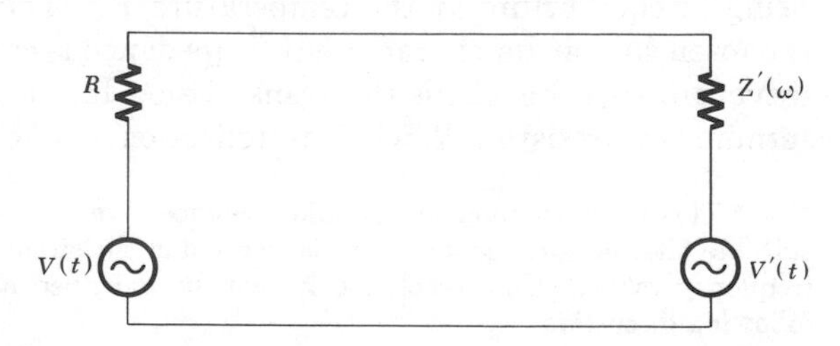

***Fig. 15·17·2** A resistor R connected to an impedance Z′(ω), both being in thermal equilibrium at the temperature T.*

emf's associated with the resistor R and the impedance Z' are represented by $V(t)$ and $V'(t)$, respectively. In the equilibrium situation illustrated in Fig. 15·17·2, the [mean power $\mathcal{P}'$ absorbed by the impedance Z' due to the random emf $V(t)$ associated with R] must be equal to the [mean power $\mathcal{P}$ absorbed by the resistance R due to the random emf $V'(t)$ associated with Z']. Thus

$$\mathcal{P}' = \mathcal{P} \tag{15·17·8}$$

Focus attention on a narrow frequency range between ω and $\omega + d\omega$. The equality (15·17·8) must be valid in any such range if it is to be valid in general.* But the frequency component $V_0(\omega)$ of the emf $V(t)$ produces in the circuit a current $I_0 = V_0/(R + Z')$; the mean power $\mathcal{P}'$ consequently absorbed by Z' is given by

$$\mathcal{P}' \propto |I_0|^2 R' = \left| \frac{V_0}{R + Z'} \right|^2 R' \tag{15·17·9}$$

Similarly, the frequency component $V_0'(\omega)$ of the emf $V'(t)$ produces in the circuit a current $I_0' = V_0'/(R + Z')$; the mean power $\mathcal{P}$ consequently absorbed by R is given by

$$\mathcal{P} \propto |I_0'|^2 R = \left| \frac{V_0'}{R + Z'} \right|^2 R \tag{15·17·10}$$

with the same constant of proportionality as in (15·17·9). Hence it follows by (15·17·8) that

$$|V_0|^2 R' = |V_0'|^2 R$$

or

$$J_+(\omega) R'(\omega) = J_+'(\omega) R \tag{15·17·11}$$

where J_+ is the spectral density of $V(t)$ and J_+' that of $V'(t)$. Hence it follows by (15·17·7) that

$$J_+'(\omega) = \frac{R'(\omega)}{R} \left(\frac{2}{\pi} kTR \right) = \frac{2}{\pi} kTR'(\omega) \tag{15·17·12}$$

This shows that the spectral density of the thermal fluctuating voltage of *any* impedance is always associated with its resistive part (at the particular frequency) in accordance with Nyquist's theorem.

Finally, it is instructive to present Nyquist's original derivation of his theorem by noting that the problem of the resistor R can be viewed as a simple one-dimensional case of black-body radiation. One need only consider an ideal (lossless) one-dimensional transmission line of great length L which is terminated on both ends by resistances R (see Fig. 15·17·3), the whole system being in equilibrium at the temperature T. The particular transmission line is chosen so that its characteristic impedance is equal to R. Then any voltage wave propagating along the transmission line is completely absorbed by the terminating resistor R without reflection. The resistor is then indeed the

* This is an argument of detailed balance. Indeed, one could always conceive that a suitable filter of pure reactance X_f is inserted in the circuit so as to pass current only in this frequency range. This reactance X_f can be imagined added to the reactance X' in the following discussion.

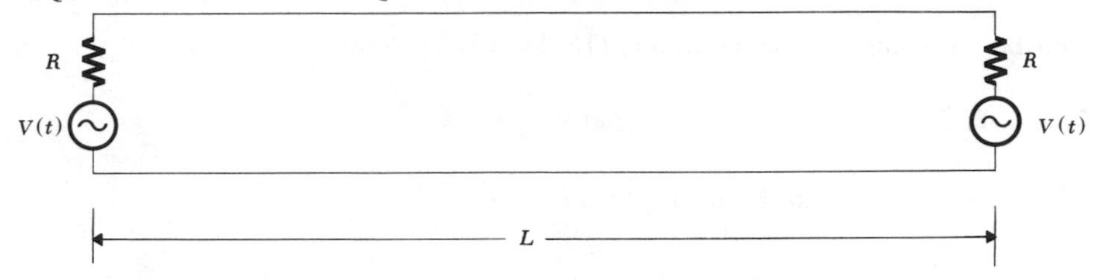

***Fig. 15·17·3** A long transmission line of length L terminated at both ends by equal resistances R equal to its characteristic impedance.*

analogue of a black body in one dimension. A voltage wave of the form $V = V_0 \exp[i(kx - \omega t)]$ propagates along the transmission line with a velocity $c' = \omega/k$. To count the possible modes, one can consider the domain between $x = 0$ and $x = L$ and impose the boundary condition $V(L_x) = V(0)$ on the possible propagating waves. Then $kL = 2\pi n$, where n is any integer, and there are $\Delta n = (1/2\pi)\, dk$ such modes *per unit length* of the line in the frequency range between ω and $\omega + d\omega$. The mean energy in each mode is given by (9·13·1) as

$$\epsilon(\omega) = \frac{\hbar\omega}{e^{\beta\hbar\omega} - 1} \to kT \qquad \text{for } \hbar\omega \ll kT \tag{15·17·13}$$

One can now resort to the familiar detailed balance argument by equating (in any small frequency range between ω and $\omega + d\omega$) the power absorbed by a resistor to the power emitted by it. Since there are $(2\pi)^{-1}(d\omega/c')$ propagating modes per unit length in this frequency range, the mean energy per unit time incident upon a resistor in this frequency range is

$$\mathcal{P}_i = c'\left(\frac{1}{2\pi}\frac{d\omega}{c'}\right)\epsilon(\omega) = \frac{1}{2\pi}\,\epsilon(\omega)\, d\omega \tag{15·17·14}$$

This is the power *absorbed* by the resistor. By the principle of detailed balance this must be equal to the power *emitted* by the resistor in this frequency range. But if the thermal emf generated by a resistor is V, this voltage sets up a current $I = V/2R$ in the line. Hence the mean power emitted down the line and absorbed by the resistor at the other end is

$$R\langle I^2\rangle = R\left\langle\frac{V^2}{4R^2}\right\rangle = \frac{1}{4R}\int_0^\infty J_+(\omega)\, d\omega$$

or $(4R)^{-1}J_+\, d\omega$ in the frequency range between ω and $\omega + d\omega$. Equating this to (15·17·14) then gives

$$\frac{1}{4R} J_+(\omega)\, d\omega = \frac{1}{2\pi}\,\epsilon(\omega)\, d\omega$$

Thus

$$J_+(\omega) = \frac{2}{\pi}\frac{\hbar\omega}{e^{\beta\hbar\omega} - 1} R \tag{15·17·15}$$

which is the correct Nyquist result if quantum-mechanical corrections are taken into account. Since at ordinary temperatures $\hbar\omega \ll kT$ up to frequencies well beyond the microwave range, (15 · 17 · 15) becomes,

for $\hbar\omega \ll kT$,
$$J_+(\omega) = \frac{2}{\pi} kTR$$

which agrees with (15 · 16 · 6) or (15 · 17 · 7).

GENERAL DISCUSSION OF IRREVERSIBLE PROCESSES

15 · 18 *Fluctuations and Onsager relations*

We shall conclude this chapter by showing that the arguments used in Sec. 15 · 7 to study the Brownian-motion problem can be extended on a more abstract level to lead to general results of very wide applicability.

Consider an isolated system A described macroscopically by n parameters $\{y_1, \ldots, y_n\}$. Imagine that the possible values of each parameter y_i are subdivided into small ranges of magnitude δy_i, and denote by $\Omega(y_1, \ldots, y_n)$ the number of states accessible to A when the parameters lie in the small range near $\{y_1, \ldots, y_n\}$. Then $S = k \ln \Omega$ is, by definition, the corresponding entropy of A. It may be helpful in this general discussion to keep in mind some specific simple examples, e.g., the situation illustrated in Fig. 15 · 18 · 1 where the single parameter y denotes the position of the weight, or the Brownian-motion problem where y denotes the velocity v of the particle.

In the equilibrium situation the probability in the representative ensemble of finding A with its parameters in the range near $\{y_1 \ldots y_n\}$ is given by

$$P(y_1, \ldots, y_n) \propto \Omega(y_1, \ldots, y_n) = e^{S(y_1, \ldots, y_n)/k} \qquad (15 \cdot 18 \cdot 1)$$

The most probable situation is that in which S has its maximum value $\tilde{S}$, and this occurs for values of the parameters $y_i = \tilde{y}_i$. Then

$$\left[\frac{\partial S}{\partial y_i}\right] = 0 \qquad \text{for all } i \qquad (15 \cdot 18 \cdot 2)$$

where the square bracket denotes that the derivatives are evaluated when

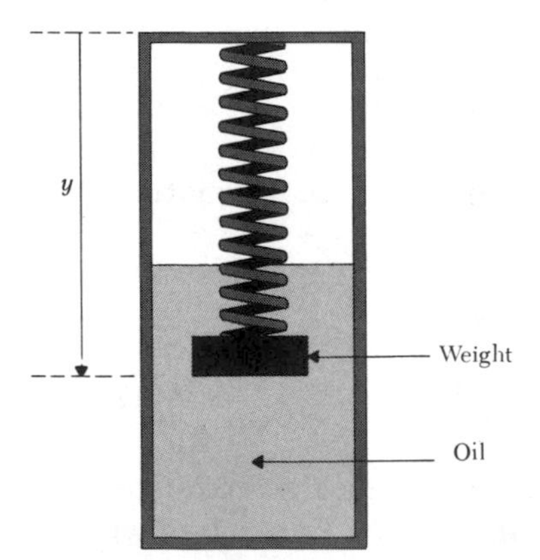

***Fig. 15 · 18 · 1** An isolated system where a weight suspended from a spring is immersed in a viscous oil.*

$y_j = \tilde{y}_j$ for all j. The fact that S is a maximum implies then, by Taylor's series expansion, that for small values of $|y_i - \tilde{y}_i|$,

$$S(y_1, \ldots, y_n) - S(\tilde{y}_1, \ldots, \tilde{y}_n) = \frac{1}{2} \sum_{ik} C_{ik}(y_i - \tilde{y}_i)(y_k - \tilde{y}_k) \quad (15 \cdot 18 \cdot 3)$$

where

$$C_{ik} = C_{ki} \equiv \left[\frac{\partial^2 S}{\partial y_i \partial y_k} \right] \quad (15 \cdot 18 \cdot 4)$$

Suppose now that one or more parameters y_i are known to be different from $\tilde{y}_i$. This situation may have arisen as a result of external intervention (e.g., clamping the weight in position), or as a result of spontaneous fluctuations in the system. If the parameters are now left free to adjust themselves, A will be in a highly improbable situation. One expects therefore that the parameters y_i will change in time until A approaches the most probable situation. Our present interest is to examine this more difficult nonequilibrium problem and to arrive at some statements about the *rate* at which the parameters change.

Purely phenomenologically one may reason as follows. Quantities such as $\dot{y}_i \equiv dy_i/dt$ and $\partial S/\partial y_i$ are very rapidly fluctuating functions of time. Let us therefore focus attention on their ensemble averages, which are slowly varying functions of time and which describe the macroscopically readily observable quantities. If $\overline{\partial S/\partial y_i} = 0$ for all i, the system is by virtue of $(15 \cdot 18 \cdot 2)$ in equilibrium; then the mean values $\bar{y}_i$ of all parameters remain unchanged; i.e., $d\bar{y}_i/dt = 0$ for all i. On the other hand, when all the quantities $\overline{\partial S/\partial y_i}$ do not vanish, the system is not in equilibrium. One expects then that

$$\frac{d\bar{y}_i}{dt} = f(\bar{Y}_1, \bar{Y}_2, \ldots, \bar{Y}_n), \qquad \text{where } Y_i \equiv \frac{\partial S}{\partial y_i} \quad (15 \cdot 18 \cdot 5)$$

and where f is some function such that $f = 0$ in the equilibrium situation when $\bar{Y}_i = 0$ for all i. If the situation is not too far removed from equilibrium, then the quantities $\bar{Y}_i$ are small and the parameters $\bar{y}_i$ change relatively slowly in time. One can then expand $(15 \cdot 18 \cdot 5)$ in a Taylor's series and obtain for the lowest order nonvanishing terms a linear relationship of the form

$$\blacktriangleright \qquad \frac{d\bar{y}_i}{dt} = \sum_{j=1}^{n} \alpha_{ij} \bar{Y}_j \quad (15 \cdot 18 \cdot 6)$$

where the "friction coefficients" α_{ij} are constants. By $(15 \cdot 18 \cdot 3)$

$$\blacktriangleright \qquad \bar{Y}_j = \overline{\frac{\partial S}{\partial y_j}} = \sum_k C_{jk}(\bar{y}_k - \tilde{y}_k) \quad (15 \cdot 18 \cdot 7)$$

If $\bar{y}_k = \tilde{y}_k$ for all k, the "driving forces" $\bar{Y}_j$ vanish; the relation $(15 \cdot 18 \cdot 6)$ then properly describes the equilibrium situation where $d\bar{y}_i/dt = 0$. In the more general nonequilibrium case, the quantities $\bar{y}_i$ must change in such a direction that the entropy S increases.

Let us now treat the situation by more detailed statistical arguments similar to those used in Sec. $15 \cdot 7$ to discuss the Brownian-motion problem.

Each parameter y_i can be regarded as a random variable which fluctuates rapidly in time. Its rate of fluctuation is given by $\dot{y}_i \equiv dy_i/dt$ and can be characterized by some correlation time τ^* which measures the time between successive maxima (or minima) of the function $\dot{y}_i$. The time τ^* is also of the order of the relaxation time necessary for the system to return to equilibrium after some sudden small disturbance away from equilibrium. This time τ^* is very short on a macroscopic scale, and our interest is in describing the behavior of y_i over time intervals very much greater than τ^*. We introduce, therefore, a time interval τ which is microscopically large in the sense that $\tau \gg \tau^*$, but which is macroscopically so small that (for each i) the ensemble average $\bar{y}_i \equiv \langle y_i \rangle$ changes only very little during such a time τ. We then consider the statistical ensemble and shall try to find the "coarse-grained" time derivative of $\langle y_i \rangle$, i.e., the value of

$$\frac{1}{\tau} \langle y_i(t+\tau) - y_i(t) \rangle = \frac{1}{\tau} \int_t^{t+\tau} \langle \dot{y}_i(t') \rangle \, dt' \tag{15·18·8}$$

In the first approximation one can assume that the value of $\langle \dot{y}_i \rangle$ at time t is the same as if the parameters $\bar{y}_j$ of A did not change in time, i.e., that $\langle \dot{y}_i(t) \rangle = 0$ as it would be in an equilibrium situation. But, to make the approximation adequate, one must take into account the fact that, when the parameters of the system change in time, the values assumed by $\dot{y}_i$ in the ensemble at any subsequent time t' do depend on how the parameters y_j themselves change in time. Let us assume that the parameters change in a small time interval τ' ($\tau' > \tau^*$) by small amounts from $y_j(t)$ to $y_j\,(t + \tau') = y_j(t) + \Delta y_j(\tau')$. Since $\tau' > \tau^*$, the system A can then reestablish internal equilibrium so that it is again equally likely to be found in any of its Ω accessible states consistent with these new values of the parameters. In the time τ' the number of accessible states Ω (or, equivalently, the entropy $S = k \ln \Omega$ of the system) therefore changes by an amount

$$\Delta S(\tau') \equiv k \ln \Omega[y_1(t+\tau'), y_2(t+\tau'), \cdots] - k \ln \Omega[y_1(t), y_2(t), \cdots] \tag{15·18·9}$$

Suppose that the system has a probability W_r of being found in some configuration r where the value of $\dot{y}_i$ is $(\dot{y}_i)_r$. When the system is in internal equilibrium this probability is proportional to the number of states accessible to the system under these circumstances. Hence one can compare the respective probabilities that the system is found in the *same* configuration r at time t and time $t + \tau'$ by writing

$$\frac{W_r(t+\tau')}{W_r(t)} = \frac{\Omega(t+\tau')}{\Omega(t)} = e^{\Delta S(\tau')/k} \tag{15·18·10}$$

or

$$W_r(t+\tau') \approx W_r(t) \left[1 + \frac{1}{k} \Delta S(\tau') \right] \tag{15·18·11}$$

where $\Delta S(\tau')$ is the entropy change of the system in the time τ'. The mean value of $\dot{y}_i$ at the time $t' = t + \tau'$ is then given by

$$\langle \dot{y}_i(t') \rangle = \sum_r W_r(t + \tau')(\dot{y}_i)_r = \left\langle \dot{y}_i \left[1 + \frac{1}{k} \Delta S(\tau') \right] \right\rangle_0$$

where the last average is to be computed with the original probability $W_r(t)$ at time t where we assumed an equilibrium situation so that $\langle \dot{y}_i \rangle_0 = 0$. Hence

$$\langle \dot{y}_i(t') \rangle = \frac{1}{k} \langle \dot{y}_i(t') \, \Delta S(\tau') \rangle_0 \tag{15·18·12}$$

Here one can expand ΔS in (15·18·9) and write

$$\Delta S(\tau') = S[y_1(t + \tau'), \ . \ . \ .] - S[y_1(t), \ . \ . \ .] = \sum_j \frac{\partial S}{\partial y_j} \Delta y_j(\tau') \tag{15·18·13}$$

Since $\tau \gg \tau'$, it is permissible to use the result (15·18·12) in the integrand of (15·18·8). A calculation analogous to that of Sec. 15·7 then gives

$$\begin{aligned}
\langle y_i(t + \tau) - y_i(t) \rangle &= \frac{1}{k} \int_t^{t+\tau} \langle \dot{y}_i(t') \, \Delta S(t' - t) \rangle_0 \, dt' \\
&= \frac{1}{k} \int_t^{t+\tau} dt' \left\langle \dot{y}_i(t') \sum_j \frac{\partial S}{\partial y_j} \Delta y_j \, (t' - t) \right\rangle_0 \\
&= \frac{1}{k} \sum_j \int_t^{t+\tau} dt' \left\langle \dot{y}_i(t') Y_j(t) \int_t^{t'} dt'' \, \dot{y}_j(t'') \right\rangle_0 \\
&\approx \frac{1}{k} \sum_j \bar{Y}_j(t) \int_t^{t+\tau} dt' \int_t^{t'} dt'' \, \langle \dot{y}_i(t') \dot{y}_j(t'') \rangle_0
\end{aligned}$$

where we have put $Y_j \equiv \partial S / \partial y_j$ and have averaged separately over this more slowly varying function. Putting $s \equiv t'' - t'$ and using the same change of variable as in (15·8·6) then gives

$$\langle y_i(t + \tau) - y_i(t) \rangle = \frac{1}{k} \sum_j \bar{Y}_j \int_{-\tau}^{0} ds \int_{t-s}^{t+\tau} dt' \, K_{ij}(s) = \frac{1}{k} \sum_j \bar{Y}_j \tau \int_{-\infty}^{0} ds \, K_{ij}(s) \tag{15·18·14}$$

where
$$K_{ij}(s) \equiv \langle \dot{y}_i(t) \dot{y}_j(t + s) \rangle_0 = \langle \dot{y}_1(0) \dot{y}_j(s) \rangle_0 \tag{15·18·15}$$

is the "cross-correlation" function relating $\dot{y}_i$ and $\dot{y}_j$ in equilibrium. Thus (15·18·14) gives, after division by τ, the coarse-grained time derivative

▶
$$\frac{d\bar{y}_i}{dt} = \sum_j \alpha_{ij} \bar{Y}_j \tag{15·18·16}$$

where

▶
$$\alpha_{ij} = \frac{1}{k} \int_{-\infty}^{0} ds \, K_{ij}(s) \tag{15·18·17}$$

Equation (15·18·16) is identical to (15·18·7), but the friction coefficients α_{ij} are now given explicitly in terms of correlation functions of the quantities $\dot{y}$ in the *equilibrium* situation. Equation (15·18·17) is a general form of the fluctuation-dissipation theorem.

Symmetry properties The fact that $\langle \dot{y}_i(t)\dot{y}_j(t+s)\rangle_0$ must be independent of the absolute value of the time t in the equilibrium situation implies that this quantity is unchanged under the substitution $t \to t - s$. Hence

$$\langle \dot{y}_i(t)\dot{y}_j(t+s)\rangle_0 = \langle \dot{y}_i(t-s)\dot{y}_j(t)\rangle_0 = \langle \dot{y}_j(t)\dot{y}_i(t-s)\rangle_0$$

or

$$K_{ij}(s) = K_{ji}(-s) \tag{15·18·18}$$

Hence this argument based on invariance under *translation* of the origin of time does *not* imply that $K_{ij}(s)$ is an even function of s unless $i = j$.

On the other hand, a physical argument based upon time *reversal* does lead to further interesting results. Imagine that the sign of the time is reversed from t to $-t$. Correspondingly, all particle velocities also reverse their signs; furthermore, if any external magnetic field **H** is present, it also reverses its sign since the currents producing it also reverse theirs. But the microscopic equations of motion are invariant under such a reversal of the sign of the time; hence, if the sense of time is reversed, all particles simply retrace their paths backward in time. (This is the property of "microscopic reversibility" already mentioned in Sec. 15·1.) Let us denote by a dagger superscript those quantities that refer to a situation where the sign of time is imagined reversed. The correlation functions describing the fluctuations in the equilibrium situation must, therefore, be such that

$$\langle \dot{y}_i(0)\dot{y}_j(s)\rangle_0 = \langle \dot{y}_i(0)y_j(-s)\rangle_0^{\dagger}$$

or

$$K_{ij}(s) = K_{ij}^{\dagger}(-s) \tag{15·18·19}$$

where the correlation functions on the right are evaluated under conditions of time reversal. Combining this with the result (15·18·18) one has the symmetry property

$$K_{ij}(s) = K_{ji}^{\dagger}(s) \tag{15·18·20}$$

or by (15·18·17)

▶ $$\alpha_{ij} = \alpha_{ji}^{\dagger} \tag{15·18·21}$$

More explicitly, imagine a general situation where an external field $\boldsymbol{H}$ is present. In the usual case where y_i and y_j are *both* displacements or are *both* velocities, the quantity $\dot{y}_i\dot{y}_j$ does not change sign under time reversal. Then (15·18·19) and (15·18·18) imply that $K_{ij}(s;\boldsymbol{H}) = K_{ij}(-s;-\boldsymbol{H}) = K_{ji}(s;-\boldsymbol{H})$ and one has

$$\alpha_{ij}(\boldsymbol{H}) = \alpha_{ji}(-\boldsymbol{H}) \tag{15·18·22}$$

On the other hand, if y_i is a displacement and y_j is a velocity, or vice versa, then $\dot{y}_i\dot{y}_j$ does change sign under time reversal. Thus

$$K_{ij}(s;\,\boldsymbol{H}) = -K_{ij}(-s;\,-\boldsymbol{H}) = -K_{ji}(s;\,-\boldsymbol{H})$$

and one has

$$\alpha_{ij}(\boldsymbol{H}) = -\alpha_{ji}(-\boldsymbol{H}) \tag{15·18·23}$$

The symmetry relations (15·18·21) (or more specifically (15·18·22) and (15·18·23)) are known as the "Onsager reciprocal relations." Together with the equations (15·18·16) they form the basis for the entire discipline of macroscopic irreversible thermodynamics. A discussion of this subject and its applications to phenomena such as thermoelectric effects would take us too far afield in this book. The interested reader is referred to the references at the end of this chapter.

Finally, it is worth pointing out that the *micro*scopic reversibility in no way contradicts the fact that in a macroscopic description all parameters always tend to change in time so as to approach some equilibrium value and thus to exhibit irreversible behavior. Consider a variable y whose fluctuations in an equilibrium situation are illustrated schematically in Fig. 15·18·2. For simplicity we assume that there is no magnetic field present and that y is some quantity, e.g., a displacement, which is invariant under time reversal. If, starting at some time $t = t_0$, one looks into the future at times $t_0 + s$ where $s > 0$, or into the past at times $t_0 - s$ where $s > 0$, the situation is then quite indistinguishable. Or, phrasing this in more picturesque terms, suppose that one took a moving picture of the fluctuating variable y. Then a spectator watching the movie run through the projector could not distinguish whether the movie is run through the projector forward or backward. This illustrates concretely the microscopic reversibility of the process under consideration. On the other hand, suppose that one *knows* that the parameter lies in the small range between y_1 and $y_1 + \delta y$, where y_1 is appreciably different from the mean value $\tilde{y}$ about which y fluctuates. Then y must lie near one of the very few peaks like A, which correspond to such improbably large fluctuations that y attains a value near y_1 (see Fig. 15·18·2). But then y, being near a maximum like A, will decrease as time goes on. (As a matter of fact, it would also decrease if one went backward in time.) The situation where y has a value near y_1 and yet *increases* would correspond to the occurrence of a peak like B, which is even larger than A; on the rising side of this peak, y would indeed *increase*. But the occurrence of peaks as large as B is enormously less probable than the already very improbable occurrence of peaks as large as A. Hence, if y is *known* to be as large as y_1, one can conclude that it will practically always *decrease* and thus approach closer to $\tilde{y}$.

Does the preceding argument, however, not contradict the fact that y has

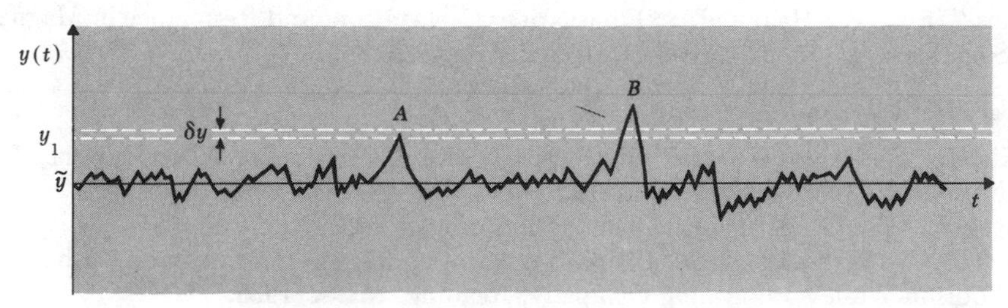

Fig. 15·18·2 Diagram illustrating the time dependence of a fluctuating parameter $y(t)$ in equilibrium.

to *increase* in the first place in order to attain a value as large as y_1 as a result of spontaneous fluctuations? The answer is that there is no contradiction because if $|y_1 - \tilde{y}|$ is at all appreciable, it would take an unimaginably long time before a fluctuation as large as this actually did occur; i.e., spontaneously occurring peaks as large as A are a fantastically rare occurrence if $|y_1 - \tilde{y}|$ is not small. Indeed, in this case the only practical hope of observing what happens when y has a value near y_1 is to bring it to this value by *external* intervention. If one then removes the external constraint, the situation will be just the same as if y had attained the value y_1 as a result of prior spontaneous fluctuations, and y will practically always *decrease* toward $\tilde{y}$.

SUGGESTIONS FOR SUPPLEMENTARY READING

Fluctuation phenomena

C. Kittel: "Elementary Statistical Physics," secs. 25–32, John Wiley & Sons, Inc., New York, 1958.

R. Becker: "Theorie der Wärme," chap. 6, Springer-Verlag, Berlin, 1955.

D. K. C. MacDonald: "Noise and Fluctuations: An Introduction," John Wiley & Sons, Inc., New York, 1962.

M. Wax (ed.): "Selected Papers on Noise and Stochastic Processes," Dover Publications, New York, 1954. (See particularly the paper of S. Chandrasekhar, reprinted from *Rev. Mod. Physics*, vol. 15, pp. 1–89 (1943), which gives an extensive discussion of Brownian-motion problems.)

C. W. McCombie: Fluctuation Theory in Physical Measurements, *Reports on Progress in Physics*, vol. 16, pp. 266–320 (1953).

Fluctuation theory in applied physics

A. van der Ziel: "Noise," Prentice-Hall, Englewood Cliffs, N.J., 1954.

J. J. Freeman: "Principles of Noise," John Wiley & Sons, Inc., New York, 1958.

W. N. Davenport and W. L. Root: "Random Signals and Noise," McGraw-Hill Book Company, New York, 1958.

Advanced discussions of the relation between fluctuation and dissipation

H. B. Callen and T. A. Welton: Irreversibility and Generalized Noise, *Phys. Rev.*, vol. 83, p. 34 (1951); also J. L. Jackson, *Phys. Rev.*, vol. 87, p. 471 (1952).

H. B. Callen: "The Fluctuation-dissipation Theorem and Irreversible Thermodynamics," in D. Ter Haar (ed.): "Fluctuations, Relaxation, and Resonance in Magnetic Systems," pp. 15–22, Oliver and Boyd, London, 1962.

Irreversible thermodynamics

C. Kittel: "Elementary Statistical Physics," secs. 33–35, John Wiley & Sons, Inc., New York, 1958.

R. Becker: "Theorie der Wärme," chap. 6, Springer-Verlag, Berlin, 1955.

J. F. Lee, F. W. Sears, D. L. Turcotte: "Statistical Thermodynamics," chap. 15, Addison-Wesley Publishing Company, Reading, Mass., 1963.

S. R. DeGroot: "Thermodynamics of Irreversible Processes," Interscience Publishers, New York, 1951.

PROBLEMS

15.1 A sphere or radius a moves with uniform speed v through a fluid of viscosity η. The frictional retarding force f acting on the sphere must then be some function of a, v, and η. (It cannot depend on the density of the fluid since the inertial properties of the fluid are of no consequence in the absence of acceleration.) Use arguments of dimensional analysis to find (except for a constant of proportionality) the dependence of the frictional force f on these parameters. Show that the result agrees with Stokes's law in (15·6·2).

15.2 W. Pospišil (*Ann. Physik*, vol. 83, p. 735 (1927)) observed the Brownian motion of soot particles of radius $0.4 \cdot 10^{-4}$ cm. The particles were immersed in a water–glycerin solution having a viscosity of 0.0278 gm cm^{-1} sec^{-1} at the temperature of 18.8°C of the experiment. The observed mean square x component of displacement in a 10-second time interval was $\overline{x^2} = 3.3 \cdot 10^{-8}$ cm^2. Use these data and the known value of the gas constant to calculate Avogadro's number.

15.3 Consider a system of particles (each having a charge e) confined within some finite volume. The particles might, for example, be ions in a gas or ions in a solid like NaCl. The particles are in thermal equilibrium at the temperature T in the presence of an electric field $\mathcal{E}$ in the z direction.

(*a*) Denote the mean number of particles per unit volume at the position z by $n(z)$. Use the results of equilibrium statistical mechanics to relate $n(z + dz)$ to $n(z)$.

(*b*) Suppose that the particles can be characterized by a diffusion coefficient D. By using the definition of this coefficient, find the flux J_D (the number of particles crossing unit area per unit time in the z direction) due to the concentration gradient calculated in (*a*).

(*c*) Suppose that the particles are also characterized by a mobility μ relating their drift velocity to the applied field $\mathcal{E}$. Find the particle flux J_μ resulting from the drift velocity produced by the field $\mathcal{E}$.

(*d*) By making use of the fact that in equilibrium the net particle flux $J_D + J_\mu$ must vanish, find a relation between D and μ. The result thus obtained constitutes a very general derivation of the Einstein relation (15·6·14).

15.4 Consider the Langevin equation

$$\frac{dv}{dt} = -\gamma v + \frac{1}{m} F'(t) \tag{1}$$

where the first term on the right is a phenomenological expression for the slowly varying part of the interaction force, whose rapidly fluctuating part is denoted by $F'(t)$. If F' is neglected, the solution of the resulting equation is $v = u \exp(-\gamma t)$ where u is a constant. In the general case where $F' \neq 0$, assume a solution of the same form with $u = u(t)$ and show that the solution of the Langevin equation gives for the velocity at time t the result

$$v = v_0 e^{-\gamma t} + \frac{1}{m} e^{-\gamma t} \int_0^t e^{\gamma t'} F'(t')\, dt' \tag{2}$$

where $v_o \equiv v(0)$.

15.5 To exploit the fact that the correlation time τ^* of the fluctuating force F' is very short, consider a time τ such that $\tau \gg \tau^*$, but which is macroscopically very short in the sense that $\tau \ll \gamma^{-1}$. The force F' is then not correlated in successive intervals of length τ (all correlations due to the slowly varying interaction force having already been explicitly absorbed in the term $-\gamma v$ of the Langevin equation). By dividing the time interval t into N successive intervals τ so that $t = N\tau$, show that in the preceding problem the solution of the Langevin equation can be written in the form

$$v - v_0 \, e^{-\gamma t} = Y \equiv \sum_{k=0}^{N-1} y_k \tag{1}$$

where $$y_k \equiv e^{-\gamma t} \, e^{\gamma \tau k} G_k = e^{-\gamma \tau (N-k)} G_k \tag{2}$$

and $$G_k \equiv \frac{1}{m} \int_0^\tau F'(k\tau + s) \, ds \tag{3}$$

Since $\tau \gg \tau^*$, the statistical properties of G_k are the same in each interval of length τ. Furthermore, the quantities y_k (or G_k) are statistically independent of each other.

15.6 Show that the results of the preceding problem can also be obtained directly from the Langevin equation (1) of Problem 15.4 by integrating the latter over the small time interval τ so as to relate the velocity v_k at time $k\tau$ to the velocity v_{k-1} at the time $(k-1)\tau$. Successive application of this result then allows one to relate v_N to v_0.

15.7 Use the results of Problem 15.5 to relate $\bar{Y}$ to $\bar{G}$, the ensemble average of G_k which is independent of k. Show that the result $\bar{G} = 0$, expected from the property that $\overline{F'} = 0$, is consistent with the equilibrium value $\bar{v} = 0$ which must be attained when $t \to \infty$.

15.8 Use the results of Problem 15.5 to relate $\overline{Y^2}$ to $\overline{G^2}$ for all times t. Show that $\overline{G^2}$ can then be determined since it is known that, when $t \to \infty$, the velocity must be given by the equilibrium Maxwell distribution, so that $\frac{1}{2}m\overline{v^2} = \frac{1}{2}kT$. Find thus the explicit value of $\overline{G^2}$. Hence find also an explicit expression for $\overline{Y^2}$ valid at all times.

15.9 The central limit theorem in its general form (see Problem 1.27) can be applied to Eq. (1) of Problem 15.5 to find for large N the probability distribution of $Y = \Sigma y_k$ since this quantity is a sum of statistically independent variables. By combining this result with the value of $\overline{Y^2}$ found in Problem 15.8, find the probability $P(v,t|v_0) \, dv$ for the velocity v after any time interval t. Show that the result thus obtained agrees with the solution (15·12·8) of the Fokker-Planck equation.

15.10 Express $\overline{G^2}$ in terms of the correlation function $K(s) = \langle F'(t)F'(t+s) \rangle$ of the random force and use the result of Problem 15.8 to rederive in this way the fluctuation-dissipation theorem (15·8·8) relating the friction constant to $K(s)$.

15.11 Integrate the Langevin equation (1) of Problem 15.4 directly over the small time interval τ to find $\Delta v \equiv v(\tau) - v_0$.

(*a*) Use this result to express $\overline{\Delta v}$ and $\overline{(\Delta v)^2}$ in terms of $\bar{G}$ and $\overline{G^2}$. Show that these moments are proportional to τ and find their explicit values by using the results of Problems 15.7 and 15.8.

(*b*) Express $\overline{(\Delta v)^3}$ and $\overline{(\Delta v)^4}$ in terms of moments of G; show that these quantities are proportional to τ^2.

(*c*) Find an explicit expression for $\overline{(\Delta v)^3}$.

15.12 Use the solution (2) of Problem 15.4 to find the velocity correlation function $\langle v(0) \, v(t) \rangle$. Express the result in terms of T, m, γ, and t.

15.13 Use the solution (2) of Problem 15.4 to calculate directly $\overline{v^2(t)}$ for any time $t \gg \tau^*$ without explicitly breaking up the range of integration of the integral into discrete intervals as was done in Problem 15.5. Use the fact that $\tau^* \ll \gamma^{-1}$ so that the correlation function $\langle F'(0)F'(s)\rangle$ is appreciable only when $\gamma s \ll 1$. Show that the result thus obtained agrees with that derived previously. Show also that this result yields immediately the general fluctuation-dissipation theorem (15·8·8) if one makes use of the requirement that $\frac{1}{2}m\overline{v^2} = \frac{1}{2}kT$ in the final equilibrium situation when $t \to \infty$.

15.14 Consider the Langevin equation (1) of Problem 15.4.

(*a*) Write F' and v in terms of Fourier integrals, and show how their Fourier coefficients must be related to satisfy the Langevin equation. Express the spectral density of the velocity in terms of the spectral density of the force $F'(t)$.

(*b*) Use the Wiener-Khintchine relations to find the spectral density of the velocity v from the known correlation function (15·10·9) of this quantity.

(*c*) By combining the preceding results of this problem, find an explicit expression for the spectral density of the force F' in terms of γ. This constitutes another derivation of Nyquist's theorem.

Appendices

A·1 *Review of elementary sums*

If $f(x)$ denotes a function of a variable x which can assume the discrete values $x_1, x_2, \ldots, x_m$, then the sum

$$f(x_1) + f(x_2) + \cdots + f(x_m) \equiv \sum_{i=1}^{m} f(x_i) \qquad \text{(A·1·1)}$$

is conveniently abbreviated by the compact notation on the right. The distributive property of addition permits one to rearrange the terms of a sum in convenient ways; e.g.,

$$\sum_{i=1}^{m} \sum_{j=1}^{n} x_i y_j = \left(\sum_{i=1}^{m} x_i\right) \left(\sum_{j=1}^{n} y_j\right) \qquad \text{(A·1·2)}$$

Here the right side is obtained by summing first over all values of y for a given value of x, and then summing the resulting products over all values of x.

A frequently occurring sum is that of a "geometric series"

$$S \equiv a + af + af^2 + \cdots + af^n \qquad \text{(A·1·3)}$$

where each term is obtained from the preceding one as a result of multiplication by f. This factor f may be real or complex. To evaluate the sum (A·1·3), multiply both sides by f to obtain

$$fS = af + af^2 + \cdots + af^n + af^{n+1} \qquad \text{(A·1·4)}$$

Subtracting (A·1·4) from (A·1·3) then yields

$$(1 - f)S = a - af^{n+1}$$

so that

$$S = a \frac{1 - f^{n+1}}{1 - f} \qquad \text{(A·1·5)}$$

If $|f| < 1$ and the geometric series (A · 1 · 3) is infinite so that $n \to \infty$, the series converges. Indeed, $f^{n+1} \to 0$ in that case so that

$$\text{for } n \to \infty, \qquad S = \frac{a}{1-f} \tag{A·1·6}$$

A · 2 *Evaluation of the integral* $\int_{-\infty}^{\infty} e^{-x^2}\, dx$

The *indefinite* integral $\int e^{-x^2}\, dx$ cannot be evaluated in terms of elementary functions. Let I denote the desired *definite* integral

$$I \equiv \int_{-\infty}^{\infty} e^{-x^2}\, dx \tag{A·2·1}$$

The following clever artifice exploits the properties of the exponential function to lead to the evaluation of I. One can equally well write (A · 2 · 1) in terms of a different variable of integration

$$I = \int_{-\infty}^{\infty} e^{-y^2}\, dy \tag{A·2·2}$$

Multiplication of (A · 2 · 1) and (A · 2 · 2) then yields

$$I^2 = \int_{-\infty}^{\infty} e^{-x^2}\, dx \int_{-\infty}^{\infty} e^{-y^2}\, dy$$

$$= \int_{-\infty}^{\infty} \int_{-\infty}^{\infty} e^{-x^2} e^{-y^2}\, dx\, dy$$

or

$$I^2 = \int_{-\infty}^{\infty} \int_{-\infty}^{\infty} e^{-(x^2+y^2)}\, dx\, dy \tag{A·2·3}$$

This is an integral extending over the entire xy plane.

Let us express the integration over this plane in terms of polar coordinates r and θ. Then one has simply $x^2 + y^2 = r^2$, and the element of area in these coordinates is given by $(r\, dr\, d\theta)$. In order to cover the entire plane, the varia-

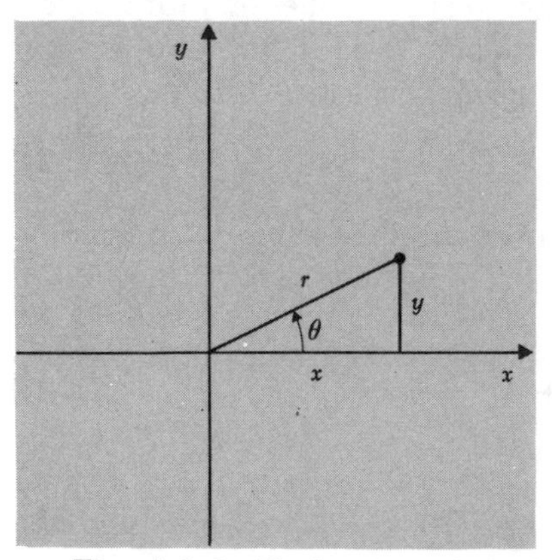

***Fig. A · 2 · 3** Evaluation of the integral (A · 2 · 3) in terms of polar coordinates.*

bles θ and r must range over the values $0 < \theta < 2\pi$ and $0 < r < \infty$. Hence (A·2·3) becomes

$$I^2 = \int_0^\infty \int_0^{2\pi} e^{-r^2}\, r\, dr\, d\theta = 2\pi \int_0^\infty e^{-r^2}\, r\, dr \qquad \text{(A·2·4)}$$

since the integration over θ is immediate. But the factor r in this integrand makes the evaluation of this last integral trivial. Thus

$$I^2 = 2\pi \int_0^\infty (-\tfrac{1}{2})d(e^{-r^2}) = -\pi[e^{-r^2}]_0^\infty = -\pi(0-1) = \pi$$

or

$$I = \sqrt{\pi}$$

Thus

$$\blacktriangleright \qquad \int_{-\infty}^{\infty} e^{-x^2}\, dx = \sqrt{\pi} \qquad \text{(A·2·5)}$$

Note that since e^{-x^2} is an even function (i.e., since it assumes the same value for x and $-x$)

$$\int_{-\infty}^{\infty} e^{-x^2}\, dx = 2\int_0^\infty e^{-x^2}\, dx$$

Hence

$$\blacktriangleright \qquad \int_0^\infty e^{-x^2}\, dx = \tfrac{1}{2}\sqrt{\pi} \qquad \text{(A·2·6)}$$

A·3 *Evaluation of the integral* $\int_0^\infty e^{-x}\, x^n\, dx$

For $n = 0$, the evaluation is trivial.

$$\int_0^\infty e^{-x}\, dx = -[e^{-x}]_0^\infty = -[0-1] = 1 \qquad \text{(A·3·1)}$$

More generally, the integral can be simplified by integration by parts. Thus, for $n > 0$,

$$\begin{aligned}\int_0^\infty e^{-x}\, x^n\, dx &= -\int_0^\infty x^n d(e^{-x}) \\ &= -[x^n e^{-x}]_0^\infty + n\int_0^\infty x^{n-1}\, e^{-x}\, dx\end{aligned}$$

Since the first term on the right vanishes at both limits, one obtains the recurrence relation

$$\int_0^\infty e^{-x}\, x^n\, dx = n\int_0^\infty e^{-x}\, x^{n-1}\, dx \qquad \text{(A·3·2)}$$

If n is a positive integer, one can apply Eq. (A·3·2) repeatedly to get

$$\int_0^\infty e^{-x}\, x^n\, dx = n(n-1)(n-2)\cdots(2)(1)$$

or

$$\blacktriangleright \qquad \int_0^\infty e^{-x}\, x^n\, dx = n! \qquad \text{(A·3·3)}$$

Actually the integral is also well defined if n is nonintegral and $n > -1$. Quite generally the "gamma function" is defined by the relation

$$\Gamma(n) \equiv \int_0^\infty e^{-x}\, x^{n-1}\, dx \tag{A·3·4}$$

By (A·3·3) it thus follows that if n is a positive integer

$$\Gamma(n) = (n-1)! \tag{A·3·5}$$

Equation (A·3·2) implies the general relation

$$\Gamma(n) = (n-1)\,\Gamma(n-1) \tag{A·3·6}$$

By (A·3·1) it follows that

$$\Gamma(1) = 1 \tag{A·3·7}$$

Note also that

$$\Gamma(\tfrac{1}{2}) = \int_0^\infty e^{-x}\, x^{-\frac{1}{2}}\, dx = 2\int_0^\infty e^{-y^2}\, dy$$

where we have put $x = y^2$. Thus (A·2·6) yields

$$\Gamma(\tfrac{1}{2}) = \sqrt{\pi} \tag{A·3·8}$$

A · 4 *Evaluation of integrals of the form* $\int_0^\infty e^{-\alpha x^2}\, x^n\, dx$

Let

$$I(n) \equiv \int_0^\infty e^{-\alpha x^2}\, x^n\, dx, \qquad \text{where } n \geq 0 \tag{A·4·1}$$

Putting $x \equiv \alpha^{-\frac{1}{2}} y$, we note that for $n = 0$

$$I(0) = \alpha^{-\frac{1}{2}} \int_0^\infty e^{-y^2}\, dy = \frac{\sqrt{\pi}}{2}\, \alpha^{-\frac{1}{2}} \tag{A·4·2}$$

where we have used (A·2·6). Also for $n = 1$

$$I(1) = \alpha^{-1}\int_0^\infty e^{-y^2}\, y\, dy = \alpha^{-1}[-\tfrac{1}{2}\, e^{-y^2}]_0^\infty = \tfrac{1}{2}\alpha^{-1} \tag{A·4·3}$$

All integrals $I(n)$ with integral values $n > 1$ can then be reduced to the integrals $I(0)$ or $I(1)$ by differentiation with respect to α considered as a parameter. Indeed, one can write (A·4·1) in the form

$$I(n) = -\frac{\partial}{\partial \alpha}\left(\int_0^\infty e^{-\alpha x^2}\, x^{n-2}\, dx\right) = -\frac{\partial I(n-2)}{\partial \alpha} \tag{A·4·4}$$

which gives a recurrence relation which can be applied as often as necessary. For example,

$$I(2) = -\frac{\partial I(0)}{\partial \alpha} = -\frac{\sqrt{\pi}}{2}\frac{\partial}{\partial \alpha}(\alpha^{-\frac{1}{2}}) = \frac{\sqrt{\pi}}{4}\, \alpha^{-\frac{3}{2}}$$

Alternatively one can put $x = (u/\alpha)^{\frac{1}{2}}$ in (A·4·1). Then $dx = \frac{1}{2}\alpha^{-\frac{1}{2}}u^{-\frac{1}{2}}\,du$, and the integral assumes the form

$$I(n) = \tfrac{1}{2}\alpha^{-\frac{1}{2}(n+1)} \int_0^\infty e^{-u}\, u^{\frac{1}{2}(n+1)}\, du$$

By virtue of the definition (A·3·4) of the Γ function, this can then be written

$$I(n) \equiv \int_0^\infty e^{-\alpha x^2} x^n\, dx = \frac{1}{2}\Gamma\left(\frac{n+1}{2}\right)\alpha^{-(n+1)/2} \tag{A·4·5}$$

Using the property (A·3·6) of the Γ function and the values Γ(0) and $\Gamma(\frac{1}{2})$ given in (A·3·7) and (A·3·8), one can apply (A·4·5) to construct a small table of integrals. Thus one obtains explicitly

$$\begin{aligned}
I(0) &= \tfrac{1}{2}\sqrt{\pi}\,\alpha^{-\frac{1}{2}} \\
I(1) &= \tfrac{1}{2}\alpha^{-1} \\
I(2) &= \tfrac{1}{2}(\tfrac{1}{2}\sqrt{\pi})\alpha^{-\frac{3}{2}} = \tfrac{1}{4}\sqrt{\pi}\,\alpha^{-\frac{3}{2}} \\
I(3) &= \tfrac{1}{2}(1)\alpha^{-2} = \tfrac{1}{2}\alpha^{-2} \\
I(4) &= \tfrac{1}{2}(\tfrac{3}{2}\times\tfrac{1}{2}\sqrt{\pi})\alpha^{-\frac{5}{2}} = \tfrac{3}{8}\sqrt{\pi}\,\alpha^{-\frac{5}{2}} \\
I(5) &= \tfrac{1}{2}(2\times 1)\alpha^{-3} = \alpha^{-3}
\end{aligned} \tag{A·4·6}$$

A·5 *The error function*

The integral $\int e^{-x^2}\,dx$ cannot be evaluated in terms of elementary functions, although we have seen in Appendix A·2 that the *definite* integral between 0 and ∞ has the simple value

$$\int_0^\infty e^{-x^2}\, dx = \frac{\sqrt{\pi}}{2} \tag{A·5·1}$$

Since the indefinite integral occurs frequently, it is useful to *define* the following function of y by the relation

$$\operatorname{erf} y \equiv \frac{2}{\sqrt{\pi}} \int_0^y e^{-x^2}\, dx \tag{A·5·2}$$

This is called the "error function." The integral can be evaluated numerically for various values of y and is tabulated in many books.* By its definition (A·5·2), erf y is a monotonically increasing function of y since the integrand in (A·5·2) is positive. Obviously erf $0 = 0$. The factor in front of the integral (A·5·2) has been chosen so that, by virtue of (A·5·1), erf $y \to 1$ as $y \to \infty$. The behavior of erf y is illustrated in Fig. A·5·1.

* See, for example, B. O. Peirce and R. M. Foster, "A Short Table of Integrals," 4th ed., Ginn and Company, Boston, 1956.

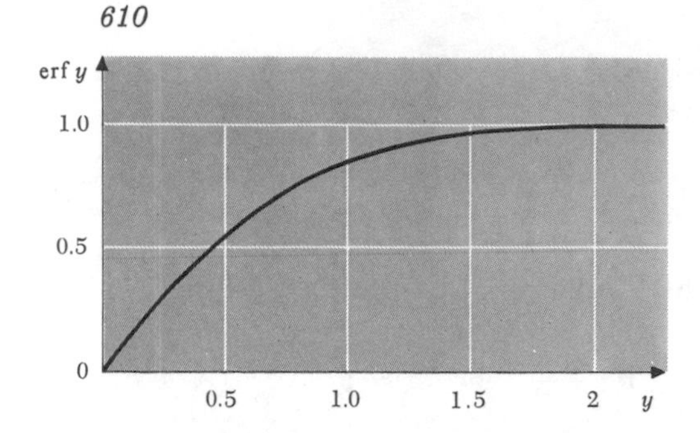

Fig. A·5·1 The error function erf y.

Series expansions The integrand in (A·5·2) can be expanded in a power series valid throughout the range of integration. Thus

$$\text{erf } y = \frac{2}{\sqrt{\pi}} \int_0^y \left(1 - x^2 + \frac{1}{2}x^4 - \cdots\right) dx$$

or

$$\text{erf } y = \frac{2}{\sqrt{\pi}} \left(y - \frac{1}{3}y^3 + \frac{1}{10}y^5 - \cdots\right) \tag{A·5·3}$$

This expression is useful when y is sufficiently small that the series converges rapidly. In the opposite limit, where $y \gg 1$, it is more convenient to write (A·5·2) in the form

$$\text{erf } y = \frac{2}{\sqrt{\pi}} \left(\int_0^\infty e^{-x^2}\,dx - \int_y^\infty e^{-x^2}\,dx\right) = 1 - \frac{2}{\sqrt{\pi}} \int_y^\infty e^{-x^2}\,dx$$

where the last integral is small compared to 1. One can express it as an asymptotic series in y^{-1} by successive integrations by parts.

$$\begin{aligned}\int_y^\infty e^{-x^2}\,dx &= -\frac{1}{2}\int_y^\infty \frac{1}{x}\,d(e^{-x^2}) = -\frac{1}{2}\left[\frac{e^{-x^2}}{x}\right]_y^\infty - \frac{1}{2}\int_y^\infty \frac{1}{x^2}e^{-x^2}\,dx \\ &= \frac{1}{2}\frac{e^{-y^2}}{y} + \frac{1}{4}\int_y^\infty \frac{1}{x^3}\,d(e^{-x^2}) \\ &= \frac{1}{2}\frac{e^{-y^2}}{y} - \frac{1}{4}\frac{e^{-y^2}}{y^3} - \cdots\end{aligned}$$

Thus,

for $y \gg 1$,

$$\text{erf } y = 1 - \frac{e^{-y^2}}{\sqrt{\pi}\,y}\left(1 - \frac{1}{2y^2}\cdots\right) \tag{A·5·4}$$

A · 6 *Stirling's formula*

The calculation of $n!$ becomes very laborious for large values of n. We should like to find a simple approximate formula by which n can be calculated in the limit when n is large.

It is very simple to derive an approximation which is good when n is *very*

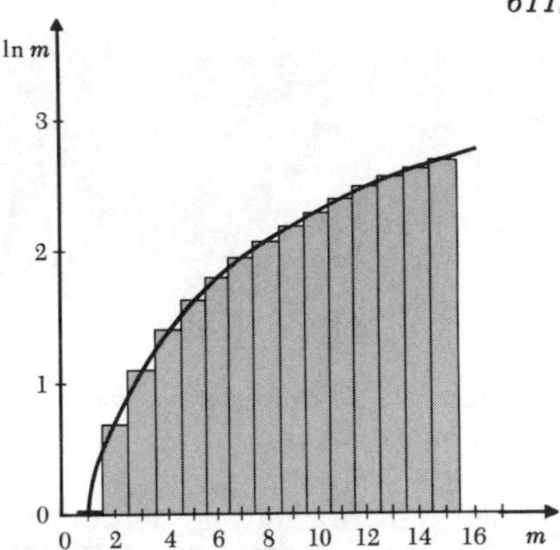

Fig. A·6·1 Behavior of ln n as a function of n.

large. By its definition

$$n! \equiv 1 \times 2 \times 3 \times \cdots \times (n-1) \times n$$

Thus
$$\ln n! = \ln 1 + \ln 2 + \cdots + \ln n = \sum_{m=1}^{n} \ln m \tag{A·6·1}$$

One can approximate this sum (given by the area under the steps of Fig. A·6·1) by an integral (the area under the continuous curve of Fig. A·6·1). This replacement by an integral is increasingly good in the range where m is large, since $\ln m$ then varies only slightly when m is increased by unity. With this approximation (A·6·1) becomes

$$\ln n! \approx \int_1^n \ln x \, dx = [x \ln x - x]_1^n$$

or

$$\ln n! \approx n \ln n - n \tag{A·6·2}$$

since the lower limit is negligible when $n \gg 1$.

In most applications of statistical mechanics the numbers of interest are so large that (A·6·2) is an adequate approximation. It is, however, readily possible to obtain a much better approximation by a method which is of very general utility. As a starting point one needs a convenient analytic expression for $n!$. The integral formula of (A·3·3) provides one such convenient expression, namely,

$$n! = \int_0^\infty x^n e^{-x} \, dx \tag{A·6·3}$$

Consider the integrand $F \equiv x^n e^{-x}$ when n is large. Then x^n is a rapidly increasing function of x, while e^{-x} is a rapidly decreasing function of x. Hence the product $F \equiv x^n e^{-x}$ is a function which exhibits a sharp maximum for some value $x = x_0$ and falls off rapidly for x appreciably removed from x_0. Let us locate the position x_0 of this maximum of the integrand F. It is equivalent and more convenient to work with the logarithm of F. (Since $\ln F$ is a mono-

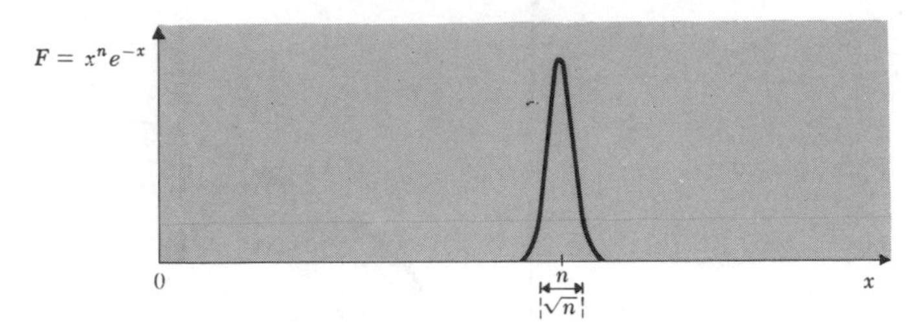

Fig. A·6·2 Behavior of the integrand $F(x) = x^n e^{-x}$ as a function of x for large values of n.

tonically increasing function of F, a maximum in $\ln F$ corresponds, of course, to a maximum of F.) To find this maximum, put

$$\frac{d \ln F}{dx} = 0$$

or

$$\frac{d}{dx}(n \ln x - x) = \frac{n}{x} - 1 = 0$$

Hence

$$x_0 = n \tag{A·6·4}$$

But, to the extent that the maximum of the integrand F is very sharp, only values of x in the vicinity of $x_0 = n$ contribute appreciably to the integral (A·6·3). Hence a knowledge of the integrand F in the vicinity of n is adequate for an evaluation of the integral. But convenient power-series expansions exist which represent F adequately in this region. Hence one is led to a convenient approximate evaluation of the integral.

To find an expression for F valid near $x = n$, write

$$x \equiv n + \xi, \qquad \text{where } \xi \ll n \tag{A·6·5}$$

and expand $\ln F$ in a Taylor's series in ξ about the point $x = n$.

> ***Remark*** We expand $\ln F$ rather than F directly because F has a sharp maximum and is therefore a very rapidly varying function of x; hence it is difficult to expand it in a power series valid over an appreciable range. On the other hand, for $F \gg 1$, $\ln F$ is a much more slowly varying function of x than F itself and can therefore readily be expanded. This comment is similar to that discussed at the beginning of Sec. 1·5.

Then

$$\ln F = n \ln x - x = n \ln (n + \xi) - (n + \xi) \tag{A·6·6}$$

But, expanding the logarithm in Taylor's series,

$$\ln (n + \xi) = \ln n + \ln \left(1 + \frac{\xi}{n}\right) = \ln n + \frac{\xi}{n} - \frac{1}{2}\frac{\xi^2}{n^2} + \cdots \tag{A·6·7}$$

Substituting this in (A·6·6), the first-order term linear in ξ vanishes, of course, since one is expanding ln F about its maximum. Thus

$$\ln F = n \ln n - n - \frac{1}{2}\frac{\xi^2}{n}$$

or

$$F = n^n e^{-n} e^{-\frac{1}{2}(\xi^2/n)} \tag{A·6·8}$$

The last exponential term shows explicitly that F has a maximum for $\xi = 0$ and that it becomes quite small when $|\xi| \gg \sqrt{n}$. If n is large, $\sqrt{n} \ll n$ and the maximum is very sharp (see Fig. A·6·2). By (A·6·8) the integral (A·6·3) then becomes

$$n! = \int_{-n}^{\infty} n^n e^{-n} e^{-\frac{1}{2}(\xi^2/n)} \, d\xi = n^n e^{-n} \int_{-\infty}^{\infty} e^{-\frac{1}{2}(\xi^2/n)} \, d\xi \tag{A·6·9}$$

In the last integral we have replaced the lower limit $-n$ by $-\infty$, since for values $\xi < -n$ the integrand is already negligibly small. By (A·4·6) this last integral equals $\sqrt{2\pi n}$. Thus (A·6·9) yields the result

$$n! = \sqrt{2\pi n}\, n^n e^{-n} \qquad \text{for } n \gg 1 \tag{A·6·10}$$

This is known as Stirling's formula. It can also be written in the form

$$\ln n! = n \ln n - n + \tfrac{1}{2} \ln (2\pi n) \tag{A·6·11}$$

If n is *very* large, $\ln n \ll n$. (For example, if $n = 6 \times 10^{23}$, Avogadro's number, then $\ln n = 55$.) In that case (A·6·11) reduces properly to the simple result (A·6·2).

Accuracy of Stirling's formula To investigate this question, one needs only to go systematically to the next higher approximation in expanding ln F. Then (A·6·7) gives

$$\ln (n + \xi) = \ln n + \frac{\xi}{n} - \frac{1}{2}\left(\frac{\xi}{n}\right)^2 + \frac{1}{3}\left(\frac{\xi}{n}\right)^3 - \frac{1}{4}\left(\frac{\xi}{n}\right)^4 + \cdots$$

so that

$$\ln F = n \ln n - n - \frac{1}{2}\frac{\xi^2}{n} + \frac{1}{3}\frac{\xi^3}{n^2} - \frac{1}{4}\frac{\xi^4}{n^3}$$

The integral (A·6·9) becomes then

$$n! = n^n e^{-n} \int_{-\infty}^{\infty} \exp\left(-\frac{1}{2}\frac{\xi^2}{n}\right) \exp\left(\frac{1}{3}\frac{\xi^3}{n^2} - \frac{1}{4}\frac{\xi^4}{n^3}\right) d\xi \tag{A·6·12}$$

Here one can make one further approximation to evaluate the integral. The factor $\exp(-\frac{1}{2}\xi^2/n)$ is the predominant one which makes the integrand negligibly small when $|\xi| > n^{\frac{1}{2}}$. Hence a knowledge of the second factor in the integrand is required only in the significant range where $\xi \lesssim n^{\frac{1}{2}}$. There this factor can be expanded in a Taylor's series, since

$$\frac{\xi^3}{n^2} \lesssim \frac{n^{\frac{3}{2}}}{n^2} = n^{-\frac{1}{2}} \qquad \text{and} \qquad \frac{\xi^4}{n^3} \lesssim \frac{n^2}{n^3} = n^{-1}$$

i.e., all these terms are much less than unity when n is large. Hence (A · 6 · 12) can be approximated by

$$n! = n^n e^{-n} \int_{-\infty}^{\infty} e^{-\frac{1}{2}(\xi^2/n)} \left[1 + \left(\frac{1}{3}\frac{\xi^3}{n^2} - \frac{1}{4}\frac{\xi^4}{n^3}\right) + \left(\frac{1}{18}\frac{\xi^6}{n^4} + \cdots\right)\right] d\xi \quad \text{(A·6·13)}$$

where we have used the expansion

$$e^y = 1 + y + \tfrac{1}{2}y^2 + \cdots$$

and have been careful to retain all terms of order $n^{-\frac{1}{2}}$ and n^{-1}. (Note that $\xi^6/n^4 \lesssim n^3/n^4 \approx n^{-1}$ is still of order n^{-1}.) Here the second integral involving ξ^3 vanishes by symmetry, since the integrand is an odd function of ξ. The remaining three integrals can be evaluated by the results of Appendix A · 4. Hence

$$n! = n^n e^{-n} \left\{\sqrt{2\pi n} + 0 - \frac{1}{4n^3}\left[\frac{3}{4}\sqrt{\pi}\,(2n)^{\frac{5}{2}}\right] + \frac{1}{18n^4}\left[\frac{15}{8}\sqrt{\pi}\,(2n)^{\frac{7}{2}}\right]\right\}$$

$$= \sqrt{2\pi n}\, n^n e^{-n}\left[1 - \frac{3}{4n} + \frac{5}{6n}\right]$$

Thus

▶ $$n! = \sqrt{2\pi n}\, n^n e^{-n}\left[1 + \frac{1}{12n} + \cdots\right] \quad \text{(A·6·14)}$$

This shows the next correction term for Stirling's formula. Thus, even when n is as small as 10, Stirling's formula is already accurate to better than 1 percent.*

A · 7 *The Dirac delta function*

The Dirac δ function is a very convenient "function" (or more exactly, the limiting case of a function) having the property of singling out a particular value $x = x_0$ of a variable x. The function is characterized by the following properties:

$$\left.\begin{array}{lll} & \delta(x - x_0) = 0 & \text{for } x \neq x_0 \\ \text{but} & \delta(x - x_0) \to \infty & \text{for } x \to x_0 \\ \text{in such a way that,} & & \\ \text{for any } \epsilon > 0, & \displaystyle\int_{x_0-\epsilon}^{x_0+\epsilon} \delta(x - x_0)\, dx = 1 & \end{array}\right\} \quad \text{(A·7·1)}$$

That is, the function $\delta(x - x_0)$ has a very sharp peak at $x = x_0$, but the area under the peak is unity. It follows that, given any smooth function $f(x)$, one

* More rigorous estimates of the maximum error committed in using Stirling's formula can be found in R. Courant, "Differential and Integral Calculus," p. 361, Interscience Publishers, New York, 1938; also in R. Courant and D. Hilbert, "Methods of Mathematical Physics, vol. I, p. 522, Interscience Publishers, New York, 1953.

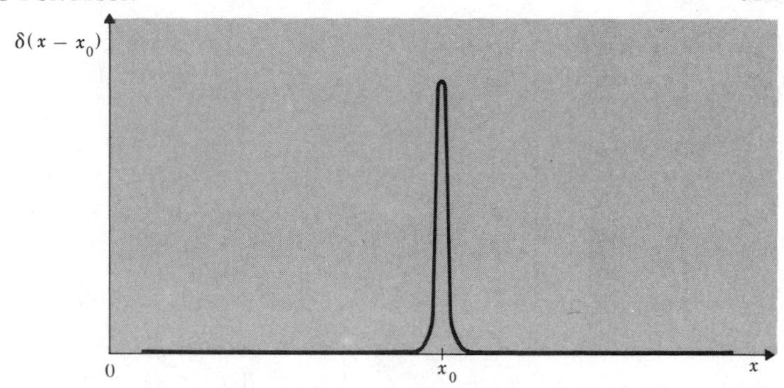

Fig. A·7·1 Schematic plot of $\delta(x - x_0)$ as a function of x.

has

$$\int_A^B f(x)\delta(x - x_0)\, dx = f(x_0) \int_A^B \delta(x - x_0)\, dx$$

since $\delta(x - x_0) \neq 0$ only when $x = x_0$, and there $f(x) = f(x_0)$. Hence

$$\int_A^B f(x)\delta(x - x_0)\, dx = \begin{cases} f(x_0) & \text{if } A < x_0 < B \\ 0 & \text{otherwise} \end{cases} \qquad (\text{A}\cdot 7\cdot 2)$$

The property (A·7·2) implies all the characteristics of (A·7·1) and can be taken as the definition of the function $\delta(x - x_0)$.

The δ function is a mathematical representation of a very common physical approximation, the "physical point." (An example is the electron considered as a point charge.) It corresponds to a finite physical quantity (e.g., an electrical charge) concentrated in a region much smaller than all other dimensions relevant in a physical discussion. Subtle questions concerned with limiting processes involving mathematical points are therefore usually irrelevant in discussions of physical problems.*

The following are examples of various analytical representations of the δ function. In all of these the positive parameter γ is taken in the limit $\gamma \to 0$. (This is a physical limit where γ is smaller than all other relevant dimensions.)

Example 1:
$$\delta(x) = \begin{cases} \dfrac{1}{\gamma} & \text{for } -\dfrac{\gamma}{2} < x < \dfrac{\gamma}{2} \\ 0 & \text{otherwise} \end{cases} \qquad (\text{A}\cdot 7\cdot 3)$$

Example 2:
$$\delta(x) = \frac{1}{\pi} \frac{\gamma}{x^2 + \gamma^2} \qquad (\text{A}\cdot 7\cdot 4)$$

Example 3:
$$\delta(x) = \frac{1}{\sqrt{2\pi}\,\gamma} e^{-x^2/2\gamma^2} \qquad (\text{A}\cdot 7\cdot 5)$$

The most convenient and important representation is, however, one involving an integral.

* The reader interested primarily in questions of mathematical rigor is referred to M. J. Lighthill, "Introduction to Fourier Analysis and Generalized Functions," Cambridge University Press, Cambridge, 1958.

Integral representation of the δ function The periodic character of the complex exponential function yields the familiar result

$$\int_{-\pi}^{\pi} e^{in\phi}\,d\phi = \begin{cases} 2\pi & \text{for } n = 0 \\ \dfrac{e^{in\pi} - e^{-in\pi}}{in} = \dfrac{[(\pm 1) - (\pm 1)]}{in} = 0 & \text{for } n \neq 0 \end{cases}$$

i.e.,

$$\frac{1}{2\pi}\int_{-\pi}^{\pi} e^{in\phi}\,d\phi = \delta_{n,0} \tag{A·7·6}$$

where the right-hand side is a shorthand notation defined by

$$\delta_{n,m} = \begin{cases} 1 & \text{if } n = m \\ 0 & \text{if } n \neq m \end{cases} \tag{A·7·7}$$

This useful symbol is called the "Kronecker delta symbol." It is obviously the analog, for discrete variables, of the definition of the Dirac δ function $\delta(x - x_0)$ for continuous variables.

To make the connection between the discrete and continuous cases explicit, choose a very large number L so that

$$x \equiv \frac{2\pi n}{L} \tag{A·7·8}$$

covers essentially all possible values of the continuous variable x as n assumes all possible integral values.* The relation (A·7·8) associates with each integer n the range of x lying between

$$\frac{2\pi}{L}\left(n - \frac{1}{2}\right) < x < \frac{2\pi}{L}\left(n + \frac{1}{2}\right)$$

i.e., a range of magnitude

$$\Delta x = \frac{2\pi}{L} \tag{A·7·9}$$

which becomes infinitesimally small as $L \to \infty$.

* The factor 2π in (A·7·8) is introduced purely for convenience so that trigonometric functions such as $\cos Nx$ (where N is any integer) remain unchanged when x changes by L.

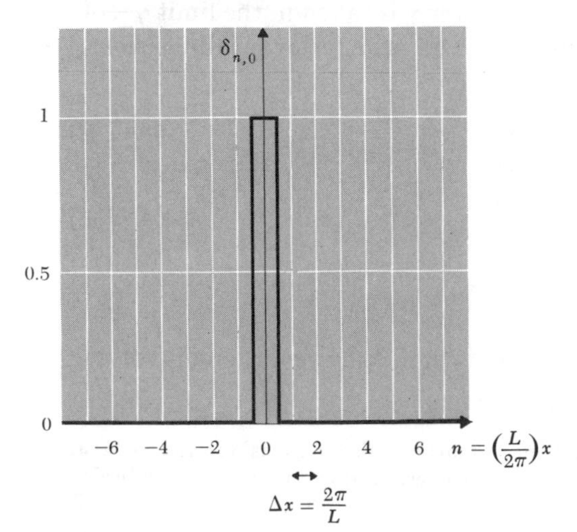

Fig. A·7·2 The Kronecker delta symbol $\delta_{n,0}$ as a function of the continuous variable $x = 2\pi n/L$.

According to the definition (A·7·7) it then follows that

$$\delta_{n,0} = \begin{cases} 1 & \text{when } -\frac{1}{2}\,\Delta x < x < \frac{1}{2}\,\Delta x \\ 0 & \text{otherwise} \end{cases}$$

By (A·7·3) one can then write for the δ function the expression

$$\delta(x) = \lim_{\Delta x \to 0} \frac{\delta_{n,0}}{\Delta x} \tag{A·7·10}$$

Hence $\delta(x) = \lim_{\Delta x \to 0} \dfrac{1}{2\pi\,\Delta x} \displaystyle\int_{-\pi}^{\pi} e^{in\phi}\,d\phi$ using (A·7·6)

$$= \lim_{L \to \infty} \frac{1}{2\pi}\left(\frac{L}{2\pi}\right) \int_{-\pi}^{\pi} e^{(iLx/2\pi)\phi}\,d\phi \qquad \text{using (A·7·8) and (A·7·9)}$$

Let
$$k \equiv \frac{L}{2\pi}\,\phi$$

so that
$$d\phi = \frac{2\pi}{L}\,dk$$

Then

$$\delta(x) = \frac{1}{2\pi} \int_{-\infty}^{\infty} e^{ikx}\,dk \tag{A·7·11}$$

This is the desired integral representation of the δ function, a very useful result.

Remark Let us verify directly that the representation (A·7·11) has the desired properties of the δ function. We see that for $x \neq 0$, the integrand in (A·7·11) is a rapidly oscillating function so that the integral vanishes; also for $x = 0$, $e^{ikx} = 1$, so that the integral approaches infinity. In more detail, introduce in the integrand the factor $e^{-\gamma|k|}$ (where γ is infinitesimally small and positive) to avoid convergence ambiguities for $|k| \to \infty$. Then

$$\begin{aligned}
\delta(x) &= \frac{1}{2\pi} \int_{-\infty}^{\infty} e^{ikx-\gamma|k|}\,dk \\
&= \frac{1}{2\pi} \int_{0}^{\infty} e^{ixk-\gamma k}\,dk + \frac{1}{2\pi} \int_{-\infty}^{0} e^{ixk+\gamma k}\,dk \\
&= \frac{1}{2\pi} \int_{0}^{\infty} e^{(ix-\gamma)k}\,dk + \frac{1}{2\pi} \int_{-\infty}^{0} e^{(ix+\gamma)k}\,dk \\
&= \frac{1}{2\pi} \left\{ \frac{[e^{(ix-\gamma)k}]_0^{\infty}}{ix-\gamma} + \frac{[e^{(ix+\gamma)k}]_{-\infty}^{0}}{ix+\gamma} \right\} \\
&= \frac{1}{2\pi} \left\{ \frac{-1}{ix-\gamma} + \frac{1}{ix+\gamma} \right\}
\end{aligned}$$

or
$$\delta(x) = \frac{1}{\pi}\frac{\gamma}{x^2+\gamma^2}, \qquad \text{where } \gamma \to 0 \tag{A·7·12}$$

This is the same as the representation (A·7·4) mentioned previously. Note that, for $x \neq 0$, $\delta(x) = \gamma/\pi x^2 \to 0$; for $x = 0$, $\delta(x) = (\pi\gamma)^{-1} \to \infty$. Also the integral of (A·7·12) is

$$\int_{-\infty}^{\infty} \frac{1}{\pi}\frac{\gamma}{x^2+\gamma^2}\,dx = \frac{1}{\pi}\left[\tan^{-1}\frac{x}{\gamma}\right]_{-\infty}^{\infty} = \frac{1}{\pi}\left[\frac{\pi}{2} - \left(-\frac{\pi}{2}\right)\right] = 1$$

as required.

Thus (A·7·6) and (A·7·11) yield the very useful results

$$\delta_{n,m} = \delta_{m,n} = \frac{1}{2\pi} \int_{-\pi}^{\pi} e^{i\phi(n-m)}\, d\phi \tag{A·7·13}$$

$$\delta(x - x_0) = \delta(x_0 - x) = \frac{1}{2\pi} \int_{-\infty}^{\infty} e^{ik(x-x_0)}\, dk \tag{A·7·14}$$

Remark Since (A·7·13) vanishes when $n \neq m$, its right side can, without affecting the equality, be multiplied by any function which reduces to unity when $n = m$. In particular, one can write the more general result

$$\delta_{n,m} = \frac{1}{2\pi} \int_{-\pi}^{\pi} e^{i\phi(n-m)}\, d\phi \cdot e^{\phi_0(n-m)}$$

or

$$\delta_{n,m} = \frac{1}{2\pi} \int_{-\pi}^{\pi} e^{(\phi_0+i\phi)(n-m)}\, d\phi \tag{A·7·15}$$

where ϕ_0 is *any* arbitrary parameter. Similarly, one can write

$$\delta(x - x_0) = \frac{1}{2\pi} \int_{-\infty}^{\infty} e^{(k_0+ik)(x-x_0)}\, dk \tag{A·7·16}$$

where k_0 is *any* arbitrary parameter.

A · 8 *The inequality ln $x \leq x - 1$*

We wish to compare $\ln x$ with x itself for positive values of x. Consider the difference function

$$f(x) \equiv x - \ln x \tag{A·8·1}$$

$$\left.\begin{array}{lll} \text{For } x \to 0, & \ln x \to -\infty; & \text{hence } f(x) \to \infty \\ \text{For } x \to \infty, & \ln x \ll x; & \text{hence } f(x) \to \infty \end{array}\right\} \tag{A·8·2}$$

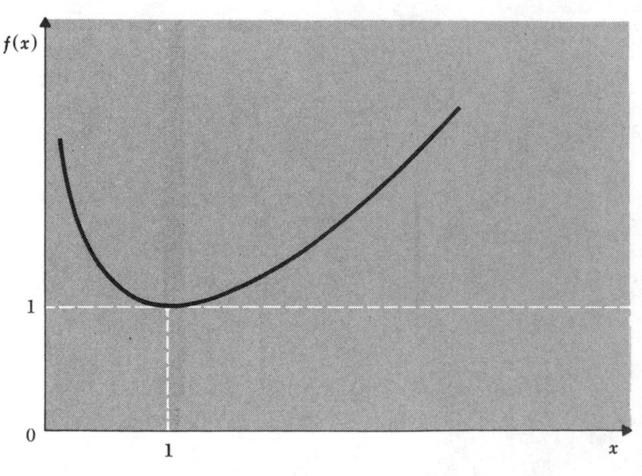

Fig. A·8·1 *The function $f(x) \equiv x - \ln x$ as a function of x.*

To investigate the behavior of $f(x)$ between these limits, we note that

$$\frac{df}{dx} = 1 - \frac{1}{x} = 0 \qquad \text{for } x = 1 \tag{A·8·3}$$

when $f = 1$. Since $f(x)$ is a continuous function of x satisfying (A·8·2) and having a single extremum given by (A·8·3), it follows that $f(x)$ must have the appearance shown in Fig. A·8·1 with a minimum at $x = 1$. Hence

$$f(x) \geq f(1) = 1 \qquad (= \text{sign if } x = 1) \tag{A·8·4}$$

or, by (A·8·1),

$$\ln x \leq x - 1 \qquad (= \text{sign if } x = 1) \tag{A·8·5}$$

A·9 *Relations between partial derivatives of several variables*

Consider three variables x, y, z, two of which are independent. Then there exists a functional relationship of the form

$$z = z(x,y) \tag{A·9·1}$$

if one considers x and y as the independent variables. Given infinitesimal changes in x and y, the corresponding change in z can then be written

$$dz = \left(\frac{\partial z}{\partial x}\right)_y dx + \left(\frac{\partial z}{\partial y}\right)_x dy \tag{A·9·2}$$

where the subscripts indicate explicitly the variables held constant in taking the partial derivatives.

Alternatively, one might wish to consider y and z as the independent variables and to use (A·9·1) to express x in terms of these; then

$$x = x(y,z) \tag{A·9·3}$$

Analogously to (A·9·2), infinitesimal changes in the variables would then be related by

$$dx = \left(\frac{\partial x}{\partial y}\right)_z dy + \left(\frac{\partial x}{\partial z}\right)_y dz \tag{A·9·4}$$

We should like to express the partial derivatives occurring in (A·9·4) in terms of the partial derivatives of (A·9·2) which involve x and y as independent variables.

Evaluation of $(\partial x/\partial y)_z$ Here one is asked to keep z constant and find the ratio of the increment dx associated with an increment dy. But if z is kept constant in (A·9·2), one has $dz = 0$, so that

$$0 = \left(\frac{\partial z}{\partial x}\right)_y dx + \left(\frac{\partial z}{\partial y}\right)_x dy$$

Thus

$$\frac{dx}{dy} = -\frac{(\partial z/\partial y)_x}{(\partial z/\partial x)_y}$$

Since z was kept constant, this yields the result

$$\left(\frac{\partial x}{\partial y}\right)_z = -\frac{(\partial z/\partial y)_x}{(\partial z/\partial x)_y} \tag{A·9·5}$$

Evaluation of $(\partial x/\partial z)_y$ Here one is asked to keep y constant and find the ratio of the increment dx associated with an increment dz. But if y is kept constant in (A·9·2), one has $dy = 0$ so that

$$dz = \left(\frac{\partial z}{\partial x}\right)_y dx$$

Thus

$$\frac{dx}{dz} = \frac{1}{(\partial z/\partial x)_y}$$

or

$$\left(\frac{\partial x}{\partial z}\right)_y = \frac{1}{(\partial z/\partial x)_y} \tag{A·9·6}$$

A · 10 *The method of Lagrange multipliers*

Suppose that it is desired to find the extremum (maximum or minimum) of the function

$$f(x_1, x_2, \ldots, x_n) \tag{A·10·1}$$

where the n variables $x_1, x_2, \ldots, x_n$ satisfy the equation of constraint

$$g(x_1, x_2, \ldots, x_n) = 0 \tag{A·10·2}$$

If f is to have an extremum for a given set of values $\{x_1^{(0)}, \ldots, x_n^{(0)}\}$, then there can be no change in f for *any* infinitesimal departure of the variables from this set; i.e.,

$$df = \frac{\partial f}{\partial x_1} dx_1 + \frac{\partial f}{\partial x_2} dx_2 + \cdots + \frac{\partial f}{\partial x_n} dx_n = 0 \tag{A·10·3}$$

Here the derivatives are to be evaluated at the (n-dimensional) "point" $\{x_1^{(0)}, \ldots, x_n^{(0)}\}$.

Furthermore, since the relation (A·10·2) is always to be satisfied, one must require that for any small departure from the extremum point

$$dg = \frac{\partial g}{\partial x_1} dx_1 + \frac{\partial g}{\partial x_2} dx_2 + \cdots + \frac{\partial g}{\partial x_n} dx_n = 0 \tag{A·10·4}$$

where the derivatives are again to be evaluated at the extremum point $\{x_1^{(0)}, \ldots, x_n^{(0)}\}$.

Now, *if* all the variables $x_1, x_2, \ldots, x_n$ *were* completely independent, then one could in (A·10·3) choose each dx to be zero except a particular one, say

dx_k. One could then immediately conclude from (A·10·3) that $(\partial f/\partial x_k) = 0$ for all k.

But all the variables $x_1, x_2, \ldots, x_n$ are *not* independent, since they are interrelated by the condition (A·10·2). Since one equation connects the variables, one of the variables, say, x_n, can by (A·10·2) be expressed in terms of the other $(n - 1)$ variables; these may then be chosen quite independently without further restrictions. In other words, one must solve (A·10·3) subject to the relation (A·10·4). The straightforward (and difficult) way of doing this is to use (A·10·4) to solve dx_n in terms of the differentials $dx_1, \ldots, dx_{n-1}$. Substituting this value in (A·10·3) would then give a linear relation involving only the $(n - 1)$ differentials $dx_1, \ldots, dx_{n-1}$ which are now completely independent. Putting all these (except dx_k) equal to zero, would immediately give an equation to be satisfied as a result of the extremum condition (A·10·3). Since $k = 1, 2, \ldots, n - 1$, there would be a total of $(n - 1)$ such equations.

The above method is perfectly correct and feasible, but it is complicated because it spoils the symmetry of the problem. A much easier procedure of accomplishing the same task is due to Lagrange. In this method one introduces a parameter λ, to be determined later, by which one multiplies the restrictive condition (A·10·4). One then adds the result to (A·10·3) to obtain

$$\left(\frac{\partial f}{\partial x_1} + \lambda \frac{\partial g}{\partial x_1}\right) dx_1 + \left(\frac{\partial f}{\partial x_2} + \lambda \frac{\partial g}{\partial x_2}\right) dx_2 + \cdots + \left(\frac{\partial f}{\partial x_n} + \lambda \frac{\partial g}{\partial x_n}\right) dx_n = 0 \qquad \text{(A·10·5)}$$

Here, of course, only $(n - 1)$ of the differentials dx_i are independent, e.g., $dx_1, \ldots, dx_{n-1}$. But the value of the parameter λ is still at one's disposal. Choose it so as to eliminate the coefficient of dx_n, e.g., choose λ so that

$$\frac{\partial f}{\partial x_n} + \lambda \frac{\partial g}{\partial x_n} = 0$$

(Note that λ is a constant characteristic of the particular extremum point $\{x_1^{(0)}, \ldots, x_n^{(0)}\}$, since the derivatives in (A·10·3) are evaluated at this point.) But with this one term in (A·10·5) eliminated, all the remaining differentials $dx_1, \ldots, dx_{n-1}$ are independent. Since any one of these can be set equal to zero, one can immediately conclude that

$$\frac{\partial f}{\partial x_k} + \lambda \frac{\partial g}{\partial x_k} = 0 \qquad \text{for } k = 1, \ldots, n - 1$$

The net result of this argument is that one can write

$$\frac{\partial f}{\partial x_k} + \lambda \frac{\partial g}{\partial x_k} = 0 \qquad \text{for } all\ k = 1, \ldots, n \qquad \text{(A·10·6)}$$

This means that, after the "Lagrange multiplier" λ has been introduced, the expression (A·10·5) can be treated *as if* all differentials dx_k were mutually independent. The awkward constraining condition (A·10·4) has thus been handled very elegantly.

Of course, the constraint has not disappeared; all one has done is to postpone the complications introduced by it to a later stage of the problem where it is more readily handled. For, after solving Eq. (A·10·6), the solutions will still be in terms of the unknown parameter λ. This parameter can then be determined by requiring that the solution satisfies the original restrictive condition (A·10·2).

The method can readily be generalized to the case when there are m equations of constraint. In that case only $(n - m)$ of the variables are independent, and the problem can be handled by introducing m Lagrange parameters, $\lambda_1, \ldots, \lambda_m$, one for each equation of constraint.

A · 11 *Evaluation of the integral* $\int_0^\infty (e^x - 1)^{-1} x^3\,dx$

Let

$$I \equiv \int_0^\infty \frac{x^3\,dx}{e^x - 1} \tag{A·11·1}$$

This can be evaluated by expanding the integrand in a series. Since $e^{-x} \leq 1$ throughout the range of integration, one can write

$$\frac{x^3}{e^x - 1} = \frac{e^{-x}x^3}{1 - e^{-x}} = e^{-x}x^3([1 + e^{-x} + e^{-2x} + \cdots])$$
$$= \sum_{n=1}^{\infty} e^{-nx}x^3$$

Hence (A·11·1) becomes

$$I = \sum_{n=1}^{\infty} \int_0^\infty e^{-nx}x^3\,dx = \sum_{n=1}^{\infty} \frac{1}{n^4} \int_0^\infty e^{-y}y^3\,dy$$

or

$$I = 6 \sum_{n=1}^{\infty} \frac{1}{n^4} \tag{A·11·2}$$

since the integral equals $3! = 6$ by virtue of (A·3·3). This series converges rapidly and can thus be easily evaluated numerically; alternatively, its exact value can also be found analytically to be

$$\sum_{n=1}^{\infty} \frac{1}{n^4} = \frac{\pi^4}{90} \tag{A·11·3}$$

Hence

▶ $$I = \frac{\pi^4}{15} \tag{A·11·4}$$

The exact value can be found in straightforward fashion by contour integration in the complex plane.* The sum (A·11·2) suggests a sum of residues at all integral values. Since the function $(\tan \pi z)^{-1}$ has simple poles at all integral values, it follows that the sum (A·11·2) can be expressed in terms of a contour integral along the path C of Fig. A·11·1*a*, i.e.,

$$\sum_{n=1}^{\infty} n^{-4} = \frac{1}{2i} \int_C \frac{dz}{z^4 \tan \pi z} \tag{A·11·5}$$

where the integrand vanishes as $z \to \infty$. Since the sum is unchanged if $n \to -n$, it can be equally well expressed in terms of an identical integral along C'. Thus (A·11·5) can be written

$$2i \sum_{n=1}^{\infty} n^{-4} = \int_C = \int_{C'} = \tfrac{1}{2} \int_{C+C'} = \tfrac{1}{2} \int_{\text{contour } b} \tag{A·11·6}$$

where the integrand is understood to be that of (A·11·5). In the last step we have completed the path of integration along the semicircles at infinity (see Fig. A·11·1*b*) since the integrand vanishes there. Since the resulting enclosed area contains no singularities except at $z = 0$, we have then shrunk this contour down to the infinitesimal circle C_0 surrounding the origin (see Fig. A.11.1*c*). But this last integral is simple. Expanding the integrand in powers of z about $z = 0$, only the term involving z^{-1} contributes to the integral while other powers of z do not. Thus one gets

$$\frac{1}{z^4 \tan \pi z} = \frac{1}{z^4[\pi z + \frac{1}{3}\pi^3 z^3 + \frac{2}{15}\pi^5 z^5 + \cdots]} = -\frac{\pi^3}{45}\frac{1}{z} + \text{other powers of } z$$

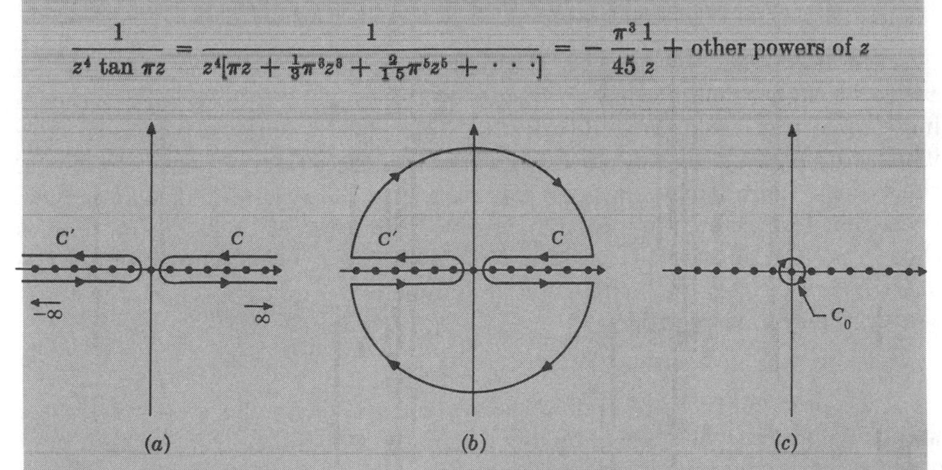

***Fig. A·11·1** (a) The contours C and C′ (closed at infinity) (b) The contours C and C′ closed by the semicircles at infinity. (c) The contour in (b) shrunk down to an infinitesimal circle C_0 about the origin.*

* This general method of summing series can be found described in P. M. Morse and H. Feshbach, "Methods of Theoretical Physics," vol. I, p. 413, McGraw-Hill Book Company, New York, 1953.

Hence (A·11·6) becomes

$$2i \sum n^{-4} = \frac{1}{2}(-2\pi i)\left(-\frac{\pi^3}{45}\right) = 2i\left(\frac{\pi^4}{90}\right)$$

which yields the result (A·11·4).

Alternatively, the integral I of (A·11·1) can be extended from $-\infty$ to ∞ after integration by parts. Thus

$$I = \frac{1}{8}\int_{-\infty}^{\infty} \frac{x^4 e^x \, dx}{(e^x - 1)^2}$$

This integral can then be evaluated directly, without series expansion, by using contour integration in a manner similar to that outlined in Problem 9.27.

A · 12 *The H theorem and the approach to equilibrium*

We consider in greater detail the situation described at the end of Sec. 2·3. Let us denote the approximate quantum states of an isolated system by r (or s). The most complete description of interest to us in considering this system is one which specifies at any time t the probability $P_r(t)$ of finding this system in any one of its accessible states r. This probability is understood to be properly normalized so that summation over all accessible states always yields

$$\sum_r P_r(t) = 1 \qquad \text{(A·12·1)}$$

Small interactions between the particles cause transitions between the accessible approximate quantum states of the system. There exists accordingly some transition probability W_{rs} per unit time that the system originally in a state r ends up in some state s as a result of these interactions. Similarly, there exists a probability W_{sr} per unit time that the system makes an inverse transition from the state s to the state r. The laws of quantum mechanics show that the effect of small interactions can to a good approximation be described in terms of such transition probabilities per unit time, and that these satisfy the symmetry property*

$$W_{sr} = W_{rs} \qquad \text{(A·12·2)}$$

The probability P_r of finding the system in a particular state r *increases* with time because the system, having originally probability P_s of being in any other state s, makes transitions to the given state r; similarly, it *decreases* because the system, having originally probability P_r of being in the given state r, makes transitions to all other states s. The change per unit time in the probability P_r can, therefore, be expressed in terms of the transition prob-

* The conditions necessary for the validity of this description in terms of transition probabilites and the symmetry property (A·12·2) are discussed more fully in connection with Eq. (15·1·3).

abilities per unit time by the relation

$$\frac{dP_r}{dt} = \sum_s P_s W_{sr} - \sum_s P_r W_{rs}$$

or

$$\frac{dP_r}{dt} = \sum_s W_{rs}(P_s - P_r) \tag{A·12·3}$$

where we have used the symmetry property (A·12·2).*

Consider now the quantity H defined as the mean value of $\ln P_r$ over all accessible states; i.e.,

$$H \equiv \overline{\ln P_r} \equiv \sum_r P_r \ln P_r \tag{A·12·4}$$

This quantity changes in time since the probabilities P_r vary in time. Differentiation of (A·12·4) then gives

$$\frac{dH}{dt} = \sum_r \left(\frac{dP_r}{dt} \ln P_r + \frac{dP_r}{dt} \right) = \sum_r \frac{dP_r}{dt} (\ln P_r + 1)$$

or

$$\frac{dH}{dt} = \sum_r \sum_s W_{rs}(P_s - P_r)(\ln P_r + 1) \tag{A·12·5}$$

where we have used (A·12·3). Interchange of the summation indices r and s on the right side does not affect the sum so that (A·12·5) can equally well be written

$$\frac{dH}{dt} = \sum_r \sum_s W_{sr}(P_r - P_s)(\ln P_s + 1) \tag{A·12·6}$$

Using the property (A·12·2), dH/dt can then be written in very symmetrical form by adding (A·12·5) and (A·12·6). Thus one gets

$$\frac{dH}{dt} = -\frac{1}{2} \sum_r \sum_s W_{rs}(P_r - P_s)(\ln P_r - \ln P_s) \tag{A·12·7}$$

But since $\ln P_r$ is a monotonically increasing function of P_r, it follows that if $P_r > P_s$, then $\ln P_r > \ln P_s$, and vice versa. Hence

$$(P_r - P_s)(\ln P_r - \ln P_s) \geq 0 \qquad (= \text{sign only if } P_r = P_s) \tag{A·12·8}$$

Since the probability W_{rs} is intrinsically positive, each term in the sum (A·12·7) must be positive or zero. Hence one can conclude that

▶
$$\frac{dH}{dt} \leq 0 \tag{A·12·9}$$

where the equals sign holds only if $P_r = P_s$ for *all* states r and s between which transitions are possible (so that $W_{rs} \neq 0$), i.e., for *all* accessible states. Thus

$$\frac{dH}{dt} = 0 \qquad \text{only if } P_r = C \text{ for all accessible states} \tag{A·12·10}$$

* Note that the relation (A·12·3) is just the "master equation" discussed in (15·1·5).

where C is some constant independent of the particular state r. The result (A·12·9) is called the "H theorem" and expresses the fact that the quantity H always tends to decrease in time.*

An isolated system is not in equilibrium when any quantity, and in particular the quantity H, changes systematically in time. Now (A·12·7) shows that, irrespective of the initial values of the probabilities P_r, the quantity H tends always to decrease as long as not all of these probabilities are equal. It will thus continue to decrease until H has reached its minimum possible value when $dH/dt = 0$. This final situation is, by (A·12·10), characterized by the fact that the system is then equally likely to be found in any one of its accessible states. This situation is clearly one of equilibrium, since any subsequent change in the probabilities P_r could only make some probabilities again unequal and thus again increase H, a possibility ruled out by (A·12·9). The final equilibrium situation is thus indeed consistent with the postulate of equal a priori probabilities.†

A · 13 *Liouville's theorem in classical mechanics*

Consider an isolated system specified classically by f generalized coordinates and momenta $\{q_1, \ldots, q_f, p_1, \ldots, p_f\}$. In a statistical ensemble of such systems let

> $\rho(q_1, \ldots, q_f, p_1, \ldots, p_f; t)\, dq_1 \cdots dq_f\, dp_1 \cdots dp_f =$ the number of systems in the ensemble which, at time t, have positions and momenta in the element of volume $(dq_1 \cdots dq_f\, dp_1 \cdots dp_f)$ of phase space lying between q_1 and $q_1 + dq_1$, q_2 and $q_2 + dq_2$, $\ldots$, p_f and $p_f + dp_f$.

Every system in the ensemble moves in time according to its classical equations of motion which are

$$\dot{q}_i = \frac{\partial \mathcal{H}}{\partial p_i}, \qquad \dot{p}_i = -\frac{\partial \mathcal{H}}{\partial q_i} \tag{A·13·1}$$

where $\mathcal{H} = \mathcal{H}(q_1, \ldots, q_f, p_1, \ldots, p_f)$ is the Hamiltonian of the system. As a result of this motion the density ρ of systems in phase space changes in time. We are interested in finding $\partial\rho/\partial t$ at a given point of phase space.

Focus attention on any given fixed element of volume of phase space located between q_1 and $q_1 + dq_1$, q_2 and $q_2 + dq_2$, $\ldots$, p_f and $p_f + dp_f$ (see Fig. A·13·1). The number of systems located in this volume $(dq_1 \cdots dq_f\, dp_1 \cdots dp_f)$ changes as the coordinates and momenta of the systems vary in

* Note that, by (6·6·24), $S = -kH$. Thus (A·12·9) expresses the fact that the entropy tends to increase.

† More elaborate and critical discussions of the H theorem can be found in R. C. Tolman, "The Principles of Statistical Mechanics," chap. 12, Oxford University Press, Oxford, 1938; also in D. ter Haar, *Rev. Mod. Phys.*, vol. 27, p. 289 (1955). The H theorem was first proved, in somewhat special form, by Boltzmann in 1872.

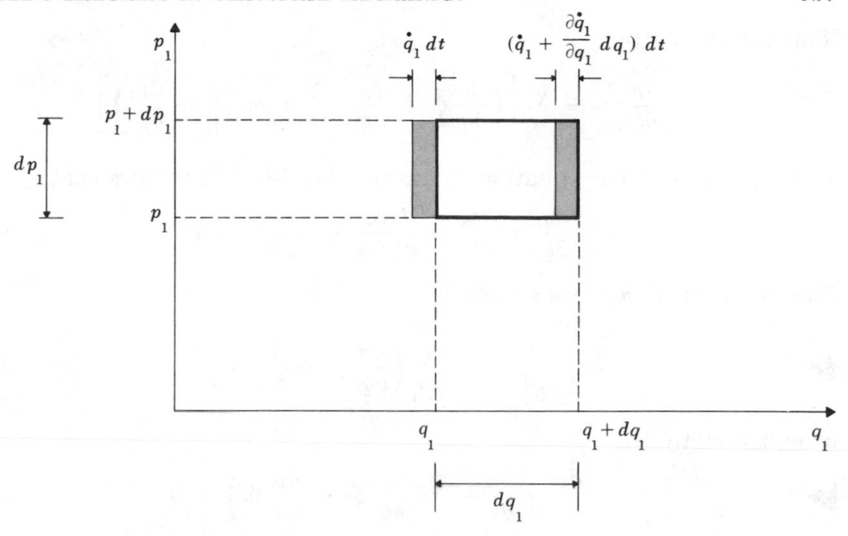

Fig. A·13·1 Diagram illustrating a fixed volume element of (two-dimensional) phase space.

accordance with (A · 13 · 1). In a time dt the change in the number of systems within this volume of phase space is given by $(\partial\rho/\partial t)\,dt(dq_1 \cdots dp_f)$. This change is due to the number of systems entering and leaving this volume $(dq_1 \cdots dp_f)$ in the time dt. But the number of systems entering this volume in time dt through the "face" $q_1 =$ constant is just the number contained in the volume $(\dot{q}_1\,dt)(dq_2 \cdots dp_f)$, i.e., it is equal to the quantity $\rho(q_1, \ldots, p_f; t)(\dot{q}_1\,dt\,dq_2 \cdots dp_f)$. The number of systems leaving through the "face" $q_1 + dq_1 =$ constant is given by a similar expression except that both ρ and $\dot{q}_i$ must be evaluated at $q_1 + dq_1$ instead of at q_1. Hence the net number of systems entering the phase-space volume $(dq_1 \cdots dp_f)$ in time dt through the "faces" $q_1 =$ constant and $q_1 + dq_1 =$ constant is given by

$$\rho\dot{q}_1\,dt\,dq_2 \cdots dp_f - \left[\rho\dot{q}_1 + \frac{\partial}{\partial q_1}(\rho\dot{q}_1)\,dq_1\right]dt\,dq_2 \cdots dp_f = -\frac{\partial}{\partial q_1}(\rho\dot{q}_1)\,dt\,dq_1\,dq_2 \cdots dp_f$$

The total net increase in time dt of the number of systems in this volume of phase space is then obtained by summing the net numbers of systems entering the volume through all the "faces" labeled by $q_1, q_2, \ldots, q_f$ and $p_1, p_2, \ldots, p_f$. Thus one obtains the relation

$$\frac{\partial\rho}{\partial t}\,dt\,dq_1 \cdots dp_f = \left[-\sum_{i=1}^{f}\frac{\partial}{\partial q_i}(\rho\dot{q}_i) - \sum_{i=1}^{f}\frac{\partial}{\partial p_i}(\rho\dot{p}_i)\right]dt\,dq_1 \cdots dp_f$$

or

$$\frac{\partial\rho}{\partial t} = -\sum_{i=1}^{f}\left[\frac{\partial}{\partial q_i}(\rho\dot{q}_i) + \frac{\partial}{\partial p_i}(\rho\dot{p}_i)\right] \qquad \text{(A·13·2)}$$

This can be written

$$\frac{\partial \rho}{\partial t} = - \sum_{i=1}^{f} \left[\left(\frac{\partial \rho}{\partial q_i} \dot{q}_i + \frac{\partial \rho}{\partial p_i} \dot{p}_i \right) + \rho \left(\frac{\partial \dot{q}_i}{\partial q_i} + \frac{\partial \dot{p}_i}{\partial p_i} \right) \right] \tag{A·13·3}$$

But, by virtue of the equations of motion (A · 13 · 1) it follows that

$$\frac{\partial \dot{q}_i}{\partial q_i} + \frac{\partial \dot{p}_i}{\partial p_i} = \frac{\partial^2 \mathcal{H}}{\partial q_i \partial p_i} - \frac{\partial^2 \mathcal{H}}{\partial p_i \partial q_i} = 0 \tag{A·13·4}$$

Hence (A · 13 · 3) reduces simply to

► $$\frac{\partial \rho}{\partial t} = - \sum_{i=1}^{f} \left(\frac{\partial \rho}{\partial q_i} \dot{q}_i + \frac{\partial \rho}{\partial p_i} \dot{p}_i \right) \tag{A·13·5}$$

or equivalently to

► $$\frac{d\rho}{dt} = \frac{\partial \rho}{\partial t} + \sum_{i} \left(\frac{\partial \rho}{\partial q_i} \dot{q}_i + \frac{\partial \rho}{\partial p_i} \dot{p}_i \right) = 0 \tag{A·13·6}$$

where $d\rho/dt$ is the *total* time derivative of $\rho(q_1, \ldots, p_f; t)$; it measures the rate of change of ρ if one moved along in phase space with the point representing a system. By (A · 13 · 1) one can also write (A · 13 · 5) as

$$\frac{\partial \rho}{\partial t} = - \sum_{i=1}^{f} \left(\frac{\partial \rho}{\partial q_i} \frac{\partial \mathcal{H}}{\partial p_i} - \frac{\partial \rho}{\partial p_i} \frac{\partial \mathcal{H}}{\partial q_i} \right) \tag{A·13·7}$$

The relation (A · 13 · 5) or (A · 13 · 6) is known as Liouville's theorem.

Suppose that at any one time t, ρ is a constant (i.e., the systems are uniformly distributed over all of phase space). Or, more generally, suppose that ρ is at time t only a function of the energy E of the system, this energy being, of course, a constant of the motion. Then*

$$\frac{\partial \rho}{\partial q_i} = \frac{\partial \rho}{\partial E} \frac{\partial E}{\partial q_i} = 0 \qquad \text{and} \qquad \frac{\partial \rho}{\partial p_i} = \frac{\partial \rho}{\partial E} \frac{\partial E}{\partial p_i} = 0$$

and Liouville's theorem (A · 13 · 5) implies that $\partial \rho / \partial t = 0$. This means that the distribution of systems over states remains then *unchanged* in time, i.e., one has an equilibrium situation. In particular, a specification such as that of the microcanonical ensemble (where ρ = constant when $E_0 < E < E_0 + \delta E$, but $\rho = 0$ otherwise) is then indeed consistent with an equilibrium situation.

In quantum mechanics an equivalent discussion can be given in terms of the density matrix ρ_{mn} instead of the classical phase-space density ρ. This discussion is actually simpler than the classical one for someone familiar with elementary quantum mechanics. The interested reader is referred to the references.†

* The same result would hold if ρ is only a function of *any* set of parameters $\alpha_1, \ldots, \alpha_m$, which are constants of the motion.

† See C. Kittel, "Elementary Statistical Physics," sec. 23, John Wiley & Sons, Inc., New York, 1958; also R. C. Tolman, "The Principles of Statistical Mechanics," chap. 9, Oxford University Press, Oxford, 1938.

Numerical constants

IN THE TABLE below the mole is defined in accordance with the modern convention according to which the isotope C^{12} is assigned the atomic mass 12.* The estimated error limits are three standard deviations applied to the last digit of the preceding column.

Physical constants

Quantity	*Value*	*Error*
Elementary charge	$e = 4.80298 \times 10^{-10}$ esu	±20
	$= 1.60210 \times 10^{-19}$ coulombs	± 7
Speed of light in vacuum	$c = 2.997925 \times 10^{10}$ cm sec^{-1}	± 3
Planck's constant,	$h = 6.6256 \times 10^{-27}$ ergs sec	± 5
$\hbar \equiv h/2\pi$	$\hbar = 1.05450 \times 10^{-27}$ ergs sec	± 7
Electron rest mass	$m_e = 9.1091 \times 10^{-28}$ grams	± 4
Proton rest mass	$m_p = 1.67252 \times 10^{-24}$ grams	± 8
Bohr magneton $e\hbar/2m_ec$	$\mu_B = 9.2732 \times 10^{-21}$ ergs gauss^{-1}	± 6
Nuclear magneton $e\hbar/2m_pc$	$\mu_N = 5.0505 \times 10^{-24}$ ergs gauss^{-1}	± 4
Avogadro's number	$N_a = 6.02252 \times 10^{23}$ mole^{-1}	±28
Boltzmann's constant	$k = 1.38054 \times 10^{-16}$ ergs deg^{-1}	±18
Gas constant	$R = 8.3143 \times 10^{7}$ ergs deg^{-1} mole^{-1}	±12
Stefan-Boltzmann constant	$\sigma = 5.6697 \times 10^{-5}$ ergs cm^{-2} sec^{-1} deg^{-4}	±29

* The values quoted are those recommended in 1963 by the Committee on Fundamental Constant of the National Academy of Sciences, National Research Council. See *Physics Today*, vol. 17, pp. 48–49 (February, 1964).

Conversion factors

Quantity		*Value*	*Error*
Triple point of water		$\equiv 273.16°K$	by definition
Celsius temperature	$X°C$	$\equiv (273.15 + X)°K$	by definition
1 atmosphere $\equiv$ 760 mm Hg		$\equiv 1.013250 \times 10^6$ dynes cm^{-2}	by definition
1 joule		$\equiv 10^7$ ergs	by definition
1 thermochemical calorie		$\equiv 4.1840$ joules	by definition
1 electron volt		$= 1.60210 \times 10^{-12}$ ergs	± 7
		$= 1.16049 \times 10^4 °K$	± 16
1 kilocalorie/mole		$= 6.94726 \times 10^{-14}$ ergs/molecule	± 32
		$= 4.33634 \times 10^{-2}$ ev/molecule	± 28

Approximate temperature T corresponding to various frequencies ν ($kT = h\nu$)

$\nu = 10^6$ cycles sec^{-1}	$\leftrightarrow$	$T = 4.80 \times 10^{-5} °K$	(radio-wave photon)
$\nu = 10^{10}$ cycles sec^{-1}	$\leftrightarrow$	$T = 0.48 °K$	(microwave photon)
$\nu = 10^{13}$ cycles sec^{-1}	$\leftrightarrow$	$T = 480 °K$	(infrared photon)
$\nu = 10^{15}$ cycles sec^{-1}	$\leftrightarrow$	$T = 4.8 \times 10^4 °K$	(visible photon)

Bibliography

This bibliography includes references to works appreciably more advanced than the present book, but is not meant to be exhaustive.

General textbooks

Allis, W. P., and M. A. Herlin: "Thermodynamics and Statistical Mechanics," McGraw-Hill Book Company, New York, 1952.

Crawford, F. H.: "Heat, Thermodynamics, and Statistical Physics," Harcourt, Brace and World, Inc., New York, 1963.

King, A. L.: "Thermophysics," W. H. Freeman & Company, San Francisco, 1962.

Lee, J. F., F. W. Sears, and D. L. Turcotte: "Statistical Thermodynamics," Addison-Wesley Publishing Company, Reading, Mass., 1963.

Morse, P. M.: "Thermal Physics," rev. ed., W. A. Benjamin, New York, 1964.

Sears, F. W.: "Thermodynamics, the Kinetic Theory of Gases, and Statistical Mechanics," 2d ed., Addison-Wesley Publishing Company, Reading, Mass., 1953.

Sommerfeld, A.: "Thermodynamics and Statistical Mechanics," Academic Press, New York, 1956.

Soo, S. L.: "Analytical Thermodynamics," Prentice-Hall Publishing Company, Englewood Cliffs, N.J., 1962.

Probability theory

Cramér, H.: "The Elements of Probability Theory," John Wiley & Sons, Inc., New York, 1955.

Feller, W.: "An Introduction to Probability Theory and Its Applications," 2d ed., John Wiley & Sons, Inc., New York, 1957.

Hoel, P. G.: "Introduction to Mathematical Statistics," 3d ed., John Wiley & Sons, Inc., New York, 1962.

Mosteller, F., R. E. K. Rourke, and G. B. Thomas: "Probability and Statistics," Addision-Wesley Publishing Company, Reading, Mass., 1961. (An elementary introduction.)

Munroe, M. E.: "The Theory of Probability," McGraw-Hill Book Company, New York, 1951.

Parzen E.: "Modern Probability Theory and Its Applications," John Wiley & Sons, Inc., New York, 1960.

Macroscopic thermodynamics

Callen, H. B.: "Thermodynamics," John Wiley & Sons, Inc., New York, 1960. (A modern sophisticated introduction.)

Fermi, E.: "Thermodynamics," Dover Publications, New York, 1957. (A good introduction.)

Guggenheim, E. A.: "Thermodynamics," 4th ed., Interscience Publishers, New York, 1960.

Kirkwood, J. G., and I. Oppenheim: "Chemical Thermodynamics," McGraw-Hill Book Company, New York, 1961.

Pippard, A. B.: "The Elements of Classical Thermodynamics," Cambridge University Press, Cambridge, 1957.

Zemansky, M. W.: "Heat and Thermodynamics," 4th ed., McGraw-Hill Book Company, New York, 1957. (A careful exposition of fundamental ideas and good introduction.)

Statistical mechanics

Andrews, F. C.: "Equilibrium Statistical Mechanics," John Wiley & Sons, Inc., New York, 1963. (A readable elementary introduction.)

Becker, R.: "Theorie der Wärme," Springer-Verlag, Berlin, 1955. (A good introduction to statistical mechanics and fluctuation phenomena.)

Brillouin, L.: "Science and Information Theory," 2d ed., Academic Press, New York, 1962. (Discusses the connection with information theory and some paradoxes such as "Maxwell's demon".)

R. Brout: "Phase Transitions," W. A. Benjamin, New York, 1965.

Chisholm, J. S. R., and A. H. Borde: "An Introduction to Statistical Mechanics," Pergamon Press, New York, 1958.

Davidson, N.: "Statistical Mechanics," McGraw-Hill Book Company, New York, 1962.

Eyring, H., D. Henderson, B. J. Stover, and E. M. Eyring: "Statistical Mechanics and Dynamics," John Wiley & Sons, Inc., New York, 1963.

Fowler, R. H.: "Statistical Mechanics," 2d ed., Cambridge University Press, Cambridge, 1955. (Treatise based on the Darwin-Fowler method.)

Fowler, R. H., and E. A. Guggenheim: "Statistical Thermodynamics," rev. ed., Cambridge University Press, Cambridge, 1949. (A less theoretical treatise than the previous one, but also based on the Darwin-Fowler method.)

Frenkel, J. I.: "Statistische Physik," Akademie-Verlag, Berlin, 1957.

Hill, T. L.: "An Introduction to Statistical Thermodynamics," Addison-Wesley Publishing Company, Reading, Mass., 1960. (An introduction oriented toward chemistry.)

Hill, T. L.: "Statistical Mechanics," McGraw-Hill Book Company, New York, 1956. (A good exposition of more advanced and modern topics.)

Huang, K.: "Statistical Mechanics," John Wiley & Sons, Inc., New York, 1963. (Treats advanced topics of interest to physicists.)

Kittel, C.: "Elementary Statistical Physics," John Wiley & Sons, Inc., New York, 1958.

Khinchin, A. I.: "Mathematical Foundations of Statistical Mechanics," Dover Publications, New York, 1949.

Landau, L. D., and E. M. Lifshitz: "Statistical Physics," Addison-Wesley Publishing Company, Reading, Mass., 1959.

Lee, J. F., F. W. Sears, and D. L. Turcotte: "Statistical Thermodynamics," Addison-Wesley Publishing Company, Reading, Mass., 1963. (A readable elementary introduction.)

Lindsay, R. B.: "Introduction to Physical Statistics," John Wiley & Sons, Inc., New York, 1941.

MacDonald, D. K. C.: "Introductory Statistical Mechanics for Physicists," John Wiley & Sons, Inc., New York, 1963.

Mayer, J. E., and M. G. Mayer: "Statistical Mechanics," John Wiley & Sons, Inc., New York, 1940.

Münster, A.: "Statistische Thermodynamik," Springer-Verlag, Berlin, 1956.

———: "Prinzipien der Statistischen Mechanik," in "Handbuch der Physik," vol. III/2, pp. 176–412, Springer-Verlag, Berlin, 1959.

Muto, T., and Y. Takagi: "Order-Disorder in Alloys," in "Advances in Solid-state Physics," vol. 1, pp. 194–282; Academic Press, New York, 1955. (Discussion of cooperative phenomena.)

Rushbrooke, G. S.: "Introduction to Statistical Mechanics," Oxford University Press, Oxford, 1949.

Schrödinger, E.: "Statistical Thermodynamics," 2d ed., Cambridge University Press, Cambridge, 1952. (A concise and clearly written book.)

Ter Haar, D.: "Foundations of Statistical Mechanics," *Rev. Mod. Phys.*, vol. 27, p. 289 (1955).

———: "Elements of Statistical Mechanics," Holt, Rinehart and Winston, New York, 1954.

Tolman, R. C.: "The Principles of Statistical Mechanics," Oxford University Press, Oxford, 1938. (A careful exposition of the fundamental concepts and a classic in the field.)

Tribus, M.: "Thermostatics and Thermodynamics," D. Van Nostrand & Company, New York, 1961. (Written by an engineer from the point of view of information theory.)

Wilks, J.: "The Third Law of Thermodynamics," Oxford University Press, Oxford, 1961.

Wilson, A. H.: "Thermodynamics and Statistical Mechanics," Cambridge University Press, Cambridge, 1957.

Kinetic theory

Chapman, S., and T. G. Cowling: "The Mathematical Theory of Non-uniform Gases," Cambridge University Press, Cambridge, 1952. (A standard advanced work.)

Guggenheim, E. A.: "Elements of the Kinetic Theory of Gases," Pergamon Press, New York, 1960. (An elementary introduction to exact transport theory.)

Hirschfelder, J. O., C. R. Curtiss, and R. B. Bird: "The Molecular Theory of Gases and Liquids," John Wiley & Sons, Inc., New York, 1954. (An advanced book.)

Jeans, J.: "An Introduction to the Kinetic Theory of Gases," Cambridge University Press, Cambridge, 1952.

———: "Kinetic Theory of Gases," Cambridge University Press, Cambridge, 1946.

Kennard, E. H.: "Kinetic Theory of Gases," McGraw-Hill Book Company, New York, 1938.

Loeb, L. B.: "The Kinetic Theory of Gases," 2d ed., McGraw-Hill Book Company, New York, 1934.

Patterson, G. N.: "Molecular Flow of Gases," John Wiley & Sons, Inc., New York, 1956.

Present, R. D.: "Introduction to the Kinetic Theory of Gases," McGraw-Hill Book Company, New York, 1958. (A good introduction.)

Sommerfeld, A.: "Thermodynamics and Statistical Mechanics," Academic Press, New York, 1956. (Chapters 3 and 5 contain a good discussion of kinetic theory.)

Waldman, L.: "Transporterscheinungen in Gasen von mittlerem Druck," in "Handbuch der Physik," vol. XII, pp. 295–514, Springer-Verlag, Berlin, 1958.

Watson, K. M., J. W. Bond, and J. A. Welch: "Atomic Theory of Gas Dynamics," Addison-Wesley Publishing Company, Reading, Mass., 1965. (An advanced modern treatment.)

Fluctuations and irreversible processes

Becker, R.: "Theorie der Wärme," Springer-Verlag, Berlin, 1955. (Chapter 6 gives a good discussion of fluctuation phenomena.)

Chandrasekhar, S.: "Stochastic Problems in Physics and Astronomy," *Rev. Mod. Phys.*, vol. 15, pp. 1–89 (1943). (A good discussion of Brownian motion and random walk problems.)

Cohen, E. D. G. (ed.): "Fundamental Problems in Statistical Mechanics," John Wiley & Sons, Inc., New York, 1962.

Cox, R. T.: "Statistical Mechanics of Irreversible Change," Johns Hopkins Press, Baltimore, Md., 1955.

Denbigh, K. G.: "Thermodynamics of the Steady State," John Wiley & Sons, Inc., New York, 1951.

DeGroot, S. R.: "Thermodynamics of Irreversible Processes," Interscience Publishers, New York, 1951.

DeGroot, S. R., and P. Mazur: "Non-equilibrium Thermodynamics," Interscience Publishers, New York, 1962.

Eisenschitz, R.: "Statistical Theory of Irreversible Processes," Oxford University Press, Oxford, 1958.

Kirkwood, J. G., and J. Ross: "The Statistical Mechanical Basis of the Boltzmann Equation," in I. Prigogine (ed.), "International Symposium on Transport Processes in Statistical Mechanics," p. 1, Interscience Publishers, New York, 1958.

Kubo, R.: "Some Aspects of the Statistical Mechanical Theory of Irreversible Processes," in University of Colorado, "Lectures in Theoretical Physics," vol. I, pp. 120–203; Interscience Publishers, New York, 1959.

McCombie, C. W.: "Fluctuation Theory of Physical Measurements," *Reports on Progress in Physics*, vol. 16, pp. 266–320 (1953).

MacDonald, D. K. C.: "Noise and Fluctuations: an Introduction," John Wiley & Sons, Inc., New York, 1962.

Prigogine, I.: "Introduction to Thermodynamics of Irreversible Processes," 2d ed., John Wiley & Sons, Inc., New York, 1962.

Rice, S. A., and H. Frisch: "Some Aspects of the Statistical Theory of Transport," *Ann. Review of Phys. Chem.*, vol. 11, pp. 187–272 (1960). (A review article on irreversible statistical mechanics.)

Ross, J.: "Contribution to the Theory of Brownian Motion," *J. Chem. Phys.*, vol. 24, p. 375 (1956). (A detailed derivation of the Fokker-Planck equation.)

Wax, M. (ed.): "Selected Papers on Noise and Stochastic Processes," Dover Publications, New York, 1954.

Zwanzig, R. W.: "Statistical Mechanics of Irreversibility," in University of Colorado, "Lectures in Theoretical Physics," vol. 3, pp. 106–141, Interscience Publishers, New York, 1961.

History

Broda, E.: "Ludwig Boltzmann," Franz Deuticke, Vienna, 1955. (In German.)

Holton, G., and D. H. D. Roller: "Foundations of Modern Physical Science," chaps. 19, 20, and 25, Addison-Wesley Publishing Company, Reading, Mass., 1958.

Koenig, F. O.: "On the History of Science and of the Second Law of Thermodynamics," in H. M. Evans (ed.), "Men and Moments in the History of Science," University of Washington Press, Seattle, 1959.

Roller, D.: "The Early Development of the Concepts of Temperature and Heat," Harvard University Press, Cambridge, Mass., 1950.

Wheeler, L. P.: "Josiah Willard Gibbs," Yale University Press, 1951 (paperback edition, 1962).

Answers to selected problems

CHAPTER 1

1.2 (*a*) 0.402; (*b*) 0.667; (*c*) 0.020 ***1.5*** (*a*) $(\frac{5}{6})^N$; (*b*) $(\frac{1}{6})(\frac{5}{6})^{N-1}$; (*c*) 6

1.6 $\overline{m} = 0$; $\overline{m^2} = N$; $\overline{m^3} = 0$; $\overline{m^4} = 3N^2 - 2N$ ***1.8*** $[(2N)!/(N!)^2][\frac{1}{2}]^{2N}$

1.11 (*a*) 0.37; (*b*) 0.08 ***1.13*** 0.003; 0.086; 0.162 ***1.15*** 85 ***1.18*** Nl^2

1.19 $(N^2V^2/R)p^2[1 + (1 - p)/Np]$ ***1.23*** (*a*) Nl; (*b*) $Nb^2/3$

1.25 (*a*) $A(\mu/a^3 + b)^{-\frac{1}{2}}\, db$, where $A = \frac{1}{2}\sqrt{a^3/3\mu}$
(*b*) $\frac{1}{2}A(\mu/a^3 - b)^{-\frac{1}{2}}\, db$, if $-2\mu/a^3 < b < -\mu/a^3$;
$\frac{1}{2}A[(\mu/a^3 - b)^{-\frac{1}{2}} + (\mu/a^3 + b)^{-\frac{1}{2}}]\, db$, if $-\mu/a^3 < b < \mu/a^3$;
$\frac{1}{2}A(\mu/a^3 + b)^{-\frac{1}{2}}\, db$ if $\mu/a^3 < b < 2\mu/a^3$

1.26 $Nb\, dx\, [\pi(x^2 + N^2b^2)]^{-1}$

1.29 $(8\pi l^3)^{-1}$ if $0 < r < l$; $(3l - r)(16\pi l^3 r)^{-1}$ if $l < r < 3l$; 0 if $r > 3l$

CHAPTER 2

2.3 (*a*) $\dfrac{dx}{\pi(A^2 - x^2)^{\frac{1}{2}}}$ ***2.4*** (*a*) $\dfrac{N!}{[(N/2) - (E/2\mu H)]![(N/2) + (E/2\mu H)]!}\dfrac{\delta E}{2\mu H}$

2.7 (*b*) $\bar{p} = \frac{2}{3}\bar{E}/V$ ***2.10*** $(\bar{p}_f V_f - \bar{p}_i V_i)/(1 - \gamma)$

2.11 (*a*) $W = 22{,}400$ joules, $Q = 11{,}800$ joules

CHAPTER 3

3.2 (*a*) $E = -N\mu H \tanh(\mu H/kT)$; (*b*) $E > 0$; (*c*) $M = N\mu \tanh(\mu H/kT)$

3.3 (*a*) $\tilde{E}/\mu^2 N = \tilde{E}'/\mu'^2 N'$; (*b*) $\mu^2 N(bN\mu H + b'N'\mu' H')(\mu^2 N + \mu'^2 N')^{-1}$;
(*c*) $NN'H(b'\mu'\mu^2 - b\mu'^2\mu)(\mu^2 N + \mu'^2 N')^{-1}$;
(*d*) $P(E)\, dE = (2\pi\sigma^2)^{-\frac{1}{2}} \exp[-(E - \tilde{E})^2/2\sigma^2]\, dE$,
where $\sigma \equiv \mu\mu' H[NN'/(\mu^2 N + \mu'^2 N')]^{\frac{1}{2}}$; (*e*) σ^2; (*f*) $\mu'/(\mu b' N^{\frac{1}{2}})$

CHAPTER 4

4.1 (*a*) $\Delta S_W = 1310$ joules deg^{-1}, $\Delta S_{res} = -1120$ joules deg^{-1}, $\Delta S_{tot} = 190$ joules deg^{-1}; (*b*) 102 joules deg^{-1}

4.3 $\Delta S = c \ln (T_f/T_i) + R \ln (V_f/V_i)$ **4.4** $2.27\, Nk$

CHAPTER 5

5.1 (*a*) $T_f = T_i(V_f/V_i)^{1-\gamma}$ **5.2** (*a*) 314 joules; (*b*) 600 joules; (*c*) 1157 joules

5.4 (*c*) 3.24 joules deg^{-1} **5.5** (*c*) $[T_0 + (mgV_0/\nu c_V A)](1 + R/c_V)^{-1}$

5.6 $\gamma = 4\pi^2\nu^2 mV_0(p_0A^2 + mgA)^{-1}$ **5.12** $\alpha T\, \Delta p/\rho c_p$

5.13 $\alpha^2 vT - vT(d\alpha/dT)$

5.14 (c) $S(L,T) = S(L_0,T_0) + b(T - T_0) - aT(L - L_0)^2$; (*e*) $C_L(L,T) = bT - aT(L - L_0)^2$

5.15 (*b*) $2l\sigma_0 x$; (*c*) $-2l\sigma x$ **5.18** (*a*) $-(1/C_V)[T(\partial p/\partial T)_V - p]$

5.19 (*c*) $(p' + 3/v'^2)(v' - \frac{1}{3}) = \frac{8}{3}T'$ **5.20** $p' = 9 - 12(\sqrt{T'} - \sqrt{3})^2$

5.21 (*b*) $T_0 \exp \left\{ \int_{\vartheta_0}^{\vartheta} [\alpha'\, d\vartheta'/(1 + \mu' C_p'/V)] \right\}$ **5.22** (*a*) $T_i/(T_i - T_0)$

5.23 (*a*) $C(T_1 + T_2 - 2T_f)$; (*b*) $T_f \geq \sqrt{T_1T_2}$; (*c*) $C(\sqrt{T_1} - \sqrt{T_2})^2$

5.25 $\exp(-10^{18})$ **5.26** $1 - (V_1/V_2)^{\gamma-1}$

CHAPTER 6

6.1 (*a*) $e^{-\hbar\omega/kT}$; (*b*) $(\hbar\omega/2)(1 + 3e^{-\hbar\omega/kT})(1 + e^{-\hbar\omega/kT})^{-1}$ **6.4** T^{-1}

6.6 (*c*) $\bar{E} = N(\epsilon_1 + \epsilon_2 e^{-(\epsilon_2-\epsilon_1)/kT})(1 + e^{-(\epsilon_2-\epsilon_1)/kT})^{-1}$

6.7 (*b*) $S = N_a k \ln (1 + 2e^{-\epsilon/kT}) + (2N_a\epsilon/T)e^{-\epsilon/kT}(1 + 2e^{-\epsilon/kT})^{-1}$

6.8 $(Nea/2) \tanh (e\mathcal{E}a/2kT)$

6.10 (*a*) $\rho(r) = \rho(0) \exp (m\omega^2 r^2/2kT)$; (*b*) $\mu = [2N_a kT/\omega^2(r_1^2 - r_2^2)] \ln [\rho(r_1)/\rho(r_2)]$

6.11 $\alpha = 0.15k/U_0$ **6.12** $(A/3)^{\frac{1}{2}}, (A/3)^{\frac{1}{2}}, \frac{1}{2}(A/3)^{\frac{1}{2}}$

CHAPTER 7

7.2 (*b*) $kT + mgL(1 - e^{mgL/kT})^{-1}$

7.3 (*a*) $2p(1 + b)^{-1}$; (*b*) $\nu R \ln [(1 + b)^2/b]$; (*c*) $\nu R \ln [(1 + b)^2/4b]$

7.5 $Na \tanh (Wa/kT)$ **7.9** (*a*) Ma/α; (*b*) kT/α; (*c*) $\sqrt{kT\alpha/g^2}$

7.10 (*b*) $\frac{3}{4}Nk$ **7.12** (*b*) 164°K **7.14** $\bar{M}_z = N_0\mu\,[\coth\,(\mu H/kT) - (kT/\mu H)]$

7.16 (*c*) $1 + \frac{1}{2}(\mu/kT)^2(H_2{}^2 - H_1{}^2)$ **7.17** 0.843

7.18 $u = (\gamma/2)^{\frac{1}{2}}\bar{v}$; 37 percent have speed less than u.

7.20 (*b*) $2\pi n(\pi kT)^{-\frac{3}{2}}\epsilon^{\frac{1}{2}}e^{-\epsilon/kT}\,d\epsilon$ **7.22** (*c*) $I_0 \exp\,[-mc^2(\nu - \nu_0)^2/(2kT\nu_0{}^2)]\,d\nu$

7.23 (*a*) 1.1×10^{18}; (*b*) 1.7×10^{11}; (*c*) 2.4×10^{-8} mm of Hg

7.24 $4V/\bar{v}A$ **7.27** $2^{(1-\sqrt{\mu_{He}/\mu_{Ne}})}$

7.28 (*a*) $\frac{1}{2}[p_1(0) + p_2(0)] + \frac{1}{2}[p_1(0) - p_2(0)]e^{-A\bar{v}t/V}$;
(*b*) $\dfrac{V}{2T}\left[p_1(0)\ln\dfrac{2p_1(0)}{p_1(0) + p_2(0)} + p_2(0)\ln\dfrac{2p_2(0)}{p_1(0) + p_2(0)}\right]$; **7.30** (*b*) $4\bar{\epsilon}_i/3$

7.31 $\bar{p}A\left[1 - \cos^3\left(\sin^{-1}\dfrac{R}{L}\right)\right] \to \dfrac{2}{3}\bar{p}A\left(\dfrac{R}{L}\right)^2$ if $R \ll L$

CHAPTER 8

8.1 $$\mathcal{P}(V,T)\,dV\,dT = \frac{1}{2\pi}\left(\frac{c_V\rho_0}{k^2T^3\kappa}\right)^{\frac{1}{2}} \exp\left[-\frac{Mc_V}{2kT_0{}^2}(T - T_0)^2 - \frac{\rho_0}{2M\kappa kT_0}\left(V - \frac{M}{\rho_0}\right)^2\right]dV\,dT$$

8.2 (*a*) 195°K; (*b*) l = 31,220 joules mole^{-1} for sublimation, l = 25,480 joules mole^{-1} for vaporization; (*c*) 5740 joules mole^{-1}

8.5 (*a*) $QRT_r/L\mathcal{V}$; (*b*) $T_0[1 - (T_0R/L)\ln\,(p_m/p_0)]^{-1}$ **8.6** $T^{-1}[(L/RT) - \frac{1}{2}]$

8.7 $(c_{p_2} - c_{p_1}) + (l/T)\{1 - [T(\alpha_2v_2 - \alpha_1v_1)]/(v_2 - v_1)\}$

8.8 $(2mgkT/abcl^2\rho_i)[(1/\rho_i) - (1/\rho_w)]$

8.11 (*a*) $\frac{5}{2}kT + mgz - kT[\ln\,(kT/p) + \frac{3}{2}\ln T + \sigma_0]$

8.13 2, for small dissociation **8.17** $K_p(T) = p\xi^3(1 - \xi)^{-2}(2 + \xi)^{-1}$

8.19 (*d*) $p/kT = e^{-\beta\eta}/ev_0$; (*f*) $L/RT_b = \ln\,(v_g/v_0)$

CHAPTER 9

9.2 $S = -k\sum_r [\bar{n}_r \ln \bar{n}_r \pm (1 \mp \bar{n}_r)\ln\,(1 \mp \bar{n}_r)]$ (upper sign for FD, lower for BE)

9.4 (*c*) $\bar{p}h(2\pi m)^{-\frac{1}{2}}(kT)^{-\frac{1}{2}}\exp\,(\epsilon_0/kT)$ **9.6** (*c*) $2/3N$ **9.7** $\frac{5}{2}R$

9.8 $\left[\dfrac{(m + M)^2}{mM}\right]\exp\left[-\dfrac{\hbar\omega}{2kT}\left(1 - \dfrac{m^{\frac{1}{2}} + M^{\frac{1}{2}}}{[2(m + M)]^{\frac{1}{2}}}\right)\right]$

9.13 (*a*) $T_0\sqrt{R/2L}$ **9.15** $p(t) = (\Delta M/rt)\sqrt{RT/2\pi\mu}$

9.16 $\overline{v_x{}^2} = 2\mu/5m$ **9.18** (*a*) $[(3\pi^2)^{\frac{2}{3}}/5](\hbar^2/m)(N/V)^{\frac{2}{3}}$ **9.22** $\chi = 3n\mu_m{}^2/2\mu$

9.23 $(2/h^3)(2\pi mkT)^{\frac{3}{2}}e^{-\Phi/kT}$ **9.24** $[4\pi me(1-r)/h^3](kT)^2e^{-\Phi/kT}$

9.26 (b) $I_2 = 4\sum_{n=0}^{\infty}(-1)^n/(n+1)^2$ **9.27** (b) $J(k) = \pi k/\sinh \pi k$

CHAPTER 10

10.2 (a) $\ln Z = (N\eta/kT) - 3N\ln(1-e^{-\Theta_D/T}) + ND(\Theta_D/T)$;
(b) $\bar{E} = -N\eta + 3NkTD(\Theta_D/T)$;
(c) $S = Nk[-3\ln(1-e^{-\Theta_D/T}) + 4D(\Theta_D/T)]$

10.3 For $T \ll \theta_D$, $\ln Z = (N\eta/kT) + (N\pi^4T^3/5\Theta_D^3)$,
$\bar{E} = -N\eta + (3\pi^4/5)(NkT^4/\Theta_D^3)$,
$S = (4\pi^4/5)Nk(T/\Theta_D)^3$;
for $T \gg \Theta_D$, $\ln Z = (N\eta/kT) + N[-3\ln(\Theta_D/T) + 1]$,
$\bar{E} = -N\eta + 3NkT$, $S = Nk[-3\ln(\Theta_D/T) + 4]$

10.4 $\bar{p} = N(\partial\eta/\partial V) + (3\gamma NkT/V)D(\Theta_D/T)$

10.7 $-\frac{1}{2}k\int_0^\infty [1 - e^{-\beta u}(1+\beta u)]4\pi R^2\,dR$

10.9 (a) $\bar{E} = -2nNS^2J$, if $T \ll T_c$;
$$\bar{E} = -\frac{5NS(S+1)NkT}{S^2+S+1}\left[1 - \frac{3kT}{2nJS(S+1)}\right], \text{ if } T \approx T_c;$$
$\bar{E} = 0$, if $T \gg T_c$;
(b) $C \to 0$, if $T \ll T_c$; $C = \dfrac{15Nk^2T}{nJ[S^2+S+\frac{1}{2}]}\left[1 - \dfrac{S(S+1)nJ}{3kT}\right]$, if $T \approx T_c$;
$C = 0$ if $T \gg T_c$

CHAPTER 11

11.2 $\Delta S = -AH_0^2/2(T-\theta)^2$ **11.3** $\Delta v = -A\theta_0\alpha H_0^2/2[T - \theta_0(1+\alpha\bar{p})]^2$

11.4 (b) $C_s - C_n = (VT/4\pi)(dH/dT)^2 + (VTH/4\pi)(d^2H/dT^2)$;
(c) $(VT_c/4\pi)(dH/dT)^2$ **11.6** (a) $\alpha = 3\gamma/T_c^2$; (b) $-\gamma T_c^2/4$

CHAPTER 12

12.3 (a) $(e\mathcal{E}/m)\tau^2$; (b) 0.757 **12.4** $\frac{1}{4}(a_1+a_2)^2$ **12.8** (a) $G = 2\pi R^3L\eta\omega/\delta$

12.9 (b) $\eta \propto T^{(s+3)/2(s-1)}$

12.10 (a) $(\pi/8)(\rho a^4/\eta L)(p_1 - p_2)$;
(b) $(\pi/16)(\mu a^4/\eta RTL)(p_1^2 - p_2^2)$

12.11 $\partial T/\partial t = (\kappa/\rho c)(\partial^2T/\partial z^2)$

12.12 $(I^2R/2\pi\kappa)\ln(b/a)$

12.15 $t \approx 3$ hours

12.16 $\mathfrak{F} = \mu n\bar{v}VL^2$, $t = M\ln 2/mn\bar{v}L^2$

CHAPTER 13

13.6 $\bar{v}_z = 0$ gives $(1/n)(\partial n/\partial z) = (1/\beta)(\partial \beta/\partial z)$,
$f = g + (gv_z\tau/T)(dT/dz)[\frac{5}{2} - \frac{1}{2}(mv^2/kT)]$

13.11 $\kappa = (\pi^2/3)(nk^2T/m)\tau_F$ *13.12* $\kappa/\sigma_{el} = (\pi^2/3)(k/e)^2T$

CHAPTER 14

14.1 $\sigma_{el} = (3\pi^{\frac{1}{2}}/8)(ne^2/n_1\sigma_{im}^{(0)})(2\mu kT)^{-\frac{1}{2}}$, where $\mu = m_1m/(m_1 + m)$

14.2 (b) $\kappa = \frac{25}{32}(k/\overline{\sigma_\eta})\sqrt{\pi kT/m}$; (c) $\kappa = \frac{75}{64}(k/\sigma_0)\sqrt{\pi kT/m}$

14.3 $\kappa/\eta = 15k/4m$

CHAPTER 15

15.7 $\bar{Y} = \sum_{k=0}^{N-1} e^{-\gamma\tau(N-k)}\bar{G} = 0$ *15.8* $\overline{G^2} = 2kT\gamma\tau/m$, $\overline{Y^2} = (kT/m)(1 - e^{-2\gamma\tau})$

15.10 $\gamma = (2mkT)^{-1}\int_{-\infty}^{\infty} K(s)\,ds$

15.11 (a) $\overline{\Delta v} = -\gamma v_0\tau$, $\overline{(\Delta v)^2} \approx \overline{G^2} = 2kT\gamma\tau/m$;
(b) $\overline{(\Delta v)^3} = -3\gamma v_0\tau\overline{G^2}$, $\overline{(\Delta v)^4} = \overline{G^4}$; (c) $\overline{(\Delta v)^3} = -6kT\gamma^2\tau^2v_0/m$

15.13 $\overline{v^2(t)} = v_0^2e^{-2\gamma t} + [(1 - e^{-2\gamma t})/2\gamma m^2]\int_{-\infty}^{\infty} K(s)\,ds$

15.14 (a) $J_{F'}(\omega) = m^2(\gamma^2 + \omega^2)J_v(\omega)$;
(b) $J_v(\omega) = (kT/\pi m)[\gamma/(\gamma^2 + \omega^2)]$;
(c) $J_{F'^{(+)}}(\omega) = (2/\pi)mkT\gamma$

Index